T0191661

Advances in Intelligent Systems and Computing

Volume 820

Series editor

Janusz Kacprzyk, Polish Academy of Sciences, Warsaw, Poland
e-mail: kacprzyk@ibspan.waw.pl

The series "Advances in Intelligent Systems and Computing" contains publications on theory, applications, and design methods of Intelligent Systems and Intelligent Computing. Virtually all disciplines such as engineering, natural sciences, computer and information science, ICT, economics, business, e-commerce, environment, healthcare, life science are covered. The list of topics spans all the areas of modern intelligent systems and computing such as: computational intelligence, soft computing including neural networks, fuzzy systems, evolutionary computing and the fusion of these paradigms, social intelligence, ambient intelligence, computational neuroscience, artificial life, virtual worlds and society, cognitive science and systems, Perception and Vision, DNA and immune based systems, self-organizing and adaptive systems, e-Learning and teaching, human-centered and human-centric computing, recommender systems, intelligent control, robotics and mechatronics including human-machine teaming, knowledge-based paradigms, learning paradigms, machine ethics, intelligent data analysis, knowledge management, intelligent agents, intelligent decision making and support, intelligent network security, trust management, interactive entertainment, Web intelligence and multimedia.

The publications within "Advances in Intelligent Systems and Computing" are primarily proceedings of important conferences, symposia and congresses. They cover significant recent developments in the field, both of a foundational and applicable character. An important characteristic feature of the series is the short publication time and world-wide distribution. This permits a rapid and broad dissemination of research results.

More information about this series at http://www.springer.com/series/11156

Sebastiano Bagnara · Riccardo Tartaglia
Sara Albolino · Thomas Alexander
Yushi Fujita

Editors

Proceedings of the 20th Congress of the International Ergonomics Association (IEA 2018)

Volume III: Musculoskeletal Disorders

 Springer

Editors
Sebastiano Bagnara
University of the Republic of San Marino
San Marino, San Marino

Riccardo Tartaglia
Centre for Clinical Risk Management
    and Patient Safety, Tuscany Region
Florence, Italy

Sara Albolino
Centre for Clinical Risk Management
    and Patient Safety, Tuscany Region
Florence, Italy

Thomas Alexander
Fraunhofer FKIE
Bonn, Nordrhein-Westfalen
Germany

Yushi Fujita
International Ergonomics Association
Tokyo, Japan

ISSN 2194-5357   ISSN 2194-5365 (electronic)
Advances in Intelligent Systems and Computing
ISBN 978-3-319-96082-1   ISBN 978-3-319-96083-8 (eBook)
https://doi.org/10.1007/978-3-319-96083-8

Library of Congress Control Number: 2018950646

This Springer imprint is published by the registered company Springer Nature Switzerland AG
The registered company address is: Gewerbestrasse 11, 6330 Cham, Switzerland

Preface

The Triennial Congress of the International Ergonomics Association is where and when a large community of scientists and practitioners interested in the fields of ergonomics/human factors meet to exchange research results and good practices, discuss them, raise questions about the state and the future of the community, and about the context where the community lives: the planet. The ergonomics/human factors community is concerned not only about its own conditions and perspectives, but also with those of people at large and the place we all live, as Neville Moray (Tatcher et al. 2018) taught us in a memorable address at the IEA Congress in Toronto more than twenty years, in 1994.

The Proceedings of an IEA Congress describes, then, the actual state of the art of the field of ergonomics/human factors and its context every three years.

In Florence, where the XX IEA Congress is taking place, there have been more than sixteen hundred (1643) abstract proposals from eighty countries from all the five continents. The accepted proposal has been about one thousand (1010), roughly, half from Europe and half from the other continents, being Asia the most numerous, followed by South America, North America, Oceania, and Africa. This Proceedings is indeed a very detailed and complete state of the art of human factors/ergonomics research and practice in about every place in the world.

All the accepted contributions are collected in the Congress Proceedings, distributed in ten volumes along with the themes in which ergonomics/human factors field is traditionally articulated and IEA Technical Committees are named:

    I. Healthcare Ergonomics (ISBN 978-3-319-96097-5).
   II. Safety and Health and Slips, Trips and Falls (ISBN 978-3-319-96088-3).
  III. Musculoskeletal Disorders (ISBN 978-3-319-96082-1).
 IV. Organizational Design and Management (ODAM), Professional Affairs, Forensic (ISBN 978-3-319-96079-1).
  V. Human Simulation and Virtual Environments, Work with Computing Systems (WWCS), Process control (ISBN 978-3-319-96076-0).

VI. Transport Ergonomics and Human Factors (TEHF), Aerospace Human Factors and Ergonomics (ISBN 978-3-319-96073-9).
VII. Ergonomics in Design, Design for All, Activity Theories for Work Analysis and Design, Affective Design (ISBN 978-3-319-96070-8).
VIII. Ergonomics and Human Factors in Manufacturing, Agriculture, Building and Construction, Sustainable Development and Mining (ISBN 978-3-319-96067-8).
IX. Aging, Gender and Work, Anthropometry, Ergonomics for Children and Educational Environments (ISBN 978-3-319-96064-7).
X. Auditory and Vocal Ergonomics, Visual Ergonomics, Psychophysiology in Ergonomics, Ergonomics in Advanced Imaging (ISBN 978-3-319-96058-6).

Altogether, the contributions make apparent the diversities in culture and in the socioeconomic conditions the authors belong to. The notion of well-being, which the reference value for ergonomics/human factors is not monolithic, instead varies along with the cultural and societal differences each contributor share. Diversity is a necessary condition for a fruitful discussion and exchange of experiences, not to say for creativity, which is the "theme" of the congress.

In an era of profound transformation, called either digital (Zisman & Kenney, 2018) or the second machine age (Bnynjolfsson & McAfee, 2014), when the very notions of work, fatigue, and well-being are changing in depth, ergonomics/human factors need to be creative in order to meet the new, ever-encountered challenges. Not every contribution in the ten volumes of the Proceedings explicitly faces the problem: the need for creativity to be able to confront the new challenges. However, even the more traditional, classical papers are influenced by the new conditions.

The reader of whichever volume enters an atmosphere where there are not many well-established certainties, but instead an abundance of doubts and open questions: again, the conditions for creativity and innovative solutions.

We hope that, notwithstanding the titles of the volumes that mimic the IEA Technical Committees, some of them created about half a century ago, the XX Triennial IEA Congress Proceedings may bring readers into an atmosphere where doubts are more common than certainties, challenge to answer ever-heard questions is continuously present, and creative solutions can be often encountered.

Acknowledgment

A heartfelt thanks to Elena Beleffi, in charge of the organization committee. Her technical and scientific contribution to the organization of the conference was crucial to its success.

References

Brynjolfsson E., A, McAfee A. (2014) The second machine age. New York: Norton.

Tatcher A., Waterson P., Todd A., and Moray N. (2018) State of science: Ergonomics and global issues. Ergonomics, 61 (2), 197–213.

Zisman J., Kenney M. (2018) The next phase in digital revolution: Intelligent tools, platforms, growth, employment. Communications of ACM, 61 (2), 54–63.

Sebastiano Bagnara
Chair of the Scientific Committee, XX IEA Triennial World Congress
Riccardo Tartaglia
Chair XX IEA Triennial World Congress
Sara Albolino
Co-chair XX IEA Triennial World Congress

Organization

Organizing Committee

Riccardo Tartaglia (Chair IEA 2018)	Tuscany Region
Sara Albolino (Co-chair IEA 2018)	Tuscany Region
Giulio Arcangeli	University of Florence
Elena Beleffi	Tuscany Region
Tommaso Bellandi	Tuscany Region
Michele Bellani	Humanfactor^x
Giuliano Benelli	University of Siena
Lina Bonapace	Macadamian Technologies, Canada
Sergio Bovenga	FNOMCeO
Antonio Chialastri	Alitalia
Vasco Giannotti	Fondazione Sicurezza in Sanità
Nicola Mucci	University of Florence
Enrico Occhipinti	University of Milan
Simone Pozzi	Deep Blue
Stavros Prineas	ErrorMed
Francesco Ranzani	Tuscany Region
Alessandra Rinaldi	University of Florence
Isabella Steffan	Design for all
Fabio Strambi	Etui Advisor for Ergonomics
Michela Tanzini	Tuscany Region
Giulio Toccafondi	Tuscany Region
Antonella Toffetti	CRF, Italy
Francesca Tosi	University of Florence
Andrea Vannucci	Agenzia Regionale di Sanità Toscana
Francesco Venneri	Azienda Sanitaria Centro Firenze

Scientific Committee

Sebastiano Bagnara (President of IEA2018 Scientific Committee)	University of San Marino, San Marino
Thomas Alexander (IEA STPC Chair)	Fraunhofer-FKIE, Germany
Walter Amado	Asociación de Ergonomía Argentina (ADEA), Argentina
Massimo Bergamasco	Scuola Superiore Sant'Anna di Pisa, Italy
Nancy Black	Association of Canadian Ergonomics (ACE), Canada
Guy André Boy	Human Systems Integration Working Group (INCOSE), France
Emilio Cadavid Guzmán	Sociedad Colombiana de Ergonomia (SCE), Colombia
Pascale Carayon	University of Wisconsin-Madison, USA
Daniela Colombini	EPM, Italy
Giovanni Costa	Clinica del Lavoro "L. Devoto," University of Milan, Italy
Teresa Cotrim	Associação Portuguesa de Ergonomia (APERGO), University of Lisbon, Portugal
Marco Depolo	University of Bologna, Italy
Takeshi Ebara	Japan Ergonomics Society (JES)/Nagoya City University Graduate School of Medical Sciences, Japan
Pierre Falzon	CNAM, France
Daniel Gopher	Israel Institute of Technology, Israel
Paulina Hernandez	ULAERGO, Chile/Sud America
Sue Hignett	Loughborough University, Design School, UK
Erik Hollnagel	University of Southern Denmark and Chief Consultant at the Centre for Quality Improvement, Denmark
Sergio Iavicoli	INAIL, Italy
Chiu-Siang Joe Lin	Ergonomics Society of Taiwan (EST), Taiwan
Waldemar Karwowski	University of Central Florida, USA
Peter Lachman	CEO ISQUA, UK
Javier Llaneza Álvarez	Asociación Española de Ergonomia (AEE), Spain
Francisco Octavio Lopez Millán	Sociedad de Ergonomistas de México, Mexico

Contents

Development of a Risk Assessment Procedure for Upper Limbs Based on Combined Use of EAWS 4th Section and OCRA High Precision Checklist 793
Enrico Occhipinti and Lidia Ghibaudo

Workshop: How to Diagnose and Treat a Work Related Musculoskeletal Disorder?

Deepak Sharan[1](✉), Joshua Samuel Rajkumar[2], and Jerrish A. Jose[2]

[1] Department of Orthopaedics and Rehabilitation, RECOUP
Neuromusculoskeletal Rehabilitation Centre, 312, 10th Block, Anjanapura,
Bangalore 560108, KA, India
deepak.sharan@recoup.in
[2] Department of Physiotherapy, RECOUP Neuromusculoskeletal Rehabilitation
Centre, 312, 10th Block, Anjanapura, Bangalore 560108, KA, India
{joshua.samuel,rajarajeshwari.b}@recoup.in

Abstract. WRMSD is the commonest occupational health condition across the world. WRMSDs can affect workers in the industrial, commercial, healthcare, hospitality, and service sectors. Occupations at high risk for WRMSDs include machine operators, parts assembly operators, manual and agricultural labourers, nurses and other healthcare professionals, computer or visual display unit users, typists, and musicians. Diagnosis of WRMSD can be controversial and challenging, especially since soft tissue evaluation requires specialised manual skills not usually provided in medical training. Symptoms of WRMSD are often poorly localised, nonspecific, and episodic. Accurate diagnosis helps determine the most appropriate treatment and protects the worker from treatments that are ineffective, unwarranted, or unnecessarily invasive (54%). The current best practices regarding investigations, treatment of WRMSD using a staged, comprehensive, multi-disciplinary protocol, and criteria for return to work will be discussed. Common WRMSDs like Myofascial Pain Syndrome, Thoracic Outlet Syndrome, Fibromyalgia Syndrome, Tendinopathies, Compressive Neuropathies and Disc Disorders will be covered.

Keywords: Work related musculoskeletal disorder · Diagnosis
Treatment

1 Introduction

Work Related Musculoskeletal Disorders (WRMSDs) are impairments of body structures such as muscles, joints, tendons, ligaments, nerves, bones or a localised blood circulation system caused or aggravated primarily by the performance of work and by the effects of the immediate environment where the work is carried out. WRMSD is the commonest occupational health condition across the world. WRMSDs can affect workers in the industrial, commercial, healthcare, hospitality, and service sectors. Occupations at high risk for WRMSDs include machine operators, parts assembly operators, manual and agricultural labourers, nurses and other healthcare professionals, computer or visual display unit users, typists, and musicians. U.S. companies spent 50 billion dollars on direct costs of WRMSDs in 2011 (source: CDC). Indirect costs can

© Springer Nature Switzerland AG 2019
S. Bagnara et al. (Eds.): IEA 2018, AISC 820, pp. 1–6, 2019.
https://doi.org/10.1007/978-3-319-96083-8_1

be up to five times the direct costs of WRMSDs (source: OSHA). The average WRMSD comes with a direct cost of almost $15,000 per worker (source: BLS). Diagnosis of WRMSD can be controversial and challenging, especially since soft tissue evaluation requires specialised manual skills not usually provided in medical training. Symptoms of WRMSD are often poorly localised, nonspecific, and episodic. Accurate diagnosis helps determine the most appropriate treatment and protects the worker from treatments that are ineffective, unwarranted, or unnecessarily invasive. In this inter-active and practical workshop, we will present the latest criteria for diagnosis and for classification of WRMSD, and demonstrate diagnostic tests for different regions of the body. The diagnosis of WRMSD of each region of the body will include history taking, physical examination, special tests, investigations, differential diagnosis and time rule. Establishing work relatedness of MSDs has significant organisational, financial and medico-legal implications, besides helping the clinician to make considered decisions about a worker's possible condition and management. Classification criteria, by con-trast, aim to identify homogenous subgroups to help ensure that future studies are comparable, and that data can be combined across industries, even countries. The current best practices regarding investigations, treatment of WRMSD using a staged, comprehensive, multi-disciplinary protocol, and criteria for return to work will be discussed. Common WRMSDs like Myofascial Pain Syndrome, Thoracic Outlet Syndrome, Fibromyalgia Syndrome, Tendinopathies, Compressive Neuropathies and Disc Disorders will be covered.

2 Definition, Classification and Diagnosis of WRMSD, Including Criteria to Determine Work-Relatedness of a MSD

Diagnosis of WRMSD can be controversial and challenging, especially since soft tissue evaluation requires specialised manual skills not usually provided in medical training. Symptoms of WRMSD are often poorly localised, nonspecific, and episodic. Accurate diagnosis helps determine the most appropriate treatment and protects the worker from treatments that are ineffective, unwarranted, or unnecessarily invasive. The diagnosis of WRMSD of each region of the body will include history taking, physical examination, special tests, investigations, differential diagnosis and time rule. The WRMSD's are classified in to two broad categories: 1. Specific (that are diagnosed with a specific cause and are usually elicited by provocative medical examination/tests) and 2. Non-Specific (Patient presents with pain, discomfort or functional impairment but no specific diagnosis or pathology can be ascertained). Non-Specific WRMSDs are diagnosed based on typical individual factors. That is Myofascial Pain Syndrome (MPS) by Simon's major and minor criteria, Fibromyalgia Syndrome (FMS) by American College of Rheumatology (ACR) criteria as mentioned below and CRPS based on Budapest criteria. Whereas, specific WRMSD's are diagnosed based on three factors: (1) Time rule: Symptoms present now or at least 4 days during the last 1 week, (2) Symptoms: Intermittent pain/ache in that region and (3) Signs: Provocation of symptoms (physical examination and special test). The commonly presenting specific conditions of WRMSD as reported by Sluiter et al. 2001, shows that it can be categorised into 5 types as tendon related, nerve related, joint related, circulatory related and pain syndromes. Various specific WRMSD's of the head, neck and upper limb include TMJ disorder, tension headache, cervicogenic head ache, tendinosis of the supraspinatus, biceps, tennis elbow, compressive neuropathies, thoracic outlet syndrome etc.

3 WRMSD of Head, Neck, Upper Limb

The prevalence of WRMSD in head, neck and upper limbs are reported to be higher, especially among dentists and surgeons. Amstrong et al. 1993, illustrated a conceptual model "cause-effect cascade" which better explains the developments of WRMSD, especially in the upper extremity. Repetitive and forceful wrist movements cause increased pressure inside the carpal tunnel which in turn increases the intra-neural pressure of the median nerve, which causes a decrease in intra-neural blood supply, which causes cell membrane instability, causing defective axonal transport and signal propagation and electrolyte destabilisation, which cause oedema that cause further increase in intra-neural pressure, and so on. The defective axonal signal propagation causes symptoms of numbness and tingling in the fingers; positive nerve compression provocation signs, and a decrease in median nerve conduction velocity over the wrist region. These symptoms can be regarded as "acute" reversible effects. Sustained high intra-neural pressure can cause chronic effects, such as a decreased function in 2-point discrimination ability and trophic changes in the thenar muscles with permanent damage, which can be observed as muscle weakness and thenar atrophy. The symptoms and dysfunction may cause decreased productivity and quality at work and a need for exposure elimination (sick-leave or work modification) and perhaps also medical care including surgery. Some of the commonest WRMSD related to head, neck and upper limb are as follows: Myofascial Pain Syndrome (MPS), Cervicogenic headaches, TMJ disorder, Thoracic outlet syndrome, tenosynovitis, tendinitis, epicondylitis, rotator cuff syndrome, carpal tunnel syndrome, cubital tunnel syndrome, Guyon canal syndrome, radial tunnel syndrome, hand/arm vibration syndrome, degenerative arthritis and adhesive capsulitis of shoulder.

4 WRMSD of Spine, Thorax, Pelvis and Lower Limbs

The back is the most frequently injured part of the body (22% of 1.7 million injuries) (NSC, Accident Facts, 1990) with overexertion being the most common cause of these injuries. However, many back injuries develop over a long period of time by a repetitive loading of the discs caused by improper lifting methods or other exertions. Lower limb WRMSD are currently a problem in many jobs, they tend to be related with disorders in other areas of the body. The epidemiology of these WRMSD has received modest awareness until now. However, there is appreciable evidence that some activities (e.g., kneeling/squatting, climbing stairs or ladders, heavy lifting, walking/standing) are causal risk factors for their development. Other causes for acute lower limb injuries are related with slip and trip hazards (HSE, 2009). Common back problems include MPS of the back and vertebral disc problems. Hip and thigh conditions include Osteoarthritis, Piriformis Syndrome, Trochanteritis, Hamstring strains, and Sacroiliac Joint Pain. Knee and lower leg conditions included Osteoarthritis, Bursitis, Beat Knee/Hyperkeratosis, Meniscal Lesions, Patellofemoral Pain Syndrome, Pre-patellar Tendinosis, Shin Splints, Infrapatellar Tendinosis, and Stress Fractures. Ankle and foot conditions include Achilles Tendinopathy, Blisters, Foot Corns, Hallux Valgus (Bunions), Hammer Toes, Pes Traverse Planus, Plantar Fasciitis, Sprained Ankle, Stress fractures, Varicose veins, and Venous disorders.

5 Principles of Management of WRMSD

The principles of management of WRMSD include: Early reporting of injuries, timely access to healthcare, job modification, light or modified duty if required, and close medical follow-up. The goals of management of WRMSD are to reduce symptoms, prevent progression, promote healing, reduce impairment and disability, and prevent future injury. The usual care for WRMSD includes medication (painkillers, sleep medication, antiepileptics), postural correction, ergonomic changes, wrist splints, complete rest, voice recognition software, physiotherapy (mainly strengthening), a wide array of complementary therapies, psychological counselling, work hardening and encouragement to "live with it" and "manage the pain." Our published research suggests that this approach is sub-optimal. It is commonly seen that if a patient with a localised musculoskeletal pain is followed over years and if the problem is not effectively treated or resolved early enough the pain starts to spread outside the region of origin of pain. When acute pain becomes chronic, it often results in missed work, disability, and significantly high cost of care. Half of the persons seeking care for pain conditions still have pain 5 years later, and up to 25% of them receive long-term disability. Utmost emphasis should be given to early, accurate diagnosis of the WRMSD by Physicians skilled in manual musculoskeletal medicine, followed by intensive, protocol based, multi-disciplinary rehabilitation with emphasis on direct therapy to muscles through counter stimulation to desensitise the soft tissues, using myofascial release and manual therapies and reduction of all contributing factors that strain the muscle(s) and heighten peripheral and central sensitisation, to prevent acute conditions from becoming chronic. Workers in chronic pain are best served by aggressive inpatient rehabilitation for several hours a day lasting several months by an experienced, multidisciplinary medical and rehabilitation team with a solid track record. This is rarely achieved by psychological approaches alone. Patients with complicated neuropathic pain often need intensive rehabilitation for up to 6 months and sometimes more. In contrast, acute onset localised WRMSDs can usually be completely and permanently resolved within a few weeks.

6 Prognosis and Criteria for Return to Work

The usual care for WRMSD includes medication (painkillers, sleep medication, antiepileptics), postural correction, ergonomic changes, wrist splints, complete rest, voice recognition software, physiotherapy (mainly strengthening), a wide array of complementary therapies, psychological counselling, work hardening and encouragement to "live with it" and "manage the pain." Our published research suggests that this approach is sub-optimal. It is commonly seen that if a patient with a localised musculoskeletal pain is followed over years and if the problem is not effectively treated or resolved early enough the pain starts to spread outside the region of origin of pain. When acute pain becomes chronic, it often results in missed work, disability, and significantly high cost of care. Half of the persons seeking care for pain conditions still have pain 5 years later, and up to 25% of them receive long-term disability. Utmost emphasis should be given to early, accurate diagnosis of the WRMSD by Physicians skilled in manual musculoskeletal medicine, followed by intensive, protocol based, multi-disciplinary rehabilitation with emphasis on direct therapy to muscles through counter stimulation to desensitise the soft tissues, using myofascial release and manual therapies and reduction of all contributing factors that strain the muscle(s) and heighten peripheral and central sensitisation, to prevent acute conditions from becoming chronic. Workers in chronic pain are best served by aggressive inpatient rehabilitation for several hours a day lasting several months by an experienced, multidisciplinary medical and rehabilitation team with a solid track record. This is rarely achieved by psychological approaches alone. Patients with complicated neuropathic pain often need intensive rehabilitation for up to 6 months and sometimes more. In contrast, acute onset localised WRMSDs can usually be completely and permanently resolved within a few weeks. If an employee's treatment plan required time away from work, the next step is to return the employee to work in a manner that will minimise the chance for re-injury. Employees returning to the same job without a modification of the work environment are at risk for a recurrence. Communication between Employee, Employee and HCP is essential to ensure prompt treatment, an expedient return to work consistent with the employee's health status and job requirements, and regular follow-up to manage symptoms and modify work restrictions as appropriate. The factors that determine return to work include type of MSD condition, severity of the MSD condition, MSD risk factors present on the job, and duration of disability and time off work. The factors that lead to successful recovery from WRMSD are expert medical assessment by a Physician with a sound track record with WRMSD, protocol based rehabilitation including intensive, skilled manual therapy and body awareness approaches, identification and correction of all predisposing factors, e.g., posture, stress, habitual deep muscle tension, medical co-morbidities, and workers take responsibility for making changes in their approach to work and other activities. The poor prognostic factors for recovery from WRMSD are the presence of Neuropathic pain; Associated medical co morbidities (Fibromyalgia, Multidirectional Instability of Shoulders, Thoracic Outlet Syndrome), often associated with Osteopenia or Osteoporosis and CRPS; Depression; Poor social and economic support; Insomnia; Pain avoidance behaviour, Kinesiophobia, catastrophising; Poor treatment compliance and Passive treatment seekers.

For assessment of severity of WRMSD and prognosis, we will describe the published and validated DEEPAK SHARAN'S Severity Score

- Duration of symptoms
- Effect on activities of daily living (ADL)
- Exertion (use of physical energy/hard work) rate during work
- Pain intensity
- Ability to control the speed of work and take breaks
- Known medical conditions
- Stress and psychological factors
- Hours of work
- Area affected
- Restricted movements of affected area
- Activity restriction
- Neuropathic pain
- Sleep disturbance

How to Perform an Ergonomic Workplace Analysis?

Deepak Sharan[1(✉)], Jerrish A. Jose[1], and Joshua Samuel Rajkumar[2]

[1] Department of Orthopedics and Rehabilitation, RECOUP
Neuromusculoskeletal Rehabilitation Centre, 312, 10th Block, Anjanapura,
Bangalore 560108, KA, India
{deepak.sharan,jerrish}@recoup.in
[2] Department of Physiotherapy, RECOUP Neuromusculoskeletal Rehabilitation
Centre, 312, 10th Block, Anjanapura, Bangalore 560108, KA, India
joshua.samuel@recoup.in

Abstract. Work related musculoskeletal disorders (WRMSD) are one of the commonest causes for morbidity among workers and are associated with significant financial and social burden. The use of a methodical, step-by-step approach and validated, simplified risk estimation tools which can be used even by non-experts in workplaces with limited resources is necessary to evaluate conditions of biomechanical overload and prevent WRMSDs. The work-related risk factors are psychosocial, physical, personal, organisational, environment etc. Identification of potential risk factors is an important component in preventing WRMSD and Ergonomic Workplace Analysis (EWA) plays a key role in it. EWA is a process where the ergonomic risk factors are evaluated using various validated tools. An EWA usually consist of 12 steps which included collection of personal information, work details, subjective analysis questionnaires, psychosocial information, vital sign assessment, metabolic assessment, assessment of work effort and fatigue, postural evaluation, other specific evaluations, environment details which is followed by analysis, recommendation and then follow up evaluations

Keywords: Ergonomic workplace analysis
Work related musculoskeletal disorder · Risk assessment

1 Introduction

Work related musculoskeletal disorders (WRMSD) are one of the commonest causes for morbidity among workers and are associated with significant financial and social burden. One of the crucial steps in prevention of WRMSD is the identification of ergonomic risk of a job by a process called Ergonomic Workplace Analysis (EWA), where the ergonomic risk factors are evaluated using various standardised and validated tools.

© Springer Nature Switzerland AG 2019
S. Bagnara et al. (Eds.): IEA 2018, AISC 820, pp. 7–11, 2019.
https://doi.org/10.1007/978-3-319-96083-8_2

2 Ergonomic Workplace Analysis

The use of a methodical, step-by-step approach and validated, simplified risk estimation tools which can be used even by non-experts in workplaces with limited resources is necessary to evaluate conditions of biomechanical overload and prevent WRMSDs. These tools are especially relevant in small and medium sized enterprises, craftwork, agriculture, and in industrially developing countries. The steps used in the toolkits include: Step one: identification of preliminary occupational hazards and priority setting via "key-enter." Step two: identification of risk factors for WRMSDs, consisting of a "quick assessment" and substantially aimed at identifying three possible conditions: acceptable/no consequences; critical/redesign urgently needed; more detailed analysis required. Step three: recognized tools for estimating risk (of WRMSDs) are used depending on the outcomes of step two. Examples of such tools include "adaptations" of the Revised NIOSH Lifting Equation, Liberty Mutual Psychophysical Tables, OCRA Checklist, etc. The goal of EWA is to identify ergonomic risk factors, quantify them, and then make measurable improvements to the workplace, ensuring that jobs and tasks are within workers' capabilities and limitations. The best approach for doing that is to make ergonomics an ongoing process of risk identification and risk reduction based on objective, scientific analysis of the workplace. However, an assessor often feels lost due to the numerous EWA Tools available. This tutorial aims at presenting a simplified overview of the popular EWA tools, merits and demerits of each and will provide recommendations for selection of tools for different tasks, based on the faculty's experience of conducting over 8,00,000 EWA in diverse industries. In this tutorial various validated tools available to assess posture, muscle effort, risk associated with repetitive action and risk associated with manual handling tasks like lifting and carrying will be discussed and will give an overview about various tools and recommend the best tool to assess a specific task identified to be risky.

3 Introduction to EWA, Steps Involved in Ergonomic Assessment, Checklists in EWA

The work-related risk factors are psychosocial, physical, personal, organisational, environment etc. Identification of potential risk factors is an important component in preventing WRMSD and Ergonomic Workplace Analysis (EWA) plays a key role in it. EWA is a process where the ergonomic risk factors are evaluated using various validated tools. An ergonomist needs to evaluate the physical work environment, psychosocial risk factors as well as various generic risk factors which leads to the development of WRMSD while performing an EWA. However, an assessor often feels lost in the myriad of EWA tools available. This tutorial will provide an overview of commonly used tools which are useful in EWA with their advantages and limitations. An EWA usually consist of 12 steps which included collection of personal information, work details, subjective analysis questionnaires, psychosocial information, vital sign assessment, metabolic assessment, assessment of work effort and fatigue, postural evaluation, other specific evaluations, environment details which is followed by analysis, recommendation and then follow up evaluations. Checklists are used as a pre-evaluation tools that could help us to know whether further detailed evaluation is needed. Common check list used are Plan för identifiering av belastningsfaktorer (PLIBEL) and Quick exposure check (QEC). Observational methods are still the most frequently used. These methods are simple, low-cost, and more flexible than other methods in field studies.

4 Subjective Analysis Tools and Psychophysiological Assessment Tools

A subjective task analysis tool is narrative, quantitative and/or checklist system that provides a standardised evaluation of a job or task. Tools are based on biomechanical and physiological information compiled and calculated providing a relative risk of injury probability and can be specific for body region and/or risk factors. Pschophysiological Assessment can be done using Questionnaires or Psychophysiological Methods. Commonly used Questionnaires include Workstyle Questionnaire, NIOSH Generic Job Stress Questionnaire, Caplan's Job Stress Questionnaire and Occupational Stress Index. The Psychophysiological Methods include Electrodermal analysis, EMG, EEG, Heart rate and Respiratory rate. Human factors methods include Task Analysis Techniques, Cognitive task analysis techniques, Process Charting methods, Human error identification and human reliability analysis techniques, Situation awareness assessment techniques, Mental workload assessment techniques, Team performance analysis techniques, Interface analysis techniques.

5 Work Effort and Fatigue Assessment Tools

Measuring the exertion level and fatigue level is important to know the physical load of a task or a job. Physical load is one of the important factor that determines the level of risk of getting MSD following the job and so identification of physical load is important in risk identification and task modification. There are various methods of assessment of physical load, among which Borg's rate of perceived exertion (RPE scale), Borg's Category Ratio Scale 10 (CR 10) and Rodger's Muscle Fatigue Analysis (MFA) are commonly used tools. RPE scale is a quantitative measure of perceived exertion during physical activity. The original scale introduced by Gunnar Borg rated exertion on a scale of 6–20. Borg then proposed a category (C) ratio (R) scale, the Borg CR10 Scale. This is used in clinical diagnosis of breathlessness and dyspnoea, chest pain, angina and musculoskeletal pain. The CR-10 scale is best suited when there is an overriding sensation arising either from a specific area of the body, for example, muscle pain, ache or fatigue in the biceps or from pulmonary responses. The Muscle Fatigue Analysis was proposed by Rodgers to assess the amount of fatigue that accumulates in muscles during various work patterns within 5 min of work. The hypothesis was that a rapidly fatiguing muscle is more susceptible to injury and inflammation. If fatigue can be minimised, so should muscle injuries. MFA is most appropriate to evaluate the risk for fatigue accumulation in tasks that are performed for an hour or more and where awkward postures or frequent exertions are present. Based on the risk of fatigue, a Priority for Change can be assigned to the task.

6 Postural Assessment Tools

Posture is the relationship between different parts of the body. Posture analysis refers to decision-making about the magnitude of a posture, relative to a convention specified in the tool or method used. For example, video can be used to record or collect body postures in the workplace. These postures can be analysed later with software to determine the angle of the body segments, as viewed on the video (NIOSH). Rapid upper-limb assessment (RULA) was developed by McAtamney and Corlett (1993) to calculate the rate of musculoskeletal loads in tasks where people have a risk of neck and upper-limb loading. Rapid entire body assessment (REBA) was developed by Hignett and McAtamney (2000) to assess the type of unpredictable working postures found in health-care and other service industries. The other posture assessment methods include Video analysis, Fishbone diagram, Goniometer, Motion capture systems, Accelerometry, Inertial measurement units, Computerised tools: Task recording and analysis on computer (TRAC), Portable ergonomic observation (PEO), Hands relative to the body (HARBO), Observational tools: Ovako working posture assessment system (OWAS), Arbeitswissenschaftliches erhebungsverfahren zur tätigkeitsanalyse (AET), Posture targeting, Posture, activity, tools and handling (PATH), Washington State ergonomic checklist, Video- och datorbaserad arbetsanalys (VIDAR), Postural loading on the upper-body assessment (LUBA), Chung's postural workload evaluation, OREGE (Outil de Repérage et d'Evaluation des Gestes), SUVA (National Institute of the Swiss Accident Insurance), Novel Ergonomic Postural Assessment Method (NERPA), Rapid Office Strain Assessment (ROSA) and EPM – TACO tool.

7 Tools to Assess Repetitive Actions

The commonly used tools are Hand activity level (HAL), Strain Index (SI), Assessment of repetitive action (ART) and Occupational Repetitive action (OCRA). The American Conference of Governmental Industrial Hygienists (ACGIH®) Threshold Limit Value® (TLV®) for Hand Activity (2001) is offered for the evaluation of job risk factors associated with musculoskeletal disorders of the hand and wrist. The evaluation is based an assessment of hand activity and the level of effort for a typical posture while performing a short cycle task. SI was developed by Steven Moore and Arun Garg, and is a method of evaluating jobs to determine if they expose workers to increased risk of developing musculoskeletal disorders of the distal upper extremity. ART was developed by Health and Safety Executive (HSE) in UK. The ART tool is designed to help risk assess tasks that require repetitive movement of the upper limbs (arms and hands). It assists in assessing some of the common risk factors in repetitive work that contribute to the development of Upper Limb WRMSD. The ART tool is intended for people with responsibility for the design, assessment, management, and inspection of repetitive work. Repetitive tasks are typically found in assembly, production, processing, packaging, packing and sorting work, as well as work involving regular use of hand tools. ART is not intended for Display Screen Equipment (DSE) assessments. Occhipinti and Colombini (1996) developed the OCRA methods to analyse workers' exposure to tasks featuring various upper-limb injury risk factors: repetitiveness, force, awkward postures and movements, lack of recovery periods, and others, defined as "additionals".

8 Tools for Assessing Lifting, Carrying, Pushing and Pulling Tasks

Low back pain affects more workers and result in higher costs to industry than any other WRMSD. The tools commonly used to assess the risk for low back pain are NIOSH (National Institute of Occupational safety and Health) Lifting Equation, Manual Handling Assessment Charts (MAC), SNOOK Tables, Key Indicator method (KIM) and Concise Back Screening Instrument (CBSI). The purpose of the 1991 Revised NIOSH (National Institute of Occupational safety and Health) Lifting Equation is to provide a means of quantifying the relative risk or acceptability of a specific lifting task, to subsequently be able to identify specific task deficiencies, and then plan for their elimination. MAC is an initial screening tool to identify high-risk manual handling activities, which incorporates a numerical scoring system to assist with prioritising interventions using a colour scheme which indicates which element of the manual handling task is high-risk. The MAC can assess single person lifting, carrying and team handling operations. Snook Tables were developed at Liberty Mutual Insurance Company. These data were developed based on the subjects' subjective responses to the loads that they believed they could handle safely. This type of study is known as "psychophysical." It can be used to find the percent of an industrial population capable of sustaining the efforts tabulated in lifting, lowering, pushing, pulling, and carrying. While the NIOSH equation establishes a recommended weight limit for lifting, the Snook tables provide guidance as to the proportion of the population that should be able to do the tasks as a regular part of daily work. KIM is a screening *method* for assessing physical workload during manual handling operations. There are forms with instructions and partially integrated calculation aids for the types of physical workload Lifting/Holding/Carrying, Pushing/Pulling, and Manual work processes. CBSI is also a screening instrument, utilised to recognise suitable cases for further assessment in the risk management of workers. Early identification of such cases could additionally prevent injuries, absenteeism, and decreased productivity through appropriate interventions.

MSDs Reducting with an Innovative Approach of Professional Gestures - Collaborating 15 Years with Faurecia Group

Eric Caulier[1][✉] and Georgette Methens-Renard[2]

[1] Université Côte D'Azur LAPCOS (EA 7278), Nice, France
[2] Charleroi, Belgium

1 Introduction

In a lot of our physical activities, our motions are usually initiated from the upper limbs and superficial muscles. Therefore, the generated power is scattered, erratic and very uptight. The perception of movement and the awareness of trajectories are weak. This way of moving causes MSDs that mainly affect the upper limbs. A lot of models for investigation of works disorders gives guidelines for upper limbs This is an important in the beginning but for us ergonomists, we have to make a real breakthrought.

2 The Master Is Pointing to the Moon Ans the Student Is Looking at the Finger

Efficiency based on « sports models » is high cost. Efficiency inspired by the "active principles" of Chinese Martial arts that we extracted is low cost.

Our many stays in China have led us to discover other ways to use the body through the practice of Chinese internal arts. In these arts, the movement comes from the internal support within the whole body. The gesture is smooth and the flow is continuous. Relaxing enhances the individual's perception of their own body map and the awareness of trajectories in accordance to the environment. The practitioner becomes again the main actor of his health (Delassus: 2011). His attitudes and thoughts, like his gestures become more fluid. This leads him to live more and more often flow experiences within which all parts of the body work in perfect synergy in a state of amplified consciousness (Csikszentmihalyi 2004).

However, in high performance sport, Philippe Fleurance (2009) studies dynamic approaches, embodied, located, enactive. It takes into account as much the subjectivity of the operator as the environment in the evolving one. The researcher highlights how the best athletes use their routines and perceptual abilities to create economic and effective behaviors. These testify to an extraordinary production of extraordinary action adapted to the dynamic situation (ibid., 7–27).

© Springer Nature Switzerland AG 2019
S. Bagnara et al. (Eds.): IEA 2018, AISC 820, pp. 12–18, 2019.
https://doi.org/10.1007/978-3-319-96083-8_3

Philippe Fleurance concludes that « *the best cognitive system is then a unified system that unites "body-mind-world" (ibid., 14).* The separation between theory and practice, action and knowledge is therefore obsolete. The expert is the one who is able to perfect and master the changes.

High performance is not achieved by strictly applying a plan of action, but opening us up to a certain number of possibilities and adapting ourselves to the contingencies. These writings renews the vision and understanding of expertise not only in sport, but also in human activity. The main characteristics of Taoism are linked to it: economy and efficiency, no distinction between theory and practice, support for change, use of the potential of the situation.

Philippe Fleurance's cognitive system strangely recalls the Taoist body of Kristofer Schipper (1997, 18).

3   Active Principles and Accessible Protocols

In China I learned many forms (sequences of codified movements) that are as much movement and gestures performed in a complete perceptual consciousness. However, over time, it is the clarity and power of each of his movements that has impressed me the most.

The sharpness and precision allowed me to implement a global and elastic force. This mastery rhymes with a keen awareness of my body and my environment.

With a particular Chinese art, I learned the transmission via a remarkable pedagogy, sober and structured. This is part of the method I developed during the 1990s, which I have been using for more than twenty years in different contexts. I have deepened and completed it at different levels: enhance of an additional series synthesizing the whole, lighting with the new paradigms (third included, systemic, enaction, biotenségrité, etc.) and new technologies (capture and analysis of the movement with 3D images) at the Numediart laboratory).

The recent arise of Motion Capture (MoCap) technologies provides new possibilities, but also new challenges in human motion analysis. Motion capture (MoCap) allows new prospects in many areas, including human computer interaction (HCI), entertainment, movement education (sport, music, etc.), and health movement analysis.

However, the exploration and analysis of a motiondatabase is a complex task, due to the high dimensionality of motion data, and the number of independent factors that can affect movements.

Secondly, in a data baset of any type of movement performance, a large number of factors can influence the way a performer moves. Factors are the variables, intra- or inter-individual, which can have an effect on movements. Inter-individual factors may be of various types, including social factors (e.g., culture, education), psychological factors (e.g., personality, emotions, state of concentration), physiological factors (e.g., gender, age, morphology, force, suppleness), or psychophysiological factors (e.g., handedness, motor skills). On the other hand, intra-individual factors are related to the performance, independently of the performer, such as the type of dance, or the purpose of a particular exercise. Many specific researches on these factors have been conducted in different contexts, ...

Many experiences of practice (experimentation in my own body), research (questioning, definition, modeling) and transmission in other domains (transfer of principles in understandable forms) gradually led me to highlight categories / active principles (flexibility, breathing, imagination, etc.) then to qualify them (soft flexibility, inverted breathing, creative imagination). These categories allowed me to grasp the fundamental contours and dynamics of the internal arts. The linking of these categories has led to the emergence of an integrative model highlighting the interactions between these active principles. Thus, the judicious dosage of tone and relaxation in the various parts of the body promotes a soft flexibility facilitating a deep breathing. These four qualities increase the perception which allows coordination more and more fine.

My approach consists of using a few simple movements in the vast repertoires of the internal arts so as to make perceptible their gestural principles (use of the whole body in a fluid and elastic way), relational (exchanges without opposition or fusion) and attentional (availability, vigilance).

The tutorials, in the practical gestures is first perform the movements in slow motion. Experienced craftsmen, knowing the importance of relaxation for the precision of the gesture, have always cultivated this clever mix of strength and softness. As you get older, the increase in perception and consciousness compensates for the decrease in muscle tone. One can think that, consciously or unconsciously, intentionally or involuntarily, the Chinese internal arts have developed from a fine observation of these phenomena. I turned these evidences into guiding principles.

Teaching these movement arts in different contexts has allowed us to highlight their "active principles". Collaborating with various university departments, as well as our own research has led us to develop a scientific approach to their operational modes.

The various «tools» of our method have been developed and refined successfully because of the Faurecia group. Faurecia is now a global leader in its three areas of business: automotive seating, interior systems and emissions control technologies. Faurecia is the world's number one supplier of seat frames and mechanisms, emissions control technologies and vehicle interiors.

At the Faurecia group, we have achieved a new approach to motion and postures in the work environment by moving softly and placing the body in front of the work station. Opening slightly the workstation and implementing into each workers to abilities to place himself more ergonomically.

The production lines are hard on the workers' bodies. Postures and repetitive gestures lead to blocked backs, tendonitis and other so-called musculoskeletal disorders, with their attendant pain, work stoppages and disabilities by inadequate posture, poor appreciation of movements, automatic and repetitive gestures.

The usual ergonomic methods, as referenced in the charter of many companies are based on giving norms and standards and asking to apply them. The results are exerted by an external adhesion, an "external forcing". Stronger membership and longer-term results can be implemented with a different approach. The one where the operator is convinced in his flesh, in his articulations of the results, the one where he feels co-constructor, participating in his well-being at work, the one where he is a full actor body, mind, emotions..

4 Breakthrough

In our method, we used the principles of the internal arts to get the operators to think for themselves about what could improve their conditions and how to adapt, to develop the consciousness of their body so that they improve on their own their gestures and postures. We consider production standards as standardized forms What is a form if not a spatial matrix (the directions of movement for example) in interdependence of a temporal structure (sequence of movements)? All production standards are characterized by this definition.

We take the actors out of their workstations to re-instate them to perceive the direction of the movement, then put them back in situation on their machines. My learning of the internal arts has tremendously structured all the work around postures and gestures and all the "basic grammar" is found in the fundamentals of ergonomics. Here are some excerpts from semi-structured interviews

1. Roots: Standing, it is a question of regaining consciousness of its support on the ground, to use more the legs which, at the beginning, are like pieces of dry wood; in the movement, it will be necessary to learn to transmit the thrust of the ground in the legs.
2. Axis: Tuck your chest slightly and stretch your back. The knees are unlocked to allow the back to bend its natural curves.
3. Space. In the matrix posture (unitary posture), we define a static space. What do the classics of the internal arts tell us? Lower your shoulders and let your elbows fall. This is exactly what the RULA evaluation method shows us. Let's go back to the biomechanical design that verifies the classics: I can only open and close my arms ergonomically when my shoulder blades rest on my rib cage. The amplitudes allowed are the ergonomic amplitudes. At the level of the posture, the alignment, the stability, the relaxation, the equilibrium are then realized
4. Continuity in the movement: it is to connect the movements without interruption. Each stop in the movement is an energetic break, a break in the flow of the transmission of the efforts of the feet (roots up to the palm). The production standard is no longer a canvas where the man and the machine fight, it becomes "the standard form of the production movement in the factory". There is no more opposition Man/ Machine but agreement Man and Machine. For this agreement to be realized, the scenography of the ballet is set up: now, the operator moves to better position himself at the joints of the lower limbs to the upper limbs. This motor power is transmitted in a correct and correct positioning of the back. The trajectories of the efforts are made without "break", without point of crawling, nothing is in force, the movement develops. At the level of external movement, there is continuity, coordination, trajectories.
5. Development of the perception and the sensation of one's body: To recover the hearing of the messages emitted by the body and to modify if necessary the gesture so that the movements remain in the elastic limits: "Sometimes when I took the carpet and me Now I do not do it anymore (M. B) My A. thinks that there will be less muscular or bony pains if you work in this way, the person and for the management of the Faurecia group".

6. The development of presence and consciousness leaves less room for inevitable automatisms, to recruit in one's own internal resources and thus to better adapt to the particular situation and the changing context. "Sometimes so much that it's going fast, and after I forget... My brain tells me ... I have to do it, I have to come back in a more integrated way ... to the gesture you made at the beginning" (MB). As for Mr. S's experience: "We do things a little robotically, we do the same things every second, every minute, every hour, every day, we work really stupidly without consequence of the acts that can happen to our body". We discovered it, we say to ourselves: "How is it that I get hurt, why can not I bend down? ... we say now, it would have been better, I would never have hurt myself this way. It would have been easier, we find the work, easy, in its own limits ... By doing some things we realize that it is thin, it goes alone!"

But how to translate these principles and make standard forms of factory production? So that the production standard becomes simple and flexible, some elements still remain to add:

1. The intention is developed ... by the techniques of the visualization, in order to manifest the spirit through the form of the body cad, the glance makes a preprogramme of the gesture to be able to feel the trajectories.
2. The unity of the person: a point that activates the whole body. Inside and outside are inextricably linked: intention, strength, form, qi and spirit (another way of talking about qi-energy when the gesture gets a global approach): body, psyche, thought, environment. "With the release gun it was very hard for the wrist". Mr C., gaplealer is grafted today to work with the whole body: "It is true, for the application of the release agent, I have already completely changed compared to before ... Now these are great movements" (It shows a large horizontal sweep in the space where the arm, the trunk, the hips rotate together from right to left and vice versa). It is no longer the small movements of the wrist.

For example, let's see how a small operator will develop continuity in his movement when he takes a cumbersome piece of a table must put it in a cart whose floor is 50 cm from the ground? How to make sure that these men and women who work do not become unemployable.

Mr L. believes as many of his colleagues that we must say "we must think and after thinking, it must come from legs to their hands."

It's not just about thinking about doing it, you still have to think about it, and even force yourself to think about it.

To return to the post after the work in the room, we found several breakings in his work sequence. The first being when taking the piece and the second when filing. Each time the operator took the piece extending his back backward (angle $> -10°$), the second, when throwing the coin onto the cart he pushed his load from his lumbar axis! So we have, without load, started his standard factory ballet: 2 steps back, while keeping the pelvis relaxed and according to his step we have adjusted the distance of the cart, the last step developing the impulse that spreads from the feet up 'in the hands. The pushing gave a slight acceleration to the carpet so that it naturally settles on the carriage. We did the same sequence back in charge to arrive at a continuity of the flow of work.

Mr L., who at first was not convinced, transmits consciously, during each work cycle, a kinetic energy with a slight acceleration in the carpet. Mr L. thinks himself and creates himself in his movement, in his intention. To keep this intention awake, the right movement ends up coming. The brakes are actually habits: "it's been 34 years that we do bad things, ... it's not obvious" And the results in all this? Mr L. acknowledges that he did not believe it at first. Today he says: "I find that I already have less pain in the joints ... and even the back. All this is true that it makes a difference".

For all, it is the same story of constant attention: "I pay more attention to my everyday gestures ... Yes I am more attentive. Although it is always the same gesture, the brain is still awake (Mr B.)". After having followed operators on automobile assembly lines, I am today asked by the world of the company to support managers in self management. This project brings me close together different theoretical models combined with my approach of the internal arts to favor:

- intuition as an expanded mode of knowledge (Petitmengin 2001)
- creativity through the transition to creative action (Caulier 2012)
- maxi-performances (Gardfield 1987)
- integrative health (Wayne 2013)
- Self-realization through the realization of a work (Caulier 2012).

It seems that we are dealing with various variations of the same approach: the entry into the state of flow. This state makes it possible to anticipate (intuition), to be creative (actions perfectly adapted to the context), to have the maximum efficiency with the minimum of effort (ease, evidence), to mobilize the whole of its resources (well-being, fulfillment) and to be oneself (feeling of fullness).

These different testimonies highlight undeniable links between the body and consciousness. They show that it is possible to increase or even modify the consciousness of one's body. These changes, in the cases, took place through a practice and extend to the outside. This gestural practice is a body technique that works on (bio) mechanical movements but also on breathing, perception, active imagination and attention. Relationships with others and interactions with the environment are important. The technique used is a specific approach to the Chinese internal arts. It focuses on the active imagination, the interface between the body and consciousness.

5 Conclusion

These modes of operation introduce, in different fields of interesting perspectives for the industry. They emphasize the importance of upstream collaboration, with the training of other key players such as R & D center development engineers, process experts and even buyers, to go even further and work in the field with respect for the human being and his potential.

18   E. Caulier and G. Methens-Renard

References

Billeter JF (2002) Leçons sur Tchouang-Tseu. Allia, Paris
Billeter JF (2012) Un paradigme. Allia, Paris
Bonardel F (1993) Philosophie de l'alchimie. PUF, Paris
Caulier E (1998) Voyage au coeur du taijiquan. Trédaniel, Paris
Caulier E (2010) Comprendre le taijiquan - Tome 1. EME, Fernelmont
Caulier E (2012) Contribution interculturelle à l'étude de modélisations de l'agir créatif contemporain. ANRT, Lille
Caulier, E (2013) Pratique de soi, empathie et approche compréhensive par le Taijiquan. Dans Bernard Andrieu et al., L'expérience corporelle. AFRAPS, Clapiers, pp 125–134
Caulier E (2015) Du geste formel à la gestuelle habitée: la voie du taijiquan. Rech Educ 13:59–71
Caulier, E (2016a) Création d'une ergonomie énactive. Dans Nicolas Burel (dir.), Corps et méthodologie - Corps vivant, corps vécu, corps décrit. L'Harmattan, Paris, pp 145–156
Caulier E (2016b) Prendre conscience du corps. UPPR, Paris
Caulier, E (2016c). Taijiquan: une voie vers la sérénité. M@gma, 14/2
Caulier E (2017) Approches traditionnelles et scientifiques du tai chi chuan. Le Livre en papier, Strépy-Bracquenies
Csikszentmihalyi M (2004) Vivre - la psychologie du bonheur. Robert Laffont, Paris
Daumal R (1972) L'évidence absurde - Écrits et notes 1. Gallimard, Paris
Dujacquier, I (2016) Être débutante en taijiquan, c'est … Espace Taiji, 101, 16
Gardfield CA (1987) Maxi-performance. Le Jour, Montréal
Hobsbawm EJ, Ranger, T (2012) L'invention de la tradition. Amsterdam, Paris
Fleurance P (2009) Introduction - Je vois la balle avec mes mains. Intellectica 52:7–27
Naëije M (2010) De l'interaction du xingyiquan et du bagua zhang sur ma pratique du taijiquan. Espace Taiji 78:15
Petitmengin C (2001) L'expérience intuitive. L'Harmattan, Paris
Schipper K (1997) Le corps taoïste. Fayard, Paris
Wayne PM (2013) The Harvard Medical School Guide to Tai Chi - 12 weeks to a healthy body, strong heart et sharp mind. Shambala, Boston

Liens

Blog: http://www.eric-caulier.be/
Site: http://www.taijiquan.be/
Page: http://ericcaulier.strikingly.com/
Numédiart: https://numediart.org/axes/
Ebook gratuit: http://www.taijiquan.be/ebook-gratuit-taichi/

Systematic Reviews as Evidence-Base for Dutch Guidelines to Assess Musculoskeletal Disorders as Occupational Disease: Examples of Shoulder, Knee and Low Back Disorders

Henk F. van der Molen$^{(\boxtimes)}$, Monique H. W. Frings-Dresen, and P. Paul F. M. Kuijer

Department: Coronel Institute of Occupational Health, Amsterdam Public Health Research Institute, Academic Medical Center, University of Amsterdam, PO Box 22700, 1100 DE Amsterdam, The Netherlands
h.f.vandermolen@amc.nl

Abstract. In The Netherlands, occupational physicians are obligated to report occupational diseases (ODs) to the Netherlands Center for Occupational Diseases (NCOD). Evidence-based occupational disease guidelines have been developed for musculoskeletal disorders. At the moment, the NCOD has 23 guidelines for work-related musculoskeletal disorders. The aim of this session is to explain the development of these criteria for the work-related assessment of subacromial pain syndrome, knee osteoarthritis and non-specific low back pain in the Netherlands. To enhance the application of these guidelines for ergonomic professionals, the exposure-criteria for the same job demands are categorized for instance in exerted hand force, movement, posture and vibration. When a job demand exceeds the exposure criteria, evidence is available that a worker has an increased risk of developing an occupational disease. For example, hand-arm elevation ($\geq 60°$ 1 h/day), hand force ($\geq 10\%$ maximum voluntary force exertion) and hand-arm vibration (>2.5 m/s^2 8 h/day) are examples of exposure criteria for the work-relatedness of subacromial pain syndrome. International comparison of OD exposure criteria and knowledge about corresponding effective preventive measures could be enhanced by scientific collaboration in defining evidence-based work-related diagnostic criteria and guidelines. The knowledge could be anchored in instruments and tools for ergonomic professionals.

Keywords: Musculoskeletal disorders · Work-relatedness · Exposure criteria Systematic reviews · Guidelines

1 Introduction

In the Netherlands, occupational physicians (OPs) are obligated to report occupational diseases (ODs) to the Netherlands Center for Occupational Diseases (NCOD). Evidence-based occupational disease guidelines [1] have been developed for musculoskeletal disorders (MSDs) by the NCOD. In addition to their instrumental assistance to assess ODs, these guidelines contain evidence-based exposure criteria for ODs that

© Springer Nature Switzerland AG 2019
S. Bagnara et al. (Eds.): IEA 2018, AISC 820, pp. 19–21, 2019.
https://doi.org/10.1007/978-3-319-96083-8_4

might also be useful for ergonomists when assessing high job demands and prioritizing preventive measures. The aim is to explain the evidence-based development of criteria to assess the work-relatedness of subacromial pain syndrome (SAPS), knee osteoarthritis and non-specific low back pain in the Netherlands.

2   Methods

In the Netherlands, an OD is defined as a disease for which the work-related fraction is >50%. OPs use a structured anamnesis and other information obtained through clinical examination and tests and exposure assessments to decide whether or not for the patient at hand an OD is present and should be reported. Exposure assessments of risk factors in the workplaces can be based upon self-reports of workers, the obligatory Risk Assessment and Evaluation of the employer, or through exposure assessment by ergonomists or other health professionals like occupational hygienists. Also the work history is taken into account when deciding about the presence of an occupational disease.

A guideline is instrumental for the assessment of the work-relatedness and consists of two parts: a clinical case definition of the disease and disease-specific exposure criteria. Exposure criteria are based on systematic reviews of the scientific literature, preferably including a meta-analyses. Exposure criteria are, if possible, defined in terms of the intensity, frequency and/or duration of specific occupational demands.

3   Results

At the moment, the NCOD has 23 guidelines for work-related MSDs. Examples are the guidelines for the work-relatedness of SAPS, knee osteoarthritis and non-specific low back pain. To enhance the application of these guidelines for ergonomic professionals, the exposure-criteria for the same job demands are grouped together, for instance in exerted hand force, movement, posture and vibration. When a job demand exceeds the exposure criteria, sufficient evidence is available that a worker has an increased risk of developing an occupational disease.

For SAPS [2], arm-hand elevation ($\geq 60°$ 1 h/day), hand force ($\geq 10\%$ maximum voluntary force exertion) and hand-arm vibration (>2.5 m/s^2 8 h/day) are examples of minimum exposure criteria.

For knee osteoarthritis [3], kneeling or squatting (>1 h/day ≥ 1 year), lifting of loads (≥ 5 kg ≥ 10 times/week ≥ 1 year), jumping (>15 times/day ≥ 10 years), climbing (> 30 stairs/day $\geq$ 10 years) are examples of minimum exposure criteria.

For non-specific low back pain [4], lifting or carrying of loads (>5 kg ≥ 2 times per minute 2 h/day or >25 kg ≥ 1 time per day), bending or twisting of the trunk 20° >2 h/day) and whole-body vibration (>0.5 m/s^2 per day) are examples of minimum exposure criteria.

4 Conclusions

International comparison of OD criteria and knowledge about corresponding effective preventive measures could be enhanced by scientific collaboration in defining evidence-based work-related diagnostic criteria and guidelines. The knowledge could be anchored in instruments and tools for ergonomic professionals.

References

1. http://www.occupationaldiseases.nl
2. Van der Molen HF, Foresti C, Daams JG, Frings-Dresen MHW, Kuijer PPFM (2017) Work-related risk factors for specific shoulder disorders: a systematic review and meta-analysis. Occu Environ Med 74:745–755. https://doi.org/10.1136/oemed-2017-104339
3. Verbeek J, Mischke C, Robinson R, Ijaz S, Kuijer P, Kievit A, Anneli Ojajärvi A, Neuvonen K (2018) Occupational exposure to knee loading and the risk of osteoarthritis of the knee: a systematic review and a dose-response meta-analysis. Saf Health Work 8:130–142
4. Miedema HS, Van der Molen HF, Kuijer PP, Koes BW, Burdorf A (2014) Incidence of low back pain related occupational diseases in the Netherlands. Eur J Pain 18(6):873–882

The Effect of Short Time Computer Work on Muscle Oxygenation in Presence of Delayed Onset Muscle Soreness

Afshin Samani[(✉)] and Ryan Godsk Larsen

Sport Sciences, Department of Health Science and Technology,
Aalborg University, Aalborg, Denmark
afsamani@hst.aau.dk

Abstract. We aimed at investigating the effect of a short time low load repetitive task on the local muscle oxygenation kinetics in presence of delayed onset muscle soreness. Computer work was investigated as a model of low load repetitive task. Nine healthy male subjects participated in an experimental protocol consisting of a rest period and two blocks, each including two maximum voluntary contractions (MVC) of isometric bilateral shoulder elevation and a computer work session with 2 or 5 min duration between the MVCs in each block. Then a set of unaccustomed eccentric exercise (ECC) of shoulder elevation was implemented to induce delayed onset muscle soreness (DOMS) in the trapezius muscle. Identical experimental blocks were performed immediately and 24 h after ECC. Local tissue saturation index (TSI) was continuously recorded over upper trapezius throughout the experiment. TSI parameters such as mean TSI at rest, during computer work as well as TSI drop, recovery and their descending and ascending slopes were computed following the MVCs. TSI drop and recovery and their corresponding descending and ascending slopes were reduced immediately after ECC ($p < 0.05$). The computer work caused an increase in TSI at rest prior to the MVCs ($p < 0.05$) and elevated oxygen consumption ($p < 0.05$). The observed changes in oxygenation kinetics seemed to appear after a strenuous exercise and they do not seem to be affected by the presence of DOMS. However, computer work as a model of low load repetitive task caused increased oxygen consumption in a following bout of muscle contraction.

Keywords: Computer work · Musculoskeletal disorders
Near infrared spectroscopy

1 Introduction

Musculoskeletal disorders (MSD) are a major global health problem and are known to be multifactorial phenomena [1]. Our understanding of mechanisms underlying development of MSD particularly in jobs with low workload intensity with repetitive patterns, such as computer work is limited [2]. Repetitive low load muscular activity may result in alter blood supply and muscle oxidative metabolism [3].

© Springer Nature Switzerland AG 2019
S. Bagnara et al. (Eds.): IEA 2018, AISC 820, pp. 22–31, 2019.
https://doi.org/10.1007/978-3-319-96083-8_5

Near infrared spectroscopy (NIRS) is a non-invasive technique which enables the investigation of muscle oxidative metabolism locally [4], and thus provides a tool to investigate the link between the disturbance of muscle tissue oxygenation and the development of MSD. Notably, contrasting results have been reported when comparing muscle oxygenation between healthy subjects and patients with MSD. In healthy adults, a decrease in muscle oxygenation has been reported in forearm muscles during isometric submaximal contractions [5] and in trapezius during a repetitive task [6] and load bearing [7]. In contrast, studies in MSD patients have found no changes in oxygenation of trapezius and wrist extensor muscles during low load repetitive tasks and submaximal isometric contractions [8, 9]. Such discrepancies may be due to a bias in patient selection, differences in experimental setup, choice of NIRS variables and body regions [10].

As an endogenous muscle pain model, delayed onset muscle soreness (DOMS) has been widely used to mimic clinical musculoskeletal pain [11]. Thus, investigating the disturbances in muscle oxygenation in presence of DOMS may improve our understanding of the underlying mechanism contributing to development of pain. The literature regarding the effect of DOMS on oxygenation kinetics is equivocal [12, 13]. Particularly, our knowledge about the effect of DOMS on the kinetics of oxygenation in upper extremity muscles during low load repetitive work with relatively high precision demand such as computer work is lacking.

Prolonged duration of computer work has been suggested as a risk factor for development of MSD in the neck and shoulder area [14]. Additionally, sustained activity of trapezius muscle over 4 min periods in half of a working day has been found to be associated with the development of neck and shoulder pain [15]. Therefore, it is interesting to investigate the short time effects of computer work on muscle oxygenation below and above the suggested time threshold (4 min) and its interaction with the presence of DOMS. Due to structural damage and edema in eccentrically exercised muscle tissue [16], we hypothesized that DOMS would slow the kinetics of muscle oxygenation. Additionally, we hypothesized that the computer work over 4 min would intensify the effects of DOMS.

2 Materials and Methods

2.1 Subjects

Nine right-handed recreationally active males (24.2 (2.0) years; 181 (9) cm; Weight 75 (6) kg), with no history of neck- and shoulder disorders, volunteered to participate in the study. All subjects abstained from strenuous exercise three days prior to and during the trials. Informed written consent was obtained from each participant, prior to initiating the experiment and conducted in accordance with the Declaration of Helsinki. The study was approved by the local ethics committee of Northern Denmark Region (N-20120036).

2.2   Experimental Procedure

The study was performed over two consecutive days. Identical procedures were performed at three stages, i.e., before, immediately after and 24 h after a standardized eccentric exercise (ECC; described below). The procedures consisted of (i) 2 min instructed rest where the subjects sat comfortably on a chair with their hands on their lap, feet on the ground, looking straight ahead, (ii) a 5 s maximum voluntary isometric contraction (MVC) of the trapezius muscle (MVC Pre), (iii) either a 2 or 5 min standardized computer mouse work session, (iv) right after step iii, step ii was repeated (MVC Post), (v) after at least 5 min rest, steps ii, iii and iv were repeated but with a different duration of standardized computer work session performed in step ii (the order of computer work duration was randomized across subjects but kept identical across the experimental stages for each subject). MVC trials were added to the protocol to increase the intramuscular pressure and disturb tissue perfusion such that the contraction-induced response of the vascular system could be investigated [17].

Maximum Voluntary Contractions. The subject gripped both sides of the seat pan while sitting on a comfortable chair, the elbows were fully extended, and the shoulders were in neutral position. A handheld dynamometer (Vernier, HD-BTA, Beaverton, OR) was placed under the subject grip to the seat pan to measure the exerted force during the MVC trial. Visual feedback of the exerted force and verbal encouragement were provided to the subjects.

Eccentric Exercise. The subjects performed the ECC for 30 min in a shoulder dynamometer [18]. The exercise intensity was set to 120% MVC measured under the shoulder dynamometer before the ECC trial started. The EEC consisted of 5 bouts of 10 repetitions (3 min of continuous activity), separated by a 3 min of rest. During each repetition, participants had to counteract the dynamometer vertical downward force as much and for as long as possible along the previously defined range of shoulder elevation [19].

Computer Work Sessions. The computer work-station was adapted individually according to guidelines (e.g. seat and desk height, full arm support) [20]. The task was composed of a cyclic mouse clicking on 6 markers to draw a graph. The time allowed for drawing a graph was 8 s and this procedure repeated with a randomly generated graph each time [19]. Before starting the preparation of the experimental setup and data recordings, the subject was familiarized with the experimental setup and the computer work.

2.3   Data Recordings

Sensory Motor Assessments. The subjects were asked to score the level of the perceived pain in the shoulder area on a visual analog scale (VAS) prior to all the three stages (i.e. before, immediately after and 24 h after the ECC). VAS was constructed on a scale from 0 'no pain' to 10 'worst imaginable pain'. Maximum force produced was measured by the handheld dynamometer during the MVC trials. The subjects' performance during computer work was evaluated by multiplying the average rate of graph drawings in the allocated time (i.e., 8 s) and the ratio of correct clicks to total number of clicks (higher number reflecting better performance).

Near Infrared Spectroscopy. Local tissue oxygen saturation in the trapezius muscle was monitored using near-infrared spectroscopy (NIRS) with a continuous wave technique (Oxymon Mk III, Artinis Medical Systems BV, Netherlands). The NIRS probe setup consisted of one receiver and three transmitters placed in a row with 3.5 cm the mean of inter-optode distances. The transmitter optode transmits NIRS beams at 860 nm and 765 nm which are mostly sensitive to oxy- and de-oxyhemoglobin (O2Hb and HHb), respectively [21]. The tissue saturation index (TSI) (%) was calculated as the ratio of O2Hb to O2Hb + HHb. The O2Hb, HHb and TSI values have been shown to be valid estimators of changes in tissue deoxygenation status representing regional imbalances between O2 delivery and O2 utilization in the tissue under the probe [22].

Before initiating the trials, the NIRS apparatus was calibrated and the TSI probe was placed over the descending part of the trapezius muscle centered on the lateral third of distance between 7th cervical vertebrae and the acromion. A wax marker was used to ensure an identical placement of the probe across the experiment stages. The TSI was continuously measured in all the aforementioned experimental steps from i through v. Prior to mounting the NIRS measurement system, ultrasound imaging (GE, Logiq S7 Expert, BCF Technology) was used to obtain an estimate of subcutaneous fat and muscle thickness at the site of NIRS measurement.

2.4   Data Analysis

From the TSI signal, we chose a 65 s window starting from 10 s before the onset of MVC trial. The choice of the window length was based on previous results [23] showing an approximately 30% drop in TSI with an approximately 0.5% per second recovery rate in tissue oxygenation of trapezius following a 70% MVC. The lowest point of the TSI in this time window was determined and the signal in the window was divided into a descending and an ascending part. The descending part of the window was fitted to a trapezoidal shape function starting with a steady level followed by a linear decay. A non-linear least square routine was applied to derive an optimal level and descending slope for the trapezoidal function [24]. A similar procedure was performed to derive the level and ascending slope of the TSI signal. The cross-correlation coefficient (CC) between fitted trapezoidal function and the experimental values was calculated and if the CC was below 0.75, that window of TSI signals was removed from the rest of the analysis (two trials from one subject and one trial from three subjects). The fitting quality was also visually inspected. The initial steady level of TSI, the amount of TSI drop and recovery from the steady levels to the corresponding value of TSI at the lowest point and the slope of the drop and recovery were calculated. TSI descending and ascending slopes would reflect the rate of oxygen usage and recovery of oxygen saturation, respectively [12]. This procedure was performed for the MVC trial before the computer work (MVC pre) and after the computer work (MVC post) during all three stages of the experiment. The mean value of TSI was also computed during the instructed rest and during the computer work sessions.

2.5 Statistical Analysis

A linear mixed model (LMM) was applied to the outcome measures, i.e., VAS, exerted force during the MVC trials, computer work performance, TSI level prior to the MVC trial, the TSI drop and recovery and their corresponding descending and ascending slopes and TSI mean during computer work and rest. The effect of delay onset muscle soreness and its interaction with the computer work on the outcomes were investigated and compared. Experimental stages (before, immediately after and 24 h), computer work duration (2 min 5 min, rest (only for the TSI mean)), contraction order (MVC pre and MVC post) were introduced as fixed factors in the LMM. Additionally, a repeated factor associated with the introduced factors was added to the model. When a significant effect was observed, a Bonferroni adjustment was performed for a pairwise comparison. We reported an index of effect size Ω^2 which indicates the partial variance described by a factor with respect to the full LMM model [25]. In all tests, P < 0.05 was considered significant. Mean values (standard deviation) were reported.

3 Results

3.1 Sensory Motor Assessments

Experimental stages had a significant effect on reported VAS ($F_{2,16}$ = 9.7, p = 0.002, Ω^2 = 0.36), MVC force ($F_{2,15.1}$ = 10.4, p = 0.001, Ω^2 = 0.3) and computer work performance ($F_{2,14.1}$ = 26.7, p < 0.001, Ω^2 = 0.21). VAS was greater 24 h after ECC compared with VAS before ECC. VAS increased from 0 (0) before, 1.1 (1.2) immediately after and 2.5 (2.2) 24 h after the ECC. For all MVC trials, the MVC force was greater 24 h after ECC compared with the MVC immediately after ECC. Table 1 shows the obtained force level in MVC trials. Computer work performance score was lowest before ECC and greatest 24 h after the ECC. The performance score increased from 1.03 (0.18) before, 1.19 (0.14) immediately after and 1.26 (0.10) 24 h after the ECC.

3.2 The Local Muscle Oxygenation

The thickness of subcutaneous fat and muscle tissue at the site of NIRS measurement were 4.6 (1.9) and 20.4 (3.4) mm, respectively. The contraction order ($F_{1,46.2}$ = 78.6, p < 0.001, Ω^2 = 0.19) had a significant effect on the TSI level prior to the MVC trials. The TSI level prior to the MVC after computer work was greater than TSI level prior to the MVC before computer work. Table 1 includes the mean and standard deviation of TSI parameters. The contraction order ($F_{1,50.2}$ = 27.1, p < 0.001, Ω^2 = 0.09) and the interaction of duration × the experimental stages ($F_{2,24.4}$ = 7.1, p = 0.004, Ω^2 = 0.12) had significant effects on the TSI drop. The TSI drop during the MVC was greater after computer work compared with before computer work. With the 2 min computer work, the TSI drop was lower immediately after the ECC compared with the drop before and 24 h after the ECC. The experimental stages ($F_{2,41.5}$ = 16.7, p < 0.001, Ω^2 = 0.15) and the contraction order ($F_{2,45.7}$ = 4.7, p = 0.03, Ω^2 = 0.01) had significant effects on the TSI recovery. The TSI recovery was lower immediately after the ECC compared with

Table 1. The mean and standard deviation of exerted force during maximum voluntary contractions (MVC) and derived parameters from the tissue saturation index (TSI). * and $ indicate significant difference between the experimental stage immediately after the eccentric exercise (ECC) on the one hand and the experimental stage before and 24 h after the ECC, respectively. # shows a significant difference between the experimental stage immediately after and 24 h after the ECC.

Outcomes	Experimental stages					
	Before		Immediately after		24 h after	
	Pre	Post	Pre	Post	Pre	Post
MVC force (N)	484 (168)	475 (163)	441 (142)	443 (154)	510 (154)#	521 (147)#
TSI level prior the MVC (%)	61.9 (8.2)	66.8 (8.9)	62.5 (11.3)	66.2 (10.3)	59.5 (7.0)	64.9 (5.9)
TSI drop (%)	23.7 (16.2)	24.5 (14.1)	17.1 (11.9)	22.5 (12.0)	23.3 (15.5)	25.0 (15.9)
TSI recovery (%)	21.6 (15.2)	17.5 (12.1)	13.3 (10.6)*$	12.6 (9.8)*$	21.1 (15.3)	18.9 (14.7)
Descending slope (%/s)	3.3 (3.0)	3.2 (2.6)	2.5 (2.5)*$	2.8 (2.3)*$	3.3 (2.6)	3.8 (2.9)
Ascending slope (%/s)	1.8 (1.8)	1.8 (1.9)	1.5 (1.5)$	1.7 (1.8)$	2.2 (1.5)	1.9 (1.3)

before and 24 h after the ECC. The recovery was greater before the computer work compared with the recovery after the computer work.

The experimental stages ($F_{2,33.9} = 9.7$, $p < 0.001$, $\Omega^2 = 0.05$) and the contraction order ($F_{1,52.8} = 6.7$, $p = 0.01$, $\Omega^2 = 0.01$) had significant effects on the descending slope of TSI. The descending slope was smaller immediately after the ECC compared with before and 24 h after the ECC. The descending slope after the computer work was greater than the descending slope prior to the computer work. The experimental stages ($F_{2,21.6} = 7.5$, $p = 0.003$, $\Omega^2 = 0.02$) had a significant effect on the ascending slope of TSI. The ascending slope was smaller immediately after the ECC compared with 24 h after the ECC. TSI mean during computer work did not change significantly compared to the rest period. Table 2 shows the mean and standard deviation of TSI mean during the computer work sessions.

Table 2. The mean and standard deviation of the tissue saturation index (TSI) across subjects during the computer work sessions.

TSI mean (%)	Experimental stages		
	Before	Immediately after	24 h after
Rest	60.6 (6.7)	65.7 (11.5)	60.9 (8.3)
2 min	65.9 (7.9)	67.5 (9.8)	64.5 (5.7)
5 min	65.7 (11.5)	65.3 (9.5)	63.5 (6.0)

4 Discussion

The TSI drop and recovery and their corresponding descending and ascending slopes were reduced immediately after ECC. The reduced drop in TSI may partly be a consequence of a decrease in the exerted force during the MVC trials, while slower recovery of TSI suggests impaired re-oxygenation of the muscle tissue following ECC. The computer work caused an increase in TSI level prior to the MVC, most likely due to exercise induced hyperemia. The computer work also elevated oxygen consumption, reflected by the greater drop in TSI after computer work compared with the drop before computer work. The duration of computer work did not result in significant different TSI responses during and after the MVC, except that TSI drop was lower immediately after the ECC compared with 24 h after and before the ECC with 2 min computer work only.

The increased VAS 24 h after the ECC may indicate some typical structural changes in the muscle or connective tissue as similar levels of VAS have been reported by [26] using a very similar ECC protocol with somewhat lower exercise intensity compared to our protocol. Despite significant sensory-motor changes in presence of DOMS, previous studies, in line with our results, have also not found a significant change in the exerted force during MVC trials [27].

Our results are in accordance to [28] as no alteration in the NIRS parameters in presence of DOMS was observed. However, as opposed to Walsh et al. (2001) [28], who have investigated the effect of DOMS on muscle oxygenation in vastus lateralis, we focused our study on oxygenation kinetics in the trapezius muscle. It is expected that ECC may damage slow twitch muscle fibers to a lesser extent compared with fast twitch fibers [29] and type II fibers constitute the major part of vastus lateralis in males [30], whereas the trapezius has a larger proportion of type I fibers [31].

In this study, DOMS did not seem to have an effect on the TSI parameters at rest which might be due to a different acute immunological response to DOMS compared with the conditions of chronic MSD [16]. Ahmadi et al. (2008) [12] have reported TSI levels at rest and during isometric contractions of the knee extensors before and after exercise induced muscle damage. Similar to our findings, they have not reported any changes in TSI at rest across the experimental stages. However, in contrast to our study, they reported faster oxygenation kinetics in presence of DOMS. Apart from a different site of recording and constant level of force across the experimental stages (discussed below), the subjects in Ahmadi et al. (2008) [12] were relatively sedentary, and the ECC protocol may therefore have had a larger impact on the muscle tissue in their participants. Interestingly, the subjects in Walsh et al. (2001) [28] consisted of recreationally active individuals, similar to our subject group, and they reported no changes in TSI parameters in presence of DOMS.

The reduced drop and recovery in TSI and their corresponding slopes immediately after the ECC is partly explained by the reduced level of exerted force during MVC trials [24]. However, eccentric exercise has been suggested to impair oxygen extraction and reduce microvascular function [32], which may also contribute to altered kinetics of muscle oxygenation in our study. Ahmadi et al. (2008) [12] kept a constant submaximal force level across the experimental stages, which may affect the reliability of the NIRS parameters at lower relative force levels [23]. Particularly, the NIRS parameters have been recommended to be assessed at maximal force level in clinical applications [33]. Therefore, the MVC trials constitute a crucial part of the experimental design enabling the assessment of the effect of computer work on tissue oxygenation kinetics with and without presence of DOMS.

The TSI level prior to the MVC trials was greater after the computer work compared with TSI level prior to the MVC trials before the computer work. This is probably due to increased muscle perfusion which is low at rest but gradually increases with exercise intensity [34]. As the muscle contraction level during the computer work is quite low, below 10% of maximum capacity [35], the intramuscular pressure cannot disturb the blood flow [17]. However, the drop in TSI was greater after the computer work, suggesting that short term computer work elevates the oxygen consumption in the trapezius muscle during the MVC trials.

In contrast to our hypothesis the duration of computer work did not result in significant changes in the TSI parameter. However, TSI drop was lower immediately after the ECC compared with 24 h after and before the ECC only with 2 min computer work. This may suggest that the abovementioned increased oxygen consumption after computer work may be intensified with the accumulated effect of 5 min computer work.

The subjects' performance during computer work increased across the experimental stages, which may be due to a learning effect. However, in a similar experimental setup, the performance did not change significantly when comparing computer work performance under experimental pain condition and baseline on the same day [36]. Along the timeline of our experiment, the subjects may have been mentally motivated to supersede their previous performance.

We conclude that despite the lack of DOMS effect on the TSI parameters, short time low intensity repetitive work such as computer work can affect the kinetics of muscle tissue oxygenation. Notably, the changes in oxygenation kinetics appear immediately after ECC and not 24 h post-exercise. This may suggest that these changes are due to the exercise bout itself and not due to structural changes in muscle tissue after eccentric exercise. Further, studies are warranted to specifically investigate the interaction of DOMS and prolonged low load repetitive tasks.

References

1. Bernard BP (1997) Musculoskeletal disorders and workplace factors, Second Edn, vol 1, pp 97–141. NIOSH Publication, US Department of Health and Human Services, Cincinnati, OH
2. Perrey S, Thedon T, Bringard A (2010) Application of near-infrared spectroscopy in preventing work-related musculoskeletal disorders: brief review. Int J Ind Ergonomics 40:180–184
3. Visser B, van Dieën JH (2006) Pathophysiology of upper extremity muscle disorders. J Electromyogr Kinesiol 16:1–16

4. Ferrari M, Muthalib M, Quaresima V (2011) The use of near-infrared spectroscopy in understanding skeletal muscle physiology: recent developments. Philos Trans A Math Phys Eng Sci 369:4577–4590

5. Brunnekreef JJ, Oosterhof J, Thijssen DH, Colier WN, Van Uden CJ (2006) Forearm blood flow and oxygen consumption in patients with bilateral repetitive strain injury measured by near-infrared spectroscopy. Clin Physiol Funct Imaging 26:178–184

6. Sjøgaard G, Rosendal L, Kristiansen J, Blangsted AK, Skotte J, Larsson B, Gerdle B, Saltin B, Søgaard K (2010) Muscle oxygenation and glycolysis in females with trapezius myalgia during stress and repetitive work using microdialysis and NIRS. Eur J Appl Physiol 108:657–669

7. Mao CP, Macias BR, Hargens AR (2015) Shoulder skin and muscle hemodynamics during backpack carriage. Appl Ergon 51:80–84

8. Elcadi GH, Forsman M, Aasa U, Fahlstrom M, Crenshaw AG (2013) Shoulder and forearm oxygenation and myoelectric activity in patients with work-related muscle pain and healthy subjects. Eur J Appl Physiol 113:1103–1115

9. Flodgren GM, Crenshaw AG, Hellstrom F, Fahlstrom M (2010) Combining microdialysis and near-infrared spectroscopy for studying effects of low-load repetitive work on the intramuscular chemistry in trapezius myalgia. J Biomed Biotechnol 2010:513803

10. Elcadi GH, Forsman M, Hallman DM, Aasa U, Fahlstrom M, Crenshaw AG (2014) Oxygenation and hemodynamics do not underlie early muscle fatigue for patients with work-related muscle pain. PLoS ONE 9:e95582

11. Graven-Nielsen T, Arendt-Nielsen L (2008) Impact of clinical and experimental pain on muscle strength and activity. Curr Rheumatol Rep 10:475–481

12. Ahmadi S, Sinclair PJ, Davis GM (2008) Muscle oxygenation after downhill walking-induced muscle damage. Clin Physiol Funct Imaging 28:55–63

13. Davies RC, Eston RG, Poole DC, Rowlands AV, DiMenna F, Wilkerson DP, Twist C, Jones AM (2008) Effect of eccentric exercise-induced muscle damage on the dynamics of muscle oxygenation and pulmonary oxygen uptake. J Appl Physiol 105:1413–1421

14. Blatter B, Bongers P (2002) Duration of computer use and mouse use in relation to musculoskeletal disorders of neck or upper limb. Int J Ind Ergonomics 30:295–306

15. Hanvold TN, Wærsted M, Mengshoel AM, Bjertness E, Stigum H, Twisk J, Veiersted KB (2013) The effect of work-related sustained trapezius muscle activity on the development of neck and shoulder pain among young adults. Scand J Work Environ Health 39:390–400

16. Smith LL (1991) Acute inflammation: the underlying mechanism in delayed onset muscle soreness? Med Sci Sports Exerc 23:542–551

17. de Ruiter CJ, Goudsmit JF, Van Tricht JA, de Haan A (2007) The isometric torque at which knee-extensor muscle reoxygenation stops. Med Sci Sports Exerc 39:443–453

18. Madeleine P, Nie H, Arendt-Nielsen L (2006) Dynamic shoulder dynamometry: a way to develop delay onset muscle soreness in shoulder muscles. J Biomech 39:184–188

19. Samani A, Holtermann A, Søgaard K, Madeleine P (2009) Effects of eccentric exercise on trapezius electromyography during computer work with active and passive pauses. Clin Biomech 24:619–625

20. Kroemer KHE, Kroemer HB, Kroemer-Elbert KE (2001) Ergonomics: how to design for ease and efficiency. Prentice-Hall, Englewood Cliffs

21. Boushel R, Piantadosi C (2000) Near-infrared spectroscopy for monitoring muscle oxygenation. Acta Physiol Scand 168:615–622

22. Van Beekvelt MC, Colier WN, Wevers RA, Van Engelen BG (2001) Performance of near-infrared spectroscopy in measuring local O(2) consumption and blood flow in skeletal muscle. J Appl Physiol 90:511–519

23. Crenshaw AG, Elcadi GH, Hellstrom F, Mathiassen SE (2012) Reliability of near-infrared spectroscopy for measuring forearm and shoulder oxygenation in healthy males and females. Eur J Appl Physiol 112:2703–2715
24. Felici F, Quaresima V, Fattorini L, Sbriccoli P, Filligoi GC, Ferrari M (2009) Biceps brachii myoelectric and oxygenation changes during static and sinusoidal isometric exercises. J Electromyogr Kinesiol 19:e1–e11
25. Xu R (2003) Measuring explained variation in linear mixed effects models. Stat Med 22:3527–3541
26. Vangsgaard S, Nørgaard LT, Flaskager BK, Søgaard K, Taylor JL, Madeleine P (2013) Eccentric exercise inhibits the H reflex in the middle part of the trapezius muscle. Eur J Appl Physiol 113:77–87
27. Binderup AT, Arendt-Nielsen L, Madeleine P (2010) Pressure pain threshold mapping of the trapezius muscle reveals heterogeneity in the distribution of muscular hyperalgesia after eccentric exercise. Eur J Pain 14:705–712
28. Walsh B, Tonkonogi M, Malm C, Ekblom B, Sahlin K (2001) Effect of eccentric exercise on muscle oxidative metabolism in humans. Med Sci Sports Exerc 33:436–441
29. Lieber R, Friden J (1988) Selective damage of fast glycolytic muscle fibres with eccentric contraction of the rabbit tibialis anterior. Acta Physiol Scand 133:587–588
30. Staron RS, Hagerman FC, Hikida RS, Murray TF, Hostler DP, Crill MT, Ragg KE, Toma K (2000) Fiber type composition of the vastus lateralis muscle of young men and women. J Histochem Cytochem 48:623–629
31. Lindman R, Eriksson A, Thornell LE (1990) Fiber type composition of the human male trapezius muscle: enzyme-histochemical characteristics. Am J Anat 189:236–244
32. Larsen RG, Hirata RP, Madzak A, Frokjaer JB, Graven-Nielsen T (2015) Eccentric exercise slows in vivo microvascular reactivity during brief contractions in human skeletal muscle. J Appl Physiol 119:1272–1281
33. Muthalib M, Millet GY, Quaresima V, Nosaka K (2010) Reliability of near-infrared spectroscopy for measuring biceps brachii oxygenation during sustained and repeated isometric contractions. J Biomed Opt 15:017008
34. Delp M, Laughlin M (1998) Regulation of skeletal muscle perfusion during exercise. Acta Physiol Scand 162:411–419
35. Jensen C, Borg V, Finsen L, Hansen K, Juul-Kristensen B, Christensen H (1998) Job demands, muscle activity and musculoskeletal symptoms in relation to work with the computer mouse. Scand J Work Environ Health 24:418–424
36. Samani A, Holtermann A, Søgaard K, Madeleine P (2009) Experimental pain leads to reorganisation of trapezius electromyography during computer work with active and passive pauses. Eur J Appl Physiol 106:857–866

Effects of Shift Work on Knee Pain and Knee Osteoarthritis Among Retired Chinese Workers

Min Zhou[1,2], Dongming Wang[1,2], Yanjun Guo[1,2],
and Weihong Chen[1,2(✉)]

[1] Department of Occupational and Environmental Health, School of Public
Health, Tongji Medical College, Huazhong University of Science
and Technology, Wuhan 430030, China
[2] Key Laboratory of Environment and Health in Ministry of Education
& Ministry of Environmental Protection, and State Key Laboratory
of Environmental Health (Incubating), School of Public Health, Tongji Medical
College, Huazhong University of Science and Technology,
Wuhan 430030, China
wchen@mails.tjmu.edu.cn

Abstract. Objectives: To evaluate the association between shift work with the
risk of knee pain and knee osteoarthritis (KOA), we studied 13,906 retired
workers from the Dongfeng-Tongji cohort.

Methods: Physical examinations and face-to-face interviews were performed.
Knee pain was diagnosed by self-reported pain or stiffness. Clinical KOA was
diagnosed from knee pain complains and clinical X-ray radiographs. Occupation
history including work content and shift work experience in each job was collected from questionnaires.

Results: The prevalence of knee pain and clinical KOA was 39.0% and 6.7%,
respectively. After adjusting for potential confounders, shift work was independently associated with elevated risk of knee pain (OR 1.24, 95% CI 1.15–
1.33) and clinical KOA (1.15, 1.01–1.32). Such associations remained stable in
stratified analyses by age, gender, BMI, work postures, or chronic diseases.
Additionally, in comparison with daytime workers, the risks increased with
prolonged duration of shift work, the ORs (95% CI) of knee pain for participants
with 1–9 years, 10–19 years, and ≥ 20 years of shift work were 1.20 (1.08–
1.33), 1.26 (1.14–1.40), and 1.26 (1.12–1.40), and ORs (95% CI) of clinical
KOA were 1.06 (0.87–1.30), 1.15 (0.94–1.40), and 1.26 (1.02–1.56). However,
the effects of shift work on knee gradually reduced with the extended duration of
leaving shift work.

Conclusions: Shift work might be independent risk factor for knee pain and
clinical KOA among the retired workers.

Keywords: Knee osteoarthritis · Knee pain · Shift work

© Springer Nature Switzerland AG 2019
S. Bagnara et al. (Eds.): IEA 2018, AISC 820, pp. 32–42, 2019.
https://doi.org/10.1007/978-3-319-96083-8_6

1 Introduction

Knee osteoarthritis (KOA) is one of the most common osteoarthritis. This disease is manifested as degenerative knee arthritis that results from the breakdown of joint cartilage and underlying bone [1]. The most common and disturbing symptom of KOA is knee pain. The results of the Global Burden of Disease Study 2016 showed that osteoarthritis caused 16.28 million of disability-adjusted life-years in 195 countries [2]. Globally, osteoarthritis affected 301.57 million people, and 14.70 million new cases were reported in 2016 [3]. In Asia, KOA is also a common disease, the prevalence ranged from 7.5% to 25.2% [4–6]. Moreover, KOA is believed to be the leading cause of lower productivity at work and resulting in a huge medical burden [7–9].

Occupational factors have long been considered as causative agents of KOA. Published literature indicated that occupational knee joint activities such as task repetition, heavy load, and awkward postures were potential risk factors for knee joint damage [10–12]. While workers are performing these occupational activities, extra force is exerted on the knee joint and gradually weaken the muscles and ligaments around the knee joint and further reduce joint stability. Shift work is another vital occupational factor recognized as a risk factor of many adverse health outcomes [13, 14]. Shift work is an employment practice designed to make use of or provide service across all 24 h of the clock each day of the week. The practice typically divides the day into shifts, set periods of time during which different groups of workers take up their posts. Shift work is one of the most common occupational exposures in the industrialized countries worldwide with 20% of the total active population working in shift schedule [15]. Shifts are thought to induce circadian rhythm and sleep disruption [16]. The homeostasis of skeletal tissue is under the control of normal circadian rhythm, and the homeostasis is broken along with the disrupted circadian rhythm [17]. Meanwhile, the inflammatory response, obesity, lipids disorder, and diabetes followed the disrupted circadian rhythms also add to the onset and progression of KOA [13, 18, 19]. However, few study has focused on the effect of shift work on knee pain or KOA.

In order to fill the knowledge gap we conducted the present study with 13,906 retired workers from the Dongfeng-Tongji cohort [20]. The objective of this study was to quantify the association of shift work with knee pain and KOA.

2 Methods

2.1 Study Population

The Dongfeng-Tongji cohort was established in 2008 and was first followed up in 2013 [20]. In the follow-up survey, questionnaires were administered by trained interviewers through face-to-face interviews to collect information on demographics, medical history, occupation history, and lifestyle. Along with the follow-up, a group of 14,438 participants, i.e. all of the participants that underwent physical examinations at the Central Hospital of the Dongfeng Motor Corporation, were firstly additionally examined knee joint condition (tenderness, range of motion, extension test, and McMurray's test [21]) and questioned about their knee health status including pain, stiffness, and history of osteoarthritis, rheumatoid arthritis, accidental injury, and surgery. We developed the present study with this group of participants using data concerning shift work experience before retirement, and knee health status during the first follow-up. After excluding individuals with missing data regarding knee health status (N = 182), knee surgery caused by accidental injury which was associated with secondary knee osteoarthritis (N = 247), history of rheumatoid arthritis (N = 103), data from 13,906 participants was finally analyzed.

2.2   Ethics Statement

This study was approved by the Medical Ethics Committee of Dongfeng General Hospital, Dongfeng Motor Corporation, and the School of Public Health, Tongji Medical College, Huazhong University of Science and Technology. All participants signed written informed consent.

2.3   Assessment of Knee Pain and Clinical KOA

Cases of self-reported knee pain was defined when an individual's one or both knee(s) met at least one of the following conditions: (1) pain within the past 12 months; (2) persistent pain within the past week; or (3) stiffness within the past 12 months.

Information regarding clinical diagnosed knee osteoarthritis cases was collected from the questionnaires and confirmed by insurance records and treatment information. Clinical KOA cases were defined only if the participant had knee complains (pain, stiffness, tenderness, or decreased range of motion) and the bilateral weight-bearing anteroposterior X-ray radiographs showed a Kellgren & Lawrence grade $\geq$ 2 in at least one knee [22].

2.4   Assessment of Occupational Factors

Occupation history including job titles, calendar years of each job for individual's full duration of employment, and job content was collected from face-to-face interviews. Based on the job held for the longest duration, the work posture was grouped into sitting, standing, squatting or kneeling, or bending if the participant accumulatively held the posture for half time or longer in each shift (i.e. $\geq$ 4 h for most cases) according to the description of job content. The occupational chemicals (e.g. industrial dust) level for each job title at workplace came from the company occupational hazard monitoring records, and was categorized into yes when the concentrations of chemicals were higher than the national occupational chemicals exposure limit.

Shift work was identified as working with a schedule involving unusual working hours as opposed to the normal daytime work schedule, i.e. from 8:00 to 17:00, for at least one year. There were 3 kinds of shift work in Dongfeng Motor Corporation: two-shifts where day work (8:00–17:30) and night work (17:30–2:30) shifts in weekly rotation; three-shifts during which 3 crews of workers succeed each other at 8:00, 16:00, and 00:00; and four shifts in which 4 crews of workers succeed each other at 8:00, 14:00, 20:00, and 2:00. The workers in any kind of shift work had to take turns to work in the early morning and at night. Shift work duration was calculated by the starting year to the ending year of shifts excluding the years without shift, and was categorized into 1–9 y, 10–19 y, and ≥ 20 y.

Years after leaving shift work was the duration from the year left the latest shift work to 2013. For participants engaged in 1–9 years of shift work, the duration of leaving shift was tertiled into 1–14 y, 15–24 y and ≥ 25 y; for those with 10–19 years of shift work, it was tertiled into 1–9 y, 10–19 y, and ≥ 20 y; and for those with ≥ 20 years of shift work, it was tertiled into 1–4 y, 5–9 y, and ≥ 10 y.

2.5 Assessment of Covariates

Body weight was measured to the nearest 0.1 kg and body height was measured to the nearest 0.1 cm with participants wearing light indoor clothing and without shoes. The body mass index (BMI) was calculated from weight in kilograms divided by body height in squared meters.

Information regarding smoking status (current, former, never), drinking status (current, former, never), and physical exercise was collected from the questionnaires. Participants who had been smoking as much as one cigarette a day for at least 6 months were considered current smokers, and those who had been drinking alcohol as often as once per week for at least 6 months were considered current drinkers. Physical exercise was defined as regular exercise of at least 20 min per day over the past 6 months [20].

Chronic diseases diagnosed by physicians were reported by the participants and confirmed by insurance records and treatment information, including hypertension, hyperlipidemia, diabetes, coronary heart disease, myocardial infarction, stroke, and cancer.

2.6 Statistical Analysis

Basic characteristics of participants were presented as mean (SD) for continuous variables and as number (percentage) for categorical variables. Logistic regressions were used to evaluate the association of shift work with knee pain and clinical KOA. The associations were further stratified analysis by age, gender, BMI, work postures, and chronic diseases, for previous researches suggested that these were risk factors for KOA [23–25]. In the multivariate models, we adjusted for age (as numerical variable), gender, BMI (as numerical variable), work postures, chronic diseases (yes/no), smoking status (current, former, never), drinking status (current, former, never), and physical exercise (yes/no). To explore the independent effect of shift work on knee pain and clinical KOA, we used multivariable logistic regression with a stepwise selection procedure, which started with potential risk factors and ended with variables that were statistically significantly associated with a p-value < 0.05. The models were conducted with daytime workers as the reference group. The statistical tests were two sided with a significance set at p < 0.05. All statistical analyses were performed using SAS 9.4 software (SAS Institute, Cary, NC).

3 Results

The basic characteristics of the participants were reported by categories of knee pain and clinical KOA (Table 1). A total of 13,906 participants (female 7,560, 54.4%) with a mean age of 64.7 ± 8.2 years old were included in the analysis. The prevalence of knee pain and clinical KOA was 39.0% and 6.7%, respectively. A total of 5,537 (39.8%) workers had ever engaged in shift work, among that 64.0% were two shifts, 34.1% were three shifts, and1.9% were four shifts.

Prolonged working years was positively associated the risk of knee pain (OR 1.16, 95% CI 1.01–1.33) and clinical KOA (OR 1.48, 95% CI 1.18–1.87) among participants in the upper tertile age group. In comparison with participants engaged in sitting work posture, the risk of knee pain was elevated among those with standing (OR 1.18, 95% CI 1.09–1.28) or bending work posture (OR 1.20, 95% CI 1.08–1.34) (Table 2).

We examined the effects of shift work on knee pain and clinical KOA (Table 2). Compared with daytime workers, the odds ratios (ORs) of knee pain and clinical KOA were 1.19 (95% CI 1.10–1.29) and 1.18 (95% CI 1.01–1.38) for shift workers, after adjusting for age, gender, BMI, smoking status, drinking status, physical exercise, and chronic diseases. Different shifts type showed similar adverse effect on knee pain, the ORs (95% CI) for two shifts, three shifts, and four shifts were 1.17 (1.06–1.28), 1.24 (1.09–1.40), and 1.96 (1.20–3.18), respectively. The risk of clinical KOA increased among the two shifts workers (OR 1.24, 95% CI 1.03–1.49), but not among three shifts or four shifts, perhaps because of the limited number of the clinical KOA cases (N = 88 and N = 3, respectively). We noted that the risk for knee pain or clinical KOA showed a rising trend with prolonged duration of shift work. After adjusting for the above potential confounders, each 5-years increase in shift work duration was associated with 4% and 3% increase in the risks of knee pain (OR 1.04, 95% CI 1.02–1.06) and KOA (OR 1.03, 95% CI 1.01–1.07), respectively. Compared with daytime worker, the ORs for participants with 1–9 years, 10–19 years, and ≥ 20 years of shift work were 1.20, 1.26, and 1.26 for knee pain, and 1.06, 1.15, and 1.26 for clinical KOA. At the meantime, the ORs of shift work on knee pain showed a trend of decreasing with the extended duration of leaving shift work, with the p values for trend were 0.01, <0.01, and <0.01 for those with 1–9 years, 10–19 years, and ≥ 20 years of ever shift work, respectively. The effect on knee pain may return to the daytime worker level after 25 years ceasing shift work for participants ever engaged in 1–9 years of shift work, while it takes 10 years for those with 10–19 years' shift work, and 5 years for those with ≥ 20 years' shift work. While the ORs of shift work on clinical KOA significantly reduced with the extended duration of leaving shift work among workers with ≥ 20 years of ever shift work (p value for trend 0.03).

The stratified analyses suggested that the association between shift work and knee pain was statistical significant in each subgroup. While the association between shift work and risk of clinical KOA was more pronounced in participants of lower tertile age (OR 1.43, 95% CI 1.07–1.90) and in those with standing work posture (OR 1.22, 95% CI 1.01–1.50). In the stepwise multivariate logistic regression analyses, shift work was independently associated with increased risk of knee pain (OR 1.24, 95% CI 1.15–1.33) and clinical KOA (OR 1.15, 95% CI 1.01–1.32) (Table 3).

Table 1. The characteristics of the study population (N = 13,906)

Variables	Total	Knee pain			Clinical KOA		
		Yes	No	p-value	Yes	No	p-value
Female gender (%)	7560 (54.4)	3516 (64.9)	4044 (47.6)	<0.01*	646 (69.7)	6914 (53.3)	<0.01*
Age (years, mean ± SD)	64.7 ± 8.2	64.9 ± 8.2	64.6 ± 8.3	0.09	66.3 ± 8.2	64.6 ± 8.2	<0.01*
BMI (kg/m², mean ± SD)	24.2 ± 3.3	24.6 ± 3.4	24.0 ± 3.2	<0.01*	25.2 ± 3.7	24.2 ± 3.3	<0.01*
Work postures				<0.01*			0.19
Sitting	4819 (34.8)	1769 (32.8)	3050 (36.1)		313 (33.9)	4506 (34.9)	
Standing	6101 (44.2)	2482 (45.9)	3619 (42.9)		431 (46.7)	5670 (43.9)	
Bending	1894 (13.7)	779 (14.4)	1115 (13.2)		124 (13.4)	1770 (13.7)	
Squatting or kneeling	1030 (7.4)	372 (6.9)	658 (7.8)		55 (6.0)	975 (7.6)	
Occupational chemicals (%)	1753 (12.6)	687 (12.7)	1066 (12.6)	0.84	119 (12.8)	1634 (12.6)	0.83
Working years (years, mean ± SD)	30.7 ± 8.7	29.5 ± 8.7	31.5 ± 8.6	<0.01*	29.6 ± 8.4	30.8 ± 8.7	<0.01*
Shift work (%)	5537 (39.8)	2301 (42.4)	3236 (38.1)	<0.01*	387 (41.8)	5150 (39.7)	0.21
Shift work duration (years)	14.5 ± 10.1	14.1 ± 9.8	14.7 ± 10.3	0.02*	14.2 ± 9.4	14.5 ± 10.2	0.54
Chronic diseases (%)	7407 (53.3)	3047 (56.2)	4360 (51.4)	<0.01*	542 (58.5)	6865 (52.9)	<0.01*
Smoking (%)				<0.01*			<0.01*
Non-Smokers	9821 (70.8)	4171 (77.1)	5650 (66.8)		744 (80.4)	9077 (70.1)	
Current-Smokers	2170 (15.6)	651 (12.0)	1519 (17.9)		78 (8.4)	2092 (16.1)	
Ex-Smokers	1886 (13.6)	590 (10.9)	1296 (15.3)		104 (11.2)	1782 (13.8)	
Drinking (%)				<0.01*			0.03*
Non-Drinkers	9469 (68.3)	3857 (71.4)	5612 (66.3)		668 (72.1)	8801 (68.0)	
Current-Drinkers	3511 (25.3)	1258 (23.3)	2253 (26.6)		203 (21.9)	3308 (25.6)	
Ex-Drinkers	887 (6.4)	287 (5.3)	600 (7.1)		55 (5.9)	832 (6.4)	
Physical exercise (%)	12520 (90.0)	4839 (89.3)	7681 (90.5)	0.02*	828 (89.3)	11692 (90.1)	0.45

BMI: body mass index. SD: standard deviation. Chronic diseases: hyperlipidemia, diabetes, hypertension, coronary heart disease, myocardial infarction, stroke, or tumor. *p-value < 0.05.

Table 2. Odds ratios (95% CI) for knee pain and clinical KOA by working years, work postures, occupational chemicals, and shift work

Variables	Knee pain	Clinical KOA
Age (T1: male < 65 years, female < 55 years)		
Working years (years)		
<35 (female: <25)	1	1
35–40 (female: 25–30)	1.04 (0.88–1.22)	1.02 (0.70–1.49)
≥ 40 (female: ≥ 30)	0.92 (0.88–1.07)	1.06 (0.75–1.50)
Age (T2: male 65–70 years, female 55–65 years)		
Working years (years)		
<35 (female: <25)	1	1
35–40 (female: 25–30)	0.88 (0.76–1.02)	0.86 (0.64–1.15)
≥ 40 (female: ≥ 30)	0.90 (0.78–1.05)	1.20 (0.90–1.59)
Age (T3: male ≥ 70 years, female ≥ 65 years)		
Working years (years)		
<35 (female: <25)	1	1
35–40 (female: 25–30)	0.96 (0.83–1.12)	0.98 (0.74–1.30)
≥ 40 (female: ≥ 30)	1.16 (1.01–1.33)*	1.48 (1.18–1.87)*
Work postures		
Sitting	1	1
Standing	1.18 (1.09–1.28)*	1.09 (0.94–1.27)
Bending	1.20 (1.08–1.34)*	1.01 (0.81–1.25)
Squatting or kneeling	0.96 (0.85–1.12)	0.81 (0.60–1.09)
Occupational chemicals		
No	1	1
Yes	1.04 (0.94–1.16)	1.05 (0.86–1.29)
Shift work		
No	1	1
Yes	1.19 (1.10–1.29)*	1.18 (1.01–1.38)*
Shift work duration (years)		
Daytime workers	1	1
1–9	1.20 (1.08–1.33)	1.06 (0.87–1.30)
10–19	1.26 (1.14–1.40)*	1.15 (0.94–1.40)
≥ 20	1.26 (1.12–1.40)*	1.26 (1.02–1.56)*
Shift work types		
Daytime workers	1	1
Two shifts	1.17 (1.06–1.28)*	1.24 (1.03–1.49)*
Three shifts	1.24 (1.09–1.40)*	1.13 (0.89–1.43)
Four shifts	1.96 (1.20–3.18)*	0.50 (0.12–2.08)

Adjusted for age (continuous), gender, BMI (continuous), smoking status, drinking status, physical exercise, and chronic diseases (hyperlipidemia, diabetes, hypertension, coronary heart disease, myocardial infarction, stroke, or tumor).
*p-value < 0.05.

Table 3. Odds ratios (95% CI) for knee pain and clinical KOA from multivariate stepwise logistic regression model

Variables	Knee pain	Clinical KOA
Shift work	1.24 (1.15–1.33)*	1.15 (1.01–1.32)*
Age (years)	1.02 (1.01–1.02)*	1.04 (1.03–1.05)*
Female gender	2.29 (2.13–2.48)*	2.43 (2.09–2.82)*
BMI (kg/m^2)	1.06 (1.05–1.07)*	1.09 (1.07–1.11)*
Chronic diseases	1.24 (1.15–1.33)*	/

The models included work postures, occupational chemicals, shift work, gender, age (continuous), BMI (continuous), chronic diseases (hyperlipidemia, diabetes, hypertension, coronary heart disease, myocardial infarction, stroke, or tumor), smoking status, drinking status, and physical exercise. *p-value < 0.05.

4 Discussion

In the present study, we not only observed adverse effects of awkward work postures on knee pain, but also identified a positive association between shift work and risk of knee pain or clinical KOA in a large group of retired workers. The risks of knee pain or clinical KOA increased with prolonged shift duration. The effects of shift work gradually reduced with extended duration of leaving the shift work.

Several cross-sectional studies showed that shift work was positively associated with musculoskeletal disorders including low back pain and shoulder pain [26–28], few study has estimated the association between shift work with knee pain or KOA. Therefore, the present study adds new information to adverse health of shift work on knee joint. Our results have notable implications for occupational health, because the high prevalence of KOA and knee pain seriously affects quality of life and reduces work productivity [27], while shift work remains as a vital component of the modern economy and the shift policies could be modified.

Although the mechanisms for the increased risk due to shift work remains unclear, several factors may be underlying the association. First, shift work may influence KOA through sleep quality. In our previous reports, shift work was associated with reduced sleep quality [29]. In the present analysis, we found that poor sleep quality was associated with increased risk of knee pain (OR 1.11, 95% CI 1.09–1.12) and clinical KOA (OR 1.10, 95% CI 1.08–1.13), meanwhile the poor sleep quality significantly mediated the total effects of shift work on knee pain and clinical KOA by 10.91% and 15.84%, respectively. Poor sleep may directly reduce recovery time of the knee problem. Second, long-term of shift work is associated with chronic circadian rhythms disturbance [30], which directly disturbs the homeostasis of cartilage, bone, and tendon [31–33]. Simultaneously, circadian misalignment has been found to result in metabolic disorders and inflammatory consequence [34, 35], which have been linked to the cartilage and bone homoeostasis [36]. Third, shift workers had different dietary habits from daytime workers, i.e. greater energy density, increased saturated fat, and decreased dietary fiber were reported among shift workers [37, 38]. High intake of fat and saturated fatty acids was associated with the onset and progression of KOA, while the dietary fiber was inversely associated with KOA worsening [39, 40].

The strengths of this study include the large sample size and detailed occupational history. We are aware of the limitations of this study. First, we did not provide knee X-ray examination for each participants although all participants could take free knee X-ray examination at any time if they have knee symptoms and the company will pay for such examinations. Second, information regarding shift work was self-reported, which may lead to misclassifications of the exposure. However, we checked the shift work information (yes or no) used in the present study with that collected at baseline, and the agreement (Kappa coefficient) was 0.76 (95% CI 0.75–0.78). Third, although information regarding physical activity was collected over the past year, the individuals may have changed their physical activity levels over time.

In conclusion, the present study showed that shift work was independently associated with risk of both knee pain and clinical KOA among large sample of retired workers. These findings have substantial implications for the prevention of KOA through the modification of shift work policies.

Acknowledgements. The authors are indebted to Dr. Stephen S. Bao at the Washington State Department of Labor and Industries for providing valuable comments. We also thank the study participants for their help. This work was supported by the National Natural Science Foundation of China [grant number 81573121]; and the Fundamental Research Funds for the Central Universities, Huazhong University of Science and Technology [grant number 2016JCTD116].

References

1. Felson DT (2006) Clinical practice. osteoarthritis of the knee. N Engl J Med 354:841–848. https://doi.org/10.1056/NEJMcp051726
2. GBD 2016 DALYs and HALE Collaborators (2017) Global, regional, and national disability-adjusted life-years (DALYs) for 333 diseases and injuries and healthy life expectancy (HALE) for 195 countries and territories, 1990–2016: a systematic analysis for the Global Burden of Disease Study 2016. Lancet 390:1260–1344. https://doi.org/10.1016/s0140-6736(17)32130-x
3. GBD 2016 Disease and Injury Incidence and Prevalence Collaborators (2017) Global, regional, and national incidence, prevalence, and years lived with disability for 328 diseases and injuries for 195 countries, 1990–2016: a systematic analysis for the Global Burden of Disease Study 2016. Lancet 390:1211–1259. https://doi.org/10.1016/s0140-6736(17)32154-2
4. Chopra A (2013) The COPCORD world of musculoskeletal pain and arthritis. Rheumatology 52:1925–1928. https://doi.org/10.1093/rheumatology/ket222
5. Fransen M, Bridgett L, March L, Hoy D, Penserga E, Brooks P (2011) The epidemiology of osteoarthritis in Asia. Int J Rheum Dis 14:113–121. https://doi.org/10.1111/j.1756-185X.2011.01608.x
6. Tang X, Wang S, Zhan S et al (2016) The prevalence of symptomatic knee osteoarthritis in China: results from the China health and retirement longitudinal study. Arthritis Rheumatol 68:648–653. https://doi.org/10.1002/art.39465

7. Agaliotis M, Fransen M, Bridgett L et al (2013) Risk factors associated with reduced work productivity among people with chronic knee pain. Osteoarthritis Cartilage 21:1160–1169. https://doi.org/10.1016/j.joca.2013.07.005

8. Palazzo C, Nguyen C, Lefevre-Colau MM, Rannou F, Poiraudeau S (2016) Risk factors and burden of osteoarthritis. Ann Phys Rehabil Med 59:134–138. https://doi.org/10.1016/j.rehab. 2016.01.006

9. Spector JT, Adams D, Silverstein B (2011) Burden of work-related knee disorders in Washington State, 1999 to 2007. J Occup Environ Med 53:537–547. https://doi.org/10.1097/ JOM.0b013e31821576ff

10. Haukka E, Ojajarvi A, Takala EP, Viikari-Juntura E, Leino-Arjas P (2012) Physical workload, leisure-time physical activity, obesity and smoking as predictors of multisite musculoskeletal pain. A 2-year prospective study of kitchen workers. Occup Environ Med 69:485–492. https://doi.org/10.1136/oemed-2011-100453

11. Solidaki E, Chatzi L, Bitsios P, Coggon D, Palmer KT, Kogevinas M (2013) Risk factors for new onset and persistence of multi-site musculoskeletal pain in a longitudinal study of workers in Crete. Occup Environ Med 70:29–34. https://doi.org/10.1136/oemed-2012-100689

12. Lau E, Cooper C, Lam D, Chan V, Tsang K, Sham A (2000) Factors associated with osteoarthritis of the hip and knee in Hong Kong Chinese: obesity, joint injury, and occupational activities. Am J Epidemiol 152:855–862

13. Pan A, Schernhammer ES, Sun Q, Hu FB (2011) Rotating night shift work and risk of type 2 diabetes: two prospective cohort studies in women. PLoS Med 8:e1001141. https://doi.org/ 10.1371/journal.pmed.1001141

14. De Bacquer D, Van Risseghem M, Clays E, Kittel F, De Backer G, Braeckman L (2009) Rotating shift work and the metabolic syndrome: a prospective study. Int J Epidemiol 38:848–854. https://doi.org/10.1093/ije/dyn360

15. Eurofound (2012) Working time and work–life balance in a life course perspective, Eurofound, Dublin

16. Linton SJ, Kecklund G, Franklin KA et al (2015) The effect of the work environment on future sleep disturbances: a systematic review. Sleep Med Rev 23:10–19. https://doi.org/10. 1016/j.smrv.2014.10.010

17. Berenbaum F, Meng QJ (2016) The brain-joint axis in osteoarthritis: nerves, circadian clocks and beyond. Nat Rev Rheumatol 12:508–516. https://doi.org/10.1038/nrrheum.2016.93

18. Potter GD, Skene DJ, Arendt J, Cade JE, Grant PJ, Hardie LJ (2016) Circadian rhythm and sleep disruption: causes, metabolic consequences, and countermeasures. Endocr Rev 37:584–608. https://doi.org/10.1210/er.2016-1083

19. Zhou M, Guo Y, Wang D et al (2017) The cross-sectional and longitudinal effect of hyperlipidemia on knee osteoarthritis: results from the Dongfeng-Tongji cohort in China. Sci Rep 7:9739. https://doi.org/10.1038/s41598-017-10158-8

20. Wang F, Zhu J, Yao P et al (2013) Cohort Profile: the Dongfeng-Tongji cohort study of retired workers. Int J Epidemiol 42:731–740. https://doi.org/10.1093/ije/dys053

21. Jackson JL, O'Malley PG, Kroenke K (2003) Evaluation of acute knee pain in primary care. Ann Intern Med 139:575–588

22. Kellgren JH, Lawrence JS (1957) Radiological assessment of osteo-arthrosis. Ann Rheum Dis 16:494–502

23. Reijman M, Pols HA, Bergink AP et al (2007) Body mass index associated with onset and progression of osteoarthritis of the knee but not of the hip: the Rotterdam Study. Ann Rheum Dis 66:158–162. https://doi.org/10.1136/ard.2006.053538

24. Eymard F, Parsons C, Edwards MH et al (2015) Diabetes is a risk factor for knee osteoarthritis progression. Osteoarthritis Cartilage 23:851–859. https://doi.org/10.1016/j. joca.2015.01.013

25. Zhang Y, Hunter DJ, Nevitt MC et al (2004) Association of squatting with increased prevalence of radiographic tibiofemoral knee osteoarthritis: the beijing osteoarthritis study. Arthritis Rheum 50:1187–1192. https://doi.org/10.1002/art.20127

26. Takahashi M, Matsudaira K, Shimazu A (2015) Disabling low back pain associated with night shift duration: sleep problems as a potentiator. Am J Ind Med 58:1300–1310. https://doi.org/10.1002/ajim.22493

27. Lipscomb JA, Trinkoff AM, Geiger-Brown J, Brady B (2002) Work-schedule characteristics and reported musculoskeletal disorders of registered nurses. Scand J Work Environ Health 28:394–401

28. Choobineh A, Soltanzadeh A, Tabatabaee H, Jahangiri M, Khavaji S (2012) Health effects associated with shift work in 12-hour shift schedule among Iranian petrochemical employees. Int J Occup Saf Ergon 18:419–427. https://doi.org/10.1080/10803548.2012.11076937

29. Guo Y, Liu Y, Huang X et al (2013) The effects of shift work on sleeping quality, hypertension and diabetes in retired workers. PLoS ONE 8:e71107. https://doi.org/10.1371/journal.pone.0071107

30. Gumenyuk V, Howard R, Roth T, Korzyukov O, Drake CL (2014) Sleep loss, circadian mismatch, and abnormalities in reorienting of attention in night workers with shift work disorder. Sleep 37:545–556. https://doi.org/10.5665/sleep.3494

31. Gossan N, Zeef L, Hensman J et al (2013) The circadian clock in murine chondrocytes regulates genes controlling key aspects of cartilage homeostasis. Arthritis Rheum 65:2334–2345. https://doi.org/10.1002/art.38035

32. Dudek M, Meng QJ (2014) Running on time: the role of circadian clocks in the musculoskeletal system. Biochem J 463:1–8. https://doi.org/10.1042/bj20140700

33. McDearmon EL, Patel KN, Ko CH et al (2006) Dissecting the functions of the mammalian clock protein BMAL1 by tissue-specific rescue in mice. Science 314:1304–1308. https://doi.org/10.1126/science.1132430

34. Scheer FA, Hilton MF, Mantzoros CS, Shea SA (2009) Adverse metabolic and cardiovascular consequences of circadian misalignment. Proc Natl Acad Sci USA 106:4453–4458. https://doi.org/10.1073/pnas.0808180106

35. Guo B, Yang N, Borysiewicz E et al (2015) Catabolic cytokines disrupt the circadian clock and the expression of clock-controlled genes in cartilage via an NFκB-dependent pathway. Osteoarthritis Cartilage 23:1981–1988

36. Aspden RM, Scheven BA, Hutchison JD (2011) Osteoarthritis as a systemic disorder including stromal cell differentiation and lipid metabolism. Lancet 357:1118–1120. https://doi.org/10.1016/s0140-6736(00)04264-1

37. Nea FM, Kearney J, Livingstone MB, Pourshahidi LK, Corish CA (2015) Dietary and lifestyle habits and the associated health risks in shift workers. Nutr Res Rev 28:143–166. https://doi.org/10.1017/s095442241500013x

38. Bonnell EK, Huggins CE, Huggins CT, McCaffrey TA, Palermo C, Bonham MP (2017) Influences on dietary choices during day versus night shift in shift workers: a mixed methods study. Nutrients 9:193. https://doi.org/10.3390/nu9030193

39. Lu B, Driban JB, Xu C, Lapane KL, McAlindon TE, Eaton CB (2017) Dietary fat intake and radiographic progression of knee osteoarthritis: data from the osteoarthritis initiative. Arthritis Care Res 69:368–375. https://doi.org/10.1002/acr.22952

40. Dai Z, Niu J, Zhang Y, Jacques P, Felson DT (2017) Dietary intake of fibre and risk of knee osteoarthritis in two US prospective cohorts. Ann Rheum Dis 76:1411–1419. https://doi.org/10.1136/annrheumdis-2016-210810

Financial Impact and Causes of Chronic MSD Cases in Malaysia Based on SOCSO Claims Record

Raemy Md Zein[1]([⊠]), Jafri Mohd Rohani[2], Norsheila Zainal Abidin[2], and Ismail Abdul Rahman[1]

[1] Ergonomics Excellence Centre, NIOSH, 81400 Senai, Johor, Malaysia
raemymdzein@gmail.com, eec@niosh.com.my
[2] Faculty of Mechanical Engineering, UTM, 81300 Skudai, Johor, Malaysia

Abstract. Musculoskeletal disorder (MSDs) are major occupational health issues among the private sector and government. This paper analyzed total direct cost being paid by the Social Security Organization of Malaysia (SOCSO) to the Malaysian workers due to musculoskeletal disease. Total direct cost consists of compensation cost, return to work and rehabilitation cost, and medical cost. The objectives of this study are (1) to determine the total direct costs incurred as a result of cases of chronic musculoskeletal injuries approved by SOCSO from 2010–2014. (2) To examine the age category imposed for the total average cost on the highest MSDs claims for the four categories setting. This study has utilize data provided by Social Security Organization of Malaysia (SOCSO) on occupational diseases and adopt top-down approach. The claims data is limited to MSD reported between 2009 until 2014. A total of 416 claims related to MSD has been analyzed. All the category recorded the highest total direct cost for types of industry, types of injury, causes of accident and type of body part. Then age claimant is identified from total average cost earns from the highest total direct cost for four categories. Manufacturing industry, strenuous movement, sprain and strain and back are recorded as high total direct cost with bear cost of RM 5,181,282.34 (n = 185), RM 7,088,839.51 (n = 264), RM 8,753,975.13 (n = 335), RM 5,526,590.69 (n = 209) respectively. Then the age 35–44 is recorded as the high total average cost for this four parameter. This study will provide basis for future studies and intervention on MSD related injuries in working environment in Malaysia.

Keywords: Musculoskeletal disorder · Direct cost · Type of injury
Cause of accident · Body part location

1 Introduction

Industrialization carried many with it many a human cost. Malaysia is one of the growing industrialized country and expected to become developed country status for the next decade [1]. Occupational diseases are always considered as most significant problems for workers, especially for less developed country [2]. Meanwhile, for the developed country such as Canada also claimed occupational diseases are common

© Springer Nature Switzerland AG 2019
S. Bagnara et al. (Eds.): IEA 2018, AISC 820, pp. 43–53, 2019.
https://doi.org/10.1007/978-3-319-96083-8_7

health problems and major contributor to disability and cost in working population [6]. This shows both developed and less developed country facing the same problems and being apprehensive to the all organizations nowadays.

A study being done in United States accounted about 29%–35% of occupational diseases and injuries are belong to Musculoskeletal Disorder [9] The others had reported a survey in United Kingdom Health and Safety Executive states musculoskeletal disorders were the most common disease and 37% of working days lost were from MSD [19]. In Malaysia itself, a total number of claims of 553 that related to MSD is recorded between years 2009 until 2014. In addition, the figure is corresponding to 25.22% from overall occupational diseases and all MSD related consist of temporary and permanent disability [18].

Generally, MSD problem is caused by the work related physical risk factors such as repetitiveness, work environment and psychosocial factors. This MSD will gave claimants experiences of pain or discomfort of the muscle, nerves and tendons region including other soft tissue [13]. Back 50.2% was most occurrence body parts injury were recorded from SOCSCO database for the period 2009-2014 followed by hip 36.5% and shoulder 2.6%.18 However, the percentage of claimants will be vary depending on the scale of sectors. There is study that shows MSD commonly occurred at lower back 48%, shoulder 13%, extremities 1 5%, knee 1 5%, ankle or foot 2% and multiple site 5% [8]. The other researcher specific the body part as lumbar spine and also was most frequent cases [8, 9].

MSD give significant to loss workdays that will affect productivity of business and after that will impact on negative economy on individual and community neither [18]. MSDs cases have incurred about US $171.7 million of productivity losses in Columbia in 2005 [16]. The productivity can be related to the efficiency between the input and output that the number output unit given the usual or less input hours due to workers being away from work due to illness and absenteeism [10]. While the study in Korea economic cost of MSD was estimated $6.89 billion, which represent 0.7 of the Korean gross domestic product in 2008 [14].

The study in the United States had differ the cost between direct cost and indirect cost [5]. Direct cost can be classified as an illness cost being compensate, medical cost, indemnity cost and others related cost during the claimants' recovery. Meanwhile indirect cost basically refer to the losses of potential output, at work or home this lead to morbidity or premature mortality and reduction of quality life [4]. In separate study, the researcher had proposed three cost categories in estimating the cost related to MSDs. The three categories includes of direct cost, indirect cost and quality of life cost [16]. Meanwhile, the other researcher defined the cost into three categories which are direct cost, indirect cost and intangible cost [14]. According to Middlesworth indirect cost is difficult to measure as need to consider loss of production time, training and compensate a replacement worker and related to absenteeism filing [12].

Currently there is little study being done on the cost associated towards total direct cost claimed in Malaysia. There are study being done at oversea on total direct cost which include of medical cost and wage compensation cost in their study but they are not include return to work and rehabilitation cost [5, 8, 9, 16]. Meanwhile, the other researcher only report compensation claims cases from SOCSO database for the period 2002 to 2006 and focusing to non-governmental employee [1].

This paper is limited to secondary data only and primary data were not included. Only total direct cost (compensation claim, medical cost and return to work and rehabilitation cost) are consider in this study. The purpose of this research was (1) to determine the total direct costs incurred as a result of cases of chronic musculoskeletal injuries approved by SOCSO from 2010–2014. (2) To examine the age category imposed for the total average cost on the highest MSDs claims for the three category setting which are types of industry, types of injury, causes of injury and body part.

2 Materials and Method

The descriptive data of occupational diseases claims were provided to the research team by Social Security Organization of Malaysia (SOCSO). The data is limited from the period 2009 until 2014 and only MSD cases is filtered in as to align with the scope of study. The chosen of MSDs cases due to highest claims during the five years period compared to miners' nystagmus, hearing impairment caused by noise, any other physical causes, agent's diseases caused vibration, and occupational vitiligo. In addition, only permanent disability MSDs cases with paid status were accounted and others type of status (i.e. claim rejected, wrong data entry, incomplete documents and no payment made) were filtered out in order to validate the data.

As to achieve the objective of this study, the cost of medical and return to work and rehabilitation were then collected at each selected branches. The selected branches were chosen by comprising the highest percentage with the others branches that accumulating 77.5% (n = 416) from the overall reported by SOCSO (537 cases). The selected branches were Melaka (n = 207), Butterwoth (n = 55), Sungai Petani (n = 49), Kuantan (n = 49), Kuala Lumpur (30) and Rawang (26). The collection data vary depending on each branches and losing of many data is possible due to lack on record keeping. The data of claimants such as name and identification number are not reporting in this study as to protect the claimants' personal data.

The critical part in collecting the cost data is the financial and distribution of cost due to confidential data. The authority permission is needing in order to have data from the SOCSO branches. The main idea is visiting the main SOCSO office which located at Jalan Ampang to get overview on how the cost is being distribute to the claimants. Return to work and rehabilitation cost only kept by the main SOCSO branches. Meanwhile, the medical cost claims is recorded by every SOCSO branches. The cost claims at each branches is collected through visits, emails and fax.

The process of descriptive data analysis is performed by analyzing on demographic characteristic such as gender, age groups, type of industry, causes of injury, location of body part injury and type of injury. In order to determine the age category, the study in Ohio is referred. The researcher had classified the age category by six categories which are 16–24, 25–34, 35–44, 45–54, 55–64 and above 65 [8]. Unrelated body part for MSD also is filtered out. All the total direct cost were calculated based on actual payment made by SOCSO to employee based on permanent disability cases for the year 2009 until 2014.

Statistical Package for the Social Science (SPSS) is tool use in order to analyze the descriptive data. SPSS is one of the most popular statistical package that can perform highly complex data manipulation. Those generation need as to manipulate the huge amount of numbers. SPSS is easier to manipulate data on distribution cost between variables setting. This is needs for the future assessment or action taken by the related agency and employers

3   Results and Discussion

3.1   Compensation Cost Claims on Type of Occupational Disease

Table 1 shows the comparison compensation cost between other diseases and MSD cases reported every year from 2009 until 2014. In the table shows the types of occupational disease being reported by the SOCSO over the year. Musculoskeletal disorder is highest cases claims among the others diseases 25.5%. The sum of other diseases include of 37 type of disease (miners' nystagmus (21.0%), diseases caused by any other physical agents (16.4%), hearing impairment caused by noise (16.3%), occupational vitiligo (6.0%), diseases caused vibration (5.8%), diseases caused by chemical agents (1.7%), any other respiratory diseases (1.2%), and the rest of 31 other diseases each counted less than 1%. From the table, the total cost incurred to the occupational diseases is increasing over the periods. For the year 2009, MSD cost is RM 1,049,700.86, year 2010 the cost increase to RM 912,485.64 even though the sample size is same. Meanwhile, for year 2011 the total cost claim is slightly increase to RM 1,791,629.86 compare to year 2009. The claimant for year 2012 is increasing and the consequent of cost is greatly increase to RM 3,274,073.07. The number of claimants is dropping to n = 101 that costing RM 2,624,981.55 for the year 2013. For the year 2014, the total cost is more costing compare for year 2012, RM 3,940,486.61. This lead to the total cost of 537 MSD claims is about RM 13,593,357.59. While for occupational disease it can been seen clearly the trend is increasing over the year. The total cost for year 2009 is RM 1,599,074.26, RM 3,722,228.27 for year 2010, RM 4,714,369.26 for year 2011, RM 6,148,457.45 for year 2012, RM 8,163,071.11 for year 2013 and the total cost is drastically increase for year 2014 with the sum up claimants n = 1567 (RM 34,465,120.24). The total up for the occupational diseases made up n = 2104 claimants with average total cost mean RM 22,841.48.

Table 1. Total cost for occupational disease from 2009 to 2014

Year	Type of diseases			
	Musculoskeletal		Other diseases	
	Compensation cost		Compensation cost	
	n	RM	n	RM
2009	40	1,049,700.86	81	1,599,074.26
2010	40	912,485.64	181	3,722,228.27
2011	74	1,791,629.86	222	4,714,369.26
2012	129	3,274,073.07	276	6,148,457.45
2013	101	2,624,981.55	355	8,163,071.11
2014	153	3,940,486.61	452	10,117,919.89
Total	537	13,593,357.59	1567	34,465,120.24

3.2   Total Direct Cost Claims on Sociodemographic

Table 2 shows the total direct average cost claims for socio demographic characteristic for period 2009 to 2014. Male workers were recorded as higher claimants with average mean cost RM 28,220.00 as compared to female workers with the average cost RM 20,600.00. The age between 35–44 years old significantly shows the highest average total direct cost claims with RM 28,264.42 which shows in percent 38.9%. This category of age also have the highest total direct cost compare to the other category with costing of RM 4,578,835.89 for that particular age. For the second higher claims on average total direct cost was at age 25–34 (RM 28,029.98) and then at age 45–54 (RM 23,934.81) follow to claimants from age 16–24 (RM 18,488.10) and the least average total direct cost was on age 55–64 (RM 15,161.27).

Table 2. Average total direct cost claimants for socio demographic characteristic

Variable	n	(%)	Total direct cost (RM)	Average total direct cost (RM)
Gender				
Male	322	77.4	9,086,979.96	28,220.00
Female	94	22.6	1,936,435.29	20,600.00
Age group				
16–24	19	4.6	351,273.94	18,488.10
25–34	138	33.2	3,868,137.92	28,029.98
35–44	162	38.9	4,578,835.89	28,264.42
45–54	86	20.7	2,058,393.53	23,934.81
55–64	11	2.6	166,773.97	15,161.27
Above 65	0	0	0	0

The age category number of average total direct cost follow the bell shape distribution. Even though the other researchers had setting different category, but the trend is still follow the bell shape [1, 5, 8, 13]. The other researcher had setting the age parameter by group them into younger and older category. The authors had set the claimants' age below 45 years old as younger and above 45 years old as older and the authors also had stated that the claims is high for younger worker [15, 18].

3.3   Total Direct Cost Claims on Chronic MSD Related to Types Industry, Causes Injury, Types Injury and Body Part

The claims for types industry were total up to 23 types of industry including manufacturing sectors, administration, services, civil, farming, education and others industry. Most of the total direct cost claims were from manufacturing industry accounted RM 5,181,282.34 and accumulated near to 50% from overall cost other than others industry. In manufacturing industry, male workers compensate approximate total direct cost RM 4,455,094.00 and female worker were costing RM 726,188.27. The average total direct cost claims for manufacturing most frequently at age 35–44 (RM 31,453.3) which is 25.5% from the others age categories. The others types industry shows a digit percentage in its distribution of total direct cost. The least of total direct cost were from supporting services to water transport industry (RM 33,842.90). All claimants from the types of industry has been summarized in Table 3.

Table 3. Total direct cost based on types industry

Types of industry	Total direct cost (RM)		
	n	Sum	%
Manufacturing	185	5,181,282.34	47.0
other services	29	988,636.54	9.0
Iron and Steel Basic industries	36	865,306.98	7.8
Personal services	33	804,805.87	7.3
Business services, exc. machinery and leasing	22	696,375.89	6.3
others industry	29	578,633.91	5.2
General contractors incl. civil engineering	12	269,201.70	2.4
Monetary institutions	9	268,786.34	2.4
Electric light and power	6	232,947.74	2.1
Social security and welfare	11	228,413.13	2.1
Palm oil	6	169,112.63	1.5
Printing, publishing and Allied Industries	4	108,351.32	1.0
Research and scientific institutes	4	92,737.48	0.8
General administration	3	89,611.99	0.8
Medical, dental and other health services	3	80,469.59	0.7
Engineering, architectural and tech services	4	74,743.16	0.7
Parts and accessories for motor vehicles	4	64,943.76	0.6
Hotels, rooming houses, camps, lodging places	4	61,986.52	0.6
Restaurants, cafes and other eating places	3	55,035.59	0.5

The cost of MSDs is different through sectors and profession. Manufacturing is most favourite parameter used by researcher because many claimants contribute high cost on manufacturing industry. There is study in Malaysia reported the manufacturing industry were leading on claimants between the range 2001–2006 [1]. This is similar to the others study reported manufacturing was frequent claims [8, 9]. Manufacturing industry is sector that produce or manufacture electrical and electronic products; chemicals, chemical products and petroleum products; wood and wood products; textiles, apparel and footwear; construction-related materials; transport equipment; and food products, beverages and others products. Manufacturing involve of mass scale and require worker to stand and sitting for prolong and need to over time in time to fulfil the production target. Besides, manufacturing sector is envisaged to expand, spurred by robust domestic demand and export-oriented industries.

Table 4 shows the type causes of injury being reported and classified by SOCSO over the periods. From the analyse, strenuous movements recorded as most frequent claims with cost RM 7,088,839.51 with similar to 64.3% and 264 claimants highest compare to over-exertion in lifting objects (59), others type of accident, over exertion in pushing or pulling objects (19), Striking against stationary objects (26) and others causes of accidents were listed in the Table 4. Strenuous movement was most occurrence at age 35–44 years with average cost RM 29,833.86. Over-exertion in lifting objects and other types of accident were recorded at age 35–44 as highest average total direct cost compare to other age category. Contrast to over-exertion in pushing or pulling object was most frequent at age 25–34 years old.

Table 4. Total direct cost for causes of injury

Causes of injury	Total direct cost (RM)		
	n	Sum	%
Strenuous movements	264	7,088,839.51	64.3
Over-exertion in lifting objects	59	1,542,937.56	14.0
Other types of accident, not classified	20	529,602.02	5.4
Over-exertion in pushing or pulling objects	19	590,446.88	5.2
Striking against stationary objects	26	574,142.82	4.8
Over-exertion in handling or throwing objects	11	357,671.00	3.2
Caught between a stationary object and a moving object	9	186,988.81	1.7
Accidents not classified for lack of data	3	57,458.40	0.5
Falls of persons on the same level	2	28,414.89	0.3
Struck by moving objects, excl. falling object	1	26,254.41	0.2
Striking against moving objects	1	21,613.80	0.2
Falls of persons from heights and into depths	1	19,045.15	0.2
Total	**416**	**11,023,415.25**	**100.0**

Most frequent types of injury experience on sprain and strain nearly to 80% claims of total direct cost. The cost incurred for sprain and strain was RM 8,753,975.13. Dislocation reported as the highest second place of claimable claims which incurred total direct cost of RM 665,204.98, followed by concussion and other internal injuries RM 653,374.58. The least cost for total direct cost was multiple injuries of different nature that cost RM 74,808.93 (7%). Total direct cost for cause injury were distributed in the Table 5. The most claimants on average total direct cost for sprain and strain were at age 35–44 years old (RM 28,086.39).Dislocation also incurred average total direct cost at age 35–44 years old (RM 36,577.49). For concussions and other internal injuries, the average total of direct cost was recorded RM 50,939.32 at age 25–34.

Sprain and strain consequent to strenuous movement and the claims similarly high. Research on American Journal of Industrial Medicine, 1989 reported strain was the most frequent claim and also dislocation about 31.9% claimant followed by cuts, lacerations, punctures, scratches, and abrasions 14.9%, and chemical burns 13.4% [11]. 73% on sprain and strain claims was recorded in Ohio and also most frequent claim [8]. Similarly to other research reported 35% is from a sprain and strain pain [20].

Table 5. Total direct cost for types of injury

Types of injury	Total direct cost (RM)		
	Count	Sum	%
Sprains and strains	335	8,753,975.13	79.4
Dislocations	20	665,204.98	6.0
Concussions and other internal injuries	25	653,374.58	5.9
Contusions and crushings	11	302,051.76	2.7
Other wounds	8	207,687.22	1.9
Other and unspecified injuries	5	137,598.08	1.2
Effects of weather, exposure"	3	117,070.52	1.1
Fractures	5	111,644.05	1.0
Multiple injuries of different nature	4	74,808.93	0.7
Total	**416**	**11,023,415.25**	**100.0**

Sprain is a stretch or tear of a ligament, a strong band of connective tissue that connect the end of one bone with another. Meanwhile, a strain is an injury to a muscle or tendons. Tendons are fibrous cords of tissue that attach muscles to the bone. Strains often occur in foot, leg (typically the hamstring) or back. Strain is more likely a simple stretch in the muscle or tendon compare to sprain. The symptom of strain include of feel the pain, muscle spasm, muscle weakness, swelling, inflammation and cramping. Typically, sprain and strain pain limiting the claimants to perform their daily job efficiently. Thus, this limiting is reducing the productivity of workers and yet will reduce the number of Malaysia gross domestic product in order to fulfill domestic and international demands.

In Table 6, back shows the highest total direct cost with RM 5,526,590.69, with half percent of claims. Hip incurred about 40% on total direct cost (RM 4,447,674.63) follow by abdomen (RM 269,894.88), shoulder (RM 202,493.48), knee (RM 141,679.05), trunk, multiple location (RM 99,965.29), hand (except fingers alone) (RM 72,697.32), fingers (RM 53,554.42), neck (RM 50,277.99), wrist (RM 41,612.15), and least total direct cost was at elbow (RM 7,434.10). Back and hip recorded the average total direct cost at age 35–44 respectively. Meanwhile abdomen recorded the total average cost at age 25–34.

Other researchers have specified the specific part on back region as lumbar spine or lower back and this body part recorded as similar most frequent MSD claim [3, 5, 8, 15]. According to Murphy and Courtney lumbar spine pain was main contributor reported in many industries and represent 25% of the cost claims [17]. In this data were not provided the specific of body part on back pain. It's probably include all of the lower backpain and upper back pain. Back pain is definitely common pain for the both industry and developed countries, with up to 50 percent of claimants suffering over the year. Back pain lead to disability adjusted life years and major of absenteeism [7].

Table 6. Total direct cost for body part

Body part	Total direct cost (RM)		
	n	Sum	%
Back	209	5,526,590.69	50.1
Hip	152	4,447,674.63	40.3
Abdomen	9	269,894.88	2.4
Shoulder	11	202,493.48	1.8
Knee	7	141,679.05	1.3
Trunk, multiple locations	6	99,965.29	0.9
Hand (except fingers alone)	6	72,697.32	0.7
Fingers	3	53,554.42	0.5
Neck	2	50,277.99	0.5
Wrist	3	41,612.15	0.4
Lower limb, unspecified location	1	34,402.13	0.3
Leg (lower leg)	2	27,040.28	0.2
Feet (except toes alone)	1	22,437.00	0.2
Ankle	1	16,516.76	0.1
Trunk and one or more limbs	1	9,145.08	0.1
Elbow	2	7,434.10	0.1
Total	416	11,023,415.25	100.0

4 Conclusion

Occupational morbidity imposes important health and economic burden on individual workers, employers, and society. The SOCSO database were provided the statistical data of occupational disease based on standardize reporting. From the analyzing the data, the chronic MSDs was determined. MSD total direct cost were reported such amount RM 11 billion for the five period back (2009–2014) and only for the 416 claimants. This tremendous cost will greatly effect on social economy. This cost incurred shows significantly there is need to concern on MSDs disease from the bottom line to the top level.

The information is bias due to unknown specific work of claimant and hazardous work untold. In addition MSDs can be noticed when the claimants are experience pain at the chronic level and exclude the factor or any causes out from working daily task. The distribution of cost also not mention the types of claimant profession and this cost will be vary depending on the wages earning and years of experiences.

The manufacturing industry is the main contributor to the economy growth and the blue collar such as operator more likely to experience the chronic MSDs disorder. At meantime, the production volume was at high most. Therefore, the workers tend to carry similar work at short time so that can achieve short tack time work plan. It might the reason of the claimant having strenuous movement and back pain at high frequency and cost due to prolong standing and too many movement because of the unsuits workplace design.

The major of total direct cost on types of industry, types of injury, causes of accident and body part injury shows the average total cost significantly on age between 35–44 years old. Special attention should be given to this range age as to benefit their quality life as this age is categorize as younger worker. The objectives of this study has been achieve.

This study is important as to provide basis for future studies and intervention on MSDs related injuries in working environment in Malaysia.

Acknowledgement. The authors would like to acknowledge the financial support from National Institute of Occupational Safety and Health Malaysia (NIOSH) and occupational diseases claims data from SOCSO.

References

1. Abas AB, Said AR, Mohammed MA, Sathiakumar N (2008) Occupational disease among non-governmental employees in Malaysia: 2002–2006. Int J Occup Environ Health 14:263–271
2. Abbas M (2015) Trend of occupational injuries/diseases in Pakistan: index value analysis of injured employed persons from 2001–02 to 2012–13. Saf Health Work 6:218–226
3. Silverstein BA, Viikari-Juntura E, Kalat J (2002) Use of a prevention index to identify industries at high risk for work-related musculoskeletal disorders of the neck, back, and upper extremity in Washington State, 1990–1998. Am J Ind Med 41:149–169
4. Baldwin ML (2004) Reducing the costs of work-related musculoskeletal disorders: targeting strategies to chronic disability cases. J Electromyogr Kinesiol 14:33–41
5. Bhattacharya A (2014) Costs of occupational musculoskeletal disorders (MSDs) in the United States. Int J Ind Ergonomics 44:448–454
6. Bultmann U, Franche RL, Hogg-Johnson S, Cote P, Lee H, Severin C, Vidmar M, Carnide N (2007) Health status, work limitations, and return-to-work trajectories in injured workers with musculoskeletal disorders. Qual Life Res 16:1167–1178
7. Connelly LB, Woolf A, Brooks P (2006) Cost-effectiveness of interventions for musculoskeletal conditions. In: Disease control priorities in developing countries, pp 963–980
8. Davis K, Dunning K, Jewell G, Lockey J (2014) Cost and disability trends of work-related musculoskeletal disorders in Ohio. Occup Med 64:608–615
9. Dunning KK, Davis KG, Cook C, Kotowski SE, Hamrick C, Jewell G, Lockey J (2010) Costs by industry and diagnosis among musculoskeletal claims in a state workers compensation system: 1999–2004. Am J Ind Med 53:276–284
10. Escorpizo R (2008) Understanding work productivity and its application to work-related musculoskeletal disorders. Int J Ind Ergon 38:291–297
11. McCurdy SA, Schenker MB, Lassiter DV (1980) Occupational injury and illness in the semiconductor manufacturing industry. Am J Ind Med 15:499–510
12. Middlesworth M (2016) Financial burden of musculoskeletal disorders (MSDs). http://ergo-plus.com/financial-burden-of-musculoskeletal-disorders-msd/
13. Nur NM, Dawal SZMd, Dahari M (2014) The prevalence of work related musculoskeletal disorders among workers performing industrial repetitive tasks in the automotive manufacturing companies. In: Proceedings of the 2014 international conference on industrial engineering and operations management Bali, Indonesia, 7–9 January 2014

14. Oh I-H, Yoon S-J, Seo H-Y, Kim E-J, Kim YA (2011) The economic burden of musculoskeletal disease in Korea: a cross sectional study. BMC Musculoskelet Disord
15. Peele PB, Xu Y, Colombi A (2005) Medical care and lost work day costs in musculoskeletal disorders: older versus younger workers. In: International congress series, pp 214–18
16. Piedrahita H (2006) Costs of work-related musculoskeletal disorders (MSDs) in developing countries: Colombia case. Int J Occup Saf Ergon 12:379–386
17. Murphy PL, Courtney TK (2002) Low back pain disability: relative costs by antecedent and industry group. Am J Ind Med 37:558–571
18. Rohani JM, Zainal AM, Johari MF, Sirat RMd (2016) Analysis of compensation cost related to musculoskeletal disorders (MSDs) against younger and older Malaysian manufacturing workers. In: Proceedings of the 2016 international conference on industrial engineering and operations management, Kuala Lumpur, Malaysia
19. A, Cheng S, Douwes J, Ellison-Loschmann L, McLean D, Pearce N (2011) Prevalence of musculoskeletal symptoms in relation to gender, age, and occupational/industrial group. Int J Ind Ergon 41:561–572
20. WorkCoverSA (2010) Statistical Review Part 1, WorkCoverSA. http://www.workcover.com/site/workcover/resources/publications.aspx. Accessed 4 Oct 2016

Musculoskeletal Symptoms in Midwives and Work-Related Contributory Risk Factors

Kubra Okuyucu[1][(✉)] , Sue Hignett[1] , Diane Gyi[1] ,
and Angie Doshani[2]

[1] Loughborough Design School, Loughborough University, Loughborough, UK
k.arslan@lboro.ac.uk
[2] University Hospitals of Leicester NHS Trust, Leicester, UK

Abstract. This paper presents an exploration of work related musculoskeletal symptoms and contributory factors in midwives. Data were collected with a survey (n = 635) and interviews (n = 15). The survey results showed that the majority of midwives (92%) reported musculoskeletal discomfort within the last 12 months, most commonly for the low back, neck and shoulders. The suggested main contributory factors were awkward working positions, increased work load with longer shifts and fewer breaks, and less support leading to defensive practice. The results of this research indicate that musculoskeletal symptoms are a problem among midwives with serious impacts. Strategies should be developed to manage risk factors to improve patient safety and staff well-being.

Keywords: Musculoskeletal disorders · Midwifery · Workload

1 Introduction

Musculoskeletal disorders (MSDs) and predisposing factors have been explored extensively for nurses; however there is relatively little literature specific to midwifery practice. Midwives were first investigated for MSDs related to tasks unique to midwifery such as supporting mother with breast feeding and delivering babies in different positions by Hignett [1], followed by Royal College of Midwives [2] and Steele and Stubbs [3]. Midwifery tasks and working conditions were reported to be physically and psychologically demanding leading to the risk of developing musculoskeletal symptoms [1, 4]. Contributory factor for MSD risk in midwifery may be because the birth process (delivery) is mother-centred; the pregnant woman is encouraged to choose the most comfortable position for her. The midwife, therefore, is expected to accommodate her chosen position, which might not always result in comfortable working positions for the midwife [1, 5]. Another reason might be that midwives have two 'loads': mother and baby. These kinds of musculoskeletal problems are also known to have a considerable impact on both staff well-being and economy [6].

A recent survey about midwives' health, safety and wellbeing showed that midwives have been affected by working demands and pressure; linked to absenteeism at work [7]. 62% of the participants (n = 1361) were absent from work, most commonly due to stress and musculoskeletal problems. This research aimed to investigate

S. Bagnara et al. (Eds.): IEA 2018, AISC 820, pp. 54–59, 2019.
https://doi.org/10.1007/978-3-319-96083-8_8

musculoskeletal symptoms and explore workload related to contributory factors among midwives in the UK to develop risk management.

2 Methods

A self-administered online survey was designed to investigate MSDs, individual and work related contributory factors. The Nordic Musculoskeletal Questionnaire [8] was used to assess the life-time, point (7 days) and period (12 months) prevalence of symptoms as well as severity, with respect to the effect on work and leisure activities during last 12 months. Additional questions asked about work modifications due to symptoms; sickness absence and MSD treatment; working characteristics including years of work experience, working hours in a week, work place, duration of a shift work, proportion of night shift and breaks; and job satisfaction. The survey was distributed through the Royal College of Midwives (RCM), the Head of Midwifery and the Consultant Midwifery networks. The survey data were uploaded into IBM SPSS Statistics 23 and analysed with descriptives to present the frequency of musculoskeletal symptoms and characteristics of the sample.

In order to explore the issues raised by the survey, semi-structured interviews were conducted with 15 midwives recruited using purposive and snowball sampling. The interview schedule was developed from the survey and relevant literature [1, 10]. Interviewees were asked about their symptoms, perceptions of contributory factors with respect to working conditions and burden, support and actions undertaken by their employer. The interview data were recorded, transcribed, and imported into NVivo10 for thematic analysis, where the text was coded, labelled and then grouped as a theme [11].

This research was approved by Loughborough University Ethics committee and further approvals were confirmed as part of the Health Research Authority (HRA).

3 Results

A total of 635 midwives responded to the survey, with 634 female respondents. More than half of the participants (n = 57%) were over 40 years of age. Of the 635 midwives, 92% reported musculoskeletal symptoms within the last 12 months, most commonly for low back, neck and shoulders. Table 1 shows the prevalence rates of reported musculoskeletal symptoms for life-time, 12 months and 7 days, and severity (impact on normal activities) in nine body parts. Just over half of respondents (51%) thought that their symptoms caused reduction in activities at work and/or leisure time.

Overall, a third of the respondents reported being hospitalised due to their symptoms, most commonly in low back, knee and shoulders. Many (58%) self-managed their symptoms with pain killers. Almost half of the respondents (45%) had to change jobs or duties because of their symptoms and 30% of the participants required sick leave due to musculoskeletal symptoms within the last 12 months.

The respondents' mean experience in midwifery was 15 years (SD = 11.10) with a range from 1 to 46 years, and 56% (n = 357) reported working full time. Over half of the respondents' work place was reported as a maternity unit in a hospital (66%), and

Table 1. Reported musculoskeletal symptoms, prevalence rates and severity (impact on normal activities) within 12 months.

Body area	Life-time (n = 635)		12 months (n = 633)		7 days (n = 627)		Severity (n = 633)	
	n	%	n	%	n	%	n	%
Neck	342	54	287	45	114	18	116	18
Shoulders	327	52	282	45	141	23	156	25
Upper back	188	30	187	30	69	11	92	15
Elbows	78	12	78	12	32	5	34	5
Wrists/hands	196	31	162	26	70	11	83	13
Low back	511	81	452	71	272	43	323	51
Hips/thighs	229	36	183	29	105	17	124	20
Knees	230	36	201	32	87	14	128	20
Ankles/feet	154	24	145	23	75	12	93	15

the remainder was based in midwife-led units in a hospital (8%), standalone midwifery units (4%), or home births (19%). A shift duration was more than 8 h for most respondents (84%). 39% of respondents also reported working more than 12 h in a shift. Of those answering the night shift question (n = 625), 65% have night shifts with 4% only working at night. 43% of respondents said that they were not given sufficient breaks during work.

Of those interviewed, 85% (n = 11) reported having musculoskeletal symptoms, most commonly in back. MSDs were mostly attributed to work related activities or aggravated by working tasks. For example, symptoms in back were thought to be due to assisting breast feeding: *'I* would *imagine that lower back pain is because of twisting and being in [an] awkward position to try to get the woman to feed'* (M11); or the positions for internal examination e.g. sitting on edge of the bed, turning and twisting to access the woman. Shoulder symptoms may be exacerbated by acting as a resistance during pushing (birthing process), quoted as: *'so they are on the bed and we sit side ways and push our shoulders – I know for certain that was the cause majority of my damage to my shoulder'* (M10). Another awkward position contributing to their symptoms was delivering in birthing pool with regular bending over the pool and stretching for examinations: *'obviously, the pools are static – you can't get them up or down, so when you are listening to foetal heart in the pool, you do a lot of bending'* (M03).

Not having enough breaks during the shifts was mentioned many times as a contributory factor for MSDs *'the midwives very rarely actually manage to get a break and if they do it is a short break constantly rushing around'* (M12). This could also result in dehydration and irregular eating patterns, which would impact on staff well-being: *'I have my breakfast at 6.00 and I don't normally have my lunch before 17.00'* (M08).

The interviewees all agreed that fewer staff and an increased work load leads to gradual exhaustion, as one interviewee said: *'I can guarantee to you that in practice, when you have got [a] busy ward, not enough members on duty, coordinators pressing you to make a space, emergencies ongoing on, the last thing you will think is your backache and how to prevent it'* (M08).

The shift hours could also contribute to MSD; all interviewees commented that this change has impacted negatively on their health: *'The hours kill me. I was much better with the shorter hours – 12 h do kill me.'* (M01). *'I think the most extreme work related that challenges midwife role is the working hours'* (M08).

With respect to support from employers, there were some positive comments but most were negative. One favourable comment was: *'I would say that organizationally, yes I think support is there. We are taught, advised and given information, and we can access support for that.'* (M12). However, some believed that a lot of training was given instead of improving working conditions, as stated: *'There is a lot of stuff could be better is not necessarily provided by the trust either, but you are expected still to give that care because like I said you can't deny woman – that choice'* (M02).

Another key theme from the interviews was concern about complaints from the patients which could result in an environment where midwives practice defensively: *'I think we tend to accept that what the patient wants the patient gets because should the patient then complain we would be seem to be fault'* (M10). Defensive practice may lead midwives to do much more than they should: *'they say 'help me move'; actually I am not meant to help to move. You feel awful by saying - if you just do it by yourself. It is not always nice to say - you do it because I am not meant to hurt myself. That makes you look not caring'* (M07).

4 Discussion

The survey showed high prevalence rates of musculoskeletal problems most frequently in low back (71%), neck (45%) and shoulders (45%). These results are in line with previous studies for Australian midwives [4, 12]. The rates are higher than the UK general population MSD prevalence rates, at 34% for neck [13] and 37% for low back [14]. Most respondents attributed their discomfort to static or awkward positions during the delivery and assisting women with breast feeding, similar to the reported physical demands of midwifery for Australian midwives [4].

The impact can be clearly seen as half of the respondents' activities were affected by MSDs. It is therefore very likely that these problems will impact on the quality of care provided and/or patient safety. Sickness absenteeism is known to result in additional work load for other staff as well as financial consequences [6].

Despite previous guidelines and redesigning of equipment e.g. birthing pool [1, 2], MSD prevalence and impact rates are still very high. The reason for this is not clear but it may be explained by the increased work load over the last 20 years. From the interviews, older interviewees discussed the changes in working characteristics as: increased work load, fewer staff and longer shift hours. Midwives recently started 12 h shifts; some might find benefits in working 12 h as they work fewer days of the week. However longer shifts with fewer breaks have been argued to lead to lower productive and opportunities for errors [15, 16].

Practicing defensively was one of the most interesting emergent findings from the interviews, which was linked with fear of complaints from the patients. Defensive practice has been previously been discussed among health professionals, linked to poor staff health, both physically and psychologically, and impact on patient safety [17–19].

5 Conclusion

This research explored MSDs and work related contributory factors among midwives in the UK. The survey found very high prevalence rates and detrimental impacts such as activity reduction and sickness absence. In the interviews, such problems were often attributed to both physical (working activities, long hours, fewer breaks) and psychosocial (defensive practice) challenges. In conclusion, midwifery working conditions put many pressure on the staff which will impact on the quality of care, patient safety and staff well-being. However, such discomforts and their impacts could be reduced by improving working conditions and developing strategies to provide the staff with a better working environment.

References

1. Hignett S (1996) Manual handling risks in midwifery: identification of risk factors. Br J Midwifery 4(11):590–596
2. Royal College of Midwives (1997) Handle with care: a midwife's guide to preventing back injury. Royal College of Midwives, London
3. Steele D, Stubbs D (2002) Measuring working postures of midwives in the healthcare setting. In: McCabe P (ed) Annual conference of the ergonomics society 2002, pp 39–44. Taylor and Francis Group
4. Long MH, Johnston V, Bogossian FE (2013) Helping women but hurting ourselves? Neck and upper back musculoskeletal symptoms in a cohort of Australian midwives. Midwifery 29(4):359–367
5. De Jonge A, Teunissen DAM, Van Diem MT, Scheepers PLH, Scheepers PLH (2008) Lagrol-Janssen ALM.: Women's positions during the second stage of labour: views of primary care midwives. J Adv Nurs 63(4):347–356
6. Boorman S (2009) NHS health and well being. Final Report. Department of Health, London
7. Royal College of Midwives (2016) Caring for you campaign: survey results RCM campaign for healthy workplaces delivering high quality care. Royal College of Midwives, London
8. Kuorinka I, Jonsson B, Kilbom A, Vinterberg H, Biering-Sørensen F, Andersson G et al (1987) Standardised nordic questionnaires for the analysis of musculoskeletal symptoms. Appl Ergon 18(3):233–237
9. Siegrist J (1996) Adverse health effects of high-effort/low-reward conditions. J Occup Health Psychol 1(1):27
10. Long MH, Bogossian FE, Johnston V (2013) Midwives' experiences of work-related shoulder musculoskeletal problems. Int J Childbirth 3(1):52–64
11. Robson C, McCartan K (2016) Real world research, 4th edn. Wiley, London
12. Long MH, Bogossian FE, Johnston V (2013) Functional consequences of work-related spinal musculoskeletal symptoms in a cohort of Australian midwives. Women Birth 26(1): 50–58

13. Palmer KT, Walker-Bone K, Griffin MJ, Syddall H, Pannett B, Coggon D et al (2001) Prevalence and occupational associations of neck pain in the British population. Scand J Work Environ Health 1:49–56

14. Papageorgiou AC, Croft PR, Ferry S, Jayson MIV, Silman AJ (1995) Estimating the prevalence of low back pain in the general population: evidence from the South Manchester back pain survey. Spine 20(17):1889–1894

15. Griffiths P, Dall'Ora C, Simon M, Ball J, Lindqvist R, Rafferty A-M et al (2014) Nurses' shift length and overtime working in 12 European countries: the association with perceived quality of care and patient safety. Med Care 52(11):975

16. Rogers AE, Hwang W-T, Scott LD, Aiken LH, Dinges DF (2004) The working hours of hospital staff nurses and patient safety. Health Aff 23(4):202–212

17. Passmore K, Leung WC (2002) Defensive practice among psychiatrists: a questionnaire survey. Postgrad Med J 78(925):671–673

18. Surtees R (2010) Everybody expects the perfect baby… and perfect labour… and so you have to protect yourself: discourses of defence in midwifery practice in Aotearoa/New Zealand. Nurs Inquiry 17(1):82–92

19. Symon A (2000) Litigation and defensive clinical practice: quantifying the problem. Midwifery 16(1):8–14

Multitask Analysis of UL Repetitive Movements by OCRA Method: Criteria and Tools

Occhipinti Enrico[⊠] and Colombini Daniela

EPM International Ergonomics School, Milan, Italy
epmenrico@tiscali.it

Abstract. The OCRA Checklist is one of the tools of the OCRA method for assessing risk associated with repetitive movements of the upper limbs. Task rotation is when a worker alternates between two or more tasks during a certain period of time (i.e.: a day, a month, a year). The multitask analysis process by the OCRA Checklist involves the following steps: 1 - Determine the full task rotation period: daily up to yearly; 2 - Identify repetitive tasks performed in the period; 3 - Use the OCRA checklist to calculate the "intrinsic" score for each task as if it was performed for the entire period; 4 - Analyse tasks and scores with respect to real time exposure and apply two alternative computational models: Time-Weighted Average and Multitask Complex. While, in the industrial sectors, tasks rotate often in a similar way every day and consequently the previous procedures could be easily applied, in some productive sectors exposure assessment is much more complex being characterized by the presence of several tasks over periods longer than a day (weekly, monthly, yearly). In general the study of these conditions are based on the use of the Checklist OCRA and on adaptations of the two multitask analysis models.

Keywords: Repetitive movements · Multitask analysis · OCRA method

1 Introduction

The OCRA method for assessing risk associated with repetitive movements of the upper limbs consists of two tools, the OCRA Checklist and the OCRA Index [1]. The tools feature different analytical details and purposes, although both are inspired by the same conceptual model. The OCRA checklist is the simpler of the two tools and is used for the initial screening of workstations [2]; the OCRA Index is a more complex and was chosen as the preferred risk assessment method by international standards relating to high-frequency repetitive manual work [3].

The OCRA checklist is a simplified tool (based on the OCRA Index) for measuring the risk of biomechanical overload of the upper limbs, which can be used both in the initial stage of estimating risk levels in a certain productive sector or later, for managing the aforesaid risk. The OCRA checklist consists of five parts that focus on the four main risk factors (lack of recovery time, frequency, force, awkward posture/stereotyped movement) and a number of additional risk factors (vibration, low temperatures, precision work, repeated impacts, etc.), and also factoring the net

© Springer Nature Switzerland AG 2019
S. Bagnara et al. (Eds.): IEA 2018, AISC 820, pp. 60–71, 2019.
https://doi.org/10.1007/978-3-319-96083-8_9

duration of repetitive jobs on the final estimate of risk. The classic analysis proposed by the OCRA checklist entails using pre-assigned scores (the higher the score, the higher the risk) to define the risk associated with each of the aforementioned factors. The sum and product of the partial values generate a final score which estimates the exposure level, featuring four different levels (green, yellow, red and purple). The calculation procedure for reaching the final result (see Fig. 1) shows how all the risk factors are included: the *lack of recovery period* factor is a multiplier to be applied, along with the *duration factor*, to the sum of the scores for the other risk factors.

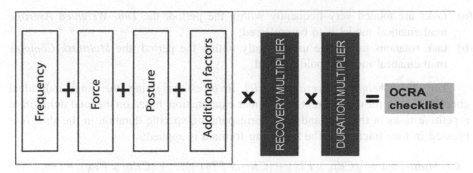

Fig. 1. OCRA checklist: final score calculation procedure

The OCRA checklist is used to describe a workstation and estimate the exposure level embedded in the task, as if this task were the only performed by a single worker for the entire duration of the shift.

Task rotation is when a worker alternates between two or more tasks during a certain period of time (i.e.: a daily shift, a week, a month, a year). While in the industrial sectors tasks rotate often in a similar way every day, in some productive sectors (agriculture, construction, cleaning, supermarket, etc.) exposure assessment is much more complex being characterized by the presence of several tasks over periods longer than a day (weekly, monthly, yearly turnover). In these special situations, such as when the worker has to perform a large number of tasks and the tasks are distributed "asymmetrically" throughout the period of time, risk assessments can become extremely complex. In general the study of those conditions are based on the use of the Checklist OCRA and on adaptations of the two multitask analysis approaches (average and complex) that will be presented.

2 General Approach for Multitask Analysis

The multitask analysis process by the OCRA Checklist involves the following steps:

STEP 1 - Determine how long it takes to complete a full task rotation: daily, weekly, monthly, yearly; STEP 2 - Analyse the work organization to identify repetitive tasks performed in the period; STEP 3 - Use the OCRA checklist to calculate the intrinsic score of repetitive tasks, analysing each task as if it were performed for the entire shift;

STEP 4 - Analyse tasks and scores (risk level) with respect to real time exposure: identify a homogeneous group, describe duration of real exposure of the homogeneous group (allocation of tasks in the period), apply computational mathematical models and interpret the results.

In general, independently from the length of the rotating period, there are two main ways of organizing worker turnover, and consequently two different calculation methods:

(a) tasks are rotated very frequently within the period: the *Time-Weighted Average* mathematical model is to be preferred;
(b) task rotations take place unfrequently within the period: the *Multitask Complex* mathematical model should be used.

The *Time-Weighted Average model* involves weighting the final individual checklist scores for the different tasks under examination, based on the total duration of repetitive tasks in the shift and their corresponding specific duration in the shift (expressed in time fractions). The following formula is applied:

$$CK\,Multi\text{-}TWA = [(CKi_1 \times FT_1) + (CKi_2 \times FT_2) + \ldots + (CKi_N \times FT_N)] * Dm_{tot} \quad (1)$$

where

$CKi_{1,2,\ldots,N}$ are the *intrinsic scores with recovery* obtained from the intrinsic OCRA checklists for the various tasks that the worker performs, calculated using the *Recovery Multiplier* corresponding to the actual distribution and duration of recovery periods in the shift while the *Duration Multiplier* is a constant = 1
$FT_{1,2,\ldots}$, FT_N represent the time fractions of duration of the various repetitive tasks versus the total duration of repetitive work
Dm_{tot} = *Total Duration Multiplier*, relative to the net duration of all repetitive tasks in the shift

The *Multitask Complex model* is based on the concept of task generating the highest overload as minimum. With this approach, the final result will be at the least equivalent to the Checklist OCRA score for the most overloading task, according its continuous duration, and at the most equal to the Checklist OCRA score for the same task applied, however (only theoretically), to the overall duration of all the repetitive tasks examined. In this case the procedure is based on the following formula:

$$CK\,Multi\text{-}MC = CKi_{1(Dm1)} + (\Delta CKi_1 \times K) \quad (2)$$

where
$$K = \frac{(CKi_{1(Dm\,tot)} * FT_1) + (CKi_{2(Dm\,tot)} * FT_2) + \ldots + (CKi_{N(Dm\,tot)} * FT_N)}{(CKi_{1(Dm\,tot)})}$$

1, 2, 3, J ..., N = repetitive tasks listed according to the score (CK) of the OCRA checklist (task 1 = the task with the highest CK score; task N = the task with the lowest CK score);

Dm_j = *Duration Multiplier* according to the actual duration of each task$_j$ in the shift

Dm_{tot} = *Duration Multiplier* for the total duration of all repetitive tasks in the shift

$CKi_{1(Dm1)}$ = the CK score of task$_1$, calculated with the *Lack of Recovery Time Multiplier* actually present in the tasks and considering Dm_1 (it is worth remembering that task$_1$ is the task with the higher risk)

$CKi_{1(Dm\ max)}$ = the CK score of task$_1$ (the highest risk task) considering Dm_{tot}

$\Delta CKi_1 = CKi_{1(Dm\ max)} - CKi_{1(Dm1)}$

FT_j = Time fraction (between 0 and 1) of each task j with respect to the total repetitive working time in the shift.

3 Daily Rotation

Considering a daily shift, two main ways of rotation among repetitive task should be considered:

(a) task rotation takes place within a period of less than 90 consecutive minutes for each task performed: the *Time-Weighted Average* mathematical model is used;
(b) task rotation takes place within a period of more than 90 consecutive minutes for each task performed: the *Multitask Complex* mathematical model is used.

3.1 The Time-Weighted Average Model

This approach and calculation model is only suitable when the task rotation rate is fairly high, for instance once every 90 min or less. The approach should also be used when different products (or models of the same product) are processed at the same workstation during the shift. In such cases, it can be assumed that higher risk exposure is somewhat offset by lower risk exposure, with the worker alternating between the two within a relatively short time frame. To obtain the Checklist score for a worker performing a job entailing repetitive tasks at several workstations, using the multitask analysis with Time-weighted Average method (*CK Multi-TWA*), the formula (1) must be applied.

3.2 The Multitask Complex Model

Conversely, if the repetitive task rotation occurs more than once every 90 min, for instance a repetitive task for all the morning and another for all the afternoon, the time-weighted average approach could underestimate the actual exposure level. This problem is particularly acute in the study of certain jobs where tasks featuring high intrinsic risk indexes alternative with lighter tasks.

In such cases it is more realistic to adopt an approach based on the task generating the highest overload as minimum. With this approach, the result will be at the least equivalent to the OCRA indicator for the most overloading task, in terms of its duration, and at the most equal to the OCRA indicator for the same task applied to the overall duration of all the repetitive tasks examined. The scores calculated using this procedure are defined as the Checklist Multitask Complex (*CK Multi-MC*). The *CK Multi-MC* computation method - see formula (2) - also take into account the Duration Multipliers that adjust the exposure level as a function of the total time spent performing repetitive tasks within a routine work shift. Table 1 shows the Duration Multipliers to use as a function of both the overall duration (in minutes) of the repetitive work (sum of the duration of each of the repetitive tasks present in the shift and included in the rotation) and of the continuous intrinsic durations of each task.

Table 1. OCRA Checklist Duration Multipliers as a function of the duration of repetitive tasks in the shift

Duration of repetitive task/s (minutes)	<1,9	1,9– 3,6	3,7– 7,4	7,5– 14	15– 29	30– 59	60– 120	121– 180	181– 240	241– 300	301– 360	361– 420	421– 480	>480
Duration Multiplier (Dm)	0,007	0,018	0,05	0,1	0,2	0,35	0,5	0,65	0,75	0,85	0,925	0,95	1	1,5

In practice, in order to calculate *CK Multi-MC*, the following steps are necessary:

- Calculate a traditional CK score for each task performed by the worker in the shift, considering its intrinsic continuous (real) duration, using the relative Dm_j as the *Duration Multiplier*. Repeat the calculation with the *Recovery Multiplier* for the entire shift. Select the worst task, which will correspond to $CKi_{1(Dm1)}$.
- Calculate the same CKi_j for each task keeping all the parameters identical except for the *Duration Multiplier*, which in this case will be considered in relation to the total duration of all the repetitive tasks in the shift (Dm_{tot}.). Thus, the respective $CKi_{j(Dm\ max)}$ will be obtained for each task. Find the highest value (i.e. the score for the worst task): this is $CKi_{1(Dm\ max)}$.
- Calculate ΔCKi_1 for task 1 (the highest risk task): $CKi_{1(Dm\ max)} - CKi_{1(Dm1)}$
- Calculate the time fraction of tasks 1, 2, 3, etc. (FT_j), dividing their respective duration (in minutes) by the total repetitive work time in the shift.
- Calculate K using the formula (3); in practice:
- *multiply each $CKi_{j(Dm\ max)}$ by its respective FT_j and add up the resulting values.*
- *divide this amount by $CKi_{1(Dm\ max)}$.*

K is usually within the range of 0 and 1.

3.3   Examples of Calculations

Consider three repetitive tasks (work stations) and their relative intrinsic checklist score; Consider one worker rotation in the shift with the corresponding durations:

-Intrinsic checklist values for real duration (as if each task lasted for the real 400 min in the shift, with Duration Multiplier (Dm_{tot}) = 0,95 and with the real Recovery Multiplier)

Task A = $CKi_{(Dm\ max)}$ = 23,8; duration in the shift = 100 min
Task B = $CKi_{(Dm\ max)}$ = 12,83; duration in the shift = 140 min
Task C = $CKi_{(Dm\ max)}$ = 8.08; duration in the shift = 160 min.

The total time assigned to repetitive work is 400 min (Dm_{tot} = 0,95) and the various time fractions for these minutes are:

Task A: FT_A = 25% (0.25)
Task B: FT_B = 35% (0.35)
Task C: FT_C = 40% (0.4)

Two rotation schemes alternating the three tasks should also be considered, where the first (scheme A) features a task rotation duration of more than 90 min and the second (scheme B) a task rotation duration of less than 90 min.

Scheme A: calculation of exposure risk for task distribution in the shift with task rotations of more than 90 min.

If the rotations are less frequent (for instance, if the tasks are performed consecutively with each lasting more than 90 min) then CK Multi-MC will be calculated using the relevant formula (2).

Based on the previous data and performing the necessary calculation, the result will be:

- $CKi_{1(Dm1)}$ = Checklist score for $task_1$ considering Dm_1 ($task_1$ being the highest risk task) = 12,5
- $CKi_{1(Dm\ max)}$ = Checklist score for $task_1$ considering IR for $task_1$ considering Dm_{tot} = 23,8
- ΔCKi_1 = (23,8 − 12,5) = 11,3
- K = [(23,8 * 0,25) + (12,8 * 0,35) + (8,1 * 0,4)]/23,8 = 0,57
- CK Multi-MC = 12,5 + (11,3*0,57) = 18,9

Scheme B: calculation of exposure risk for task distribution in the shift with task rotations of less than 90 min.

Applying the The Time-Weighted Average model (CK Multi-TWA) and taking into account the total duration of repetitive work (Dm_{tot} = 0,95), the formula (1) results in:

- CK Multi-TWA = [(23,8*0,25) + (12,8*0,35) + (8,1*0,4)] = 13,01

This value represents the exposure level for a worker alternating between three work stations based on the durations indicated but with rotation frequencies of more than one every 90 min.

4 Rotations Over Periods Longer Than a Day (Weekly, Monthly, Yearly)

The next step is to define a set of procedures and criteria for estimating risk in more complex situations, where workers perform multiple tasks variously distributed, in qualitative and quantitative terms, over periods longer than a day (weekly, monthly, yearly). In these cases organizational analysis becomes more and more relevant.

4.1 Identification of the Homogeneous Group and of the Tasks Performed

The first step is to identify, in the organization (plant, department, etc.), the homogeneous group of workers, that's to say the group of workers that performs the same tasks, in the same workplaces and with similar durations (or time patterns) during the selected period that can be a week, a month or a year depending on the way the tasks are rotated; note that a homogeneous group may sometimes be made up of just one person, if no other workers perform the same tasks qualitatively and quantitatively. Secondly, a complete list of the repetitive task performed by the group during the period is needed.

4.2 Analysis of Each Individual Task Using the OCRA Checklist to Calculate the Intrinsic Score

All the tasks performed by the group should be analyzed using the OCRA Checklist to calculate the "intrinsic" score for each task (right and left arm). Calculating the intrinsic risk score for a certain task means evaluating each task as if it is the only one performed by the worker all the time (i.e. for the whole shift and the whole period). To estimate the intrinsic Checklist Score (CKi), reference is made to a shift constant scenario featuring:

- 430/460 net minutes of repetitive work (modal value = 440, *Duration Multiplier* = 1)
- one 30-min meal break and two 10-min breaks (*Recovery Multiplier* = 1,33).

Often it is better to have a complete analysis (by the Ocra Checklist) of all the repetitive tasks performed in the organization, regardless of who performs them and than attribute specific tasks to a specific worker or homogeneous group of workers.

4.3 Exposure Time Constants

Before tackling the organizational analysis of the homogeneous group, Table 2 reports the exposure constants that will be referred to for calculating the exposure time prevalences for the homogeneous group with respect to the various tasks. In this type of non-daily analysis, it is not uncommon to come across task distribution patterns that are irregular and erratic. It has been found to be useful to adopt as reference several exposure constants representing the typical exposure level for the industry.

Table 2. Exposure time constants

Hours/day constant	8	Hours/month constant	160
Minutes/day constant	440	Days/month constant	20
Days/week constant	5	Months/year constant	11
Minutes/week (440 min * 5 days) constant	2200	Days/year constant	220
Weeks/month constant	4	Hours/year constant	1760

4.4 Organizational Analysis: Typical Day, Task Assignment, Task Duration and Calculation of Proportional Task Duration Using Constants

This section focusing on the organizational analysis, needs to be broken down into its constituent parts.

Description of a Typical Working Day. As for a normal OCRA Checklist, the analysis will discover the shift duration, number and duration of pauses, and duration of non-repetitive tasks in order to obtain the net duration of repetitive work *(Duration Multiplier)* and score for the lack of recovery times *(Recovery Multiplier)*. This analysis could be more or less detailed depending if the period of time considered is a week, a month or a year. In the first case we need to have data day by day, in the last we may consider one "typical or modal" shift for all the year as reported for example in Table 3.

Table 3. Description of a *typical or modal working day* with respect to annual exposure: duration of shift, pauses, non repetitive tasks, net repetitive tasks duration.

DESCRIPTION OF A TYPICAL WORKING DAY	
OVERALL SHIFT DURATION (min)	420
Nr. of actual breaks (recovery periods) during the shift, with a duration of at least 8 minutes (except meal break)	3
Overall duration of all actual breaks (excluding meal break) in minutes	30
Actual duration of meal break if included in shift duration (min)	
Nr. of other breaks (i.e. meal break out of working time; travel time to/from different company locations) that last at least 30 minutes.	1
Total duration of non-repetitive tasks (e.g.: cleaning, fetching supplies, etc.) in minutes	30
ESTIMATED NET DURATION OF REPETITIVE WORK (min)	360

Distribution of Tasks and Calculation of Their Proportional Duration During the Period. This step involves assigning the tasks performed by the homogeneous group (or individual exposed worker) qualitatively and, most of all, quantitatively. This part of the analysis is the most difficult one since it is needed to know how much time is spent, during the period (week, month, year) in performing the different tasks. The level of detail in the analysis depends from the period examined: it regards presence and duration of different tasks in single days for analysis of a week, single days or weeks for analysing months, single months for analysing a year. The level of detail depends also from how much information about these elements could be derived from production data. Extreme accuracy is not required for the proportional assignment of tasks: the employer, or even the members of the homogeneous group, should be able to provide this information.

The final aim of the step is to compute the total number of hours worked by each homogeneous group on each task in the representative period. This will produce their proportion with respect to the total number of hours worked and also with respect to the constant for the representative period (refer to Table 2). This brings us to the main findings that can be used to convert a defined period (week or month or year) into minutes spent in performing each task in a *fictitious day*. This conversion is preferably referred to the proportions with respect to the constants for the representative period, since these proportions are computed with respect to a fixed reference and thus take into account possible conditions of part-time or overtime work in single elements of the period. Table 4 shows an example of this final step.

Table 4. Example of calculation of minutes worked per a representative week: preliminary data for obtaining the *fictitious day*, representative of the week

	MONDAY	TUESDAY	WEDNESDAY	THURSDAY	FRIDAY	SATURDAY	SUNDAY	TOTAL MINUTES WORKED PER TASK IN A WEEK	Duration of FICTITIOUS WORKING DAY in minutes, representative of a week	Duration Multiplier for FICTITIOUS DAY
AAA	4.50	9.00		9.00	9.00	68.00		100		
BBB	4.50	9.00		9.00	9.00	68.00		100		
CCC	4.50	9.00		9.00	9.00	68.00		100		
DDD	2,25	4.50		4.50	4.50	68.00		84	METHOD OF CALCULATION = 1,600/5 (constant days worked per week)	
EEE	2,25	4.50		4.50	4.50	68.00		84		
FFF	13.50	27.00		27.00	27.00	0.00		95		
GGG	90.00	180.00		180.00	180.00	0.00		630		
HHH	9.00	18.00		18.00	18.00	0.00		63		
III	4.50	9.00		9.00	9.00	0.00		32		
LLL	45.00	90.00		90.00	90.00	0.00		315		
TOT	180	360	0	360	360	340	0	1600	320	0.925

4.5 Definition of the Fictitious Day and Application of the Models for Calculating the Final Score

For calculating the final exposure scores using the models and the formula that have been presented for daily rotations, it has been necessary to convert the data relative to the distribution of the tasks in the observed period into a *fictitious working day*, representative of the whole period. In particular, the duration in minutes of each task and the overall duration of all tasks in the fictitious day have been derived from the complex study of work organization performed in previous steps. These data allow now, together with those regarding the "intrinsic" OCRA Checklist scores for each task (see Sect. 4.2), to compute the final OCRA Checklist scores using both the

Time-Weighted Average and the Multitask Complex model and their corresponding formula (1) and (2). Both mathematical models have been adjusted for calculating weekly monthly or yearly multi-task exposure by OCRA Checklist.

In order to complete the estimation of a OCRA Checklist score for *fictitious working day* representative of the observed period, certain procedures are required that involve recalculating the intrinsic Checklist scores (CKi, i.e. estimated for tasks lasting 8 h with one meal break and two 10-min breaks) to generate the new *adjusted intrinsic Checklist scores*, or CKic, reflecting the actual organizational conditions in the homogeneous group.

The following parameters must be used:

- *Recovery Multiplier*: this value is derived from the organizational data describing the presence and distribution of breaks tasks on a *modal day*; in the example of Table 3, the *Recovery Multiplier* is 1,12.
- *Total Duration Multiplier (Dm$_{tot}$)*: this value is derived from the overall duration of all the repetitive tasks considered in the *fictitious working day* representative of the whole period; it is calculated using the criteria in Table 1; in the example of Table 4, Dm$_{tot}$ is 0,925

Table 5. Example of calculation of monthly final Checklist score using the Checklist Multitask Complex model

ACTIVE TASKS	INTRINSIC OCRA CHECKLIST VALUES (8 hours per shift, 2 breaks, 1 meal)	% VS. TOTAL HOURS WORKED PER MONTH	FICTITIOUS MINUTES PER TASK IN FICTITIOUS DAY	Partial Duration Multiplier for each task of the fictitious day representative of the month	Checklist values calculated for each task of the fictitious day (month) CONSIDERING ITS ACTUAL INTRINSIC DURATION	Total Duration Multiplier for the fictitious day (month) and Recovery Multiplier	Checklist values (calculated for each task of the fictitious day (month) CONSIDERING ITS TOTAL DURATION (i.e. 312.5 minutes)
AAA	5.3	4.86%	15.20	0.200	1,0		4.4
BBB	6.7	4.86%	15.20	0.200	1.2	**0.925** Duration Multiplier for 312.5 minutes of total duration the fictitious day representative of the month	5.6
CCC	8.0	4.86%	15.20	0.200	1.4		6.7
DDD	10.6	3.87%	12.10	0.100	1.0		8.9
EEE	11.3	3.87%	12.10	0.100	1.0		9.4
FFF	13.3	5.95%	18.60	0.200	2.4		11.1
GGG	14.6	39.8%	123.98	0.650	8.6		12.2
HHH	16.0	3.97%	12.40	0.100	1.4		13.3
III	22.6	5.10%	15.95	0.200	4.1		18.9
LLL	31.9	22.96%	71.74	0.500	14.4		26.6
TOTAL		100%	312.5				
Maximum score for most overloading task, based on its partial duration CKic$_{1(Dm1)}$					14.4	**1.2** Modal Recovery Multiplier	
Maximum score for most overloading task, based on total duration CKic$_{1 (Dm tot)}$					26.6		
Δ CKic$_1$						26.6-14.4 = 12.2	
K						0.549	
Final Checklist Score (Multitask Complex model) CK Multi-MC = CKi$_{1(Dm1)}$ + (ΔCKi$_1$ x K)						**21.1**	

Moreover for completing the use of the two formula, the following parameters should be also considered:

- *Partial Duration Multiplier (Dm_J):* this value is derived from the individual duration of each single repetitive task considered in the *fictitious working day*; it is calculated using the criteria in the Table 1.
- *Fraction of time of each task* (FT_J): this value is derived from the ratio between duration of any (each) single task in the fictitious day and the overall duration of all the repetitive tasks considered in the *fictitious working day*.

Table 5 show synthetically how to derive data for calculating the final Checklist score (Multitask Complex model) in a month scenario with 10 repetitive tasks.

5 Final Remarks

When using the OCRA Index and Checklist to assess exposure to several repetitive manual tasks featuring potential biomechanical overload of the upper limbs, the traditional approach has been to refer to calculation models based on the time-weighted average concept. However, this approach is unsuitable to certain applications if not actually misleading, such as when there is high continuous exposure for approximately half the shift, followed by slight exposure for the rest of the shift. In such cases, the weighted average does not reflect the peak continuous exposure of the half shift.

Based on multiple-task load lifting analyses already reported in the literature [4] and tested in practice, calculation models of the OCRA Checklist have been tested for analyzing multiple repetitive tasks, based on the task generating the highest overload as minimum concept. Based on experience acquired with different applications, the hypothesis, still to be confirmed by further research, has even been put forward that the calculation model based on the *time-weighted average* is still valid even if the task rotations are daily and generally fairly frequent (at least every 90 min), or when long lasting tasks are broken down into sub-tasks. Conversely, if exposure to repetitive tasks entails much less frequent daily rotations (over 90 min) or non-daily rotations (i.e. weekly, monthly or annually), it might be more effective to use the new calculation methods presented here.

The mathematical models suggested here could also be applied to scenarios where the task rotation is annual, monthly or weekly after converting the organizational data for the period into a *fictitious working day*. For the time being, the authors prefer to observe the outcome of the "longer periods" risk assessment, applying both mathematical models, since there is still little epidemiological data on the prevalence of UL-WMSDs linked to risk evaluation studies. Based on preliminary clinical findings, the Multitask Complex model appears to be more predictive of adverse health effects, at least for full-time exposures. Further studies and data are needed to confirm these preliminary findings.

References

1. Colombini D, Occhipinti E (2016) Risk analysis and management of repetitive actions - a guide for applying the OCRA system (OCccupational Repetitive Action). CRC PRESS - Taylor & Francis, Boca Raton and New York
2. ISO: ISO TR 12295 (2014) Ergonomics - application document for International Standards on manual handling (ISO 11228-1, ISO 11228-2 and ISO 11228-3) and evaluation of static working postures (ISO 11226). ISO, Geneva, Switzerland
3. ISO: ISO 11228-3 (2007) Ergonomics - manual handling - handling of low loads at high frequency. ISO, Geneva, Switzerland
4. Waters T, Lu ML, Occhipinti E (2007) Using the NIOSH lifting equation for evaluating sequential manual lifting tasks. Ergonomics 50(11):1761–1770

Application Study: Biomechanical Overload in Agriculture

Daniela Colombini[✉]

EPM International Ergonomics School, Milan, Italy
daniela.colombini@fastwebnet.it

Abstract. This project has the goal of defining the basic criteria for producing a guide (Technical Report-TR) for the specific use of the ISO 11228 series, of ISO 11226 and of ISO TR 12295 [4–8] in the agricultural sector. Specifically, the project is aimed at providing the potential users with additional information on how to use existing standards in a world widespread working sector as agriculture where, also if with different characteristics, biomechanical overload is a relevant aspect, WMSDs occurrence is very high and where specific preventive actions are needed.

One of the main goals is therefore to provide all users, and particularly those who are not experts in ergonomics, with criteria and practical procedures for a correct risk analysis. Method as checklist OCRA, NIOSH, carrying, push and pull evaluation are used, applied to annual multitask analysis, through the use of original software by EPM IES, free download from www.epmresearch.org

Keywords: Agriculture · Biomechanical overload · Multitask analysis
Repetitive movements · Manual lifting · OCRA method · TACOs method
Pushing-pulling

1 Purpose and Justification

Agriculture is by far the biggest working sector in the world. It is estimated that 2.6 billion people or 40% of the world's population are farmers. Agriculture is one of the most hazardous sectors in both the developing and the developed worlds. WMSDs are caused mainly by manual handling, heavy physical work, awkward postures and repetitive movements. Increasing attention is being drawn to the application of practical actions in agricultural settings to help reduce work-related accidents and illness and WMSDs in particular. ISO 11226, ISO 11228 series and, more recently, ISO TR 12295 could be useful for this specific scope: experiences of application of these standards have been performed in different parts of the world, but rarely in agriculture.

2 Scope of Proposed Project and Normative References

This project has the goal of defining the basic criteria for producing a guide (Technical Report-TR) for the specific use of the ISO 11228 series, of ISO 11226 and of ISO TR 12295 [4–8] in agricultural sector. Specifically, the project is aimed at providing the

S. Bagnara et al. (Eds.): IEA 2018, AISC 820, pp. 72–83, 2019.
https://doi.org/10.1007/978-3-319-96083-8_10

potential users with additional information on how to use existing standards in a world widespread working sector as agriculture where, also if with different characteristics, biomechanical overload is a relevant aspect and where preventive actions are needed.

One of the main goals is therefore to provide all users, and particularly those who are not experts in ergonomics, with criteria and procedures:

- to identify the situations in which they can apply the standards of the ISO 11228 series and/or ISO 11226 and ISO TR 12295 in different agricultural contexts (*key-enter level*)
- to provide a *quick assessment* method (according to the criteria given in the relative standard) to easily recognize activities that are "*certainly acceptable*" or "*certainly critical*". If an activity is "not acceptable because critical" it is necessary to proceed as soon as possible with the subsequent improvement actions. Where the quick assessment method shows that the activity risk falls between the two exposure conditions, then it is necessary to refer to the detailed methods for risk assessment set out in the relevant standard. This scope and approach is illustrated in the flowchart in Fig. 1.

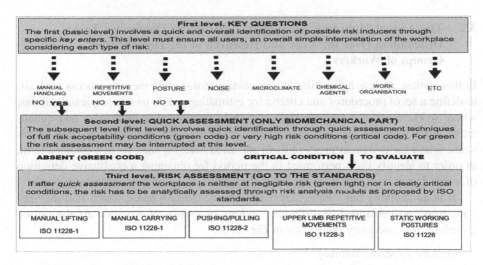

Fig. 1. The different risk assessment levels approach, present in ISO 11226 series for the part regarding biomechanical overload estimation [4–8]

3 General Outline of the Working Process in Annual Analysis in Agriculture: Qualitative Tasks Distribution Over the Year and Research of the Homogeneous Groups of Workers

3.1 Introduction and General Structure of a Multi-task Analysis

Task rotation is when a worker alternates between two or more tasks during a certain period of time. In special situations, such as in agriculture, where the worker has to perform a large number of tasks and the tasks are distributed "asymmetrically" throughout the shift, risk assessments can become extremely complex. This is why it is necessary to carry out a thorough preliminary study of how the work is organized. At any rate, the risk analysis process involves a number of steps, listed further on.

The first step consists in defining the time required to complete the task rotation schedule. This is the *macro-cycle time*, which may be: daily, weekly, monthly, yearly. In agriculture, the cycle of alternating the tasks, it is most often an annual cycle, or depending on the weather, multi-annual (2 or 3 yearly harvests).

3.2 The Annual Multi-task Analysis in Agriculture: Qualitative Tasks Distribution Over the Year and Research of the Homogeneous Groups of Workers

In this contest, such as agriculture, it is necessary, for starting the risk process analysis, to define a set of procedures and criteria for estimating risk in more complex situations, where workers perform multiple tasks variously distributed in qualitative and quantitative terms over the year (annual cycle). The general risk evaluation process entails steps, where the two preliminary of them are listed here: (a) analysis of work on a farm, in order to identify tasks performed in the period for obtaining a qualitative definition of work during each month of the year; (b) identification of homogeneous group/s.

It is an anything but simple matter to identify farming tasks, which may be very numerous and performed by different workers or groups of workers. At the outset, therefore, it is necessary (as shown in Fig. 2) to:

- identify a specific farming;
- break down the growing activities work into MACRO-PHASES and PHASES: all of the relevant tasks must be identified;
- list all the tasks required annually to grow and harvest the crop, regardless of who performs them: the allocation of tasks to workers (either on an individual basis or as a homogeneous group for risk exposure).

The same operation obviously can be carried out in several different ways; each method should be viewed as a separate task and listed accordingly. It is important to note that all the tasks performed on the farm over the year must be evidenced, including preparing the soil, applying fertilizers and disinfectants and other seemingly ancillary activities, regardless of who performs them.

MACRO PHASES	PHASES	TASKS	JAN	FEB	MAR	APR	MAY	JUNE	JULY	AUG	SEPT	OCT	NOV	DEC
SOIL PREPARATION	SOIL PREPARATION	Plowing (tractor)												
		Installing irrigation system												
	SOW SEEDS/PLANT SEEDLINGS	Planting (manual)											40%	10%
		Planting (mechanical)											20%	90%
PRUNING	DRY PRUNING	Pruning large branches with chainsaws												
		Pruning with manual shears										50%	5%	
		MMH of large branches										40%	30%	
	GREEN PRUNING	Pruning with manual shears	70%	70%	60%	60%	60%							
		MMH of small branches	20%	20%	30%	30%	30%							
HARVESTING	HARVESTING	Manual harvesting on ground							45%	45%	45%			
		Manual harvesting on ladder							35%	35%	35%			
		MMH of ladder							10%	10%	10%			
TREATMENTS	SOIL FERTILIZING	Preparing machine to apply fertilizer												
		Driving tractor												
		Composting (manual)												
	CROP TREATMENT	Disinfection (manual)												
		Disinfection (tractor)												
OTHER TASKS	PUSH PULL	Push/pull trolley with large branches	5%	5%	5%	5%	5%					5%	5%	
		Push/pull trolley with small branches	5%	5%	5%	5%	5%					5%		
		Push/pull trolley with large branches							10%	10%	10%			

Fig. 2. Example of semi-quantitative description of tasks (pruning and harvesting tasks) per month among homogeneous group of workers NO. 1.

The next step is to assign tasks to an *individual worker* or *group of workers exposed to the same risk*, to identify *homogeneous groups*. For each type of growing, tasks will be assigned to different groups of workers. When *tasks of the same nature, duration* and *side* are assigned to the same group of workers, we may speak of a *homogeneous group in terms of risk exposure*. A homogeneous group may sometimes be made up of just one person, if no other workers perform the same tasks qualitatively and quantitatively. For instance, typically (as presented in Fig. 2), a single group of workers may be assigned the job of actually growing a crop (tasks may include pruning, harvesting, etc.: homogeneous group NO. 1), whilst other workers prepare and disinfect the soil, apply fertilizers and so on (homogeneous group NO. 2). The assignment of the tasks to a homogeneous group (or individual worker) even just qualitatively (or semi-quantitative as here), is absolutely necessary before proceeding with any level of risk evaluation. The Fig. 2 shows (as example) all the tasks, subdivided by macro phases and phases that characterize the whole crop, highlighting the tasks actually performed by homogeneous group NO. 1 during the entire year, broken down into each month.

4 Pre-mapping of Danger and Discomfort

4.1 Foreword

One of the latest developments being pursued by the World Health Organization (WHO) and other international organizations (ILO, ISO), in relation to preventing work-related diseases and disorders, concerns the creation of toolkits.

The main aim is to rapidly but accurately identify the presence of possible sources of risk, using instruments that can easily be used by accident prevention officers, occupational physicians, business owners, workers, trade union representatives and security services. However, this objective also reflects the criteria set forth in ISO/TR 12295 with respect to the risk of biomechanical overload.

Against this backdrop, the "problem" of WMSDs must be considered together with other occupational "hazards" (be they physical, chemical, or other), for the more general purposes of prevention. Aim here is to suggest a methodology and some simple tools for bringing together various parties to undertake a preliminary *mapping of discomfort/danger* (i.e. to identify *risk sources in the work cycle*) in the work place.

The tool does not pretend to replace the standard risk evaluation process, but to support such a process in order to identify hazardous situations in the work place, based on which to single out emerging problems that need to be submitted to a full risk assessment (in the appropriate order of priority). The procedure presented here demands a cooperative approach towards assessing and managing risk, as it also entails interviews with workers.

In accordance with the recommendations of the WHO three main criteria underpin the methodology:

- *globality*: a global approach towards assessing the worker's discomfort, due to either the task or the work place;
- *simplicity*: the methodology consists in an easy to use model for collecting data;
- *priority-setting*: the results obtained automatically via dedicated software and depicted clearly in bar graphs will not only help to identify problems but also offer a scale of priorities for conducting subsequent assessments (Fig. 3).

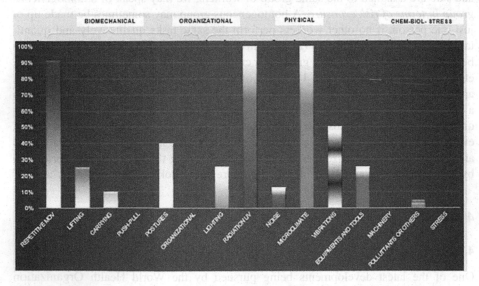

Fig. 3. Final summary results generated automatically by the software, depicting histograms for all possible risks presented by homogeneous group NO. 1 full time (example in Fig. 2)

4.2   The Pre-mapping Model

Please note that the pre-mapping model should be given to interview for homogeneous groups of workers. The operation involves two levels of intervention before presented in Fig. 1. The methodology consists in an easy to use computer-based model for collecting data (Excel spreadsheet): the EPMIES- **ERGOCHECK premapEN** () MULTIYEAR software can be free downloaded in Italian or English from www. epmresearch.org.

The model provides a general preliminary overview of all the main risk factors that may be present, regardless of the size of the manufacturing facility, and is underpinned by the basic tenets of ergonomics, entailing a global interpretation of the worker's discomfort deriving from the task or the work place [2]. The results of the pre-mapping exercises carried out via *key enters* and *quick assessments* can also be summarized graphically to more comprehensively define the "PRE-MAP" and corrective action priorities. Figure 3 shows an example of a summary of the results obtained from "pre-mapping" of the homogeneous group NO. 1 full time: the figure shows the final summary results generated automatically by the software, depicting histograms for all possible risks. These histograms are merely descriptive scores, to be used to "rank" events from the best to the worst. The scores do not reflect an analysis or assessment of risk: they are simply descriptive scales designed to help not only to identify problems but also to set priorities for the analyses and evaluations that will have to be undertaken to adopt immediate measures to reduce risk, especially for conditions defined as "critical".

5 Analytical Study of Working Process in Annual Multi-task Analysis: The Typical Working Day and the Quantitative Tasks Distribution

To switch from pre-mapping of discomfort and dangers (level 1 and 2), the actual assessment phase of the risk (Level 3), whatever the risk to be analyzed, it is necessary to deepen the organizational studies with data no longer just only qualitative but also quantitative. Three phases are necessary.

- Phase a – Description of a typical working day
- Phase b – Estimation of the total number of hours worked every month of the year
- Phase c – Assignment of tasks to a homogeneous group (or individual worker) and calculation of their proportional duration in each individual month.

Before going on to complete the organizational analysis of the risk-exposed worker or homogeneous group of workers, listed below are the exposure constants (Table 1) to which reference is made for calculating exposure time prevalence to various tasks and also for reconstructing the *fictitious working day* [2]. that will be representative of the whole year (see below). It has been found to be useful to adopt several exposure constants representing the typical exposure level for the industry.

Obtained the duration of tasks in each month, we obtain the critical figure enabling the final risk to be evaluated: the *total number of hours worked per year on each task* by each member of the homogeneous group and the proportion of these hours to both the total number of hours worked and to the constant 1,760 h/year

Table 1. Exposure time constants essential to build the fictitious working day, representative of the work actually carried out during the year

HOURS / DAY CONSTANT	8	HOURS / MONTH CONSTANT	160
MINUTES/ DAY CONSTANT	440	DAYS / MONTH CONSTANT	20
DAYS /WEEK CONSTANT	5	MONTHS / YEAR CONSTANT	11
MINUTES / WEEK (440 min * 5 days) CONSTANT	2200	DAYS / YEAR CONSTANT	220
WEEKS / MONTH CONSTANT	4	HOURS / YEAR CONSTANT	1760

6 Annual Multitask Risk Assessment of Biomechanical Overload for Upper Limbs

To arrive at a final risk index with use of the OCRA method for multi-analysis tasks, it is necessary to proceed by the following successive steps:

– *Phase a* – Analysis of each individual task using the OCRA checklist to calculate the intrinsic score and prepare the "basic tasks risk evaluation" for each crop.
– *Phase b* – Application of mathematical models: preliminary preparation of "fictitious working day" representative of the whole year and of every month of the same year.

Two models are proposal for calculating the final exposure risk index: one based on the *Time-Weighted Average* and the other on the *Multitask Complex*, which has based on the most overloading task (calculated with respect to its actual duration), as the minimum exposure score that must be increased versus the score of the other tasks, taking their relative durations into account.

In order to apply them to annual and monthly exposure, as mentioned before, it has been necessary to convert the data relative both to the individual months and to the year into a *fictitious working day, representative first of each month of the year and then of the full year* [2].

Figure 4 compares, month-by-month, the checklist OCRA risk index obtained in two homogeneous groups (**4a**-homogeneous group working full time, eleven months; **4b**-homogeneous group working on a crop for half of the year, but with very low risk for three out of the six months), using Multitask Complex model, as obtained automatically using the Excel template. Discrepancy between the scores obtained is quite remarkable. The interpretation of the results for the attribution of risk could be, in some case, problematic. It goes without saying that health surveillance findings proving the exposure risk index scores would be most welcome but, in these cases, with workers exposed to risk for only half of the year, it might be difficult to attribute a disease or disorder to occupational factors when the worker's activities in the other half of the year are unknown.

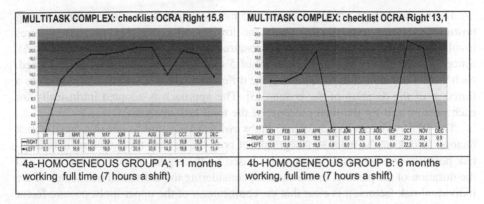

4a-HOMOGENEOUS GROUP A: 11 months working full time (7 hours a shift)	4b-HOMOGENEOUS GROUP B: 6 months working, full time (7 hours a shift)

Fig. 4. a, b – Risk index scores (for homogeneous group A, full time but working eleven months/year and B, full time but working seven months/year) plotted by month over the whole year using Multitask Complex formula.

7 The TACOS Method: Contents and Criteria for Back and Lower Limbs Posture Analysis

As a general approach towards identifying and describing postures and posture duration, the following general rules were followed:

- postures should not be identified and described for each part of the body and subsequently aggregated using ergonomic evaluation tools (RULA; REBA; OWAS);
- *overall postures for different body segments* (various standing, sitting, squatting postures, etc.) have been defined using sketches and simple descriptions (Fig. 5);
- postures must be identified by task, and each task will thus be defined by the posture (s) characterizing it.
- the duration of postures in each task can be readily measured with the help of pie charts depicting different risk scores; stopwatches are seldom required.

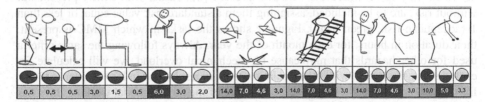

Fig. 5. Some examples of the analysed postures, proposed by TACOs method, divided for body areas and their specific scores defined on the basis of the incongruity of the posture and its duration.

The figures help to group postures into 5 main categories: standing postures, sitting postures, postures primarily involving the lower limbs, complex (mixed) postures involving many parts of the body, and postures involving the cervical spine (Fig. 5). Since the maximum score is 14, a series of 5 categories has been arbitrarily defined, each with a different colour, indicating the different degrees of awkwardness, also as a function of the duration of the posture [3]. The approach is adopted intrinsically for each posture and for the final outcome of the total task analysis.

The OCRA system analyses upper limbs awkward postures as one of a range of risk factors. It looks at the main joint segments of both the left and the right arm, defines at-risk postures and assigns different scores based on the region of the limb involved and the duration of the posture. In this method, considering the individual risk scores of the individual risk factors, it is clear that the contribution of the upper limbs posture factor to the final task score is about 50%.

As always, to start these studies with multitask exposure, you must have the intrinsic risk assessments of each of them (as if each task only lasted a standard shift). This is present both for the OCRA checklist, as well as for intrinsic postural risk, relative to the spine and lower limbs, obtained by the TACOs method.

Having the intrinsic risk scores and the duration of each task available for each task in the year, it becomes possible, with the same calculation models used for the OCRA checklist, to estimate the risk indexes for spine and lower limbs, both the Time-Weighted average and the Multitask Complex. The global result is presented also divided into four major areas) standing rachis, spine in sitting posture, lower limbs, head-neck. The final result illustrates also the proportional distributions of the different postures of the spine and lower limbs in their entirety [3].

8   Annual Multitask Risk Assessment of Manual Material Handling

In order to study the annual exposure risk for manual lifting of loads, it is necessary, as for other factors, to start from the quantitative organisational studies already set out before. Starting from Fig. 1, as a starting point, dedicated to identifying the tasks performed during the year involving upper limbs repetitive tasks, we now have to activated only tasks where the MMC is present [8]. Tasks, with MMH present but without risk, are to be included, assigning them a standard risk value equal or less than 1 (acceptable risk, green band). Figure 1 shows the tasks in which MMC is present, their duration in hours, for each month of the year. Always following the same criteria used for other hazards that may cause biomechanical overload, we will calculate for each task, intrinsic risk, with the calculation techniques defined in ISO standards [5, 8]: for the analysis a day, defined as representative of the annual modal working days has to be used The tasks performed may be characterized by manual lifting of loads of type *mono task,* or *composite task* or *variable task.* There are no rotations between tasks of *sequential task* type (typical of work done on assembly lines or workbenches, with turn-over every few hours on 2–3, maximum 4 tasks in a shift) since, in general in agriculture, the operators perform the same tasks for several days, alternating their tasks mainly with the change of season. For this reason, when calculating the intrinsic indexes for MMH in agriculture, always use the frequency/duration multiplier for *long duration* [1, 5]. The intrinsic values are calculated separately for adult male, adult female, younger/older male, younger/older female [1, 5]. Now having the intrinsic risk indices and the proportional duration of each task, both within each month of the year and throughout the year, becomes possible calculate risk indices through the reconstruction of *fictitious working days*, representative of each month and of the year.

This procedure is the same as used for calculating exposure to repetitive movements with OCRA method and awkward postures of the spine (without load lifting) and lower limbs with TACOS method

One examples of application are proposed in Fig. 6, concerning MMH risk assessment in the homogeneous NO. 1 group, full time contract (7 h per shift), where the manual handling, when present, is less then all shift period: here only the Multitask Complex has to be used, the only suggested in NIOSH method.

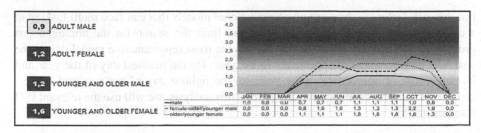

Fig. 6. The MMH risk indices (homogeneous NO. 1 full-time) calculated using the Multitask Complex model for each month of the year and the year after.

9 Annual Multitask Risk Assessment of Pushing and Pulling

In order to study the annual exposure risk for manual pushing and pulling it is necessary, as for other factors, to start again from the quantitative organisational studies already set out.

Starting from Fig. 1, as a starting point, dedicated to identifying the tasks performed during the year, we now have to activated only tasks where the PUSHING and/or PULLING are present [6]. Always following the same criteria used for other hazards that may cause biomechanical overload, we will calculate for each task, intrinsic risk [8]. Now having the intrinsic risk indices, hours and percentages of the duration of each task, both within each month of the year and throughout the year, becomes possible through reconstruction of *fictitious working days*, representative of each month and year, calculate risk indices. The procedure is the same as used for calculating exposure to repetitive movements and awkward postures of the spine (without load lifting) and lower limbs and MMH.

An example of application is proposed in Fig. 7 concerning PUSHING and/or PULLING risk assessment in the homogeneous NO. 1 group, full time contract (7 h per shift), The calculation model, is only the Multitask Complex. as for MMH.

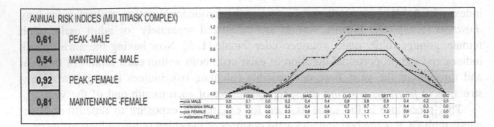

The table in the figure:

ANNUAL RISK INDICES (MULTITASK COMPLEX)	
0,61	PEAK -MALE
0,54	MAINTENANCE -MALE
0,92	PEAK -FEMALE
0,81	MAINTENANCE -FEMALE

	JAN	FEBB	MAR	APR	MAG	GIU	LUG	AGO	SETT	OTT	NOV	DIC
pick MALE	0,0	0,1	0,0	0,2	0,4	0,4	0,8	0,8	0,8	0,4	0,2	0,0
maintenance MALE	0,0	0,1	0,0	0,2	0,4	0,4	0,7	0,7	0,7	0,4	0,3	0,0
pick FEMALE	0,0	0,2	0,0	0,3	0,6	0,6	1,2	1,2	1,2	0,6	0,3	0,0
maintenance FEMALE	0,0	0,2	0,0	0,2	0,7	0,7	1,1	1,1	1,1	0,7	0,5	0,0

Fig. 7. The PUSHING/PULLING risk indices (relative to the homogeneous group NO. 1 full-time) calculated using the Multitask Complex model for each month of the year and the year after.

10 Risk Assessment of Carrying

For this risk factor, as there are no computational models that can face multi-tasking at a cycle different from the daily one, you can limit the search for the homogeneous group and then evaluate for it the risk for: (a) the most representative modal day of the year and its percentage of appearance in the year; (b) the heaviest day of the year and its percentage of appearance in the year; (c) the lightest day of the year and its percentage of appearance in the year. To address this analysis, we will use the relevant ISO standard [5, 8]

11 Conclusion

This document discusses the complex procedures for addressing the biomechanical overload in agriculture, which includes the study of the risk of repetitive upper limb movements, manual lifting and awkward postures including lower limb and spine studies. Criteria and measurement methods already present in the ISO standards dedicated to biomechanical [4–8] have been proposed and adapted to the characteristics of agricultural work that involves a more frequent annual cycle organization, with exposure to more work tasks, which diversify qualitatively and quantitatively over the course of the year. Given the clear complexity of the risk assessment, it is suggested to start with a first qualitative and simple analysis (using the *key questions* and *quick assessment*) but extended not only to biomechanical risk factors but to all risks, so as to obtain a sort of *global risk pre-mapping*, which points out the presence of discomforts and dangers and with what priorities should be addressed the future more precise risk assessments. A simple tool to deal with this first phase of analysis is available. In all situations where the staff is unable to complete the evaluation phase of the real risk (small companies, etc.), they will obtain, by pre-mapping, at least one document that indicates the potential risk factors present and priority. This paper also discusses how to conduct the real risk assessment level, illustrating strategies for applying risk calculation methods (OCRA, NIOSH, TACOs, PUSHING and/or PULLING) adapted to analysis in agriculture, all published in specific manuals produced by the author [1–3]. Here too, a simple tool in available.

References

1. Colombini D, Occhipinti E, Alvarez CE, Waters T (2012) Manual Lifting, A Guide to Study of Simple and Complex Lifting Tasks. CRC Press, Taylor & Francis, New York
2. Colombini D, Occhipinti E (2016) Risk Analysis and Management of Repetitive Actions: A Guide for Applying the OCRA System (Occupational Repetitive Actions). CRC Press, Taylor & Francis, New York
3. Colombini D, Occhipinti E (2018) Working Posture Assessment, The TACOs Method, Time-Based Assessment Computerized Strategy. CRC, Taylor & Francis, New York
4. ISO 11226, Ergonomics — Evaluation of static working postures
5. ISO 11228-1, Ergonomics — Manual handling — Part 1: Lifting and carrying
6. ISO 11228-2, Ergonomics — Manual handling — Part 2: Pushing and pulling
7. ISO 11228-3, Ergonomics — Manual handling — Part 3: Handling of low loads at high frequency
8. ISO/TR 12295, Ergonomics — Application document for International Standards on manual handling (ISO 11228-1, ISO 11228-2 and ISO 11228-3) and evaluation of static working postures (ISO 11226). ISO 12100, Safety of machinery—General principles for design

A Study on Effects of Muscle of Lower Limb Associated with Whole-Body Vibration

Shih-Yi Lu[1]([✉]) [iD], Xiang-An Cheng[1], Yen-Hui Lin[1],
and Cheng-Lung Lee[2]

[1] Chung Shan Medical University, Taichung 40201, Taiwan, ROC
sylu@csmu.edu.tw
[2] Chaoyang University of Technology, Taichung 41349, Taiwan, ROC

Abstract. Long-term exposure to whole-body vibration in the workplace will increase the chances of lower back pain, spinal disc herniation, and other diseases. However, many studies have also indicated that vibration stimulation, often used in physical therapy, clinical treatment, and muscle strength training, has a variety of positive effects for the human body.

A commercially available electric vibrating machine was chosen, and 20 subjects were recruited to statically stand on the vibration platform with knee flex at different angles (0°, 60°, 90°), and to dynamically stand (squatting and rising) on the platform, while being exposed to different vibration frequencies (0 Hz, 20 Hz, 35 Hz, 50 Hz). The experiment used surface electromyography to assess the effects of posture and frequency on the neuromuscular activation. Each subject was asked to rate the perceived exertion on three monitored muscles (gastrocnemius, rectus femoris, vastus lateralis).

The results showed that the knee flex angle had a significant effect on the muscles of the lower limbs, especially the thigh muscles, which were regarded as the support for body weight, and that the most obvious impact on the lower limb muscles was on the calf muscles during whole-body vibration. Surface EMG signals detected in dynamic posture were generally higher than those in static posture; however, through the subjective perception and assessment of subjects, we found that the scores of the two situations were quite close. The study results showed that the higher-frequency vibration activated muscles more easily. However, excessive fatigue would also result in injuries, a cause for caution and care.

Keywords: Whole-body vibration · EMG · Muscle fatigue · Lower limb

1 Introduction

Many studies have found that the occupational injuries caused by the unpleasant occupational exposure to systemic vibration in humans include dizziness, discomfort and lower back pain, and occupational injuries such as degenerative or prominent disc disease [1–4]. In today's workplace, exposure to whole-body vibration is identified as one of the major risk factors impacting physical health. Prolonged exposure is a major concern for employers and workers in a given exposed population. Interestingly, however, whole-body vibration has often been used in recent years as a physical

© Springer Nature Switzerland AG 2019
S. Bagnara et al. (Eds.): IEA 2018, AISC 820, pp. 84–91, 2019.
https://doi.org/10.1007/978-3-319-96083-8_11

therapy and in other clinical rehabilitation; for professional athletes using vibration stimulation combined with resistance training applied to muscle strength training, it can be called "whole-body-vibration training" (WBVT) [5, 6]. WBVT is a new type of neuromuscular training developed in recent years. It was initially applied to astronauts in zero G mainly to slow down the loss of bone-mass density and calcium, and then as the relevant research papers became more developed and their applications became widespread, the body vibration instrument successfully entered the mass market.

The EU's 2010 survey data on employees and self-employed workers showed the hazard factors at work. Of all the factors that account for more than 20% of the working time for exposure to hazard factors, vibration accounts for 22.8% of the total, and male workers account for most of the exposure to these factors [7]. These data show that laborers, especially men, are highly exposed in the workplace to the hazards of vibration; this may be because heavily physical tasks involve whole-body vibration. In 1974, NIOSH estimated that about 8 million workers (about 9% of the U.S. workforce) were exposed to vibration hazards, 80% of which were attributable to the vibration of the whole body. Excessive exposure to vibration may easily lead to muscle fatigue. In particular, when working under full-body vibration, laborers' improper use of vibration equipment or excessive exposure to vibration can easily cause musculoskeletal pain and even occupationally unrecoverable diseases.

The use of surface electromyography (sEMG) to measure and analyze the myo-electric signals of the muscles reveals the state of muscle-force exertion, the timing of muscle fatigue, and the recruitment of motor units in the muscle spindles [8, 9]. The EMG signal is the inevitable physiological signal in muscle contraction. Through the surface electrode patch attached to the human skin, the sum of the moving potential of this part of the motor unit is measured. The original signal is collected through the transmission line to the amplifier to enhance the original EMG signal, and then the filter is used to exclude noise. The EMG signal after treatment can then be used as exper-imental data.

In recent years, the related literatures have still been missing information on the effect of whole-body vibration on the muscles of the lower limbs. The existence of vibration or changes in body posture, and how the frequency of vibration affects the lower limb muscles are interesting to explore. Therefore, this study used the EMG signal analysis of different knee-bending angles with different vibration frequencies to explore the changes in muscle EMG.

2 Materials and Methods

2.1 Subjects

This study recruited 20 healthy males for the experiment. These subjects were ordinary college students without muscular skeletal diseases in the lower limbs, and they had never suffered from any associated, relevant diseases. Their ages, heights, and weights were 23.1 ± 1.2 (years), 174.1 ± 4.6 (cm), and 71.1 ± 9.8 (kg), respectively. The study procedures were approved by the Chung Shan Medical University Hospital's Institutional Review Board, and informed consents were obtained from all subjects.

2.2   Apparatus

This study aimed to probe the effects of different vibration frequencies and postures on the muscle groups in the lower limbs after a WBV exposure. In the experiment, an electric vibrator (BH-YT18, Taiwan) and a surface EMG measuring system (Zebris Medical GmbH, Germany) were adopted. The sEMG was often used to obtain myoelectric signals from active muscles to determine local muscle fatigue (Cifrek et al. 2009). The sEMG measurements from three muscles in the subjects' lower limbs using disposable skin surface electrodes were obtained using a computer-based data acquisition system. The raw sEMG signals were amplified with a gain of 5,000, lowpass filtered at 500 Hz, and digitized at a rate of 1,000 Hz.

2.3   Experimental Process

First, the subjects were informed of the purposes and specific procedures of the experiment, and they were required to sign the consent letter. In the early stage of the experiment, the muscle groups in the gastrocnemius, rectus femoris, and vastus lateralis muscles were scrubbed with medical alcohol. The first step was to measure the maximum voluntary contraction (MVC) of the muscles and the signals of static, uncontracting muscles when the subjects relaxed their legs. Then, the maximum force exertion was measured for 3 to 5 s, and the numerical values of the EMG signals were collected for 3 s. These numbers were taken as the denominators of standardization of the follow-up measurements of MVC% (percentage of maximum voluntary contraction). The subjects stood on the vibrator (BH-Y18) and grabbed the handle of the machine. The experimental situations were random combinations of four postures with knee flexion of 0° (standing straight), knee flexion of 60° (squatting), knee flexion of 90° (deep squatting) and dynamic (squatting and rising); four vibration frequencies (0 Hz, 20 Hz, 35 Hz, 50 Hz) for a total of 16 sessions. Meanwhile, the physiological EMG signals were collected. Each experimental session lasted for 5 min, and there was a 10-min interval between the two experimental sessions. Figure 1 shows the static postures.

2.4   Statistical Analysis

The study was interested in evaluating the effects of vibration exposure. The approach considered was to measure the MVC% of a sample of participants before and after vibration exposure, and to analyze the differences using a paired sample t-test. IBM SPSS Statistics were used to determine whether there were any statistically significant differences between the means of the independent groups.

3   Results

3.1   Effects of Vibration on Muscle Electromyography

As shown in Table 1, the presence or absence of whole-body vibration in the gastrocnemius and rectus femoris muscles had a significant effect on myoelectric responses

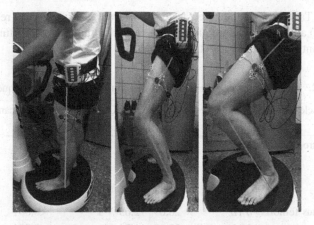

Fig. 1. Static posture (left: knee flexion of 0°; middle: knee flexion of 60°; right: knee flexion of 90°).

in either static or dynamic postures. In addition, in static postures, the presence or absence of whole-body vibration in the vastus lateralis muscle could have a significant effect on myoelectric responses; however, in dynamic postures, the application of whole-body vibration had no significant effect on myoelectric responses.

Table 1. The effect of vibration on the tested muscles in different postures.

Muscles	Postures	Changes[a] (MVC%)		Significance
		Mean	Standard deviation	
Gastrocnemius	0°	−7.52	8.12	<0.001
	60°	−3.40	5.83	<0.001
	90°	−3.93	5.31	<0.001
	Dynamic	−2.17	8.40	0.05
Rectus femoris	0°	−3.15	4.97	<0.001
	60°	−3.28	6.33	<0.001
	90°	−3.61	7.62	0.001
	Dynamic	−3.00	10.61	0.033
Vastus lateralis	0°	−5.94	8.04	<0.001
	60°	−3.79	8.20	0.001
	90°	−7.29	8.57	<0.001
	Dynamic	−1.29	9.72	0.307

[a]Changes = MVC (no vibration) − MVC (with vibration)

3.2   Effects of Dynamic Posture on Muscle Electromyography

Using a squatting-rising motion of dynamic posture (knee angle flexion from 0° to 90°, then to 0° as a cycle) compared to static muscle activation (knee flexes to 90°), the paired t-test was applied to examine the difference between the two situations.

As shown in Table 2, for the three tested muscles (gastrocnemius, rectus femoris, lateral femoral) at all vibration frequencies, the muscles in dynamic posture statistically had a significantly higher MVC% than the muscles in static posture.

Table 2. The effect of postures on the tested muscles at different frequencies.

Muscles	Frequencies	Changes[a] (MVC%)		Significance
		Mean	Standard deviation	
Gastrocnemius	0 Hz	−9.90	8.00	<0.001
	20 Hz	−11.26	12.01	<0.001
	35 Hz	−5.30	5.55	<0.001
	50 Hz	−7.85	8.05	<0.001
Rectus femoris	0 Hz	−20.09	12.65	<0.001
	20 Hz	−20.00	11.76	<0.001
	35 Hz	−20.69	9.99	<0.001
	50 Hz	−17.72	8.98	<0.001
Vastus lateralis	0 Hz	−18.52	12.79	<0.001
	20 Hz	−11.52	11.44	<0.001
	35 Hz	−15.31	10.10	<0.001
	50 Hz	−10.74	12.50	0.001

[a]Changes = MVC (static) − MVC (dynamic)

3.3 Joint Analysis of EMG Spectrum and Amplitude (JASA)

The mean scores of subjective perception exertion evaluation of 20 subjects was measured. It showed that the most prone to fatigue situations were static posture with knee flex of 90° and dynamic posture while exposed to 50 Hz vibration. The results seemed to imply that the higher-frequency vibration activated muscles more easily. Therefore, these two situations were selected to observe fatigue. Luttmann and Jager [10] pointed out that in order to assess muscle strength and fatigue properly, it is necessary to consider both the strength and spectral properties of the myoelectric signal. The "Joint analysis of EMG spectrum and amplitude (JASA)" was used to evaluate muscle fatigue. It is known that muscle fatigue causes a reduction in the median frequency (MDF) and an increase in the electrical activity (EA). The judgment method is to make the situation into JASA figures. In the figure, where the two trend lines (MVC%, MDF) intersect, it can be considered that the muscle has entered a fatigue state. As shown in Fig. 2, the trend lines of the gastrocnemius muscle in the top two figures did not quite conform to the fatigue rule, especially at static posture, whereas the middle and bottom in Fig. 2 show that both the rectus femoris muscle and the vastus lateralis muscle presented a perfect crossover trend line, so it can be speculated that the rectus femoris muscle and the vastus lateralis muscle may show a state of fatigue as the experimental time progresses.

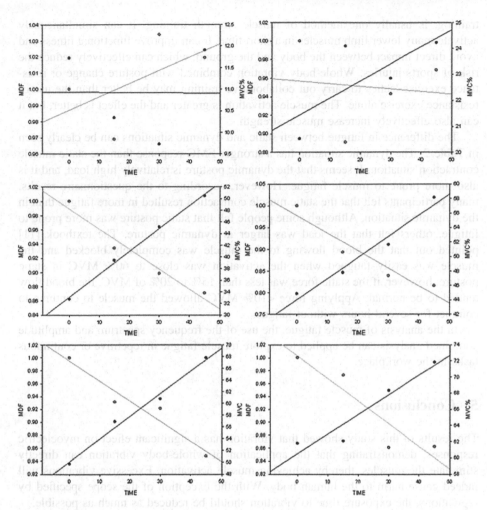

Fig. 2. Joint analysis of the EMG spectrum and amplitude (top left: gastrocnemius with static posture; top right: gastrocnemius with dynamic posture; middle left: rectus femoris with static posture; middle right: rectus femoris with dynamic posture; bottom left: vastus lateralis with static posture; bottom right: vastus lateralis with dynamic posture).

4 Discussions

The study results showed that the knee flex angle had a significant effect on the muscles of the lower limb, especially the thigh muscles, which were regarded as the support for body weight, and the most obvious impact on the lower limb muscles was on the calf muscles during whole-body vibration. The proximity of these muscles to the vibration source might be the reason for this. As shown in Table 1, the presence or absence of vibration in different postures caused a significant change in the myoelectric signal, revealing that the vibration significantly affected the myoelectric responses of the leg muscles. In addition, the knee flexion movement, similar to the deep squat in weight training, is usually one method of muscle strength training. It can simultaneously activate many lower limb muscles in a short time. It can improve functional fitness and avoid direct impact between the body and the ground, which can effectively reduce the risk of sports injuries. Whole-body vibration combined with posture change or resistance exercise tactics to carry out collaborative training may be better than the use of resistance exercise alone. The muscle activation is greater and the effect is better, but it can also effectively increase muscle strength.

The difference in fatigue between static and dynamic situations can be clearly seen in Table 2. The dynamic situation has a stronger EMG response than the static muscle contraction situation. It seems that the dynamic posture is relatively high load, and it is also more prone to muscle fatigue. However, according to the questionnaire results, many participants felt that the static muscle contraction resulted in more fatigue than in the dynamic situation. Although some people felt that static posture was more prone to fatigue, others felt that the load was larger in dynamic posture. The textbook [11] pointed out that the blood flowing to the muscle was completely blocked and the muscle was easily fatigued when the activation was close to 60% MVC in static posture; however, if the static force was less than 15% to 20% of MVC, the blood flow tended to be normal. Applying force <10% MVC allowed the muscle to continue to contract for several hours without fatigue.

In the analysis of muscle fatigue, the use of the frequency spectrum and amplitude combined analysis can be applied to explore muscle fatigue in repetitive or continuous tasks in the workplace.

5   Conclusions

The results of this study showed that vibration has a significant effect on myoelectric responses, demonstrating that the application of whole-body vibration can directly stimulate the muscles, thereby achieving muscle activation. Excessive vibrations will indeed cause harm to the human body. With the exception of the scope specified by regulations, the exposure time to vibration should be reduced as much as possible.

References

1. Bovenzi M (1996) Low back pain disorders and exposure to whole-body vibration in the workplace. Semin Perinatol 20(1):38–53
2. Seidel H (1993) Selected health risks caused by long term whole body vibration. Am J Ind Med 23:589–604
3. Bovenzi M, Hulshof CTJ (1999) An updated review of epidemiologic studies on the relationship between exposure to whole-body vibration and low back pain. Int Arch Occup Environ Health 72:351–365
4. Lings S, Leboeuf-Yde C (2000) Whole-body vibration and low back pain: a systematic, critical review of the epidemiological literature 1992–1999. Int Arch Occup Environ Health 73:290–297

5. Prisby RD, Lafage-Proust MH, Malaval L, Belli A, Vico L (2008) Effects of whole body vibration on the skeleton and other organ systems in man and animal models: what we know and what we need to know. Ageing Res Rev 7:319–329
6. Cardinale M, Wakeling J (2005) Whole body vibration exercise: are vibrations good for you? Br J Sports Med 39(9):585–589
7. Agnès PT, Greet V, Gijs H, Maija LY, Isabella B, Jorge C (2012) Fifth European Working Conditions Survey. Publications Office of the European Union, Luxembourg, Eurofound
8. Cifrek M, Medved V, Tonković S, Ostojić S (2009) Surface EMG based muscle fatigue evaluation in biomechanics. Clin Biomech 24:327–340
9. Chesler NC, Durfee WK (1997) Surface EMG as a fatigue indicator during FES-induced isometric muscle contractions. J Electromyogr Kinesiol 7(1):27–37
10. Luttmann A, Jager M (1996) Joint analysis of spectrum and amplitude (JASA) of electromyograms applied for the indication of muscular fatigue among surgeons in urology. In: Advances in occupational ergonomics and safety, vol 1, pp 523–528
11. Kroemer KH, Grandjean E (1997) Fitting the task to the human. A textbook of occupational ergonomics, 5th edn. CRC Press, Boca Raton

Work Postural and Environmental Factors for Lower Extremity Pain and Malalignment in Rice Farmers

Manida Swangnetr Neubert[1,2(✉)], Rungthip Puntumetakul[2,3],
and Usa Karukunchit[2,4]

[1] Program of Production Technology, Faculty of Technology, Khon Kaen
University, Khon Kaen 40002, Thailand
manida@kku.ac.th
[2] Research Center in Back, Neck, Other Joint Pain and Human Performance,
Khon Kaen University, Khon Kaen 40002, Thailand
[3] Division of Physical Therapy, Faculty of Associated Medical Sciences,
Khon Kaen University, Khon Kaen 40002, Thailand
[4] Faculty of Physical Therapy, Saint Louis College, Bangkok 10120, Thailand

Abstract. In many Southeast Asian Countries, most tasks of the rice cultivation rely heavily on manual labor and require prolonged working in muddy terrain. Due to the lack of comprehensive ergonomic assessment and interventions, a series of studies were conducted to examine work postural and environmental factors contributing to lower extremity (LE) pain and malalignment in Thai rice farmers. This paper evaluates the collective results of our previous studies, which can be divided into two stages. The initial stage included a survey of pain perception and physical examination of LE alignment to specify the most problematic LE parts in a large group of 250 farmers, revealing that farmers generally perceived elevated hip pain. However, physical examination identified a high prevalence of foot pronation and knee valgus. A subsequent detailed analytical stage, conducted to identify factors of LE pain on a smaller group of 30 farmers, included two-stage ergonomic risk assessment and investigation of effects of muddy work terrain in different laboratory settings. The ergonomic assessment results indicated the planting process to pose the highest risk for LE injury, specifically leading to perception of knee pain induced by motion and posture factors, and foot pain induced by force exertion. Experiential results showed muddy ground to induce significantly higher force on knees and higher levels of knee and ankle muscle exertion. The findings suggest that further development of interventions should focus on reducing awkward posture and muscular exertion due to mud resistive force, particularly for knee and foot during the planting process.

Keywords: Lower extremity pain · Risk assessment
Muddy work environment · Muscle activity · Rice cultivation process

© Springer Nature Switzerland AG 2019
S. Bagnara et al. (Eds.): IEA 2018, AISC 820, pp. 92–102, 2019.
https://doi.org/10.1007/978-3-319-96083-8_12

1 Introduction

Agricultural work has been known to pose ergonomic hazards due to requiring repetitive force exertion and awkward postures [1, 2]. In Thailand and many Southeast Asian Countries, the rice production industry generates a significant market volume of agricultural products [3]. With the increase in production demands and concomitant lack of technological interventions for the majority of the workforce, farmer health and safety has become a key factor to ensure high efficiency and productivity of the agricultural sector. The rice cultivation process involves multiple stages, including field preparation, seeding, planting, nursing and fertilization, and harvesting [4]. Most tasks rely heavily on strenuous manual efforts and require prolonged working in muddy terrain. Such viscous muddy environment causes farmers to preferably perform tasks without footwear. As a result of these extreme work conditions, a previous study indicated a high prevalence of musculoskeletal disorder (MSD) in the lower extremities (LE) among Thai rice farmers (10.29–41.16%) [5]. LE MSDs have been found to cause chronic leg pain and may promote the development of LE malalignment [6, 7]. Clearly, such detrimental health conditions severely impact the working capability and require costly medicine and healthcare service.

Although epidemiological studies of the prevalence of LE pain among rice farmers have already been reported [5], there is still a lack of comprehensive ergonomics assessment and intervention programs for rice farmers. Therefore, we conducted a series of studies to examine work postural and environmental factors contributing to LE pain and malalignment in Thai rice farmers. This paper aims to present a comprehensive assessment of the collective results of our previous studies, which can be divided into two stages (see Fig. 1). An initial stage included a survey of pain perception and physical examination of LE alignment to specify the most problematic LE parts. The subsequent detailed analytical stage, conducted to identify factors of LE pain, included two-stage ergonomic risk assessment and investigation of effects of muddy work terrain in different laboratory settings. The objectives of this study were to specify the most problematic LE parts and to identify factors of LE pain of rice farmers. The implications of this research are anticipated to benefit the development of work guidelines and interventions for rice farmers that minimize specified risk factors of the LE parts, of which pain and malalignment are of most concern.

2 Methods and Results of Stage I

2.1 Participants

Two hundred and fifty (250) experienced rice farmers (108 male and 142 female farmers with age between 20–59 years) in rice farming villages located in Sila district, Khon Kaen Province, were recruited to participate in the study. Participants were required to have at least one year of experience in rice cultivation. One participant was excluded from the study due to the following exclusion criteria: (1) chronic leg and foot pain within two weeks prior to testing, such as gouty arthritis, rheumatoid arthritis, or ankylosing spondilitis; (2) previous or current injury to the lower extremities or; (3) any

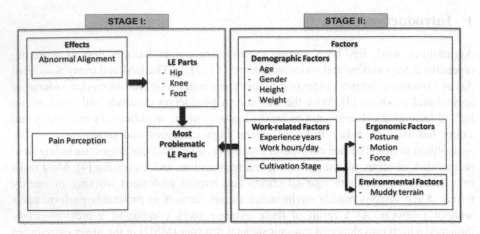

Fig. 1. Overall study framework

previous history that would affect lower extremity alignment, such as a fracture and/or surgery.

2.2 Examination of LE Malalignments

LE alignment measurements examined in our study [8] included pelvic angle, femoral antetorsion, quadriceps (Q) angle, tibiofemoral angle, genu recuvatum, tibial torsion, rearfoot angle, and medial longitudinal arch angle. All measurements were repeated 3 times by a single examiner who had excellent test-retest reliability on all lower extremity measures (ICC range of 0.89–0.98). Participants were asked to wear shorts that exposed lower limb parts for testing. Abnormal ranges of alignment were subsequently classified into malalignment characteristics of hip (anterior pelvic tilt, posterior pelvic tilt and excessive femoral antetorsion), knee (genu valgus, genu varus, knee hypertension and tibial torsion) and foot (pronate and supinate foot). The details of LE alignment measurement and classification methods were described in [8].

The average of three LE alignment measurements was used for analyses. The findings (see Table 1) indicated foot and knee malalignment to be common among Thai rice farmers. The prevalence of LE malalignments of rice farmers was found to be highest for foot pronation (36.14%), followed by knee valgus (34.94%). The highest malalignment characteristics of hip was found in terms of anterior pelvic tilt (30.52%). No posterior pelvic tilt and supinate foot conditions were observed in this study.

2.3 Survey of LE Pain Perception

LE pain investigation was conducted to compare pain perceived by rice farmers for each cultivation step. Rice farmers were asked to rate the Thai version of the Visual Analog Scale (VAS) using a body chart region modified from the Standardized Nordic Questionnaire [9] for LE part discomfort before and after each farming stage for the

Table 1. Prevalence of LE malalignment

Malalignment characteristics	Number	%
Anterior pelvic tilt	76	30.52
Excessive femoral antetorsion	70	28.11
Knee valgus	87	34.94
Knee varus	77	30.92
Knee hyperextension	28	11.24
External tibial torsion	53	21.29
Foot pronation	90	36.14

complete cultivation cycle. The VAS for pain rating scales ranged from 0 to 10 points, where 0 represents no pain and 10 was intolerable pain.

Results of perceived pain ratings for each LE parts for rice farmers in each cultivation steps are shown in Fig. 2. Analysis of Variance (ANOVA) demonstrated that rice farmers perceived significantly greater LE pain during performance of rice cultivation tasks than before the cultivation activity. Pain rating results further indicated that farmers perceived the highest LE pain during the planting task. Results of paired t-test comparing each LE part revealed that, in general, farmers perceived higher pain in the hip, as compared with knee and foot.

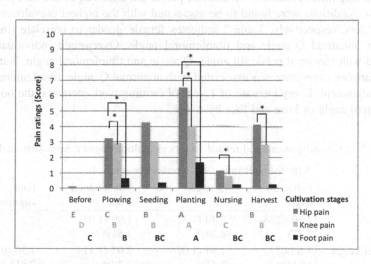

Fig. 2. Results of LE pain for each cultivation stage (Mean with the asterisk (*) and different letters are significantly different at alpha = 0.05)

2.4 Problematic LE Parts

Although rice farmers experienced less knee and foot pain during cultivation activities, malalignment examination results indicated higher prevalence in foot pronation and knee valgus. Excessive foot pronation may also originate from other malalignment conditions, including excessive internal femoral rotation, external tibial rotation or anterior pelvic tilt position [10, 11], which might contribute to hip muscle pain [12]. Therefore, all LE parts were still considered for further investigation in subsequent detailed analyses.

3 Methods and Results of Stage II

3.1 Participants

The same set of 250 participants was used to identify risk factors of LE abnormal alignments. For detailed analyses of ergonomic risk and environmental factors, a smaller group of 30 farmers was randomly selected to participate in the experiments.

3.2 Risk Factor Identification of LE Malalignments

Analysis of demographic factors (age, gender and body mass index (BMI)) and work-related factors (average working hours per day and years of work experience) for LE malalignments was conducted for 250 rice farmers in our previous study [8]. The objective of the analysis was to investigate whether common malalignment conditions among farmers were caused by performing farming activities. This paper presents only specific malalignment conditions of: anterior pelvic tilt, knee valgus and foot pronation, since these conditions were found to be associated with the highest prevalence of hip, knee and foot, respectively. Table 2 indicates female gender to correlate (increased risk) with abnormal Q angle and tibiofemoral angle. Overweight individuals were associated with abnormal pelvic tilt angle, Q angle and tibiofemoral angle. Number of years of farming experience was associated with abnormal Q angle, tibiofemoral angle, and foot alignment. Every increase of 1 year of farming work increased the odds ratio for abnormal angle of knee and foot by 4–6%.

Table 2. LE malalignments and related factors in multiple logistic regression analysis

Factors	Adjusted odds ratio (95% confidence interval)			
	Pelvic tilt angle	Q angle	Tibiofemoral angle	Foot alignment
Sex: Female	1.70 (0.88–2.37)	2.40 (1.32–4.39)*	1.90 (1.03–3.51)*	
BMI: Overweight	8.6 (4.7–16.15)*	1.94 (1.07–3.55)*	2.05 (1.11–3.78)*	1.56 (0.87–2.81)
Age	1.03 (0.99–1.07)			
Daily working hours		1.18 (0.91–1.51)	1.23 (.95–1.58)	
Years of experience		1.04 (1.01–1.08)*	1.04 (1.01–1.08)*	1.06 (1.03–1.1)*

Note: * indicates a significance at p < 0.05.

3.3 Ergonomic Risk Factor Identification of LE Pain

The two-stage ergonomic risk assessment included first-stage worst case risk assessment and the second-stage risk assessment for 30 individual participants based on the highest-risk subtask (see detailed in [13])). The assessments were conducted independently by 3 expert analysts based on direct and video-based observation during task performance in each stage of the rice cultivation. The evaluation method was based on the modified Rapid Upper Limb Assessment (RULA) [14] tool with an extension to cover whole body parts for ergonomic risks in terms of improper posture, motion and force. In line with the previous analysis of pain, ANOVA results also indicated the planting process to pose the highest ergonomic risk for LE injury in rice farmers ($p < 0.0001$).

Subsequent multiple linear regression analyses were conducted to identify ergonomic risk factors associated with LE pain during the planting process. Unfortunately, hip pain did not establish a linear relationship with risk factors ($p > 0.05$, $R^2 < 0.3$). Therefore, a more detailed investigation of the complex relationship between these factors using a non-linear modelling approach will be required in future studies. For other significantly associated factors of perceived pain, results showed knee pain perception to be significantly induced by motion and posture factors ($p = 0.017$ and 0.008, respectively); while foot pain perception was primarily caused by force exertion ($p = 0.014$).

3.4 Environmental Factor Identification of LE Pain

A unique work condition encountered in rice cultivation results from most tasks being typically performed with bare feet in muddy terrain. Such working environments might lead to increasing both the force acting on joints as well as the muscular force requirements in the LEs of farmers, which increase due to force resulting from body weight and mud viscosity [15]. The environmental factor of muddy work terrain was investigated in a group of 30 participants during the simulated planting setup on rigid and muddy work surface. A specific posture, involving lifting one foot off the work surface, was selected due to strong association with tensile viscous force. A 3D Static Strength Prediction Program (3DSSPP; Center of Ergonomics, University of Michigan ref) was used to estimate tensile force loading on the LE joints. Muscle activity was captured using electromyography (EMG) during a 10-session simulated task. Both investigations were conducted for only one side of the legs. Details of 3DSSPP analysis and muscle activity analysis along with laboratory settings were described in [16] and [17], respectively. According to 3DSSPP computation results, the knee joint was found to be exposed to the greatest force increase due to mud resistance (see Fig. 3). Results of repeated measure ANOVA of EMGs (see Fig. 4) showed muddy ground to induce significantly higher levels of biceps femoris (BF; knee flexor) and gastrocnemius (GA; ankle plantar flexor) muscle exertion, as compared with rigid ground.

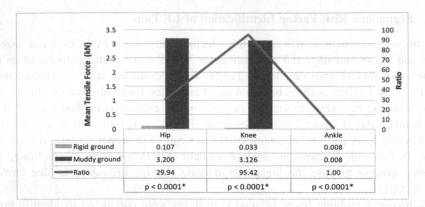

Fig. 3. Results of force on LE joints for muddy and rigid ground conditions.

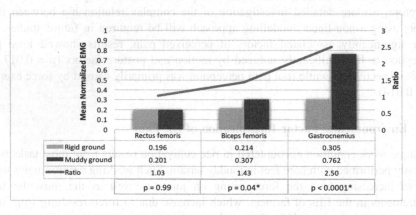

Fig. 4. Results of LE muscle activity for muddy and rigid ground conditions.

4 Discussion

The study found the highest prevalence of LE malalignments among rice farmers in foot pronation, followed by knee valgus. The highest malalignment characteristics for the hip was found in terms of anterior pelvic tilt. These conditions might be associated to each other. Excessive foot pronation may be a result of knee valgus or anterior pelvic tilt position [10, 11]. Also, abnormal knee alignment may be caused by hip, ankle, or foot malalignment [18, 19]. Detailed analysis of risk factors showed this specific type of hip and knee malalignments to be associated with individual factors of BMI. Overweight individuals have a high risk for developing abnormal pelvic tilt angle, Q angle and tibiofemoral angle, since an increase in joint loading can produce injury to weight-bearing joints of the LEs [20, 21]. Knee malalignment was also found to be associated with gender. Female individuals were found prone to increased risk of abnormal Q angle and tibiofemoral alignment. Knee malalignment in females may result from wider pelvic width and amplified internal rotation of the hips, as compared with males [22]. Moreover, female individuals on average have lower knee muscle strength, which may lead to excessive abnormal loading on the knee joint, in turn leading to knee malalignment [23, 24].

The work-related factor of years of farming experience was found to be associated with knee and foot malalignments. However, hip malalignment could not be proved to relate to rice farming experience. Since most farming tasks are performed in awkward postures with repetitive LE motion and involving heavy loads, lifting or carrying excessive weight may overload muscles and tendons of knee and foot [1, 21]. Exposure to such working conditions was found to be associated with degeneration of joints, which is a risk factor for malalignment [1]. Repetitive awkward postures, such as stooping and twisting, are associated with chronic knee musculoskeletal symptoms. As a result, increased abnormal knee loading may lead to injury, such as osteoarthritis (OA) and patellofemoral pain syndrome, and might develop into knee malalignment [18, 19]. Moreover, farming activities also involved working on slippery uneven surface of muddy terrain. Such environmental conditions may be challenging for farmers to control leg alignment and may lead to increased foot and knee pain arising from postural instability and fatigue [25]. Years of exposure to such risk factors could lead to progressive abnormal biomechanical function and result in increasing risk of LE malalignments [18, 19].

Contrary to the malalignment examination, pain analysis results showed that farmers perceived higher pain in the hip, as compared with knee and foot. As mentioned previously, hip pain due to malalignment may originate from foot pronation and knee valgus [12]. However, analyses using linear models could not establish strong relationships between ergonomic risk factors and hip pain. With regard to the the cultivation stages, the study showed that rice farmers perceived the highest LE pain during the planting task. Detailed analyses indicated that, for the planting stage, perceived knee pain was significantly induced by motion and posture factors, while foot pain perception was primarily caused by force exertion. During the planting activities, the typical farming work posture constitutes a combination of squatting, forward bending and rotation of the trunk, assumed for several hours during a work shift. These activities were found to induce knee discomfort, especially in the hamstring muscle group, as a result of maintaining stooped postures leading to overuse and fatigue of affected knee muscles [26]. With respect to foot pain perception induced by the force factor, performing planting work while carrying heavy load on a muddy walking surface was found to create excessive loading on trunk and foot muscles [1].

To further investigate the force factor due to the adverse environmental conditions of muddy work terrain, a simulated rice planting task was conducted to compare the force acting on joints as well as the muscular force requirements to the LEs of farmers on muddy vs. rigid working surface. Experiment results revealed BF and GA muscle activities to significantly increase when participants were working on muddy ground in comparison with the rigid ground condition. Participants utilized higher activity levels of a certain set of BF and GA muscles to compensate for mud viscous force. An increase in BF and GA activity, which assisted knee flexion, was in line with results from the biomechanical model estimation of tensile force loading on LE joints, indicating the knee is exposed to the highest force increase due to mud viscosity. Substantially higher GA activity might result from a higher propulsion force from plantar flexion required for backward gait initiation. Such high GA activity requirements also complement the result of foot pain perception induced by excessive force. Prolonged repetition of GA exertion might lead to muscle fatigue and increase of risk of structural malalignment of the feet [25], which correlates well with the high prevalence of abnormal alignment found in foot condition.

5 Conclusion and Future Research

The collected results of our studies revealed the knee and foot to be priority LE parts that need to be addressed due to the high prevalence of knee valgus and foot pronation. The environment condition of muddy work terrain also increased force loading on the knee and foot joints and muscles due to adverse effects of ground viscous force on the legs. Although the hip was found to be the most problematic part in terms of pain perception, there is currently no evidence of relationships between hip pain and rice cultivation activities. Moreover, hip pain might be attributed to abnormal alignments of foot and knee. Based on these findings, future research should emphasize on developing self-care programs, personal protective equipment and/or assistive devices to prevent knee and feet injury during rice planting task performance. More specifically, the development of protective and assistive interventions should focus on reducing awkward posture and repetitive motion of the knee, and on alleviating forceful exertion of knee and foot during work on muddy terrain. Self-care programs should be designed to develop self-awareness of improper work postures, excessive load and repetitive movement, and to implement exercises that focus on strengthening the hamstring group and gastrocnemius muscles.

Regarding future research in our group, a particular focus is set on designing customized footwear with medical wedging at various places underneath the foot to reduce eversion. Such footwear is anticipated to have beneficial impact on preventing knee arthritis attributable to knee valgus as well as soft tissue and joint injuries of the foot and ankle secondary to foot pronation. In addition, the use of skin-comfort materials will be investigated where the footwear contacts the skin. The resulting footwear designs are expected to be useable alongside with adaptive knee braces, which can be manually or automatically locked to prevent or support specific motions and work postures. Such a design concept may also be applicable for farmers, who demonstrate excessive foot pronation and knee valgus due to the extreme work conditions in the field, by incorporating materials that may help reduce mud viscous force for footwear shells.

References

1. Reid CR, Bush PM, Karwowski W, Durrani SK (2010) Occupational postural activity and lower extremity discomfort: a review. Int J Ind Ergon 40(3):247–256
2. Fathallah FA (2010) Musculoskeletal disorders in labor-intensive agriculture. Appl Ergon 41 (6):738–743

3. Ministry of Commerce of Thailand. Thailand trading report (agricultural products). http://www.ops3.moc.go.th/menucomen/export_topn_re/report.asp. Accessed 18 Apr 2018
4. Mokkamul P (2006) Ethnobotanical study of rice growing process in northeastern, Thailand. Ethnobot Res Appl 4:213–222
5. Puntumetakul R, Siritaratiwat W, Boonprakob Y, Eungpinichpong W, Puntumetakul M (2011) Prevalence of musculoskeletal disorders in farmers: case study in Sila, Muang Khon Kaen, Khon Kaen province. J Med Technol Phys Ther 23(3):297–303
6. Osborne A, Blake C, Fullen BM, Meredith D, Phelan J, McNamara J, Cunningham C (2012) Risk factors for musculoskeletal disorders among farm owners and farm workers: a systematic review. Am J Ind Med 55(4):376–389
7. Woolf AD, Pfleger B (2003) Burden of major musculoskeletal conditions. Bull World Health Organ 81(9):646–656
8. Karukunchit U, Puntumetakul R, Swangnetr M, Boucaut R (2015) Prevalence and risk factor analysis of lower extremity abnormal alignment characteristics among rice farmers. Patient Prefer Adher 9:785–795
9. Saetan O, Khiewyoo J, Jones C, Ayuwat D (2010) Musculoskeletal disorders among northeastern construction workers with temporary migration. Srinagarind Med J (SMJ) 22 (2):165–173
10. Khamis S, Yizhar Z (2007) Effect of feet hyperpronation on pelvic alignment in a standing position. Gait Posture 25:127–134
11. Barwick A, Smith J, Chuter V (2012) The relationship between foot motion and lumbopelvic-hip function: a review of the literature. Foot 12:224–231
12. Chuter VH, Janse de Jonge X (2012) Proximal and distal contributions to lower extremity injury: a review of the literature. Gait Posture 36:7–15
13. Neubert MS, Karukunchit U, Puntumetakul R (2017) Identification of influence demographic and work-related risk factors associated to lower extremity pain perception among rice farmers. Work J Prev Assess Rehabil 58:489–498
14. McAtamney L, Corlett EN (1993) RULA: a survey method for the investigation of work-related upper limb disorders. Appl Ergon 24(2):91–99
15. Tropea C, Yarin AL, Foss JF (2007) Springer handbook of experimental fluid mechanics, vol 1. Springer science & business media, Heidelberg
16. Juntaracena K, Swangnetr M (2016) Effects of muddy work terrain on force of rice farmer lower extremity joints during rice planting process. In: Proceedings of the 4th IIAE international conference on industrial application engineering, Beppu, Japan
17. Juntaracena K, Neubert MS, Puntumetakul R (2018) Effects of muddy terrain on lower extremity muscle activity and discomfort during the rice planting process. Int J Ind Ergon 66:187–193
18. Daneshmandi H, Saki F, Shahheidari S, Khoori A (2011) Lower extremity malalignment and its linear relation with Q angle in female athletes. Procedia Soc Behav Sci 15:3349–3354
19. Shultz SJ, Nguyen A, Levine BJ (2009) The relationship between lower extremity alignment characteristics and anterior knee joint laxity. Sports Health 1(1):54–60
20. Viester L, Verhagen EA, Hengel KMO, Koppes LL, van der Beek AJ, Bongers PM (2013) The relation between body mass index and musculoskeletal symptoms in the working population. BMC Musculoskelet Disord 14(1):238
21. Messing K, Tissot F, Stock S (2008) Distal lower-extremity pain and work postures in the Quebec population. Am J Public Health 98(4):705–713
22. Nguyen AD, Shultz SJ (2007) Sex differences in clinical measures of lower extremity alignment. J Orthop Sports Phys Ther 37(7):389–398
23. Keogh E, Herdenfeldt M (2002) Gender, coping and the perception of pain. Pain 97:195–201

102   M. S. Neubert et al.

24. Scerpella TA, Stayer TJ, Makhuli BZ (2005) Ligamentous laxity and noncontact anterior cruciate ligament tear: a sex-based comparison. Orthopedics 28:656–660
25. Nguyen AD, Shultz SJ (2009) Identifying relationships among lower extremity alignment characteristics. J Athletic Train 44:511–518
26. Meyer RH, Radwin RG (2007) Comparison of stoop versus prone postures for a simulated agricultural harvesting task. Appl Ergon 38(5):549–555

Multitask Analysis of Whole Body Working Postures by TACOs: Criteria and Tools

Marco Tasso[(⊠)]

EPM International Ergonomics School, Milan, Italy
prevenzione.tasso@gmail.com

Abstract. Work-related musculoskeletal disorders (WMSDs) are very common in different professional profiles due to the awkward postures maintained during their work.

TACOs (Timing Assessment Computerized Strategy) method enables to assess this risk starting from work organizational study to identify the various tasks making up the job and it offers criteria for calculating scores even in highly complex scenarios where individuals perform multiple tasks sometimes over cycles lasting longer than one day (weekly, monthly, yearly).

Keywords: Biomechanical overload · Risk assessment
Occupational awkward postures · TACOs method · Musculoskeletal disorders
Multitask analysis

1 A New Approach to Assess Working Postures Risk

Scientific literature and International standards does not suggest any definitive recommendations concerning how to manage multi-task situations in terms of different postures and length of exposure [1, 2, 5–7].

TACOs (Timing Assessment Computerized Strategy) method offers new criteria for approaching this specific risk even in highly complex scenarios.

The strategy of the method allow to identify postures that need to be analysed within each task and to define scores for these postures depending on their duration, applying time-integration principles and calculating final risk scores.

In short, TACOs is a Timing Assessment method, that is an analysis and assessment strategy, delivering criteria for:

- identifying postures that need to be analysed and defining intrinsic scores for these postures depending on their duration;
- identifying the interval and tasks to be analysed;
- timing assessment of postures within each task;
- organizational study of turnover (i.e. task rotation) in order to define the duration of exposure to postures over multiple tasks;
- applying time-integration principles and calculating final risk scores via simple tools.

This strategy is applied to the study of physically tiring postures overlooked by the *OCRA method* [4] (that has to be use to study upper limb awkward postures) and by the

© Springer Nature Switzerland AG 2019
S. Bagnara et al. (Eds.): IEA 2018, AISC 820, pp. 103–111, 2019.
https://doi.org/10.1007/978-3-319-96083-8_13

Revised NIOSH Lifting Equation [3] (that has to be use to study spine posture during manual lift of loads).

For these postures (spine without lifting and lower limbs), additional criteria and forms proposed for are required:

- lower limbs;
- low back for standing postures;
- low back for sitting postures;
- sitting postures with pedal;
- neck-head postures;
- complex and integrated postures, for specific tasks (on ladder, on bed, carrying on the head, etc.)

The recommended analytical process always begins with an organizational study in order to identify the various tasks making up the job and to assess their duration in the work cycle. Then, within each task identified, separately assess the intrinsic risk scores both with regards to complete biomechanical overload of the upper limbs (OCRA) or the spine, if manual lifting is involved (NIOSH), and to the postural risk scores for the spine (without manual lifting) and the lower limbs.

At the end the data for each job needs to be integrated in order to calculate the final exposure index, based on percentages assigned to each task, using specific mathematical models.

2   The TACOs Method: Criteria for Back and Lower Limbs Posture Analysis

The proposed postures result from various methods reported in the literature and in international standards [1, 2, 5–7]; they have been simplified and/or regrouped to ensure a clear understanding.

As to the criteria adopted for assigning scores, it should be noted that the exposure scores simply rank postures from the least uncomfortable (generally "neutral") to the most awkward.

Scores are assigned starting from the most comfortable posture (e.g. seated upright, reclining against a back rest – Fig. 1):

- score 0.5 when the worker spends 1/3 of the time (approx. 20–40%)
- score 1.5 when the worker spends half the time (approx. 40–60%)
- score 3 when the worker spends 3/3 of the time (approx. 60–100%)

When postures are frequently changed, such as moving from standing to sitting with a back rest, at least every hour, the risk score will be 0.5 (no risk) for durations of 1/3, 2/3 and 3/3 of the time (Fig. 2).

As the discomfort caused by the posture increases, the scores increase accordingly, starting from a duration of 1/3 of the time: the other scores start from this figure and eventually they will double or triple their score. More tiring or unusual postures receive a score even if they are adopted for short periods (approx. 1/10 of the time).

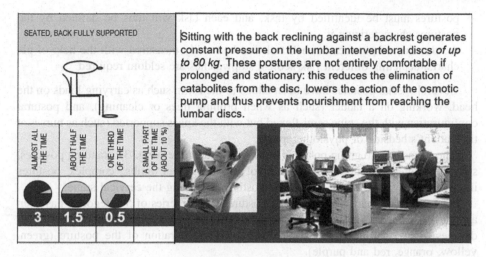

Fig. 1. Different exposure scores defined on the basis of posture and its duration

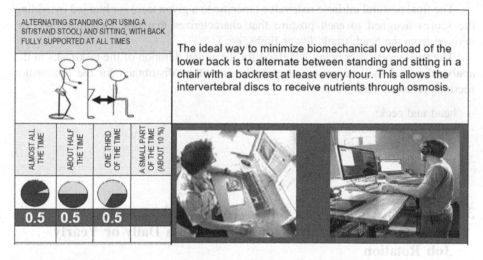

Fig. 2. Same exposure scores when posture changes at least every hour

In other words, these descriptive exposure scores are adjusted on the basis of the posture awkwardness and duration of the task.

As a general approach on identifying postures and on posture duration, some general rules are proposed:

- postures should not be identified and described for each part of the body and subsequently aggregated;
- whole body postures (various standing, sitting, squatting postures, etc.) have been defined using sketches and simple descriptions;

- postures must be identified by task, and each task will thus be defined by the postures characterizing it;
- the duration of postures in each task can be readily measured with the help of pie charts, depicting different risk scores; stopwatches are seldom required.

Several special postures have been added to the form, such as carrying loads on the head, working on a ladder (such as when pruning trees or cleaning), and postural configuration with the spine semi-flexed but with the knees supported (such as nurses at the patient's bedside or physiotherapists).

The forms help to group postures into 5 main categories (Fig. 3): standing postures, sitting postures, postures primarily involving the lower limbs, complex postures involving many parts of the body, and postures involving the cervical spine.

Since the maximum score for each posture is 14, a series of 5 categories has been arbitrarily defined (Table 1), each with a different colour, indicating the different degrees of awkwardness, also as a function of the duration of the posture (green, yellow, orange, red and purple).

This approach is adopted intrinsically for each posture and for the outcome of the total task analysis.

The final postural intrinsic risk value (specific to a given task) is obtained by adding the scores assigned to each posture that characterizes that task (whether they are derived from the head, trunk, lower limbs, etc.).

The final score, after being weighted for the actual duration of the task/tasks in the analysed period, will be presented globally as well as distributed for the four major sections reported here:

- head and neck;
- spine in the standing position;
- spine in sitting position;
- lower limbs.

3 Example: Analysis of a Single-Task Job Performed on a Full-Time or Part-Time Basis with a Daily or Yearly Job Rotation

An example of single task analysis with TACOs method is reported below.

Figure 4 shows the result of the risk assessment for biomechanical overload of the upper limbs, which considers all the risk factors required to estimate the final risk level. The risk level was quite high, with scores in the purple zone for the right side (very high risk) and in the red zone for the left side (high risk).

Figure 5 shows the TACOs method application to an additional timing analysis of the postures: in this example low back, lower limbs and head/neck postures.

The example shows two evenly alternating awkward back postures (the lower limbs are not subject to awkward postures):

Fig. 3. Some examples of the analysed postures, proposed by TACOs method, divided for body areas and their specific scores

Table 1. The five levels characterized by a color (green, yellow, orange, red and purple), each representing a different degree of posture risk, also in relation to duration.

Zone	TACOs	Risk classification
Green	Up to 0.55	Acceptable
Yellow	0.56–2.00	Borderline or very slight
Red-low	2.1–3.9	Slight
Red-medium	4.00–8.00	Medium
Red high	Sup. 8.00	High

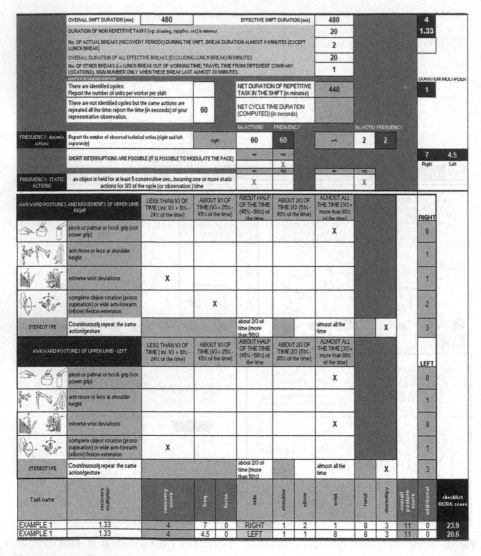

Fig. 4. *Example 1*. Full-time, exposure to 440 min of net repetitive work/shift involving the upper limbs: application of the OCRA Checklist method to assess the risk of biomechanical overload of the upper limbs

- back completely flexed (with work area below knee height), with a score of 7 for a duration of 50%;
- back semi-flexed (between 30° and 60°), with a score of 5 for a duration of 50%.

The sum of the corresponding scores produces the total intrinsic postural risk score associated with that particular task.

When the networking time is equal to the constant of 480 min/day, the total postural risk value coincides with the intrinsic value (in this case equal to 15).

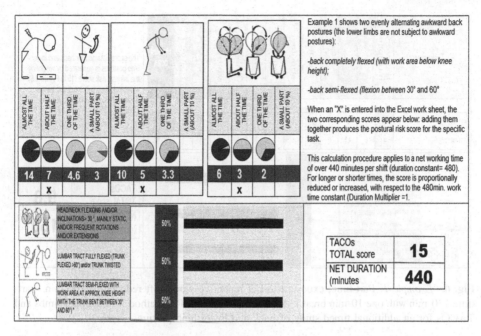

Fig. 5. *Example 1.* Full-time, exposure to 440 min of net repetitive work/day: application of the TACOs method for an additional timed study of back and lower limb postures and automatic calculation criteria for the final back and lower limb index.

In this example, however, the networking time is 440 min per shift. For longer or shorter times, the score will be proportionally reduced or increased, with respect to the 480 min work time constant.

Figure 6 shows the indexes obtained with the OCRA checklist and TACOs recalculated for the same task, but on a part-time basis (i.e. 170 min in the shift, or 160 min net).

The awkward postures are the same, but the duration of the work in the shift is limited to three hours.

In our calculation model the scores decrease considerably, since the final scores are weighted versus pre-defined exposure constants. The calculations are included in Fig. 6.

In studying the task in terms of the risk of biomechanical overload, whether the job is full-time or part-time, it is always necessary to match the scores obtained with the OCRA checklist (upper limbs) with those obtained with the TACOs method (back and lower limbs). If the task had involved manual lifting of loads weighing more than 3 kg, we would have also matched the score using the NIOSH formula.

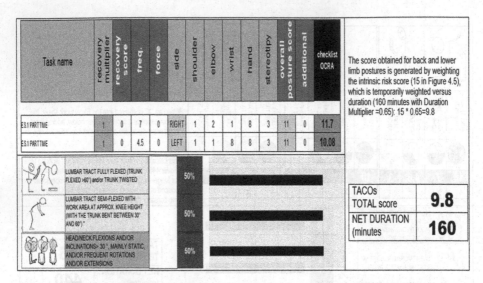

Task name	recovery multiplier	recovery score	freq.	force	side	shoulder	elbow	wrist	hand	stereotipy	overall posture score	additional	checklist OCRA
E.S.I PARTTIME	1	0	7	0	RIGHT	1	2	1	8	3	11	0	11.7
E.S.I PARTTIME	1	0	4.5	0	LEFT	1	1	8	8	3	11	0	10.08

The score obtained for back and lower limb postures is generated by weighting the intrinsic risk score (15 in Figure 4.5), which is temporarily weighted versus duration (160 minutes with Duration Multiplier =0.65): 15 * 0.65=9.8

LUMBAR TRACT FULLY FLEXED (TRUNK FLEXED >60°) and/or TRUNK TWISTED	50%
LUMBAR TRACT SEMI-FLEXED WITH WORK AREA AT APPROX. KNEE HEIGHT (WITH THE TRUNK BENT BETWEEN 30° AND 60°) °	50%
HEAD/NECK FLEXIONS AND/OR INCLINATIONS> 30 °, MAINLY STATIC, AND/OR FREQUENT ROTATIONS AND/OR EXTENSIONS	50%

TACOs TOTAL score	9.8
NET DURATION (minutes	160

Fig. 6. *Example 1.* Part-time, exposure to net repetitive work/shift reduced to 160 min (part-time 170 min with one 10-min break). Application of the OCRA method for the upper limbs and TACOs for an additional timed study of back and lower limb postures.

4 Conclusions

This paper discusses TACOs (Timing Assessment Computerized Strategy) method, a new approach for assessing awkward postures risk even in highly complex scenarios.

Starting from an analytic organizational study in order to identify the various tasks making up the job and to assess their duration in the work cycle, this strategy offers criteria to define postures and related scores.

TACOs method considers the following body parts overlooked by the OCRA method (that has to be use to study upper limb awkward postures) and the Revised NIOSH Lifting Equation (that has to be use to study spine posture during manual lift of loads):

- lower limbs;
- low back for standing postures;
- low back for sitting postures;
- sitting postures with pedal;
- neck-head postures;
- complex and integrated postures, for specific tasks (on ladder, on bed, carrying on the head, etc.)

The final postural intrinsic risk value (specific to a given task) is obtained by adding the scores assigned to each posture that characterizes that task (whether they are derived from the head, trunk, lower limbs, etc.).

The final score, based on the duration of the task/tasks in the analysed period, is presented globally as well as distributed for the four major sections reported here:

- head and neck;
- spine in the standing position;
- spine in sitting position;
- Lower limbs

References

1. CEN (European Committee for Standardization) (2003) EN 1005-2. Safety of machinery - Human physical performance - Part 2: Manual handling of machinery and component parts of machinery. CEN Management System, Bruxelles, Belgium
2. CEN (European Committee for Standardization) (2005) EN-1005-4: Safety of machinery – human performance – Part 4: Evaluation of working postures and movements in relation to machinery. CEN Management System, Bruxelles, Belgium
3. Colombini D, Occhipinti E, Alvarez-Casado E, Waters T (2012) Manual lifting, a guide to study of simple and complex lifting tasks. CRC Press, Taylor & Francis, Boca Raton, New York
4. Colombini D, Occhipinti E (2016) Risk analysis and management of repetitive action - a guide for applying the OCRA system (Occcupational Repetitive Action). CRC Press, Taylor & Francis, Boca Raton, New York
5. Karhu O, Kansi P, Kuorinka I (1977) Correcting working postures: a practical method for analysis. Appl Ergon 8(4):199–201
6. ISO, 2000. ISO 11226. Ergonomics — Evaluation of static working postures. ISO, Geneva, Switzerland
7. McAtamney L, Corlett EN (1993) RULA: a survey method for the investigation of work-related upper limb disorders. Appl Ergon 24(2):91–99

Repetitive Manual Tasks Risk Assessment Among Supermarket Workers: Proposal of an Analysis Model

S. Tello-Sandoval[1(✉)], E. Alvarez-Casado[1], and D. Colombini[2]

[1] Centro de Ergonomia Aplicada (CENEA), Barcelona, Spain
sonia.tello@cenea.eu
[2] Research Unit and International School of Ergonomics of Posture and Movement (EPM), Milan, Italy

Abstract. Introduction: Although the risk of upper limbs repetitive movements of cashier workers is clear in the scientific literature, there are still very few studies providing a comprehensive risk assessment for all supermarket jobs regarding musculoskeletal conditions due to repetitive movements.
Methods: An organizational analysis of 5 supermarkets was carried out onsite to determine the duration and content of the shifts, i.e. to identify which tasks entail a potential biomechanical overload and how they are assigned to homogeneous groups of workers. A total of 67 tasks were identified and analyzed. Each individual task carried out by supermarket staff was analyzed and filmed. The videos were later assessed using the OCRA checklist, for both the right and left arm, to calculate the intrinsic task risk.
Result: Based on the evaluations, risk of upper limb musculoskeletal disorders is often "medium" and sometimes "high" (e.g. deli counter and butcher) among full-time workers; for part-time workers risk is often medium/low red or borderline (yellow).
Discussion: The analysis of the exposure to repetitive movements should be performed based an exhaustive organizational analysis. This kind of study will contribute to manage this risk and improving working conditions in an effective way. More research is needed to identify the best mathematical model to compute the weekly risk score.

Keywords: Work related upper limb disorders · Repetitive tasks
Risk assessment · Supermarket

1 Introduction and Magnitude of the Issue

1.1 Description of Work Centers

Large scale retailing is a modern type of trade where goods are sold through a chain of large grocery stores, supermarkets, hypermarkets and so forth.

Outlets or points of sale are generally classified by channel according to the amount of space (in square meters) allocated to sales, without calculating common areas, parking lots, etc., and on the assortment of goods on sale, as following:

© Springer Nature Switzerland AG 2019
S. Bagnara et al. (Eds.): IEA 2018, AISC 820, pp. 112–121, 2019.
https://doi.org/10.1007/978-3-319-96083-8_14

- Large-scale retailing can be broken down into the following sales channels:
- Hypermarket: retail premises measuring over 2,500 m^2;
- Supermarket: retail premises measuring between 400 and 2,500 m^2;
- Convenience store: retail premises measuring between 100 and 400 m^2;
- Discount store: this type of store does not stock branded products.
- Cash and carry: a store open only to wholesalers;
- Traditional retail stores: stores that sell products made by many different brands, in premises measuring up to 100 m^2;
- Self-service specialist drug stores: stores selling home and personal care products.

1.2 Epidemiological Data

In the literature there are numerous reports stressing the close link between upper limb disorders, primarily carpal tunnel syndrome (CTS) and cervical pain, and the job of supermarket cashiers. Conversely, there are few studies on spinal conditions among supermarket workers due to prolonged standing.

In particular, Bonfiglioli [1, 2] observed increases in CTS among cashiers as opposed to teachers. In these studies, 40% of cashiers were found to suffer from bilateral CTS whilst 18.4% of teachers had bilateral CTS, 8.2% on the right side and 7.7% on the left.

This study [2] indicates that 31% of full-time cashiers and 19.3% of part-time cashiers have CTS, versus 16.3% of teachers.

Another study [9] narrows down the type and location of upper arm disorders among supermarket cashiers. The problems primarily affect the shoulders (right 19.2%, left 15.6%), followed by the wrists (right 11.7%, left 7.2%). The conditions include CTS (right 10.5%, left 12%), epicondylitis (right and left 11.7%), other elbow disorders (right 5.8%, left 15.6%) and De Quervain syndrome (right and left 5.8%).

Margolis [7] compares musculoskeletal disorders among supermarket workers in different positions, including cashiers, deli counter staff, stock clerks, fresh produce staff, and office workers. The study draws a distinction between conditions affecting the neck, upper limbs in general and specifically the wrist and the hand. The higher risk positions are cashiers for the neck (7%) and stock clerks for wrist-hand (8,1%).

In 2005, a Japanese study [6] found an association between musculoskeletal symptoms and workers assigned to cold stores.

1.3 Problem Description

There are still very few studies providing a comprehensive comparison between risk assessments and the results of clinical assessments for all supermarket jobs regarding musculoskeletal conditions due to manual load handling and/or repetitive movements among supermarket workers in general. So far it would seem that only two such studies have been published in the literature:

The first is by Osorio [8], where the authors clearly document the presence of carpal tunnel syndrome among workers highly exposed to strenuous and repetitive movements, especially among cashiers, butchers, pastry cooks and bakers.

The second is by Draicchio [5], who used the Revised Niosh Lifting Equation (RNLE) and other methods to measure risk due to manual materials handling in the fresh produce section of a medium sized store.

The lack of data on supermarkets and the challenges involved in assessing risk confirm the need to undertake more systematic research into risk due to biomechanical overload and the specific damage that it causes.

We will next be suggesting an effective and specific methodology for analyzing risk based on more recent experiences.

2 Method

2.1 Risk Analysis Procedure

Most of the traditional methods for analyzing risk among workers exposed to multiple repetitive tasks tend to focus on daily exposure; in supermarkets, however, exposure may vary considerably in terms of type and duration. There may be both daily and weekly variations.

So before applying models for analyzing exposure risk for weekly multitask schedules, it is first necessary to conduct an organizational analysis and examine the duration and content of the shifts, i.e. to identify which tasks entail a potential biomechanical overload and how they are assigned to homogeneous groups of workers.

As usual, the aim is to create an easy-to-use "tool" (in the form of software or spreadsheets) for automatically estimating exposure risk with respect to the various tasks.

This tool represents a "best practice" for use by numerous professionals, including accident prevention officers and occupational health experts, insurance companies and so on, who are involved in preventing and managing risk due to biomechanical overload, as well as for medico-legal purposes, such as achieving an objective risk assessment in order to recognize occupational disorders.

The data presented here have been developed based on investigations conducted in Milan, Italy, as part of a partnership between the city's local health units and the Occupational Health Centre (CEMOC) at the Clinica del Lavoro "Luigi Devoto" in Milan. The aim here was to assess the risk of biomechanical overload in the upper limbs of supermarket workers.

2.2 Data Collection

CEMOC, the occupational health center at the Clinica del Lavoro in Milan, conducted a more systematic analysis. The study began with visits to several small, medium and large supermarket chains in Milan and ended with interviews with management to accurately identify the relevant operational areas and tasks.

During the initial visit, each individual task carried out by supermarket staff was analyzed and filmed. The videos were later assessed using the OCRA checklist [4], for both the right and left arm, in order to calculate the intrinsic task risk (i.e. as if the worker performed the same task for the whole 8-h shift, with one meal break and two breaks lasting at least 8–10 min each). A total of 67 tasks were identified and analyzed.

3 Results

3.1 Intrinsic Risk Level for Every Operation Area

The various operational areas are listed below, along with the risk factors and intrinsic OCRA Checklist results.

Cashier

There are various types of checkout counters fitted with different types of scanners: flat (horizontal), vertical or hand-held. Cashiers also have different job descriptions, which may include bagging merchandise for customers. At the supermarkets in our study, the checkout counters had horizontal scanners and the cashiers did not bag merchandise (as is generally the case in Italy).

To collect data on work cycles (i.e. average duration and operations performed), with one customer representing one work cycle, 25 customers were analyzed at 4 different checkouts to obtain a representative sample.

Four sub-tasks were identified for each cycle/customer, and the OCRA Checklist was used to measure the risk and respective mean duration of every sub-task for each representative average customer/cycle.

Based on the results of the analysis the following information emerged: the average duration of each customer/cycle, average number of items bought, average number of "heavy" items, average waiting time within the customer/cycle.

The average duration of a customer/cycle is approx. 2.5 min. In percentage terms, this is the duration of the relative sub-tasks: scanning merchandise (63%); payment by cash or credit/bank card (19%); scan customer's loyalty card (4%); wait for payment (14%).

These were the intrinsic OCRA checklist values corresponding to each sub-task making up the average customer/cycle:

- 10.6 – scan customer loyalty card;
- 21.9 – scan items;
- 13.3 – payment;
- 0 – wait for payment.

Given the percentage duration of the above sub-tasks, the time-weighted average representing the intrinsic risk for this job (featuring a net duration of 420–440 min, with two 10-min breaks and a meal break) is 16.1.

Cashiers obviously spend some time waiting for customers (the main component of total unsaturation); daily average waiting times will vary throughout the day.

With regards to the different days of the week, the highest saturation levels are reported on Thursday, Friday and Tuesday; the lowest is on Monday.

Besides this information, other research involving a large supermarket chain has confirmed the average customer/cycle scores: here the average unsaturation rate is estimated to be between 17% and 19%.

The distribution of the unsaturation rate (per customer/cycle) is random and non-scheduled and the customer waiting time is variable (but seldom over 8 consecutive minutes), therefore the unsaturation neither modifies nor reduces the score for the lack of recovery time risk factor. Consequently, there are no recovery times within the cycle in the "customer at checkout" cycle (though they are often mistakenly calculated!).

Stock Clerk

Stock clerks perform many tasks that involve moving merchandise using devices such as 4-wheeled carts and manual or electric forklifts, and stocking/filling shelves manually.

The merchandise may weigh anything from a few grams, in the case of sale signs and labels, to 12 kg, such as detergents and crates of bottles. Heavier weights are quite rare. The items are moved to and from different heights.

For heavier items weighing more than 3 kg, it will also be necessary to assess manual load handling risk (using the Revised Niosh Lifting Equation –RNLE- suggested by the ISO standard 12228-1).

This job encompasses the following tasks:

1. move crates/boxes (6–12 kg) manually to shelves;
2. stock shelves with items weighing up to 2 kg;
3. fill and transport cardboard boxes weighing between 6 and 12 kg using 4-wheeled carts;
4. open cardboard boxes/items;
5. position items weighing between 6 and 12 kg along aisles;
6. position special offer signs;
7. stock shelves with bottled water;
8. return all unwanted items left at checkout counter to appropriate shelf locations.

The risk assessment for the various tasks performed by stock clerks leads to OCRA Checklist scores ranging from 16 to 20 (RED). Collecting unwanted items left at the checkout counter is associated with slight risk (11, YELLOW), whilst stocking shelves with bottled mineral water is high risk (27, PURPLE).

Stock clerks use both upper limbs in their work therefore the values calculated for the left side are the same as for the right side.

Grocery Clerk

Tasks include stocking and replenishing refrigerated grocery counters with fresh food including cheese and dairy, salami, ham, etc.

Generally speaking, grocery counter tasks are handled not by stock clerks but by deli staff. The usual stock clerk risks (as described above) are present, in addition to the additional very high risk associated with taking items out of refrigerators – here, the OCRA Checklist score may be as high as 25.5 (PURPLE).

Butcher

Butchers perform numerous tasks that involve cutting entire sides of meat suspended on hooks, slicing meat, portioning chickens, etc., as well as preparing trays. In most of the supermarkets in the study, the meat arrives partly pre-cut; in this case the butchers do not have to handle and cut heavy loads of meat such as whole sides.

The butchers are also in charge of replenishing and stocking the meat counter. The tasks performed by the butchers in our study include the following:

1. halve chickens using a cleaver and place on trays;
2. remove chicken breasts and place on trays;
3. remove chicken thighs and place on trays;
4. prepare and clean pieces of meat;
5. cut large portions of meat into smaller ones and place on trays;
6. slice meat into steaks using electric slicer and place on trays;
7. cut pork chops using a knife and place on trays;
8. cut through backbones using a knife to make chops and slice meat using a knife;
9. cut through backbones using cleaver to make chops and slice meat using a knife;
10. pack trays of pre-cut meat;
11. pack chicken breasts in cardboard boxes;
12. pack sausages on trays;
13. wrap, weigh and price meat trays;
14. restock meat counter;
15. restock meat fridge;
16. cut half sides suspended from ceiling hook;
17. debone femurs.

The 17 tasks been analyzed using the OCRA Checklist for the left and right upper limb. The task associated with the highest risk is cutting sides of meat suspended from hooks and deboning the femur, both of which require the use of considerable force, especially deboning "prosciutto" hams. Workers are also obliged to adopt awkward postures with the arm held above shoulder height (PURPLE).

Other tasks that are high-risk (PURPLE) due to the frequency with which they are performed include packing meat cuts on trays, packing chicken breasts in cardboard boxes and wrapping full trays; the OCRA Checklist scores for these tasks is 24. Other tasks for supermarket butchers range from slight to medium, with OCRA Checklist scores of between 12 and 21.

Comparing the checklist values for the left and right side, it can be seen that the dominant hand experiences the highest biomechanical overload.

For standard meat cutting operations, the left side may be at higher risk than the right side depending on the position of the left hand. When cutting larger pieces of meat, while the right hand is holding the knife (grip), the left hand is holding the meat (pinch or palmar grasp).

Often butchers wear a metal mesh glove or protective safety gloves. These gloves are indispensable for preventing accidents but make it harder to grip objects and may become slippery, requiring even more force.

Deli Counter

The deli counter performs numerous tasks that also differ in terms of risk: from deboning ham to slicing and packaging deli meats and cheeses.

The main tasks carried out by deli counter staff are the following:

1. cut hard cheeses into large pieces using a knife;
2. pack and wrap cheese portions (low machine);

3. pack and wrap cheese portions (tall machine);
4. open large packages of cheese;
5. weigh and price cheeses;
6. prepare prosciutto hams for slicing on electric slicer;
7. wrap prosciutto hams;
8. move and position deli meats and cheeses on shelves (weight: over 8–10 kg);
9. slice deli meats on manual slicer and prepare (slice, package, weigh, label);
10. slice deli meats on electric slicer and prepare (slice, package, weigh, label);
11. cut small rounds of soft and semi-soft cheeses;
12. debone prosciutto hams;
13. move and position smaller deli meats and cheeses on shelves;

Certain tasks, such as deboning ham and cutting hard cheeses manually, require peak force and are extremely high-risk (OCRA Checklist scores between 23 and over 30).

When shelves are taller than the deli staff, lifting and lowering deli meats weighing over 8–10 kg may subject the shoulders to extreme biomechanical overload (OCRA Checklist score: 43.5 - PURPLE).

The other deli counter operations are classified as RED (medium risk, scores between 14 and 20), except for moving and positioning smaller deli meats and cheeses (weighing less than 3 kg) on shelves, which has an OCRA Checklist score of 7 (GREEN).

Deli counter tasks may be carried out at the front or in the back of the store. The risk score will be identical in either case. The only difference is the net duration of exposure, which is higher for back-of-store tasks since there is no unsaturation time, due to waiting for customers for example.

Fresh Produce
In the fresh produce section, workers perform a variety of tasks, such as:

1. package lettuce or other fruit and vegetables;
2. pack and wrap watermelon;
3. price fruit and vegetables;
4. position price tags;
5. move and position hampers on display units;
6. position fruit and/or vegetables (individually or in small packs in display cases);
7. position pre-bagged fruit and/or vegetables;
8. close plastic crates for fresh produce.

Workers in the fresh produce section of supermarkets are at very high risk with respect to repetitive movements of the upper limbs when moving and positioning hampers full of fruit or vegetables on display units; the OCRA Checklist score is 25.5 (PURPLE). While some tasks are at negligible or no risk (i.e. positioning individual pieces of fruit or vegetables or small groups on display units), most tasks are classified RED (OCRA Checklist scores: between 14.5 and 22).

Fish Counter

Workers at the fish counter perform only a few tasks and all are in the fish section:

1. prepare and slice large fish;
2. cut and bag fish;
3. prepare the fish counter display with crates and individual fish;
4. package fish slices;
5. weigh and sell fish at the counter.

In certain instances, the tasks entail different levels of risk for the left and right arms:

– packing fish on trays puts the right arm at greater risk primarily due to both the high frequency of movements and stereotypy (OCRA Checklist score: 24 - PURPLE). The other tasks feature OCRA Checklist values of between 14 and 19 (RED).
– the left arm is generally at only slight risk (OCRA checklist scores: 10.5-YELLOW), however, preparing the fish counter with crates and individual fish and cutting large fish into slices are tasks classified as RED (OCRA Checklist scores: 17 and 18).

Bakery

The following operations are performed in the bakery section:

1. load frozen bread into oven;
2. sell small loaves at bakery counter;
3. package bread.

All these operations are classified as RED. The highest risk task is packaging bread, due to the high frequency of action and stereotypy. The OCRA Checklist scores are between 12 and 20.5.

Cleaning

Most of the cleaning tasks involve cleaning the shelves and rotisserie; the risk is due mainly to high frequency of action and stereotypy.

The OCRA Checklist scores are between 19.5 and 22.5 for the dominant arm (RED); cleaning shelves presents no risk for the non-dominant arm.

Merchandise Handling Using Pallet Jacks

These operations require assessment due to the presence of manual materials handling rather than to evaluate risk for the upper limbs.

The manual handling of empty pallets and the use of inadequate pallet jacks may, however, also place a strain on the upper limbs and therefore need assessing.

The OCRA Checklist values are between 8 and 20.5 for the dominant side (YELLOW/RED); the non-dominant side is at slight risk (YELLOW, risk score 8.5).

4   Example of Risk Assessment for a Homogeneous Group of Workers

The exposure risk is highly variable in terms of the shift duration and contents, i.e. which tasks make up the turnover both qualitatively and quantitatively; we now offer an example of risk assessment of cashiers.

In our example, cashiers worked a 380-min shift over 6 days a week. They did not perform any other tasks, except for returning unwanted items from the checkout (10 min a day).

The risk of biomechanical overload of the upper limbs was assessed over one full-time 380-min shift with one 20-min break.

The work was over a 6-day week; therefore, the weekly risk was 17 for the right side and 16 for the left. Since every day of the week featured the same exposure and the main job of the cashiers was to man the registers (their secondary tasks being minimal), the Time-Weighted Average and the Multitask Complex method [3] both produced the same risk indexes.

However, the exposure index calculated with the time-weighted average method per day (15.6 on the right side and 15.8 with the Multitask Complex method) was slightly higher than per week (16.9 and 17). This slight difference is negligible.

5   Discussion

Considering the risk assessment results outlined so far, the following observations can be made with reference to open issues that still require further research.

5.1   Intrinsic Risk Scores

Tasks are performed in much the same way in the various supermarkets: the intrinsic risk values assigned to the various tasks can, for the time being, be regarded as representative of all supermarkets (thus providing non-experts with libraries containing lists of pre-assessed tasks and making it easier for them to assess risk).

5.2   Duration and Content of Weekly Shifts

In terms of assessing risk in the various supermarkets, the differences stem primarily from the way the work is organized (i.e. duration of shift(s) over the week, duration and distribution of breaks, task assignment and turnover). All these aspects must be analyzed in the utmost detail in each individual situation.

5.3   Weekly Risk: Mathematical Models and Their Predictive Accuracy

The decision as to whether to use the Time-Weighted Average or the Multitask Complex formula to calculate the final weekly risk score should consider more statistical data looking at the association between the risk indexes and the clinical data deriving from on-site assessments, also to test which method produces the most predictive results.

References

1. Bonfiglioli R et al (2005) Carpal tunnel syndrome among supermarket cashiers. G. Ital. Med. Lav. Ergon. 27(1):106–111
2. Bonfiglioli R et al (2007) Relationship between repetitive work and the prevalence of carpal tunnel syndrome in part-time and full-time female supermarket cashiers: a quasi-experimental study. Int Arch Occup Environ Health 80(3):248–253
3. Colombini D, Occhipinti E (2017) Risk analysis and management of repetitive actions: a guide for applying the OCRA system (occupational repetitive actions), 3rd edn. Taylor & Francis Group, CRC Press, Boca Raton
4. Colombini D, Occhipinti E, Alvarez-Casado E (2013) The revised OCRA checklist method, 1st edn. Factor Humans, Barcelona
5. Draicchio F et al (2007) Biomechanical risk assessment of manual material handling in vegetables and fruit departments of supermarkets. G. Ital. Med. Lav. Ergon. 29(3):573–575
6. Inaba R et al (2005) Subjective symptoms among female workers and winter working conditions in a consumer cooperative. J Occup Health 47(5):454–465
7. Margolis W, Kraus JF (1987) The prevalence of carpal tunnel syndrome symptoms in female supermarket checkers. J Occup Med 29(12):953–956
8. Osorio AM et al (1994) Carpal tunnel syndrome among grocery store workers. Am J Ind Med 25(2):229–245
9. Panzone I et al (1996) Repetitive movement of the upper limbs: results of exposure evaluation and clinical investigation in cash register operators in supermarkets. Med Lav 87(6):634–639

Neck Postures During Smartphone Use in University Students and Office Workers: A Field Study

Grace Szeto[1(✉)], Daniel To[1], Sharon Tsang[1], Arnold Wong[1],
Jay Dai[1], and Pascal Madeleine[2]

[1] Department of Rehabilitation Sciences, The Hong Kong Polytechnic
University, Hung Hom, Kowloon, Hong Kong
grace.szeto@polyu.edu.hk
[2] Physical Activity and Human Performance Group, SMI, Department of Health
Science and Technology, Aalborg University, Aalborg, Denmark

Abstract. Office workers and university students are known to suffer from neck pain as they are frequent users of electronic devices. The present study utilized inertial motion sensors to examine real-time spinal kinematics in office workers and university students for 3 h in their natural working/studying environment. Office workers (10 males, 10 females) and university students (11 males, 11 females) were recruited by convenience sampling. Their mean ages were significantly different (Workers: 40.8 ± 8.5 years; Students: 21.5 ± 2.6 years). Five inertial motion sensors (Noraxon myoMotion™) were attached firmly onto the occipital protuberance, and the spinous processes of C6, T3, T12, and the sacrum, respectively, yielding angular displacements of the cervical, upper thoracic, lower thoracic and lumbar segments. The mean postural angles and the number of variations (zero crossings per minute) were analyzed. Self-reported neck pain score was higher in office workers (4.1 ± 3.7 on a 0–10 numeric scale) than that in students (2.3 ± 2.0, $P = 0.076$). Students adopted significantly greater degrees of lumbar flexion compared with office workers. Conversely, office workers tended to adopt slightly larger neck postural angles than students. Similarly, there was no significant between-group difference in zero crossings for the different spinal segments. Multivariate regression analyses showed that interaction of posture and duration of smartphone use were factors significantly contributing to musculoskeletal symptoms in students ($P < 0.001$), while age and gender were significant risk factors for symptoms in office workers ($P = 0.001$).

1 Introduction

1.1 Increasing Popularity of Electronic Devices

Smartphones have become an essential everyday instrument of communication and entertainment in people's lives worldwide. Recent reports suggested that individuals may spend up to 4 h per day using their smartphones [1]. Given the current trends with the public fascinated by new developments in multitouch smartphone technology, it is anticipated that there will be increasing reports of musculoskeletal complaints related to

S. Bagnara et al. (Eds.): IEA 2018, AISC 820, pp. 122–125, 2019.
https://doi.org/10.1007/978-3-319-96083-8_15

intensive use of these devices [2]. The multitouch smartphones and tablet devices may also induce different musculoskeletal demands due to their different screen sizes and different weights, and the different task context may also affect the physical workloads. The habit of adopting sustained neck flexion posture in using various electronic devices would inevitably lead to neck-shoulder pain and other musculoskeletal disorders. Office workers and university students are more susceptible to such symptoms since they are frequent users of multiple electronic devices.

1.2 Research on Biomechanical Measures Associated with Electronic Device Use

In the past few years, there has been a number of studies published examining the spinal kinematics and muscle activity in using handheld electronic devices. These were mainly laboratory based studies that involved standardized postures and standardized tasks. Typically the participants may be asked to perform a text typing task on the smartphone or tablet computer for a short duration such as a few minutes [3–6]. Kietrys et al. [3] determined the effect of screen size on cervical posture for a very brief texting task (about 10 s only). Other studies have compared different body positions or tilt angles of placing the electric devices [4–6]. Only two laboratory studies in the literature have estimated the range of neck flexion angles from 33–46° among most smartphone tasks [5, 6]. In our recent study [7], we reported on the spinal kinematics when participants performed smartphone texting with one hand versus two hands, but they were instructed to remain seated and resting their backs against the chair. These studies may not be accurate reflections of people's natural postures and movements when they are in their usual work or study environment.

With the advances in technology, wearable motion sensors provide a way to monitor people's real time postures and movements. These light-weight inertial motion sensors consisting of accelerometers, gyroscopes and magnetic sensors, can be applied on the human body in an unobtrusive manner and able to record movement data for long periods such as hours. The present study utilized inertial motion sensors to examine the real-time postures and movements in the cervical and thoracic spine of office workers and university students in their natural working/studying environment.

2 Methods

2.1 Study Design and Participants

This is a quasi-experimental cross-sectional study. Office workers (10 males, 10 females) and university students (11 males, 11 females) were recruited by convenience sampling. Their mean ages were significantly different (Workers: 40.8 + 8.5 years; Students: 21.5 + 2.6 years). We recruited mainly those who have had at least 1 year's experience in using touchscreen smartphones and they had to spend on the average at least 2 h per day on their smartphones as well as using a desktop or laptop computer regularly at their workplace or study setting.

2.2 Study Procedures

Five inertial motion sensors (Noraxon myoMotion™) were attached by double-side tapes onto the subject's occipital protuberance, and the spinous processes of C6, T3, T12, and the sacrum, respectively. The relative changes in positions between motion sensors were recorded as the angular displacements of the cervical, upper thoracic, lower thoracic and lumbar segments, respectively.

In this field study, each participant wore the sensors for 3 h during which they performed their usual work/study routine (e.g., using computers and smartphones) at their usual workplaces. The kinematics signals were recorded using a data-logger worn on a waist belt. The postural angles of the cervical and thoracic spine, as well as the number of times that participants varied their postures (zero crossings per minute) were analyzed.

3 Results

The mean duration of smartphone use in both groups was 0.9 h during the 3-h data collection. Self-reported neck pain score was higher in office workers (4.1 ± 3.7 on a 0–10 numeric scale) than that in students (2.3 ± 2.0, P = 0.076).

Students adopted postures with significantly greater degrees of lumbar flexion [median (interquartile range) = 34.6 (26.9–51.2)] (P < 0.001) compared with office workers [9.9 (6.0–17.1)] (P < 0.001). Conversely, office workers tended to adopt slightly larger neck postural angle [6.3 (3.3–10.8)] than students [4.3 (2.5–10.9)] (P = 0.093). Similarly, there was no significant between-group difference in zero crossings for the different spinal segments (P values: Cx: 0.515; UTx: 0.619; LTx: 0.696).

Multivariate regression analyses showed that interaction of posture and duration of smartphone use were factors significantly contributing to musculoskeletal symptoms in students (P < 0.001), while age and gender were significant risk factors for symptoms in office workers (Age: P = 0.001; Gender: P < 0.001).

4 Discussion

While neck-shoulder pain is common among office workers and university students, little is known regarding the relation between neck/lumbar postures and pain. The present field study utilized wearable sensors to bridge this knowledge gap and revealed important correlations. This approach opens a new horizon for ergonomic research.

Past research on neck postures and kinematics have only been reported based on laboratory studies involving short duration of smartphone use from less than 1 min [3, 4] to up to 10 min of texting tasks [5, 6]. The tasks performed were usually standardized and may not reveal the individuals' natural posture variations in the workplace. The future trend in research would involve field studies using the wearable sensors to gain insight about people's postural habits over the whole working day and how this is related to their experience of musculoskeletal symptoms. In fact, it would be ideal to utilize the smartphone to collect such data as well as using wearable sensors.

References

1. Salesforce 2014 (2014) Mobile behavior report. Salesforce, Singapore. http://www.exacttarget.com/system/files_force/deliverables/etmc-2014mobilebehaviorreport.pdf?download=1&download=1. Accessed 20 Nov 2016
2. Xie Y, Szeto G, Dai J (2017) Prevalence and risk factors associated with musculoskeletal complaints among users of mobile handheld devices: a systematic review. Appl Ergon 59:132–142
3. Kietrys DM, Gerg MJ, Dropkin J, Gold JE (2015) Mobile input device type, texting style and screen size influence upper extremity and trapezius muscle activity, and cervical posture while texting. Appl Ergon 50:98–104
4. Lee S, Kang H, Shin G (2015) Head flexion angle while using a smartphone. Ergonomics 58:220–226
5. Ning X, Huang Y, Hu B, Nimbarte AD (2015) Neck kinematics and muscle activity during mobile device operations. Int J Ind Ergon 48:10–15
6. Kim MS (2015) Influence of neck pain on cervical movement in the sagittal plane during smartphone use. J Phys Ther Sci 27:15–17
7. Xie Y, Madeleine P, Tsang SMH, Szeto GPY (2017) Spinal kinematics during smartphone texting – a comparison between young adults with and without chronic neck-shoulder pain. Appl Ergon 68:160–168

Handheld Mobile Devices—How Do We Use Them at Work? A University Case Study

Abdullah Alzhrani[1(⊠)], Margaret Cook[1], Kelly Johnstone[1], and Jolene Cooper[2]

[1] School of Earth and Environmental Sciences,
The University of Queensland, Brisbane, Australia
A.alzhrani@uq.edu.au
[2] Health, Safety, and Wellness,
The University of Queensland, Brisbane, Australia

Abstract. The use of handheld mobile devices (tablets and smartphones) is common among the general population. It is acknowledged that the use of handheld mobile devices exposes users to the recognized ergonomic risk factors of duration, repetition, and awkward and static postures; however, the nature of this exposure is currently poorly defined. This cross-sectional survey collected information about the use of smartphones and tablets concerning: type of devices, duration of use, duration of work-related tasks, environmental settings, hand-grip and tablet position, and musculoskeletal discomfort. Three hundred and ninety-eight (398) university employees and research students responded to the survey. The survey results highlighted that the use of smartphones differs from the use of tablets, with differences noted for postures adopted, duration of use, location of use, tasks undertaken, environmental settings, and hand-grip and tablet position. The results emphasized the potential ergonomic risks that workers are exposed to during the use of smartphones and tablets especially in relation to posture, duration, and environmental settings.

Keywords: Smartphone · Tablet · Musculoskeletal discomfort

1 Introduction

The use of handheld mobile devices (HMD), i.e., tablets and smartphones, is common among the general population. Previous studies have shown that the use of handheld mobile devices is associated with several adverse health effects (Lanaj et al. 2014; Lemola et al. 2014; Sarwar and Soomro 2013), including musculoskeletal disorders (Berolo et al. 2011; Blair et al. 2015). It is acknowledged that the use of handheld mobile devices exposes users to the recognized ergonomic risk factors of duration, repetition, and awkward and static postures; however, the nature of this exposure is currently poorly defined. With the increasing use of handheld mobile devices in the workplace, the nature of work and the work environment is changing (Honan 2015), and a better understanding of exposure is critical for the health of workers. This research aims to explore workers' use of handheld mobile devices and to explore key ergonomic risk factors associated with the use of these devices.

© Springer Nature Switzerland AG 2019
S. Bagnara et al. (Eds.): IEA 2018, AISC 820, pp. 126–137, 2019.
https://doi.org/10.1007/978-3-319-96083-8_16

2   Methods

This cross-sectional survey used an online-based questionnaire to collect self-reported information about the use of smartphones and tablets among the University of Queensland (UQ) employees and Higher Degree by Research (HDR) students.

2.1   Participants

The survey targeted all UQ staff and HDR students who used a smartphone and/or tablet for work use only or both for work and personal use. The survey was administered online using UQ's Checkbox survey tool. An introductory email was sent to UQ OHS network members, and they were requested to forward the survey invitation to UQ staff and HDR students within their schools, institution, and other organizational divisions and units. Additionally, a request for participation was advertised through the UQ website and UQ social media sites. The participants' consent was obtained prior to participation. The survey was administered from the 4th September–31st October 2017. The study was approved by the ethical officer at the School of Earth and Environmental Sciences, the University of Queensland [SEES number 201705-02].

2.2   Questionnaire

The questionnaire construction was primarily based on previously developed questions reported in the literature (Balakrishnan et al. 2016; Berolo et al. 2011; Blair et al. 2015; Cornell University Ergonomics Web 2017; Eapen et al. 2010; Goldfinch et al. 2011; Hedge et al. 1999; Hegazy et al. 2016; Kim and Kim 2015; Shan et al. 2013; Stalin et al. 2016). The survey was modified and enhanced based on feedback obtained from six school managers and key informant staff from the UQ library and Information and Technology Services. Further, the questionnaire was reviewed by three experts in the field of ergonomics and occupational health and safety. The first question 'what type/s of devices do you regularly use (at least once per week) for work or other purposes?' was used as a qualifying question. Those who use the smartphone and/or tablet for 'work use only' or for 'both work and personal use' were included and those who use the smartphone and/or tablet for 'personal use only' or 'do not use the device' were excluded.

The questionnaire consisted of six domains, which were:

(1) Demographic information: gender, age, handedness, role at UQ, and height;
(2) HMD information: type (tablet or smartphone), ownership, number of devices used, length of use in years, daily duration of use, and location of use;
(3) Duration of work-related tasks: reading, typing, emailing, web browsing, calling/watching/listening, organizing tasks, document creation, course-related use, social networking, and specific use;
(4) Environmental setting: use while sitting at a desk, sitting with no desk, standing at a desk, standing with no desk, walking, lying on sofa/bed, and lying or sitting on the floor/ground;

(5) hand-grip and tablet position: (a) for tablet: one hand holds and the other operates, both hands hold and operate, tablet is placed in the lap during use, tablet is placed flat on a supporting surface during use, and tablet is placed tilted on a support surface during use, (b) for smartphone: one hand holds and operates, one hand holds and the other hand operates and both hands hold and operate;

(6) Musculoskeletal discomfort/pain/ache during the last seven days in the neck, back, shoulders, arms, and wrists/hands.

The IBM SPSS statistics 24 was used for data analysis.

3   Results and Discussion

3.1   Demographics

A total of 398 respondents competed the survey. The respondents had a mean age of 39 years (SD ± 12yrs), with females representing 64% of the sample, which is a true reflection of the overall UQ population with 61.5% females in the UQ population (Table 1). The responses comprised of academic staff (26.7%), professional staff (45.4%), and HDR students (27.9%). Comparing to the overall UQ population, the responses were a good representation of academics with a similar population of academics in the survey. Professional staff were overrepresented amongst survey respondents, while HDR students were underrepresented (Table 2).

Table 1. A comparison between survey respondents and UQ population by gender.

Gender	Survey respondents	UQ population
Male	35.6%	38.5%
Female	64.4%	61.5%

Table 2. A comparison between survey respondents and UQ population by Role at UQ.

Role at UQ	Survey respondents	UQ population
Academic/Research	26.7%	25.1%
Professional	45.4%	34.8%
HDR	27.9%	40.1%

3.2   HMD Use

Smartphones were utilized by 97.5% of the respondents. The majority used their smartphone for both work and personal use (72.1%), whilst 1.8% used their smartphone for work only, and 23.6% used their smartphone for personal use only. Tablets were used by fewer respondents (61.5%), with 5% using their tablet for work only, 25.9% using their tablet for personal use only, and 30.4% using their tablet for both work and personal use (Table 3).

Smartphones are more common than tablets. Berolo et al. (2011), who conducted a study on a sample of university staff and students, reported that 98% of the participants used smartphones, which is similar to our findings (97.5%). Seventy six point five (76.5%) of the participants, who are university staff and students, in Blair et al. (2015) used a tablet, which is higher than our findings (61.5%). This variation can be attributed to the sample variation as Blair et al. had 22.8% participation from staff compared to 72.1% in our study.

In our study, around 49.6% (n = 121) of the tablet users utilized their tablet for both work and personal purposes, and 8.2% (n = 20) used tablets for work only (Table 2). This was found to be less than the percentages reported in (Stawarz et al. 2013) where 74% used their tablet for work and entertainment equally, and 24% used their tablet for work mainly.

The majority of smartphone (72.1%) and tablet (30.4%) users operated HMDs for work and personal purposes, which makes the separation of work use from personal use quite difficult. In addition, these results showed that workers may use HMDs at work for personal purposes and use HMDs at home for work purposes.

Table 3. Type of use for smartphone and tablet.

Device	Type of use			
	Work only	Personal only	Both work and personal use	Don't use this device
Smartphone	7 (1.8%)	94 (23.6%)	287 (72.1%)	10 (2.5%)
Tablet	20 (5.0%)	103 (25.9%)	121 (30.5)	153 (38.5%)

The availability of UQ purchased mobile devices appears to have increased over the last three years, with 41% of the respondents having a UQ owned smartphone for three years or less and 58% of respondents having a UQ owned tablet for 3 years or less (Table 4). This reflects the increased dependency on smartphone and tablet use for work purposes.

Table 4. Smartphone and tablet length of use in years.

	Length of use in years				
	0–1 Year (n (%))	1–3 years (n (%))	3–5 years (n (%))	5–10 years (n (%))	More than 10 years (n (%))
UQ smartphone	7 (11.5)	18 (29.5)	12 (19.7)	16 (26.2)	8 (13.1)
personal smartphone	10 (5.1)	32 (16.2)	37 (18.8)	91 (46.2)	27 (13.7)
UQ tablet	9 (15.0)	26 (43.3)	16 (26.7)	7 (11.7)	2 (3.3)
Personal tablet	8 (10.1)	22 (27.8)	25 (31.6)	19 (24.1)	5 (6.3)

Regardless of the ownership, 46.7% and 49.2% of smartphone and tablet users, respectively, were required to use the device for work as part of their role. For those who were provided with a UQ purchased HMD, 90.6% reported they were required to use the provided smartphone for work and 76.2% were required to use the provided tablet for work. For those who were not provided with a UQ purchased HMD, some respondents noted that they were required to use their personal HMD for work use, including smartphones (29%) and tablets (13.8%) (Table 5). These findings indicate that the workplace provided HMDs for some of its employees when HMDs were required as part of their role, but not all of the time.

Whether a HMD is required for work or not, workers may think the device is important for work. Stawarz et al. (2013) findings show that 35% of respondents see the device as necessary for work or at least useful (59%).

Table 5. The requirement for smartphone and tablet use at work.

	Is smartphone/tablet use required at work?	
	Yes (%)	No (%)
Smartphone		
Regardless of ownership	121 (46.7)	138 (53.3)
UQ owned	58 (90.6)	6 (9.4)
Personal	65 (29.0)	159 (71.0)
Tablet		
Regardless to ownership	59 (49.2)	61 (50.8)
UQ owned	48 (76.2)	15 (23.8)
Personal	12 (13.8)	75 (86.2)

Smartphones were used for a longer duration in a typical day than tablets (Table 6). The average total time of use in a day for the smartphone was 3.63 (SD + 3.45) hours per day and for the tablet was 1.81 (SD + 1.65) hours per day. On average, the duration of work-related use of smartphones and tablets were 1.65 (SD + 2.01) and 1.02 (SD + 1.23) hours per day for work. Comparing to other studies, Berolo et al. (2011) reported 4.65 h (SD + 5.67) of daily smartphone use. The reported tablet use was less than 3 h per day for the majority of respondents in other studies (Blair et al. 2015; Chiang and Liu 2016).

In addition to a longer duration of use, Müller et al. (2015) reported that the incidence of smartphone use was four times higher than tablets and that the smartphone was used on a daily basis, unlike the tablet. This could be explained by the fact that there are more smartphone users, as well as the smaller size of the smartphone, which makes the smartphone highly portable.

Most respondents reported using their smartphone in multiple locations, including in their work office (79.8%), in non-office workplaces (69.8%), in public places (82.8%), on transportation (80.9%), at home out of the bedroom (81.7%), and at home

Table 6. Smartphone and tablet average duration of use in a typical day (hours).

	Number of cases	Mean	Std. Deviation
Smartphone			
Total duration of use	257	3.63	3.45
Work duration of use	205	1.65	2.01
Tablet			
Total duration of use	125	1.81	1.65
Work duration of use	118	1.02	1.23

in the bedroom (68.3%) (Table 7). Compared to the general population, Müller et al. (2015) findings showed that most incidences of the smartphone use occurred outside of the home (56.5%), but only 19% of the use took place at work. It should be noted that Müller et al. asked their participants to report smartphone use incidents, while we asked our participants to report whether they use HMDs regularly at each location.

Tablet use was more restricted, with the most popular location being at home out of the bedroom (63.9%), followed by non-office workplace locations (54.6%), with the least popular location being public places (33.6%) (Table 7). These findings are supported by Stawarz et al. (2013), who surveyed workers and found that the most common location of use was at home (87%), followed by meeting rooms, commute and the office. On the other hand, Müller et al. 2015 found that the most common locations of use among the general population were at home (73.3%), with most of the use located in the bedroom (35.1%). These findings show the differences between tablet use for work purposes and personal or entertainment purposes. In addition, the findings show that HMD users tend to use the smartphone outside the home more than inside the home and use the tablet inside the home more than outside the home.

Table 7. Smartphone and tablet location of use.

Location of use	YES (%)	NO (%)
Smartphone		
At work—Office	210 (79.8)	53 (20.2)
At work—Non-office workplace	183 (69.8)	79 (30.2)
In public—Public places	217 (82.8)	45 (17.2)
In public—Transportation vehicles	212 (80.9)	50 (19.1)
At home—In bedroom	179 (68.3)	83 (31.7)
At home—Outside of bedroom	214 (81.7)	48 (18.3)
Tablet		
At work—Office	58 (47.5)	64 (52.5)
At work—Non-office workplace	67 (54.9)	55 (45.1)
In public—Public places	41 (33.6)	81 (66.4)
In public—Transportation vehicles	45 (36.9)	77 (63.1)
At home—In bedroom	63 (51.6)	59 (48.4)
At home—Outside of bedroom	78 (63.9)	44 (36.1)

3.3   Duration of Work-Related Tasks

The use patterns differed between the smartphone and the tablet. For the smartphone, the most common three tasks performed were emailing (94.4%), web browsing (82.8%), and organizing tasks (72.0%). For the tablet, the most common three tasks were web browsing (75.0%), emailing (73.1%), and reading (67.0%). A higher percentage of the respondents reported using the smartphone for typing (69.4%) than the tablet (51.9%). Table 8 shows the types and the estimated duration of each task.

Other studies found emailing and web browsing to be the most common tasks undertaken using a tablet (Stawarz et al. 2013). Among the general population (Müller et al. 2015), a diary study revealed that text messages, emails, and phone calls are common tasks for the smartphone, while the tablet tasks included email, playing games, social networking, and browsing.

Table 8. Types and duration of tasks performed on the smartphone and tablets.

	I don't do this task (%)	0–1 h (%)	1–2 h (%)	2–3 h (%)	3 h or more (%)
Smartphone					
Reading	140 (60.9)	74 (32.2)	11 (4.8)	3 (1.3)	2 (0.9)
Typing	71 (30.6)	112 (48.3)	32 (13.8)	8 (3.4)	9 (3.9)
Emailing	13 (5.6)	169 (72.5)	35 (15.0)	11 (4.7)	5 (2.1)
Web browsing	40 (17.2)	151 (64.8)	26 (11.2)	8 (3.4)	8 (3.4)
Calling, watching, or listening	108 (46.8)	100 (43.3)	14 (6.1)	3 (1.3)	6 (2.6)
Organizing tasks	65 (28.0)	150 (64.7)	8 (3.4)	6 (2.6)	3 (1.3)
Document creation	110 (47.2)	109 (46.8)	10 (4.3)	3 (1.3)	1 (0.4)
Course/teaching-related use	209 (92.1)	16 (7.0)	1 (0.4)	1 (0.4)	0 (0.0)
Social networking for work	113 (48.7)	85 (36.6)	20 (8.6)	6 (2.6)	8 (3.4)
Use of job specific apps	172 (74.5)	53 (22.9)	5 (2.2)	1 (0.4)	0 (0.0)
Tablet					
Reading	35 (33.0)	61 (57.5)	7 (6.6)	2 (1.9)	1 (0.9)
Typing	50 (48.1)	50 (48.1)	2 (1.9)	2 (1.9)	0 (0.0)
Emailing	28 (26.9)	68 (65.4)	5 (4.8)	3 (2.9)	0 (0.0)
Web browsing	26 (25.0)	64 (61.5)	11 (10.6)	3 (2.9)	0 (0.0)
Calling, watching, or listening	67 (65.0)	31 (30.1)	4 (3.9)	1 (1.0)	0 (0.0)
Organizing tasks	50 (48.1)	53 (51.0)	1 (1.0)	0 (0.0)	0 (0.0)
Document creation	73 (70.9)	26 (25.2)	3 (2.9)	1 (1.0)	0 (0.0)
Course/teaching-related use	87 (86.1)	10 (9.9)	3 (3.0)	0 (0.0)	1 (1.0)
Social networking for work	65 (64.4)	33 (32.7)	2 (2.0)	1 (1.0)	0 (0.0)
Use of job specific apps	56 (54.4)	42 (40.8)	2 (1.9)	1 (1.0)	2 (1.9)

3.4   Environmental Settings

With respect to the environmental settings, the majority of smartphone users reported that they usually (or always) used their smartphone while they were sitting on a chair/sofa without a desk/table (43.3%), sitting on a chair at a desk/table (37.1%), standing without a desk/table/bench (37%), lying on a bed/sofa (33.6%), walking (24.1%), sitting or lying on the floor (14.5%), and standing with a desk/table/bench (13.7%) (Table 9). The findings indicated the vast variety of postures that may be assumed during smartphone use.

The majority of tablet users reported they usually (or always) used their tablet while sitting on a chair either with or without a desk/table (38.9% and 43.4%). The tablet was found to be used less frequently in the rest of the environmental settings (Table 9).

These findings indicate that the neck, back, and upper extremities may not be in an optimal posture or supported while using HMDs. For instance, respondents reported that they use HMDs in an environmental setting such as lying on a bed/sofa. This requires them to assume and maintain non-neutral postures which expose their neck and upper extremity to ergonomic risks. Furthermore, the findings suggests that HMD users are in stationary status during HMDs use as 24.1% and 3.9% of smartphone and tablet users, respectively, indicated that they usually (or always) use their HMD while walking compared to the rest who do so less frequently.

Table 9. Frequency of smartphone and tablet use in six environmental settings.

Environmental settings	Always (%)	Usually (%)	Sometimes (%)	Seldom (%)	Never (%)
Smartphone use while					
Sitting on a chair at a desk/table	10 (4.2)	78 (32.9)	118 (49.8)	26 (11.0)	5 (2.1)
Sitting on a chair/sofa WITHOUT desk/table	11 (4.7)	91 (38.6)	124 (52.5)	7 (3.0)	3 (1.3)
Standing with a desk/table/bench	5 (2.1)	27 (11.6)	82 (35.2)	61 (26.2)	58 (24.9)
Standing WITHOUT a desk/table/bench	9 (3.8)	78 (33.2)	117 (49.8)	21 (8.9)	10 (4.3)
Walking	10 (4.2)	47 (19.9)	111 (47.0)	56 (23.7)	12 (5.1)
Lying on a bed/sofa	15 (6.4)	64 (27.2)	92 (39.1)	40 (17.0)	24 (10.2)
Sitting or lying on the floor/ground	7 (3.0)	27 (11.5)	62 (26.4)	53 (22.6)	86 (36.6)
Tablet use while					
Sitting on a chair at a desk/table	5 (4.6)	37 (34.3)	44 (40.7)	19 (17.6)	3 (2.8)
Sitting on a chair/sofa WITHOUT desk/table	2 (1.9)	44 (41.5)	40 (37.7)	9 (8.5)	11 (10.4)
Standing with a desk/table/bench	0 (0.0)	9 (8.7)	23 (22.1)	32 (30.8)	40 (38.5)
Standing WITHOUT a desk/table/bench	2 (1.9)	7 (6.6)	35 (33.0)	24 (22.6)	38 (35.8)
Walking	1 (1.0)	3 (2.9)	15 (14.3)	26 (24.8)	60 (57.1)
Lying on a bed/sofa	4 (3.7)	12 (11.1)	50 (46.3)	17 (15.7)	25 (23.1)
Sitting or lying on the floor/ground	1 (0.9)	2 (1.9)	19 (17.9)	24 (22.6)	60 (56.6)

3.5 Handgrip and Tablet Position

The smartphone users frequently (always or usually) reported holding the device with one hand and using the other hand to operate (51.3%), with just 31.1% using the same hand to both hold and operate the smartphone. For the tablet, 49% frequently (always or usually) reported holding the tablet with one hand and operating the tablet with the other hand, compared to just 9.9% who frequently used both hands to hold and operate the tablet (Table 10).

With regards to the tablet position, fewer participants reported that they frequently (always or usually) place the tablet on their lap during use (15.4%), compared to 37.5% who do sometimes, and 47.1% who do so less frequently (seldom or never). The majority of the tablet users reported they sometimes place the tablet flat on a support surface during use (44.1%), while 21.6% reported doing so frequently, and 34.3% doing so less frequently. Fewer participants of the tablet users reported that they place the tablet tilted on a support surface (30.4%), compared to 36.3% who reported doing so sometimes, and 33.3% reported doing so less often (Table 10).

Table 10. Hand-grip and tablet position.

	Always (%)	Usually (%)	Sometimes (%)	Seldom (%)	Never (%)
Smartphone use while					
The same hand holds the smartphone and performs a task	6 (2.6)	65 (28.5)	77 (33.8)	49 (21.5)	31 (13.6)
One hand holds the smartphone and the other performs the task	24 (10.5)	93 (40.8)	68 (29.8)	28 (12.3)	15 (6.6)
Both hands hold the smartphone and perform the task	11 (4.8)	45 (19.8)	47 (20.7)	61 (26.9)	63 (27.8)
Tablet use while					
One hand holds the tablet and the other hand performs a task	8 (7.7)	43 (41.3)	27 (26.0)	13 (12.5)	13 (12.5)
Both hands hold the tablet and perform the task	0 (0.0)	10 (9.9)	21 (20.8)	31 (30.7)	39 (38.6)
The tablet is placed in my lap while performing the task	1 (1.0)	15 (14.4)	39 (37.5)	28 (26.9)	21 (20.2)
The tablet is placed FLAT on a support surface while performing the task	6 (5.9)	16 (15.7)	45 (44.1)	25 (24.5)	10 (9.8)
The tablet is placed on a support surface and is TILTED while performing the task	4 (3.9)	27 (26.5)	37 (36.3)	18 (17.6)	16 (15.7)

3.6 Musculoskeletal Discomfort

Eighty point one percent (80.1%) of the participants reported they experienced musculoskeletal discomfort during the last seven days in at least one body region. The highest three body regions respondents reported having musculoskeletal discomfort in were the neck (64.7%), back (53.8%), and dominant shoulder (38.8%) (Table 11). The neck (2.44) and the back (1.91) received the highest discomfort ratings, as reported on a 10-point scale (Table 12). This discomfort interfered with the ability to work in less than 10% of respondents (Table 13).

These are similar to other findings reported in previous studies. Berolo et al. (2011) reported that 84% of their participants reported pain of any severity in at least one body region. In addition, Berolo et al. shows that the neck (68%), upper back (62%), and

Table 11. Discomfort prevalence.

Discomfort experienced in the last seven days in	Never (%)	1–2 times last week (%)	3–4 times last week (%)	Once every day (%)	Several times every day (%)
Neck	101 (35.3)	111 (38.8)	30 (10.5)	17 (5.9)	27 (9.4)
Back	132 (46.2)	92 (32.2)	26 (9.1)	16 (5.6)	20 (7.0)
Dominant shoulder	175 (61.2)	65 (22.7)	26 (9.1)	7 (2.4)	13 (4.5)
Non-dominant shoulder	206 (72.0)	45 (15.7)	20 (7.0)	4 (1.4)	11 (3.8)
Dominant arm	216 (75.5)	50 (17.5)	11 (3.8)	3 (1.0)	6 (2.1)
Non-dominant arm	242 (84.6)	27 (9.4)	10 (3.5)	1 (0.3)	6 (2.1)
Dominant wrist/hand	196 (68.5)	53 (18.5)	17 (.9)	9 (3.1)	11 (3.8)
Non-dominant wrist/hand	238 (83.2)	32 (11.2)	9 (3.1)	1 (0.3)	6 (2.1)

Table 12. Musculoskeletal discomfort rates by body regions.

	Number of cases	Mean	Std. Deviation
Neck	285	2.44	2.50
Back	285	1.91	2.37
Dominant shoulder	284	1.38	2.18
Non-dominant shoulder	286	1.00	1.93
Dominant arm	285	0.81	1.76
Non-dominant arm	285	0.51	1.51
Dominant wrist/hand	286	1.17	2.15
Non-Dominant wrist/hand	286	0.50	1.40

Table 13. Discomfort interference with the ability to work

	Yes (%)	No (%)
Neck	25 (9.1)	250 (90.9)
Back	18 (6.5)	261 (93.5)
Dominant shoulder	10 (9.4)	96 (90.6)
Non-dominant shoulder	6 (7.7)	72 (92.3)
Dominant arm	8 (12.1)	58 (87.9)
Non-dominant arm	5 (11.6)	38 (88.4)
Dominant wrist/hand	13 (15.3)	72 (84.7)
Non-dominant wrist/hand	7 (15.6)	38 (84.4)

right shoulder (52%) were the highest three body regions affected by musculoskeletal pain. Blair et al. (2015) also reported that the neck and the upper back/shoulder areas are the most affected regions by musculoskeletal symptoms during the use of the tablet.

4 Conclusion

This cross-sectional survey targeted workers and HDR students who used smartphones and tablets for work and/or for work and personal use. A total of 398 participants responded to the survey.

Smartphone and tablets are used for work and personal purposes by most of the respondents. Smartphones are more common among the participants than tablets. The availability of employer-owned devices appeared to have increased during the last three years. Almost half of the smartphone and tablet users were required to use HMDs for work regardless of the ownership.

Variations between smartphone and tablet use were seen with the hours of daily use, most common location of use, and tasks undertaken. Smartphones were used on average 3.6 ($\pm$3.5) hours per day, with a heavy emphasis on use of email in public places; this provides sufficient opportunity for exposure to risks of repetition and duration. Tablets were used less frequently (1.8 $\pm$ 1.7) hours per day, with a heavy emphasis on web browsing conducted in the home environment.

Although there were large variations in use patterns, both smartphones and tablets were most frequently used whilst seated without a work surface with the user holding the device with one hand and interfacing with their other hand. In addition, the HMDs were used regularly at different locations such as public places, home, and workplaces, which indicated the use of HMDs at these locations may not provide appropriate environmental settings that encourage the assumption of optimal postures. This highlights that postures of the neck, back, and upper extremity may be compromised, particularly if positions are held statically for long periods of time.

Around 84% of the respondents reported having musculoskeletal discomfort during the last seven days in any body region, with the neck (64.7%), back (53.8%), and dominant shoulder (38.8%) being the most affected by musculoskeletal discomfort.

Although we used self-reported data to explore the use of HMDs among workers, future research will focus on measuring the use of HMDs objectively to provide more accurate data and eliminate any self-report bias. In addition, investigating the HMDs use with a focus on smartphones use outside the home and tablet use at home would be beneficial to highlight ergonomic risks, regardless to the use for work or personal purposes.

Ergonomics intervention and research to reduce the exposure to HMD is warranted, given the high prevalence rates of musculoskeletal symptoms and the high use of HMD.

References

Balakrishnan R, Chinnavan E, Feii T (2016) An extensive usage of hand held devices will lead to musculoskeletal disorder of upper extremity among student in AMU: a survey method. Int J Phys Educ Sport Health 3:368–372

Berolo S, Wells RP, Amick BC (2011) Musculoskeletal symptoms among mobile hand-held device users and their relationship to device use: a preliminary study in a Canadian university population. Appl. Ergon. 42:371–378. https://doi.org/10.1016/j.apergo.2010.08.010

Blair B, Gama M, Toberman M (2015) Prevalence and risk factors for neck and shoulder musculoskeletal symptoms in users of touch-screen tablet computers. University of Nevada

Chiang HY, Liu CH (2016) Exploration of the associations of touch-screen tablet computer usage and musculoskeletal discomfort. Work 53:917–925. https://doi.org/10.3233/WOR-162274

Cornell University Ergonomics Web (2017) Musculoskeletal discomfort questionnaires [WWW Document]. http://ergo.human.cornell.edu/ahmsquest.html. Accessed 21 Aug 2017

Eapen C, Kumar B, Bhat AK (2010) Prevalence of cumulative trauma disorders in cell phone users. J Musculoskelet Res 13:137–145. https://doi.org/10.1142/S0218957710002545

Goldfinch S, Gauld R, Baldwin N (2011) Information and communications technology use, e-government, pain and stress amongst public servants. New Technol Work Employ 26:39–53. https://doi.org/10.1111/j.1468-005X.2010.00256.x

Hedge A, Morimoto S, McCrobie D (1999) Effects of keyboard tray geometry on upper body posture and comfort. Ergonomics 42:1333–1349. https://doi.org/10.1080/001401399184983

Hegazy A, Alkhail B, Awadalla N, Qadi M, Al-Ahmadi J (2016) Mobile phone use and risk of adverse health impacts among medical students in Jeddah, Saudi Arabia. Br J Med Med Res 15:1–11. https://doi.org/10.9734/BJMMR/2016/24339

Honan M (2015) Mobile work: ergonomics in a rapidly changing work environment. Work 52:289–301. https://doi.org/10.3233/WOR-152164

Kim H-J, Kim J-S (2015) The relationship between smartphone use and subjective musculoskeletal symptoms and university students. J Phys Ther Sci 27:575–579. https://doi.org/10.1589/jpts.27.575

Lanaj K, Johnson RE, Barnes CM (2014) Beginning the workday yet already depleted? Consequences of late-night smartphone use and sleep. Organ Behav Hum Decis Process 124:11–23. https://doi.org/10.1016/j.obhdp.2014.01.001

Lemola S, Perkinson-Gloor N, Brand S, Dewald-Kaufmann JF, Grob A (2014) Adolescents' electronic media use at night, sleep disturbance, and depressive symptoms in the smartphone age. J Youth Adolesc 44:405–418. https://doi.org/10.1007/s10964-014-0176-x

Müller H, Gove JL, Webb JS, Cheang A (2015) Understanding and comparing smartphone and tablet use. In: Proceedings of the Annual Meeting of the Australian Special Interest Group for Computer Human Interaction on – OzCHI 2015. ACM Press, New York, pp 427–436. https://doi.org/10.1145/2838739.2838748

Sarwar M, Soomro TR (2013) Impact of smartphone's on society. Eur J Sci Res 98:216–226

Shan Z, Deng G, Li J, Li Y, Zhang Y, Zhao Q (2013) Correlational analysis of neck/shoulder pain and low back pain with the use of digital products, physical activity and psychological status among adolescents in Shanghai. PLoS One 8:e78109. https://doi.org/10.1371/journal.pone.0078109

Stalin P, Abraham SB, Kanimozhy K, Prasad RV, Singh Z, Purty AJ (2016) Mobile phone usage and its health effects among adults in a semi-urban area of Southern India. J Clin Diagn Res 10:LC14-LC16. https://doi.org/10.7860/JCDR/2016/16576.7074

Stawarz K, Cox AL, Bird J, Benedyk R (2013) I'd sit at home and do work emails: how tablets affect the work-life balance of office workers. In: CHI 2013 Extended Abstracts on Human Factors in Computing Systems on - CHI EA 2013. ACM Press, New York, p 1383. https://doi.org/10.1145/2468356.2468603

Investigation of Sensitivity of OWAS and European Standard 1005-4 to Assess Workload of Static Working Postures by Surface Electromyography

Tobias Hellig[(⊠)] [iD], Alexander Mertens [iD], and Christopher Brandl [iD]

Chair and Institute of Industrial Engineering and Ergonomics,
RWTH Aachen University, Aachen, Germany
t.hellig@iaw.rwth-aachen.de

Abstract. The present study investigates the sensitivity of the Ovako Working Posture Analyzing System (OWAS) and European Standard 1005-4 for an assessment of work load of static working postures. Therefore a comparison of these methods with surface electromyography (EMG) is conducted. For this purpose muscle activity of eight muscles is captured in a laboratory study (n = 24) during 16 different static working postures. The results are compared with risk assessment categories of OWAS and European Standard 1005-4. A repeated–measures analysis of variance revealed a significant increase of muscle activity with increasing back angles and shoulder angles. However, this increase of muscle activity and the associated increase of musculoskeletal injury risk are not represented by OWAS and European Standard 1005-4 to the same extent. Thus, for an investigation of static working postures European Standard 1005-4 is more recommendable to identify musculoskeletal injury risk, since the high variance of muscle activity in the investigated working postures is represented better by the spread of three zones of European Standard 1005-4.

Keywords: Working posture · OWAS · Muscle activity

1 Introduction

Today musculoskeletal disorders (MSD) are a common problem in many industrialized countries. The World Health Organization stated in 2003, that prevalence of MSDs will more than double in 2020 compared with the prevalence in the year 2000 [1]. The accuracy of this prediction is shown by the fact that in Germany and other industrialized countries MSDs are the main reason for health-related sick leave and early retirement in 2015 [2]. Investigations have identified awkward working postures as one major risk factor for the development of MSDs [3, 4]. To influence the development of MSDs positively a reduction of work load has to be achieved. To reduce work load at first an analysis and assessment of work load is elementary [5]. Therefore a wide range of ergonomic analysis and assessment methods are known. These methods can be classified into subjective methods, observation based methods and objective methods [6]. Subjective methods like interviews and questionnaires can be used to collect data

S. Bagnara et al. (Eds.): IEA 2018, AISC 820, pp. 138–147, 2019.
https://doi.org/10.1007/978-3-319-96083-8_17

on physical and psychosocial factors of workplace exposure. Observation based methods like Owako working posture analyzing system (OWAS) [7] or European Standard 1005-4 have been developed to analyze and assess exposure to work load by systematic observations of various exposure factors. Objective methods, i.e. sensor based capture of muscle activity like surface electromyography (EMG), have been developed to derive exposure data directly from the subject. Systematic reviews of those methods can be found at Winkel and Mathiassen [6], David [8] and Takala et al. [9]. Each of those numerous methods entail both advantages and disadvantages. The use of observation based methods is inexpensive and can be done without any influence on the working person [10]. However, some studies have proofed the reliability of observations based methods to be a problem [9, 10]. Besides this, objective methods are able to capture highly accurate data on a wide range of exposure variables [9]. However, the use of objective methods is expensive and a considerable effort of capturing and interpretation of data is required. The choice of methods used for evaluation of work load depends upon the workplace and tasks as well as the kind of investigation which can be research-orientated or practice-orientated [8, 10]. Due to a limited operational capability of objective methods in practice observation based methods like OWAS are used mostly [11]. However, observation based methods sometimes suffer from validity and reliability [9]. To investigate reliability and validity it has become habitual to compare the results of observation based methods and objective methods [8, 12, 13].

In the context of the stress-strain concept [14] an objective evaluation of work load is conducted by the measurement of appropriate physiological or biochemical exposure variables. The selection of exposure variables depends on the one hand on the kind of work load and on the other hand on the limiting factor of the human organism [15]. Since our focus in this paper is on static working postures, which is one major risk factor for the development of MSDs, an evaluation of exposure has to be based on the measurement of muscle activity [5]. Especially during static working postures muscle activity represents the limiting factor of the human body since the internal muscle pressure causes an interruption of blood flow and thus an oxygen shortage of the muscle occurs [16].

To investigate sensitivity of OWAS and European Standard 1005-4, which is very similar to International Standard 11226, to assess work load of static working postures a comparison of these methods with objective measurements of EMG is carried out.

2 Method

2.1 Participants

For the current investigation of sensitivity of OWAS and European Standard 1005-4 to assess workload of static working postures a laboratory study was conducted. Therefore 24 test persons aged between 20 and 28 (M = 24.6, SD = 1.99) were recruited. The test persons were excluded from this study if they reported symptoms of musculoskeletal injuries during the last twelve months. Informed written consent was obtained prior to participating in this study.

2.2   Experimental Design

During the study test persons assumed and hold 16 different one minute lasting static working postures. The variation of working postures was achieved by a systematic increase of ventral inclination of the back and shoulder flexion. All assumed working postures with their corresponding back inclination angle and shoulder flexion angle are illustrated in Fig. 1. The investigated working postures were chosen to cover the entire range of motion of the back and the shoulder. The period of one minute was determined to investigate a static working posture and additionally to ensure an appropriate total time of the experiment.

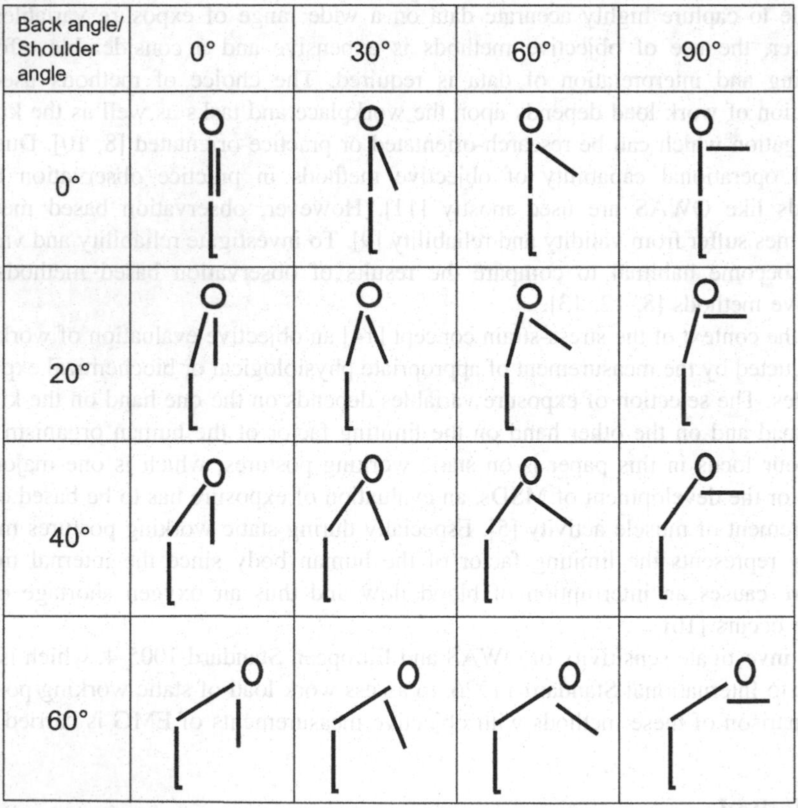

Back angle/ Shoulder angle	0°	30°	60°	90°
0°				
20°				
40°				
60°				

Fig. 1.   Different types of investigated working postures

During the one minute isometric contraction phase, muscle activity of the eight following muscles was measured: left and right trapezius pars descendes (LUT, RUT), left and right trapezius pars ascendens (LLT, RLT), left and right anterior deltoideus (LAD, RAD) and left and right erector spinae (LES, RES). These muscles were selected due to exposure of static working postures on the musculoskeletal system. In order to eliminate influence of muscle fatigue, a one minute rest period was hold after each working posture, and a five minute rest period was hold after five working postures. Additionally the order of working postures was randomised between the test persons.

Test persons were required to rate perceived exertion (RPE) regarding the whole body for each investigated working posture. RPE were obtained according to Borg's Category Ration scale (CR-10) [17].

2.3 Procedure

The experimental procedure was divided into two main parts: (1) preparation of test persons and (2) investigation of work load during 16 static working postures.

In the first part of the experimental procedure the sensor and electrode placement was conducted according to the European Recommendations for Surface Electromyography (SENIAM) [18]. Furthermore values of maximal voluntary contraction (MVC) were measured in order to normalise the EMG signal. In the second part of the experimental procedure test persons assumed the 16 different working postures. Continuous EMG signals were captured during the entire contraction phase. During the one minute static procedure working postures were monitored with a self-written software to calculate back angle and shoulder angle using Microsoft Kinect V2 for capturing of working postures.

2.4 Data Recording and Processing

Surface Electromyography

For the investigation in this study a surface electromyography device (Desktop DTS Receiver, Noraxon, Scottsdale, AZ, USA) was used to capture bilateral muscle activity of eight muscles. Ag/AgCl self-adhesive 8–shaped dual electrodes (dimensions of adhesive: 4×2.2 cm; diameter of the two circular adhesives: 1 cm; inter-electrode distance: 1.75 cm) were applicate on the skin according to SENIAM standards [18]. Signals were amplified with a gain of 1000 V/V, input impedance of 100 MΩ and a common mode rejection ratio of 100 dB. Signals were sampled during muscle contractions with a sampling frequency of 1500 Hz and digitally band-pass filtered (10–500 Hz) with a first-order high-pass filter. Signals were recorded using the biomechanical analysis software MyoResearch 3.8 (Noraxon, Scottsdale, AZ, USA). EMG data were recorded during the 60-s contraction periods. Root mean square (RMS) amplitude was calculated with an overlapping moving window of 100 ms. In order to compare EMG data for different conditions of working postures, RMS values were normalized relative to the values of initial MVC. To minimize the effects of initial movements of the body parts capturing of EMG signals was started once a test person had assumed the corresponding working posture.

Observation Based Assessment of Working Postures According to OWAS and European Standard 1005-4

In addition to the objective measurement of muscle activity, the 16 working postures were evaluated using OWAS and European Standard 1005-4. OWAS assigns one of four action categories (AC) to each working posture, depending on the degree of exposure.

AC 1 characterizes a working posture of minimal exposure level. In ascending order AC 2 to AC 4 represent an increasing exposure level and increasing risk of musculoskeletal injury.

Due to the lack of knowledge on interaction effects of working postures European Standard 1005-4 only evaluates postures of single body parts e.g. bending trunk forward is evaluated independent from postures of arms and legs [19]. Postures of the back and the arm are evaluated depending on the degree of exposure. European Standard 1005-4 assigns one of four zones to each posture of the back as well as to each posture of the arm. Analogous to OWAS Zone 1 characterises a posture of an acceptable exposure level. Zone 2 to Zone 3 characterise an increasing exposure level from conditionally acceptable to not acceptable.

Data Analysis

In order to investigate the sensitivity of OWAS and European Standard 1005-4 to assess work load of static working postures a comparison of these methods with objective measurements of EMG was carried out. A repeated–measures analysis of variance (MANOVA) was used. Significance was accepted at $p < .05$. All statistical analyses were conducted using IBM SPSS Statistics 22.

3   Results

3.1   Repeated-Measures MANOVA

A repeated-measure MANOVA revealed significant effects of back angle and shoulder angel on muscle activity. Using Pillai's trace, back angle was found to have a significant effect on muscle activity, $V = 1.31$, $F(21, 195) = 7.20$, $p < .001$. Using Pillai's trace, shoulder angle was found to have a significant effect on muscle activity, $V = 1.44$, $F(21, 195) = 8.57$, $p < .001$. Using Pillai's trace, back angle and shoulder angle were found have a significant interaction effect on muscle activity, $V = 1.00$, $F(72, 1656) = 3.47$, $p < .001$. Due to a significant interaction effect of back angle and shoulder angle on muscle activity an evaluation of exposure based on an isolated consideration of back angle or shoulder angle is inadmissible.

3.2   EMG Amplitude

Figure 2 contains averaged normalized muscle activity of the muscles in the left and right side of the body as well as the classification of investigated working postures according to OWAS and European Standard 1005-4. As evident from Fig. 2 muscle activity of the eight investigated muscles (LUT, RUT, LLT, RLT, LAD, RAD, LES, RES) increases with increasing back angles and shoulder angles. Overall an increasing shoulder angle increases muscle activity of LUT and RUT. Increasing back angles were found to raise muscle activity in LUT and RUT reasonably. Muscle activity of LLT and RLT is increased considerably by increasing shoulder angles. Apart from this increasing back angles cause increasing muscle activity in LLT and RLT. An increase in shoulder angle contributes to a considerable increase in muscle activity of LAD and RAD whereas a decreasing muscle activity in LAD and RAD could be recognized as a consequence of an increasing back angle. Increasing shoulder angles result in an increase of muscle activity of LES and RES. Furthermore increasing back angles are associated with increasing muscle activity in LES and RES.

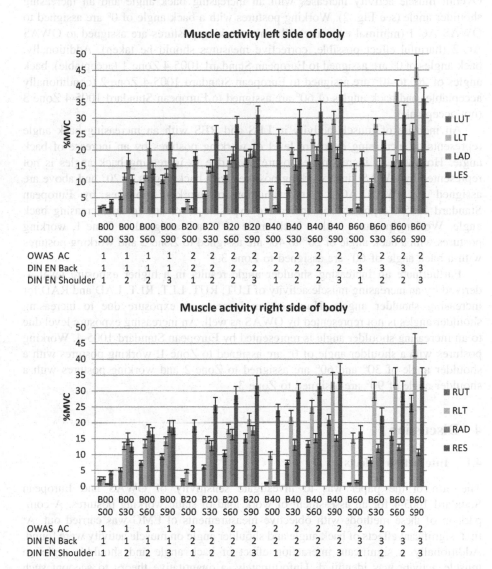

Fig. 2. Averaged muscle activity and standard error of the mean, B: ventral inclination angle of the back, S: Shoulder flexion angle, Evaluation of working posture according to OWAS and European Standard 1005-4

3.3 Comparison of Muscle Activity, OWAS and European Standard 1005-4

Overall muscle activity increases with an increasing back angle and an increasing shoulder angle (see Fig. 2). Working postures with a back angle of 0° are assigned to OWAS AC 1 (minimal exposure) all other working postures are assigned to OWAS AC 2 (harmful effect possible, corrective measures should be taken). Additionally, back angles of 0° are assigned to European Standard 1005-4 Zone 1 (acceptable), back angles of 20° to 40° are assigned to European Standard 1005-4 Zone 2 (conditionally acceptable) and back angles of 60° are assigned to European Standard 1005-4 Zone 3 (not acceptable).

An increase of muscle activity in LES and RES with an increasing back angle represents an increasing exposure level of working postures by an increase of back angle. However, an increasing exposure level due to increasing back angles is not represented by OWAS since working postures with back angles of 20° and above are assigned to OWAS AC 2. An evaluation of working postures by European Standard 1005-4 represents an increasing exposure level due to an increasing back angle. Working postures with a back angle of 0° are assigned to Zone 1, working postures with a back angle of 20° to 40° are assigned to Zone 2 and working postures with a back angle of 60° are assigned to Zone 3.

Furthermore an increasing shoulder angle results in a higher exposure level as derived by an increasing muscle activity of LUT, RUT, LLT, RLT, LAD and RAD for increasing shoulder angels. However, an increase in exposure due to increasing shoulder angles is not represented by OWAS as well. An increasing exposure level due to an increasing shoulder angle is represented by European Standard 1005-4. Working postures with a shoulder angle of 0° are assigned to Zone 1, working postures with a shoulder angle of 30° and 60° are assigned to Zone 2 and working postures with a shoulder angle of 90° are assigned to Zone 3.

4 Discussion

4.1 Interaction Effects

The aim of the paper was to investigate sensitivity of OWAS and European Standard 1005-4 for an assessment of work load of static working postures. A comparison of these methods with objective measurements of EMG was carried out. At first, significant effects of back angle and shoulder angle on muscle activity were found. Additionally, a significant interaction effect of back angle and shoulder angle on muscle activity was identified. Unfortunately, a quantitative theory to account such interaction effects has not yet been formulated [10, 19]. This current lack of knowledge may lead to an inaccurate evaluation of exposure of working postures [20]. However, an isolated evaluation of exposure in a single body segment, i.e. back or arm, is possible using existing methods for a quantification of work load.

4.2   Comparison of EMG, OWAS and European Standard 1005-4

A descriptive comparison of averaged muscle activity of working postures with evaluations of working postures by OWAS and European Standard 1005-4 was carried out. Muscle activity of LES and RES shows an increasing course for an increase in back angle and shoulder angle. As can be seen in Fig. 2, a back angle of 20° reveals in levels of muscle activity of more than 20% MVC for LUT and RUT. Even levels of muscle activity of more than 15% MVC during static working postures cause an interruption of blood flow since the internal muscle pressure exceeds the blood pressure and thus an oxygen shortage of the muscle occurs [16, 21]. Therefore a musculoskeletal injury risk due to a back angle of 20° and more is evident. Such injury risk states are not represented by OWAS since working postures with a back angle of 20° and more are assigned to OWAS AC 2. Compared with OWAS European Standard 1005-4 shows a higher sensitivity for a muscular skeletal injury risk due to static working postures with a back angle of 20° and more. However, European Standard 1005-4 assigns working postures with a back angle of more than 60° to Zone 3 (not acceptable). For an evaluation of static working postures neither OWAS nor European Standard 1005-4 represent critical exposure levels of static working postures. Nevertheless, a back angle between 20° and 60° causes high exposures, which is proofed by an investigation of muscle activity using EMG.

As displayed in Fig. 2 an increase in muscle activity of LUT, RUT, LLT, RLT, LAD and RAD is related to an increase of shoulder angle. These muscles are employed to carry out a rotation of the scapula, which is necessary for a shoulder flexion [22]. Already a shoulder angle of 30° is accompanied by a muscle activity of more than 20% MVC in the muscles of the shoulder which causes a musculoskeletal injury risk since the internal muscle pressure exceeds the blood pressure and thus an oxygen shortage of the muscle occurs [16, 21]. This risk is not represented by an evaluation of static working postures by OWAS. Working postures investigated in this study are assigned to OWAS AC 1 and AC 2, which indicates only a minimal harmful effect. As well European Standard 1005-4 assigns shoulder flexion angles of 90° to Zone 3 (not acceptable) but flexion angles of less than 90° are assigned to Zone 1 (acceptable) and Zone 2 (conditionally acceptable). Also investigation of muscle activity in shoulder muscles leads to conclude neither OWAS nor European Standard 1005-4 represent critical exposure levels of static working postures.

The results of this study indicate a low sensitivity of OWAS and European Standard 1005-4 to assess workload of static working postures. The variance of muscle activity revealed in this study provides compelling evidence for this conclusion. Muscle activity in OWAS AC 1 varies between 2% MVC and 22% MVC whereas muscle activity in OWAS AC 2 varies between 2% MVC and 45% MVC. Therefore OWAS AC 1 and AC 2 include working postures of low exposure and high exposure. Hence, a higher sensitivity for representing a musculoskeletal injury risk is indicated by European Standard 1005-4 compared to OWAS.

4.3   Limitations

This study was conducted in a controlled laboratory setting. A generalising of our findings is possible only to a limited extent. Especially work in real work systems is characterised by dynamic muscle contractions which were not in the focus of this investigation. Therefore no conclusions can be drawn about sensitivity of OWAS and European Standard 1005-4 for an investigation of dynamic muscle contractions. Furthermore for an investigation of static working postures based on EMG an examination of skeletal muscles should be taken into account. However, during this study an investigation of skeletal muscles was not possible due to the necessity of invasive examination methods, e.g. needle electrodes. Furthermore, the higher sensitivity of European Standard 1005-4 compared to OWAS stated in this work is limited to the working postures investigated in this study.

5   Conclusion

Nevertheless, this study could proof a need for further developments of ergonomic analysis and assessment methods, since the two investigated methods OWAS and European Standard 1005-4 proofed only low sensitivity for a representation of exposure of static working postures. However, for an investigation of static working postures European Standard 1005-4 is more recommendable to identify musculoskeletal injury risk, since the high variance of muscle activity in the investigated working postures is represented better by the spread of three zones of European Standard 1005-4. Nevertheless, using European Standard 1005-4 a slightly higher effort in evaluating working postures is necessary than using OWAS, since every single body part has to be evaluated independent by European Standard 1005-4. Additionally, this study revealed significant interaction effects of back angle and shoulder angle on muscle activity. Since interaction effects are not part of ergonomic analysis and assessment methods future work is needed to increase reliability and validity of analysis results of physical work.

Acknowledgement. The research is carried out within the "Smart and Adaptive Interfaces for INCLUSIVE Work Environment" project, funded by the European Union's Horizon 2020 Research and Innovation Program under Grant Agreement N723373. The authors would like to express their gratitude for the support given.

References

1. World Health Organisation (WHO) (2003) The burden of musculoskeletal conditions at the start of the new millennium
2. Robert Koch-Institut (2016) Gesundheit in Deutschland – die wichtigsten Entwicklungen. Gesundheitsberichterstattung des Bundes. Gemeinsam getragen von RKI und Destatis. RKI, Berlin
3. Widanarko B, Legg S, Stevenson M, Devereux J, Eng A, Mannetje AT, Cheng S, Pearce N (2012) Gender differences in work-related risk factors associated with low back symptoms. Ergonomics 3:327–342. https://doi.org/10.1080/00140139.2011.642410

4. Roman-Liu D (2014) Comparison of concepts in easy-to-use methods for MSD risk assessment. Appl Ergon 3:420–427. https://doi.org/10.1016/j.apergo.2013.05.010
5. Schlick C, Luczak H, Bruder R (2010) Arbeitswissenschaft. Springer, Heidelberg
6. Winkel J, Mathiassen SE (1994) Assessment of physical work load in epidemiologic studies: concepts, issues and operational considerations. Ergonomics 6:979–988. https://doi.org/10.1080/00140139408963711
7. Karhu O, Kansi P, Kuorinka I (1977) Correcting working postures in industry: a practical method for analysis. Appl Ergon 4:199–201. https://doi.org/10.1016/0003-6870(77)90164-8
8. David GC (2005) Ergonomic methods for assessing exposure to risk factors for work-related musculoskeletal disorders. Occup Med 3:190–199. https://doi.org/10.1093/occmed/kqi082
9. Takala E-P, Pehkonen I, Forsman M, Hansson G-A, Mathiassen SE, Neumann WP, Sjøgaard G, Veiersted KB, Westgaard RH, Winkel J (2010) Systematic evaluation of observational methods assessing biomechanical exposures at work. Scand J Work Environ Health 1:3–24. https://doi.org/10.5271/sjweh.2876
10. Li G, Buckle P (1999) Current techniques for assessing physical exposure to work-related musculoskeletal risks, with emphasis on posture-based methods. Ergonomics 5:674–695. https://doi.org/10.1080/001401399185388
11. Brandl C, Mertens A, Schlick CM (2017) Ergonomic analysis of working postures using OWAS in semi-trailer assembly, applying an individual sampling strategy. Int J Occup Saf Ergon JOSE 1:110–117. https://doi.org/10.1080/10803548.2016.1191224
12. Li G, Buckle P (1999) Evaluating change in exposure to risk for musculoskeletal disorders. In: Health and safety executive, Sheffield
13. Viikari-Juntura E, Rauas S, Martikainen R, Kuosma E, Riihimäki H, Takala E-P, Saarenmaa K (1996) Validity of self-reported physical work load in epidemiologic studies on musculoskeletal disorders. Scand J Work Environ Health 4:251–259. https://doi.org/10.5271/sjweh.139
14. Rohmert W (1986) Ergonomics. Appl Psychol 2:159–180. https://doi.org/10.1111/j.1464-0597.1986.tb00911.x
15. Rohmert W (1983) Formen menschlicher Arbeit. In: Rohmert W, Rutenfranz J (eds) Praktische Arbeitsphysiologie. Thieme Verlag, Stuttgart
16. Rohmert W (1962) Untersuchungen über Muskelermüdung und Arbeitsgestaltung, Aachen
17. Borg G (1990) Psychophysical scaling with applications in physical work and the perception of exertion. Scand J Work Environ Health 1:55–58. https://doi.org/10.5271/sjweh.1815
18. Hermens HJ (1999) European recommendations for surface ElectroMyoGraphy. Roessingh Research and Development, Enschede
19. European Standard 1005-4:2009-01: Safety of machinery - Human physical performance - Part 4: Evaluation of working postures and movements in relation to machinery
20. Lim C-M, Jung M-C, Kong Y-K (2011) Evaluation of upper-limb body postures based on the effects of back and shoulder flexion angles on subjective discomfort ratings, heart rates and muscle activities. Ergonomics 9:849–857. https://doi.org/10.1080/00140139.2011.600777
21. Rohmert W (1983) Statische Arbeit. In: Rohmert W, Rutenfranz J (eds) Praktische Arbeitsphysiologie. Thieme Verlag, Stuttgart
22. Perotto A, Delagi EF (2005) Anatomical guide for the electromyographer. Charles C Thomas, Springfield

Low Back Pain (LBP) and Physical Work Demands

F. Serranheira[1(✉)], M. Sousa-Uva[1], F. Heranz[2], F. Kovacs[3],
and A. Sousa-Uva[1]

[1] Public Health Research Center, NOVA National School of Public Health,
Lisbon, Portugal
serranheira@ensp.unl.pt
[2] Lisbon, Portugal
[3] Unidad de la Espalda Kovacs, Hospital Universitario HLA-Moncloa,
y Red Española de Investigadores en Dolencias de la Espalda, Madrid, Spain

Abstract. Low back pain (LBP) is a common occupational health complaint
and an important public health concern. Analyzing the association between
physical demands at work and occupational outcomes can be useful for
improving LBP prevention. In this study, workers filled out a questionnaire
gathering data on socio-demographic and work-related characteristics, general
health, LBP (episodes in the last 12 months, pain severity and intensity), and
other occupational hazards related with physical demands (DMQ). 735 workers
answered the questionnaire (male n = 359). They worked in different sectors.
507 (69%) reported LBP in the last year. The highest proportion of subjects with
>6 episodes of LBP per year was found among public services (31.8%) and the
lowest among administrative working in offices (10.3%). Most workers reported
having sedentary-type work (39%), 34% a low/moderate physical intensity one,
and 27% a highly physically demanding one. Results of logistic regression
showed that, after adjusting for age, gender and sector: sedentary work (vs. high
work intensity) was associated with a lower likelihood of having 3 to 6 LBP
episodes per year (OR = 0.4; 95%CI 0.2–0.8), and >6 LBP episodes per year
(OR = 0.5; 95%CI 0.3–0.9); low/moderate work intensity (vs. high work
intensity) was also associated with a lower likelihood of having 3 to 6 LBP
episodes per year (OR = 0.5; 95%CI 0.3–0.9) and >6 LBP episodes per year
(OR = 0.6; 95%CI 0.3–1.0). Findings suggest that occupational high physical
demands are associated with a higher likelihood of presenting LBP. For
Occupational Health Services these results may contribute to design and assess
better LBP prevention programs.

Keywords: Ergonomics · Occupational health · Occupational disorders

F. Heranz—Clinical Investigator, Occupational Doctor.

S. Bagnara et al. (Eds.): IEA 2018, AISC 820, pp. 148–153, 2019.
https://doi.org/10.1007/978-3-319-96083-8_18

1   Introduction

Work-related low back pain (WRLBP) often presents as a result from a biomechanical requirement (mediated by physical demands on the workplace, individual, and psychosocial, including organizational factors) involving a pain response to tissue stimulation [1–3].

In general, the relationship established when there is an applied load on the tissues and their physiological and biomechanical tolerance (including elastic components) determines the risk of work-related musculoskeletal disorders (WRMSDs) [4]. Thus, whenever the external demands of the work impose individual responses with physical load (in any of the anatomical tissues of the lower back) below the tissue tolerance, the risk of injury is reduced. The opposite is also valid, identifying high risk of low back pain (LBP) and WRMSDs, at least for a certain proportion of the population. Despite the expected relationship between physical work and LBP in recent studies indicate that only 11 to 18% of all LBP episodes are related to occupational exposure, namely with physical work demands [5]. The present study tries to analyze possible associations between occupational physical demands and LBP episodes in workers from different Portuguese occupational backgrounds.

2   Materials and Methods

An observational, analytical and prevalence study was carried out in the context of health surveillance in Occupational Health and Safety Services. Four companies from the Lisbon region (designated as companies A, B, C and D) participated in the study to identify the influence of different factors (individual, professional, psychosocial and clinical) on the prevalence of LBP.

A questionnaire, after informed consent, was applied to a group of workers in each of the companies to collect socio-demographic data, professional background, general health, LBP episodes in the last 12 months, pain intensity, and data related to physical work requirements (DMQ - [6]).

The results presented here are related to the association between physical workload and LBP episodes. They were analyzed using SPSS statistical software through univariate and bivariate statistical analyzes using the independence test of chi- square. The association between the number of LBP episodes and the work requirements was quantified using multinomial regression. For all statistical tests a significance level of 0.05 was set.

3   Results

A public hospital, two large private companies, and a public company participated in the study. 745 workers answered the questionnaire. The population studied was predominantly female (51.4%), with a mean age of 43 years, mainly with a secondary school degree education or a technical school degree (38.1%). An average profession period of 16 years and 40 h per week work was observed.

Regarding the type of schedule practiced, 11.6% of respondents work in shifts and 21.0% have night work.

Manual work is the most frequent work observed (52.3%). Almost two-thirds (60.5%) of the respondents work in the private sector.

Regarding the classification of labor requirements, 38.7% of workers perform sedentary work; 34.2% work with low/moderate physical requirements; and 27.1% have intense work with high physical demands.

The most frequent professional requirements are (i) "sitting more than 50% of the workday" (57.1%); (ii) "standing more than 50% of the working day" (49.9%), (iii) "frequent rotation of the body" (40.3%), (iv) "or trunk" (35.5%), (v) "manual mobilization of loads" (32.6) and the less frequent professional requirement is (vi) "exposure to vibrations if transmitted to the whole body (2.1%).

About 70% of self-reported participants (Table 1) have experienced at least one to two episodes of LBP in the past year. Of these, most describe that the pain lasted from 2 to 14 days (56.2%).

Table 1. LBP description

Variables	Categories	n	Frequency (%)
LBP episodes in the last year	0	218	29.7
	1–2	222	30.3
	3–6	131	17.9
	>6	162	22.1
Pain habitual duration	≤ 1 day	166	32.9
	2–14 days	284	56.2
	15–30 days	27	5.4
	>30 days	28	5.5
Duration of the longest episode	≤ 1 day	117	23.5
	2–14 days	286	57.6
	15–30 days	54	10.9
	>30 days	40	8.0
Low back pain goes down the leg	Yes	216	43.6
	No	279	56.4
Low back pain lying down (sciatica)	Yes	264	37.2
	No	445	62.8
Low back pain when lifting from bed	Yes	271	38,1
	No	440	61,9
LBP when in a sitting position	Yes	337	47,7
	No	370	52,3

Low back pain was irradiated (sciatica) in 43.6% of respondents. Among those who reported having suffered one or more episodes of LBP in the last year, 49.6% reported feeling pain lying down and 51.4% when they got up; 64.6% when sitting and 86.2% need to change, often, the trunk posture to be more comfortable.

The presence of more than 6 episodes of LBP was more common in company D in the last year, and it was also in company D that the majority of workers reported the presence of intense physical work ($\chi2 = 29.7$, $p = 0.001$) (Tables 2 and 3).

Table 2. LBP in the last 12 mounts

		None	1–2	3–6	>6
A	n	24	33	28	26
	%	21.60%	29.70%	25.20%	23.40%
B	n	39	34	14	10
	%	40.20%	35.10%	14.40%	10.30%
C	n	118	112	58	73
	%	32.70%	31.00%	16.10%	20.20%
D	n	35	40	30	49
	%	22.70%	26.00%	19.50%	31.80%

Table 3. Work physical demands

		Sedentary work	Light/moderate work	Intense work
A	n	21	42	45
	%	19.4%	38.9%	41.7%
B	n	90	5	3
	%	91.8%	5.1%	3.1%
C	n	125	145	80
	%	35.7%	41.4%	22.9%
D	n	39	47	63
	%	26.2%	31.5%	42.3%

From the multivariate analysis it is emphasized that sedentary work, in relation to light/moderate or intense work, presents a decreased probability (protecting factor) of having at least 3–6 episodes (or more than 6) of LBP in the last year (Table 4).

Table 4. Elements of the work situation that influence the occurrence of LBP

Physical demands	Categories	LBP episodes in the last year	OR *	CI 95% Min.	Max.	p value
Sedentary work		1–2	0,8	0,5	1,3	0,325
		3–6	0,4	0,2	0,8	**0,008**
		>6	0,5	0,3	0,9	**0,026**
		None	Reference category			
Low/moderate intensity work		1–2	0,7	0,4	1,2	0,248
		3–6	0,5	0,3	0,9	**0,023**
		>6	0,6	0,3	1	**0,043**
		None	Reference category			
High intensity work			Reference category			

4   Discussion

Several studies carried out essentially in the last two decades on the etiology of LBP emphasize individual factors, mainly psychosocial variables, and assign them a predominant role, devaluing occupational hazards such as the physical demands at work [7, 8]. This model, which attributes low back pain a psychosocial nature, has focused almost exclusively on individual variables, underestimating occupational risk factors and the resulting incapacity [9–12].

On the other hand, studies, such as Oliv et al. [13], report that elder workers not exposed to physically demanding jobs during their life time occupations have now better health levels, namely less presence of back pain, than those exposed to high physical demands at work.

Many queries about the etiology of LBP and the workplace physical risk factors role remain and should be placed on the agenda, facing the results of the present study, that suggest the risk of LBP to higher among workers with physical demands, compared to sedentary workers or lower effort requirements workers, in terms of force application and load mobilization.

These findings occurring in a considerable number of workers need to be addressed in further studies on the association between physical demands at work and LBP.

There are probably other numerous risk factors taking influence in the increased occurrence of LBP, since its etiology is based on a multifactor matrix.

Thus, both psychosocial factors and other occupational factors should be valued in different weights according to the context of this study. As underlined before there is a need for greater knowledge in this field, in order to contribute for a better prevention of LBP.

5   Conclusions

The results of this study found a significant association between work with high physical demands and the frequency of at least three episodes of LBP in the last year compared to sedentary work.

Despite the recognized influence of individual factors, exposure to occupational hazards may contribute to increased episodes of pain in the lower back. In this context, the need to develop knowledge of the interdependencies between the physical demands at work and LBP is highlighted aiming at healthy and safer workplaces.

Acknowledgments. Thanks to the support of the Authority for Working Conditions (ACT – Autoridade para as Condições de Trabalho), in particular by funding project 027ESC/13: "Chronic low back pain and Work". It is also appreciated the collaboration of the companies and their Occupational Health Services, and in particular of all workers and occupational doctors who participated in this study.

References

1. N.R.C./I.O.M. (The National Research Council/Institute of Occupational Medicine) (2001) Musculoskeletal disorders and the workplace: low back and upper extremities. National Academy Press, Washington
2. Serranheira F, Sousa-Uva A (2016) Lesões musculoesqueléticas, Fatores individuais e Trabalho: interações e interdependências (1ª parte). Segurança 232:20–24
3. Serranheira F, Sousa-Uva A (2016) Lesões musculoesqueléticas, Fatores individuais e Trabalho: interações e interdependências (2ª parte) Segurança 233:20–24
4. Serranheira F, Uva A, Lopes F (2008) Lesões músculo-esqueléticas e trabalho: alguns métodos de avaliação do risco. Cadernos Avulso 5. Lisboa: Sociedade Portuguesa de Medicina do Trabalho
5. Marras WS (2012) The complex spine: the multidimensional system of causal pathways for low-back disorders. Hum Factors 54(6):881–889
6. Hildebrandt VH, Bongers PM, van Dijk FJ, Kemper HC, Dul J (2001) Dutch Musculoskeletal Questionnaire: description and basic qualities. Ergonomics 10;44 (12):1038–1055
7. Waddell G, Burton AK, Main C (2003) Screening to identify people at risk of long-term incapacity for work. A conceptual and scientific review. Disabil Med 3(3):72–83
8. Hartvigsen J, Lings S, Leboeuf-Yde C, Bakketeig L (2004) Psychosocial factors at work in relation to low back pain and consequences of low back pain; a systematic, critical review of prospective cohort studies. Occup Environ Med 61(1):e2
9. Schultz IZ, Crook JM, Berkowitz J, Meloche GR, Milner R, Zuberbier OA (2008) Biopsychosocial multivariate predictive model of occupational low back disability. In: Handbook of complex occupational disability claims. Springer, London, pp 191–202
10. Schultz IZ, Law AK, Cruikshank LC (2016) Prediction of occupational disability from psychological and neuropsychological evidence in forensic context. Int J Law Psychiatry 49:183–196
11. Schultz IZ, Crook J, Fraser K, Joy PW (2000) Models of diagnosis and rehabilitation in musculoskeletal pain-related occupational disability. J Occup Rehabil 10(4):271–293
12. White MI, Wagner SL, Schultz IZ, Murray E, Bradley SM, Hsu V et al (2015) Non-modifiable worker and workplace risk factors contributing to workplace absence: a stakeholder-centred synthesis of systematic reviews. Work 52(2):353–373
13. Oliv S, Noor A, Gustafsson E, Hagberg M (2017) A lower level of physically demanding work is associated with excellent work ability in men and women with neck pain in different age groups. Saf Health Work 8(4):356–363

The Revised ISO Standard 11228-1 on Manual Lifting, Lowering and Carrying: Special Focus on Extensions of the Revised NIOSH Lift Equation and a Strategy for Interpretation

Robert R. Fox[✉]

Chair, US TAG to ISO TC159/SC3, General Motors Company,
Warren, MI, USA
robert.r.fox@gm.com

Abstract. This short paper summarizes a number of the revisions to the ISO 11228-1 standard on manual lifting, lowering and carrying related to the use and extensions of the Revised NIOSH Lifting Equation (RNLE). The extensions allow the RNLE to be applied to more complex multiple lifting tasks that are commonly found in industry and warehousing. Also reviewed is a recommend approach to interpreting the calculated result (LI, CLI, SLI, VLI) of the RNLE and its extensions.

Keywords: Revised NIOSH Lifting Equation · Manual lifting tasks
ISO standards

1 Introduction

The NIOSH Lifting Equation, specifically the 1991 Revised NIOSH Lifting Equation or RNLE (Waters et al. 1993) is used worldwide by occupational ergonomists, industrial engineers and safety specialists to assess and (re)design manual lifting tasks and industrial systems that involve either single or repetitive lifting. Studies have shown that the RNLE is the most commonly used ergonomics assessment tool by practicing ergonomists globally (Dempsey et al. 2005). The ISO 11228-1 standard, "Ergonomics – Manual handling-Part 1: Lifting and carrying" was first publish as an ISO standard in 2003 and was directed at practitioners to use in assessing lifting tasks. ISO 11228-1 was subjected to periodic systematic reviews and in 2014 the decision was made by the TC159/SC3 Working Group 4 (WG4 manual handling and strength) to revise the standard. A WG4 writing group was formed and an extensive literature review performed on the RNLE which formed the basis of the revision. While the revision of ISO 11228-1 introduces a number of new multipliers and extensions of the RNLE, the focus in this short paper will be on the extensions to variable and sequential lifting tasks and to how the RNLE can be interpreted.

As originally devised, the RNLE assessed lifting tasks that either involved: (1) single-task manual lifting (the task characteristics do not vary from lift to lift) or; (2) multiple-task lifting which consists of a small set (less than 10) of repetitive lifting

© Springer Nature Switzerland AG 2019
S. Bagnara et al. (Eds.): IEA 2018, AISC 820, pp. 154–158, 2019.
https://doi.org/10.1007/978-3-319-96083-8_19

tasks. The RNLE calculates a *Recommended Mass Limit* (RML, also referred to as the *Recommended Weight Limit* or RWL) which discounts the mass constant of the RNLE given the measured and assessed variables of the task (horizontal distance, vertical height, distance moved, frequency of lift, asymmetry and coupling). The RML is then divided by the actual weight of the lift or lower to calculate the Lift Index (LI). For multiple lifting Waters et al. 1993 devised a method to calculate a Composite Lift Index (CLI).

It should be noted that lifting tasks in industry have continued to undergo changes as industrial and business practices, technology and philosophies have changed in the decades since the RNLE was first published. For one, many manual handling tasks have become more varied in industry with a greater variety of sizes and weights of objects handled. Work practices, besides introducing more variety in the objects handled, may also incorporate job rotation where the worker's handling tasks vary during the rotation. Marras et al. (2009) noted that manufacturing increasingly involves work where employees perform a variety of tasks as part of their job which may involve rotation through different work stations throughout the day.

2 Addressing Variable Lifting Tasks

Over a number of year, Dr. Thomas Waters of NIOSH, working with various US and European collaborators, developed a number of extensions of the RNLE that allowed it to be applied to a wider range of lifting task categories other than the single and limited multiple lifting tasks. These include the following:

(1) Sequential manual lifting tasks where a worker rotates between different tasks or workstations during a shift. For each task or workstation the worker performs a different series of lifting tasks which may consist of single or multiple lifting tasks.
(2) Variable-task manual lifting tasks where all of the lifts are highly variable. These types of jobs include warehousing, baggage handling and small lot material delivery in manufacturing operations.

Annex F of the revised ISO 11228-1 presents a number of methods developed by Waters and others to address different aspects of multiple task lifting including the sequential lifting tasks and the variable lifting tasks. To address the issue of sequential manual lifting tasks, a method utilizing a Sequential Lifting Index (SLI) was devised (Waters et al. 2007). The method is similar to the Composite Lifting Index (CLI) of the RNLE and involves the individual single or multiple lift analysis of the lifting jobs occurring within a rotation cycle. The procedure then involves the calculation of a Maximum Lifting Index (LImax) using the overall duration time category for each task, an ordering of the tasks, calculation of a time fraction with respect to 480 min for each task and the determination of a new LI for each task based upon its individual actual continuous duration. The SLI is calculated starting from the LI of the most strenuous task (highest LI) considering its actual duration and adding a fraction (called "K") of the difference between the LImax and LI of the same task. K is the ratio between the weighted average of the LImax values multiplied by the task time fraction and LImax of the most strenuous task.

Highly variable lifting tasks involving potentially the lifting of a wide variety of weights with different lift parameters posed a unique challenge to the use of the RNLE. The Variable Lift Index (VLI) was developed specifically to address these lifting situations (Waters et al. 2015, 2009). The VLI procedures involve collecting lifting parameter data on a subset of the lifting tasks to be analyzed. The Frequency Independent Lifting Index (FILI) is calculated for each of the tasks selected for analysis and then each lift is fitted into one of a set number of FILI categories. The number of categories may range from six to nine, depending upon the sample size and range of the FILI values. The representative values for each FILI category and the corresponding frequency of lifts in each category are then inputs into the Composite Lifting Index (CLI) equation of the RNLE to obtain the VLI for a variable manual lifting job. The revised ISO 11228-1 provides methods for composite lifting tasks to calculate the CLI for greater than 10 subtasks and up to 30 subtasks using a simplified aggregation of the lift parameters into a small number of LI categories.

The revised ISO 11228-1 provides detailed discussion as well as examples of the above SLI and VLI methods. It also provides decision tables (see the figure below) that can assist the user/practitioner in selecting the appropriate method. It should also be noted that the revised ISO 11228-1 also provides guidance on applying the RNLE to one-handed lifting, multiple-person lifting and the analysis of lifting tasks that exceed 8 hours per day (Fig. 1).

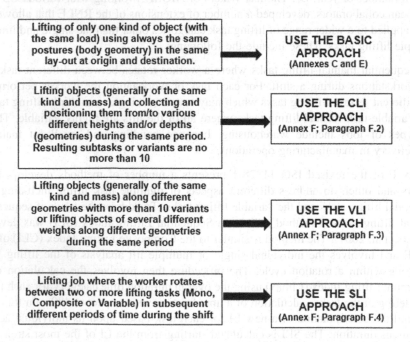

Fig. 1. Decision table for the appropriate lifting approach (Taken from the Revised ISO 1122-1: *Ergonomics – Manual handling – Part 1: Lifting, lowering and carrying*)

3 Interpreting the LI, CLI, SLI and VLI in the Revised ISO 11228-1

Since the Waters et al. (1993) publication of the RNLE, the interpretation of the LI and CLI have been problematic in regards to the risk if back injury inferred from the LI or CLI score. A LI or CLI value of 1.0 was taken as a design goal based upon the assumption that a value of 1.0 met the biomechanical, psychophysical and physiological parameters defined originally (Waters et al. 1993, 1994). However, the relationship between those parameters and actually back injury risk was not established. The LI and CLI have been described as indices of *exposure* to the physical stressors of a lifting task (Garg et al. 2014), not as risk thresholds for illness/injury. "Exposure" is defined as the extent to which a person is subject to a change or to specific circumstances (e.g. task conditions including frequency, force, heights or depths reached, duration etc.). The amount of that exposure can be expressed in quantitative terms, and can refer to the extent to which a person is subjected to a hazard (potential source of harm) or combination of hazards. "Risk" in contrast is defined as the combination of the probability of occurrence of harm, (e.g. likelihood of having an adverse health outcome such as back pain) and the severity of that harm.

Since the original publications, there have been numerous studies on the RNLE and industrial lifting tasks and the relationship between the LI/CLI scores and a variety of injury and illness outcomes. Many of these studies revealed that the associations are neither simplistic nor clear-cut. The ISO 11228-1 revision writing group felt that more specific advice should be provided in the revision on how the LI, CLI, SLI and VLI should be interpreted. To that end the research literature on the RNLE conducted over the past several decades was reviewed including very recent studies by NIOSH and other researchers. The intent of the review was, based upon the findings of the literature, to develop a classification of exposure and risk that could be employed by users of the standard (or of the RNLE in general) to make informed interpretations and decisions about the illness/injury likelihood of lifting tasks. To that effect a table of risk and exposure given ranges of the LI/CLI was developed and presented in Annex D of the revised standard. An abbreviated version of that table with exposure level/risk implication and recommended actions is provided below (Table 1):

Table 1. Interpretation of Lifting Index (m_A/RML values)

Lifting index value	Exposure level/Risk Implication	Recommended actions
LI ≤ 1,0	Very low	None in general for the healthy working population
1,0 < LI ≤ 1,5	Low	Pay attention to low frequency/high load conditions and to extreme or static postures
1,5 < LI ≤ 2,0	Moderate	Redesign tasks and workplaces according to priorities to reduce the LI, followed by analysis of results to confirm effectiveness
2,0 < LI ≤ 3,0	High	Changes to the task to reduce the LI should be a high priority
LI > 3,0	Very high	Changes to the task to reduce the LI should be made immediately

It is the intent of a number of the WG4 experts to provide a detailed explanation and description of the literature review and how the exposure level/risk implications and recommended actions were arrived at in a future journal paper.

4 Conclusion

The revised ISO 11228-1 standard on manual lifting, lowering and carrying provides users of the standard and occupational ergonomics practitioners a number of useful extensions of the RNLE that greatly enhance its applicability to various types of lifting tasks and situations found currently in industry. Furthermore, it offers users an improved basis for interpreting the results of the RNLE and its extensions based upon a rigorous evidenced-based review of the literature of the validation of the RNLE.

References

Dempsey PG, McGorry RW, Maynard WS (2005) A survey of tools and methods used by certified professional ergonomists. Appl Ergon 36:489–503

Garg A, Boda S, Hegmann KT, Moore JS, Kapellusch JM, Bhoyar P, Thiese MS, Merryweather A, Beckow-Schaefer G, Bloswick D, Malloy EJ (2014) The NIOSH lifting equation and low-back pain, part 1: association with low-back pain in the backworks prospective cohort study. Hum Factors 56(1):6–28

Marras WS, Cutlip RG, Burt SE, Waters TR (2009) National Occupational Research Agenda (NORA) future directions in occupational musculoskeletal disorder health research. Appl Ergo 40:15–22

Waters T, Occhipinti E, Colombini D, Alvarez-Casado E, Fox R (2015) Variable Lifting Index (VLI): a new method for evaluating variable lifting tasks. Hum Factors. https://doi.org/10.1177/001872081561225

Waters T, Occhipinti E, Colombini D, Alvarez E, Hernandez A (2009) The Variable Lifting Index (VLI): a new method for evaluating variable lifting tasks using the Revised NIOSH Lifting Equation. In: Proceedings 17th IEA world conference, Beijing, China

Waters TR, Lu M-L, Occhipinti E (2007) New procedure for assessing sequential manual lifting jobs using the revised NIOSH lifting equation. Ergonomics 50(11):1761–1770

Waters TR, Putz-Anderson VP, Garg A, Fine LJ (1993) Revised NIOSH equation for the design and evaluation of manual lifting tasks. Ergonomics 36(7):749–776

Waters TR, Putz-Anderson V, Garg A (1994) Applications manual for the revised NIOSH lifting equation. DHHS (NIOSH) Publication 94-110. National Institute for Occupational Safety and Health, Centers for Disease Control and Prevention

The Influence of Physiological Breaks and Work Organization on Musculoskeletal Pain Index of Slaughterhouse Workers

Roberta Schwonke Martins, Fernando Gonçalves Amaral(✉) ⓘ, and Marcelo Pereira da Silva ⓘ

Universidade Federal do Rio Grande do Sul, Porto Alegre
Rio Grande do Sul, Brazil
amaral@producao.ufrgs.br

Abstract. Beef cattle and the meat processing chain are of great economic importance in Brazil, since they are responsible for creating thousands of jobs. Nevertheless, the issue of health and safety of the sector's workers is a major problem given the high number of accidents at work and workers in sick leave, as a consequence of working at low temperatures, doing repetitive tasks, and carrying weights. In light of that, this article aims at identifying the association between musculoskeletal pains and discomforts and the implementation of a system of break periods and task rotations in the slaughter sector of a cattle slaughterhouse. The method employed consisted of direct observation, interviews with workers, and the application of ergonomic analysis methods. Results show that the lack of guidance in the implementation of task rotation still generates complaints, discomforts, and sick leaves. Finally, we developed a functional task rotation plan grounded on the analysis of work stations' risks.

Keywords: Slaughterhouse workers · Physiological breaks · OCRA Musculoskeletal pains and discomfort

1 Introduction

The growth of Brazilian slaughterhouse sector in the last decade raised the competitivity of companies, increasing production and reducing costs. The intensification and the acceleration of the work pace, increase in control, time pressure, and the mental and physical demands of workers are observed in this context [4, 19]. It is estimated that 80% of the slaughterhouse work tasks are done manually and in standing position. Slaughterhouses are humid places, with loud noises, and low temperatures. Slaughtering procedures occur sequentially, and the working speed is not determined by the worker, rather it is determined by the daily production goal. Additionally, sharp objects are manipulated with steady, sudden, and repetitive motions. An analysis of such conditions under the ergonomic optics reveals the possibility of developing musculoskeletal pathologies [1, 4, 8, 10, 11, 15–18, 20].

In this context, improvements in slaughterhouses were necessary in order to increase production without causing risks to workers and without causing losses to the

S. Bagnara et al. (Eds.): IEA 2018, AISC 820, pp. 159–168, 2019.
https://doi.org/10.1007/978-3-319-96083-8_20

company due to work overload. Therefore, companies became concerned about the factors that directly influence such conditions as: the environment, working hours, workstations, tasks, work organization, remuneration, nutrition, and wellbeing, with the aim to provide a healthy and comfortable environment [8, 20].

In Brazil, since 2013, a regulation of safety and health at work that specifically addresses slaughterhouses, and meat and derivatives processing companies has gone into effect (Regulatory Norm 36 – NR36). Its implementation involves changes in work routines, equipment, work environment, in addition to the definition of breaks and planned workstation rotation [2].

The implementation of breaks for psychophysiological recovery during working hours is an important measure, which allows workers' muscular recovery and rest to reduce fatigue. Fatigue is considered the main factor for productivity reduction, since it reduces the organism's capacity and diminishes the quality of the executed task [9, 19]. Planned and functional workstation rotation are also important for reduction of fatigue and musculoskeletal pain, and that is done through activity switching as a means to alternate postures and motions [14, 21]. This scheme is also indicated by NR36, although it lacks evidences to prove its efficacy [12, 19].

In this context, this study aims to analyze the influence of (physiological) breaks and job rotation (work organization) on the reduction of musculoskeletal discomfort and pain index in workers from the slaughter sector in a Brazilian cattle slaughterhouse.

2 Methods

We conducted an exploratory descriptive research through direct observations, interviews with workers and the application of ergonomic analysis methods. We analyzed a cattle slaughterhouse located in southern Brazil, composed of 60 workers, who work in fixed 8-h shifts.

Data and information was collected during weekly visits to the company in January 2017. Initially, we selected a sector based on its working conditions and the types of workers' complaints. Afterwards, an interview was conducted with one of the workers from the selected sector. The aim was to comprehend how the process worked before NR36 going into effect (before 2013), and the improvements made after its implementation. Aspects addressed were namely: task description; data about the workers' profile such as age and sex, working hours, company service length, length of time executing the task, practice of physical activity, other simultaneous professional activities, workstation and sector identification, information regarding the workstation, history of possible accidents at work, and information about breaks/pauses and workstation rotations, as well as data about workers' sicknesses. Then, through the information provided by the company, we analyzed the current chart of breaks and workstation rotations.

To identify musculoskeletal pain and discomfort complaints, we used the pain diagram of Corlett and Manenica [7]. Based on the diagram, we asked the sectors' workers to indicate the degree of discomfort for each of the body parts. The discomfort index was classified into eight levels ranging from zero (no discomfort) to seven (extreme discomfort). Complaints for the activity above level three were flagged as requiring immediate attention [9].

Afterwards, we conducted a task analysis to understand work organization by observing the space designated for workers and furniture, the types of tools and work equipment (shape, dimensions, and handling). Finally, we applied the Occupational Repetitive Actions (OCRA) method, which gathers quantitative data on the level of workers' exposure, considering frequency of actions, excessive force exertion, stereotyped and/or awkward upper limb motion and posture, and the lack of resting periods [5, 6, 13].

3   Results

For this study, we selected the slaughter sector since it demands greater physical efforts and muscle overload. Additionally, the levels of complaint and discomfort are high in this sector, since around 90% of the workers reported some type of discomfort during task execution. The process is segmented and comprises seven workstations in sequence: stunning, bleeding, skinning, evisceration, beheading, carcass cut, and carcass washing.

We interviewed the workers responsible for the bleeding task, which is when big blood vessels from the animals' neck are cut with a knife. The results from the interviews indicated that before NR36 implementation, "work was heavier and more tiresome", "the day was long", "at the end of the morning, everyone was already exhausted", and "many workers took painkillers during their tasks".

According to NR36, physiological breaks for recovery should be applied in tasks that involve repeatability and muscular overload, and the breaks should respect the intervals presented in Table 1.

Table 1. Work time organization

Working hours	Tolerance time for break application	Break time
up to 6 h 00	up to 6 h 20	20 min
up to 7 h 20	up to 7 h 40	45 min
up to 8 h 48	up to 9 h 10	60 min

The analysis of company's break organization indicates three ergonomic breaks, accounting for 45 min throughout a working period of 7h20 min. The breaks are taken outside of the work environment, and the company provides seats and drinking water. At first glance, the company complies with the regulation of NR36. Nevertheless, an analysis of the break distribution (Table 2) chart shows that the break before lunchtime is poorly placed, since this time is already compensated by the one-hour break for lunch.

Table 2. Slaughter sector's break chart

Clock in (shift start)	Change of clothes	1st break	Industry work	2st break	Return from the break	Return from the break	Change of clothes	Total worked time
06:55	07:00	08:40	08:55	10:15	10:25	10:50	10:55	04:00
07:00	07:05	08:45	09:00	10:20	10:30	10:55	11:00	04:00
07:05	07:10	08:50	09:05	10:25	10:35	11:00	11:05	04:00
07:10	07:15	08:55	09:10	10:30	10:40	11:05	11:10	04:00
07:15	07:20	09:00	09:15	10:35	10:45	11:10	11:15	04:00

Clock in (return from lunch)	Change of clothes	Industry work	3st break	Industry work	Change of clothes	Clock in (shift end)	Total worked time
12:05	12:10	13:55	14:15	16:00	16:05		04:00
12:10	12:15	14:00	14:20	16:05	16:10		04:00
12:15	12:20	14:05	14:25	16:10	16:15		04:00
12:20	12:25	14:10	14:30	16:15	16:20		04:00
12:25	12:30	14:15	14:35	16:20	16:25		04:00

Regarding the organization of activities, the company does not have a workstation rotation plan that switches the demand between muscle groups, or that reduces monotony and alternates cadency. Furthermore, workers only change workstations when team members are absent. Often, depending on the number of animals to be slaughtered, workers from the slaughter sector also assist in deboning animals.

To identify work-related musculoskeletal pains and discomforts, 15 workers from the sector responded to the Corlett and Manenica [7] questionnaire. The results indicated that 93,3% of the employees reported some shoulder discomfort, 86,7% reported arm discomfort, 60% reported feeling forearm discomfort, 60% reported hand discomfort, 66,7% reported neck discomfort, 60% reported discomfort in upper and middle back, and 73,3% in lower back. In the other segments, the percentage of discomfort was low, as Fig. 1 shows. Thus, we found that the body areas with the greatest indices of pain and discomfort were shoulders, arms, and lower, and upper back, respectively.

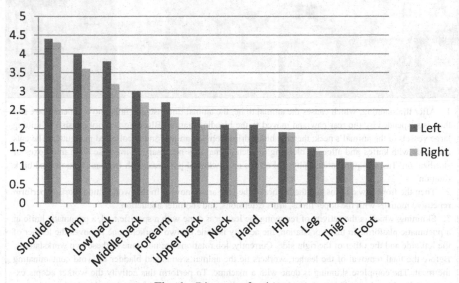

Fig. 1. Diagram of pain areas

We analyzed workstations of the bleeding sector by direct observations and through videos recorded during a work shift. After a brief description of how the work is organized, the main task characteristics are presented in Fig. 2.

The results from the OCRA method evidenced shoulder overload associated to the high motion rate. Out of the nine workstations analyzed, only the Carcass Sawing station is within the risk level limit of the method, as shown in Table 3.

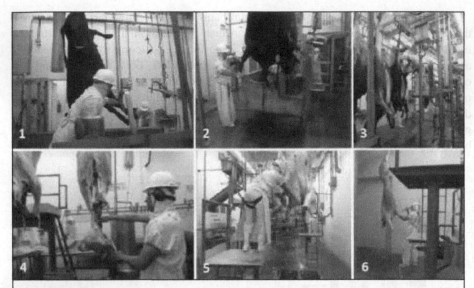

1. After the stunning, which causes the animal to be, the animal unconscious falls into a concrete box and then is suspended by the rear paw and moved to the bleeding station. Bleeding activity consists of cutting large vessels of the animal's neck, then the blood drips from the animal and is collected in a gutter. The cuts are made with knifes and after the bleeding of each animal it is necessary to immerse them into boxes of sterilization. Tasks require intense physical efforts and awkward postures, with frequent flexions and back rotation

2. Then the front paws, horns and the leather of the face are removed. In this workstation workers perform repetitive motions with the upper limbs, arms extensions, and shoulder abduction.

3. Skinning, which is the activity of removing the leather is done with a specific tool, a pneumatic knife in a pneumatic platform. In this workstation, the activity is done by two workers, one removes the leather of the left side and the other of the right side. Currently, job rotations are not done in these two workstations. Before the total removal of the leather, workers tie the animal's anus and bladder to avoid contaminating the meat. The complete skinning is done with a machine. To perform this activity the worker adopts extreme and awkward postures, performs large amplitude motions and repetitive technical actions.

4. Before opening the animal's carcass, the workers manually close the esophagus, displace the trachea, and remove the head. The worker performs repetitive motions with the upper limbs, shoulder abduction and flexions and constant ulnar deviations in the hand that grabs the knife, which causes fatigue and musculoskeletal discomfort.

5. The evisceration happens in a fixed platform. The viscera are manually removed and fall into a bench for inspection. Then the carcass is sawed into smaller portions using a suspended chainsaw. The activity is performed in a permanent standing position, with frequent inclinations of the trunk. After, residues and the fat excess are manually removed with a knife performing continuous motions with the hands and arms, inducing shoulder and back flexion due to the distance between the carcass and the platform.

6. Finally, the half carcasses are washed and taken to refrigeration through a suspended rail. Continuous hand and arm motion, and moderate physical efforts are done to lift and transport the boxes in the viscera separation in a predominantly standing posture.

Fig. 2. Workstations work organization

Table 3. OCRA checklist for the right (R) and left (L) side

Number	Line or homogenous area	Recovery multiplier	N hours without recovery	Frequency	Force	Side	Shoulder	Elbow	Wrist	Hand	Stereotype	Total posture	Complementary	Checklist value
P1	Bleeding	1,12	2	9	2	R	12	2	4	3	0	12	1	17,47
P2	Horn, paws and face leather removal	1,12	2	9	2	R	24	3	4	3	1,5	25,5	1	27,30
P3	Skinning	1,12	2	9	6	R	8	8	4	3	3	11	1	19,66
P4	Rectum removal	1,12	2	9	2	R	8	3	3	2	1,5	9,5	1	15,65
P5	Esophagus closure	1,12	2	9	2	R	12	3	4	3	0	12	1	17,47
P6	Evisceration	1,12	2	9	4	R	8	2	3	1	0	8	1	16,02
P7	Carcass sawing	1,12	2	2,5	2	R	8	2	1	1	0	8	1	9,83
P8	Fat removal and cleaning	1,12	2	9	2	R	12	2	4	2	1,5	13,5	1	18,56
P9	Toilet	1,12	2	2,5	1	R	24	1	1	1	0	24	1	20,75
P1	Bleeding	1,12	2	0	1	L	24	1	1	1	0	24	1	17,47
P2	Horn, paws and face leather removal	1,12	2	9	2	L	8	2	1	1	1,5	9,5	1	27,30
P3	Skinning	1,12	2	9	2	L	2	3	2	1	3	6	1	19,66
P4	Rectum removal	1,12	2	9	2	L	8	1	1	1	1,5	9,5	1	15,65
P5	Esophagus closure	1,12	2	9	1	L	12	2	1	1	0	12	1	17,47
P6	Evisceration	1,12	2	9	4	L	8	2	1	1	0	8	1	16,02
P7	Carcass sawing	1,12	2	0	2	L	8	1	1	1	0	8	1	9,83
P8	Fat removal and cleaning	1,12	2	9	1	L	12	1	4	1	1,5	13,5	1	18,56
P9	Toilet	1,12	2	0	1	L	24	1	1	1	0	24	1	20,75

Workstation 1	Rotation	Workstation 2
Bleeding	→	Toilet
Horn, paws and face leather removal	→	Bleeding
Rectum	→	Carcass cleaning
Esophagus	→	Rectum
Evisceration	→	Esophagus
Sawing	→	Horn, paws and face leather removal
Carcass cleaning	→	Evisceration
Toilet	→	Sawing

Fig. 3. Job rotation program

4 Conclusions

After the identification of the breaks and main discomforts related to the activities, we identified that even after the implementation of NR36 the pain and muscular discomfort levels in the dorsal area and shoulders were still high. The total break time complies with the regulation; however, their distribution could be improved by anticipating the break immediately before lunch in one hour.

The OCRA method application showed high sickness risk in almost all the workstations, mainly related to the frequency of motions and awkward postures of the upper limbs. From the data gathered we developed a job rotation scheme, as depicted in Fig. 3. The objective of this scheme was to alternate the demanded muscle groups. To reach efficacy, such change needs to be implemented with operational training courses for the different activities. Also, the company's Ergonomics Committee must monitor the reduction of worker's pain complaints in the sector. The appropriate time to alternate between activities is every two hours.

However, the mere application of the norm's recommendations, without the adequate workstation modifications and the definition of break and activities, is not sufficient to make work safer and more comfortable. Moreover, other recommended measures to protect occupational health are:

(a) To implement 20-min rest breaks after every 1 h and 40 min of continuous work, for workers in activities located in artificially cold environments, which count as working hours;
(b) To implement break periods between 10 min and 20 min. The breaks must be distributed in order to not occur during the first working hour, mealtime, or the last working hour. Breaks must occur outside the working environment, in places that provide thermal and acoustic comfort, benches or chairs and drinking water;
(c) To allocate a compatible number of workers according to the production needs;

(d) Regarding productivity demands, the company must respect the individual production capacity of each worker along with the demanded quality of the product;
(e) To guarantee that the equipment of the workstations meets the anthropometric specifications, at least of 95% of the users, providing facilities that deliver good work posture, visualization, and operation conditions;
(f) To guarantee workers' participation in the evaluation and diagnosis of the overall working conditions, definition of opportunities for improvements, planning and implementation of methods, tools, procedures, and modifications in the workstations.

The objective of this study was to analyze the influence of breaks and job rotation on the reduction of musculoskeletal discomfort and pain index in workers from the slaughter sector of a Brazilian cattle slaughterhouse. The data found indicates that these practices alone are not enough to reduce the complaints and associated biomechanical risks. Further research considering other related factors and other possible improvements are necessary.

References

1. Arvidsson I, Balogh I, Hansson GA, Ohlsson K, Åkesson I, Nordander C (2012) Rationalization in meat cutting e Consequences on physical workload. Appl Ergon 43:1026e1032
2. BRASIL (2013) Ministério do Trabalho e Emprego - MTE. NR 36 - SEGURANÇA E SAÚDE NO TRABALHO EM EMPRESAS DE ABATE E PROCESSAMENTO DE FRANGOS E DERIVADOS. Disponível em. http://trabalho.gov.br/images/Documentos/SST/NR/NR36.pdf. Acesso em: Abril 2017
3. Buzanello MR, Moro ARP (2012) Association between repetitive work and occupational cold exposure. Work 41:5791–5793
4. Christensen H, Sgaard K, Pilegaard M, Olsen HB (2000) The importance of the work/rest pattern as a risk factor in repetitive monotonous work. Int J Ind Ergon 25:367–373
5. Colombini D (1998) An observational method for classifying exposure to repetitive movements of the upper limbs. Ergonomics 41:1261–1289
6. Colombini D (2014) Método OCRA para análise e prevenção do risco por movimentos repetitivos: manual para a avaliação e a gestão do risco. Edição do autor, Curitiba, PR
7. Corlett EN, Manenica J (1980) The effects and measurement of working postures. Appl Ergon 11(1):7–16
8. Gandon LFM, Ferraz RRN, Pavan LMB, Zaions APDRE (2017) Reduction of accidents at work and absenteism with ergonomic improvements in a South Brazilian fridge company. Revista Eletrônica Gestão & Saúde. Rev. Gestão & Saúde (Brasília), January 2017, vol 08, no 01, pp 92–113. ISSN 1982-4785
9. Iida I (2016) Ergonomia: projeto e produção, 3ª edn. Editora Blucher, São Paulo
10. Kaka B, Idowu OA, Fawole HO, Adeniyi AF, Ogwumike OO, Toryila MT (2016) An analysis of work-related musculoskeletal disorders among Butchers in Kano Metropolis, Nigeria. Saf Health Work 7:218e224
11. Kasaeinasab A, Jahangiri M, Karimi A, Tabatabaei HR, Safari S (2017) Respiratory disorders among workers in slaughterhouses. Saf Health Work 8:84e88

12. Leider PC, Boschman JS, Frings-Dresen MH, Van der Molen HF (2015) Effects of job rotation on musculoskeletal complaints and related work exposures: a systematic literature review. Ergonomics 58:18–32
13. Occhipinti E (1998) OCRA: a concise index for the assessment of exposure to repetitive movements of the upper limbs. Ergonomics 41:1290–1311
14. Pinetti ACH, Buczek MR (2015) Ergonomic work analysis in a Brazilian poultry slaughterhouse cutting room. In: Proceedings 19th Triennial Congress of the IEA, Melbourne, 9–14 August 2015
15. Ramos E, Reis DC, Tirloni AS, Moro ARP (2015) Thermographic analysis of the hands of poultry slaughterhouse workers exposed to artificially cold environment. Proc Manuf 3:4252–4259
16. Reis DC, Ramos E, Moro ARP, Reis P (2016) Upper limbs exposure to biomechanical overload: occupational risk assessment in a poultry slaughterhouse. In: Advances in physical ergonomics and human factors. Advances in intelligent systems and computing, vol 489
17. Soares ACC (2004) Estudo Retrospectivo de Queixas Músculos-esqueléticas em Trabalhadores de Frigorífico. 2004. 75f. Dissertação (mestrado em Engenharia de produção) – Programa de Pós-Graduação em Engenharia de Produção, Universidade de Santa Catarina, Florianópolis. Disponível em. https://repositorio.ufsc.br/bitstream/handle/123456789/86820/225307.pdf. Acesso em abril 2017
18. Takeda F (2010) Configuração Ergonômica do Trabalho em Produção Contínua: o Caso de Ambiente de Cortes em Abatedouro de Frangos. [Dissertação Mestrado], Programa de Pós-Graduação em Tecnologia, Universidade Tecnológica Federal do Paraná. Ponta Grossa
19. Tirloni AA, Reis DC, Ramos E, Moro ARP (2018) Evaluation of bodily discomfort of employees in a slaughterhouse. In: Advances in physical ergonomics and human factors. Advances in intelligent systems and computing, vol 602
20. Vilanova MAS, Dengo CS, Fumagalli, LAW (2016). Absenteeism in Company of Poultry Slaughterhouse area with Emphasis on Ergonomics. Rev Ciênc Juríd Empres, Londrina, vol 17, no 2, pp 142–150
21. Vogel K, Karltun J, Eklund J, Engkvist IL (2013) Improving meat cutters' work: changes and effects following na intervention. Appl Ergon 44:996e1003

An Investigation of the Maximum Acceptable Weight of Lift by Indonesian Inexperienced Female Manual Material Handlers

Ardiyanto Ardiyanto[✉], Dhanaya A. Wirasadha,
Novi W. Wulandari, and I. G. B. Budi Dharma

Department of Mechanical and Industrial Engineering,
Universitas Gadjah Mada, Yogyakarta, Indonesia
ardiyanto@ugm.ac.id

Abstract. The purpose of this study was to investigate the maximum acceptable weight of lift (MAWL) by Indonesian inexperienced female manual material handlers. Twenty-one females who were selected based on their physical activity categories voluntarily participated in this study. The participants were asked to determine their MAWLs using the psychophysical method for the lifting tasks at two different lifting frequencies: one lift/5 min and 4 lifts/min. The participants' heart rate was also recorded to be utilized for investigating the physiological responses while lifting the obtained MAWLs. The obtained MAWLs were also compared to the MAWLs of other study populations, using published data. As the results, two-way analysis of variance test result revealed no significant effect of physical activity category on MAWL (p = 0.890). On the other hand, a significant effect on MAWL for the lifting frequency was observed at p < .001. The comparison of MAWLs obtained in this study to other studies showed that no significant differences on MAWL were found between Indonesian and Chinese inexperienced manual handlers for the lifting tasks at one lift/5 min. The significant differences on MAWL were observed between Indonesian inexperienced and American experienced female manual handlers. However, no significant differences were observed for the lifting tasks at 4 lifts/min. Furthermore, the physiological data analysis revealed that the energy expenditures required for lifting at one lift/5 min was significantly less than the physiological limit, while those of lifting at 4 lifts/min was significantly higher than the limit. The results of this study might be utilized as consideration for determining the safety lifting limit for Indonesian inexperienced female manual material handlers.

Keywords: Lifting · Maximum acceptable weight of lift
Indonesian inexperienced female manual material handlers
Psychophysical approach

1 Introduction

Many working activities in the developing country industries are still dominated by manual material handling (MMH) activities such as lifting, lowering, carrying, pushing and pulling [1, 2]. On the other hand, MMH activities are considered as a risk factor for

S. Bagnara et al. (Eds.): IEA 2018, AISC 820, pp. 169–178, 2019.
https://doi.org/10.1007/978-3-319-96083-8_21

musculoskeletal disorders such as low back pain [3, 4]. In addition to the adverse health effects caused by these activities, MMH also associates with economic losses since it represents one of the largest sources of claims and cost [5].

In order to prevent the workers from adverse health effects caused by MMH activities, the National Institute for Occupational Safety and Health (NIOSH) developed a lifting equation which can be utilized to determine the recommended weight limit for various MMH activities. The equation first developed in 1981 [6] and revised in 1991 to accommodate broader lifting tasks. The NIOSH lifting equation is developed based on three criteria: biomechanics, physiology, and psychophysics [7].

The psychophysical criterion are determined using the maximum acceptable weight of lifts (MAWLs). The MAWL refers to the amount of weight a person chooses to lift under a specific condition for a defined period [7]. It typically measured by asked the workers lift as hard as they can without straining themselves or without becoming unusually tired, overheated, weakened or out of breath' [8].

Although the equation may be protective to the working population, various data that become the provision of the equation, particularly for the psychophysical criterion, were determined based on the experienced manual material handlers as the subjects [7]. On the other hand, novice or inexperienced manual material handlers perform MMH activities differently to the experienced manual material handlers. Novice manual material handlers have different lifting strategies that influence their balance on performing MMH activities [9]. Also, the spinal loads of the inexperienced handlers are higher than experienced handlers during the identical lifting condition [10].

In addition to the fact that it was developed based on experienced manual material handlers, this equation might also not consider the racial differences. The research to construct these equations was conducted in Europe and North America thus make these equations were not probably directly applicable to other populations [11–13]. On the other hand, this equation is widely used as the tool for evaluating MMH activities in many developing countries such as Indonesia [14, 15] and Malaysia [16, 17].

This study was performed to investigate the MAWLs of Indonesian inexperienced female manual material handlers. The specific objectives of this study were as follows: (1) to investigate the MAWLs by Indonesian inexperienced female manual handlers, (2) to compare the obtained MAWLs to the MAWLs by other populations using data of published studies, and (3) to investigate the physiological responses of the participants while lifting the obtained MAWLs. The results of this study might be useful in determining the adjustment of the NIOSH lifting equation which is designed for the Indonesian inexperienced manual material handlers.

2   Methods

2.1   Participants

Twenty-one college female students without any experiences in manual materials handlings voluntary participated in this study. All the participants were free from musculoskeletal disorders in the upper extremity, neck, back, and the lower extremity at the beginning of the study (self-reported). Also, the participants were not in the menstrual period during the data collection sessions. The summary of participants' anthropometry is given in Table 1.

Before joining the study, potential participants were screened by filling out the Global Physical Activity Questionnaire [18] to determine their physical activity categories. Since we wanted to investigate three physical activity categories, including low, moderate, and high, the potential participants fitted with these categories were followed up. The potential participants who were willing to participate in this study were selected based on their time availability. The number of participants in each group was balanced (n = 7).

Two hours before the data collection sessions, the participants were instructed to avoid smoking, consuming alcoholic or carbonated beverages. They were also requested to wear comfortable clothing and their daily shoes during the data collection sessions. In order to familiarize with the experimental procedures, one-hour training sessions were introduced one day before the first data collection session.

Table 1. Summary of participants' anthropometry

Physical activity category	Age (years)	Weight (kg)	Height (cm)	Hand grip strength (kg)	Back strength (kg)
Low (n = 7)	20.71 (0.95)	50.8(5.5)	155.4(5.9)	22.5(5.5)	41.29(13.5)
Moderate (n = 7)	20.29 (1.25)	49.3(5.9)	158.4(4.9)	22.6(5.9)	34.79(10.5)
High (n = 7)	18.72 (0.49)	56.1(5.0)	156.9(4.5)	21.0(5.0)	39.43(10.2)
Total	19.90 (1.26)	52.1(5.9)	156.89 (5.0)	22.0(3.1)	38.5(10.5)

2.2 Experimental Design

The lifting tasks were performed during optimal lifting condition based on the NIOSH lifting equation [7]. The optimal lifting condition means that all multipliers in NIOSH lifting equation have the optimal value. Hence, the value of the horizontal distance, the vertical distance, the lifting distance, and the asymmetric angle were 25 cm, 75 cm, 25 cm, and was 0^0. Furthermore, this study applied two lifting frequencies, i.e., one lift/5 min (0.2 lifts/min) and 4 lifts/min. The first frequency is the lowest frequency in NIOSH lifting equation, while the second frequency was selected because psychophysical criterion is reliable for determining the maximum acceptable weight of lifts until 4 lifts/min frequency [19]. The order of the lifting frequencies was randomized.

2.3 Equipment

The weights were contained in a 40 cm wide × 30 cm deep × 30 cm high wooden box. The container was equipped with handles on each side which were categorized as good coupling based on the NIOSH lifting equation. Also, it had false bottoms able to hold up to 6 kg to minimize the visual cues. A motorized machine was utilized to lower the box after each lift performed by the participants. The anthropometric and strength measurements were performed using a chair-shaped anthropometer, a handgrip dynamometer, and a back muscle dynamometer (Takei Scientific Instruments, Japan). The heart rate data were recorded using a heart rate monitor (Polar Electro, Finland).

2.4 Procedure

After arriving at the data collection site, the participants were asked to have strength and anthropometric measurements. Following this step, the participants were requested to perform a warming up activity by moving their upper extremities for 5 min. Subjects were then asked to perform manual lifting tasks from waist height (75 cm) to chest height (100 cm). In determining the maximum acceptable weight of lift (MAWL), a psychophysical methodology as described by Ciriello and Snook [20] and Ciriello et al. [21] was applied in this study.

The participants were requested to lift the weights with randomized initial weights continuously for 40 min using their preferable lifting technique. They were encouraged to make as many weight adjustments until the maximum weight that they felt they could lift continuously for an 8-h period without "straining themselves or without becoming unusually tired, weakened, overheated, or out of breath" [22]. The weight adjustments were conducted by adding several kilograms of pebbles as requested by the subjects. The minimum weight for each adjustment was 1 kg. The final weight where no more weight adjustment requests would be the MAWL. The 40 min data collection session was conducted twice with two different lifting frequencies. Thirty minutes break were given to the subjects before switching to the lifting tasks at another frequency.

The participants were requested to return at the data collection site at least one day after the first data collection session. During the second data collection session, they were asked to lift the MAWLs determined from the first session for one hour at the two lifting frequencies. The participants' heart rate was continuously recorded during the last 5 min. The heart rate average of the last 5 min was utilized for analysis. The second data collection session was aimed to investigate the physiological responses while lifting the obtained MAWLs by examining the energy expenditure which was calculated based on the participants' heart rate. Keytel et al.'s equation [23] was utilized to predict the energy expenditure from the recorded heart rate. The predicted energy expenditures were compared to the energy expenditure limit for 1-h lifting task which is 3.7 kcal/min [7].

2.5 Data Analysis

Two-way analysis of variance was performed to determine the effect of lifting frequency and physical activity category on MAWL. The descriptive statistics of the MAWL were summarized in terms of mean, standard deviation (SD), and percentile values. Furthermore, the differences among MAWL of the Indonesian and those of other studies were examined by performing one-way analysis of variance or student's t-tests. If significant differences were observed, Tukey HSD tests were performed to follow-up the results of the one-way analysis of variance. One sample t-tests were conducted to determine statistically significant differences existed between energy expenditure from the participants in the study and the energy expenditure limit according to Waters et al. [7]. This study applied 5% as the significance level of the statistical tests.

3 Results and Discussion

3.1 Effect of Lifting Frequency and Physical Activity on MAWL

Figure 1 shows the average MAWLs for the two lifting frequencies by the three physical activity categories. A two-way analysis of variance revealed a statistically significant main effect on MAWL by lifting frequency, $F(1, 36) = 16.17$, $p < .001$. On the other hand, there was no statistically significant main effect by physical activity category, $F(2, 36) = .11$, $p = .890$. There was also no statistically significant interaction between physical activity category and lifting frequency, $F(1, 36) = .21$, $p = .815$.

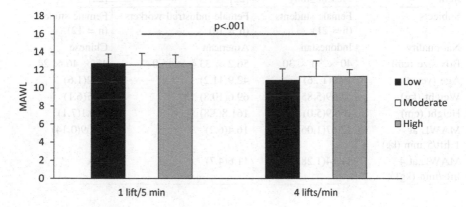

Fig. 1. Maximum acceptable weights of lifts (kg) for the two lifting frequencies by the three physical activity categories

3.2 MAWL Comparison with Other Study Populations

Due to no significant effect of physical activity category, the MAWL data for the further analysis were combined. Table 2 presents the descriptive statistics of the MAWL data in terms of mean, standard deviation, 25th percentile, 50th percentile, and 95th percentile. The comparison between the MAWLs obtained in this study and the MAWLs of other studies is presented in Tables 3 and 4.

The average MAWL at one lift/5 min was 12.67(1.06) kg. Meanwhile, the 5th percentile MAWL at this frequency was 11 kg (Table 2). This value might represent the loads that could be lifted by 75% of the participants. The one-way analysis of variance (ANOVA) revealed significant differences on MAWL at one lift/5 min for three different study populations at $p < .05$ (Table 4). The Tukey HSD test indicated significant differences between the MAWLs of Indonesian and American (p < .05). The average MAWL of Indonesian females was 23% lower than the average MAWL of American females. On the other hand, although the average MAWL of Indonesian females was 8% higher than the average MAWL of Chinese females. No significant difference was observed between these study populations. Furthermore, the student's t-test revealed no significant differences on MAWL at 4 lifts/min between Indonesian and American (p = .6749).

Table 2. Means, standard deviations, and percentile values of the MAWL for the two lifting frequencies

Frequency	Mean	SD	5th	50th	95th
1 lift/5 min (N = 21)	12.67	1.06	11	12	14
4 lifts/min (N = 21)	11.14	1.28	9	11	13

Table 3. Characteristics of comparison samples from other published studies

Item	Present study	[22]	[24]
Subjects	Female students (n = 21)	Female industrial workers (n = 10)	Female students (n = 12)
Nationality	Indonesian	American	Chinese
Box size (cm)	40 × 30 × 30	56.2 × 33.4 × 16.0	45 × 46 × 24
Age (years)	19.9(1.26)	42.9(11.2)	21.2(1.6)
Weight (kg)	52.09(5.85)	69.6(10.8)	51.5(6.1)
Height (cm)	156.9(5.0)	161.8(3.8)	160.1(7.1)
MAWL at 1 lift/5 min (kg)	12.67(1.06)	16.4(6.2)	11.69(0.14)
MAWL at 4 lifts/min (kg)	11.14(1.28)	11.6(4.7)	N/A

Table 4. Comparison of the present study's MAWL to other study populations

Frequency	Indonesian vs American	Indonesian vs Chinese
1 lift/5 min	0.0075*	0.6484
4 lifts/min	0.6749	N/A

* Significant at p < .05

3.3 Analysis of Physiological Responses for the Obtained MAWLs

The participants' energy expenditures for the two lifting frequencies are shown in Fig. 2, while the results of one sample t-tests are presented in Table 5. The energy expenditure of the lifting tasks at 1 lift/5 min was statistically significantly lower than the energy expenditure limit, t(20) = 2.806, p = 0.011. On the contrary, the energy expenditure of lifting at 4 lifts/min was statistically significantly higher than the limit, t(20) = 30.765, p = < .001.

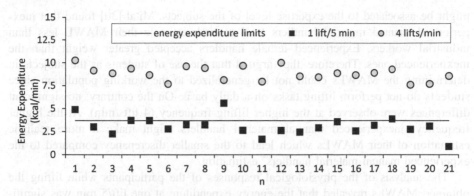

Fig. 2. Energy expenditure for the two lifting frequencies

Table 5. Mean, standard deviations, and summary of the one sample t-tests of the energy expenditure for the two lifting frequencies

Frequency	Mean (kcal/min)	SD (kcal/min)	t-statistics	p-value
1 lift/5 min (N = 21)	3.086	1.003	2.806	0.011*
4 lifts/min (N = 21)	8.808	0.761	30.765	<0.001**

* Significant at $p < .05$
** Significant at $p < .001$

3.4 Discussion

This study found no significant effect of physical activity category on MAWL. Although physical activity categories have a positive correlation to muscle mass [25], other factors besides the muscle mass such as the lifting postures might also influence the lifting capability. Plamondon et al. [26] found that better lifting postures, particularly on the lumbar spine and knees can reduce the external back load which correlate to the ability to lift the higher loads. Therefore, the physical activity category might not be the only factor that influence the lifting capability of a person.

The results of this study were in line with the results of previous studies that the MAWL decreased with the increase in the lifting frequency [24, 27, 28]. Wu [27] found that there was a significant effect of lifting frequency on heart rate. Hence, higher lifting frequency might demand greater energy consumption which influences the subjects in deciding their MAWLs. Also, the higher lifting frequency might make subjects come up with psychological barriers in believing that the lifting tasks were more tiring than the lower frequency. Thus, they tend to limit their self by choosing lower weights for the given lifting tasks [29].

The comparison between the MAWL data of this study and those of other studies showed that the Indonesian subjects in this study had significantly lower MAWLs compared to the American experienced manual material handlers for lifting at one lift/5 min. Also, no significant differences on MAWL were observed between the Indonesian and the Chinese students. The differences on MAWL at the lower frequency might be associated to the expertise level of the subjects. Mital [30] found that inexperienced manual material handlers tend to underestimate their MAWL less than industrial workers. Experienced female handlers accepted greater weight than the inexperienced ones. Therefore, they argued that the use of students as the subjects in determining the MAWLs could not be generalized to the working population since students do not perform lifting tasks on a daily basis. On the contrary, no significant differences were observed at the higher lifting frequency (4 lifts/min). At the higher frequency, inexperienced manual material handlers might make a more realistic estimation of their MAWLs which lead to the smaller discrepancy compared to the experienced manual material handlers' estimation.

The analysis of the physiological responses of the participants while lifting the obtained MAWLs revealed that the energy expenditure at one lift/5 min was significantly lower than the energy expenditure limit used as the physiological criterion during the development of NIOSH lifting equation. On the other hand, those of lifting at 4 lifts/min was significantly greater than the limit. These findings confirm the results of Legg and Myles [31] that there was no evidence of cardiovascular, metabolic or subjective fatigue for the soldiers who lifted and lowered their MAWLs at the low frequency (5 lifts/2 min). Significant differences were observed when the soldiers lifted and lowered higher loads. Although the analysis of the physiological responses has shown that the obtained MAWLs created significantly different energy expenditures compared to the energy expenditure limit, another approach such as biomechanical analysis should be performed. The biomechanical criterion of the NIOSH lifting equation determines that the MMH activities should also not create greater or equal to 3.4 kN compression force at the L5/S1 disc [7].

4 Conclusion

This study was performed to investigate the maximum acceptable weight of lift by Indonesian inexperienced female manual material handlers. As the results, the MAWLs were significantly affected by lifting frequency. On the contrary, no significant differences on MAWL were observed for the three physical activity categories. The results of this study also indicated that the 5th percentile MAWL for the lifting tasks at one lift/5 min was 11 kg. Furthermore, the comparison of MAWLs obtained in this study to other studies revealed that no significant differences were observed between Indonesian and Chinese inexperienced manual material handlers for the lifting tasks at one lift/5 min. On the other hand, Indonesian inexperienced female manual handlers had significantly lower MAWLs than American experienced manual material handlers for the lifting tasks at one lift/5 min. However, no significant differences were observed for the lifting tasks at 4 lifts/min among them. Moreover, the analysis of the physiological responses of the participants while lifting the obtained MAWLs revealed that the energy expenditure for the lifting tasks at one lift/5 min was significantly lower than the NIOSH lifting equation's physiological limit, while those of lifting at 4 lifts/min was significantly higher than the limit.

Acknowledgment. The authors gratefully acknowledge the funding support from Department of Mechanical and Industrial Engineering, Universitas Gadjah Mada on this project. We also thank Dr. Titis Wijayanto for letting us using the heart rate monitor during the data collection process. Also, the authors would like to acknowledge the support of Lembaga Pengelola Dana Pendidikan Republik Indonesia in the form of a Travel Grant, which enabled the author to attend this conference.

References

1. Maiti R (2008) Workload assessment in building construction related activities in India. Appl Ergon 39:754–765
2. Deros BM, Daruis DDI, Basir IM (2015) A study on ergonomic awareness among workers performing manual material handling activities. Proc Soc Behav Sci 195:1666–1673
3. Smedley J, Egger P, Cooper C, Coggon D (1995) Manual handling activities and risk of low back pain in nurses. Occup Environ Med 52:160–163
4. Kuiper JI, Burdorf A, Verbeek JH, Frings-Dresen MH, van der Beek AJ, Viikari-Juntura ER (1999) Epidemiologic evidence on manual materials handling as a risk factor for back disorders: a systematic review. Int J Ind Ergon 24:389–404
5. Dempsey PG, Hashemi L (1999) Analysis of workers' compensation claims associated with manual materials handling. Ergonomics 42:183–195
6. National Institute for Occupational Safety and Health (1981) Work practices guide for manual lifting. US Department of Health and Human Services, Cincinnati, OH
7. Waters TR, Putz-Anderson V, Garg A, Fine LJ (1993) Revised NIOSH equation for the design and evaluation of manual lifting tasks. Ergonomics 36:749–776
8. Snook SH, Ciriello VM (1991) The design of manual handling tasks: revised tables of maximum acceptable weights and forces. Ergonomics 34:1197–1213
9. Authier M, Lortie M, Gagnon M (1996) Manual handling techniques: comparing novices and experts. Int J Ind Ergon 17:419–429
10. Marras WS, Parakkat J, Chany AM, Yang G, Burr D, Lavender SA (2006) Spine loading as a function of lift frequency, exposure duration, and work experience. Clin Biomech 21:345–352. https://doi.org/10.1016/j.clinbiomech.2005.10.004
11. Evans WA (1990) The relationship between isometric strength of Cantonese males and the US NIOSH guide for manual lifting. Appl Ergon 21:135–142
12. Lee KS, Park HS, Chun YH (1996) The validity of the revised NIOSH weight limit in a Korean young male population: a psychophysical approach. Int J Ind Ergon 18:181–186
13. Maiti R, Ray GG (2004) Determination of maximum acceptable weight of lift by adult Indian female workers. Int J Ind Ergon 34:483–495
14. Widodo L, Sukania IW, Kristiani C (2016) Workload analysis of the container unloading process worker. In: Proceeding of 9th international seminar on industrial engineering and management. ER-1-ER-7
15. Huda LN, Matondang R (2018) The lean ergonomics in green design of crude palm oil plant. IOP Conf Ser Mater Sci Eng 309:012109. https://doi.org/10.1088/1757-899X/309/1/012109
16. Deros BM, Daruis DDI, Rosly AL, Aziz IA, Hishamuddin NS, Hamid NHA, Roslin SM (2017) Ergonomic risk assessment of manual material handling at an automotive manufacturing company. PressAcad Proc 5:317–324
17. Shy LH (2008) Ergonomic intervention to reduce the risk of musculoskeletal disorders (MSDS) for manual materials handling tasks. Project report. UTeM Malaysia

18. Armstrong T, Bull F (2006) Development of the world health organization global physical activity questionnaire (GPAQ). J Public Health 14:66–70
19. Karwowski W, Yates JW (1986) Reliability of the psychophysical approach to manual lifting of liquids by females. Ergonomics 29:237–248
20. Ciriello VM, Snook SH (1983) A study of size, distance, height, and frequency effects on manual handling tasks. Hum Factors 25:473–483
21. Ciriello VM, Snook SH, Hughes GJ (1993) Further studies of psychophysically determined maximum acceptable weights and forces. Hum Factors 35:175–186
22. Ciriello VM (2007) The effects of container size, frequency and extended horizontal reach on maximum acceptable weights of lifting for female industrial workers. Appl Ergon 38:1–5
23. Keytel LR, Goedecke JH, Noakes TD, Hiiloskorpi H, Laukkanen R, van der Merwe L, Lambert EV (2005) Prediction of energy expenditure from heart rate monitoring during submaximal exercise. J Sports Sci 23:289–297
24. Wu S-P (1999) Psychophysically determined infrequent lifting capacity of Chinese participants. Ergonomics 42:952–963
25. Elizabeth E, Vitriana V, Defi IR (2016) Correlations between muscle mass, muscle strength, physical performance, and muscle fatigue resistance in community-dwelling elderly subjects. Int J Integr Health Sci 4:32–37
26. Plamondon A, Larivière C, Delisle A, Denis D, Gagnon D (2012) Relative importance of expertise, lifting height and weight lifted on posture and lumbar external loading during a transfer task in manual material handling. Ergonomics 55:87–102. https://doi.org/10.1080/00140139.2011.634031
27. Wu S-P (1997) Maximum acceptable weight of lift by Chinese experienced male manual handlers. Appl Ergon 28:237–244
28. Ciriello VM (2003) The effects of box size, frequency and extended horizontal reach on maximum acceptable weights of lifting. Int J Ind Ergon 32:115–120
29. Snook SH, Ciriello VM (1974) Maximum weights and work loads acceptable to female workers. J Occup Environ Med 16:527–534
30. Mital A (1987) Patterns of differences between the maximum weights of lift acceptable to experienced and inexperienced materials handlers. Ergonomics 30:1137–1147
31. Legg SJ, Myles WS (1985) Metabolic and cardiovascular cost, and perceived effort over an 8 hour day when lifting loads selected by the psychophysical method. Ergonomics 28:337–343

The Effect of the Lower Extremity Posture on Trunk While Sitting

Sangeun Jin[1]($\boxtimes$) (iD), Seulgi Kim[1], and Seong Rok Chang[2]

[1] Pusan National University, Busan 46241, Republic of Korea
sangeunjin@pusan.ac.kr
[2] Pukyong National University, Busan 48513, Republic of Korea

Abstract. The goal of this study was to investigate the interactions between upper extremity and lower extremity in sitting postures. Ten healthy participants were recruited from the university population, and were asked to sit on a chair with six different lower extremity postures (2 trunk-thigh angles, 3 knee angles). The head, trunk and lower extremity postures were captured by using fourteen motion trackers, and used to calculate the head flexion angle, thoracic flexion angle, lumbar flexion angle, pelvic flexion angle, and shoulder angle. Results showed the effects of changes in the trunk-thigh angle and the knee angle on all dependent measures. First, the bigger trunk-thigh angle (135°) showed better lumbar lordosis, smaller head flexion angle, and less rounded shoulder as compared to the 90° trunk-thigh angle. In addition, the current study revealed that the bigger the trunk-thigh angle, the better the whole trunk postures including the better lumbar and thoracic lordosis, less flexed head, and less rounded shoulder. Second, the bigger knee flexion angle negatively influenced on all trunk posture measures, suggesting that the extended knee joint tightens hamstring muscles and pulls the pelvis backward, and finally results in the bad trunk posture such as less lordotic lumbar and thoracic postures, more head flexion angle and more rounded shoulder angle. On this basis, the office chair should be designed under the consideration of the lower extremity postures such as the position of foots, the knee supporter, and the seat inclination angle.

Keywords: Sitting · Lower extremity · Trunk kinematics

1 Introduction

The sitting posture is the most common posture observed in everywhere of our life. A study revealed that three out of four workers are sitting while working [1]. The result may be highly related with an increase of office works in modern society. Previous studies showed that the prolonged sitting and overuse of the smart devices such as PC and smartphone could result in the increase of pain or discomfort symptoms on the lower back, upper back, and neck region [2–5]. An investigation of the ideal sitting postures is worthwhile regarding the increase of back pain patients, and the results should be linked to the development of a proper chair that can induce the correct posture.

© Springer Nature Switzerland AG 2019
S. Bagnara et al. (Eds.): IEA 2018, AISC 820, pp. 179–186, 2019.
https://doi.org/10.1007/978-3-319-96083-8_22

The sitting posture has been of interest in the area of spine biomechanics and have largely contributed to the knowledge of the sitting kinematics and kinetics [6–9]. Previous studies investigated various body postures (e.g., lordortic lumbar, slumped sitting, sitting with crossed legs, thigh angle etc.) and seat designs (e.g., seat height, seat inclination, seat contour, seat width, seat length, seat-back inclination, seat-back lumbar supporter, seat-back height etc.) in both symptomatic and asymptomatic participants [10]. However, those studies are limited in that they concentrated on the lumbar region and neglected the significant interactions between adjoining body parts.

Only some previous studies investigated an effect of hamstring length on the pelvic angle and lumbar lordosis [7, 9]. Keegan [9] was the first researcher captured lumbar spine and lower extremity posture in various sitting postures with the x-ray in lateral side. However, he suggested that the lower extremity postures including feet location and seat tilting angle are least important factors. In 1992, Bridger et al. [7] quantitatively investigated the lumbar and pelvic postures while standing and sitting, focusing on the effect of hip muscle length on the lumbar lordosis. Total 9 standing and sitting postures were observed, and the lumbar lordosis and pelvic tilting angle were measured with an inclinometer. Results showed that the sitting posture lost the lumbar lordosis with a backward tilting of the pelvis, suggesting that the posterior lower extremity muscles played a major role to pull back the pelvis and finally affect on the lumbar posture. Consequently, the significant connection between lower extremity and trunk should be considered as a significant factor when designing an office chair.

Previous literature provided some anatomical evidences for the spine-pelvis-leg system [11–15]. First, the trunk, hip and lower extremity muscles are largely attached to the pelvis and give significant biomechanical influence during trunk flexion and extension [11–13]. Second, the lumbardorsal fascia covering the paravertebral musculature and iliac spine passively links the trunk and lower extremity [14, 15]. Those two linkage are known as an indirectly connection between lower extremity and trunk, and showed biomechanically significant interaction while trunk flexion and extension. In addition to these indirect linkage, the psoas major muscles and its nearby muscles have a direct connection between trunk and lower extremity, originated from the transverse processes of $L1 \sim L5$ and inserted into lesser trochanter of the femur [16]. In summary, ample evidence in functional anatomy for the significant biomechanical linkage from the lower extremity to the trunk suggest the needs of a quantitative measurement of the effect of various lower extremity posture on the trunk postures including lumbar, thoracic, cervical and neck postures. On this basis, the goal of this study was to quantitatively investigate the effect of knee and thigh angle on whole spine and shoulder posture.

2   Method

2.1   Participants

Total ten male subjects participated in the current study from the university population without any chronic back problems and pain (average age 25.0 (SD 1.95) years, height 174.3 (SD 4.9) cm, and weight 66.4 (SD 7.5) kg). The experimental protocol used in the current study was approved by the institutional review board of the Pusan National University, and the written informed consent was provided to each participant before starting the experiment.

2.2 Experimental Design and Protocol

The experimental design included two independent variables such as two trunk-thigh angles (90° and 135°) and three knee angles (45°, 90° and 180°). All independent variables were fully crossed and replicated by 5 times ($2 \times 3 \times 5$ = total 30 trials). During thirty experimental trials, the participants were asked to sit on a chair (no backrest and arm supporter) with six different sitting postures while gazing straight ahead with a bare hand. Each sitting trials continued for 10 s without any request or restriction on the trunk posture. All thirty trials are fully randomized to remove unexpected sequential noise. The head, trunk and lower extremity postures were captured by using a camera-based motion capture system at the sampling frequency of 120 Hz (Model Prime 13, OptiTrack, Naturepoint, OR). Fourteen reflective markers were attached over the skin at C7, T6, T12, L3, L5, S2 spinous process, right and left tragus, right and left acromion, right and left posterior superior iliac spine (PSIS) and right and left fibula. Each marker position was used to calculate the head flexion angle, lumbar flexion angle, pelvic flexion angle, and rounded shoulder angle. Hence, four dependent variables were employed to show the effect of different lower extremity postures.

2.3 Data Processing and Analyses

The raw motion capture data were transferred into the MATLAB (MathWorks, version R2014B) and used to calculate the four dependent variables. The angles were simply the average during the 10 s trial. First, the lumbar flexion angle was calculated by the angle between two lines developed by three markers on T12, L3 and L5. Second, the pelvic flexion angle was calculated by the forward tilting angle of a rigid body, taken from the three markers comprising the markers on S2 and right and left PSIS. Third, the head flexion angle was captured by the forward inclination angle of a rigid body, developed by three markers including markers on C7 and right and left tragus [17]. Fourth, the rounded shoulder angle was calculated by the angle of the arcs between the left acromion-C7 line and the C7-right acromion line. Then, all four dependent variables were standardized by the neutral standing posture, recorded before starting thirty experimental trials.

The processed data were then used for the statistical analyses conducted with SAS® and Minitab®. Model adequacy checking was performed before performing ANOVA and the post-hoc test using the Bonferroni method. A p-value less than 0.05 was the standard level for significance.

3 Results

The ANOVAs showed significant effects of trunk angle and knee angle in all five dependent variable (see Table 1, all p-value < 0.001). Also, all four dependent variables showed significant interactions between trunk and knee angle. First, the pelvic flexion angle was tilted to backward while sitting with all six postures (about $22° \sim 37°$), and the observation was more prominent under the 180° knee flexion condition (i.e., straightened knee), especially in the smaller trunk angle (Fig. 1).

Second, the lumbar flexion angle also showed similar trend in which the bigger trunk angle and the smaller knee angle showed better lumbar lordosis. For an example, the sitting posture with 135° trunk angle and 45° knee angle showed most similar level of lumbar lordosis with the standing posture (Fig. 2).

Table 1. Table 1.

Independent variables	Dependent variables			
	Head flexion angle	Rounded shoulder angle	Lumbar flexion angle	Pelvic tilting angle
Trunk angle	p < 0.001	p < 0.001	p < 0.001	p = 0.001
Knee angle	p < 0.001	p < 0.001	p < 0.001	p < 0.001
Trunk × Knee angle	p = 0.015	p = 0.043	p = 0.031	p = 0.046

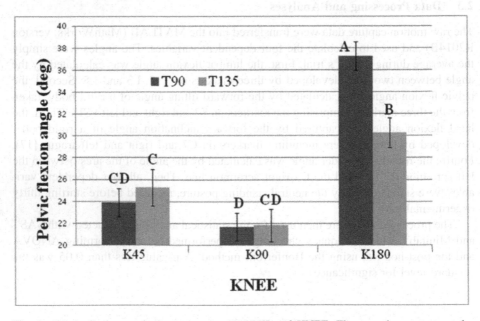

Fig. 1. Pelvic flexion angle for comparing TRUNK and KNEE. The error bars represent the standard error of the sample mean. The same letter indicate that they are not statistically significantly different.

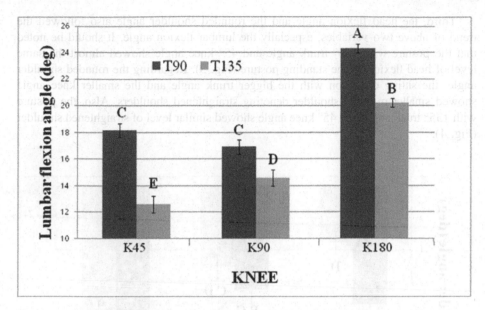

Fig. 2. Lumbar flexion angle for comparing TRUNK and KNEE. The error bars represent the standard error of the sample mean. The same letter indicate that they are not statistically significantly different.

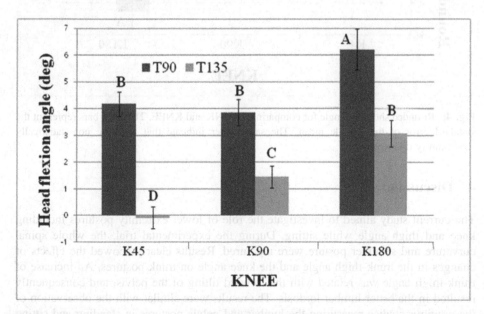

Fig. 3. Head flexion angle for comparing TRUNK and KNEE. The error bars represent the standard error of the sample mean. The same letter indicate that they are not statistically significantly different.

Third, the head flexion angle and the rounded shoulder angle also followed the trend of above two variables, especially the lumbar flexion angle. It should be noted that the posture with 135° trunk angle and 45° knee angle showed almost the same level of head flexion as the standing posture (Fig. 3). Regarding the rounded shoulder angle, the sitting condition with the bigger trunk angle and the smaller knee angle showed smaller rounded shoulder denoting straightened shoulders. Also, the posture with 135° trunk angle and 45° knee angle showed similar level of straightened shoulder (Fig. 4).

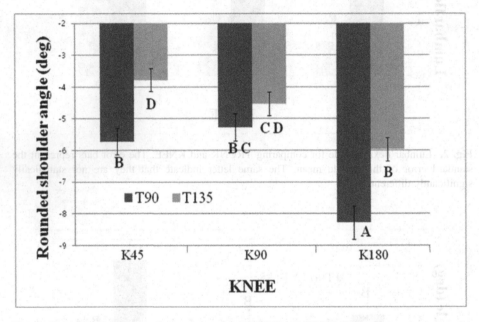

Fig. 4. Rounded shoulder angle for comparing TRUNK and KNEE. The error bars represent the standard error of the sample mean. The same letter indicate that they are not statistically significantly different.

4   Discussion

The current study aimed to investigate the role of lower extremity postures including knee and thigh angle while sitting. During the experimental trial, the whole spinal curvature and shoulder posture were measured. Results clearly showed the effects of changes in the trunk-thigh angle and the knee angle on trunk postures. An increase of trunk-thigh angle was related with the forward tilting of the pelvis, and consequently resulted in the better lumbar lordosis. The results were similar with the observation of the previous studies measuring the lumbar and pelvis postures in standing and sitting [7, 18]. The significant changes in the lumbar region depending on the lower extremity posture suggested a strong interaction and the significant role of the psoas major muscle, lumbardorsal fascia, and other hamstring muscles to control the lumbar posture while sitting. As discussed earlier, those direct and indirect linkage between trunk and lower extremity dominantly control the lumbar lordosis and showed that the bigger the trunk-thigh angle and the smaller the knee angle, the better the lumbar postures which is most similar to the upright posture. It means that the bigger trunk angle may give a chance to have longer hamstring and psoas length (i.e., no passive tension on the muscles), and the smaller knee angle may also provide no pulling force (i.e., passive tension) of the hamstring muscles. All in all, the current study clearly revealed that the lumbar posture is directly influenced by the changes in trunk and knee angle while sitting.

Meanwhile, the current study also revealed the significant changes in the head flexion angle and the rounded shoulder angle according to the lower extremity posture. The trend was exactly same with the lumbar and pelvic flexion angle by confirming better posture (i.e., similar to the upright posture) in the bigger trunk-thigh angle and the smaller knee angle. Intuitively, it may be impossible to have a direct influence of the lower extremity on the head flexion and the rounded shoulder angle. However, the changes in the center of mass to forward while sitting with the smaller knee angle and bigger trunk angle posture could effect on the head and shoulder posture by stretching shoulders and pulling chin. Additionally, the better lumbar lordosis could result in the better thoracic kyphosis and cervical lordosis, and finally induce the best head flexion and rounded shoulder angle. In summary, the changes in the lower extremity posture also significantly provide a guidance on the upper trunk posture including head flexion.

5 Conclusion

All dependent variables denoted the significant influence of the lower extremity posture on the low back, upper back, and head postures. The results confirmed a significant influence of the lower extremity posture on lumbar, thoracic and cervical vertebrae while sitting. On this basis, the office chair should be designed under the consideration of the lower extremity postures such as the position of foots, the knee supporter, and the seat inclination angle.

Acknowledgement. This research was supported by the Basic Science Research Program through the National Research Foundation (NRF) of Korea funded by the Ministry of Education (NRF-2017R1D1A1B03035470).

References

1. Callaghan JP, McGill SM (2001) Low back joint loading and kinematics during standing and unsupported sitting. Ergonomics 44:280–294
2. Karlqvist L, Tornqvist EW, Hagberg M, Hagman M, Tommingas A (2002) Self-reported working conditions of VDU operators and associations with musculoskeletal symptoms: a cross-sectional study focussing on gender differences. Int J Ind Ergon 30(4–5):277–294
3. Marcus M, Gerr F, Monteilh C, Ortiz DJ, Gentry E, Cohen S, Edwards A, Ensor C, Kleinbaum D (2002) A prospective study of computer users: II. Postural risk factors for musculoskeletal symptoms and disorders. Am J Ind Med 41: 236–249

4. Demunter C (2005) The digital divide in Europe. http://epp.eurostat.cec.eu.int/
5. Berolo S, Wells RP, Amick BC (2011) Musculoskeletal symptoms among mobile hand-held device users and their relationship to device use: a preliminary study in a Canadian university population. Appl Ergon 42(2):371–378
6. Bendix T, Biering-Sorenson F (1983) Posture of the trunk when sitting on forward inclining seats. Scand J Rehabil Med 15(4):197–203
7. Bridger RS, Orkin D, Henneberg M (1992) A quantitative investigation of lumbar and pelvic postures in standing and sitting: interrelationships with body position and hip muscle length. Int J Ind Ergon 9:235–244
8. Curran M, O'Sullivan L, Dankaerts W, O'Sullivan K (2015) Does using a chair backrest or reducing seated hip flexion influence trunk muscle activity and discomfort? A systematic review. Hum Factors 57(7):1115–1148
9. Keegan JJ (1953) Alterations of the lumbar curve related to posture and seating. J Bone Joint Surg 35(3):589–603
10. Harrison DD, Harrison SO, Croft AC, Harrison DE, Troyanovich SJ (1999) Sitting biomechanics part I: review of the literature. J Manipulative Physiol Ther 22(9):594–608
11. Cailliet R, Helberg LA (1981) Organic musculoskeletal back disorders. In: Stolov WC, Clowers MR, (eds) Handbook of severe disability. U.S. Department of Education, RAS
12. Leinonen V, Kankaanpaa M, Airaksinen O, Hanninen O (2000) Back and hip extensor activities during trunk flexion/extension: effects of low back pain and regabilitation. Arch Phys Med Rehabil 81(1): 32–37
13. Sihvonen T (1997) Flexion relaxation of the hamstring muscles during lumbar-pelvic rhythm. Arch Phys Med Rehabil 78(5):486–490
14. Snijders CJ, Vleeming A, Stoeckart R (1993) Transfer of lumbosacral load to iliac bones and legs: part 2: loading of the sacroiliac joints when lifting in a stooped posture. Clin Biomech 8 (6):295–301
15. Vleeming A, Pool-Goudzwaard AL, Stoeckart R, van Wingerden JP, Snijders CJ (1995) The posterior layer of the thoracolumbar fascia. its function in load transfer from spine to legs. Spine 20(7):753–758
16. https://en.wikipedia.org/wiki/Psoas_major_muscle. Accessed 26 Apr 2018
17. Psihogios JP, Sommerich CM, Mirka G, Moon SD (2001) A field evaluation of monitor placement effects in VDT users. Appl Ergon 32(4):131–325
18. Eklund J, Liew M (1991) Evaluation of seating : the influence of hip and knee angles on spinal posture. Int J Ind Ergon 8(1):67–73

Occupational Diseases of the Musculoskeletal System – The Situation in Germany

Elke Ochsmann[✉]

Luebeck University and University Hospital, 23563 Luebeck, Germany
elke.ochsmann@uksh.de

Abstract. The reporting and acceptance of a musculoskeletal occupational disease in Germany is based on an extensive procedure.

An admissible occupational disease has to be named on the list of occupational diseases defined by a medical specialist committee of the Federal Ministry of Labour and Social Affairs. This committee defines the scientific guidelines for reporting doctors. For most of the occupational diseases these guidelines include information about occupational exposures or defined occupational groups. For the acceptance of an occupational disease, more elaborate criteria have to be fulfilled.

In Germany, there are currently fourteen occupational musculoskeletal diseases. Three more occupational diseases are currently being discussed by the medical specialist committee. Disc related diseases of the lumbar spine are the occupational diseases for which there are the most reports. The scientific guidelines include tables with loads and forces. The criteria for acceptance include duration of exposure, frequency of exposure, and kind of exposure (heavy loads, trunk inclination). Only few of the initially reported cases are finally compensated.

The identification and compensation of occupational diseases in Germany started as soon as 1885 and developed into a complex process. New input by international comparisons will be beneficial for the future discussion.

Keywords: Occupational disease · Musculoskeletal disorders · Germany

1 Introduction

Any medical doctor or any employer in Germany has to report a disease, to either the relevant Social Accident Insurance or one of the institutions of the Governmental Employment Protection, if he or she suspects that this disease is an occupational disease, Employees or Statutory Health Insurances may also to report a suspected occupational disease. Scientific guidelines are available for reporting doctors or employers to verify the relevance of the suspected occupational disease. The report of a suspected occupational disease sets in motion an extensive process of verification, involving players from Governmental institutions and Social Accident Insurances, and leading to the acceptance or rebuttal of an occupational disease.

© Springer Nature Switzerland AG 2019
S. Bagnara et al. (Eds.): IEA 2018, AISC 820, pp. 187–190, 2019.
https://doi.org/10.1007/978-3-319-96083-8_23

2 Methods

An occupational disease stands in a medico-legal context and only comes into effect by being added to the list of occupational diseases defined by a medical expert committee of the Federal Ministry of Labour and Social Affairs. Legally speaking, the term "occupational diseases" refers to diseases mentioned in Annex 1 of the Ordinance of Occupational Diseases ("Berufskrankheitenverordnung") that insured persons suffer from due to an activity warranting insurance protection in accordance with section 2, 3 or 6 of Book Seven of the German Social Code (SGB VII). Since 1885, statutory accident prevention has served to protect workers in Germany from work-related risks that are directly related to their insured activity [1].

Relevant exposure criteria are pre-defined in the scientific guidelines by this medical expert committee. In some cases, these criteria are refined by other expert commissions, e.g. of the Social Accident Insurances. This more detailed information is used e.g. for the appraisal of the occupational disease by (independent, medical) experts (e.g. [2]). For most of the occupational diseases the guidelines or criteria or recommendations include information about occupational exposures or defined occupational groups, which are at least necessary to nurture the suspicion that the respective disease has an occupational background.

The process to an acceptance or admission of an occupational disease starts with the notification about the disease of the Social Accident Insurance or the Statutory Occupational Safety and Health Inspectorate, in most cases by the attending physician. Theses two institutions inform each other about the notification. The Social Accident Insurance then starts the verification process. Health and safety officers are requested to judge occupational exposures in the individual case and examine the relevance of these exposures for the relevant occupational disease. These judgments are sometimes difficult as in many cases there were only scarce documentations of exposure measurements in the past, which can be relied on for these individual appraisals. If the relevant occupational exposures can be verified, the verification of the notified disease and of causality between exposure and disease is judged by medical experts of the Social Accident Insurance, sometimes independent experts are consulted. The case is then presented to the Statutory Occupational Safety and Health Inspectorate and re-evaluated there. The final decision about the acceptance of the occupational disease is made by the Social Accident Insurance.

For compensatory reasons, in Germany, there is a strong interest in the strict differentiation between "occupational diseases" and "diseases of the general population". While the former are managed by the Statutory Accident Insurances, the latter are managed mostly by the Statutory Health Insurances.

3 Results

There are currently twelve occupational musculoskeletal diseases in the Annex of the German Ordinance of Occupational Diseases which can be reported by doctors or employers (Table 1) [3]. Three more occupational diseases are currently being discussed by the medical specialist committee of the Federal Ministry of Labour and

Social Affairs. Table 1 also shows the reporting of suspected cases of occupational diseases.

Table 1. Occupational diseases of the musculoskeletal system in Germany [1, 3]

No. of the occupational disease on the legal list	Rough description	No. of reported cases in 2016
2101	Diseases of the tendon sheath or tendon or muscle insert	688
2102	Diseases of the meniscus	1003
2103	Diseases caused by working with vibrating tools	420
2105	Chronic diseases of a bursa	344
2106	Diseases caused by pressure on nerves	71
2107	Avulsion fracture of vertebral body	1
2108	Disc related diseases of the lumbar spine caused by heavy lifting or carrying or working in awkward postures	4759
2109	Disc related diseases of the cervical spine caused by heavy lifting on one shoulder	692
2110	Disc related diseases of the lumbar spine caused by whole body vibration	158
2112	Osteoarthrosis of the knee caused by long kneeling	1385
2113	Disease caused by pressure on nervus medianus in the Carpal tunnel	1009
2114	Thenar- and Hypothenar-hammer syndrome	48

Disc related diseases of the lumbar spine because of carrying heavy loads are the occupational diseases for which there are the most reports in the musculoskeletal area (approximately 12.500 cases were reported in 2000, approximately 4.800 cases were reported in 2016). In 2016, 443 of these were accepted and 275 of these accepted occupational diseases were compensated by the Social Accident Insurance. In the following paragraph, the criteria for acceptance of disc related diseases are depicted exemplarily.

For acceptance, the existence of the occupational disease (which equates to the ICD-codes M42, M47.9 or M51) has to be medically verified. In order to standardize this process, medical assessment criteria were defined. These refer to diagnostic criteria like location of disc lesions, radiological alterations, and pain intensity. Competing diagnosis have to be considered. Apart from this medical verification process, the occupational exposure requirements have to be fulfilled. These include at least 10 years of lifting or carrying heavy loads or working in extreme trunk flexion. Heavy loads for men are defined as at least 15 kg (age: 15–17 years), 25 kg (age 18–39 years) and 20 kg (age: 40 years and more). Suspected weights are lower for women. These weights have to be carried or lifted regularly during most work shifts. Extreme trunk flexion is defined as working in workspaces which are lower than 100 cm or working in a posture which is flexed from the upright position by 90° and more. Typical jobs where these criteria apply are weight carriers (transport sector), personnel in the health care sector (nursing personnel) and persons working in deep mines or working in the construction sector (e.g. as brick layers/masons). Finally, the causality between occupational exposure and medical diagnosis has to be discussed, regarding competing factors. To do so, the age of onset of the disease is taken into account in order to differentiate between disc related diseases of the general population and occupational disc related diseases. The severity of the disease at a certain age is considered. Also, the distribution of lesions is taken into account.

For the case of disc related diseases, there exists a special feature, namely that the occupation has to be given up, so that the occupational disease can be accepted and compensated.

All in all, the process of acceptance of the occupational disease is rather complex; relatively few of the initially suspected occupational diseases are finally being accepted, even less are compensated.

4 Conclusion

The identification and compensation of occupational diseases in Germany started as soon as 1885 and developed into a complex process. New input by international comparisons will be beneficial for the future discussion.

References

1. BAuA (Federal Institute for Occupational Safety and Health). https://www.baua.de/EN/Service/Legislative-texts-and-technical-rules/Legislative-texts-and-technical-rules_node.html. Accessed 30 May 2018
2. Bolm-Audorff U, Brandenburg S, Brüning T et al (2005) Medizinische Beur teilungskriterien zu bandscheibenbedingten Berufskrankheiten der Lenden wirbelsaule (I) - Konsensempfehlungen zur Zusammenhangsbegutachtung der auf Anregung des HVBG eingerichteten interdisziplinären Arbeitsgruppe, vol 7. Trauma Berufskrankh, pp 211–252
3. DGUV   (2017).   http://www.dguv.de/medien/inhalt/mediencenter/dguv-newsletter/2017/bilder-okt/dguv_statistiken_deutsch_web.pdf. Accessed 30 May 2018

Motor Control with Assistive Force During Isometric Elbow Flexion

Satoshi Muraki[1]([✉]), Keisuke Hayashi[2], Nursalbiah Nasir[3],
and Ping Yeap Loh[1]

[1] Faculty of Design, Kyushu University, Fukuoka, Japan
muraki@design.kyushu-u.ac.jp
[2] Graduate School of Design, Kyushu University, Fukuoka, Japan
[3] Faculty of Mechanical Engineering, Universiti Teknologi MARA,
Shah Alam, Malaysia

Abstract. In the modern society, an assistive/powered suit has been developed to enhance the limb and trunk movements by mechanical force. The effective output of assistive products needs cooperation between the users, that is, human beings and the machine. The present study investigated the motor control of external forces that assist with physical exertion. Sixteen adult male participants performed isometric elbow flexion under two conditions of submaximal workload (20% and 40% of the maximal voluntary contraction) and four levels (0%, 33%, 67%, and 100%) of assistive force. The electromyographic (EMG) activity of the agonist and antagonist muscles (biceps and triceps, respectively) and rating of perceived exertion decreased with increased levels of assistive force under both workload conditions. At the lower level of assistance (33%), the EMG amplitude of the biceps was near the expected amplitude, which denotes that the participants made good use of the assistive force. However, at the higher level of assistance (100%), it was far from the expected values at both workload levels. These results suggest that the effectiveness of assistive force changes according to the level of workload and assistive force, and that various human physiological regulations and motor control would be required during cooperative work with assistive force.

Keywords: Assistive technology · Electromyogram · Force release

1 Introduction

In the modern society, the development of assistive technology has advanced rapidly. For example, an assistive/powered suit has been developed to enhance the limb and trunk movements by mechanical force. Such products will be used widely in human daily living and manual working in the near future. On the other hand, the effective output of assistive products needs cooperation between the users, that is, human beings and the machine. Users have to release their force skillfully to use the mechanical force from the machine. In addition, combinations of the level of human exertion and machine assistance greatly vary to properly control human motion.

© Springer Nature Switzerland AG 2019
S. Bagnara et al. (Eds.): IEA 2018, AISC 820, pp. 191–194, 2019.
https://doi.org/10.1007/978-3-319-96083-8_24

With this background, our previous study [1] produced simple experimental devices to provide workload and assistive force, and investigated muscle activity against external assistive force at various force levels while a voluntary muscle exercise was performed at various workloads. The results suggested that although the assistive force during isometric muscle contraction relieves the exertion of the agonist muscle that accompanies the decrease in perceived exertion, their assistive effects are influenced by various human physiological and anatomical factors.

The present study used different tasks of lifting, number of conditions for the assistance level, and period of assistance, as compared with those in our previous study [1], and investigated the motor control of external forces that assist with physical exertion in more detail.

2   Methods

2.1   Participants

Sixteen right-handed young adult men (age: 22.1 ± 1.1 years) participated in this study. All the participants were right-handed and had no current physical disorders. Before the experiments, informed consent was obtained from all the participants. This study was approved by the ethics committee of the Faculty of Design, Kyushu University.

2.2   Experiment Setup

The experiment setup was designed as shown in Fig. 1. This setup was modified from that used in the study of Nasir et al. [1]. The participants sat on a chair with their body straightened, while the right arm was positioned at 90° elbow flexion and the palm was placed in the supine position. A strap was secured at the wrist to connect the forearm to the tension sensor (Takei Scientific Instruments Co., T.K.K. 1269f) that acted as workload and tension measurement. The tension value was displayed to the participant through the monitor.

Another strap was firmly secured at the middle of the length of the radius to support the forearm and connect the forearm to the assist load at the other end of the fixed pulley system.

2.3   Experiment Condition and Procedure

The participants performed isometric elbow flexion under two conditions of submaximal workload (20% and 40% of the maximal voluntary contraction) and four levels of assistive force (0%, 33%, 67%, and 100%, as theoretically expected) for 30 s (10 s without assistive force and 20 s thereafter with assistive force). Each participant performed eight trials that consisted of the above-mentioned workload and assistive force in counterbalanced order.

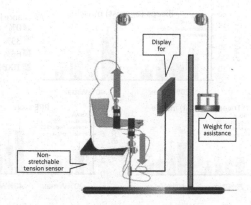

Fig. 1. Experimental setup for isometric elbow flexion and application of assistive force.

2.4 Measurement

During the task, the surface electromyography (EMG) activity of the agonist and antagonist muscles (biceps and triceps, respectively), measured using an analog-to-digital converter with a bio-amp (ADInstrument, PowerLab 16/30, ML880, Australia); the tension of the elbow flexion, measured using above-mentioned tension sensor; and the rating of perceived exertion (RPE, Borg's CR-10) were assessed. The mean values of the rectified EMG (mEMG) before and during the assist load application periods were calculated and normalized as a percentage of mEMG at maximal voluntary contraction ($\%MVC_{EMG}$).

2.5 Statistical Analysis

Two-way repeated-measures analysis of variance (2×4 factorial) was conducted to evaluate the influences of the levels of workload and assist load. A post hoc pairwise Bonferroni-corrected comparison was used to examine the mean differences in the factors. Statistical significance was accepted at p values of <0.05.

3 Results and Discussion

As the tension of knee flexion and mEMG were attained at steady state after a few seconds of assistance, the $\%MVC_{EMG}$ during the latter half of assistance (10 s) was obtained. The $\%MVC_{EMG}$ of the agonist and antagonist muscles decreased with increased levels of assistive force under both workload conditions (Fig. 2). In addition, reduction of RPE was almost consistent with that of $\%MVC_{EMG}$ of the agonist muscle. Some previous studies reported a significant reducing effect of the assistive device on muscle efforts using EMG [2–4]. The results of the present study also supported the reducing effect on muscle effort and subjective exertion.

At the lower level of assistance (33%), %MVC decreased by 23.7% at 20% workload and by 28.6% at 40% workload, as compared with that without assistance,

194   S. Muraki et al.

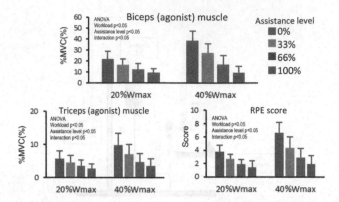

Fig. 2. %MVC$_{EMG}$ in the biceps and triceps for 10 s (20th–30th s) during assistance and RPE score

which was not far from the expected reduction rate (33%). This result implies that the participants made use of the assistive force well, that is, that the effect of assistance on reducing muscle efforts are relatively high. On the other hand, at the higher level of assistance (100%), although %MVC$_{EMG}$ was expected to be near 0%, it was far from the expected values. This phenomenon was observed at both workload levels. These results suggest that some unique motor control against assistive force occurred, especially at higher assistance level.

In conclusion, assistive force relieved the exertion in the agonist and antagonist muscles during isometric elbow flexion, although various human physiological regulations and motor control would be required during cooperative work with assistive force.

Acknowledgment. This work was supported by JSPS KAKENHI (Grant No. JP 15K14619 and 17H01454).

References

1. Nasir N, Hayashi K, Loh PY, Muraki S (2017) The effect of assistive force on the agonist and antagonist muscles in elbow flexion. Mov Health Exerc 6(2):35–52
2. Bellar D, LeBlanc N, Judge LW (2014) Examination of muscle activity with an elastic hamstring assistance device. Health Res Rev 1(1):20–24
3. Gams A, Petric T, Debevec T, Babic J (2013) Effects of robotic knee exoskeleton on human energy expenditure. IEEE Trans Biomed Eng 60(6):1636–1644
4. Tang Z, Zhang K, Sun S, Gao Z, Zhang L, Yang Z (2014) An upper-limb power-assist exoskeleton using proportional myoelectric control. Sensors (Basel) 14(4):6677–6694

A Focus on Dynamic Work Rather Than Sit or Stand Postures

David Caple(✉)

David Caple & Associates Pty Ltd., Ivanhoe East, Australia
david@caple.com.au

Abstract. The introduction of sit-to-stand workstations has resulted in the workplace community debating if standing is now the alternative to sitting at work. This debate is not well founded when the benefits of frequent changes in posture and using blood flow to maintain physical and cognitive health is the key outcome required. The term "dynamic work" should be the focus of the debate to encourage workers to change their working postures and activities to generate greater blood flow and dissipate static muscle loading. This should be discussed within a broader health debate in relation to the prevention of cardio vascular and obesity for sedentary workers. This is part of a focus on general health and wellbeing at work. Dynamic work is not about sitting or standing; it includes both of these postures as well as a broader understanding of the job design. A review of workers activities in post occupancy assessments of new office fitouts indicated that approximately 50% of workers to do not adjust their sit-to-stand workstations but leave it at a seated height at all times. These workers explained that their job design requires them frequently to change posture including getting up and walking around. They achieve the benefits of dynamic work through their multi task job design. Approximately 40% indicated that they will use the adjustment mechanism to change the workstation height between sitting and standing during the day however they were less likely to use the adjustment mechanism if it was noisy or slow to adjust. The actual timing and duration of the sitting or standing depended on their physical, cognitive and psychosocial needs of the worker. Findings also revealed that approximately 10% of workers preferred to stand at their workstations all the time. This cohort of people included workers with injuries who indicated that prolonged sitting aggravated their condition. It also included workers who regularly sat in meetings and preferred to stand when they returned to their workstation for a change in posture. There were also injured workers who use the sit-to-stand adaptor units on their fixed height workstations if they were recovering from an injury. Evidently the design of some adaptor units presented MSD risks due to the amount of force and the elevation of the shoulders required to adjust these unit. In relation to training it was found that the introduction of sit-to-stand workstations required specific guidance on how to achieve dynamic work. The training should also develop skills to relieve lower back and lower limb fatigue from prolonged standing. This is particularly important when the hands are in front of the body's centre of gravity to work on a keyboard and mouse on the work surface. Further research is required pre and post relocation to sit-to-stand workstations to assess the health and wellbeing impact on the workers as well as the impact on their productivity and cognitive demands.

© Springer Nature Switzerland AG 2019
S. Bagnara et al. (Eds.): IEA 2018, AISC 820, pp. 195–199, 2019.
https://doi.org/10.1007/978-3-319-96083-8_25

Keywords: Dynamic work · Static load · Physical demands
Sit-to-stand workstations

1 Introduction

Adjustability within office workstations has been provided in Australia since the 1980s. This was introduced as part of the prevention strategy for musculoskeletal disorders associated with computer work. These disorders were known as Repetition Strain Injuries or "RSI". The adjustment mechanisms enabled the user to set the height of the computer keyboard and monitor on their workstation to suit their anthropometry requirements. These workstations were designed for seated applications using a task chair.

During the last 10 years, engineering innovations have enabled cost-effective mechanisms and designs that enable the user to adjust the workstation surface from sitting to standing heights. This includes height adjustable work surfaces that adjust over a range from sitting to standing height, as well as adaptors that can be placed on top of fixed height workstations allowing a smaller range of height adjustments.

The introduction of these user height adjustable workstations was based on the assumption that by providing these features and options, users would achieve greater variation in postures during their working day. Lin et al. (2016) found that users of sit-to-stand workstations did not set the heights of the workstations correctly to suit their anthropometric needs. Robertson et al. (2013) provided training on how to use the sit-to-stand workstations and found that this group had less discomfort than those who were not trained. Bao and Lin (2018) compared different periods between sitting and standing to assess the relative benefits. They found that there was no specific ratio that produced consistent benefits although longer periods of standing produced less muscle fatigue.

2 Objective

To assess if users of adjustable workstations utilise the adjustability mechanism and alternate between sitting and standing postures during their working day.

3 Methodology

1. Observational studies were conducted across 10 large office fit outs in Melbourne and Sydney, Australia. These incorporated new office designs involving banking, insurance, and government services.
2. Interviews and small group discussions were conducted with ergonomics advisors and safety professionals within each workplace. This provided details of the ergonomics training and support provided to staff with specific physical needs.
3. Consultation with the architects, designers and manufacturers of the projects where sit-to-stand workstations where being introduced, was conducted to understand the design principles integrated into their brief.

4 Findings

1. The ratio of sit-to-stand workstations in the fit outs of these large buildings involving between 200 and 10,000 points of work (work stations) ranged from 20% to 100% of the supplied workstations.
2. The introduction of sit-to-stand desktop adaptors onto fixed height workstations was primarily associated with workplace modifications as a result of reported musculoskeletal discomfort, or injuries to specific staff members. The lower cost of the desktop adaptor was the main reason for their use rather than replacing the desk with a sit-to-stand workstation.
3. The company's internal ergonomist and health and safety professionals provided training for the users of these sit-to-stand workstations. This advice generally focussed on the musculoskeletal benefits of alternating posture between sitting and standing on a regular basis during the working day. The principles behind this advice were derived from their understanding that prolonged sitting is a risk factor due to static loading, particularly involving the neck, shoulder, and back musculature resulting in associated discomfort. The educational material provided by these professionals focussed on the biomechanical benefits of standing as an alternative to, or together with sitting, during the working day.
4. The data aggregated from the interviews and observations on those workers sitting and standing as a pattern of work changed during the first three months using these workstations. After initial training the majority of users tried the mechanism at various times to see if they preferred to work in a sitting or standing posture. Where individual support was provided by the trainer to have a go, many tried it for the first time. Without this support they indicated that they were happy to use it seated. Once they had tried it there was an increase in using the workstation when standing. After three months the aggregated data is provided in the following Table 1.

Table 1. Observation outcomes of postures for staff using sit-to-stand adjustable workstations.

Posture	Proportion of staff observed
1. Only sitting with the workstation adjusted to clerical height	50%
2. Only standing	10%
3. Regularly adjusting the workstation during each day between sitting and standing heights	40%

5. Further detailed analysis was obtained about the decisions to use the height adjustment mechanism on the workstation during the consultation sessions. The following table provides details of the reasons given by staff as to why they chose a particular working posture at their workstation (Table 2).

Table 2. Examples of reasons given for choosing sitting or standing heights.

Posture	Reasons for the preferred posture
1. Only use the workstation when seated on a task chair	1.1 "I don't feel any discomfort from working in a seated posture and see no reason to adjust it to standing height" 1.2 "I frequently need to get up from my workstation to walk to other areas of the office as part of my job. Hence, I don't see a need to then stand up at my workstation to relieve any posture discomfort" 1.3 "I find prolonged standing tiring and prefer to sit"
2. I only use my workstation when it is adjusted to standing height	2.1 "I spend most of my day sitting in meetings and like to stand up when I return to my workstation to have a change of posture" 2.2 "I have a musculoskeletal condition which is aggravated by prolonged sitting. By standing at my workstation, I feel less tired and less discomfort during the day" 2.3 "I like to stand up and walk around a lot. The standing workstation suits my working style"
3. Alternate between sitting and standing heights	3.1 "I feel less tired by changing between sitting and standing height during the day" 3.2 "I like to sit in the morning and stand most of the afternoon to relieve posture discomfort and tiredness" 3.3 "I like to stand when doing my emails and making phone calls but sit when I need to concentrate on report writing"

6. The training provided on the postures for the standing workstations by the companies was found to be non-existent or focused on the health benefits of standing rather than sitting. The observation of the standing users indicated that they were placing static load on their back and shoulders as their arms reached forwards to the keyboard and mouse. By training them to position their feet to balance a leaning forward posture this static load was relieved.

5 Conclusion

It was evident from observing users of sit-to-stand workstations and consulting with them about their reasons for using, or not using, the option to change their posture that there are multiple factors involved in their decision. The factors in determining if the mechanism is used include any existing musculoskeletal disorder where the changing of postures assists to relieve discomfort. It also includes their consideration of their range of postures provided by work duties undertaken away from their workstation. If they mainly sit doing other work tasks they are more likely to stand at their workstation. The nature of the task is also relevant in their decision. They are more likely to sit for tasks requiring concentration such as report writing or stand for emails or making phone calls where visual distraction from activity around the work area has less impact on completing their tasks. The introduction of the standing option should be accompanied by training on how to balance the posture when leaning forwards to work at a computer.

The introduction of sit-to-stand workstations provides the user with choice to change their working posture based on their individual needs. There was not one common frequency of when and why the workstation was adjusted.

References

Bao SS, Lin J (2018) An investigation into four different sit-stand workstation use schedules. Ergonomics 61(2):243–254

Lin M, Catalano P, Dennerlein J (2016) A psychophysical protocol to develop ergonomic recommendations for sitting and standing workstations. Hum Factors 58(4):574–585

Robertson M, Ciriello V, Garabet A (2013) Office ergonomics training and a sit-stand workstation: effects on musculoskeletal and visual symptoms and performance of office workers. Appl Ergon 44(1):73–85

Cognitive and Psychosocial Assessment of Sit or Stand Workstations

David Caple[(✉)]

David Caple & Associates Pty Ltd., Ivanhoe East, Australia
david@caple.com.au

Abstract. The biomechanical benefits of changing posture from sitting to standing are evident from the research on prevention of musculoskeletal disorders. However, the decision for workers who choose to sit or stand is also related to the cognitive and psychosocial needs in their work area. Research in office areas who have provided sit-to-stand workstations showed that the choice to stand in a team environment where other team members are sitting is often influenced by the team communication requirements. The workers indicated that maintaining eye contact with each other while discussing issues influenced their decision to adopt a standing position as a group and if all members were sitting then the decision to sit rather than stand was the general outcome. Further, where adjustment was available in both the workstation and chair design, the workers made necessary adjustments to maintain similar eye heights rather than to suit their individual anthropometry needs. Their ability to see and hear each other was assessed as more important from a cognitive and psychosocial perspective in these team settings than anthropometry. In individual settings such as focus rooms or enclosed offices where the worker is alone, they were more likely to adjust their workstation based on their physical requirements. If they spend much of their day in meetings where they are sitting, they use the time back at their workstation to stand whilst processing emails. However if they stand and walk a lot in their work, they choose to sit at the workstation as a break. Furthermore, it was found that the activity undertaken by the worker also influenced their decision to sit or stand. For example, when performing activities requiring moderate levels of concentration such as talking on the phone or writing emails, they were more likely to stand than if they were performing a task requiring focused attention such as typing a report. Thus for tasks requiring greater concentration, their preferred working position was sitting, as they were less likely to be distracted by others moving around the office. It is evident that the cognitive and psychosocial demands of the workplace will influence the utilization of sit-to-stand workstations as much or even more than the physical demands.

Keywords: Cognitive demands · Psychosocial · Workplaces

S. Bagnara et al. (Eds.): IEA 2018, AISC 820, pp. 200–204, 2019.
https://doi.org/10.1007/978-3-319-96083-8_26

1 Objective

The objective of this project was to assess the factors that influenced workers decisions to adjust their workstation to sitting or standing heights. The research was to assess how the physical, cognitive and psychosocial factors in the ergonomics study interacted in the choice of work setting arrangements.

2 Conceptual Framework

Height adjustable workstations have been utilized in Australia since the mid 1980's. The original workstations for seated computer work incorporated a work surface height range from 570 mm to 800 mm to accommodate the anthropometric requirements for the shortest to the tallest workers. In the 1990's the engineering of workstations enabled sit-to-stand adjustable settings. These adjusted between 850 mm and 1200 mm to enable the shortest to the tallest workers to stand and work at a computer. Since the early 2000's the option of adjusting a workstation from sitting at 610 mm through to standing at 1200 mm or higher were introduced. These workstations are adjustable using an electric motor or a hydraulic mechanism. Desktop adaptors that sat on a fixed height workstation were also introduced in the early 2000's. These adaptors enabled the computer to be adjusted from the 720 mm seated workstation height up to 1200 mm for a standing height and were mainly used by staff undergoing rehabilitation who were unable to maintain a sitting position as their primary working posture. Hence the provision of workstation adjustability was mainly to reduce biomechanical load from static sedentary postures and to accommodate the shortest to tallest percentiles of the population.

Robertson et al. (2017) studied the impact of psychosocial risk factors on the training and use of computer workstations. They found that opportunities for flexible work practices and training on workstation use resulted in less discomfort for the group of workers. The benefits of using sit-to-stand workstations to reduce discomfort were also discussed by Hedge and Ray (2004).

The sociotechnical model of safety proposed by Carayon et al. (2015) recognizes the multi-layered approach to understanding safety from the environmental, organizational and work systems perspective. The understanding of why users choose to work in a sitting or standing posture is an example of this system in practice.

3 Technical Approach

This project reviewed a sample of users of adjustable height workstations and asked them why and when they adjusted their workstation. All of the users were from large companies or government offices and had been provided with ergonomics training when the workstations were introduced. This training provided them with the principles of how to adjust their workstation to suit their anthropometric requirements. The training also covered the biomechanical principles to relieving static load discomfort of the shoulders and back whilst undertaking computer work.

Observations were made of the work areas in these large offices and those staff seen to be standing at their workstation were noted for follow up discussions. Focus groups were arranged with those staff members who had been provided with the adjustable workstations and questions in relation to when and how often they adjusted their workstation were discussed.

4 Proof of Findings

An initial review was undertaken in the 1990's when a large Call Centre was refitted with height adjustable workstations. They were fitted with electric motors to provide heights between 570 mm and 800 mm for seated height adjustment only. A post occupancy review of the 100 workers in this Call Centre found that 95% of the workers had their workstation within 20 mm from the previous fixed 720 mm height. This was a shift work environment and the workers "hot desked" between workstations on their respective shifts. All the workers had participated in the ergonomics training when the new workstations were introduced.

The focus groups with the workers identified that their primary determinant in setting their workstation height was their ability to easily see and hear their colleagues in their team. They wanted to set their sitting height to maintain eye contact at a consistent eye height level with their colleagues to assist in easy communications with each other. They identified that their ability to work together as a team was more important to them than their individual anthropometric needs. Thus they traded off the physical ergonomics needs for the psychosocial needs associated with working as a team. It was observed that the shortest workers often had adjusted their chairs to the highest levels to enable them to maintain this similar eye height with their colleagues. They were aware that the option of lowering the chair so their feet rested on the floor and lowering their workstation to provide a supported working posture was included in the ergonomics training but they still preferred to sit higher to be an "equal part of the team".

In reviewing the sit-to-stand workstations that are now provided in office designs the option of adjusting the postures by changing the workstation height regularly during the day is possible. When reviewing why the majority of workers chose to leave their workstation at a sitting height, a number of reasons were identified. These included:

- There were many opportunities during the day to stand up and move around the office as part of their normal work tasks. Hence the "dynamic work" requirement to generate blood flow from movement of the body was achieved through their job design. These workers found that sitting was "more relaxing" and they did not feel the need to also stand at their workstation.
- Workers who need to concentrate on their tasks with high cognitive demands such as data analysis or report preparation found that sitting enabled them to focus on the task. When they were standing at their workstation they felt more distracted by staff members walking past or sitting beside them.

- Workers who spent much of their day sitting in meetings were more likely to demonstrate a standing working position when they returned to their workstation. The different posture was found to relieve their static discomfort from prolonged sitting during meetings. The standing posture was found to suit tasks such as checking emails and talking on the phone, allowing them to move around. They would then move to a seated posture for longer periods of work to avoid prolonged standing.
- Workers who regularly changed between sitting and standing at their workstation during the day mainly included those who had existing injuries and found prolonged sitting and prolonged standing aggravated their condition. The frequency and duration of sitting and standing varied greatly. Some would diligently adjust between sitting and standing at regular intervals during the day. Others would sit most of the morning and stand for part of the afternoon when they felt physically or cognitively fatigued. Feedback from workers on their preferred frequency of sitting or standing indicated that they were not supportive of software driven reminders that send them a reminder message at pre-determined intervals to change their posture. Workers indicated that they would rather "listen to their body" and make this decision based on their individual needs.
- The design of the adjustment mechanism was also an influence on how often workers adjusted their workstation. Those workstations with a crank handle feature was seen as a disincentive to make adjustments. These crank handles often require over 120 rotations of the handle to complete the height adjustment from sit-to-stand and vice versa. The desktop adaptors that required force to grip the adjustment levers at each end of the device to raise or lower the device was also a potential risk factor to aggravate the existing injuries.

Where workers were provided with sit-to-stand workstations in team clusters it was observed that an individual worker was less likely to adjust their workstation to standing height if all their colleagues were at the lower sitting height. They expressed that standing along in the team may have been disruptive to their colleagues when they talked on the telephone and team discussions were compromised if one member was standing.

5   Impact of Findings

On the basis of these findings it is evident that the provision of sit-to-sit or sit-to-stand height adjustable workstations does not result in workers setting the heights based only on their anthropometric requirements. Those workers with injuries or discomfort associated with their working postures were the most likely to use the adjustment to set the workstation height based on their anthropometric needs. The majority of the workers do not use the adjustment mechanism at all and leave the workstation at a standard height around 720 mm. The rationale behind why others adjust their workstation heights between sitting and standing positions is associated with either the cognitive demands of their work tasks and/or the psychosocial requirements working in collaboration with their team colleagues.

The training that is provided on ergonomics when the sit-to-stand workstations are introduced should take a holistic approach and discuss the impact of the physical, cognitive and psychosocial aspects of the work in determining when and how the mechanism is used.

References

Carayon P, Hancock P, Leveson N, Noy I, Sznelwar L, Van Hootegem G (2015) Advancing a sociotechnical systems approach to workplace safety – developing the conceptual framework. Ergonomics 58(4):1–17

Hedge A, Ray EJ (2004) Effects of an electronic height-adjustable work surface on computer worker musculoskeletal discomfort and productivity. In: proceedings of the human factors and ergonomics society 48th annual meeting, New Orleans, LA, pp 1091–1095, September 2004

Robertson MM, Huang YH, Lee J (2017) Improvements in musculoskeletal health and computing behaviours: effects of a macro ergonomics office workplace and training intervention. Appl Ergon 62:182–196

Validation and Comparison of Three Positioning Protocols of Inertial Measurement Units for Measuring Trunk Movement

Liyun Yang[1,2(✉)] 🆔, Dennis Borgström[2] 🆔,
and Mikael Forsman[1,2] 🆔

[1] Division of Ergonomics, KTH Royal Institute of Technology,
141 57 Huddinge, Sweden
liyuny@kth.com

[2] Institute of Environmental Medicine, Karolinska Institutet,
171 77 Stockholm, Sweden

Abstract. Postures and movements of the trunk are of ergonomic concern when evaluating the risks at work. Technical measurement methods can be used for measurements of trunk movements for long duration with high accuracy, and are therefore increasingly used in practice and research. However, currently there is no standardized protocol for the sensor placement for trunk measurement. Three placement protocols of inertial measurement units (IMUs), including placement on C7, T4 and sternum (St), in combination with S1 spinous process, were compared with an optical motion capture (OMC) system. Four subjects performed a movement test including forward to backward bending, sideward bending and twisting of the trunk, and a symmetrical lifting task. Root-mean-square differences (RMSDs) and Pearson's correlation were calculated between the two systems. For the movement tests, the RMSDs of the forward inclination at the 10th, 50th and 90th percentiles from the three IMUs were all smaller than 7.3°. Larger differences were shown for C7 of the sideward inclination at 90th percentile (10.8°). Also for the twisting, larger differences were shown, especially for C7-S1 and T4-S1 (RMSD = 16.5° and 19.8°). For the lifting tests of forward inclination, St had the smallest differences compared to OMC (RMSDs < 4.1°), while slightly larger errors were found for C7 and T4 at the 90th percentile (RMSDs = 8.1° and 8.2°). Different positioning protocols seem to have a slightly different effect on the measurement accuracy of trunk movement. Considerations should be taken when comparing results across studies applying different protocols.

Keywords: Trunk motion · Inertial sensor · Postural assessment

1 Background

Work in non-neutral or awkward trunk postures is a critical risk factor for low back pain (NIOSH 1997; Coenen et al. 2016). Accelerometer (ACC) or inclinometer has been validated and used in several studies for monitoring truck motion (Dahlqvist et al. 2016; Afshari et al. 2014). The disadvantages of using ACC for trunk motion measurement

© Springer Nature Switzerland AG 2019
S. Bagnara et al. (Eds.): IEA 2018, AISC 820, pp. 205–211, 2019.
https://doi.org/10.1007/978-3-319-96083-8_27

include lower accuracy during dynamic motion (Amasay et al. 2009) and inability to detect the trunk rotation. Inertial Measurement Units (IMUs), which integrate accelerometer, gyroscope and magnetometer, offers the opportunity to improve the measurement accuracy and monitor axial rotation of the trunk. However, the magnetometer in the IMU often suffers from magnetic field disturbances created by ferromagnetic objects (Robert-Lachaine et al. 2017), which are commonly found in industrial settings.

With the rapid development of sensor technology and data analytics, an increase of the use of these technical measurement tools for assessing working postures and evaluating work environment has been seen. However, there is a lack of standardized protocol of positioning the sensors on the trunk. Different positions have been used by different research groups, which vary from (i) on the sternal notch and the L5/S1 (Robert-Lachaine et al. 2017; Schall et al. 2015); (ii) on the upper thoracic spine (Straker et al. 2010; Trask et al. 2014), with various description such as "between the media borders of the scapulae" (Wahlström et al. 2016), or on the chest or back at the level of the T6 (Teschke et al. 2009); and (iii) on the level of the C7-T1 spinous process (Dahlqvist et al. 2016; Hansson et al. 2010), or the T1-T2 (Korshoj et al. 2014).

The definition of the trunk angle used in ergonomic and epidemiology studies is vague and various terms have been used to describe the trunk motion, such as flexion, forward bending, inclination, lateral bending, lateral flexion, rotation or twisting. In practice for assessing working postures, the trunk is usually considered as a rigid segment for simplification. When determining the trunk posture, two referencing systems are of relevance: the joint angle of the trunk relative to the pelvis, and the inclination angle of the trunk relative to the gravitational field (Delleman et al. 2004).

In this study, the forward and lateral inclination, which is relative to gravity, and twisting, which is relative to the pelvis, were looked into as they are more commonly used in previous studies. Three positioning protocols of the IMUs were chosen: on the sternum (referred to as "St"), T4, or C7 for measuring the trunk forward and lateral inclination. The relative angle between these three sensors and another sensor on the S1 was calculated for measuring twisting.

The objective of this study was to evaluate these three positioning protocols using IMUs for measuring trunk posture and movement in three dimensions by comparing with an optical motion capture (OMC) system.

2   Methods

Four subjects (2 male and 2 female) were recruited to perform the tests in an optical motion lab, equipped with Elite 2002 (BTS Bioengineering, Italy). Four IMU sensors (LPMS-B2, LP Research, Japan) were used and placed on the sternum, C7, T4 and S1. Reflexive markers were placed on acromion process and greater trochanter on both sides, and on the back at the level of C7, T4, L2 and L5/S1. The subjects were instructed to stand straight in the beginning as a reference posture. A test including maximal movement in three planes were performed in the order of forward to backward bending, return to neutral position, left to right lateral bending, return to neutral position, left to right twisting of the trunk and lastly return to neutral position (example of one subject shown in Fig. 1). Then symmetric lifting test was performed when the subject lifted a box from floor to table and back to floor for three times, following a metronome (20 bpm).

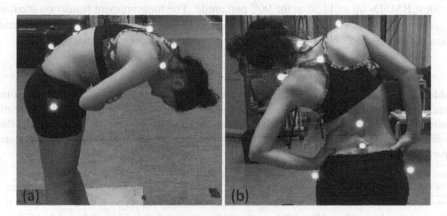

Fig. 1. One subject performing the (a) forward bending and (b) lateral bending in the movement test. Reflexive markers were placed on acromion process, C7, T4, L2, S1 spinous process and greater trochanter; four IMUs were placed on the sternum, C7, T4 and S1 spinous process.

Trunk angle in the aforementioned three dimensions were calculated from the IMUs and OMC. For the OMC, a vector between C7 and S1 was created to calculate the trunk forward inclination and lateral inclination compared to gravity. To calculate the twisting, a vector between both acromion processes and a vector between the greater trochanters were created and compared. For the IMUs, the built-in sensor fusion algorithm was used and the Euler angles from the IMUs on the sternum, C7 and T4 were retrieved to calculate the forward and lateral inclination respectively. For the twisting, the relative angles between the IMUs on the sternum, C7, T4 and S1 were calculated.

Pearson correlation coefficients and root mean square differences (RMSDs) of the selected percentiles (10th, 50th, and 90th) of the three trunk posture angles were calculated between the OMC and the IMUs in three positioning protocols. RMSDs of the median angular velocity of the trunk movements in three dimensions were also calculated.

3 Results

The RMSDs of trunk angles during the movement tests between the OMC in for forward and lateral inclination and IMUs placed on C7, T4, Sternum (St) were presented in Table 1. Besides, the RMSDs of trunk twisting between the OMC and IMUs on C7, T4, and St, relative to the IMU on S1 were also presented in Table 1. For forward inclination, all three IMUs had similar RMSDs compared to the OMC

(RMSDs < 6.7°). While when examining the measured angles along the timeline, the IMU on C7 was found to have a tendency of underestimating the peak value during forward inclination (see Fig. 2). For sideward inclination, the St IMU showed the highest accuracy at all percentiles. For twisting, the IMUs had generally worse accuracy, where the T4-S1 had the largest RMSDs up to 19.8° and the St-S1 had the smallest RMSDs up to 11.3° at the 90th percentile. The measurement results on median angular velocity between OMC and IMUs had generally high accuracy in forward and sideward inclination, while a slightly lower accuracy was found in twisting.

Table 1. The mean root-mean-square differences (RMSDs) at the 10th, 50th and 90th angular percentiles and the median angular velocity, and the Pearson's correlation of the trunk measurement in three dimensions during the movement tests between the OMC and IMUs positioned on C7, T4, Sternum (St), and L5/S1. The criterion values from the OMC are given separately in brackets.

Mean RMSDs from OMC	Forward inclination				Sideward inclination				Twisting			
	C7	T4	St	(OMC)	C7	T4	St	(OMC)	C7-S1	T4-S1	St-S1	(OMC)
10th (°)	1.0	1.3	2.2	(−33.0)	0.6	0.6	0.4	(−35.0)	1.0	2.8	1.3	(−39.4)
50th (°)	3.4	4.3	4.6	(−0.3)	4.6	3.2	2.7	(−0.3)	7.7	8.3	6.4	(1.7)
90th (°)	6.3	7.3	6.7	(76.6)	10.8	7.7	6.1	(34.1)	16.5	19.8	11.3	(39.9)
Median angular velocity (°/s)	0.8	0.3	0.3	(42.1)	0.2	0.2	0.2	(2.9)	3.4	1.2	1.6	(6.5)
Pearson's correlation	0.997	0.999	0.999	-	0.997	0.998	0.998	-	0.960	0.990	0.991	-

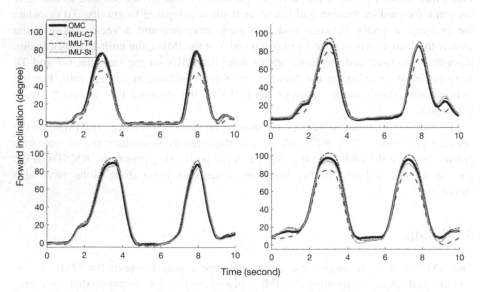

Fig. 2. Examples of the trunk forward inclination measured by the optical motion capture (OMC) system and the inertial measurement units (IMUs) placed on C7, T4 and sternum (St), during lifting tests from four subject.

Table 2. The mean root-mean-square differences (RMSDs) at the 10th, 50th and 90th percentiles and at the median angular velocity, and the Pearson's correlation of the trunk forward inclination measurement between the OMC and the IMUs positioned on C7, T4, and Sternum (St), respectively, during symmetric lifting tests. The criterion values from the OMC are given separately in the most right column.

Percentiles	C7	T4	St	OMC
	Mean RMSDs from OMC			
10th (°)	0.1	0.1	0.1	1.1
50th (°)	0.4	1.0	0.9	11.5
90th (°)	8.1	8.2	5.2	80.5
Median angular velocity (°/s)	5.2	4.4	4.2	52.9
Pearson's correlation	0.992	0.994	0.997	-

During the lifting tests, the IMUs had good accuracy at the 10th and 50th percentile of forward inclination compared to the OMC, with RMSDs < 1.0° for the three positions. Larger errors were found at the 90th angular percentile, especially for the IMU on C7 (RMSD = 8.1°) and T4 (RMSD = 8.2°) (Table 2).

4   Discussion and Conclusion

The different positioning protocols of IMUs had different effects on the measurement of trunk postures and movements compared to the standard measurement.

The underestimation by using IMUs placed on C7 for assessing forward inclination during lifting tests might be caused by the extended neck angle during the task. When examining the video recordings from the tests, it was found that participants usually raised their head during lifting the box from floor to table in order to see forward. This neck posture had an effect on the inclination angle measured by the IMU placed on C7. This effect might be commonly found in real work scenarios when the workers raise or rotate their heads.

Magnetic disturbances were observed in the tests. For postural tests, the twisting angle measured by IMUs placed on Sternum and S1 had the smallest RMSD. However, when examining the data, a drift in the twisting angle was distinct over the measured time. For lifting test, due to the presence of the ferromagnetic material of the table close to the subjects, the magnetometer of the IMU was not able to monitor the transverse rotational movement.

The limitation of this study includes that only a short period of measurement data and few tasks were compared. The number of subjects was also limited. The calculation of trunk angles from the OMC system was based on a simplified model, and combined movements in multiple planes were not included.

The forward and lateral trunk inclinations are more often used by researchers when assessing real work settings. The reasons are partly due to the long tradition of using questionnaires and observational methods for assessing exposure at work, which have reported the inclination angle other than the joint angle of the trunk relative to pelvis.

Existing recommendations and action limits based on these epidemiology studies can be found on trunk inclination, whereas little exists on the joint angle. However, the joint angles are also important dimensions to consider when analyzing the trunk posture and load as it is related to the passive strain of the muscle and other tissues (Delleman et al. 2004). Labaj et al. (2016) reported significant between group differences among daycare workers when assessing the trunk joint angle, but not in trunk inclination. Information might be lost if the analysis was only focused on trunk inclinations. While at the same time, more resources (e.g. two IMUs) are required to monitor the trunk posture of both the joint and inclination angle.

In conclusion, three different positioning protocols of IMUs for the trunk posture measurement were compared with the criterion measurement. Differences in the measurement accuracy were found among them. Hence, consideration should be made when comparing results from different studies when different protocols were used for sensor positioning.

References

Afshari D, Motamedzade M, Salehi R, Soltanian AR (2014) Continuous assessment of back and upper arm postures by long-term inclinometry in carpet weavers. Appl Ergon 45(2):278–284. https://doi.org/10.1016/j.apergo.2013.04.015

Amasay T, Zodrow K, Kincl L, Hess J, Karduna A (2009) Validation of tri-axial accelerometer for the calculation of elevation angles. Int J Ind Ergon 39(5):783–789. https://doi.org/10.1016/j.ergon.2009.03.005

Coenen P, Douwes M, van den Heuvel S, Bosch T (2016) Towards exposure limits for working postures and musculoskeletal symptoms – a prospective cohort study. Ergonomics 59(9):1182–1192. https://doi.org/10.1080/00140139.2015.1130862

Dahlqvist C, Hansson G-Å, Forsman M (2016) Validity of a small low-cost triaxial accelerometer with integrated logger for uncomplicated measurements of postures and movements of head, upper back and upper arms. Appl Ergon 55(July):108–116. https://doi.org/10.1016/j.apergo.2016.01.013

Delleman NJ, Haslegrave CM, Chaffin DB (2004) Working postures and movements: tools foe evaluation and engineering. CRC Press, Boca Raton

Hansson G-Å, Balogh I, Ohlsson K, Granqvist L, Nordander C, Arvidsson I, Åkesson I et al (2010) Physical workload in various types of work: part II. neck, shoulder and upper arm. Int J Ind Ergon 40(3):267–281. https://doi.org/10.1016/j.ergon.2009.11.002

Korshoj M, Skotte JH, Christiansen CS, Mortensen P, Kristiansen J, Hanisch C, Ingebrigtsen J, Holtermann A (2014) Validity of the Acti4 software using ActiGraph GT3X+accelerometer for recording of arm and upper body inclination in simulated work tasks. Ergonomics 57(2):247–253. https://doi.org/10.1080/00140139.2013.869358

Labaj A, Diesbourg T, Dumas G, Plamondon A, Mercheri H, Larue C (2016) Posture and lifting exposures for daycare workers. Int J Ind Ergon 54:83–92. https://doi.org/10.1016/j.ergon.2016.05.003

NIOSH, National Institute for Occupational Health and Safety (1997) Musculoskeletal disorders and workplace factors: a critical review of epidemiologic evidence for work-related musculoskeletal disorders of the neck, upper extremity, and low back

Robert-Lachaine X, Mecheri H, Larue C, Plamondon A (2017) Effect of local magnetic field disturbances on inertial measurement units accuracy. Appl Ergon 63:123–132. https://doi.org/10.1016/j.apergo.2017.04.011

Schall MC, Fethke NB, Chen H, Gerr F (2015) A comparison of instrumentation methods to estimate thoracolumbar motion in field-based occupational studies. Appl Ergon 48 (May):224–231. https://doi.org/10.1016/j.apergo.2014.12.005

Straker L, Campbell A, Coleman J, Ciccarelli M, Dankaerts W (2010) In vivo laboratory validation of the physiometer: a measurement system for long-term recording of posture and movements in the workplace. Ergonomics 53(5):672–684. https://doi.org/10.1080/00140131003671975

Teschke K, Trask C, Johnson P, Chow Y, Village J, Koehoorn M (2009) Measuring posture for epidemiology: comparing inclinometry, observations and self-reports. Ergonomics 52(9): 1067–1078. https://doi.org/10.1080/00140130902912811

Trask C, Mathiassen SE, Wahlström J, Forsman M (2014) Cost-Efficient assessment of biomechanical exposure in occupational groups, exemplified by posture observation and inclinometry. Scand J Work Environ Health 40(3):252–265. https://doi.org/10.5271/sjweh.3416

Wahlström J, Bergsten E, Trask C, Mathiassen SE, Jackson J, Forsman M (2016) Full-shift trunk and upper arm postures and movements among aircraft baggage handlers. Ann Occup Hyg 60(8):977–990. https://doi.org/10.1093/annhyg/mew043

Work-Related Musculoskeletal Disorders and Risk Factors: A Cross-Sectional Study Among Chinese Flight Baggage Handlers

Jingjing Wang[1], Yang Cao[1], Xianning Jin[1], Nazhakaiti Maimaiti[1], Lihua He[1(✉)], Zhongbin Zhang[2], Zhongxu Wang[3], and Wei Zhang[3]

[1] Department of Occupational and Environmental Health, School of Public Health, Peking University Health Science Center, Beijing 100191, China
alihe2009@126.com
[2] China Academy of Safety Science and Technology, Beijing 100012, China
[3] National Institute of Occupational Health and Poison Control, Beijing 100050, China

Abstract. Objective: The aim of this study was to investigate prevalence of work-related musculoskeletal disorders (WMSDs) as well as the contribution of personal and ergonomic factors to the occurrence of low back pain among Chinese flight baggage porters, in order to provide them with valuable suggestions for intervention.

Methods: Cluster sampling was conducted among flight baggage porters in an airport and a self-developed questionnaire was distributed to 550 flight baggage porters to collect information on musculoskeletal symptoms and relevant factors. Prevalence of work-related musculoskeletal disorders in different parts of the body were calculated. Univariate and multivariate logistic regression analysis were carried out to evaluate the influence of individual and ergonomic factors on the occurrence of regional musculoskeletal symptoms in the past 12 months.

Results: The results showed that the prevalence of lower back pain, neck pain, hand/wrist pain and shoulder pain among these porters were 62.7%, 38.1%, 35.2% and 32.4%, respectively. The results of logistic regression suggested that factors as "smoking behavior", "bending amplitude of back", "turn around frequently", "turn and bend at the same time frequently", "exercise after work" and "labor intensity" were statistically correlated with low back pain, whose ORs were 1.699, 0.762, 1.716, 8.267, 3.645, 2.508, respectively.

Conclusion: Airport porters are a group of workers at high risk of WMSDs. Some habitual and postural factors were recognized to be associated with low back pain. It is recommended to reduce labor intensity, train on proper postures, as well as develop good habits to prevent low back pain.

Keywords: Work-related musculoskeletal disorders · Low back pain
Ergonomics · Prevalence · Risk factor

S. Bagnara et al. (Eds.): IEA 2018, AISC 820, pp. 212–218, 2019.
https://doi.org/10.1007/978-3-319-96083-8_28

1 Introduction

In China, there were more than 8566 thousand airlines in 2015, which carried about 915 million passengers and 14 million tons of baggages [1]. These baggages were usually handled by flight baggage porters in the airport. After checked in, the luggage will be placed in conveyor belt, then the conveyor belt transfers it to the sorting area. Subsequently, the luggage will be sorted out and transferred to the truck. Finally, it will be transported to the cabin and be placed in the luggage compartment by the porter piece by piece. In these process, airport porters are required to transfer the heavy baggage frequently and keep sustained periods of stooping, squatting, bending and constant trunk flexion in the constrained compartment. Consequently, the majority of them are exposed to poor ergonomic conditions, which makes them prone to musculoskeletal disorders.

The US Department of Labor defined work-related musculoskeletal disorders (WMSDs) as injuries or disorders of the muscles, nerves, tendons, joints, cartilage and spinal discs associated with exposure to risk factors in the workplace [2]. These disorders are caused by a series of factors, which makes it difficult for researchers to fully elucidate the etiology. Hazards recognized in Bruno's study included several biomechanical, psychosocial and individual factors [3]. Data from the U.S. Bureau of Labor Statistics in 2015 showed that WMSDs were the most important parts of workers' compensation, which accounted for at least one third of labor time losses [4]. These diseases will not only affect the quality of life, but also impose a major economic burden because of compensation and lost wages.

Researchers have emphasized the serious problem of WMSDs in different occupational population [5–7]. However, the prevalence of WMSDs within flight baggage porters has not received sufficient attention, and few studies have focused on the ergonomic issues among Chinese population. Therefore, the aim of this study was to investigate prevalence of WMSDs as well as the contribution of personal and ergonomic factors to the prevalence of low back pain among Chinese flight baggage porters, in order to provide them with valuable suggestions for intervention.

2 Methods

2.1 Instrumentation

The study was a cross-sectional questionnaire survey. The questionnaire was designed by our research group and has been done with reliability and validity test [8]. The information collected included general information, musculoskeletal symptoms and work-related ergonomic factors. In the first domain, information concerning gender, age, vocation, length of employment, Body Mass Index (BMI), education, marriage, monthly income, smoking behavior and drinking behavior etc. was collected. The second domain captured information on musculoskeletal symptoms experienced in the past 7 days or 12 months in four body regions: neck, shoulder, upper back, hand or wrist, which are the most commonly-studied and vulnerable parts. The design of this domain was in accordance with the Standardized Nordic Musculoskeletal

Questionnaire (NMQ) [9]. The domain of work-related ergonomic factors involved items measuring postural factors, psychosocial factors as well as other factors concerning work environment and occupational health. A case in the study refers to anyone who suffered from symptoms of discomfort, numbness, pain or limitation of movement in the musculoskeletal system that occurred at any time during the past 7 days or 12 months, which lasted for at least 24 h and can't get relief after rest.

2.2 Sampling and Data Collection

The survey was carried out among a group of flight baggage porters in an airport. Cluster sampling was conducted in working groups and finally 550 people were investigated. The questionnaire was completed under the guidance of trained investigators and went through strict quality control. Of the 550 questionnaires sent to subjects, 523 were returned and valid, yielding an response rate of 95.1%. Porters who have worked in the airport for at least 1 year were recruited while those who had musculoskeletal injuries caused by sources other than workplace or other diseases which will affect the process of WMSDs were excluded. At last, 383 subjects were included.

2.3 Data Analysis

Data was analyzed using SPSS 22.0 for Windows. Descriptive statistics were performed to calculate the prevalence of musculoskeletal symptoms in each part. For the most affected part, univariate and multivariate logistic regression analysis were carried out to evaluate the influence of individual and ergonomic factors on the occurrence of regional musculoskeletal symptoms in the past 12 months. The stepwise backward removing method was used for variable selection. The inclusion level is 0.05, the elimination level is 0.10 and the significant level is 0.05. Adjusted odd ratios (ORs) with 95% confidence intervals (95% CI) were obtained as measurement of association.

3 Methods

3.1 Demographic Characteristics

Among the subjects, there were luggage trailer driver (n = 85, 19.8%), porters of narrow-bodied planes (n = 218, 50.8%), porters of wide-bodied planes (n = 39, 9.1%) and baggage sorting workers (n = 41, 9.6%). Their average age was 34.5 (SD: 6.5) years old and most of them had a length of employment for less than 3 years (n = 245, 64.0%), which indicated a high level of mobility. Because of particular occupational characteristics, the subjects were all men and their monthly income concentrated in the group of 3000–5000 RMB. 41.0% (n = 176) of them felt that they had good health, 48.3% (n = 207) felt fine about their health, while 10.8% (n = 46) felt their health status were bad or very bad.

3.2 WMSDs Prevalence

According to the descriptive statistics, the most common symptoms appeared in the lower back (n = 240, 62.7%), neck (n = 146, 38.1%), hand or wrist (n = 135, 35.2%) and shoulder (n = 124, 32.4%). Detailed information on different occupations was presented in Table 1.

Table 1. The prevalence of WMSDs in different body regions among flight baggage handlers.

Occupation	N	Region			
		Neck	Shoulder	Hand or wrist	Low back
Luggage trailer drivers	85	30 (35.3)	25 (29.4)	21 (24.7)	47 (55.3)
Porters of narrow-bodied planes	218	82 (37.6)	75 (34.4)	85 (39.0)	154 (70.6)
Porters of wide-bodied planes	39	10 (25.6)	8 (20.5)	6 (15.4)	10 (25.6)
Baggage sorting workers	41	24 (58.5)	16 (39.0)	23 (56.1)	29 (70.7)
Total	383	146 (38.1)	124 (32.4)	135 (35.2)	240 (62.7)

3.3 Risk Factors Analysis

Therefore, low back pain was selected as an outcome variable for univariate logistic regression analysis. In the univariate logistic regression analysis, 13 variables were found to be statistically significant, which were "smoking behavior", "exercise after work", "bending amplitude of back", "turn around frequently", "turn and bend at the same time frequently", "repeat the same posture frequently", "labor intensity", "repetitive work", "job stress", "personnel shortage", "pre-job training", "payment level" and "promotion system", detailed information see Table 2. In order to control confounding, these 13 variables were included in the multivariate logistic regression analysis. Results of risk factors with statistical significance in the multivariate logistic regression analysis were presented in Table 3.

4 Discussion

In our survey, total prevalence of low back pain among different kinds of porters is 62.7%, with the prevalence in porters of narrow-bodied planes and baggage sorting workers being the two highest, which were 70.6% and 70.7%, respectively. This was similar to the reported rate in Eva L. Bergsten's study [10], but rather high compared with many of other vocations [11–13]. Therefore, WMSDs, especially low back pain, among airport porters should be a matter of concern.

Different prevalence among various kinds of porters may be explained by the fact that they operate differently. The common height of narrow body airliner cabin is 123 cm, which is far below the height of the workers, so workers are forced to work in constrained space for long time with long period of bending. And the conveyor belt of the sorting workshop is below the knee, thus the workers need to bent and turned around frequently at work. Therefore, these two kinds of porters are more prone to low

Table 2. Results of univariate logistic regression

Variable	Wald χ^2	P value	OR	95% CI	
				Lower	Upper
Smoking behavior	7.311	0.007*	1.428	1.103	1.848
Exercise after work	6.357	0.012*	0.782	0.646	0.947
Bending amplitude of back	18.382	<0.001*	2.146	1.514	3.042
Lifting weight	0.403	0.526	1.084	0.845	1.389
Turn around frequently	9.297	0.002*	5.900	1.885	18.465
Turn and bend at the same time frequently	17.847	<0.001*	4.442	2.224	8.872
repeat the same posture frequently	10.728	0.001*	2.786	1.509	5.142
Work overtime frequently	1.454	0.228	0.770	0.504	1.177
Regular shift	1.521	0.217	0.676	0.363	1.259
Rest at the prescribed time	0.612	0.434	0.845	0.555	1.288
Labor intensity	18.303	<0.001*	3.128	1.855	5.274
Subjective Working Posture Comfort	1.376	0.241	0.706	0.394	1.263
Space size of working environment	2.241	0.134	0.721	0.470	1.106
Repetitive work	7.287	0.007*	2.555	1.293	5.049
Control of work progress	1.485	0.221	0.770	0.507	1.171
Job stress	8.460	0.004*	1.978	1.249	3.132
Work rhythm	0.009	0.922	1.021	0.674	1.547
Personnel shortage	4.829	0.028*	0.551	0.323	0.938
Get along well with colleagues	0.586	0.444	1.331	0.640	2.766
Sense of social responsibility	2.731	0.098	1.898	0.888	4.059
Pre-job training	4.472	0.034*	2.575	1.072	6.189
Payment level	7.084	0.008*	0.476	0.276	0.822
Promotion system	6.260	0.012*	0.585	0.385	0.890
Temperature change at work	0.083	0.773	0.922	0.533	1.597

*$P < 0.05$, with statistic significance.

Table 3. Results of factors with statistical significance in multivariate logistic regression

Variable	Wald χ^2	P value	OR	95% CI	
				Lower	Upper
Smoking behavior	12.284	<0.001	1.699	1.263	2.286
Exercise after work	6.559	0.010	0.762	0.618	0.938
Bending amplitude of back	7.784	0.005	1.716	1.174	2.507
Turn around frequently	6.620	0.010	8.267	1.654	41.318
Turn and bend at the same time frequently	10.314	0.001	3.645	1.655	8.025
Labor intensity	9.933	0.002	2.508	1.416	4.441

back pain. In contrast, wide-body aircraft carriers use mechanized large luggage containers or conveyor belts to carry luggage, besides the work space is more spacious, so porters of wide-bodied planes are less affected. For luggage trailer drivers, they sit for long time and undertake part of the handling task in narrow body aircraft. Their prevalence was 55.3%.

The results of multivariate logistic regression analysis demonstrated that such postures as "bending amplitude of back", "turn around frequently", "turn and bend at the same time frequently" were associated with low back pain among the porters, which was in agreement with YUAN's study [14]. Adverse posture may produce static load on the body. When a person works in poor posture for long duration, he will need to devote more strength to finishing the same intensity of task, which will in turn increase the muscle loading and compressive stress on the vertebral disc [15].

Airport transportation industry is characterized by its heavy labor intensity. To avoid flight delays, porters are required to load and unload the cargo as quickly as possible, which will increase job stress and workload. According to previous study, workers will develop anxiety and tiredness in a long-term stressful environment, which may lead to the decrease of self recovery and even the psychosomatic diseases. In addition, our study also indicated that lifestyle had an important impact on the occurrence of low back pain. Smokers are 1.699 times more likely to develop LBP than nonsmokers, and those who exercise regularly after work will reduce the risk to 0.762 times than usual. Regular smoking will affect the quality of sleep, then the worker can not get enough recovery at night, thereby reducing work efficiency [16].

Our study was not only an alert for relevant administration, to concern about the problem of WMSDs, especially low back pain, among flight baggage porters, but also provided them with valuable hints on how to prevent such issue. It is meaningful to develop a healthy lifestyle. Less smoking and regular exercise are recommended. More importantly, the management departments are suggested to provide working posture training as well as good working environment and equipment which meet with ergonomic principles. Moreover, it is required to adjust the work organization and schedule and reduce the work intensity.

There are some limitations in the study which should be acknowledged. For example, there is no unified case definition of WMSDs worldwide, which may affect the comparability of results among studies. Besides, cross-sectional study may not be able to prove causality and the questionnaire survey in the study will probably cause recall bias. The aforementioned limitations indicated that our results should be interpreted with caution and further research on the mechanism and progress of WMSDs is warrant.

References

1. Statistics from China Civil Aviation Authority. http://news.carnoc.com/list/340/340998.html
2. Barbe MF, Barr AE (2006) Inflammation and the pathophysiology of work-related musculoskeletal disorders. Brain Behav Immun 20(5):423–429
3. Da CB, Vieira ER (2010) Risk factors for work-related musculoskeletal disorders: a systematic review of recent longitudinal studies. Am J Ind Med 53(3):285

4. The U.S. Bureau of Labor Statistics. https://www.bls.gov/news.release/osh2.nr0.htm
5. Abledu JK, Abledu GK (2012) Multiple logistic regression analysis of predictors of musculoskeletal disorders and disability among bank workers in Kumasi, Ghana. Ergonomics 4(2):556–559
6. Cheng HYK, Cheng CY, Ju YY (2013) Work-related musculoskeletal disorders and ergonomic risk factors in early intervention educators. Appl Ergon 44(1):134–141
7. Roquelaure Y, Ha C, Rouillon C et al (2009) Risk factors for upper-extremity musculoskeletal disorders in the working population. Arthr Rheumatism 61(10):1425–1434
8. Yang C, Wang J, Zhang Wei et al (2017) Assessment of reliability and validity on WMSDs questionnaire applied in porters. Chin J Ind Med 2:87–93
9. Kuorinka I, Jonsson B, Kilbom A, Vinterberg H, Rensen FB (1987) Standardised nordic questionnaires for the analysis of musculoskeletal symptoms. Appl Ergon 18(3):233–237
10. Bergsten EL, Mathiassen SE, Vingård E (2015) Psychosocial work factors and musculoskeletal pain: a cross-sectional study among swedish flight baggage handlers. Biomed Res Int 2015(1):798042
11. Min GK, Kim KS, Ryoo JH et al (2013) Relationship between occupational stress and work-related musculoskeletal disorders in korean male firefighters. Ann Occup Environ Med 25 (1):9
12. Chen ZL, Zhao Y, Gui-Ling YI et al (2016) Survey on the prevalence of musculoskeletal disorders among workers in an electronic company. Ind Health Occup Dis
13. Cho TS, Jeon WJ, Lee JG et al (2014) Factors affecting the musculoskeletal symptoms of korean police officers. J Phys Ther Sci 26(6):925
14. Yuan ZW, Cui Y, Xiang-Rong XV et al (2016) Present incidence of WMSDs and postural load in medical staff of obstetrics and gynecology. Chin J Ind Med
15. Anderson CK, Chaffin DB, Herrin GD (1986) A study of lumbosacral orientation under varied static loads. Spine 11(11):456–462
16. Sabanayagam C, Shankar A (2011) The association between active smoking, smokeless tobacco, second-hand smoke exposure and insufficient sleep. Sleep Med 12(1):7–11

Comparison of Lift Use, Perceptions, and Musculoskeletal Symptoms Between Ceiling Lifts and Floor-Based Lifts in Patient Handling

Soo-Jeong Lee$^{(\boxtimes)}$ and David Rempel

University of California, San Francisco, San Francisco, CA 94143, USA
Soo-Jeong.lee@ucsf.edu

Abstract. Lifting equipment can reduce the risk of injury from patient handling, but its use has been far from optimal. This study examined frequency of lift use, perceptions about lift use and injury risk, and musculoskeletal symptoms by the type of available lifts (ceiling lifts vs. floor lifts only). The study analyzed data from a pooled sample of 389 California registered nurses who participated in two cross-sectional surveys in 2013 and 2016. Nurses who performed patient handling tasks and had patient lifting devices were included in the data analysis: 23% had ceiling lifts and 77% had floor lifts only. Lift use was more frequent among nurses with ceiling lifts than nurses with floor lifts only (use $\geq 50\%$ of the time needed: 48% vs. 35%, p = 0.003). Perceptions about lift use were significantly more positive among nurses with ceiling lifts, in regard to safety for workers, safety and comfort for patients, and ease of use, access and storing; however, perceptions about time burden and injury risk were not significantly different. After controlling for survey year, the prevalence of major work-related musculoskeletal symptoms (moderate or severe symptoms that either occurred at least monthly or lasted one week or more) was significantly lower among nurse with ceiling lifts than those with floor lifts only for low back pain (OR = 0.52, 95% CI 0.30 0.89) and shoulder pain (OR = 0.59, 95% CI 0.35–0.99). The findings suggest that ceiling lifts are superior to floor-based lifts in multiple aspects, including better acceptance and use by nurses in patient handling, as well as being associated with reduced musculoskeletal symptoms, particularly in the low back and shoulders.

Keywords: Ceiling lift · Lift equipment · Lift use · Musculoskeletal symptoms Patient handling

1 Background

Patient handing is a major cause of musculoskeletal injuries and symptoms among healthcare workers [1]. Since lifting equipment was introduced as an effective measure to reduce the risk of injury from manual patient handling [2–5], lift use has been stressed as a key component of safe patient handling. However, the use of lifts in practices has been far from optimal [6]. Various types of lifting equipment are currently available. Among those, ceiling lifts were more recently introduced to overcome

© Springer Nature Switzerland AG 2019
S. Bagnara et al. (Eds.): IEA 2018, AISC 820, pp. 219–222, 2019.
https://doi.org/10.1007/978-3-319-96083-8_29

problems with floor-based lifts. Studies have reported that ceiling lifts have biomechanical and biopsychological benefits compared to floor-based lifts in performing patient handling tasks [7–11]. However, little research has examined the level of lift use in patient handling practices between ceiling lifts and floor lifts. The purpose of this study was to compare frequency of lift use, perceptions about lifts and injury risk, and musculoskeletal symptoms by the type of available lifts among nurses in California.

2   Methods

2.1   Sample

This study pooled data from two cross-sectional surveys of California nurses conducted in 2013 and 2016 in order to increase statistical power. The survey samples were selected randomly from lists of registered nurses with an active license from the California Board of Registered Nursing (BRN). Among 526 respondents in the 2013 survey (response rate: 26%) and 592 respondents in the 2016 survey (response rate: 20%), 593 nurses (285 in 2013; 308 in 2016) met the initial inclusion criteria of (1) being currently employed in healthcare settings for at least 3 months and (2) performing patient handling duties. Of those, the final study sample consisted of 389 nurses, excluding the following subjects who: (1) did not have lifting equipment (n = 120 in 2013; n = 69 in 2016); (2) did not provide lift type information (n = 11); and (3) were employed in neonatal units (n = 4).

2.2   Measures

Demographic and job characteristics included age, gender, race/ethnicity, body mass index (BMI), job tenure in nursing, type of workplace, type of unit, job title, and work status. Respondents were asked if they had mechanical lifting equipment on their unit. Among those who answered yes, those who had a ceiling lift (with or without other types of lifts) were classified as "ceiling lifts group" and all others were classified as "floor lifts only group." The frequency of lift use was assessed by the question "When you lift or transfer a physically dependent patient, how often do you use a lifting device?" Perceptions about lifts were asked with seven items (1 = strong disagree to 5 = strongly agree): Using a mechanical lift is (1) easy, (2) comfortable for patient, (3) safe for workers, (4) safe for patients, (5) time-consuming, (6) It is easy to access equipment, and (7) It is easy to store equipment. Risk perception was assessed by asking about the likelihood of having a musculoskeletal injury within a year from patient handling performed using a mechanical lift (1 = extremely unlikely to 6 = extremely likely). For musculoskeletal symptoms, respondents were asked whether they had pain, aching, stiffness, burning, numbness, or tingling in the lower back, neck, shoulders, or hands/wrists in the past 12 months, with subsequent questions on work-relatedness and intensity, duration, and frequency of the symptom [12, 13]. Work-related pain was defined as symptoms caused or made worse by work. Major pain was defined as moderate or severe symptoms that either occurred at least every month or lasted for one week or longer.

2.3 Data Analysis

Chi-square tests were used to compare proportions of categorical variables and Student's t-tests were used to compare means of continuous variables. First, demographic and job characteristics were compared between the 2013 and 2016 survey samples. After pooling the 2013 and 2016 data, sample characteristics and study variable were compared between nurses with ceiling lifts and nurses with floor lifts only. For musculoskeletal symptoms, logistic regressions were performed after adjusting for survey year and produced odds ratios (ORs) and 95% confidence intervals (CIs).

3 Results

The study sample consisted of 389 nurses with the mean age of 45.1 years (SD 12.1) and the mean job tenure in nursing of 16.6 years (SD 12.0). The majority of the sample were female (87.6%) and non-Hispanic White (49.9%) or Asian (30.6%) and worked full time (75.0%) as staff nurses (89.7%) in hospitals (89.2%). There were no significant differences between the 2013 (n = 156) and 2016 (n = 233) samples. In the pooled sample, 91 nurses (23.4%) had ceiling lifts and 298 nurses (76.6%) had floor lifts only.

Lift use was significantly more frequent among nurses with ceiling lifts (p = 0.003): 47.7% of nurses with ceiling lifts used it more than 50% of the time needed while 34.6% among nurses with floor lifts only. Perceptions about lifts were significantly more positive among nurses with ceiling lifts than nurses with floor lifts only, in regard to ease of use, access, and storing and safety for workers and patients (all results p < 0.001) and comfort for patients (p = 0.006). However, perceptions about time burden and risk perception of patient handling injury were not different between the two groups. For musculoskeletal symptoms, nurses with ceiling lifts reported less symptoms in the low back, neck, shoulders, and hands/wrists than nurses with floor lifts only. After controlling for survey year, the difference was significant for major work-related low back pain (OR = 0.52, 95% CI 0.30–0.89), shoulder pain (OR = 0.59, 95% CI 0.36–0.96), work-related shoulder pain (OR = 0.59, 95% CI 0.35–0.99), and major work-related musculoskeletal symptoms (OR = 0.56, 95% CI 0.34–0.91).

4 Discussion

This study compared ceiling lifts and floor lifts in regard to lift use, perceptions about lifts and injury risk, and musculoskeletal symptoms among California nurses and provides supporting evidence of significant benefits of ceiling lifts over floor lifts. Our study demonstrated that lift use was significantly higher among nurses with ceiling lifts than nurses with floor lifts only. This finding can be explained by more positive perceptions about lift use among nurses with ceiling lifts. These nurses rated significantly higher for ease of lift use, access, and storing, patient safety and comfort, and worker safety. Although nurses with ceiling lifts used a lift more frequently, they

222   S.-J. Lee and D. Rempel

perceived it very time consuming as similar as nurses using floor lifts, suggesting the need for interventions to reduce this barrier. Our study findings also suggest that provision of ceiling lifts can contribute to a greater reduction of the risk of work-related musculoskeletal symptoms compared to floor lifts. Particularly we found significant reductions for moderate or severe symptoms in the low back and shoulders that last longer or occurs more frequently. In conclusion, the study findings suggest that ceiling lifts are superior to floor-based lifts in multiple aspects, including better acceptance and use by nurses in patient handling, as well as being associated with reduced work-related musculoskeletal symptoms in the low back and shoulders. Healthcare facilities should make ceiling lifts available for more nurses in order to promote safe patient handling practices and prevent employee injuries.

References

1. Gomaa AE et al (2015) Occupational traumatic injuries among workers in health care facilities - United States, 2012–2014. MMWR Morb Mortal Wkly Rep 64(15):405–410
2. Evanoff B et al (2003) Reduction in injury rates in nursing personnel through introduction of mechanical lifts in the workplace. Am J Ind Med 44(5):451–457
3. Yassi A et al (2001) A randomized controlled trial to prevent patient lift and transfer injuries of health care workers. Spine 26(16):1739–1746
4. Li J, Wolf L, Evanoff B (2004) Use of mechanical patient lifts decreased musculoskeletal symptoms and injuries among health care workers. Inj Prev 10(4):212–216
5. Alamgir H et al (2008) Efficiency of overhead ceiling lifts in reducing musculoskeletal injury among carers working in long-term care institutions. Injury 39(5):570–577
6. Lee SJ, Lee JH (2017) Safe patient handling behaviors and lift use among hospital nurses: a cross-sectional study. Int J Nurs Stud 74:53–60
7. Alamgir H et al (2009) Evaluation of ceiling lifts: transfer time, patient comfort and staff perceptions. Injury 40(9):987–992
8. Rice MS, Woolley SM, Waters TR (2009) Comparison of required operating forces between floor-based and overhead-mounted patient lifting devices. Ergonomics 52(1):112–120
9. Marras WS, Knapik GG, Ferguson S (2009) Lumbar spine forces during manoeuvring of ceiling-based and floor-based patient transfer devices. Ergonomics 52(3):384–397
10. Santaguida PL et al (2005) Comparison of cumulative low back loads of caregivers when transferring patients using overhead and floor mechanical lifting devices. Clin Biomech 20(9):906–916
11. Waters TR et al (2012) Ergonomic assessment of floor-based and overhead lifts. Am J Safe Patient Handl Mov 2(4):119
12. Trinkoff AM et al (2002) Musculoskeletal problems of the neck, shoulder, and back and functional consequences in nurses. Am J Ind Med 41(3):170–178
13. Lee SJ, Lee JH, Gershon RR (2015) Musculoskeletal symptoms in nurses in the early implementation phase of California's safe patient handling legislation. Res Nurs Health 38(3):183–193

An International Survey of Tools and Methods Used by Certified Ergonomics Professionals

Patrick G. Dempsey[1]([⊠]) [iD], Brian D. Lowe[2], and Evan Jones[2]

[1] Pittsburgh Mining Research Division, National Institute for Occupational
Safety and Health (NIOSH), Pittsburgh, PA, USA
pbd8@cdc.gov
[2] Division of Applied Research and Technology, National Institute
for Occupational Safety and Health (NIOSH), Cincinnati, OH, USA

Abstract. A survey of certified ergonomics professionals was conducted in 2017 by NIOSH researchers to understand the types of basic tools, observational methods, and direct measurement methods used by ergonomics professionals. This survey served to update findings from a previous survey of Certified Professional Ergonomists (CPEs) in the United States that was reported in 2005. The 2017 survey was expanded to include ergonomists certified in Australia, Canada, New Zealand, Great Britain, and Ireland. The 2005 survey content was used with the addition of technologies that were not available in 2005 (e.g., mobile devices and mobile applications) and tools likely to be of use by ergonomists outside the U.S. Overall, the participation rate was 34% (405 of 1,192 surveys that reached recipients) which was lower than the response rate of 53% for the 2005 survey. This may have been related to differences in the receipt format (postal versus internet) between the two surveys. The results for U.S. ergonomists were similar across both surveys for the most part, but there were also a number of differences. Differences across country/region were pronounced for some items, but similar for basic tools and several popular assessment tools. Overall, the results suggest that ergonomists gravitate towards inexpensive and efficient tools and methods. The strengths and limitations are discussed with suggestions for future research.

Keywords: Survey · Tools · Professional Ergonomists

1 Introduction

Surprisingly, there is little formal information available in the literature from quantitative or qualitative studies on the equipment, tools, and techniques used by practicing ergonomists on a regular or occasional basis. There are quite a few assessment tools reported in textbooks and journals (e.g., Stanton and Young 1998; Takala et al. 2010). Understanding which equipment, tools, and techniques are used frequently has the potential to provide insight into ergonomics factors such as perceived usability of the tools, types of problems prevalent in workplaces and other environments, as well as the types of equipment that could form the basis of an ergonomics 'toolkit' for practicing ergonomists. Such an analysis is central to the work of many practicing ergonomists

© Springer Nature Switzerland AG 2019
S. Bagnara et al. (Eds.): IEA 2018, AISC 820, pp. 223–230, 2019.
https://doi.org/10.1007/978-3-319-96083-8_30

that analyze various occupations. The findings provide information about the tasks, tools, and equipment that form ergonomists' work systems. In spite of the idea of using ergonomics to optimize ergonomics practice having intuitive appeal, the irony is that little progress has been made understanding the job demands, capabilities, and limitations of ergonomist populations.

Shorrock and Williams (2016) described what they considered the three critical fundamental constraints of ergonomics methods based on their experience: accessibility, usability, and contextual constraints. Accessibility referred to practitioner access to journals, software, and intellectual property. Usability referred to whether tools were useful and usable, which should be a given for tools developed by our discipline (unfortunately, *it is not*). Finally, contextual constraints refer to organizational characteristics and the influence of stakeholders within organizations where ergonomics is practiced. This is multidimensional and includes factors that ultimately lead to the organization believing and approving given methods, and feeling that the results will be relevant and helpful. Although the current survey did not assess these constraints, it is possible to infer, or at least hypothesize about how the results reflect the constraints.

Dempsey et al. (2005) conducted a survey of Certified Professional Ergonomists® (CPEs) certified by the Board of Certification in Professional Ergonomics (BCPE) in the United States to gather information on the types of basic tools, direct and observational measurement techniques, and software used by practitioners. The goal was to use data to support the recommended contents of a toolkit for ergonomists that would reflect the tools used by a broad sample of practitioners. The toolkit was envisioned to include actual tools such as tape measures and cameras, software, direct measurement devices such as instrument tools, and observational tools commonly used in workplaces. The survey was developed within the context of a workers compensation insurance carrier; thus, the survey was inherently biased towards what might be termed industrial or occupational ergonomics as the insurer had considerable business in industrial and service sectors. Overall, the results of that survey suggested that tools used by a high percentage of ergonomists are tools that are accessible and usable, with simpler tools and assessment methods having higher use rates.

In order to examine secular trends in tool use by ergonomists as well as extend the 2005 survey internationally, a revised survey was sent to ergonomics practitioners in the U.S., Canada, Australia, New Zealand, Great Britain, and Ireland via a web survey. Countries that are predominantly English speaking with a professional certification body were considered for inclusion in the survey. The authors hypothesized that there may be geographic variation in tool use, particularly for tools required by regulations or guidance provided by health and safety inspectorates or regulators. Several tools (e.g. Health and Safety Executive's (HSE) Manual Handling Assessment Charts (MAC)) were added to the survey to better reflect tools in use outside the United States. In addition to the previously administered survey questions, respondents were also asked to report use of items not in existence at the time the previous survey was administered, such as smart devices (phones and tablets) and apps developed for use on smart devices.

2   Methods

2.1   Survey

The 2005 survey first had to be updated to reflect changes in technology since the previous survey, as well as include additional tools to reflect the addition of countries beyond the U.S. The authors used an approach similar to that used in Dempsey et al. (2005). Once additional questions to reflect smart devices and associated applications were developed, the authors asked 9 CPEs in the U.S. to serve as reviewers. They were asked to provide feedback about questions, possible missing questions or tools, or any other input they felt was pertinent. Several newer tools were added to reflect their feedback and input, which was generally specific and detailed. The authors also asked several international colleagues to review the survey for other potential additions. Unlike the previous survey administered via postal questionnaire, the current survey was administered using Survey Monkey. The 2005 survey was introduced to the CPEs by a letter mailed through the postal service, and this was sent approximately a week before the survey was mailed. Reminder letters were also mailed. These two functions were performed via e-mail for the current study.

2.2   Survey Recipients

The U.S. BCPE was the first organization contacted to request assistance in the form of a list of e-mail addresses of current CPEs. The BCPE agreed to participate in the survey as in 2005 and provided the list. It should be noted that Certified Human Factors Professionals® and Certified User Experience Professionals® were not included to maintain consistency with the 2005 survey. The authors then contacted the Canadian College for the Certification of Professional Ergonomists (CCCPE), and they agreed to participate subject to requirements of anti-spam laws. Inviting Canadian ergonomics professionals required two steps to accommodate these laws. The CCCPE first had to ask members if they were willing to participate in the survey, and if so, were they able to share the contact information with the investigators. The investigators then invited the ergonomists via e-mail. The remaining participants were recruited using information publicly available on respective websites. The Centre for Registration of European Ergonomists (CREE) website listed registrants in Great Britain and Ireland, and email addresses were obtained from the website (https://www.eurerg.eu/about-ergonomics/) in early 2017. The Human Factors and Ergonomics Society of Australia and the Board for Certification of New Zealand Ergonomists websites provided contact information for Australia and New Zealand, respectively.

Based on the contact information provided by the certifying bodies or obtained as noted above, the investigators sent e-mail invitations to 1,221 individuals, with 29 invitations returned due to invalid e-mail addresses, resulting in a total across countries of 1,192 delivered invitations.

3 Results

3.1 Respondent Characteristics

There were 432 Survey Monkey response entries. Two participants declined partici-pation at the informed consent phase. Fourteen duplicate entries (based on e-mail addresses) were removed. Eleven entries with no responses were also removed. The remaining 405 responses were included in the analysis.

The overall participation rate was calculated as 34% (405/1,192), using the above defined completion numbers and successful invitations as the numerator and denomi-nator, respectively. Participation rates by countries were: Australia/NZ, 43%; USA, 36%; Great Britain/Ireland, 23%; Canada, 22.0%. Australia and New Zealand were combined, as were Great Britain and Ireland, due to the geographic proximity and the small numbers of responses. Graduate degrees (Masters or Doctoral) were held by 87% of respondents.

3.2 Survey Results

Table 1 provides the percentage of ergonomists responding positively to using the basic tools listed. The U.S. results for the current study were somewhat similar to the 2005 findings. In terms of Tablet/Smart Device use, which was a new survey item, the results across regions were similar and in the 50–55% range.

Table 1. Percentage of users reporting use of basic tools.

Basic tool	Dempsey et al. (2005)	Present study (2017)			
	USA (n = 308)	USA (n = 304)	Canada (n = 53)	GB/Ireland (n = 10)	Australia/NZ (n = 38)
Tape measure	95.8%	95.7%	98.1%	100.0%	94.7%
Digital video camera	96.1%	92.1%	92.5%	80.0%	81.6%
Digital still camera	86.7%	90.1%	88.7%	90.0%	92.1%
Stopwatch	88.3%	78.0%	79.2%	90.0%	73.7%
Laptop	81.8%	84.2%	77.4%	80.0%	73.7%
Tablet/smart device	–	55.6%	50.9%	50.0%	50.0%
Light meter	69.8%	59.5%	62.3%	60.0%	73.7%
Sound level meter	54.2%	57.9%	39.6%	40.0%	55.3%
Goniometer (joint angles)	62.3%	64.8%	58.5%	30.0%	44.7%
Spring gauge	72.4%	55.3%	66.0%	50.0%	60.5%
Scale (load cell)	68.8%	73.4%	79.2%	30.0%	50.0%
Slip meter	21.4%	17.1%	0.0%	10.0%	18.4%

Table 2 provides the percentages of users that responded positively to using the observational tools listed. In general, these results were more variable. There was some notable geographic variation, such as higher use of the Health Safety Executive (HSE) tools in GB/Ireland as would be expected. Use of the NIOSH lifting equation was high across countries, as was the use of Rapid Upper Limb Assessment (RULA) and Rapid Entire Body Assessment (REBA). RULA, REBA, and Strain Index all showed large increases in the two sets of U.S. results.

Table 2. Percentage of users reporting use of observational assessment tools.

Assessment tool	Dempsey et al. (2005)	Present study (2017)			
	USA (n = 308)	USA (n = 304)	Canada (n = 53)	GB/Ireland (n = 10)	Australia/NZ (n = 38)
RULA – Rapid Upper Limb Assessment	51.6%	78.6%	86.8%	70.0%	84.2%
REBA – Rapid Entire Body Assessment	17.9%	68.4%	67.9%	70.0%	73.7%
OWAS – Ovako Working Posture Analysis System	21.4%	25.7%	24.5%	30.0%	47.4%
PATH – Posture, Activity, Tools and Handling	9.1%	12.8%	9.4%	0.0%	5.3%
Biomechanical or digital human modelling	73.4%	69.7%	67.9%	30.0%	50.0%
Body Discomfort Map (e.g. Corlett and Bishop Map)	55.5%	61.5%	60.4%	50.0%	65.8%
JCQ – Job Content Questionnaire	29.5%	21.1%	13.2%	30.0%	34.2%
Psychophysical Material Handling Data	73.1%	78.3%	90.6%	30.0%	68.4%
NIOSH Lifting Equation	83.1%	88.5%	88.7%	80.0%	73.7%
Energy Prediction Model	43.5%	33.2%	30.2%	0.0%	28.9%
ACGIH® Threshold Limit Value® (TLV®) for Lifting	–	55.9%	66.0%	0.0%	21.1%
Health Safety Executive (HSE) Manual handling assessment charts (MAC)	–	17.8%	20.8%	80.0%	47.4%
Psychophysical Upper Extremity Data (e.g. "Snook and Ciriello Tables")	37.3%	54.9%	54.7%	30.0%	55.3%
Strain Index	39.3%	63.5%	66.0%	0.0%	34.2%
Concise Exposure Index (OCRA)	–	11.2%	22.6%	10.0%	15.8%
Health Safety Executive (HSE) Assessment of Repetitive Tasks (ART tool)	–	9.9%	9.4%	70.0%	23.7%
Muscle fatigue equations	–	17.8%	17.0%	10.0%	5.3%

Table 3 presents the percentage of respondents reporting use of the direct measurement tools listed. Grip dynamometers, pinch dynamometers and push/pull sensors were amongst the most widely used direct measurement equipment in the 2005 survey, as well as in the 2017 survey for respondents from the U.S. and Canada. Trunk and wrist goniometers and electromyography had a low percentages of users, as was the case for the 2005 survey. These tend to be technically complex and relatively more expensive than the other tools, which may explain the differences.

Table 3. Percentage of users reporting use of direct measurement tools.

Direct measurement tool	Dempsey et al. (2005)	Present study (2017)			
	USA (n = 308)	USA (n = 304)	Canada (n = 53)	GB/Ireland (n = 10)	Australia/NZ (n = 38)
Lumbar motion monitor (LMM)/other trunk electrogoniometer	16.6%	20.4%	1.9%	0.0%	21.1%
Electronic wrist goniometer	18.5%	14.8%	3.8%	10.0%	2.6%
Grip dynamometer	67.2%	68.4%	62.3%	20.0%	52.6%
Pinch dynamometer	52.6%	61.8%	50.9%	0.0%	18.4%
Instrumented hand tools (for force measurement)	31.8%	32.6%	28.3%	10.0%	23.7%
Heart rate monitor	40.9%	38.5%	37.7%	80.0%	52.6%
Push/Pull force sensors	60.7%	59.5%	64.2%	30.0%	57.9%
Electromyography	31.2%	30.6%	17.0%	20.0%	21.1%

4 Discussion

A notable difference between the current survey and the 2005 results is that the overall participation rate was lower in the present survey. Although the non-U.S. responses do not have a comparison group, the U.S. response rate decreased from 53% to 36%. This was disappointing, and the authors believe this may be due to e-mail invitations being used rather than a postal survey. Many professionals are inundated with e-mail, and this may lead to e-mails such as survey invitations not being priorities. A postal survey has the advantage of possibly having a physical presence on a desk as a reminder.

The use of basic tools indicated that still images and video are widely used in all regions, as are tape measures and laptops. In terms of secular trends, the two sets of U.S. data showed similar percentages, although several observational tools noted earlier increased in use (RULA, REBA, and Strain Index). This may be due to increasing popularity or the fact that familiarity increased due to the length of time since publication increased. The most widely used tool in the 2005 study was the NIOSH Lifting Equation, and the reported use among CPEs increased 5%.

In terms of newer technology, the use of "Tablet/Smart Device" was fairly consistent in the 50–55% range across regions. Usage may increase in the future, particularly if more applications become available. Ergonomics is a relatively small profession, and extensive software is not available. However, the proliferation of mobile applications may lead to increased availability of applications that make analyses more efficient.

In terms of geography, the results of U.S. and Canada were, broadly speaking, the most similar of the regions. This is likely due to the proximity and the fact that a number of companies have facilities and personnel in both countries (e.g. automotive manufacturers). A topic for future research is obtaining a better understanding of how ergonomists come to learn and use tools. Formal university courses, colleagues in the same organization, or professional networking are potential avenues.

4.1 Limitations

There are several limitations to note. The first is that the sample sizes for the non-U.S. countries were fairly small, and the response rates for several of these countries were low. The small sample sizes were the result of the number of ergonomics professionals in each of the organizations. The results need to be interpreted with this in mind. That said, there is little information available in the literature on tool use and the data reported. Dempsey et al. (2005) chose to study CPEs so that the results reflected a certain level of credentials. There are likely many individuals that are not credentialed from the bodies cited in Sect. 2.2 using ergonomics tools.

A second limitation is that the results do not provide insight into differences in tool use across industry sectors. Future analyses will include stratifying the responses by industry sector that respondents reported if there are sufficient sample sizes within sectors. For instance, mining and manufacturing have several distinctive characteristics, such as work in restricted postures and highly repetitive upper extremity-intensive work, respectively, that may influence tool use. More detailed sector-specific research may be needed to clarify potential differences in how ergonomists function within sectors.

The final limitation is that, like the 2005 survey, the survey items were slanted towards what is often termed occupational or industrial ergonomics. The survey was largely based on the 2005 version, so this limitation carried forward. Future studies can examine a broader range of tools.

5 Disclaimer

The findings and conclusions in this report are those of the authors and do not necessarily represent the official position of the National Institute for Occupational Safety and Health, Centers for Disease Control and Prevention. Mention of any company or product does not constitute endorsement by NIOSH.

References

Dempsey PG, McGorry RW, Maynard WS (2005) A survey of tools and methods used by certified professional ergonomists. Appl Ergon 36:489–503

Shorrock ST, Williams CA (2016) Human factors and ergonomics methods in practice: three fundamental constraints. Theor Issues Ergon Sci 17:468–482

Stanton N, Young M (1998) Is utility in the mind of the beholder? A study of ergonomics methods. Appl Ergon 29(1):41–54

Takala EP, Pehkonen I, Forsman M, Hansson G-Å, Mathiassen SE, Neumann WP, Sjøgaard G, Veiersted KB, Westgaard RH, Winkel J (2010) Systematic evaluation of observational methods assessing biomechanical exposures at work. Scand J Work Environ Health 36(1): 3–24

Assessment of Muscular Strength for Male and Female Backpacking Task

Shui Cheng Tian[1,2], Ying Chen[1,2(✉)], Kai Way Li[3], and Hong Xia Li[4]

[1] College of Safety Science and Engineering,
Xi'an University of Science and Technology, Xi'an 710054, Shaanxi, China
254004093@qq.com
[2] Key Laboratory of Western Mines and Hazard Prevention,
Ministry of Education of China, Xi'an 710054, Shaanxi, China
[3] Department of Industrial Management, Chung Hua University,
Hsin-Chu, Taiwan
[4] College of Management, Xi'an University of Science and Technology,
Xi'an 710054, Shaanxi, China

Abstract. Backpacking tasks are common. Back carrying, like many other manual materials handling (MMH) tasks, could result in muscular fatigue. Assessment of muscular strength recovery upon muscular fatigue is helpful to control musculoskeletal injuries and rest allowance management of the workplace. This research investigated back muscular fatigue for backpacking tasks of different loads via analyses of muscular strength decrease, recover, heart rate and subjective rating of physical exertion. Twelve adult participants (6 males and 6 females) were requested to carry a bag with 0%, 12.5%, and 25% of their body weights on their backs and walked on treadmill until they could no longer walk under three different speeds (2,4, and 6 km/hour) and two different ramp angles (0 and 10°). After the walk, the decrease of back muscular strength percentages (MVC) were measured in male and female subjects. In addition, the subjective rating on body fatigue was measured using the Borg RPE. It was found that back strength decreased after the carrying tasks, female subjects were more resistant to muscular fatigue than male subjects. The predictive model of RPE were been determined in Regression Equation, between the RPE score and MVC% are correlation coefficient. The RPE of results after the back-packing task were consistent with those of the muscular strength data.

Keywords: Manual materials handling (MMH) · Musculoskeletal injury
Muscular fatigue · MVC · Subjective rating

1 Introduction

Musculoskeletal-related issues are the most common injury [1]. Work-related musculoskeletal incidences have occasionally occurred. Annual incidence and prevalence ranges of Work-related musculoskeletal disorders (WRMSDs) were increased, and the highest prevalence among a miscellaneous group of workers [2]. The symptoms of Work-related musculoskeletal disorders (WRMSDs) influence the health of professional group workers [3]. Musculoskeletal disorders (MSDs) are defined as injuries and

© Springer Nature Switzerland AG 2019
S. Bagnara et al. (Eds.): IEA 2018, AISC 820, pp. 231–242, 2019.
https://doi.org/10.1007/978-3-319-96083-8_31

disorders to muscles, nerves, tendons, ligaments, joints, cartilage, and spinal discs, and the majority of MSDs have been caused by overexertion [4]. Overexertion or unnatural results in pain and discomfort of body segments of the Manual materials handling (MMH) accidents [1]. Manual materials handling (MMH) is a principal source of compensated work injuries and invest the height, distance, frequency, size, weight, and the differences in worker sex, age, and heart rate via the lifting, lowering, pushing, pulling, carrying and walking tasks [5].

Backpacking tasks are prevalent in manual materials handling (MMH) tasks. Over 40 million students use backpacks on a regular basis in the United States [6]. Back carrying, like many other manual materials handling (MMH) tasks, could result in back and knee musculoskeletal injuries. 79% of manual materials handling affect the lower back injuries, which are much reduced by the job design [7]. The back 45% of the body of injuries is due to the manual materials handling (MMH) accidents [1]. Load carriage in a long preceding period may produce the skeletal discomfort and injuries [8], and females reported hip discomfort to be significantly greater than males [9]. It is convenient and comfortable in the manual materials handling (MMH) tasks, backpacking task is worth concerning. Investigating issues of MMH tasks have been considered some contributors of musculoskeletal injuries [10].

The physical capability may be assessed a certain body segment of the muscular strength. The muscular strength decreases when the force exertion lasts for a period of time due to muscular fatigue, which could present directly or indirectly musculoskeletal injuries [11]. The decrease of the muscular strength for the backpacking tasks resulted in muscular fatigue, including the differences of gender [13], the differences of age [15], the increase of heart rate [16]; the effect of loads [13, 17, 18], influence of walking ways [19], carrying methods [20], gaits and postures [21–23], load placements [24], walking speeds [25] and the prolonged walking [26].

Muscular fatigue is defined as "any exercise-induced reduction in the ability to exert muscle force or power, regardless of whether or not the task can be sustained" [27]; "any exercise-induced reduction in capacity to generate force or power output" [28]; or alternatively the "reduction in the ability to exert force in response to voluntary effort" [12]. Based on these definitions, muscular fatigue is a common phenomenon and is regarded as a synonym of the reduction of muscular strength" for physical activities. The cause muscle fatigue is specific to the task being performed, and the development of muscle fatigue is typically quantified by the decline in the maximal force or power capacity of muscle strength and the sustained of maximum contractions after the onset of muscle fatigue [29]. Measurements of the reduction the maximum voluntary contraction (MVC) are direct assessments of muscular fatigue. Ma et al. [30] (2009) proposed a new simple muscle fatigue dynamic model in mathematics that described the maximum voluntary contraction (MVC) of muscle strength.

In this paper, a simple predictive models of the Borg RPE on fatigue model was proposed. Further analysis based on the muscle fatigue dynamic model of Ma in mathematics is carried out using mathematical regression method to determine the fatigue rate of the muscular strength for male and female backpacking task. It is about propose a mathematical parameter to describe the decrease of muscular strength capacity in the specific time for adults in different genders. Different significance of the muscular strength, and the Borg RPE of the back muscular strength will be discussed in male and female backpacking task.

2   Method

A back-packing experiment of back muscular strength capacity was conducted in the laboratory to assess the decrease of back muscular Strength for male and female backpacking task. The temperature and humidity were 21.1 °C (±2.4) and 65.2% (±9.7), respectively. After understanding the request of the backpacking task, human subjects will carry a bag and walk on a treadmill until they can no longer walk.

2.1   Subjects

Twelve college students, including six males and six females, were recruited in the study as human subjects. All adult participants were healthy, and none of them had a history of musculoskeletal disorders (WRMSDs). Before joining the experiment of muscular strength, all subjects had read and signed a consent form. The trial was completed with the help of instructors during the whole process of the experiment. Their age, body weight, stature, and BMI are shown in Table 1.

Table 1. Age, body mass, stature, and BMI of the subjects.

Variable	Male (n = 6)			Female (n = 6)		
	Mean	SD	Range	Mean	SD	Range
Age (yrs)	21.00	±1.41	(19–23)	22.33	±3.44	(19–28)
Stature (cm)	172.17	±6.05	(162–177)	166.83	±3.76	(162–173)
Body mass (kg)	68.33	±13.43	(44–80)	55.67	±3.33	(50–60)
BMI[a] (kg/m$^{2)}$	23.29	±4.04	(16.77–26.65)	20.00	±1.09	(18.37–21.72)

BMI[a] = body mass index.

2.2   Apparatus

There are a lot experiment apparatus to complete the tasks, including a treadmill, back strength dynamometer (TAKEI inc., 20–300 kgf), a sphygmomanometer, a bag, dumbbells and Borg RPE scale. The inclination and the speed of the treadmill were adjusted according to the requirement. Two slope levels of the walking task were tested on the treadmill. The first one was the flat condition and the second one was the 10° uphill inclination. Three walking speeds were adjusted on the treadmill: 2 km/h, 4 km/h and 6 km/h. The back strength dynamometer is an isometric strength measurement unit, which has a handle, a chain and a load cell included (see Fig. 1). The distance from the handle to above the platform was 38 cm, and they had been collected with a S-shape hook. This unit could measure the isometric back muscular strength.

The sphygmomanometer was used to test the heart rate of subjects before trial, after walking with the bag. Three load levels of dumbbells had been tested: 0; 12.5%, and 25% of body weight. The different weight of dumbbells were prepared to load in a bag for backpacking task of the subjects on treadmill, and the bag should been carried with both shoulders. A Borg RPE rating scale $(6 \sim 20)$ was employed to measure the subjective rating for perceiving body fatigue of the subject at the end of walking tasks. Table 2 shows the physical subjective rating burden of Borg RPE scale [31].

Fig. 1. Back strength dynamometer

2.3   Muscular Strength Test

All subjects were available volunteers to participate in the experiment, and they had no strenuous exercises within 24 h before the trial. Before each experiment, the temperature, humidity, the heart rate, the isometric back composite strengths will be measured. The heart rate (HR) of the subjects at the beginning, at the end of the walk will be recorded. For isometric back strength (see Fig. 2), the subject stood the platform and grasped the handle 38 com above the platform with bending his/her waist and pulled the handle upward with maximum force of the back muscles. The maximum forces measured before the trial will be the pre-test strength and after walking task will be the post-test strength. The pulled upward with maximum force measured in three times that will be the maximum voluntary contraction (MVC) of body muscles. After the pre-test strength test, the subjects carried a bag with 0, 12.5%, and 25% of their body weights on the back and walked on treadmill until they could no longer walk under three different speeds (2,4, and 6 km/h) and two different ramp angles (0 and 10°) (see Fig. 3). At the beginning of walk, the needed speed level had been controlled on the treadmill by instructor, in case the hurts were occurred on subjects for high speed. Each subject had taken the experiment for backpacking task for only once per day.

Table 2. Borg RPE scale

RPE	Individual physical feels	Corresponding heart rate values
20	Maximal exertion	Maximum
19	Extremely hard	195
18		
17	Very hard	170
16		
15	Hard	150
14		
13	Somewhat hard	130
12		
11	Light	110
10		
9	Very light	90
8		
7	Extremely light	70
6	No exertion at all	60

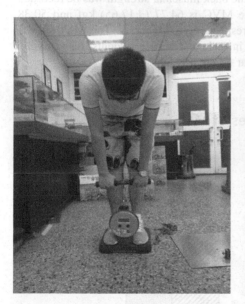

Fig. 2. Isometric back strength measurement

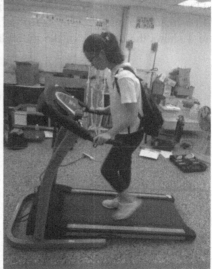

Fig. 3. Subjects carried a bag on the treadmill

2.4   Data Analysis

The trial was performed using a randomized completely block design, and each adult student was considered a block. The experiments were randomly arranged within each other. The experiment conditions included three loads (0, 12.5%, and 25% of their body weights), three speeds (2,4, and 6 km/h) and two different ramp angles (0 and 10°) conditions. A total of 216 back strengths (F(t))(12 subjects × 3 loads × 3 speeds × 2 slops) were recorded. Descriptive statistics, analysis of variance (ANOVA), RPE predicted models and regression analyses were performed in the load condition of male and female subjects. Regression analyses were performed for the data for each subject using Eq. (3). The muscular fatigue parameter of the muscular strength, or k, was calculated using Eq. (4). The predictive equation of RPE of the back muscular fatigue were presented. Statistical analyses were performed using the SAS 8.1 software.

3   Results

3.1   The Decrease and Recovery of the Back Strength

Maximum voluntary contraction (MVC) of the back muscular strength was be recorded in every member before the every test. The MVC is 64.77 (±17.65) kgf and 50.38 (±7.11) kgf in male and female students, respectively. After tests, The MVCafter is 54.74 (±16.80) kgf and 42.73 (±6.76) kgf in male and female students, respectively. Figure 4 shows the decrease of the muscular strength results.

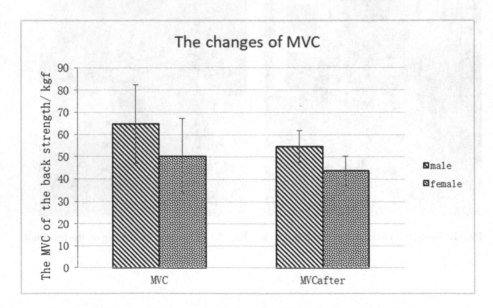

Fig. 4. The MVC results of the back muscular strength before tests and after tests.

The decrease percentage of the back muscular strength was adapt in back-packing tasks by Eq. 1:

$$MVC(\%) = (MVC - MVC_{after}) \times 100\%/MVC \qquad (1)$$

Where $MVC\%$ is the decrease percentage of the back muscular strength. The result shows that $MVC\%$ is 15.49%, 15.18% in male and female students, respectively. An ANOVA was conducted for the decrease muscular strength. The results showed that gender ($\rho < 0.0001$), load ($\rho < 0.0001$), and speed ($\rho < 0.0001$) were all significant factors for $MVC\%$ after back-packing tasks. The slops factors were not significant. Then Second and third order interactions of various factors were not significant. Further studying, $MVC\%$ of the male and female subjects is 18.89%, 18.15%, 14.69%, 13.95%, 13.22% and 10.85% in 0, 12.5%W and 25%W loads conditions, respectively. $MVC\%$ of the male and female subjects is 15.91%, 18.97%, 20.15%, 13.49%, 13.31% and 12.63% in 2, 4 and 6 km/h speeds conditions in Fig. 5, respectively.

3.2   The Predicted Model of RPE

After every test of back-packing task and an hour rest, a Borg RPE rating scale ($6 \sim 20$) (see Table 2) was employed to measure the subjective rating for perceiving body fatigue of the subject. After back-packing task, an ANOVA result of RPE shows that load ($\rho < 0.0001$), speed ($\rho < 0.0001$) and D% ($\rho < 0.0001$) were all significant factors. Then Second and third order interactions of various factors were not significant in load and speed conditions. The Pearson's correlation coefficient between the RPE score and D% was 0.66 ($\rho < 0.0001$).

RPE was been affected by many factors, like gender, BMI, load, ramp and speed. According to the principle of balanced BMI number in different levels. BMI was been divided into 3 Levels: L(19.27 $\pm$ 1.27), N(21.65 $\pm$ 3.30) and H(24.99 $\pm$ 1.92). After further regression analysis, the predictive equation of RPE in Regression Eq. (2), we have:

$$RPE = 1.41 \times G + 0.41 \times B + 0.12 \times L + 0.04 \times R + 0.54 \times v \qquad (2)$$

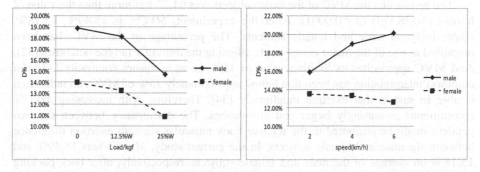

Fig. 5. The $MVC\%$ results of the back muscular strength in load and speed conditions.

$R^2 = 0.98$, where G, B, L, R and v are gender, BMI (kg/m^2), load(kg), ramp (°) and speed (km/h). An P Value shows that gender ($\rho < 0.0001$), BMI ($\rho < 0.0001$), load ($\rho < 0.0001$) and speed ($\rho < 0.0001$) were all significant factors in the model of RPE in Fig. 6.

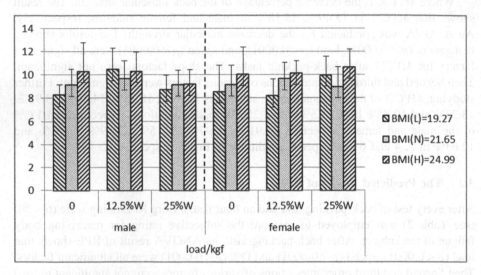

Fig. 6. The RPE in gender, BMI and load conditions.

4   Discussion

Backpacking tasks are common. A simple back-packing task was carried out by each subject under three loaded, three speeds and two ramp conditions in this study. This back-packing task, like many other manual materials handling (MMH) tasks, results in muscular fatigue. Muscular fatigue may be examined via studying the decrease of the back muscular strength and the perceive rating of RPE. Mathematical modeling of the muscular strength and the RPE are meaningful by male and female subjects in the job design for manual materials handling tasks.

The results that the MVC of the male subjects was 64.77 kgf more than the value of female (50.38 kgf) ($\rho < 0.0001$). After the experiment, $MVC\%$ is 15.49%, 15.18%, respectively, in male and female students. The percentage of the MVC, has been identified as one of the major components affecting the muscular fatigue outcomes [32]. And MVC approaches its limiting value of 15%, so as to ensure continuity [33]. An common interpretation has been that below approximately 15% of MVC, an individual is able to sustain that exertion indefinitely [34]. Therefore, with increasing $MVC\%$ recommends increasingly larger rest allowances. The discrepancy between the two genders might be attributed to the relative back muscular strength exertion difference between the male and female subjects. In the current study, $MVC\%$ was 15.49% and 15.18% on average of the male and female subjects, respectively, after back-packing tasks. These percentages were slightly higher than 15% and could be ranked as "low".

The male subjects were applying a relatively higher strength than the females during the trial. The decrease of the back muscular strength in male subjects was clipping. The interpretation and implication of our data, therefore, are only valid within those $MVC\%$ ranges. Both loads ($\rho < 0.0001$) and speeds condition ($\rho < 0.0001$) affected the back muscular strength significantly. Occupational shoulder loads in the investigated range (10–20%MVC) should be constructed with consideration to the specific endurance properties of shoulder muscles, and the large variability between individuals in endurance capacity [35]. $MVC\%$ of the back muscular strength was, therefore, somewhat different the male and female subject in different conditions. The decreasing trend of the muscular strength for the 25%W kg condition was more obvious than that of the 12.5%W kg and 0 kg conditions for male and female subjects, as was shown in Fig. 5. With the increasing of loads and speeds, $MVC\%$ and $R\%$ of the female subjects were decreased, while $MVC\%$ of male subjects was decrease in loads conditions and was increase in speeds conditions.

The Borg RPE scale was far more helpful than the surface EMG in the evaluation of fatigue [34]. The Borg RPE rating scale was obtained at the end of each back-packing task and an hour break [36]. The Pearson's correlation coefficient between the RPE score and $MVC\%$ was 0.66 ($\rho < 0.0001$), indicating a positive correlation between the RPE and D%. The Borg RPE scale A to some extent could reflect the degree of $MVC\%$. During the trial, load ($\rho < 0.0001$) and speed ($\rho < 0.0001$) were all significant influence in the RPE score. The BMI is commonly used to indicate thinness and fatness of a person [37]. High BMI implies high percentage of body fat or muscle mass. Over 62% (48/77) of our objects were within the normal BMI range (18.5–24.99 kg/m^2) suggested by the World Health Organization (WHO 2000). This percentage is slightly higher than the national percentage (58.9%) of Chinese population (BMI database, WHO). BMI ($\rho < 0.0001$) was significant factor in the predictive equation of RPE. And $R^2 = 0.98$, indicating effectiveness of the predictive equation of RPE.

There are limitations of the study. First, three different loads (0, 12.5%, and 25% of their body weights) were chosen as the conditions of the back-packing task. In real life, the loads of a bag are uncertain or more heavier. Li [15] (2004) study that backpacks loaded with 0%, 10%, 15%, and 20% of each subject's body weight, and a load heavier than 15% of body weight resulted in muscular fatigue. This paper presents three load conditions are limitations. Second, three different speeds (2,4, and 6 km/hour) were chosen, There are some speed levels in our daily life, and the different paces are adjusted. This paper presents three uniform walking speeds are also limitations. Additionally, the small sample size is also one of the limitations in male and female subjects of this study. In future study, enriching sample size and increasing the number of subjects in different ages. In this way, it is helpful to improve RPE and the extent of the predictive equation of RPE of muscular strength.

5   Conclusion

Backpacking tasks with heavier loads could result in muscular fatigue, and muscular fatigue could give rise to the decrease of the muscular strength. The purpose of this study is to make sure the decrease of the muscle strength and the predictive model of

RPE in male and female subjects. In an ANOVA study, we found that gender ($\rho < 0.0001$), load ($\rho < 0.0001$) and speed ($\rho < 0.0001$) were all significant for $MVC\%$ after back-packing tasks. By comparing the values of $MVC\%$, the results indicated that females were more resistant to muscular fatigue than males in the back-packing task. Gender ($\rho < 0.0001$), BMI ($\rho < 0.0001$), load ($\rho < 0.0001$) and speed ($\rho < 0.0001$) were all significant factors in the predictive model of RPE in Regression Equation. The Pearson's correlation coefficient between the RPE score and $MVC\%$ was 0.66 ($\rho < 0.0001$). After further regression analysis, the predictive equation of RPE were been determined. After back-packing tasks, it was found that the judgment coefficient R square values (0.503) in female subjects was larger than that (0.486) of male subjects. It is helpful to quantify physical fitness needs and reasonably arrange the strength design on Workplace in study the influence of the decrease of the muscular strength in backpacking tasks. In this way, the incidences of Work-related musculoskeletal disorders (WRMSDs) diseases are reduced by back-packing tasks.

References

1. Hughes P, Ferrett E (2015) International health and safety at work: for the NEBOSH international general certificate in occupational health and safety. Routledge, London and New York
2. Da Costa JT, Baptista JS, Vaz M (2015) Incidence and prevalence of upper-limb work related musculoskeletal disorders: a systematic review. Work 51(4):635–644
3. Serranheira F, Sousa-Uva M, Sousa-Uva A (2015) Hospital nurses tasks and work-related musculoskeletal disorders symptoms: a detailed analysis. Work 51(3):401–409
4. Maier M, Ross-Mota J (2000) Work-related musculoskeletal disorders
5. Dempsey PG, Hashemi L (1999) Analysis of workers' compensation claims associated with manual materials handling. Ergonomics 42(1):183–195
6. Dahl KD, Wang H, Popp JK, Dickin DC (2016) Load distribution and postural changes in young adults when wearing a traditional backpack versus the BackTpack. Gait Posture 45:90–96
7. Snook SH (1978) The ergonomics society the society's Lecture 1978. The design of manual handling tasks. Ergonomics 21(12):963–985
8. Stevenson JM, Bryant JT, Reid SA, Pelot RP, Morin EL, Bossi LL (2004) Development and assessment of the Canadian personal load carriage system using objective biomechanical measures. Ergonomics 47(12):1255–1271
9. Birrell SA, Haslam RA (2009) Subjective skeletal discomfort measured using a comfort questionnaire following a load carriage exercise. Mil Med 174(2):177–182
10. Waters TR, Dick RB, Davis-Barkley J, Krieg EF (2007) A cross-sectional study of risk factors for musculoskeletal symptoms in the workplace using data from the General Social Survey (GSS). J Occup Environ Med 49(2):172–184
11. Armstrong TJ, Buckle P, Fine LJ, Hagberg M, Jonsson B, Kilbom A, Viikari-Juntura ER (1993) A conceptual model for work-related neck and upper-limb musculoskeletal disorders. Scand J Work Environ Health, 73–84

12. Chaffin DB, Andersson G, Martin BJ (1999) Occupational biomechanics. Wiley, New York, pp 91–130
13. Chow DH, Kwok ML, Au-Yang AC, Holmes AD, Cheng JC, Yao FY, Wong MS (2005) The effect of backpack load on the gait of normal adolescent girls. Ergonomics 48(6):642–656
14. Hong Y, Brueggemann GP (2000) Changes in gait patterns in 10-year-old boys with increasing loads when walking on a treadmill. Gait Posture 11(3):254–259
15. Li JX, Hong Y (2004) Age difference in trunk kinematics during walking with different backpack weights in 6-to 12-year-old children. Res Sports Med 12(2):135–142
16. Hong Y, Li JX, Wong ASK, Robinson PD (2000) Effects of load carriage on heart rate, blood pressure and energy expenditure in children. Ergonomics 43(6):717–727
17. Birrell SA, Haslam RA (2010) The effect of load distribution within military load carriage systems on the kinetics of human gait. Appl Ergon 41(4):585–590
18. Johnson RF, Knapik JJ, Merullo DJ (1995) Symptoms during load carrying: effects of mass and load distribution during a 20-km road march. Percept Mot Skills 81(1):331–338
19. Hall M, Boyer ER, Gillette JC, Mirka GA (2013) Medial knee joint loading during stair ambulation and walking while carrying loads. Gait Posture 37(3):460–462
20. Hong Y, Li JX (2005) Influence of load and carrying methods on gait phase and ground reactions in children's stair walking. Gait Posture 22(1):63–68
21. Chansirinukor W, Wilson D, Grimmer K, Dansie B (2001) Effects of backpacks on students: measurement of cervical and shoulder posture. Aust J Physiotherapy 47(2):110–116
22. Devroey C, Jonkers I, De Becker A, Lenaerts G, Spaepen A (2007) Evaluation of the effect of backpack load and position during standing and walking using biomechanical, physiological and subjective measures. Ergonomics 50(5):728–742
23. Hong Y, Cheung CK (2003) Gait and posture responses to backpack load during level walking in children. Gait Posture 17(1):28–33
24. Bobet J, Norman RW (1984) Effects of load placement on back muscle activity in load carriage. Eur J Appl Physiol 53(1):71–75
25. Charteris J (1998) Comparison of the effects of backpack loading and of walking speed on foot-floor contact patterns. Ergonomics 41(12):1792–1809
26. Hong Y, Li JX, Fong DTP (2008) Effect of prolonged walking with backpack loads on trunk muscle activity and fatigue in children. J Electromyogr Kinesiol 18(6):990–996
27. Bigland-Ritchie BWJJ, Woods JJ (1984) Changes in muscle contractile properties and neural control during human muscular fatigue. Muscle Nerve 7(9):691–699
28. Vøllestad NK (1997) Measurement of human muscle fatigue. J Neurosci Methods 74 (2):219–227
29. Enoka RM, Duchateau J (2008) Muscle fatigue: what, why and how it influences muscle function. J Physiol 586(1):11–23
30. Ma L, Chablat D, Bennis F, Zhang W (2009) A new simple dynamic muscle fatigue model and its validation. Int J Ind Ergon 39(1):211–220
31. Borg G (1985) An introduction to Borg's RPE-scale. Mouvement Publications, Ithaca, New York
32. Sato H, Ohashi J, Iwanaga K, Yoshitake R, Shimada K (1984) Endurance time and fatigue in static contractions. J Hum Ergol 13(2):147
33. Elahrache K, Imbeau D (2009) Comparison of rest allowance models for static muscular work. Int J Ind Ergon 39(1):73–80

34. Garg A, Hegmann KT, Schwoerer BJ, Kapellusch JM (2002) The effect of maximum voluntary contraction on endurance times for the shoulder girdle. Int J Ind Ergon 30(2):103–113

35. Mathiassen SE, Åhsberg E (1999) Prediction of shoulder flexion endurance from personal factors. Int J Ind Ergon 24(3):315–329

36. Yi CN, Tang F, Li KW (2017) Muscular strength decrease and maximum endurance time assessment for a simulated truck pulling task, 1–8

37. Zhang Z, Li KW, Zhang W, Ma L, Chen Z (2014) Muscular fatigue and maximum endurance time assessment for male and female industrial workers. Int J Ind Ergon 44 (2):292–297

Interdisciplinary Association Between Biomechanical Analysis and Occupational Psychology: Challenges and Procedures

Adriana Savescu[1(✉)] and Pascal Simonet[2,3]

[1] Laboratoire Physiologie-Mouvement-Travail, INRS, Vandoeuvre-lès-Nancy,
France
adriana.savescu@inrs.fr
[2] Aix Marseille Université, ADEF EA 4671, 13248 Marseille, France
pascal.simonet@univ-amu.fr
[3] Equipe Psychologie du travail et Clinique de l'activité, Centre de Recherche
sur le Travail et le Développement EA 4132, CNAM, Paris, France

Abstract. For preventing musculoskeletal disorders (MSDs), various types of
intervention may be conducted, and in particular participatory ergonomics
interventions. One of the key success factors in such interventions is the active
involvement of the workers. The aim of this summary is to show how biomechanical analysis became part of occupational psychology methodology in order
to help the workers (gravediggers in this case) to analyse their movements or
"gestures" finely with a view to developing them. The construction of this
interdisciplinary methodology in which the results of the biomechanical analysis
were used by the workers as means for observing themselves, for comparing
themselves with one another, and sometimes even for trying out different
strategies for carrying out the gestures enabled gravediggers to be active in
developing the gesture and thus in making progress towards preventing MSDs.

Keywords: Interdisciplinary method · Clinic of activity · Biomechanics
Musculoskeletal disorders

1 Introduction

In France and globally, musculoskeletal disorders (MSDs) remain the most widespread
occupational diseases. One of the approaches implemented in the field for reducing
MSD risks, known as participatory ergonomics intervention, requires the involvement
of various people from the company (at all hierarchical levels) in the process of
identifying the problem and of proposing solutions [1–3]. This type of intervention
may take various forms [4] and, in order to contribute to its success, various conditions
must be satisfied, in particular worker must be involved "because he has unique insight
into his task, work, activity" [5]. That is why it is necessary to pay particular attention
to the way in which they should be mobilised in this type of intervention [6]. Positive
impacts, both on the productivity and on the comfort of workers have been identified as
a result of it being implemented [5]. In order to be fully involved, the workers must
have the possibility of being actors in the construction of the proposed solution (be able

© Springer Nature Switzerland AG 2019
S. Bagnara et al. (Eds.): IEA 2018, AISC 820, pp. 243–249, 2019.
https://doi.org/10.1007/978-3-319-96083-8_32

to take part in the decision-making and decision-taking) and the possibility of having positive feedback on the new work situation ("see" or "feel" the benefits of the future work situation). However, difficulties remain in initiating and implementing this type of intervention, due, in particular, to achieving actual involvement of the workers in the solution-seeking process when faced with the identified problem whereupon, most often, fatalism predominates. The expression "everyone works the same way" mentioned in the literature [7], too often crystallises the fatalistic feeling when faced with MSDs. The impossibility of imagining that it is possible to do things differently would appear to be one of the constants in certain situations that generate work-related MSDs [8]. Under such conditions, how can we involve the workers and make them take an active part in intervention for preventing MSDs? The path chosen by occupational psychology was the one followed by a methodology implementing self-confrontations [3, 9], cross-confrontations [10], and interdisciplinary actions [6, 11, 12]. To contribute to the methodological support and to analyse the movements or "gestures" of the workers precisely, while involving them actively, an association between occupational psychology methodology and biomechanics was constructed and implemented.

2 Objective

The intervention research described in this summary was in response to a request from a preventive medicine unit of a major French city, and relates to preventing MSDs in gravediggers. The objective was to show how biomechanical analysis could become part of occupational psychology methodology in order to help the workers (gravediggers in this case) to analyse their gestures finely with a view to developing them. Thus, in an interdisciplinary methodology, the challenge for biomechanics was not to achieve a precise diagnostic of the work situation. Based on these results and by using self-confrontation and cross-confrontation, the challenge was to work with the gravediggers on the variability of their gestures and their colleagues' gestures so as not just to accept as a fatality of the trade that "when you're a gravedigger you are condemned to hurt yourself".

3 Methodology

The methodology implemented took into account the context of the intervention and it comprised 5 successive stages.

3.1 Context of the Interdisciplinary Action

In digging a grave, 1.5 m to 2 m deep, the gravedigger has to remove the soil by throwing it backwards, constraining him to carry a tool loaded with soil at arm's length so as to chuck the soil out of the grave, with his back to the storage point. This movement or "gesture" is problematic:

– for the occupational physician who establishes a link between it being done repeatedly and possible occurrence of upper limb MSDs [13];

- for the manager who wants to ban this gesture to avoid occupational accidents and time off sick; and
- for the gravediggers who would like to avoid doing it but who have to do it when circumstances so dictate.

3.2   Methodological Framework for the Interdisciplinary Action

The methodological framework for the interdisciplinary action comprises various stages:

Presenting a Protocol for Recording the Muscular Strain on Gravediggers and Calling for Volunteers for Gathering Data. The objectives of this stage were: to start a think through debate between the gravediggers about the backward chuck gesture and the conditions under which it was carried out, to propose to them a biomechanical protocol with a view to responding to their demand for probing deeper into this gesture and for understanding it, and to mobilise them for the biomechanical data collection phase. The procedures for gathering and analysing biomechanical data were presented (using visual material). The gravediggers were asked to express themselves about the conditions under which the backward chuck gesture was carried out, and to say whether they would volunteer to take part in gathering biomechanical data.

Gathering and Analysing Biomechanical Data. Biomechanical analysis concerned the activity of certain shoulder and back muscles. It was conducted in order to highlight the intra-individual and inter-individual variability of the backward chuck gesture and in order to enable the gravediggers to "visualise" it, so as to address it in their discussions. The activity of 4 shoulder and back muscles on both sides was recorded (by surface electromyography) in 8 gravediggers. The recordings took place in real work situations. Video recordings were also taken. A backward chuck gesture was defined by the research team: the start of the gesture began when the work tool as filled with soil was no longer in contact with the soil in the grave, and ended when the soil was chucked out of the grave with a backward chuck movement. Analysis of the biomechanical data, conducted gesture-by-gesture for each gravedigger, made it possible to identify the gestures that put the most and least strain on each gravedigger. Those results were presented in the form of graphs and diagrams accompanied by the corresponding video recordings.

Presenting the Results of the Analysis to the Gravediggers and Emergence of the Initial Discussions. The objectives of this stage were to bring the results of the biomechanical analysis to the gravediggers' knowledge, with the conditions under which they were obtained being recalled, and to ask the gravediggers to express what they felt about the results. A presentation was given that recalled the conditions under which the data was gathered and analysed. The gravediggers became collectively acquainted with the way in which the biomechanical data had been gathered and with the results (cf. preceding stage). They were asked to comment on their way of doing the work and on the way their colleagues did it.

Organising Self-confrontations and Cross-Confrontations. The challenge for this stage was to make the results of the biomechanical data analysis accessible both to the occupational psychologist researcher and to the gravediggers. Associated with the video recordings, the results were used as tools for analysing their gestures during self-confrontations and cross-confrontations conducted at and away from the work station. Firstly, self-confrontations were conducted using the following scheme: self-confronted gravedigger/biomechanical results (graphical representations and video)/researchers. The instruction given to the gravedigger was to comment on what he saw himself doing or not doing in the video and in relation to the biomechanical results. Then, cross-confrontations were conducted using the following scheme: two self-confronted gravedigger/technical material for each of them/researchers. Pairs of self-confronted gravediggers were set up as follows: the larger the differences in the results of the muscular activities between two gravediggers, the more the researchers tried, with their consent, to put them together for these cross-confrontations. The instruction given to each of them was to comment, in turn, on the video of the colleague who was present, each of them having previously undergone a self-confrontation. For the two series of confrontations (self-confrontations and cross-confrontations), the gravediggers' comments, the exchanges between them, and the exchanges with the researchers were recorded on video.

Multimodal Analysis of the Cross-Confrontations. This stage involved cross-analysis of the wording used by the gravediggers and of the gestures carried out in confrontations on the basis of the video recordings of the self confrontations and of the cross-confrontations. It made it possible, in particular, to highlight how each gravedigger was able to appropriate the results of the biomechanical analysis so as to use them as a means of exchanging about their respective ways of carrying out the backward chuck gesture and of discussing them between themselves.

4 Results

The context proposed by the interdisciplinary methodology was different from the context of the real activity and enabled the gravediggers to examine the backward chuck gesture by identifying the criteria for carrying it out (Fig. 1). Thus, the gesture that was a source of pathology and to be analysed became a topic for discussion, enriched by various different ways of carrying it out (intra-individual and inter-individual variability), by going from one of its contexts to the other, through the successive stages presented in the methodology part.

It was through the aforementioned contexts that this gesture became more available for discussion between the workers and ultimately more available, for each of them, as a means to imagine doing the gesture otherwise with a view to developing their work activity. Thus, this throwing backward gesture went from the status of risk gesture to the status of debate topic gesture to be studied in detailed manner, of gesture carried out and recorded, of gesture defined by the researchers, of gesture discussed during self-confrontations and cross-confrontations, of gesture deconstructed into various elements, of gesture enriched with new knowledge, and finally to the status of gesture that some of them could imagine doing differently.

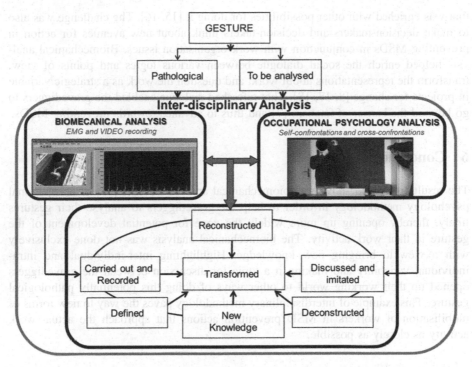

Fig. 1. Functional migration of the gesture.

This intervention research dealt a blow to fatalistic acceptance and contributed to setting up a dual dynamic for preventing MSDs:

- between gravediggers, on the ways of doing these gestures while re-thinking how they used their bodies in this activity; and
- between occupational risk preventers and supervisory staff who proposed and set up training in trade gestures that were better suited to the realities of the trade of gravedigger with a view to preventing MSDs, with the gravediggers taking a more active part in the decision-making and decision-taking process.

5 Discussion

In general, biomechanics gives elements of biological plausibility or, in the context of ergonomic study steered by ergonomics, gives metrological information that is useful for understanding or transforming the work [14]. In this study, based on an interdisciplinary methodology, the results of the biomechanical analysis were used by the workers as a means to observe themselves, to compare themselves with one another, and sometimes even to try out different strategies for doing gestures more efficiently while also not injuring themselves. Thus, through their active involvement [5], the problematic (and possibly pathological) gesture was transformed into a trade gesture that was enriched with other possibilities for doing it [15, 16]. The challenge was also to make decision-makers and decision-takers think about new avenues for action in preventing MSDs in conjunction with work organisation issues. Biomechanical analysis helped enrich the social dialogue between various logics and points of view, transform the representations about work, and question the work as a strategic variable in projects for change [8, 17, 18]. That refreshed dialogue enabled the gravediggers to go beyond the barrier of fatalism [7] and thus to advance towards preventing MSDs.

6 Conclusion

The results presented show how biomechanical analysis became part of occupational psychology methodology in order to help the gravediggers to analyse their gestures finely, thereby opening up more widely the zone for potential development of the gesture in their work activity. The biomechanical analysis was not done exclusively with a view to bringing new knowledge. Highlighting inter-individual and intra-individual variability and making it a topic for discussion between the gravediggers opened up their working world to other ways of doing this potentially pathological gesture. This example of interdisciplinary methodology paves the way to new forms of mobilisation of workers in MSD prevention actions that approach the actual work activity as closely as possible.

References

1. Roquelaure Y (2016) Promoting a shared representation of workers' activities to improve integrated prevention of work-related musculoskeletal disorders. Safety Health Work 7(2):171–174
2. Van Eerd D et al (2016) Dissemination and use of a participatory ergonomics guide for workplaces. Ergonomics 59(6):851–858
3. Kuorinka I (1997) Tools and means of implementing participatory ergonomics. Int J Ind Ergon 19(4):267–270
4. Haines H et al (2002) Validating a framework for participatory ergonomics (the PEF). Ergonomics 45(4):309–327
5. Vink P, Koningsveld EAP, Molenbroek JF (2006) Positive outcomes of participatory ergonomics in terms of greater comfort and higher productivity. Appl Ergon 37(4):537–546
6. Kloetzer L, Quillerou-Grivot E, Simonet P (2015) Engaging workers in WRMSD prevention: Two interdisciplinary case studies in an activity clinic. WORK J Prev Assess Rehabil 51(2):161–173
7. Brunet M, Riff J (2009) Analyse et exploitation de la variabilité gestuelle dans la prévention des TMS. In: PISTES
8. Coutarel F et al (2015) Marge de manœuvre situationnelle et pouvoir d'agir: des concepts à l'intervention ergonomique. Le Travail Humain 78(1):9–29
9. Mollo V, Falzon P (2004) Auto- and allo-confrontation as tools for reflective activities. Appl Ergon 35(6):531–540
10. Clot Y et al (2000) Entretiens en autoconfrontation croisée: une méthode en clinique de l'activité. PISTES 2

11. Savescu A et al (2010) Biomechanical metrology: a support in occupational controversies. In: PREMUS, Angers, France
12. Simonet P et al (2011) La pluridisciplinarité au service de la prévention des TMS: quand l'association entre psychologie du travail et biomécanique devient, pour les professionnels, support d'analyse des gestes de métier. INRIA, Paris
13. Van Trier M et al (2010) Prévention durable des TMS chez des fossoyeurs de la ville de Paris. In: 31e Congrès National de Médecine et Santé au Travail, Tooulouse, France
14. Aptel M, Vézina N (2008) Quels modèles pour comprendre et prévenir les TMS? Pour une approche holistique et dynamique. In: 2 ième Congrès francophone de recherche sur les TMS, Montréal, Canada
15. Fernandez G (2004) Développement d'un geste technique. Histoire du freinage en Gare du Nord. CNAM, Paris
16. Simonet P, Caroly S (2008) Le développement des automatismes comme conception du geste professionnel pour une prévention durable des TMS. In: 43ème congrès de la Société d'Ergonomie de Langue Française, Ajaccio, France
17. Clot Y, Simonet P (2015) Pouvoirs d'agir et marges de manœuvre. Le Travail Humain 78(1):31–52
18. Simonet P, Clot Y (2014b) Qualité du travail, santé et clinique de l'activité. Méthode pour l'action, in Encyclopédie médicochirurgicale, Pathologie professionnelle et de l'environnement, Masson E, (ed), Paris

Impacts of Typing on Different Keyboard Slopes on the Deformation Ratio of the Median Nerve

Ping Yeap Loh[1]([✉]) [iD], Wen Liang Yeoh[2], and Satoshi Muraki[1]

[1] Department of Human Science, Faculty of Design, Kyushu University,
Fukuoka 815-8540, Japan
py-loh@design.kyushu-u.ac.jp
[2] Department of Human Science, Graduate School of Design,
Kyushu University, Fukuoka 815-8540, Japan

Abstract. Carpal tunnel syndrome is a symptomatic compression neuropathy of the median nerve as it travels through the wrist. Several factors such as wrist angle and finger posture cause a change of the intra-carpal tunnel pressure. Carpal tunnel syndrome is one of the most commonly reported work-related musculoskeletal disorders. Computer users are at higher risk of upper extremity musculoskeletal symptoms and work-related musculoskeletal disorders, since time spent on the computer is associated with a higher incidence of musculoskeletal disorders. Objective: to investigate the impact of typing at two keyboard slopes (0° and +20°) on the median nerve deformation ratio. Fifteen healthy young men (24.8 ± 2.3 years) were recruited to type using both 0° and +20° inclined keyboards. The participants performed four 30-min blocks of computer typing at 0° and +20° keyboard inclinations. The left wrist median nerve was examined with an ultrasound machine after each 30-min typing block. Two-way repeated analysis of variance was performed to examine any differences in the deformation ratio of the median nerve cross-sectional area. The four time blocks and two keyboard slope conditions (0° and +20° inclination) were used as factors. Continuous typing activity causes a significant increase in the median nerve cross-sectional area deformation ratio ($p < 0.05$). Ultrasonography examination of the median nerve following computer typing can be used to generate absolute measurements and deformation ratios. These measurements help provide a better understanding of the impact of typing tasks on the median nerve.

Keywords: Carpal tunnel · Carpal tunnel syndrome · Computer ergonomics

1 Introduction

Advancements in technology have enhanced the use of computer systems in the quotidian professional environment. Although computer technology improves certain work performance and output measures, it also increases the risk of work-related musculoskeletal disorders. Studies showed that approximately 20% of computer users experience musculoskeletal symptoms involving muscles and joints [1, 2]. These

S. Bagnara et al. (Eds.): IEA 2018, AISC 820, pp. 250–254, 2019.
https://doi.org/10.1007/978-3-319-96083-8_33

work-related musculoskeletal disorders can have significant negative impacts, causing disabilities among workers that lead to socioeconomic burdens [3–5].

Carpal tunnel syndrome (CTS) is one of the most commonly reported work-related musculoskeletal disorders. CTS affects the median nerve at the wrist; increased pressure within the carpal tunnel leads to decreased function of the median nerve [6]. Occupational and work-related biomechanical factors are highly associated with workplace CTS occurrence [7–10]. Physical biomechanical stress can elevate the intra-carpal pressure and lead to an external compression of the median nerve within the carpal tunnel [11, 12].

It should also be noted that wrist posture and grip force exertion conditions affect the median nerve's morphological characteristics [13–16]. These studies suggest that wrist flexion-extension and force exertion contribute to the deformation of the median nerve at the level of the carpal tunnel. In addition, previous studies suggest that computer typing activities result in acute swelling of the median nerve at the wrist [17, 18]. However, the deformation ratio of the median nerve after prolonged keyboard typing was not well described. Therefore, the main objective here, was to investigate the impact of typing on the median nerve deformation ratio.

2 Methods

This study was approved by the Ethics Committee of the Faculty of Design, Kyushu University. Fifteen healthy right-handed young men (24.8 ± 2.3 years) were randomized to perform computer typing on both $0°$ and $+20°$ inclined keyboards. The participants were required to perform four 30-min blocks of computer typing for each $0°$ and $+20°$ keyboard condition. The left wrist median nerve was examined at pisiform level after each typing block with an ultrasound machine (LOGIQ e).

Median nerve cross-sectional area (MNCSA, mm^2) was analyzed with ImageJ [19] using the tracing method (Fig. 1). The MNCSA deformation percentage was calculated as (MNCSA after typing – MNCSA before typing)/(MNCSA before typing).

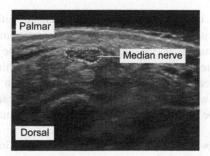

Fig. 1. Quantification of the median nerve cross-sectional area by tracing method.

To examine differences in the MNCSA deformation ratio, two-way repeated analysis of variance (ANOVA, 4×2 factorial design) was performed. The four time blocks and two keyboard slope conditions (0° and +20° inclination) were used as factors.

3 Results

Typing on a computer for 30 min led to acute swelling of the median nerve; continuous typing activity caused an increased MNCSA deformation ratio (Fig. 2). Each typing block had a significant effect on the MNCSA deformation ratio (p < 0.05). Although keyboard slope did not have a statistically significant effect, the deformation ratio for typing at +20° (approximately 20%) was larger than typing at 0° (approximately 14%).

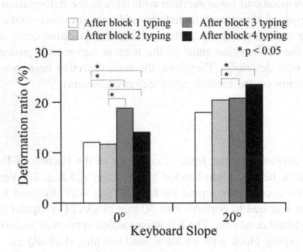

Fig. 2. Deformation ratio of the median nerve cross-sectional area after typing

4 Discussion

Biomechanical stress in the workplace may be associated with work-related musculoskeletal disorders. Wrist and finger movements during computer keyboard typing induce muscle fatigue and are also associated with acute shape changes of the median nerve at the carpal tunnel. In addition, this study indicated that the MNCSA deformation percentage is greater at higher keyboard slopes when compared to 0° slope. Typing on a keyboard with greater wrist extension could cause higher amounts of compression stress of the median nerve within the carpal tunnel.

Acknowledgement. This work is supported by JSPS KAKENHI Grant Numbers JP16J01964 and JP18K17969.

References

1. Baker NA, Cham R, Cidboy EH, Cook J, Redfern MS (2007) Kinematics of the fingers and hands during computer keyboard use. Clin Biomech (Bristol, Avon) 22:34–43. https://doi.org/10.1016/j.clinbiomech.2006.08.008
2. Sauter SL, Schleifer LM, Knutson SJ (1991) Work posture, workstation design, and musculoskeletal discomfort in a VDT data entry task. Hum Factors 33:151–167. https://doi.org/10.1177/001872089103300203
3. Bureau of Labor Statistics: Spotlight on statistics: Older workers (2008)
4. Daniell WE, Fulton-Kehoe D, Chiou LA, Franklin GM (2005) Work-related carpal tunnel syndrome in Washington State workers' compensation: temporal trends, clinical practices, and disability. Am J Ind Med 48:259–269. https://doi.org/10.1002/ajim.20203
5. Foley M, Silverstein B, Polissar N (2007) The economic burden of carpal tunnel syndrome: long-term earnings of CTS claimants in Washington State. Am J Ind Med 50:155–172. https://doi.org/10.1002/ajim.20430
6. American Academy of Orthopedic Surgeons: Clinical practice guidelines on diagnosis of carpal tunnel syndrome (2007). http://www.aaos.org/research/guidelines/CTS_guideline.pdf
7. Harris-Adamson C, Eisen EA, Kapellusch J, Garg A, Hegmann KT, Thiese MS, Dale AM, Evanoff B, Burt S, Bao S, Silverstein B, Merlino L, Gerr F, Rempel DM (2015) Biomechanical risk factors for carpal tunnel syndrome: a pooled study of 2474 workers. Occup Environ Med 72:33–41. https://doi.org/10.1136/oemed-2014-102378
8. Bao SS, Kapellusch JM, Merryweather AS, Thiese MS, Garg A, Hegmann KT, Silverstein BA, Marcum JL, Tang R (2016) Impact of work organizational factors on carpal tunnel syndrome and epicondylitis. J Occup Environ Med 58:760–764. https://doi.org/10.1097/JOM.0000000000000790
9. Bao SS, Kapellusch JM, Garg A, Silverstein BA, Harris-Adamson C, Burt SE, Dale AM, Evanoff BA, Gerr FE, Hegmann KT, Merlino LA, Thiese MS, Rempel DM (2015) Developing a pooled job physical exposure data set from multiple independent studies: an example of a consortium study of carpal tunnel syndrome. Occup Environ Med 72:130–137. https://doi.org/10.1136/oemed-2014-102396
10. Dias JJ, Burke FD, Wildin CJ, Heras-Palou C, Bradley MJ (2004) Carpal tunnel syndrome and work. J Hand Surg Br Eur 29:329–333. https://doi.org/10.1016/j.jhsb.2004.03.002
11. Uchiyama S, Itsubo T, Nakamura K, Kato H, Yasutomi T, Momose T (2010) Current concepts of carpal tunnel syndrome: pathophysiology, treatment, and evaluation. J Orthop Sci 15:1–13. https://doi.org/10.1007/s00776-009-1416-x
12. Werner RA, Andary M (2002) Carpal tunnel syndrome: pathophysiology and clinical neurophysiology. Clin Neurophysiol 113:1373–1381. https://doi.org/10.1016/S1388-2457(02)00169-4
13. Loh PY, Muraki S (2015) Effect of wrist angle on median nerve appearance at the proximal carpal tunnel. PLoS ONE 10:e0117930. https://doi.org/10.1371/journal.pone.0117930
14. Loh PY, Nakashima H, Muraki S (2016) Effects of grip force on median nerve deformation at different wrist angles. PeerJ 4:e2510. https://doi.org/10.7717/peerj.2510
15. Loh PY, Nakashima H, Muraki S (2015) Median nerve behavior at different wrist positions among older males. PeerJ 3:e928. https://doi.org/10.7717/peerj.928
16. Loh PY, Muraki S (2014) Effect of wrist deviation on median nerve cross-sectional area at proximal carpal tunnel level. Iran J Publ Health 43
17. Loh PY, Yeoh WL, Nakashima H, Muraki S (2017) Impact of keyboard typing on the morphological changes of the median nerve. J Occup Health 59. https://doi.org/10.1539/joh.17-0058-oa

18. Toosi KK, Impink BG, Baker NA, Boninger ML (2011) Effects of computer keyboarding on ultrasonographic measures of the median nerve. Am J Ind Med 54:826–833. https://doi.org/10.1002/ajim.20983

19. Schneider CA, Rasband WS, Eliceiri KW (2012) NIH image to ImageJ: 25 years of image analysis. Nat Methods 9:671–675. https://doi.org/10.1038/nmeth.2089

Musculoskeletal Complaints in a Sample of Employees in a Tertiary Hospital: An Exploratory Preliminary Pilot Study

M. C. R. Fonseca[1(✉)], F. P. F. M. Ricci[1], L. M. Gil[1], N. C. Silva[1],
E. C. O. Guirro[1], R. R. J. Guirro[1], E. R. C. Lopes[1], L. R. Santos[1],
R. I. Barbosa[2], A. M. Marcolino[2], V. R. Castro[1], T. M. Fifolato[1],
H. Nardim[1], L. Mauad[1,2], and K. S. Ferreira[1]

[1] University of Sao Paulo, Ribeirao Preto, Brazil
marisaregistro@gmail.com
[2] Federal University of Santa Catarina, Araranguá, Brazil

Abstract. Introduction: Musculoskeletal disorders can cause impact in quality of life. This exploratory preliminary study aimed to describe musculoskeletal complaints in a sample of employees in a tertiary hospital.

Methods: A university public hospital has 6114 employees that work in an emergency and outpatient units. There are several occupations with predominance of nurses, computer users, cleaning and cooking staff. A sample of ten workers was assessed by, socio-demographic sheet, The Nordic body pain map, SF-36 questionnaire, and Work Ability Index Questionnaire. Data were described in terms of frequencies by SPSS™.

Results: The sample had predominance of women (80%), mean age of 49.3 years old, all have children, working as nurses, administrative and general services. 50% has less than 5 years of experience, most have two work shifts, few are active in terms of exercise routine, 70% use medication for pain, do not have a sitting or standing working position preference. Nordic questionnaire showed predominance in pain complaints in back, neck, shoulder and knee. SF-36 questionnaire presented mean scores for subdomains: functional capacity 68.56, physical aspects 68.75, pain 42.72, general health status 67.37, vitality 66.72, social aspects 81.25, emotional aspects 70.83 and mental health 67.2. The work ability index questionnaire the majority of the scores was classified as moderate (28–36).

Conclusions: This study showed that besides high scores for SF 36 for these sedentary women, most use pain medication for musculoskeletal complaints and was identified necessity improvement of the work ability. There is need for future study focused on ergonomics and exercises.

Keywords: Musculoskeletal disorders · Hospital · Pain · Exercises

S. Bagnara et al. (Eds.): IEA 2018, AISC 820, pp. 255–260, 2019.
https://doi.org/10.1007/978-3-319-96083-8_34

1 Introduction

Work Musculoskeletal disorders (WMD) have an impact on workers' quality of life and decrease labor productivity [1, 2]. WMDs can increase the number of employees out of work [3, 4]. In addition to high expenses and treatments, workers with injuries may be discriminated by the boss, or on the part of co-workers who feel overwhelmed by the fact that the "sick" colleague often has restrictions or needs to move away frequently [5].

WMDs are usually related to the exposure and the combination of a variety of risk factors, including individual (age, gender), biomechanical/physical (posture, load and repetition of movement) and psychosocial/organizational factors (pressure at work or time, lack of social support and job satisfaction) [6]. WMDs are considered one of the most worrying health problems for almost all workers, as they can lead to different degrees of functional disability [5]. Hospital workers are part of a class of professionals who are frequently affected by WMDs, as well as psychiatric diseases [7].

The concept of ability to work can be expressed as how well or how it will be, a worker at this time or in the near future, and how capable he can perform his work in function of the demands, his state of health and the physical capacities and mental [2].

Currently, there are several tools to evaluate work capacity, such as the Work Ability Index Questionnaire (WAIQ) developed in Finland in the 1980s and 1990s [8]. WAIQ makes it possible to evaluate and detect early changes, to predict the incidence of incapacity of aging workers and to be used as a tool to subsidize information. Its results can be used at individual levels, where it is possible to identify workers with impairment of work capacity and adopt supportive measures, as well as at the collective level, where it allows the identification of a general profile of capacity for work, functional capacity and all the factors that affect them [8].

Other well-known questionnaires cover a varied of subdomains that include the individuals' quality of life assessment and symptoms of musculoskeletal disorders, such as the generic SF-36 questionnaire [9] and the Nordic questionnaire [10, 11], which is composed of 4 questions regarding a pain/discomfort body map.

This study aimed to describe the musculoskeletal complaints in a sample of employees of a tertiary hospital in a preliminary exploratory pilot study.

2 Methodology

This transversal and observational study was carried out at the Rehabilitation Center of Ribeirao Preto Medical School of University of Sao Paulo, Brazil. This University Hospital has two units, one located on the university campus, where they perform outpatient consultations, hospitalizations and scheduled surgeries, and another unit located in downtown of Ribeirao Preto city, the Urgency/Emergency Unit.

This University Hospital has a total of 6114 employees and 911 residents and graduates. From these total employees, 4861 work at the campus hospital and 1253 in the Urgency/Emergency unit. All of these employees, residents and graduates are entitled to medical care in several specialties, where they are diagnosed and, if necessary, referred through an interconsultation request for physiotherapy.

The study included workers from both units, who were referred from October 2017 to December 2017 for the musculoskeletal physiotherapy service specialized in the care of these employees. The criteria for inclusion of the sample were: to have an employment relationship with the Hospital, to have a medical diagnosis, to be referred to the physiotherapy service of the Rehabilitation Center by means of an interconsultation request. We excluded from the sample workers who did not want to answer the questionnaires, workers who were referred due to fracture and those who had already retired and left the hospital.

During the study period, 69 evaluations were scheduled, but 13 were missed the appointment, 46 were excluded, 38 were not accepted for responding to the questionnaires, 4 were referred due to fracture and 4 were retired (Fig. 1- Flowchart). A sample of 10 employees was evaluated by means of a socio-demographic evaluation form, and three questionnaires, all applied in the form of an interview, in addition to the physical examination.

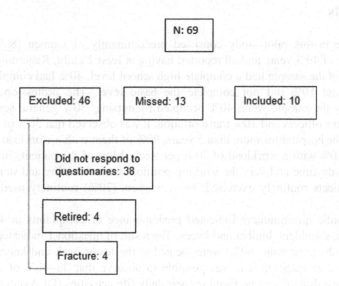

Fig. 1. Flowchart of the research participants

The socio-demographic data sheet collected information related to: age, gender, marital status, number of children, education, occupation and time on this occupation, medication use, physical activity level, smoking, and alcoholism use. This questionnaire was prepared by the researchers in order to collect general information about the patient sample. Three questionnaires were applied. The Nordic questionnaire corresponds to a body map divided into 9 body segments (neck, shoulders, upper back, elbows, wrists/hands, lower back, hips/thighs, knees, ankles/feet), each segment regarding musculoskeletal disorders. The first 3 questions referred to the period of the last 12 months and the last one refers to the last 7 days. The first question concerned about the sensations of pain, tingling and numbness; the second concerns the prevention of domestic and leisure activities related to symptoms; the third examined the demand for consultations with health professionals and the fourth evaluated the existence of some complaint or problem in the last week [10, 11]. The SF-36 questionnaire is generic tool that evaluates the quality of life through 11 questions that comprise 8 subdomains: functional capacity, physical aspects, general health status, vitality, social aspects, emotional aspects and mental health [9, 12]. The score of each subdomain ranges from 0 to 100. Values closer to 100 indicate a better quality of life. The Work Ability Index Questionnaire (WAIQ) is divided into 7 domains: current ability to work compared to best-of-life, ability to work against physical demands, number of current diagnosed illnesses, estimated loss of work, diseases, self-prognosis of work ability and mental resources, that assess the ability to work according to the different types of physical and mental requirements. The final score varies from 1 to 49 points, which classify the capacity at low (1–27), moderate (28–36), good (37–43) and optimal (44–49) [13], all described in terms of frequencies and means by the SPSS™ software.

3 Results

The sample in this pilot study consisted predominantly of women (80%) with an average age of 49.3 years and all reported having at least 1 child. Regarding education level, 50% of the sample had a complete high school level, 40% had completed higher education and 10% did not complete the basic level. The professional positions occupied by the sample were: 40% occupied by nursing, 30% general services, 20% administrative officers and 10% transportation. It was observed that 50% of the patients worked in the hospital for more than 5 years, 90% of them with a workload of 40 h per week and 10% with a workload of 30 h per week. Most of the patients in the sample work in the daytime and vary the working position between sitting and standing. Only 30% of patients routinely exercised. Most workers (70%) routinely used pain relief medication.

The Nordic questionnaire indicated predominance of complaints in 4 body segments: neck, shoulders, lumbar and knees. The result of question 1 indicated that in the last 12 months most pain (60%) were located in the lower back and knees. From the data obtained in question 2 it was possible to observe that 33–40% of the workers reported some difficulty in performing their daily life activities (DLAs) due to knee or lumbar pain. Among these workers, only 30% sought care from a health professional for this reason. Among those workers who had cervical pain and were therefore prevented from performing their DLAs in the last 12 months, only 20% sought a health professional for this reason. Through the analysis of question 4 it was possible to observe that the most affected parts of the body in the last 7 days were: shoulders, lumbar and hips/thighs.

The SF-36 questionnaire presented results divided among the subdomains: functional capacity 68.56, physical aspects 68.75, pain 42.72, general health status 67.33, vitality 66.72, social aspects 81.25, emotional aspects 70.83, and mental health 67.20. The Work Ability Index questionnaire presented general scores classified as moderate (28–36).

Theses finding related to Nordic Questionnaire showed at least one affected body segment are in accordance to other studies [14, 15]. A significant number of workers with pain or discomfort in the neck, knees, back and shoulders did not seek the help of health professionals, which may be related to individual perception about pain, unavailability of time for health care and also the self-medication mainly in individuals of the nursing sector. The result of the pain subdomain found in the SF36 questionnaire reinforces the idea that the worker's view on pain should be considered.

The impediment to performing daily life activities, which affected 33–40% of this sample, affected the performance of work, decreased the productivity or quality of services and could have an impact on the individual's personal life, with the need for interventions in various fields.

The score range (28–36) obtained from the Work Ability Index questionnaire was classified as moderate, and suggested actions to improve work capacity [8, 13]. The vitality and mental health subdomains of the SF-36 questionnaire scored below 70, and reflected the importance of the workers' biopsychosocial balance.

The future challenge is to find and bring new strategies for improving workers' health related to preventive actions in the need to be effective, in the conjunction of availability of financial resources in institutions.

This sample was composed of only 10 patients, however the application of these questionnaires has been introduced in the routine of evaluation of this specialized physiotherapy service, in order to obtain a more detailed and significant sample. From this pilot study we could draw a profile of the employees of this hospital and start a routine evaluation of work musculoskeletal risk factors. Then, propose future strategies to reduce these factors, by means of prevention with education, ergonomic and exercises approach.

4 Conclusion

This sample showed that besides high scores for SF 36 for these sedentary women, most use pain medication for musculoskeletal complaint with predominance of back, neck and shoulder and was identified necessity improvement the ability of work. Physical/biomechanical and psychosocial/organizational risk factors assessment by self-report questionnaires and ergonomic observational tools are essential, to define effective preventive measures for quality of life improvement of these workers. This pilot study showed the viability for a future intervention study including ergonomics and exercises.

References

1. David GC (2005) Ergonomics methods for assessing exposure to risk factors for work-related musculoskeletal disorders. Occup Med 55:190–199
2. Chiasson ME, Imbeau D, Major J, Aubry A, Delisle A (2015) Influence of musculoskeletal pain on workers' ergonomic risk-factor assessments. Appl Ergon 49:1–7
3. Mcdonald M, Dibonaventura M, Dacosta US (2011) Musculoskeletal pain in the workforce: the effects of back, arthritis, and fibromyalgia pain on quality of life and work productivity. J Occup Environ Med 53(7):765–770

4. Andersen LL, Clausen T, Burr H, Holtermann A (2012) Threshold of musculoskeletal pain intensity for increased risk of long-term sickness absence among female healthcare workers in eldercare. PLoS ONE 7(7):e41287

5. Walsh IAP, Corral S, Franco R, Canetti EEF, Alem MER, Coury HJCG (2004) Capacidade para o trabalho em indivíduos com lesões músculo-esqueléticas crônicas. Rev Saúde Pública 238:149–156

6. Li G, Buckle P (1999) Current techniques for assessing physical exposure to work-related musculoskeletal risks, with emphasis on posture-based methods. Ergonomics 42(5):674–695

7. Bianchessi DLC, Tittoni J (2009) Trabalho, saúde e subjetividade sob o olhar dos trabalhadores administrativo-operacionais de um hospital geral, público e universitário. Rev Saúde Coletiva 19:969–988

8. Martinez MC, Oliveira MRD, Latorre FMF (2009) Validity and reliability of the Brazilian version of the work ability index questionnaire. Rev Saúde Pública 43(3):525–532

9. Jenkinson C, Coulter A, Wright L (1993) Shortform 36(SF36) health survey questionnaire: normative data for adults of working age. BMJ 306:143740

10. Kuorinka I, Jonsson B, Kilbom A, Vinterberg H, Biering-Sørensen F, Andersson G, Jørgensen K (1987) Standardised nordic questionnaires for the analysis of musculoskeletal symptoms. Appl Ergon 18(3):233–237

11. Pinheiro FA, Troccoli BT, Carvalho CV (2002) Validity of the nordic musculoskeletal questionnaire as morbidity measurement tool. Rev Saúde Pública 36(3):307–312

12. Campolina AG, Bortoluzzo AB, Ferraz MB, Ciconelli RM (2011) Validation of the Brazilian version of the generic six-dimensional short form quality of life questionnaire (SF-6D Brazil). Ciencia Saúde Coletiva 16:1311–1330

13. Silva Junior SHA, Vasconcelos AGG, Griep RH, Rotenberg L (2011) Validity and reliability of the Work Ability Index (WAI) in nurses' work. Cad Saúde Pública 27(6):1077–1087 Rio de Janeiro

14. Isosaki M, Cardoso E, Glina DMR, Alves ACC, Rocha LE (2011) Prevalência dos sintomas osteomusculares entre trabalhadores de um serviço de nutrição hospitalar em São Paulo. SP Rev Bras Saúde Ocup 36(124):238–246

15. Skotte JH, Essendrop M, Hansen AF, Schibye B (2002) A dynamic 3D biomechanical evaluation of the load on the low back during different patient-handling tasks. J Biomech 35(10):1357–1366

Comparing the Strain Index and the Revised Strain Index Application in the Dairy Sector

Federica Masci[1]([✉]), Stefan Mandic-Rajcevic[1], Giovanni Ruggeri[2],
John Rosecrance[3], and Claudio Colosio[1]

[1] Department of Health Sciences of the University of Milan and International
Center for Rural Health of Santi Paolo e Carlo Hospital, Milan, Italy
Federica.masci@unimi.it
[2] University of Milan, Milan, Italy
[3] Department of Environmental and Radiological Health Sciences,
College of Veterinary Medicine and Biomedical Sciences,
Colorado State University, Fort Collins, CO, USA

Abstract. Repeated forceful efforts, high muscular loads in the wrist flexor and extensor muscles and awkward hand positions during milking can contribute to wrist and hand diseases. Thus, evaluating the risk of biomechanical overload of milking parlor workers' wrist with an accurate method is fundamental. Aims of the study were: (1) evaluating the wrist's biomechanical overload in the milking routine subtasks: pre-dipping, wiping/stripping, attacking, post-dipping; (2) comparing the results obtained from Strain Index and Revised Strain Index methods application; (3) evaluating the quality of the approach, the concordance and the possible difficulties in applying and interpreting the two evaluation methods.

3172 task cycles were evaluated on a population of fourteen male milking parlor workers using either SI or RSI.

The results showed that both those index have not exceeded the limit value for pre-dipping and post-dipping tasks, whilst the wiping/stripping resulted to be the most risky for milking parlor workers for both RSI and SI methods. The RSI responded to the objectives of increasing the accuracy and the precision of the evaluation.

Keywords: RSI · Risk assessment · MSD

1 Introduction

Muscle skeletal disorders (MSDs) are reported to be one of the most important issues in agricultural sector and specifically in the dairy farms [1–4].

The milking routine involves four tasks: (1) pre-dipping of the udders for sanitization; (2) udders' wiping and stripping (2–3 squirts of milk from each teat; (3) attachment of the milking cluster and (4) post-dipping of the udders for the final sanitization.

In Northern Italy's dairy farms, there are three types of milking parlors: herringbone, parallel and rotary. Obviously, the milking routine organization can differ from farm to farm due to the parlor system, its configuration and farm management.

© Springer Nature Switzerland AG 2019
S. Bagnara et al. (Eds.): IEA 2018, AISC 820, pp. 261–268, 2019.
https://doi.org/10.1007/978-3-319-96083-8_35

Despite the introduction of mechanized systems, milking tasks represent very strenuous activity both in small and in large herd size loose housing dairy farms [4, 5]. In particular pronation/supination of the wrists, milking cluster lifting [6, 7], repetitive motions [7, 8], dorsiflexion and radial deviation [9] are the main cause of the wrist/hand biomechanical overload.

The prevalence of MSDs reported in Italy in the agricultural sector and in dairy sector lead necessarily the physicians to adopt preventive strategies.

In this frame, it is evident the need of having accurate tools adequate for performing a complete risk assessment of hand/wrist district of milking parlor workers, in order to define the exposure scenarios.

In literature, there are several job analysis methods useful to assess the risk of biomechanical overload of upper extremities that is cause of an increased risk of developing MSDs [10]. In particular, the Strain Index (SI) tool was introduced in 1995 from Moore and Garg to evaluate a job's level of risk for developing a disorder of hand, wrist, forearm, or elbow [11]. The analyst evaluates six task variables (intensity of exertion, duration of exertion, exertions per minute, hand/wrist posture, speed of work, and duration of task per day). The product of the categorical "multiplier" assigned for each variable evaluated, produces a Strain Index score. The comparison of this index with a gradient identifies the risk level of the analyzed task.

In 2016, the method was revised in order to improve its accuracy and precision and address some limitations. The Revised Strain Index (RSI), conceptually similar to the SI, brings about three differences: omits the "speed of work" variable, replace the "duty cycle" with the duration per exertion and uses continuous variables and multipliers instead of categorical. These changes are supposed to provide a greater discrimination of risk predictions [12].

The present study aimed to assess the risk of wrist' biomechanical overload of a sample of milking parlor workers when performing their milking routine of pre-dipping, wiping/stripping, attacking, post-dipping, by using the Strain Index and Revised Strain Index methods. In addition, we wanted to compare the results obtained, in order to evaluate the quality of the approach, the concordance and possible difficulties in applying and interpreting the two evaluation methods.

2   Materials and Methods

A group of 40 male milking parlor workers was recruited in 21 dairy farms located in the Lombardy region (Italy), representative of the 3 milking parlors configurations in Italy: herringbone, parallel and rotary. We selected the sample from the lists of workers in the health surveillance program of the International Center for Rural Health (ICRH) of the *Santi Paolo e Carlo University Hospital* and by the company CSM services in Milan. Eligibility criteria were at least 3 years of work experience as milking parlor worker.

The first 60 min of the working shift were recorded, and Kinovea video analysis was used to analyze the 4 subtasks typical of the milking routine, as they are performed in Italy: pre-dipping, wipe/stripping, milking cluster attaching, post dipping.

The project was approved by the Ethical Committee of the Hospital. Before the project had started, the workers signed an informed consent to participation.

The risk of biomechanical overload was assessed both with SI and RSI, two methods which take into account quantitative and qualitative variables, identify a risk coefficient and determine if the work activity can be considered dangerous for the hand-wrist district.

In particular, for each subject 5 working cycles were observed and analyzed considering the 4 subtasks above mentioned, for a total of 3172 assessments. For simplicity, only the data referring to the dominant hand were processed for a total of 584 tasks.

The work cycles evaluated using the SI method were classified in 4 ranges, according to the final SI score obtained [11]:

- low risk: SI score < 3
- medium-low risk: SI score between 3 and 6.1
- medium-high risk: SI score between 6.2 and 13.5
- high risk classification: SI score above 13.5.

As regards the use of the RSI, scores lower than 10 were not considered risky, while scores greater than 10 were associated to the presence of biomechanical overload for the district under study [12].

2.1 Statistical Analysis

Data processing and statistical analyzes were performed using the programming language "R: A Language and Environment for Statistical Computing" - version 3.3.2 (2016-10-31) - "Sincere Pumpkin Patch", useful both for statistical processing and for graphical representation [13].

3 Results

The study involved a total of 40 milking parlor workers of which only 33 of the 40 milking sessions examined could be considered, due to the low quality of some videos, which did not allow analyzing and evaluating the working routine with clarity and precision.

Figures 1 and 2 show the distribution of work cycles evaluated using both the RSI and SI methods for each of the four different subtasks considered. On the ordinate axis the numbers of the analyzed tasks are shown while the calculated RSI values are indicated on the abscissa axis.

A total of Seventy pre-dipping tasks, 160 wiping/stripping tasks, 239 attaching tasks and 115 post dipping tasks were evaluated.

Study results show that low risk scores were assigned to the pre-dipping and post-dipping subtasks both with SI and RSI method, with exception of some cases were a medium-low risk index was assigned using the RSI. Conversely, the wiping/stripping task result to expose the majority of the workers to medium-high and high risk of

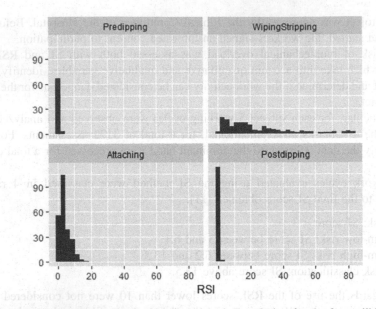

Fig. 1. Number of tasks analyzed and RSI assigned. Description for the four milking routine subtasks (pre-dipping, wiping stripping, attaching, post-dipping)

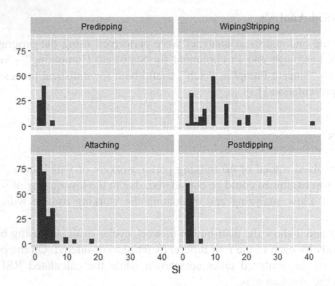

Fig. 2. Number of tasks analyzed and SI assigned. Description for the four milking routine subtasks (pre-dipping, wiping stripping, attaching, post-dipping)

biomechanical overload, considering both the risk assessment methods. To the attaching tasks are often assigned medium and low-level risk scores.

We also compared the concordance between the SI method and the RSI method based on the risk classification, as shown in Table 1. For the low risk classification, there is a greater concordance between the methods, equal to 99.2%. The concordance decreases for tasks classifies in the low to medium risk range (89.05% of concordance). The lowest agreement between the two methods is for the medium to high risk, where it is equal to 73.74%. Lastly, for the high risk classification there is a concordance equal to 87.88%.

Consequently, for each subtasks, the same comparison was made analyzing the specific SI and RSI scores (Tables 2, 3, 4 and 5).

As for the pre-dipping, Table 2 shows no medium-high or high risk classification was assigned to any of the tasks evaluated. The two methods have the maximum agreement (100%) for low and medium-low risk assessments.

Table 1. Comparison between RSI e SI and risk classification for the total of 584 tasks evaluated.

1995 Strain Index (SI)				
	SI ≤ 3.0	3.0 < SI ≤ 6.1	6.1 < SI ≤ 13.5	SI > 13.5
	Low risk	Low-medium risk	Medium-high risk	High risk
N° tasks (N)	251	201	99	33
RSI: Mean Value	1.82	4.08	30.97	41.94
RSI: Standard Deviation	1.83	4.58	54.52	23.78
RSI: Range	0.42–16.55	0.52–24.63	3.09–448.58	5.5–83.49
RSI: Median	1.23	1.55	17.07	37.17
RSI: Interquartil range	0.66–2.26	0.84–6.31	9.14–31.91	27.33–56.45
Risk classification, % of concordance	99.2%	89.05%	73.74%	87.88%

Table 2. Comparison between RSI e SI and risk classification on a total of 70 tasks of pre-dipping analyzed.

1995 Strain Index (SI) PRE-DIPPING				
	SI ≤ 3.0	3.0 < SI ≤ 6.1	6.1 < SI ≤ 13.5	SI > 13.5
	Low risk	Low-medium risk	Medium-high risk	High risk
N° tasks (N)	25	45	0	0
RSI: Mean Value	0.66	0.93	–	–
RSI: Standard Deviation	0.17	0.24	–	–
RSI: Range	0.47–0.96	0.64–1.78	–	–
RSI: Median	0.63	0.87	–	–
RSI: Interquartil range	0.51–0.81	0.77–1.01	–	–
Risk classification, % of concordance	100%	100%	–	–

Table 3. Comparison between RSI e SI and risk classification on a total of 160 tasks of wiping/stripping analyzed.

1995 Strain Index (SI) WIPING/STRIPPING SUBTASK				
	SI ≤ 3.0	3.0 < SI ≤ 6.1	6.1 < SI ≤ 13.5	SI > 13.5
	Low risk	Low-medium risk	Medium-high risk	High risk
N° tasks (N)	2	44	86	28
RSI: Mean Value	1.42	8.01	34.51	48.02
RSI: Standard Deviation	0.28	6.08	57.70	20.40
RSI: Range	1.22–1.61	1.17–24.63	3.09–448.58	23.99–83.49
RSI: Median	1.42	5.17	18.65	42.85
RSI: Interquartil range	1.32–1.51	3.39–10.86	12.58–34.72	30.84–66.83
Risk classification, % of concordance	100%	68.18%	82.56%	100%

Table 3 show the two methods have the highest percentage of agreement for the low and high risk assessment classification, while for medium-low risk and medium-high the percentages of concordance were respectively 68.18%, and 82.56%.

As for the attaching subtask, the lowest level of agreement between the SI and the RSI method were reported. As shown in Table 4, in fact, for a medium and high risk level, there was a concordance of 15.38% for the first and 20% for the second. The agreement between the two methods for the low and medium-low risk classification was definitely higher: respectively 98.74% and 87.1% of concordance.

Table 4. Comparison between RSI e SI and risk classification on a total of 239 tasks of attaching analyzed.

1995 Strain Index (SI) ATTACHING				
	SI ≤ 3.0	3.0 < SI ≤ 6.1	6.1 < SI ≤ 13.5	SI > 13.5
	Low risk	Low-medium risk	Medium-high risk	High risk
N° tasks (N)	159	62	13	5
RSI: Mean Value	2.43	6.16	7.53	7.89
RSI: Standard Deviation	2.05	3.40	2.47	1.97
RSI: Range	0.42–16.55	0.52–14.33	4.75–13.02	5.5–10.44
RSI: Median	1.97	6.55	6.90	8.33
RSI: Interquartil range	1.31–2.99	3.14–8.27	5.81–8.3	6.36–8.82
Risk classification, % of concordance	98.74%	87.1%	15.38%	20%

In similarity with the pre-dipping results, for the post-dipping subtask, there were no cases of cycles considered at medium or high risk and both low and medium-low risk evaluations had the highest percentage of concordance (Table 5).

Table 5. Comparison between RSI e SI and risk classification on a total of 115 tasks of post-dipping analyzed.

1995 Strain Index (SI) POST-DIPPING				
	SI ≤ 3.0	3.0 < SI ≤ 6.1	6.1 < SI ≤ 13.5	SI > 13.5
	Low risk	Low-medium risk	Medium-high risk	High risk
N° tasks (N)	65	50	0	0
RSI: Mean Value	0.76	0.88	–	–
RSI: Standard Deviation	0.20	0.21	–	–
RSI: Range	0.43–1.36	0.61–1.62	–	–
RSI: Median	0.75	0.83	–	–
RSI: Interquartil range	0.61–0.88	0.75–0.94	–	–
Risk classification, % of concordance	100%	100%	–	–

4 Conclusions

Among the four typical subtasks, both SI and RSI methods lead us to consider the wiping/stripping as the riskiest activity for the wrist' biomechanical overload of those workers.

In general, the methods resulted to be in agreement for the assignment of low and medium-low risk assessment classification of the tasks (RSI < 10 for and SI ≤ 6.1), while the concordance decreased for the medium-high and high classification, with some exception. In particular, for the wiping/stripping subtask we noticed an 100% of agreement also in the high risk classification. The main differences between the methods are related to the most complex tasks of the milking routine: wiping/stripping and attaching. This funding confirms the major revision of the method affect the risk assessment in the way it requires the observer a more precise evaluation of the characteristics of the efforts, before appointing a specific multiplier to the variables.

As for the comparison between SI and RSI, we may conclude that RSI seems to be more accurate and precise than the SI, since the possibility of assigning a continuous multiplier rather than a categorical one, to each of the five evaluated variables. In addition, the replacement of the "duty cycle" with "duration per exertion" in RSI facilitated the evaluation procedures.

Conversely, evaluating the wrist/hand posture with RSI resulted to be difficult for the observer, since lack of instructions nowadays available to give a sound assignment of the multiplier in particular cases we faced during our assessment. The changes introduced by the revision of the method should be supported by a more accurate description of the assessment procedures when the effort requires the worker to make both flexion and extension movements.

Further studies focusing on the RSI method application are desirable to bridge this gap and to get a better understanding of the model procedures.

These findings will also allow researcher to a make a more accurate risk assessment evaluation of the work activities in agriculture.

References

1. Raffi GB, Lodi V, Malenchini G, Missere M, Naldi M, Tabanelli S, Violante FS, Minak GJ, D'elia V, Montesi M (1966) Cumulative trauma disorders of the upper limbs in workers on an agricultural farm. Arh Hig Rada Toksikol 47(1):19–23
2. Nonnenmann MW, Anton DC, Gerr F, Yack H (2010) Dairy farm worker exposure to awkward knee posture during milking and feeding tasks. J Occup Environ Hyg 7(8):483–489
3. Pinzke S (2003) Changes in working conditions and health among dairy farmers in southern Sweden. A 14-year follow-up. Ann Agric Environ Med 10(2):185–195
4. Patil A, Rosecrance J, Douphrate D, Gilkey D (2012) Prevalence of carpal tunnel syndrome among dairy workers. Am J Ind Med 55(2):127–135
5. Masci F, Mixco A, Brents CA, Murgia L, Colosio C, Rosecrance J (2016) Comparison of upper limb muscle activity among workers in large-herd U.S. and small-herd Italian dairies. Front Publ Health 4:141. section Occupational Health and Safety
6. Stal M, Pinzke S, Hansson G-A (2003) The effect of workload by using a supporting arm in parlour milking. Int J Ind Ergon 32:121–132
7. Stal M, Pinzke S, Hansson G-A, Kolstrup C (2003) Highly repetitive work operations in a modern milking system. a case study of wrist positions and movements in a rotary system. Ann Agric Environ Med 10:67–72 Ann The effect of workload by using a support arm in parlour milking
8. Pinzke S, Stal M, Hansson G (2001) Physical workload on upper extremities in various operations during machine milking. Ann Agric Environ Med 8:7
9. Stål M (1999) Upper extremity musculoskeletal disorders in female machine milkers-an epidemiological, clinical and ergonomic study [doctoral thesis]. University of Lund, Lund, Sweden
10. Garg A, Kapellush JM (2011) Job analysis techniques for distal upper extremity disorders. Rev Hum Factors Ergon 7(1):149–196. https://doi.org/10.1177/1557234X11140386
11. Moore JS, Garg A (1995) The strain index: a proposed method to analyze jobs for risk of distal upper extremity disorders, pp 443–458
12. Garg A, Moore JS, Kapellusch JM (2017) The revised strain index: an improved upper extremity exposure assessment model. Ergonomics 60(7):912–922. http://dx.doi.org/10.1080/00140139.2016.1237678
13. R Development Core Team (2012) R: A language and environment for statistical computing. R Foundation for Statistical Computing. Vienna, Austria

Human Factors Related to the Use of Personal Computer: A Case Study

Fabíola Reinert[1]([✉]) [iD], Raoni Pontes Caselli[1],
Antônio Renato Pereira Moro[1] [iD], Leila Amaral Gontijo[1] [iD],
and Marcelo Gitirana Gomes Ferreira[2] [iD]

[1] Universidade Federal de Santa Catarina,
Caixa Postal 476, Florianópolis, SC, Brasil
fabiola.reinert@gmail.com
[2] Universidade Estadual de Santa Catarina,
Me. Benvenuta 1907, Florianópolis, SC, Brasil

Abstract. The aim of this study was to determine and compare the ergonomic requirements in the use of desktop computers and laptops. For comparison, postures in the sagittal and transversal planes involved while performing a typing and editing task on a desktop and laptop computers were verified. Thus, a case study was conducted with a Brazilian male of medium height, who works around 20 h per week in a computer. Nine spherical markers of 2.5 cm in diameter were joined up with duct tape on the right side of the participant and the positions adopted by the subject were analyzed in the use of a desktop and a laptop computer, through direct observation and filming, using the videography technique. Among the most significant results, the use of desktop computers compared to laptops allows a more upright posture of the head and neck, as well as a more neutral posture of the shoulders, elbows and wrist. It is concluded that the use of desktop computers has lesser biomechanical demands since it showed angles closer to a neutral position and consequently having a lower risk of musculoskeletal complaints.

Keywords: Postural angles · Desktop computer · Laptop computer

1 Introduction

The use of the computer is commonly related to the appearance of musculoskeletal complaints [1–3]. Generally, non-neutral postures are considered to be harmful [4]. Gerr et al. [5] states that the postures adopted in the use of the desktop computer are different from those adopted in the use of portable computers.

Portable computers can be used in a variety of postures when not attached to the workstation [4], however, increasing their portability also increases the exposure to potential risk factors for musculoskeletal complaints in relation to desktop computers [6]. In particular, since the display and keyboard are connected, the screen height is usually lower than recommended [6, 7].

Laptop computer users reported more postural constraints and higher neck muscle activity than desktop computers. The complaints of ocular and musculoskeletal

S. Bagnara et al. (Eds.): IEA 2018, AISC 820, pp. 269–276, 2019.
https://doi.org/10.1007/978-3-319-96083-8_36

discomfort as well as difficulty of typing are also greater during the work with portable computers [8]. In comparison to desktop computers, laptop computers results in increased neck flexion and head inclination [9, 10] and increased neck extension activity [5, 11].

Considering the ergonomics requirements involved in the use of personal computers, the objective of this study was to determine if there are risk factors associated to the evolution of computers, verifying the postures and body angulations involved in the use of desktop computers and laptops.

2 Method

2.1 Study Participant

The subject analyzed was a male, 29 years old and works around 20 h per week with a computer, and currently uses a laptop to work. The individual measures 1.70 m in height and has 62 kg, considering that according to Iida [12] the average height of a Brazilian is 1.70 m and the weight is 60 kg, the subject was considered as a median standard. The subject had no musculoskeletal disorders prior to the study.

2.2 Experimental Protocol

In order to carry out the study, three different situations were considered: (No. 1) use of the desktop computer coupled to the workstation, (No. 2) use of the laptop coupled to the workstation, (No. 3) use of the laptop in an armchair available in the workplace. A desktop computer of LG brand, and a laptop ASUS 546C was used.

The method of approach was documental analysis and direct observation using videography technique. For the videography was used a Nikon L110 camera positioned 1 m of the participant and parallel to the sagittal plane, and a Fujifilm S4200 camera, positioned 0.5 m of the participant and parallel to the transverse plane. The workstation consisted of a 69 × 170 cm table and an adjustable chair, and the subject was instructed to adjust the chair to a comfortable height. Nine spherical markers of 2.5 cm in diameter were joined up with duct tape on the right side of the participant. To better identify the postural angles, the markers were located in: side margin of the eye; behind the ear; spinous process of C7; acromioclavicular joint; lateral epicondyle of the humerus; ulnar styloid process; head of the fifth metacarpal; midpoint of greater trochanter of the femur; and computer screen center in the sagittal plane.

From the nine markers and the horizontal and vertical reference lines, the postural angles (Fig. 1) were defined as the following, also used by Castellucci and Benitez [7]:

(1) Angle of view: angle formed by the line going from the lateral margin of the eye to the center of the screen, with respect to the horizontal line. The horizontal reference line is designated zero. Below this, the angle value is negative and higher than this, positive.

(2) Head inclination: angle formed by the line going from the lateral margin of the eye to behind the ear with respect to the horizontal reference line. Zero is the horizontal line. Below this, the angle value is negative and higher than this, positive.

(3) Neck flexion: angle forming by the line going from behind the ear to the spinous process of C7, with respect to the vertical reference line. Zero is the vertical line. Prior to this the angle value is positive and posterior to this, negative.

(4) Craniocervical angle: angle formed by the line going from the outer margin of the eye to behind the ear and the line from behind the ear to the spinous process of C7.

(5) Cervicothoracic angle: angle formed by the line going from behind the ear to the spinous process of C7 and the line from the spinous process of C7 to the midpoint of the major trochanter of the femur.

(6) Shoulder flexion: angle formed along the line from the acromio-clavicular joint to the lateral epicondyle of the humerus, with respect to the vertical reference line. Zero is the vertical line. Prior to this the angle value is positive and posterior to this negative.

(7) Elbow flexion: angle formed by the line from the acromio-clavicular joint to the lateral epicondyle of the humerus, and the line of the lateral epicondyle of the humerus to the styloid process of the ulna. The value of this angle is progressively positive and counterclockwise.

(8) Wrist flexion-extension: angle formed by the line from the lateral epicondyle of the humerus to the styloid process of the ulna and the line from the styloid process of the ulna to the head of the fifth metacarpal. The first line is called zero. Below this the angle value is negative and higher than this, positive.

(9) Trunk inclination: angle formed by the line from the spinous process C7 to the midpoint of the greater trochanter of the femur, with respect to the horizontal reference line.

(10) Wrist Radial-ulnar deviation: angle formed by the line from the lateral epicondyle of the humerus to the styloid process of the ulna and the line from the styloid process of the ulna to the head of the fifth metacarpal in the transverse plane. The first line is called zero. Below this the angle value is negative and higher than this, positive.

The work activities of the studied individual, involving typing and text editing, were recorded for approximately 15 min, and the intermediate 5 min were analyzed. The approaches were conducted inside the working environment during the practice of the activity, in the morning of three different days, before any interaction with the computer that day. For image analysis, the softwares used were Kinovea and Corel Draw while for data collection the software used was Microsoft Excel.

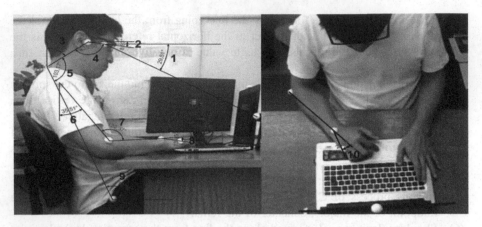

Fig. 1. Images of the subject in the sagittal and frontal plane and the identification of the anatomical reference points, with the respective angles of interest: (1) angle of view, (2) head inclination, (3) neck flexion, (4) craniocervical angle, (5) cervicothoracic angle, (6) shoulder flexion, (7) elbow flexion, (8) wrist flexion-extension, (9) trunk inclination, (10) wrist radial-ulnar.

3 Results

The results of the postures assumed during condition No. 1 (Desktop use), condition No. 2 (Laptop use), and condition No. 3 (Laptop use in the armchair) (Fig. 2) can be verified in Table 1. Sixty images of the videos of each condition were analyzed to generate the necessary data for the study of postural angles.

According to the results obtained, it can be observed that the experimental Condition No. 1, in relation to the other two conditions, presented a lower value in the: angle of view (condition No. 2: difference of 7.31°, condition No. 3: difference of 15.67°), head inclination (condition No. 2: difference of 11.32° and condition No. 3: difference of 30.87°), neck flexion (condition No. 2: difference of 10.37° and condition No. 3: difference of 27.67°), shoulder flexion (condition No. 2: difference of 13.01° and condition No. 3: difference of 15.4°), elbow flexion (condition No. 2: difference of 15.73° and condition No. 3: difference of 36.63°), wrist flexion-extension (condition No. 2: difference of 5.62° and condition No. 3: difference of 20.51°) and wrist radial-ulnar deviation (condition No. 2: difference of 5.83° and condition No. 3: difference of 9.62°). The craniocervical and cervicothoracic angles were higher in the experimental condition No. 1 (Craniocervical angle: condition No. 2: difference of 10.02° and condition No. 3: difference of 12.13°; cervicothoracic angle: condition No. 2: difference of 13.04° and condition No. 3: difference of 17.87°) while the trunk inclination was greater in condition No. 2 (condition No. 1: difference of 11.85° and condition No. 3: difference of 15.78°).

These values demonstrate that the Condition No. 1 performed with a desktop computer, compared to other conditions, translates into a more upright posture of the head and neck and a more neutral posture of the shoulders, elbows and wrists. It is also noted that the Condition No. 2, performed with a laptop on a work table, presented

Fig. 2. Pictures of the subject in the three different situations analyzed: above, condition No. 1 (desktop use), in the middle, condition No. 2 (laptop use) and below, condition No. 3 (laptop use in the armchair).

values closer to a neutral posture compared to the Condition No. 3, performed with a laptop in an armchair, in all postural angles except the craniocervical angle, the cervicothoracic angle and the trunk inclination.

The results found in this study were similar to those obtained in other studies [7], demonstrating that the most impacting factors on postural angles between the use of a

Table 1. Summary of the positions analysis assumed in the three conditions.

Postural angles	Condition No. 1(°)		Condition No. 2(°)		Condition No. 3(°)	
	Average	SD	Average	SD	Average	SD
1	−21,18	5,51	−28,49	6,83	−36,85	5,50
2	1,75	7,21	−9,57	9,06	−29,12	7,69
3	47,42	10,43	57,79	8,95	75,09	8,53
4	142,86	8,39	132,84	6,12	130,73	8,12
5	117,40	7,41	104,36	6,19	130,73	8,12
6	24,78	9,84	37,79	12,83	40,18	12,58
7	106,79	16,99	122,52	14,97	143,42	16,43
8	−8,12	11,08	−13,74	12,32	−28,63	12,30
9	102,34	8,92	114,19	9,67	98,41	8,44
10	18,52	10,92	24,35	9,19	28,14	9,34

desktop computer and a laptop are the height of the computer screen and use of external components (keyboard and mouse). Castellucci and Benitez [7] obtained the following values for angle of view, head inclination, neck flexion, craniocervical angle, cervicothoracic angle, shoulder flexion, elbow flexion, wrist flexion-extension and the trunk inclination, respectively: −40,3°; −4.64°; 61.42°; 147.13°; 113.44°; 20.87°; 94.66°; −11.59°; 94.23° in an experiment with laptop coupled to a workstation, obtaining better results using the laptop with external components as keyboard and mouse.

In this study, we obtained −9.57° for head inclination in the use of the laptop and 1.75° in the use of the desktop computer, similar to other studies that obtained the values −9.8° for head inclination in the use of the laptop and 1.75° in the use of the desktop computer [9], also 57.4° for neck flexion in the laptop and 50° in the desktop computer, while we obtained 57,79° and 47,42°, respectively. A study with laptops obtained −12.7° for wrist flexion-extension [1], and another study [3] found −12.79° for wrist flexion-extension and 39.52° for shoulder flexion, compared to −13.74° and 37.79° obtained in this case study.

Regarding the values found for angle of view, the normal line of sight is between 10 and 15° below the horizontal line [13]. This normal line of sight is the resting position of the eyes, and it is recommended that the objects to be viewed are at this limit. Also, in the posture seated with the trunk erect, people prefer to visualize objects at 20° below the horizontal line [12]. In this study, in all conditions, the angle of view exceeds 20° below the horizon line, however, in the case of the desktop computer, it exceeds only 1.18°, and for the laptop this value is higher, surpassing in around 8° in condition No. 2 and 16° in condition No. 3.

The neck flexion should be up to 20° [12]. Above 30° the pain in the neck begins to appear. Again, in all conditions the value exceeds the ideal, and in condition No. 1 the value is around 20° above the recommended, but in condition No. 2 and No. 3 this value exceeds the recommended in 33° and 51° respectively. The recommended trunk inclination is between 90° and 120° [12] and all conditions are within the recommended value.

A wrist ulnar deviation greater than 20° is significantly associated with musculoskeletal injuries [5]. In the case of the desktop computer (condition No. 1), the ulnar deviation of the wrist is 18.52°, however in the case of the laptop, the ulnar deviation is greater than 20° in both condition No. 2 (24.35°) and condition No. 3 (28.14°), being condition No. 3 the most problematic.

Even using the same working configuration, there are divergent values between the use of the laptop and the desktop computer, where the desktop computers requires less biomechanical demands since it showed angles closer to a neutral position, and consequently having a lower risk of musculoskeletal complaints.

4 Conclusion

The objective of this study was to determine and compare the ergonomic differences in the use of desktop computers and laptops, verifying the sagittal and transverse plane postures involved during the execution of a typing and editing task in both equipment. The analysis of the average postural angles assumed during the execution of tasks with a desktop computer demonstrated a more upright posture of the head and neck and a more neutral posture of the shoulders, elbows and wrist compared to the postural results during the use of the laptop.

The obtained values are similar to the researched literature, evidencing that the use of the laptop in comparison to the desktop represents a greater risk of musculoskeletal complaints. The external components of the analyzed computers such as monitor, keyboard and mouse stand out, by allowing a greater flexibility of spatial configuration and facilitating the personal adjustment of the user to more neutral postures. Although laptops allow integration with external components, the mobility factor is a strong influence in the choice of a suitable place for its use. As noted in the case study, its use in non-appropriate places such as armchairs results in less neutral postures and a greater risk of complaints.

Thus, it is concluded that although laptops allow greater mobility, they would not be adequate for periods of continuous use, and in this case the use of desktop computers is indicated. This study analyzed the postures in the laptop without considering external components such as mouse and keyboard, therefore, it is recommended as future studies the postural analysis in laptops with the help of external components, to verify the influence of the display of laptops in assuming less neutral postures.

References

1. Donoghue M, O'reilly D, Walsh M (2013) Whist postures in the general population of computer users during a computer task. Appl Ergon 44:42–47
2. Aydeniz A, Gursoy S (2008) Upper extremy musculoskeletal disorders among computer users. Turk J Med Sci 38:235–238
3. Conlon CF, Rempel DM (2005) Upper extremity mononeuropathy among engineers. J Occup Environ Med 47:1276–1284

4. Gold J, Driban J, Yingling V, Komaroff E (2012) Characterization of posture and comfort in laptop users non-desk settings. Appl Ergon 43:392–399
5. Gerr F, Monteilh CP, Marcus M (2006) Keyboard use and musculoskeletal outcomes among computer users. J Occup Rehabil 13(3):265–277
6. Asundi K, Odell D, Luce A, Dennerlein J (2012) Chances in posture through the use of simple inclines with computers placed on standard desk. Appl Ergon 43:400–407
7. Castellucci I, Benitez LZ (2011) Postura, disconfort y produtividad durante la ejecución de tareas de mecanografía en computadores personales portátiles tipo netbook, con y sin modificaciones ergonómicas. In: Proceedings of the 9th international conference on occupational risk prevention, ORP 2011, Santiago, pp 1–9
8. Jonai H, Villanueva M, Takata A, Sotoyama M, Saito S (2002) Effects of the liquid crystal display tilt angle of a notebook computer on posture, muscle activities and somatic complaints. Int J Ind Ergon 29:219–229
9. Straker L, Jones KJ, Miller J (1997) A comparison of the postures assumed when using laptop computers and desktop computers. Appl Ergon 28(4):263–268
10. Sommerich CM, Starr H, Smith CA, Shivers C (2002) Effects of notebook computer configuration and task on user biomechanics, productivity, and comfort. Int J Ind Ergon 30:7–31
11. Seghers J, Jochem A, Spaepen A (2003) Posture, muscle activity and muscle fatigue in prolonged VTD work at different screen height settings. Ergonomics 46(7):714–730
12. Iida I (2005) Ergonomia – Projeto e Produção. Edgar Blucher, São Paulo
13. Chaffin DB, Andersson GB, Martin BJ (1999) Occupational biomechanics, 3rd edn. Wiley, New York

Injury Claims from Steep Slope Logging in the United States

John Rosecrance[(⊠)] and Elise Lagerstrom

Colorado State University, Fort Collins, CO 80523, USA
John.Rosecrance@colostate.edu

Abstract. Logging is one of the most dangerous occupations in the United States. Although logging injuries within the U.S. have decreased over the last 20 years, injuries in this sector continue to exceed the rate of total recordable cases for all U.S. industries combined. **Methods:** Workers' Compensation injury claim data from two workers' compensation providers, which cover companies' active in the logging industry of Montana and Idaho were obtained. All injury and fatality claims occurring from July 2010 to June 2014 were obtained from companies in the logging industry (NAICS 113). Injury claim data from each company contained information on demographics, variables related to the time, type and source, of injury, as well as the cost associated with each injury claim. **Results:** A total of 801 workers' compensation claims were analyzed for the time period July 2010 to June 2015. The most common nature of injury were sprain/strain injuries, followed by contusions, and lacerations. Inexperienced workers (>6 months experience) accounted for over 25% of claims. Fatalities had the highest median claim cost ($274,411 USD). **Conclusions:** Injury prevention efforts in the logging industry within the Intermountain region should be focused on early training, engineering controls, and administrative controls; all designed to promote a culture and climate of safety, communication, and shared responsibility. The results of this project were used as the basis and justification for the development and implementation targeted safety interventions addressing the specific safety issues associated with logging in the Intermountain region.

Keywords: Logging · Injuries · Fatalities · Costs · Workers' compensation Claims

1 Introduction

Logging is one of the most dangerous occupations in the United States. According to the Bureau of Labor Statistics (BLS), in 2014, the national occupational fatality rate for all industries was 3.3 workers per 100,000 FTE (Bureau of Labor Statistics 2017). During that same period, the occupational fatality rate for forestry workers was 92 per 100,000 FTE, 28 times higher than all industries combined (United States Department of Labor 2015). Though logging injuries within the Untied States have decreased over the last 20 years, from 8.9 per 100 FTE in 1994 (United States Bureau of Labor Statistics 2015) to 5.1 per 100 FTE in 2014 (United States Bureau of Labor Statistics

© Springer Nature Switzerland AG 2019
S. Bagnara et al. (Eds.): IEA 2018, AISC 820, pp. 277–282, 2019.
https://doi.org/10.1007/978-3-319-96083-8_37

2015), injuries in this sector continue to exceed the rate of total recordable cases for all industries combined (3.2 per 100 FTE) (United States Bureau of Labor Statistics 2015).

Within the logging industry, there are different risks based on job task and logging system used. The forestry and logging industry subsector is divided into the following job tasks by the United States Department of Labor: fallers, supervisors, logging equipment operators, and truck drivers (United States Department of Labor 2015). The logging system used determines the number of workers in each of these different job tasks as well as the configuration of equipment. There is a wide range in the possible configurations of machines and processes used in harvesting (i.e. logging system). The logging system of choice is dependent on the terrain and region of the harvest, as well as the resources of the contracted logging company.

The variety of work methods within the logging industry likely contributes to the differences in injury and fatality rates among professional loggers in the United States. Job task, logging system type, and degree of mechanization result in differences in the prevalence, type, and severity of occupational injuries and fatalities (Myers and Fosbroke 1994). Some research has suggested that increasing mechanization of the industry may decrease the number of injuries and fatalities associated with logging (Bordas et al. 2001). For example, Myers and Fosbroke (1994) found a significant difference in the fatality rate of general logging labor in comparison to machinery operators (Myers and Fosbroke 1994). General logging labor (i.e. fellers, limbers, buckers, and choke setters) had an occupational fatality rate of 371.8 per 100,000, while machinery operators had an occupational fatality rate of 48.5 per 100,000 (Myers and Fosbroke 1994). Further, an analysis of injury claims data before and after mechanization indicated a significant decrease in injury claims when companies transitioned from conventional (i.e. chainsaw) logging to feller-buncher tree harvesting (Bell 2002).

The objective of this study was to conduct an analysis of workers' compensation data to better understand the nature and cost of injuries and fatalities in the logging industry of the Idaho and Montana intermountain region of the U.S. The goal of the analysis was to contribute to the development of logging safety interventions in the Intermountain West region.

2   Methods

Injury claim data were obtained from two workers' compensation providers (Associated Loggers Exchange of Idaho and Montana State Fund), which cover companies active in the logging industry of Montana and Idaho. All injury and fatality claims occurring from July 2010 to June 2014 were obtained from companies in the logging industry. Injury claim data from each company contained information on demographics, variables related to the time, type, and source of injury, as well as the cost associated with each injury claim. All personal identifiers were removed from the claims data by the workers' compensation providers.

Frequency statistics were calculated for demographic and injury claim variables. Chi-squared tests were performed to determine if there was a significant difference in the distribution of variables relating to the timing of claims including day of the week, month of the year, season, and fiscal year. Chi-square tests were performed to determine if there was a significant difference in the number of claims by age group, level of experience, job type, data source, state, claim type, and body part injury. Fisher's Exact Test was used in place of chi-square analysis for cell counts with fewer than five observations. We implemented a Kruskal-Wallis test to determine if differences exist in the mean age and length of experience by job title (sawyer/hooker, equipment operator, truck driver, supervisor/owner, mill operator, and other).

To accurately compare year to year workers compensation costs, all claim values were adjusted to 2015 dollars using the CPI inflation calculator provided by the Bureau of Labor Statistics (Bureau of Labor Statistics 2016). Descriptive statistics (median and range values) for age, length of experience, and claim values were calculated by job type. Kruskal-Wallis tests were performed to determine if significant differences existed in age and length of experience based upon job task. Fisher's exact test was performed to determine if there was a significant difference in the distribution of incident type and nature of injury by job type. Due to the association between nature of injury and incident type, a two-way factorial analysis of variance (ANOVA) was performed to determine if the significant differences seen in the ANOVAs for both nature of injury and incident type by adjusted total claim value were due to the effect of the factors independently, interaction, or due to confounding. As healthcare data costs were highly skewed, we log transformed data for analysis (Malehi et al. 2015).

Because workers' compensation claims data did not follow a normal distribution, a Wilcoxon-Mann-Whitney test was used to determine if there was a significant difference between the median claim value based on level of experience (<6 months v. ≥ 6 months). Kruskal-Wallis tests were performed to determine if significant differences existed in the median claim values based upon age group, job task, nature of injury, or injury type. Significance for all tests was based on alpha of 0.05.

3 Results

Eight hundred and one workers' compensation claims were analyzed for the period July 2010 to June 2015 (Table 1). Workers between ages 50–59 had the highest proportion of claims (27%) and the mean length of employment at the time of injury was 74.52 months (SD = 105.84). Approximately 26% of claims occurring to employees with less than 6-months of experience (Fig. 1).

Chi-square tests indicated significant differences in the number of claims by day of the week ($\chi 2$ = 234.89, df = 6, p < .0001), month of the year ($\chi 2$ = 67.12, df = 11, p < .0001), season ($\chi 2$ = 54.43, df = 3, p < .0001), and fiscal year ($\chi 2$ = 23.34, df = 4, p = .0001). Mondays accounted for the greatest proportion of claims (21.22%), claims decreased by day of the week with only 14 workers' compensation claims occurring on Sundays (1.75%).

The highest number of claims occurred in the months of July (10.5%), August (10.5%), and September (10.7%), while the lowest number of claims occurred in March (4.0%) and April (3.25%). The spring season only accounted for 111 claims while the autumn, summer, and winter seasons accounted for 240, 233, and 217 claims respectively.

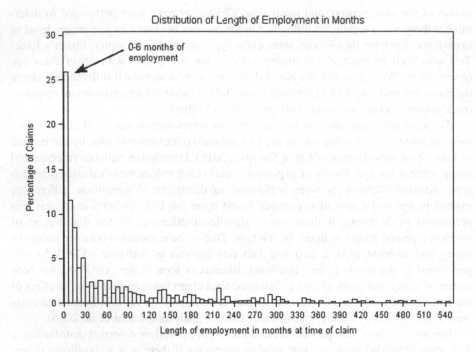

Fig. 1. Distribution of the worker's length of employment in months at the time of workers' compensation claim.

Claim characteristics of incident type and nature of injury varied significantly based up job type. Fisher's exact test indicated a significant difference in the type of incidents that occur based upon job task ($p < .0001$), as well as a significant difference in the nature of injury based upon job task ($p < .0001$). The most common incident type for sawyers and hookers was struck by injuries (51.3%), followed by falls (19.9%). The incident type for other job tasks was different than that of the sawyers and hookers, with falls being a more prominent incident type than struck by injuries for equipment operators, supervisors or owners, and truck drivers. The incident type distribution for equipment operators mirrored that of truck drivers with falls being the most prevalent accident type followed by overexertion and struck by injuries. Two out of three fatality claims recorded during this period occurred to sawyers/hookers, with the third occurring to a truck driver. Across all job types, sprain/strains were the most common nature of injury.

There was no significant difference in the median cost of the total claim values based upon age group ($\chi 2 = 8.45$, df = 5, p = 0.13), experience level ($\chi 2 = 3.18$, df = 1, p = 0.0745) or job task ($\chi 2 = 10.38$, df = 5, p = 0.0651). However, median total claim cost was significantly different by incident type ($\chi 2 = 32.23$, df = 9, p = 0.0002). Equipment overturns, caught between, and overexertion type claims had the highest median total claim values at \$5,961 (range = \$539–\$1,155,171), \$5,409 (range = \$138–\$184,587) and \$3,551 (range = \$0–\$320,773), respectively. Medical costs associated with animal bites and insect stings (median = $303, range = $122–$1,250) had the lowest claim costs.

There were significant differences in median total claim costs by the nature of injury (χ^2 = 129.99, df = 7, p < .0001). Fatalities had the highest median total claim cost at $274,411 (range = $170,762–$412,613), followed by multiple injuries ($17,138, range = $538–$1,155,171) and fractures/dislocations ($11,466, range = $0–$1,021,543). Contusions/abrasions had the lowest median claim cost with a median adjusted claim value of $810 (range = $0–$565,549).

4 Discussion

The logging industry in the U.S. Intermountain West region of Idaho and Montana presents a challenge for occupational injury and fatality prevention. Due to the terrain, remoteness of the region, and availability of emergency services, logging injuries and fatalities in this region are an ongoing challenge. While there are many similarities between the results of the present study and the work of other researchers throughout the country and world, there are also differences due to the unique attributes of the region, which contribute to differences in harvesting practices, techniques, and occupational culture.

Our findings support the findings of Laschi and associates (Laschi et al. 2016) that workdays on a Monday were determined to have the highest percentage of accidents, as well as the work of Albizu-Urionabarrenetxea and associates (Albizu-Urionabarrenetxea et al. 2013) that most injuries occurred during the spring and fall seasons. Laschi and associates attributed the finding that Mondays have the highest injury rate over other days of the week to workers' lower attention levels and carefulness on the day after a weekend (Laschi et al. 2016). Our work also supports previous research in that falling and processing job tasks were found to contribute to a disproportional number of injuries within the industry. Others have reported that the majority of workers' compensation claims were attributed to struck-by injury types (58%), followed by falls (14%) (Lefort et al. 2003). The results of the present study determined that struck-by injuries were attributed to 35% of all claims, while falls were responsible for 25% of claims in the Intermountain West region. The relatively higher prevalence of falls in the Intermountain West region may be due to the steep terrain, or due to wet or frozen conditions present during the autumn and winter seasons. While many previous studies focus on analysis and reporting of incident types (struck-by, fall, caught between, etc.), the results in the present study suggest that incident type and nature of injury (sprain/strain, laceration, fracture) are associated and the effect seen in the cost of claim can be attributed to nature of injury rather than incident type.

The high number of injuries occurring to inexperienced, young workers (Fig. 1) is highly concerning considering that more than 25% of all claims occurred to workers in their first six months on the job. A literature review by Albizu-Urionavarrenetxea and associates suggested that the primary reason many studies find a correlation between lack of experience and injury rate is due to inexperienced workers' lack of awareness of the possible occupational risks and hazards (Albizu-Urionabarrenetxea et al. 2013). Adequate training and apprenticeship programs should be in place to provide the background knowledge and skill set required for safely preforming essential job functions (Albizu-Urionabarrenetxea et al. 2013).

5 Conclusions

The results of the present research provides the background for developing focused injury prevention strategies aimed at reducing injuries and fatalities in the logging industry. The present study provides a more comprehensive understanding of logging injuries, their nature and costs. The information obtained from this study can be used to direct future intervention and training strategies. Injury prevention efforts in the Intermountain West region should focus on training related to safe work methods (especially for inexperienced workers), the development of a safety culture and safety leadership, as well as implementation of applicable engineering controls.

References

Albizu-Urionabarrenetxea P, Tolosana-Esteban E, Roman-Jordan E (2013) Safety and health in forest harvesting operations. Diagnosis and preventive actions. A review. For Syst 22(3):392–400

Bell JL (2002) Changes in logging injury rates associated with use of feller-bunchers in West Virginia. J Safety Res 33(4):463–471

Bordas RM, Davis GA, Hopkins BL, Thomas RE, Rummer RB (2001) Documentation of hazards and safety perceptions for mechanized logging operations in East Central Alabama. J Agric Safety Health 7(2):113–123

Bureau of Labor Statistics (2016) CPI Inflation Calculator

Bureau of Labor Statistics (2017) Logging workers had highest rate of fatal work injuries in 2015. The Economics Daily

Laschi A, Marchi E, Foderi C, Neri F (2016) Identifying causes, dynamics and consequences of work accidents in forest operations in an alpine context. Saf Sci 89:28–35. https://doi.org/10.1016/j.ssci.2016.05.017

Lefort J, Albert J, de Hoop CF, Pine JC, Marx BD (2003) Characteristics of injuries in the logging industry of Louisiana, USA: 1986 to 1998. Int J For Eng 14(2)

Malehi AS, Pourmotahari F, Angali KA (2015) Statistical models for the analysis of skewed healthcare cost data: a simulation study. Health Econ Rev 5(1):11. https://doi.org/10.1186/s13561-015-0045-7

Myers JR, Fosbroke DE (1994) Logging fatalities in the united states by region, cause of death, and other factors — 1980 through 1988. J Safety Res 25(2):97–105. https://doi.org/10.1016/0022-4375(94)90021-3

United States Bureau of Labor Statistics (2015) Employer-Reported Workplace Injuries and Illnesses- 2014. http://www.bls.gov/iif/oshsum.htm#14Summary_News_Release

United States Department of Labor (2015) Industries at a Glance: Forestry and Logging: NAICS 113. http://www.bls.gov/iag/tgs/iag113.htm

Evaluating the Effectiveness of Estimating Cumulative Loading Using Linear Integration Method

Rong Huangfu$^{(\boxtimes)}$, Sean Gallagher, Richard Sesek, Mark Schall, and Gerard Davis

Industrial and Systems Engineering Department, Auburn University,
Auburn, AL 36849, USA
rong.huangfu@auburn.edu

Abstract. Exposure to cumulative loading is a significant risk factor in the development of musculoskeletal disorders (MSD). To better understand the dose-response relationship, it is critically to quantify the cumulative exposure. Different integration methods have been used in estimating cumulative loading (force or torque). The general objective of the integration methods has been to sum the independently calculated task exposure. Each task was calculated by multiplying the magnitude of the task loading times the task duration which is the "area under the loading curve". An assumption of this linear integration model is that long-time exposure to low forces will result in a similar level of damage as relatively short time exposure to high forces. To evaluate the effectiveness of this model, three loading groups of eccentric exercise with the same "area under the loading curve" were performed by thirty participants (ten in each group). Maximum isometric voluntary contractions (MIVC) and relaxed elbow angle (REA) were collected before, immediately after, and 2, 4, 8 days after the exercise. The REA and MIVC changes after the eccentric exercise were significantly impacted by the loading group. It suggests that estimating cumulative loading using linear integration method may underestimate the impact of high force loading in terms of cumulative muscle damage.

Keywords: Linear integration · Cumulative load · Eccentric exercise

1 Introduction

Exposure to cumulative loading is a significant risk factor in the development of musculoskeletal disorders (MSD) [1–4]. Waters et al. (2006) summarized the integration methods used to calculate cumulative loading over time [4]. The general objective of the integration methods has been to sum the independently calculated task exposure. Each task was calculated by multiplying the magnitude of the task loading times the task duration which is the "area under the loading curve".

An assumption of this linear integration model is that long-time exposure to low forces will result in a similar level of damage as relatively short time exposure to high forces. However, based on the cadaver study of motion segment fatigue failure conducted by Brinckmann et al. (1988) Jäger et al. (2000) argued that the impact on

© Springer Nature Switzerland AG 2019
S. Bagnara et al. (Eds.): IEA 2018, AISC 820, pp. 283–288, 2019.
https://doi.org/10.1007/978-3-319-96083-8_38

injurious response from doubling of force was higher than doubling of exposure time [5, 6]. More recently, a literature review conducted by Gallagher and Heberger (2013) revealed a consistent force-repetition interaction in terms of MSD risk among different types of biological tissue [7]. Implications of this analysis are that low-force loadings will result in a low rate of tissue damage and high-force loadings will result in a more rapid progression of damage in accordance with fatigue failure theory, which would tend not to support the linear integration approach in assessing cumulative loading [7].

To better understand the efficacy of the linear integration approach of estimating exposure to cumulative loading, an experiment with three loading conditions using eccentric exercise of the elbow flexors of the non-dominant arm was performed. Each condition had the same area under the curve by design but differed in terms of loading level and number of cycles performed. Measures of transient muscle damage were observed to evaluate if significant differences between loading condition existed.

2 Methods

Thirty healthy males were participated in this study. Written informed consent was signed by every participant prior to participating in this study. Institutional Review Board approval of all study procedures was obtained from Auburn University prior to commencing study activities.

Three loading groups of eccentric exercise (high force-low repetition [HL], medium force-medium repetition [MM] and low force-high repetition [LH]) were designed to keep the same work volume (area under the curve) while differentiate the level of force and repetitions (see Table 1).

Table 1. Three levels of eccentric exercises

Types	Moment (% MIVC)	Repetition
HL	120% (High)	1 * 30 (Low)
MM	60% (Medium)	2 * 30 (Medium)
LH	30% (Low)	4 * 30 (High)

Each participant was randomly assigned to one of the three levels of eccentric exercise with the elbow flexors of the non-dominant arm using the Biodex System 4 (Biodex Medical Systems, Shirley, NY, USA). Ten days prior to the eccentric exercise, three maximum isometric voluntary contractions (MIVC) were performed by each participant. The maximum torque generated from the MIVCs for each participant was recorded using Biodex. MIVCs and relaxed elbow angle (REA) measurements were performed prior, immediately after the exercise and 2, 4, 8 days after the exercise. A summarized activities list is provided in Table 2 (Fig. 1).

Each repetition of eccentric exercise started with elbow flexed at approximately 60 degrees elbow excluded angle (Fig. 2(a)) and ended with the 180° elbow included angle (elbow fully extended) (Fig. 2(b)) with an angular velocity of 60°/s. Participants were instructed to exert force counter to the dynamometer arm motion to reach their

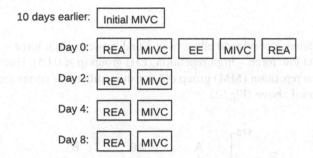

10 days earlier: | Initial MIVC |

Day 0: | REA | | MIVC | | EE | | MIVC | | REA |

Day 2: | REA | | MIVC |

Day 4: | REA | | MIVC |

Day 8: | REA | | MIVC |

Fig. 1. Summary of experiment activities: relaxed elbow angle (REA), maximum isometric voluntary contractions (MIVC) and eccentric exercise (EE)

assigned target level (30%, 60% or 120% of their own MIVC torque value) and maintain that level of exertion through each repetition. Biodex machine would bring the arm back to the starting position at 10°/s after reaching fully extended position. Participants were instructed to relax while still holding the handle during this 12 s rest period and prepared for the next cycle to start. Visual feedback of the exertion level was provided through the whole exercise session on a LCD screen in front of participant.

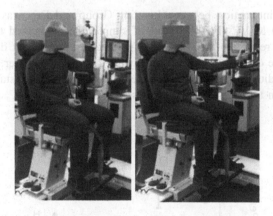

Fig. 2. Start (a) and end (b) position of eccentric exercise

Statistical analysis was performed using Minitab. A series of repeated measures ANOVAs were performed to evaluate the impact of loading conditions on the changes of MIVC torque and relaxed elbow angles on follow-up days. In the case of significant differences found in ANOVA, post hoc analyses were performed using Tukey's HSD test. Type I error rates were set at 0.05 for all tests.

3 Results

Significant difference of relaxed elbow was found between High force – low repetition (HL) group and low force – high repetition (LH) group (p < 0.05). However, medium force – medium repetition (MM) group did not show difference compared with the two groups mentioned above (Fig. 3).

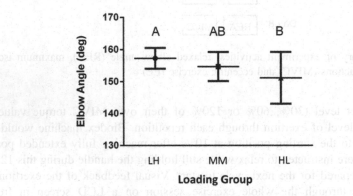

Fig. 3. Relaxed elbow angle by loading group

A significant interaction between Loading Group and Day was observed for the percent change in MIVC (Fig. 4). Tests of simple effects showed that the MIVC of high force – low repetition (HL) group experienced a significant decrease in MIVC compared with the other two groups on Days 0, 2, and 4. However, low force – high repetition and medium force – medium repetition group were not statistically different from each other in terms of MIVC changes (p > 0.05).

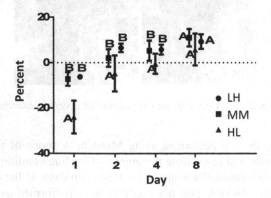

Fig. 4. Percent change in MIVC by day and loading group

4   Discussion

This study evaluated the linear integration method (area under the loading curve) of estimating cumulative loading using an eccentric exercise. Three loading groups were designed to have the same work volume (area under the loading curve) which would suggest the same cumulative loading for all three groups based on linear integration method. However, the relaxed elbow angle and the changes in MIVC observed in this study, were significantly impacted by the loading group. This result suggests the linear integration method in estimating cumulative loading may underestimate the impact of high force loading in terms of cumulative muscle damage.

According to Gallagher and Schall Jr. (2017), the development of musculoskeletal disorders may be the results of a fatigue failure process. This fatigue failure theory suggests that higher force loading has an exponentially higher damage per cycle loading instead of a linear relationship [8]. The findings from this study showed agreement with results of this study in that the linear combination of force and repetition did not appear to result in the same cumulative damage.

5   Conclusion

The relaxed elbow angle and the changes in MIVC were significantly impacted by the loading groups which were designed has the same work volume (area under the curve). This result suggests that the linear integration method of calculating cumulative loading may underestimate the impact of high force loading in terms of cumulative muscle damage.

Acknowledgement. The authors wish to thank Chad Abrams and Steve Kisor from Rehab-Works for their support during the data collection. Thanks to Aimee Sacks, Heather Murphree, Jackie McCoy, Jean Evans, Joy Shirley, Juan Barnes, Laura Allen, Lurlene Buck from the University Clinic for sharing their expertise and valuable time in this study. This research was supported by the pilot research project from the National Institute for Occupational Safety and Health (NIOSH) Sunshine Education and Research Center.

References

1. Coenen P, Kingma I, Boot CRL, Twisk JWR, Bongers PM, Van Dieën JH (2013) Cumulative low back load at work as a risk factor of low back pain: a prospective cohort study. J Occup Rehabil 23:11–18. https://doi.org/10.1007/s10926-012-9375-z
2. Kumar S (1990) Cumulative load as a risk factor for back pain. Spine (Phila Pa 1976) 15:1311–1316. https://doi.org/10.1097/00007632-199012000-00014
3. Norman R, Wells R, Neumann P, Frank J, Shannon H, Kerr M, Beaton DE, Bombardier C, Ferrier S, Hogg-Johnson S, Mondloch M, Peloso P, Smith J, Stansfeld SA, Tarasuk V, Andrews DM, Dobbyn M, Edmonstone MA, Ingelman JP, Jeans B, McRobbie H, Moore A, Mylett J, Outerbridge G, Woo H (1998) A comparison of peak vs cumulative physical work exposure risk factors for the reporting of low back pain in the automotive industry. Clin Biomech 13:561–573. https://doi.org/10.1016/S0268-0033(98)00020-5

4. Waters T, Yeung S, Genaidy A, Callaghan J, Barriera-Viruet H, Deddens J (2006) Cumulative spinal loading exposure methods for manual material handling tasks. Part 2: methodological issues and applicability for use in epidemiological studies. Theor Issues Ergon Sci 7:131–148. https://doi.org/10.1080/14639220500111392
5. Brinckmann P, Biggemann M, Hilweg D (1988) Fatigue fracture of human lumbar vertebrae. Clin Biomech 3:iS1–iiS23. https://doi.org/10.1016/s0268-0033(88)80001-9
6. Jäger M, Jordan C, Luttmann A, Laurig W, Dolly Group (2000) Evaluation and assessment of lumbar load during total shifts for occupational manual materials handling jobs within the Dortmund Lumbar Load Study – DOLLY. Int J Ind Ergon 25:553–571. https://doi.org/10.1016/s0169-8141(99)00043-8
7. Gallagher S, Heberger JR (2013) Examining the interaction of force and repetition on musculoskeletal disorder risk. Hum Factors J Hum Factors Ergon Soc 55:108–124. https://doi.org/10.1177/0018720812449648
8. Gallagher S, Schall MC Jr (2017) Musculoskeletal disorders as a fatigue failure process: evidence, implications and research needs. Ergonomics 60:255–269. https://doi.org/10.1080/00140139.2016.1208848

Capacity Index for Work, Psychosocial Risk of Work and Musculoskeletal Symptomatology in Workers of a Meat Processing Industry in Portugal

Inês Alessandra Xavier Lima[1,2(✉)], Antonio Renato Pereira Moro[1], and Teresa Patrone Cotrim[3]

[1] Programa de Pós-Graduação em Engenharia de Produção da Universidade Federal de Santa Catarina, Florianópolis, Brazil
inesaxlimal@gmail.com
[2] Universidade do Sul de Santa Catarina, Florianópolis, Brazil
[3] Faculdade de Motricidade Humana da Universidade de Lisboa, Lisbon, Portugal

Abstract. Objective: to identify the work ability perception, the psychosocial risk factors and musculoskeletal symptomatology among workers of a meat processing industry. Methods: the sample consisted of 74 subjects, 53 female (71,6%) and 21 male (28,4%), with a mean age of 41,9 years old ($\pm$ 11,3). Data collection was based on a survey including the Portuguese Version of the Work Ability Index (WAI) and of the Copenhagen Psychosocial Questionnaire II (COPSOQ II) and of the Nordic Questionnaire. Results: concerning to work ability, the average value found was 39,38 ($\pm$ 5,97), considered as good work ability. As to psychosocial risk factors, job insecurity presented critical results and favourable results were found in the scales: role clarity, rewards, social community at work, self-efficacy, work-family conflict and offensive behavior. As referring to self-reported musculoskeletal symptomatology, the region with the highest frequency of complaints was the cervical region with a prevalence of 40,5% (n = 30) of pain/discomfort, with average pain perception of 2,58 ($\pm$ 3,45) points in the last 12 months. The Spearman correlation test between the variables age and cervical pain (rho = 0,433; p = 0) and between age and low back pain (rho = 0.240; p = 0.040) show that there is a relationship between these variables. Conclusion: The results of this study point to the need for preventive actions in the meat processing industry, such as those related to work organization (rotation of tasks, adjustment of the production rhythm) and to programs for the prevention of work related musculoskeletal disorders, especially among the women.

Keywords: Musculoskeletal symptomatology · Work ability
Psychosocial risk

S. Bagnara et al. (Eds.): IEA 2018, AISC 820, pp. 289–295, 2019.
https://doi.org/10.1007/978-3-319-96083-8_39

1 Introduction

Some occupations characteristically imply a higher exposure to stress-related contexts, which may represent a greater risk potential to employees' well-being.

Meat processing industries workers are pointed out such as high-risk cases for adverse health effects, and this category of work is characterized by staff turnover, absenteeism and disciplinary actions. These characteristics are explained by the physical requirement and monotonous nature of the work, besides the inherent danger by the use of cutting tools, long work shifts in wet and cold environment, repetitive movements and the need for accelerated production (Victor and Barnard 2016).

From this work context, this study aimed to identify the perception of work capacity, psychosocial risk factors and musculoskeletal symptomatology in workers of a meat processing industry in Portugal.

2 Materials and Methods

The cross-sectional study had a sample of 74 individuals. The procedures were approved by the local Ethics Committee in Research with Human Beings, according to the Helsinke Declaration, and the data collection was carried out in May 2016.

The procedures for conducting the study were on-site observation and the instruments used were the Portuguese versions of the Work Ability Index (WAI) (Silva et al. 2011), the Copenhagen Psychosocial Questionnaire II (COPSOQ) (Silva et al. 2012) and the Nordic Musculoskeletal Questionnaire (NMQ) (Mesquita et al. 2010), which were applied in working hours and individually, after the signing of a term of science and consent.

The collected data were organized, tabulated and analyzed from descriptive statistics (frequency, mean, and standard deviation). To investigate the correlation between the variables, the Spearman correlation test (with $p < 0.05$) was used, using the statistical package Statistical Package for the Social Sciences® (SPSS), version 23).

3 Results

3.1 Sociodemographic Characteristics

The research was carried out in an industry with 460 employees, of which 210 were directly involved in the processing of pig meat, distributed in the sectors of Evisceration (60 workers), Slicing (130 workers) and Cutting (20 workers).

The sample of this analysis consisted of 74 (seventy-four) workers, 35.23% of the population studied, of which 53 (71.6%) were female and 21 (28.4) were males, with a mean age of 41.93 years ($\pm$ 11.38) and predominantly married status (46–62.2%).

Tables 1 and 2 present the results regarding the distribution of interviewed workers by sector and function, respectively.

Table 1. Distribution of interviewed workers by sector.

Sector	N	%
Evisceration	33	44,6
Slicing/Cutting	41	55,4
Total	74	100,0

Table 2. Distribution of interviewed workers by function.

Function	N	%
Eviscerator	20	27,0
Cutter	4	5,4
Packer	39	52,7
Production assistant	11	14,9
Total	74	100,0

The work process in these sectors is carried out predominantly by hand form and standing posture, in fixed shifts and in an environment with a temperature ranging from 0–6/8 °C.

3.2 Work Ability

As for the work capacity of the 74 employees, the mean value was 39.38 (± 5.97), considered as "Good" capacity for work. By category of the WAI, 36 workers (48.6%) presented "Good" capacity for work.

When analyzing the WAI by sector of work, it was verified that in the Evisceration sector, as in the Slicing/Cutting sector, the highest frequency was in the "Good" working capacity category, with a frequency of 15 and 21 responses respectively.

3.3 Psychosocial Risks

The COPSOQ includes scales whose scores range from 1 to 5. Tertiles 2.33 and 3.66 are used as a reference to define whether a subscale is a favorable or risk factor for health (Silva et al. 2011). In these scales there are variations in values, where the critical value is the lowest (Work-Family Conflict, Sleeping Problems, Burnout, Stress, Depressive Symptoms and Cognitive Requirements) while in others the critical value is the highest (Work satisfaction and Influence at work).

In this study, only the variable Labor Insecurity presented critical values (3.94 ± 1.19), while the others presented a favorable (7) and intermediate (21) classification. The favorable results were found in the scales: labor transparency (4.26, ± 0.68), rewards (3.73, ± 0.82), social community at work (3.94, ± 1.73), self-efficacy (3.89 ± 0.87), work-family conflict (2.17, ± 0.78) and offensive behavior (1.23, ± 0.48) (see Table 3).

Table 3. Classification of subscales of psychosocial risk factors.

Subscale	Average value	Classification
1. Quantitative demands	2,4054	Intermediate
2. Work pace	3,4459	Intermediate
3. Cognitive demands	3,1126	Intermediate
4. Emotional demands	2,9459	Intermediate
5. Influence at work	2,3527	Intermediate
6. Possibilities for development	3,1528	Intermediate
7. Predictability	3,4795	Intermediate
8. Role clarity	4,2658	Favorable
9. Rewards/recognition	3,7387	Favorable
10. Labor conflicts	2,8423	Intermediate
11. Peer social support	3,2072	Intermediate
12. Social support from supervisor	2,9213	Intermediate
13. Sense of community	3,9406	Favorable
14. Quality of leadership	3,5405	Intermediate
15. Horizontal confidence	2,6441	Intermediate
16. Trust regarding management	3,5571	Intermediate
17. Justice and respect	3,4475	Intermediate
18. Self-efficacy	3,8985	Favorable
19. Meaning of work	3,8288	Favorable
20. Workplace commitment	3,1892	Intermediate
21. Job satisfaction	3,3527	Intermediate
22. Job insecurity	3,9459	Health risk
23. Self-rated health	2,9459	Intermediate
24. Work/family conflict	2,1712	Favorable
25. Sleeping problems	2,3986	Intermediate
26. Burnout	2,6781	Intermediate
27. Stress	2,4257	Intermediate
28. Depressive symptoms	2,3514	Intermediate
29. Offensive behavior	1,2331	Favorable

3.4 Self-reported Musculoskeletal Symptoms

Regarding self-reported musculoskeletal symptoms, the three regions with the highest frequency of worker complaints were the cervical spine, lumbar spine and right shoulder.

Concerning the cervical spine, 30 workers (40.5%), 24 women and 6 men, reported pain/discomfort in this segment, with an average pain intensity of 2.58 (± 3.45) in the last 12 months, and of 1.82 (± 3.02) in the week prior to collection. Concerning the lumbar spine, the frequency of complaints was 28 workers (37.8%), 18 women and 10 men, with an average pain intensity of 2.39 (± 3.47) in the last 12 months, and 1.64 (± 3.22) in the week prior to collection. Regarding the right shoulder, the complaint of pain in the segment was 26 workers (35.1%), 16 women and 10 men, with an average pain intensity of 2.27 (± 3.4) in the last 12 months, and 1.74 (± 3.19) in the week prior to collection. The classification of pain intensity in the three segments could be considered "light".

Concerning the distribution of pain complaints (cervical, lumbar and right shoulder) by age of workers (category 1 = 24 to 33 years, category 2 = 34 to 43 years, category 3 = 44 to 53 years and category 4 = 54 to 63 years), there was a prevalence of complaints in categories 3 and 4, that is, among workers aged 44 or over (see Fig. 1).

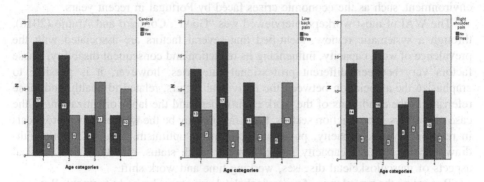

Fig. 1. Distribution of cervical pain, low back pain and right shoulder pain frequency in the last 12 months, by age.

Concerning the distribution of the complaint of pain (cervical, lumbar and right shoulder) by function of the workers, it was observed that there was prevalence of complaint in the "Packer" function (see Fig. 2).

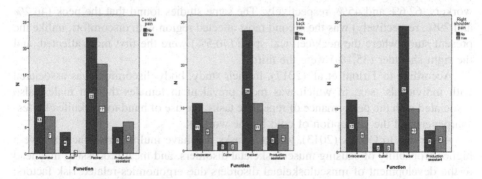

Fig. 2. Distribution of cervical pain, low back pain and right shoulder pain frequency in the last 12 months, by function performed by the worker.

The results of the Spearman correlation test between the variables age and cervical pain (rho = 0,433; p = 0) and between age and low back pain (rho = 0.240; p = 0.040) show that there is a relationship between these variables, characterizing As the age increases, so does the frequency of painful complaints in these body segments.

Although the ability to work was considered "good" by the majority of the sample, workers who did not report musculoskeletal symptoms presented better WAI.

4 Discussion

As for psychosocial risks, the variable that presents itself as critical (labor insecurity) is not necessarily negative, since labor insecurity can be influenced by the national environment, such as the economic crises faced by Portugal in recent years.

The WAI of most workers interviewed was "Good". Cordeiro and Araújo (2016), through a systematic review, identified that several factors are associated with the prevalence of work capacity, influencing its reduction and consequent disability. These factors vary between different professional categories; however, it is possible to emphasize the association between the individual factors, related to health, and those referring to the conditions of the work environment and the labor organization. In the case of workers in production sectors, these factors may be the worker's own prognosis in relation to work capacity, physical requirement, optimism, age, sex, work withdrawal, pain, physical capacity index, general health status, quality of life, clinical aspects of musculoskeletal diseases, working time and work shift.

Regarding the prevalence of musculoskeletal symptoms by body segment, Bot et al. (2005), Tirloni et al. (2012) and Reis et al. (2016) corroborate with these results. Bot et al. (2005) obtained data from the second Dutch national survey of general practice (391.294 patients at the start of the survey), the results showed that the most commonly reported complaint was neck symptoms, followed by shoulder symptoms, unlike the present study where lumbar spine (37,8%) were the second most affected. Tirloni et al. (2012), that evaluated 290 workers of a poultry slaughterhouse in Brazil, and Reis et al. (2016), reporting the shoulder region as the most affected in poultry slaughterhouse workers, 62.6% and 45%, respectively. The same studies found that the neck (46.2% and 29%, respectively) was the second most affected region with discomfort, unlike the present study where the neck/cervical spine (40,5%) were the first most affected, and the right shoulder (35,1%) were the third.

According to Tirloni et al. (2017), in their study, body discomfort was associated with individuals' sex, in which was more prevalent in females than in males; also associated with the performance of repetitive tasks, the use of hand tools (knifes/knifes-sharpener) and the perception of cold by the workers.

According to OSHA (2013), jobs and tasks that have multiple risk factors have a higher probability of causing musculoskeletal disorders, and meat processing may lead to the development of musculoskeletal disorders due ergonomics-related risk factors: repetition, forceful exertion, awkward and static posture, vibration, and cold temperatures allied to these risk factors.

5   Conclusion

In conclusion, the majority of workers classified their capacity to work as "good", identified most of the psychosocial risks as intermediaries and reported physical discomfort in the cervical spine, lumbar spine and right shoulder, respectively, and was more prevalent in females. Finally, it was found that there was a relationship between age and cervical pain and between age and low back pain, and that workers who did not report musculoskeletal symptoms showed better WAI.

The results of this study point to the need for preventive actions in the meat processing industry, such as those related to work organization (rotation of tasks, adjustment of the production rhythm) and to programs for the prevention of workrelated musculoskeletal disorders, especially among the women.

It is noteworthy that the study was conducted in only one meat processing industry, which makes it impossible to generalize the results. Therefore, it is suggested to carry out new studies with larger samples and with more than one industry involved.

References

Bot SD, van der Waal JM, Terwee CB, van der Windt DA, Schellevis FG, Bouter LM, Bekker J (2005) Incidence and prevalence of complaints of the neck and upper extremity in general practice. Ann Rheum Dis 64(1):118–123

Cordeiro TMSC, Araújo TM (2016) Capacidade para o trabalho entre trabalhadores do Brasil. Rev Bras Med Trab 14(3):262–274

Mesquita C, Ribeiro J, Moreira P (2010) Portuguese version of the standardized Nordic questionnaire: cross cultural and reliability. J Publ Health 18(5):461–466

OSHA (2013) Occupational Safety and Health Administration. Prevention of musculoskeletal injuries in poultry processing. https://www.osha.gov/Publications/OSHA3213.pdf. Accessed 30 Apr 2018

Reis DC, Moro ARP, Ramos E, Reis PF (2016) Upper limbs exposure to biomechanical overload: occupational risk assessment in a poultry slaughterhouse. In: Advances in physical ergonomics and human factors, 7th edn. Springer, Florida

Silva CF, Amaral V, Pereira A, Bem-haja P, Rodrigues V, Pereira A, Alves A (2011) Índice de capacidade para o trabalho - Portugal e os Países Africanos de Língua Oficial Portuguesa, 2nd edn. Análise Exacta, Coimbra

Silva CF, Amaral V, Pereira A, Bem-haja P, Pereira A, Rodrigues V et al (2012) Copenhagen Psychosocial Questionnaire, COPSOQ - Portugal e países africanos de língua oficial portuguesa, 1st edn. Editora Exact, Coimbra

Tirloni AS, Reis DC, Ramos E, Moro ARP (2017) Association of bodily discomfort with occupational risk factors in poultry slaughterhouse workers. Revista DYNA 84(202):49–54

Tirloni AS, Reis DC, Santos JB, Reis PF, Barbosa A, Moro ARP (2012) Body discomfort in poultry slaughterhouse workers. Work 41:2420–2425

Victor K, Barnard A (2016) Slaughtering for a living: a hermeneutic phenomenological perspective on the well-being of slaughterhouse employees. Int J Qual Stud Health Wellbeing 11:1–13

Biomechanical Methodology for Evaluating Seat Comfort During Long Term Driving According to the Variation of Seat Back Angle

Dong Hyun Kim(iD), Seohyun Kim, Sung Chul Kim, Sung Hyun Yoo, Young Jin Jung, and Han Sung Kim$^{(\boxtimes)}$(iD)

Department of Biomedical Engineering, Yonsei University,
Won-ju, Gangwon 26493, Korea
hanskim@yonsei.ac.kr

Abstract. The aim of this study is to suggest a biomechanical methodology for evaluating seat comfort during long term driving with a use of driving stimulator. Recent modern car seat has become increasingly complexed and sophisticated with ongoing seat discomfort problems. Consumers' demands for better seat comfort have also increased but most of them still choose car seats in the wrong way. Previous researches have recommended quantifying overall car seat discomfort based on questionnaire, EMG, and FEM etc. However, it is difficult to find successful objective evaluation methods in both in literature or in actual situations. Six males were recruited to participate in a laboratory study. All participants performed three trials separated by three different seat back angles (87°, 97°, and 107°) and each trial was conducted on different days. Each trial was consisted of 120 min of continuous driving on the driving simulator. Participants were recorded using Kinect v2 with a full body and pressure at both seat pan and back were measured to allow the investigator to analyze their postures. The results for each seat condition were analyzed through unpaired t-test in order to acquire statistical significance ($P < 0.05$). As expected, the overall discomfort rating increased over time, especially for neck and low back. Also, at 87° and 107°, the overall discomfort rating was higher than 97°. In addition, over time, we found an increase in pressure in both seat back and seat pan. These results suggest that a new biomechanical method could be an alternative for evaluating car seat comfort for long-term driving.

Keywords: Seat comfort · Biomechanics · Kinect v2 · Lumbar moment

1 Introduction

Seat comfort is one of the most important factors in selecting a car. In addition, customers' demand for reliable vehicle seat has increased although low-priced vehicles have been constantly released. In the advanced countries, drivers are also interested in keeping their body with a healthy spine [1]. However, ergonomics of seating has received less attention in the biomechanics of automobile than it does in office and factory.

© Springer Nature Switzerland AG 2019
S. Bagnara et al. (Eds.): IEA 2018, AISC 820, pp. 296–302, 2019.
https://doi.org/10.1007/978-3-319-96083-8_40

Previous studies have suggested that driving posture is a key factor determining drivers' comfort while driving since most drivers felt discomfort due to their inappropriate driving posture [2]. The inappropriate driving posture could cause biomechanical loads to spinal column [3]. Biomechanical loads on sitting posture which are classified into musculoskeletal loads and contact loads could cause lumbar disk degeneration, herniation and muscle spasm [4].

A number of researchers conducted various methods to objectify driver's seat discomfort which reduced biomechanical loads of lumbar region. Researchers suggested that an ideal driver's posture to reduce load on L5/S1 might be a balanced posture minimizing passive loads [5]. However, any equipment has not been found to measure the lumbar moment in real time as the car seat.

The aim of this study was to measure the changes in lumbar moment caused by changing driving posture with seat back angles.

2 Method

2.1 Sample and Design

All 6 participants, consisting of 6 males, from the local area and student population of Yonsei University were recruited to participate in a laboratory study. Participants have never suffered from musculoskeletal disorders and any kind of orthopedic surgery around the lumbar region. All participants should have acquired driver's licenses at least 2 years before the experiment. The participants were between the ages of 25 and 28. Their weight range was 74.9 ± 4.51 kg and the height range was 1.79 ± 0.24 m.

Each trial consisted of 120 min continuous driving on the driving simulator housed at Yonsei University. After measuring the subject's height and weight, the subjects were given 10 min to adjust to the simulator. The subject was then adjusted to the preferred seat distance. The seat back angle was set to 87°, 97° and 107° with respect to the floor (Fig. 1).

Each trial was conducted on different days considering fatigue of the subjects, since the experiment was carried out for 120 min at each seat back angle. Participants were required to provide subjective discomfort ratings verbally every 30 min via the use of questionnaire. Participants were trained in the use of the questionnaire prior to participation in the study. The questionnaire includes the 9 points discomfort 0 to 10 scale which was modified ISO 2631-1.

2.2 Equipment

A production car seat, Avante MD model (Hyundai, Korea), was provided by Hyundai. A rig consisted of the seat, steering wheel and pedals was used to control the driving simulator (Fig. 3). This rig replicated dimensions from SAE standard (J1100, J4002) and a production steering wheel was used. Participants were conducted simulator, Euro Truck Simulator 2 (SCS Software, Czech), the route throughout the drive on the simulator, via the use of audio prompts from built-in navigation in the game.

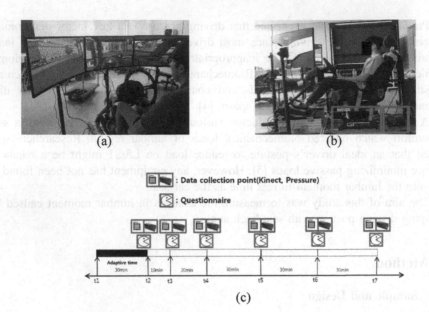

Fig. 1. (a): the experimental procedure, (b): configuration for long-term driving simulator, and (c): schematic diagram of the process of the experiment.

Two pressure-mapping mats (Xsensor™, Canada) were located over the surface of the seat back and the seat pan at a sampling rate of 30 Hz. And two depth sensor-based Kinect v2 at a sampling rate of 30 Hz were located on the left side of the participants. The pressure distribution was analyzed through using XSENSOR X3 Pro v7.0 (Xsensor™). Movements captured through Kinect camera were analyzed by iPimotion capture studio software (iPisoft, Russia) with the Biomech add-on (iPisoft).

2.3   Evaluation of Seat Comfort

Lumbar flexion moment and Extension moment were calculated by vector cross product. (1) Flexion Moment arm was calculated by vector algebra between L5 and upper body COM which was collected by Kinect. Based on anatomical basis, we calculated extension moment arm by considering that the length of the center of gravity at L5-S1 (LowerSpine) is 6 cm.

$$M_{Lumbar} = \Delta r \cdot F \tag{1}$$

Net moment was calculated as the difference between Flexion moment and Extension moment.

$$\text{Net moment} = \text{Flexion moment} - \text{Extension moment} \tag{2}$$

Our study started with the assumption that moments caused by the upper body would cause back pain. Therefore, the reaction force generated by the lower body had to be eliminated. In this respect, participants were instructed to lie flat on the floor. This indicates that a reaction force from the pressure matt placed on seat pan should present pure vertical force of lower body. This reaction force was set to zero in order to remove gravity force by participant's lower body weight.

After that, the reaction force appeared in the experiment was supposed to be caused only by the upper-body, and it was possible to obtain Flexion and Extension moment from L5/S1 region.

3   Results

3.1   Questionnaire Survey

In this study, the subjects were asked about the status of each condition at every 0, 30, 60, 90 and 120 min. The initial start was 5 points. If it was more convenient than the previous situation, the value was subtracted point(s), and if it was uncomfortable, the value was added to point(s). There was no size limit for each situation.

The highest score is 10 points while the lowest score is 0 points. 10 points indicate extremely uncomfortable condition. All six subjects complained of discomfort in most items as the driving time continued. The highest discomfort score appeared at the low back and then at the neck (Fig. 2).

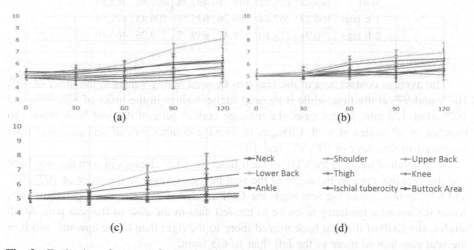

Fig. 2. Evaluation of seat comfort score when seat back angle when the seat back angle is (a): 87°, (b): 97°, and (c): 107°. (d): Legend of questionnaire items in all charts.

The score of low back discomfort decreased as seat back angle changed in the order of 87°, 107° and 97° and in the case of the neck, it is in the order of 107°, 97° and 87°. As a result of the questionnaire, it was found that the inconvenience of seat back was less at 97°, which is similar to the results of the previous study.

3.2   Pressure at Seat Pan and Back

The mean pressure, average contact area, COP of the seat back and seat pan was measured using a pressure matt. The average pressure of the seat back appeared to be generally lower than other angles at 87°. At 97°, the average pressure tends to increase compared to the first, whereas at 107° it tends to decrease slightly. The average pressure of the seat pan was measured in the order of 87°, 97° and 107° as opposed to the seat back, and the average pressure was higher than that of the seat back. Also, in all cases, the average pressure tended to increase as compared to the first, at 0 min. The changes in mean pressure of the seat back angle over time increased in the order of 87°, 97° and 107°. The average contact area of the seat back showed higher values in the order of 97°, 87°, and 107° at the first, while it showed higher values in the order of 107°, 97°, and 87° after 120 min. The average contact area of the seat back tended to increase at all angles. Changes in average contact area of seat back over time increased in the order of 107°, 97°, and 87° (Table 1).

Table 1. Reaction force calculated by product of average contact area and average pressure at seat back and seat pan.

Time	Seat back reaction force (N)			Seat pan reaction force (N)		
	87°	97°	107°	87°	97°	107°
Start	143.73	175.72	146.71	497.78	401.80	385.57
1 h later	166.23	207.44	189.26	625.35	506.43	471.12
2 h later	170.91	228.10	228.42	649.72	533.26	463.80

The average contact area of the seat pan showed higher values in the order of 87°, 107°, and 97° at the first, while it showed higher values in the order of 87°, 97°, and 107° after 120 min. In this case, the average contact area of the seat back tended to increase at all angles as well. Changes in average contact area of seat pan over time increased in the order of 97°, 87° and 107°.

As a result of tracking the COP of each time, the COP changes in both the seat back and the seat pan raise as the angle of the seat becomes larger in the order of 107°, 97° and 87°. In the case of the seat back, the COP tended to move to the upper right side, while it showed a tendency to move to the left side in the case of the seat pan. At all angles, the COP of the seat back moved more to the right than to the upward, and it of the seat pan moved more to the left than to the front.

3.3   Lumbar Moment

Extension moment and flexion moment over time were observed as follows. Extension moment showed an increasing tendency at 87° and 97°, but showed a slight decrease at 107°. The flexion moment was increased as the seat back angle increased. In addition, net moment was increased as the seat back angle increased (Fig. 3).

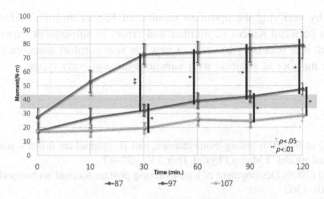

Fig. 3. Net Lumbar moment according to time for each seat back angle.

4 Conclusion

The purpose of this study was to evaluate the subjective opinion of the driver's discomfort during long - term driving according to the seat back angle of the car seat, by objectively evaluating seat comfort through questionnaires and evaluation from a new biomechanical method.

As time goes on, the distance between the body and the steering wheel became shorter, and the subjects' upper body was seen to be bent forward. Based on this, it can be deduced that the lordosis of the subject occurred when the body was forward bent over time. Also, the results show that subjects took a posture that their bodies were closely attached to the car seat overtime. Based on the motion analysis, it can be confirmed that the pelvis was tilted backward due to long period of sitting, the upper body was closely attached to the seat back and the lower body moved forward. These series of posture changes were appeared to cause pain and the larger the back angle, the larger the moment value appears.

According to previous studies, experiences that take more than two hours to make sure the right seat comfort evaluation, however this results showed that takes 30 min was enough to evaluation seat comfort. Based on these results, it is expected that it will be possible to present a suitable driving posture for drivers with an objective seat comfort.

At all angles, the COPs in both seatback and seat pan have moved in different directions. This might be due to the fact that participants did not use their left foot during operation in automatic gear shifting but constantly stepped their right foot on excel and brake pedals. In addition, as the angle of the seat increased, the changes in COP coordinate became larger. As the angle of the seat increases, the driver corrected the posture several times.

A new method for evaluating the comfort of a car seat using a questionnaire and a pressure mat is presented. It is expected that it will be easier, cheaper and quicker to evaluate the comfort of car seat than conventional methods. However, two Kinect v2 used in the experiment were analyzed and some errors were found. This will be compensated by verifying the optimum number of Kinect through performing correlation analysis between Kinect v2 number and errors in subsequent experiments.

In addition, it is necessary to verify a new car seat comfort assessment method by increasing the number of subjects with various ages and body types.

References

1. Harrison DD et al (2000) Sitting biomechanics, part II: optimal car driver's seat and optimal driver's spinal model. J Manip Physiol Ther 23(1):37–47
2. Hirao A et al (2006) Development of a new driving posture focused on biomechanical loads. SAE, 2006-01-1302
3. Bhise VD (2012) Ergonomics in the automotive design process, 1st edn. CPC Press, NewYork
4. Pope MH et al (2002) Spine ergonomics. Ann Rev Biomed Eng 4:49–68
5. Zenk R et al (2012) Technical note: spine load in automotive seating. Appl Ergon 43:290–295

The Effects of Chair Inclination, Arm Support and Touch-Typing on Shoulder and Arm Muscle Activity in Computer Work

Erwin M. Speklé[1,2]([✉]), Bas H. M. van der Doelen[3],
and Jaap H. van Dieën[4]

[1] Arbo Unie OHS, P.O. Box 791, 2003 RT Haarlem, The Netherlands
erwin.spekle@arbounie.nl
[2] Department of Public and Occupational Health, VU Medical Center,
P.O. Box 7057, 1007 MB Amsterdam, The Netherlands
[3] Flokk, P.O. Box 467, 8000 AL Zwolle, The Netherlands
[4] Department of Human Movement Sciences, VU Amsterdam,
van der Boechorststraat 9, 1081 BT Amsterdam, The Netherlands

Abstract. Aims. In an attempt to reduce the risk of work related musculoskeletal disorders (WRMD) in computer users, while maintaining comfort and functionality, different chair designs have been developed. This study aimed to determine the effect of chair inclination, arm support and touch-typing on muscle activation in the upper body.

Methods. A randomized balanced design was used to test twenty subjects performing a three-minute computer task, consisting of both typing and mouse activities, with three chair seat conditions (inclination of $-14°$, $0°$ or dynamic) and three types of arm support (none, chair, table). The angle between the seat pan and the back support was fixed, where the angle of the back support relative to the seat pan was consistent at a 93 angle. The nine resulting conditions were tested using a randomized balanced design with a 5-min familiarization period prior to the experiment and a 10-min break half-way through the measurements.

Unilateral surface electromyography (EMG) was recorded from nine upper neck and shoulder muscles at the right side. Exposure variation analysis (EVA) using normalized EMG was performed to analyse load and characteristics of muscle activity in both the amplitude and time domain. Prior to the experiment, maximal voluntary contractions (MVC) and standard isometric contractions (SIC) were recorded.

A mixed design repeated measures analysis of variance (ANOVA) was performed with chair inclination (-14, 0, dynamic) and arm support (with, without, table) as within subject variables and typing skill (touch-typing vs. non-touch-typing) as between subject variable.

Results. A chair inclination of $-14°$ resulted in lower muscle activity with higher variation in muscle activity for the trapezius ascendens and transversus muscles, whereas higher muscle activity with higher variation in activity was found for the other shoulder muscles.

Significant main effects of arm support on muscle activity were found for several muscles. Muscle activity significantly was lower in the table support condition compared to the without support condition. For several muscles, the

© Springer Nature Switzerland AG 2019
S. Bagnara et al. (Eds.): IEA 2018, AISC 820, pp. 303–304, 2019.
https://doi.org/10.1007/978-3-319-96083-8_41

table support condition resulted in a significantly lower muscle activity compared to the chair support condition.

No differences in muscle activity were found between touch-typists and non-touch typists. %MVC, APDF were similar for all chair inclination and support conditions for all muscles. Touch-typists were more productive.

Conclusions. It is concluded that: (1) A backward chair inclination of −14°, as recommended in the literature, decreases the muscle activity of the muscles depressing and retracting the shoulder, whereas activity of the muscles elevating the arm and shoulder increases. The variation in overall shoulder muscle activity increases with a backward chair inclination; (2) The use of arm support or the use of the table as an arm support resulted in lower and more dynamic muscle activity; (3) No differences in muscle activity were found between touch-typists and non-touch-typists, but touch-typists were more productive.

Keywords: Musculoskeletal disorders · Chair · Ergonomics

Implementation of the Ergonomic Principles: In the Regulations and at the Workplace

IEA 2018 Symposium: Comparison of Ergonomic Rules Between Countries

Pascal Etienne[✉]

Federation of European Ergonomics Societies (FEES),
Société d'ergonomie de langue française (SELF), Paris, France
pascal.etienne0@orange.fr

Abstract. The paper present the ergonomic principles spread in the French speaking countries by the «activity ergonomics» based on the Ergonomic Work Analysis which develops a "bottom up" problem building approach.

The different places where the traces of the ergonomics principles implementation in France are identified: legislations, collective agreements... In the French Labour Code are located the regulations, passed from the 70's to the 90's, issued mainly from the European directives transposition, which provides for to adapt the work to human. Such requirements constitute what the jurisprudence has defined as a so called "result obligation" in the field of OHS.

These principles are only partly implemented in the real world. And with the recent transformations in the work, such as the development of an economy based on services, marked by digitalization, artificial intelligence, robotization, uberization, one states that the work situations produce new risks, In front of new stakes, the issue of the principles updating is posed and the answers given recently in the laws are questioned.

Keywords: Ergonomic principles · Implementation · Social model
Work changes

1 Introduction

With the aim to present and to discuss the issue of the ergonomic principles implementation, we have to face several issues: what do we understand with ergonomics and the ways of production of these principles, its institutionalization in rules and juridical devices and its actual implementation in the real world. It is also necessary to envisage in which measure these principles allow to ensure or are an obstacle to «creativity in practice», the motto of this IEA Congress.

The plan of the presentation is as follows:

1. The principles developed by the activity ergonomics
2. The institutionalization of the principles in the public space
3. The implementation in the companies

© Springer Nature Switzerland AG 2019
S. Bagnara et al. (Eds.): IEA 2018, AISC 820, pp. 305–315, 2019.
https://doi.org/10.1007/978-3-319-96083-8_42

4. The changes in the work and in the French social model: what is at stake and which consequences for ergonomics?
5. Conclusion

2   The Principles Developed by the Activity Ergonomics

The principles developed mostly in the French speaking countries by the activity ergonomics (also called "French speaking ergonomics") are known: in order to reach the aims of well-being and performance which are stated by ergonomics, the Ergonomics Work Analysis develop a «bottom up» problem building approach, which takes into consideration the operators words as meaningful among a social group in action. Such approach takes also into account the industrial relations, the company model and its organization. The Ergonomics Work Analysis, such as developed by Wisner [1] seems also the way to identify the scientific sources where one needs to draw theories, methods and knowledges allowing to solve the problem (or problems) which characterizes the situation. It allows to come closer to the way how one operator or the operators build the problem before to solve it.

The Ergonomics Work Analysis is interested in the possibilities and the modalities of the operators and workers representative expression on the work and in the work. Therefore ergonomists, such Laville and Teiger, have underlined [2] that the workers expression fulfils several functions: it allows them to contribute to their knowledge elaboration, to objective the individually resented problems and to valorize the knowledge of each one confronted with these of the others, that transform the relation to the individual and collective work.

For some years, the stake of the development through the activity of the individuals, the collectives and the organizations has been newly discovered and brought out, in connection with the organizations changes and the demands addressed to the ergonomists as stated by Daniellou [3] and the "constructive ergonomics" stream is asserted by Falzon [4].

The demand changes addressed to ergonomics are characterized by the importance of the musculo-skeletal disorder issues, followed by the psycho-social type diseases, with the emerging concept of the impeached activity, such as described by Clot [5].

As for the constructive ergonomics, it stresses on the development of the operators skills, which are at the same time knowledges, operating modes linked to the task and knowledge of these very operators, for example on their risk mastery. In such approach, the development is also thought as a way of the ergonomics intervention: the ergonomist is an agent of a participatory and developmental approach.

3 The Institutionalization of the Principles in the Public Space

The institutionalization of the ergonomics principles has been achieved progressively and incompletely. In the field of work, more than in other activity fields, there is a long path between the emerging of the scientific disciplines, its acknowledgment as human rights and its practical implementation. Such an acknowledgment depends on various factors: on the relevance of the notions, the strength of the social movements which support it, the skill of the standing actors, and the availability of the institutions for the transcription into laws or regulations. We will shortly sketch how the ergonomics principles have been recognized in the public space and how this institutionalization is a mean to insure the implementation of these principles in the field. In our presentation, we will identify the various times and places where the tracks of these principles issuing from ergonomics (or corresponding to the ergonomic requirements) are located, in France: in the laws, the regulations, the collective agreements, ... but we will lack of space to deal with the standards.

In the Labour Code are the regulations, which have been successfully agreed from the seventies to the nineties, which stem from the social debate in France and from the European Directives that foresee to "fight the risk at the source" or to "adapt the work to the human". Such requirements, completed by provisions related to the workers representation in the occupational health and safety field and other related to the workers expression right, in addition to these concerning the machines and personal protective equipment design, constitute what the jurisprudence has defined a "safety result obligation".

We will sum up the main steps below:

The Consequences of the Post 1968 Social and Political Debate
In the seventies, with some legal provisions and collective agreements, in which ergonomists have contributed in the frame of the commissions of the public planning Administration [6], one states the emerging of ergonomics notions sustained by a socio-technical approach with three aspects: the law which set up a national Agency dedicated to the improvement of the working conditions [7] and in the companies with more than 300 employees a Commission for the improvement of the working conditions. Following this legislation, in the year 1975, a collective agreement has been adopted [8], which highlights an approach of the work based on principles inspired of those of ergonomics. This agreement is not limited to the OHS matters but deals broadly with the work organization, with the working time organization, the incentive wages, health and safety and the role of the managers. The preamble of the agreement states that the improvement of the working conditions is one of the main concerns of the time and the article 5 summarizes in a few lines numerous debates on the expected changes and the organizational models to be built. It is the matter of task enrichment, operating cycle expansion. It is clearly stated that autonomous teams of workers allow the improvement of the workers participation in the organization and the enhancement of their initiative capabilities.

Eventually in 1976 the law [9] completes the building: it provides for the integrated safety notion, as soon as the workplaces or the equipment design stance is established. This law brings out the safety obligation notion addressed to the employer.

The Emergence of «New Rights» in the Years 1980

At the beginning of the years 1980, after the political change in 1981, one sees the emergence of «new rights» in the field of work: in 1982, the laws and regulations currently called «lois Auroux» are a new step forward with a better acknowledgement of the workers representatives role in the companies and their ability to collective bargaining: in particular, one states the consecration of a specialized workers representation in the OHS field. For the employees, the law provides a right to express themselves on the working conditions, and to be withdrawn out of a dangerous work situation. Such provisions pave the way of new possibilities for the implementation of the ergonomic principles.

The laws that are actually adopted allow the spreading in the companies of the right for the workers to express their views on the working conditions [10] and give the workers representative in the OHS field with the creation of the health and safety and working conditions committees real training, enquiries and intervention tools [11]. Such means have been completed with the progressive transposition, at the beginning of the nineties, of the so called European «framework Directive» [12] with the possibility for the workers representatives to make appeal to experts coming from outside the companies.

The Transposition of the Principles Coming From the European Directives in the Nineties

The beginning of the nineties is marked by the transposition into the French law of innovative European directives which include the ergonomic principles in the risk prevention strategy that one must proceed in the design of the workplaces or in the design of the main equipment used by the operators for their work, such as machines or personal protective equipment. Such provisions are integrally transposed in the French Labour Code, with the articles 4121-1 and L. 4121-2, in particular [13].

The «framework health and safety» Directive defines the prevention as (art. 3d) «all the steps or measures taken or planned at all stages of work in the undertaking to prevent or reduce occupational risks». In the article 6, the Directive refers explicitly to ergonomics (the adaptation of the work to the individual), it sets up a hierarchy between the measures to adopt in terms of general prevention principles. Such a vision of general planned prevention in design is assured in the point d) of this article which prescribes the compliance with the ergonomic principles: «adapting the work to the individual, especially as regards the design of workplaces, the choice of work equipment and the choice of working and production methods, with a view, in particular, to alleviating monotonous work and work at a predetermined work-rate and to reducing their effect on health.» Such prescription supposes an adequate level of ergonomic knowledge and practice, as explained by Maggi [14].

For the design of the main devices used by the operators for their work, such as machines, another directive, the Directive 2006/42/EC [15] provides for in its Annex I, transposed in the Labour Code under the article R. 4312-1 [16], the principles of safety integration (point 1.1.2 (a) - Principles of safety integration). Therefore "machinery must be designed and constructed so that it is fitted for its function, and can be operated, adjusted and maintained without putting persons at risk when these operations are carried out under the conditions foreseen but also taking into account any reasonably foreseeable misuse thereof. The aim of measures taken must be to eliminate any risk throughout the foreseeable lifetime of the machinery including the phases of transport, assembly, dismantling, disabling and scrapping". In addition, the point 1.1.6 - Ergonomics of the said annex refers explicitly to ergonomics, stating that "under the intended conditions of use, the discomfort, fatigue and physical and psychological stress faced by the operator must be reduced to the minimum possible, taking into account ergonomic principles (...)."

Likewise, for the personal protective equipment (PPE), the directive 89/686/EEC [17], transposed into the French Labour Code at the article R. 4312-1 above mentioned, provides for in its Annex II (1.1.1. Ergonomics) [18] that «PPE must be so designed and manufactured that in the foreseeable conditions of use for which it is intended the user can perform the risk-related activity normally whilst enjoying appropriate protection of the highest possible level.»

The Agreements of the Years 2000: A Step Forward?
Some consider however that these regulations reflect a «technical» vision of work, in which the working conditions are faced only on the physical and material modes [19]. There would exist no rule concerning the processes, or the work organization. One state that just in time work, toyotism, matrix-type organization, lean management... are out of the regulations scope» [20]. Therefore, with the impulsion of the European social partners one states at different levels a new negotiation issue, linked with the occupational mental health. That are: the framework agreement on work-related stress in 2004 [21] and the framework agreement on harassment and violence at work, in 2007 [22]. These agreements adopted at the European level have been transposed through negotiations in France [23, 24].

In addition, new negotiation obligations have arisen in the legal system: the annual obligation to negotiate at the company level on the quality of working life since 2015 [25]; the triennial obligation to negotiate at the branch level on the taking into account of the penibility. These issues are linked with the occupational mental and physical health issues, the mental health dimension being still globally integrated in other negotiations (like the one on the working time planning). Some collective agreements stipulate as well the promotion of the notion of «quality of the working life», which is an ambiguous notion for a part, but which give new levers to ergonomic interventions.

The Solidification in the Jurisprudence of a "Safety Result Obligation"
In the year 2000, progresses in the jurisprudence of the supreme courts (the Court of Cassation and the Council of State) succeed in the definition of a «safety result obligation» for the employers [26] and a corresponding obligation for the State [27], which improves the action capabilities of the various stakeholders in the OHS field, and especially the ergonomists.

In such a way of thinking, the employer bears a general obligation to prevent the risks to damage the physical and mental health of the employees, whatsoever the project he envisages to perform could be. Such wording replaces the engineering language. The combination of both factors allows a renewed concept of prevention to come out. To that last extent, the new reference to the mental health in the Labour Code constitutes another lever. In 2002, still with the impulse of the European law, the Code is enriched with the «mental» health notion. The present article L.4121-1 of the Labour Code, mentioned above [13], provides now for that «the employer takes the necessary measures in order to insure the safety and to protect the physical and mental health of the workers».

In the same logic the recent *"Snecma"* decision from the Court of Cassation [28] specifies that the employer shouldn't put in danger the employees: "the employer is complied towards his staff with a result safety obligation which imposes him to take the necessary measures in order to ensure the safety and to protect the worker health ... and is forbidden, in the achievement of his management power, to take measures having for aim to (or resulting in) jeopardize the health and safety of the employees.»

In this new logic, the necessary measures expected from the employer are no more predetermined by regulatory provisions but, according to the risk assessment context (or the "working situations", the ergonomists would say), the employer implements the general prevention principles. So the nature of the envisaged measures must be in accordance with the design of the preventive measures that are defined in the law.

For the State, the story is also changed. Actually, the Council of State specifies that "it is the public authorities in charge of the occupational risks to keep informed of the dangers that the workers may face in their professional **activity** [*underlined by us*], taking into account the products and the substances that they handle or they are in contact with, and to decide, according to the scientific knowledge, if needed with the support of additional studies or enquiries, the more appropriate measures in order to limit, and, if possible to suppress, these dangers."

It is an obligation which mentions expressly the professional **activity** and which is translated is three articulated parts. The public authorities have to: 1 – be kept informed on the dangers; 2 – conduct the relevant studies in order to enrich the available knowledge; 3 – take the legal, regulatory and practical measures in order to fulfill such general obligation to ensure the health and safety on the workplaces.

These legal decisions have had several concrete results, such as the setup of a governmental agency dedicated to health at work (the ANSES) or a more specific governmental planning in this field: the so called "Health and Work Plans".

4   The Implementation of the Ergonomic Principles in the Companies

The data we have at hand show that the ergonomics principles are still only partly implemented in the field. Taking into account the recent changes in the working conditions, where we state too often employment status resulting in growing precariousness for new categories of workers, an increasement of the work intensification, uncertainty is present concerning the concrete implementation of these principles. The main professional disease remains in France the musculo-skeletal disorders (MSD) with more than of 40 000 new occurrences each year [29].

The statements made, for example in this domain, illustrate the difficulty and the richness of the extended implementation in the companies of the ergonomic principles mentioned above.

Achievements in the design of the actions and in the prevention are made in the field. For example, in the domain of the MSD prevention, the notion of room for maneuvre rises and the tools to evaluate it are devised, which are based on the activity model centered on the person, described by Vezina [30]. The operators benefits from points of support or operative rooms for maneuvre, allowing them to perform the necessary controls through the rooms for debate on the work.

Several components of the work situation are identified: the work context represented by the global work organization, the employers requirements and expectations, the material and organizational means that the company puts at the disposal of the workers, the individual characteristics, the work activity and the adjustment process represented by the action strategies deployed and the worker behavior and, in the end, the effects of the work situation in terms of health and performance.

In the MSD prevention, ergonomists stress also on the major components of a socio-technical driving of investment projects. According to Caroly, and alii [31] they are:

- the taking into account of the previous reality, in particular, what concerns the population in the scope of the intervention,
- a reflection on the future work, based on a precise knowledge of the "cost" for the workers to reach the present production goals,
- links between the technical devices, the organization, the training and the human resources management,
- the social dialogue development on the work in the company, which fosters the feedback, the expression of the encountered difficulties and the discussion of the solutions.

The development of the social dialogue on the issues such as the health all along life and the exclusion prevention, the quality of work, the company projects and their strategies, is based on an information circulation and original rooms for debate in the company.

In this perspective, one may lean on the statement made by Dugué in 2013 [32], according to which the prerogatives and the action field of the health and safety and working conditions committees (CHSCT) have changed consistently with the changes in the work, for example in enlarging its competence field to the work organization issues; in pointing that its mission deal with the physical and mental workers health protection; in allowing it to act upstream the changes project; in fostering its association with the occupational risk prevention processes, in the frame of a pluridisciplinary approach.

The health and safety and working conditions committee (CHSCT) becomes a major player for the takeover of the health at work and working conditions improvement issues. It has acquired a reinforced legitimacy and a better social visibility, beside all the players, and particularly the employees. It is an essential booster force in order that the health at work stakes are taken into account in the day to day decision processes and in the strategic choices of the companies. Therefore the activity ergonomists have underlined that such a capacity should be preserved in the legal projects [33].

5 The Changes in the Work and in the French Social Model: What Is at Stake and Which Consequences for Ergonomics?

The recent developments of the French social model and the stakes linked with the change of the work lead the ergonomists to start a reflection about the updating of the ergonomic principles in the regulations and to intervene in the public debate several times: which opportunities, which risks are present? Such a reflection is particularly timely in France, where the public authorities have launched comprehensive reflection on the issues of the health at work [34] and on the company governance, following a first series of measures reforming the Labour Code.

The Recent Developments of the French Social Model

The changes in the French social model since the years 2015/2018 aim clearly to combine flexibility and safety. They challenge only very partially the stakes of the work changes, but they are such that they could alter significantly the principles mentioned above. Let us say straight away the main lines of the reforms:

- the break of the work contract is made easier through a simplification of the procedures and the indemnities layoff with the creation of a collective conventional break;
- in the limit of a restricted «public social order», the negotiated rules prevail, even if they are less favorable to the workers and the collective agreements at the company level may derogate in various fields to the branch agreements;
- the workers representation and the negotiation are reduced with the setup of an unique representative body and the deletion of the instance dedicated to the health at work.

The measures which are being adopted are a step forward on the avenue of the so called "flexisecurity". At first the flexibility in the name of competitiveness and fair employment, and after the counterpart in terms of reinforced transitional rights and professional paths securisation. But the flexibility broadly prevails on safety and the structural reform with "business friendly" aims is going forward; even if one calls the social partners to find out in the new frame of the reform balanced compromises at the branch and company level.

In our oral presentation, we will analyze the main proposals of the mission conducted by the Member of Parliament Lecocq about the reform of health at work, which conclusions are expected by June 2018 and we will indicate the main proposals made by the French speaking ergonomist community [35].

The Changes Operating in the Work

The changes operating in the work present opportunities but also risks. In addition, with the recent work transformations (such as the services economy development, coined by the growing digitalization, the artificial intelligence, the robotization, the "uberization", …) on state that the work situations produce new risks:

- the automatization, with the risks of an impoverished human model and its reliability, of a skills polarisation and a depreciation of the low considered tasks; those which are in a way of automatization or those of service activities which are considered as poorly gratifying;
- The robotization with the risk for the operators of room for manoeuvre reduction, the risk of a cognitive assistance being too directive, the issue of the lack of common operative referential between the human and the robot;
- The support development via the connected things with the risks of dehumanization and the loss of meaning at work for the operators who are at the interface between the devices and the final users.

6   Conclusion

In front of these new constraints and opportunities, in the academic institutions, in the ergonomic societies and in the public space, the issue of the updating of the ergonomics principles, as they are transcribed in the legislation and the regulations is very acute. It is, for example, the matter of:

- to re-question the models of a prescriptive prevention: because we intend to favour the general principles aiming to adapt the work to the human, it means to consider the operators not as passive actors of the prevention, but mostly able to associate in the professional acting the preventive gesture and the productive one. It means to foster the operational agreements more than the technical regulations and to emphasise the discussion spaces on the work more than the top down managerial directions;
- to highlight the practical individual and collective intelligence of the operators, to take into account their experience and to favor their expression through places of debate on the work is a determinant key for the sustainability of the occupational risks;
- to incite the companies to give again the operators a power of action and means to achieve a well done work – for the health and the performance- with the acknowledgment of the moments to speak about work legitimation and the related human means, in order to be able to think, to discuss and to act on the work.

References

1. Wisner A (1996) Ergonomics work analysis. Ergonomics 38(3):595–605
2. Teiger C, Laville A (1991) L'apprentissage de l'analyse ergonomique du travail, outil d'une formation pour l'action. Travail et Emploi 47(1):53–62
3. Daniellou F (2015) L'ergonomie en quête de ses principes. Octares, Toulouse
4. Falzon P (2014) Constructive ergonomics. CRC Press, Boca Raton

5. Clot Y (2008) Travail et pouvoir d'agir. PUF, Paris
6. Christol J (2016) Entretien in Christol-Souviron et alii, Performances humaines et techniques. Octarès, Toulouse
7. Loi 73-1195 du 27 décembre 1973, article L. 4642-1 du Code du travail. https://www.legifrance.gouv.fr
8. Accord-cadre du 17 mars 1975 sur l'amélioration des conditions de travail. http://www.intefp-sstfp.travail.gouv.fr
9. Loi 76-1106 du 6 décembre 1976. https://www.legifrance.gouv.fr
10. Loi n° 82-689 du 4 août 1982 relative aux libertés des travailleurs dans l'entreprise. https://www.legifrance.gouv.fr
11. Loi 82-1097 du 23 décembre 1982relative aux CHSCT. https://www.legifrance.gouv.fr
12. https://eur-lex.europa.eu/legal-content/EN/TXT/PDF/?uri=CELEX:31989L0391
13. https://www.legifrance.gouv.fr
14. Maggi B (2003) Etude du travail et action pour le bien-être, in De l'agir organisationnel. Octares, Toulouse
15. 2006/42/EC Directive on machinery. https://eur-lex.europa.eu/legal-content/EN
16. Article R. 4312-1, Annexe I. https://www.legifrance.gouv.fr
17. 89/686/EEC Directive on personal protective equipment. https://eur-lex.europa.eu/legal-content/EN
18. Article R. 4312-1, Annexe II. https://www.legifrance.gouv.fr
19. Verkindt PY (2014) Les mots de la prévention au travail, in La prévention des risques au travail, Semaine sociale Lamy, supplément, 1655, 8 décembre 2014
20. Emmanuelle Lafuma et Cyril Wolmark, «Le lien santé-travail au prisme de la prévention. Perspectives juridiques» , Perspectives interdisciplinaires sur le travail et la santé [En ligne], 20-1 | 2018, mis en ligne le 01 novembre 2016, consulté le 17 février 2018. http://journals.openedition.org/pistes/5560. https://doi.org/10.4000/pistes.5560
21. https://osha.europa.eu/en/legislation/guidelines/Frameworkagreementonwork-relatedstress. 8 Oct 2004
22. https://osha.europa.eu/en/legislation/guidelines/Framework-agreement-on-harassment-and-violence-at-workl. 26 Apr 2007
23. Accord du 17 juin 2011 relatif au harcèlement et à la violence au travail. www.journal-officiel.gouv.fr/publications/bocc/
24. Aaccord du 2 juillet relatif au stress au travail. www.journal-officiel.gouv.fr/publications/bocc/
25. LOI n° 2015-994 du 17 août 2015 relative au dialogue social et à l'emploi. https://www.legifrance.gouv.fr/
26. Cour de Cassation, arrêts amiante (2002). https://www.courdecassation.fr/jurisprudence_2/chambre_sociale_576/arret_n_1158.html
27. Conseil d'État, Assemblée, 03/03/2004, 241152, Publié au recueil Lebon. https://www.legifrance.gouv.fr/affichJuriAdmin.do?oldAction=rechJuriAdmin&idTexte=CETATEXT000008176294&fastReqId=484622050&fastPos=1
28. Arrêt SNECMA Cass. soc. 5 mars 2008, n° 06-45.888. https://www.legifrance.gouv.fr/affichJuriJudi.do?oldAction=rechJuriJudi&idTexte=JURITEXT000018234005
29. Dossier (2017) pour une prévention durable des TMS en entreprise, Aubret-Cuvelier, Agnès. www.INRS
30. Proceedings of the SELF-ACE 2001 (2001) Conference – Ergonomics for changing work. http://docplayer.fr/619339-La-pratique-de-l-ergonomie-face-aux-tms-ouverture-a-l-interdisciplinarite.htmlVézina

31. La-Prévention durable des TMS Quels freins? Quels leviers d'action? (2008) Caroly, S. et alii, direction générale du travail. https://halshs.archives-ouvertes.fr/halshs-00373778/document
32. Dugué B (2012). https://www.anact.fr/rapport-detude-les-chsct-entre-dispositifs-et-pratiques
33. https://ergonomie-self.org/2017/11/03/l'ergonomie-face-aux-reformes-du-code-du-travail/
34. https://ergonomie-self.org/2018/05/04/contribution-de-la-self-a-la-mission-lecocq/
35. https://ergonomie-self.org/2018/05/04/positionnement-de-la-self-dans-le-cadre-de-la-mission-lecocq/

Neck Disorder Influenced by Occupational Reward Type: Results from Effort-Reward Imbalance Model Based on IPWS

Seyed Abolfazl Zakerian[1], Saharnaz Nedjat[1], Saeedeh Mosaferchi[1],
Hadi Ahsani[2], Fateme Dehghani[1], Mahdi Sepidarkish[1],
and Alireza Mortezapour Soufiani[1,3(✉)]

[1] Tehran University of Medical Sciences, Tehran, Iran
a-mortezapour@student.tums.ac.ir
[2] Iran University of Medical Sciences, Tehran, Iran
[3] Hamadan University of Medical Sciences, Hamadan, Iran

Abstract. Effort-Reward Imbalance model is known as one of the survey method of occupational stress and also as an effective element on health condition according to its parameters. Due to types of rewards generally, and reward subscales in this model specifically, each one can have a distinctive effect on health perception, current study is aimed at determination of the most effective reward subscale for managing work-related neck disorder in industries.

All of workers who participated in IPWS study (N = 1126), were entered in the statistical analysis stage. After completing personal and organizational information, they responded to Van Vegchel et al. Effort-Reward Imbalance and also Dutch questionnaires for their musculoskeletal disorders. Chi-square and t-test comparisons were performed and the final regression model was presented with a significance level of 0.05.

The mean (Standard deviation) age of workers and musculoskeletal disorders prevalence in neck were 33.21 (7.63) years and 34 percent respectively. Also in workers with neck pain, odds ratio between effort and monetary reward, between effort and respect reward, and between effort and security reward in their jobs were 1.35, 2.07 and 1.32 respectively. After elimination of confounders in final regression model, significant correlation was remained only between effort and job respect reward.

According to high prevalence of musculoskeletal disorders in neck and also large amount of effort-reward imbalance in Iranian workers, implementing interventions are recommended. Based on results of present study, it is suggested that main intervention must be focused on respect and esteem reward in jobs.

Keywords: Neck disorder · Effort-reward imbalance · Psychosocial factor Musculoskeletal complaint

© Springer Nature Switzerland AG 2019
S. Bagnara et al. (Eds.): IEA 2018, AISC 820, pp. 316–325, 2019.
https://doi.org/10.1007/978-3-319-96083-8_43

1   Introduction

Effort – Reward Imbalance is a specialized term, according to it, a conceptual model has been introduced to study stress in workplaces [1]. This model is considered as a cross-connection between sociological theories and psychosocial work factors [2]. It is based on social interactions theory [3, 4] that occupational rewards are paid for job efforts after work contraction between employee and employer. This model is one of the most valid methods of studying occupational stress [6], which is used for prediction wide range of health condition like cardiovascular disease [5], musculoskeletal disorders [6], and asthma [7] in various studies in developed [8] and even in developing countries [9]. Different components of ERI are worker effort, reward paid to him [7, 10]. Usually effort does during a Reciprocal social interaction (in the workplace and with employer) according to work contract [3]. In various studies effort considered synonymous with job demands [11]. Rewarding to worker will be done through 3 different transmission systems, including money, social support or respect to worker and career development chance or job security [1, 12]. According to main hypothesis of this model, imbalance between effort and reward (from total of three transmission systems), can cause perception of injustice in workplace and stress as a result; If the stress is not controlled, it can have many negative consequences [13]. Reward component in this model and also occupational rewards in general can impresses upon worker perception of job condition and his/her health consequently [1, 15]. Heijden et al. studied the role of social support from direct supervisor as a job reward, after controlling confounders stated that social support can be an effective predictor for determination of turnover and some health related parameters in nurses [16]. In another study, researchers studied the influence of occupational reward on health staff services quality and also some their health related parameters, as a result increasing wage can lead to job satisfaction and accordingly improving services quality in most participants [17]. Results of Hampton et al. study showed that occupational reward can effects on job satisfaction [18]; also various studies have already shown the influence of job satisfaction on mental and physical health of worker subsequently [19]. Siegrist demonstrated in his study [20] that individual understanding of being not useful in social roles (like workplaces) can have a negative effect on his health that also in other studies were noted [21, 22]. In effort- reward imbalance model, reward paid to worker using 3 different transmission systems, each of these systems creates different perception in worker from received reward. This different perception can have different health consequences [1]. In 2010 a longitudinal study was designed and implemented in some European countries to evaluate the impact of effort-reward imbalance on nurse's turnover, results during one-year follow up showed that about 8 percent of nurses had to leave their job that after elimination of confounders, it was related to effort- reward imbalance in them [23]. Another finding of this study, was stated that reward component (in all 3 transmission systems) can have a significant impact on job turnover that this point also mentioned in some studies [1, 24]. Another study has been expressed that wage level of workers can use for prediction of their health related parameters [22]. Or Mayhew showed that economic pressure on workers can have negative impacts on his health related parameters [25]. Van Vegchel is studying health parameters in medical staff using effort - reward imbalance model, results showed that in medical staff, unlike wage which had not a significant relationship with health related parameters, two other transmission systems (respect and security of job) had significant impact on health parameters [10]. Know due to high prevalence of work related neck disorders in developing countries like Iran [26], and given that no study was implemented to evaluate the separate effect of different rewards, with regards to ERI model, on work related musculoskeletal disorders, the aim of current study is to survey separate effect of 3 transmission systems including wage, respect in job and job security-as job rewards- on neck disorders at Iranian workers.

2   Materials and Methods

The current study was conducted by results of Iran Psychosocial Workplace Survey (IPWS), which has been done in 2015 and 2016. All stages of current study were approved by Tehran University of Medical Science Ethics Committee according to contract No. IR. TUMS.SPH.REC.1395.950 and participants had ensured about confidentiality of their personal information.

2.1   Data Gathering Tools

Demographic Information Questionnaire
By literature review related to the objectives of the current study and according to effective parameters in relation between dependent and independent variables, researchers gathered demographic parameters of the industry staff that could confound the main purpose of study, and another their personal information by questionnaire.

Musculoskeletal Disorders Questionnaire
Part of the Dutch Musculoskeletal Disorders Questionnaire that its reliability and validity was studied in Hildebrandt et al. study [27], were used to gathering information about musculoskeletal disorders questionnaire. This questionnaire is used only in one study within the country [28]. This part of Dutch questionnaire includes some questions about musculoskeletal disorders prevalence in nine body regions at 3 different time periods.

Van Vegchel et al. Effort- Reward Imbalance Questionnaire
Up to now, different versions of Effort- Reward Imbalance are used at various studies and with different purposes, and under consideration the strengths and weaknesses of each, the questionnaire presented by Van Vegchel et al. was used in this study [10]. Questionnaire's validity and reliability had been reported 0.7 to 0.92. This tool has been localized within the country in Oreyzi et al. study [29] and validity and reliability of its different component was presented 0.72 to 0.89. This scale includes 8 questions for effort and 10 questions in 3 sub scales of wage, respect in job, and job security for reward scale. Given that there was no need to measure over-commitment to work in current study, by considering the instructions in references in order to avoid mistakes in

final scoring, this scale was not measured in this study. Answer to questions were marked as a five point Likert in effort scale and as a four point Likert in reward scale.

2.2 Statistical Analysis

Sample Characteristics were summarized as proportions for dichotomous variables and mean with standard deviation for continuous variables. Dichotomous variables of efforts and rewards were created as following: first average scores on efforts (8 items) were divided on average scores on financial reward (2 items). Second the ratio dichotomized by coding ERI > 1 as 1 ("participants with imbalance effort- financial reward") and ERI ≤ 1 as 0 ("participants with balance effort- financial reward"). Finally, for other two rewards components (i.e., esteem, and security), an effort–reward imbalance indicator was computed in a similar vein. Comparison of continuous and dichotomous variables between two groups (with and without cervical musculoskeletal disorder) was done using Student's t-test and Chi square test, respectively. Two logistic regression models were constructed for determining the association between effort–reward imbalance and cervical musculoskeletal disorder. First, crude relationships between cervical musculoskeletal disorder and effort–reward imbalance indicator were studied separately. Second, multiple logistic regression analysis was used to study the association between effort–reward imbalance indicators and cervical musculoskeletal disorder all together. Independent variables entered to the model were age, body mass index (BMI), length of work, time of work (day/week), gender, marital status, educational level, shift work, contract status, having second job, manpower shortage, smoking status Results are presented as odds ratios with 95% confidence intervals. Model building was down based on the Hosmer–Lemeshow recommendation. The statistical analysis was done by running Stata version 13 (Stata, College Station, TX, USA).

3 Results

3.1 Study Population

In 2015 and 2016, 1126 subjects were invited to participate in the IPWS study. 1116 (90.1% males and 9.9% females) answered the self-administered questionnaire, that is, a 99.11% response rate. The overall mean ± (SD) age of the participants was 33.21 ± (7.63) years and the majority were Contractual employee 885 (79.3%) and married 822 (73. 65%). The average work tenure was 7.24 ± (6.16) with average working days/week (SD) of 6.06 ± (0.8).

3.2 Neck Musculoskeletal Disorder Distribution and Correlation with Other Measures of Working Conditions

Among existing participants, 388 cases reported having cervical musculoskeletal disorder. The prevalence of participants with cervical musculoskeletal disorder was 34.48% (CI95%: 31.70–37.27). The mean (SD) age of employees with and without cervical musculoskeletal disorder was 34.25 ± (7.69) vs. 32.65 ± (7.51) respectively; (P < 0.001). Participants with cervical musculoskeletal disorder were significantly more likely to be female (16.76% vs. 6.60%, p < 0.001) and to have shift work (39.18% vs. 31.46%, p < 0.001), and official contract (23.45% vs. 15.24%, p < 0.001). Also cases had a significantly (p = 0.014) higher education level compared to control group (36.40% vs. 19.78%, p < 0.001). Demographic variables of two groups are described in Table 1.

Table 1. Comparison of demographic characteristics of two groups

Variables		Participants with neck disorder (n = 388)	Participants without neck disorder (n = 728)	P-value
Age (year)		34.25 ± (7.69)	32.65 ± (7.51)	0.001
Body Mass Index (BMI)		25.40 ± (3.79)	25.39 ± (4.00)	0.962
Length of work (years)		8.09 ± (6.37)	6.80 ± (6.04)	0.002
Time of work (day/week)		5.99 ± (0.91)	6.09 ± (0.74)	0.054
Gender	Male	323(83.24)	680(93.40)	<0.001
	Female	65(16.76)	48(6.60)	
Marital status	Unmarried	73(18.81)	172(23.62)	0.198
	Married	294(75.77)	527(72.39)	
	Others	21(5.42)	29(3.99)	
Education level	Diploma	141(36.40)	144(19.78)	<0.001
	Low literate	247(63.60)	584(80.22)	
Shift work	Yes	152(39.18)	229(31.46)	<0.001
	No	236(60.82)	499(68.54)	
Contract status	Official	91(23.45)	111(15.24)	0.001
	Contractual	297(76.55)	617(84.76)	
Second job	Yes	29(7.47)	67(9.20)	0.319
	No	359(92.53)	661(90.80)	
Manpower shortage	Yes	208(53.60)	360(50.55)	0.135
	No	180(46.40)	368(49.45)	
Smoking status	Current smoker	44(11.34)	102(14.01)	0.214
	Ex-smoker	44(11.34)	64(8.79)	
	Non-smoker	300(77.32)	562(77.20)	

3.3 Associations Between Cervical Musculoskeletal Disorder and Effort–Reward Imbalance Indicators

In a univariate logistic regression analysis, imbalance between effort- financial reward (OR = 1.35, 95% CI: 1.01–1.79, P = 0.036), imbalance between effort- esteem reward (OR = 2.07, 95% CI: 1.59–2.69, P < 0.001) and imbalance between effort- security reward (OR = 1.32, 95% CI: 1.02–1.71, P = 0.001) showed a significant association with CVD. As shown in Table 2, after adjusting for covariates imbalance effort- esteem reward remained statistically significantly associated with CVD (OR = 2.35, 95% CI: 1.61–3.42, P < 0.001). But, the association between imbalance effort- financial reward (OR = 1.27, 95% CI: 0.83–1.95, P = 0.261) and CVD as well as imbalance effort-security reward (OR = 1.23, 95% CI: 0.88–1.71, P = 0.217) and CVD was attenuated and became non-significant after adjustment for covariates. Also we found the strong associations between Education levels and CVD (OR = 2.08, 95% CI: 1.40–3.07, P < 0.001). Also Education levels (OR = 2.08, 95% CI: 1.40–3.07, P < 0.001), gender (OR = 0.35, 95% CI: 0.19–0.64, P = 0.001) and shift work (OR = 1.44, 95% CI: 1.01–2.06, P = 0.043) were other significant risk factors. There was no significant association between the prevalence of multiple pregnancy and other variables.

Table 2. Association Between cervical musculoskeletal disorder and its predictors

Variables		Unadjusted OR	95% CI	Adjusted OR[a]	95% CI
Imbalance effort - financial reward		1.35	1.01–1.79	1.27	0.83–1.95
Imbalance effort - security reward		1.32	1.02–1.71	1.23	0.88–1.71
Imbalance effort - esteem reward		2.07	1.59–2.69	2.35	1.61–3.42
Age (year)		1.02	1.01–1.04	1.01	0.98–1.04
Body Mass Index (BMI)		1.00	0.96–1.03	0.97	0.93–1.01
Length of work (years)		1.03	1.01–1.05	1.02	0.98–1.06
Time of work (day/week)		0.85	0.73–0.99	0.92	0.74–1.14
Gender	Male	0.36	0.24–0.53	0.35	0.19–0.64
	Female	1	Referent	1	Referent
Marital status	Unmarried	1	Referent	1	Referent
	Married	1.31	0.96–1.78	1.07	0.67–1.71
	Others	1.88	0.71–4.96	1.11	0.34–3.64
Educational level	Diploma	2.34	1.77–3.08	2.08	1.40–3.07
	Low literate	1	Referent	1	Referent
Shift work	Yes	1.35	1.03–1.76	1.44	1.01–2.06
	No	1	Referent	1	Referent
Second job	Yes	0.79	0.50–1.25		
	No	1	Referent	1	Referent
Contract status	Official	1.72	1.26–2.35	0.69	0.38–1.23
	Contractual	1	Referent	1	Referent
Co-worker shortage	Yes	1.20	0.94–1.55	1.04	0.74–1.46
	No	1	Referent	1	Referent
Smoking status	Current smoker	0.81	0.55–1.19	0.76	0.46–1.25
	Ex-smoker	1.30	0.86–1.96	1.09	0.63–1.88
	Non-smoker	1	Referent	1	Referent

a. Adjusted for age, body mass index (BMI), length of work, time of work (day/week), gender, marital status, educational level, shift work, contract status, having second job, manpower shortage, smoking status.

4 Discussion

The aim of current study was to compare effect of various Occupational rewards with regards to Effort- Reward Imbalance model parameters in order to prediction of neck disorders among Iranian workers. Workers involved in current study were reported neck disorder with odds ratio of about 1.27 to 2.35 in case of understanding existence imbalance between effort and reward in their workplace. Aligned with results of this study, other researchers reported that imbalance between effort and reward has a connection with musculoskeletal disorders generally and in neck specially [6, 29]. Peter et al. [30] surveyed the effect of the same model on 1300 workers' health parameters and confirm results of this study, they concluded that imbalanced workers, also have reported musculoskeletal disorders in their body regions with odds ratio about 2. There is a reasonable agreement between the results of a study on 3600 male workers and 1700 female workers [31], and results of current study in terms of odds ratio between effort- reward imbalance with musculoskeletal disorders. Another result of this study was different effect of various components of occupational rewards in workers who have been reported neck disorder (Table 2). This generally means different occupational rewards (monetary reward or wage, reward as dignity and respect in job, and reward as having security and upgrade opportunity in job) can cause pain in neck with different odds ratios. Same as results of recent study, Van Vegchel et al. [10] surveyed health service workers and concluded that various components of reward have different effects on these worker's health. With study the impact of different components of effort- reward imbalance. Contradictory with present study, researchers have reported that financial and also security and job upgrade opportunity reward component are more effective than respect in work as a reward among those workers, perhaps reason of difference between current study results with this study is related to the differences in occupational group in two studies (industrial workers participated in current study and nurses) and each job position while doing their job, and another reason can be attributed to working structures and the importance of human resources between 2 developed and developing countries. Generally, it can be concluded that there is a fairly good consensus on different effects of various reward components about most dependent parameters (job performance, job satisfaction, …), but until now there is not this level of certainty about reward different components impact on health parameters. As the results show, workers involved in current study will experience more neck disorder, if they understand imbalance between effort and reward in the type of respect in work with higher odds ratio in comparison with imbalance between effort and financial or job security rewards. This impact increased after elimination of confounding variables, so that imbalance between effort and reward in type of getting respect in job only maintained its significant relationship with neck pains. By studying health workers, Van Vegchel et al. [10] were concluded that in case of imbalance between effort and respect in job as occupational rewards, participants will complain from their health problems with greater chance. Therefore, unlike the difference in target group results of these 2 studies can considered consistent with each other. On the other hand, getting respect in job from the manager, supervisor and co-workers can be considered as getting social support for worker. according to the Maslow motivation theory [32], this factor is a higher level needs than getting money and security (in job) and with this interpretation, it is a more important factor in individual perception of his health. According to shortage in literatures, accurate conclusion about compare the impact of different types of occupational rewards on health parameters like neck pain was not possible. Other results of present study showed that female gender, having low level of education and shift work can consider as a risk factor for neck disorder. Aligned with this results [33], Yue et al. concluded that prevalence of musculoskeletal disorders in female teachers is more than male teachers [34]. these results also confirmed by longitudinal studies [35]. Consisted with current study, Trinkoff surveyed health services workers in a longitudinal study and was reported that working time generally and having shift work in workplace can cause musculoskeletal disorders in workers by improving occupational stress [36]. also similar results could be found in Fredriksson et al. study [37]. In an agreement with present study, the participants in Luo et al. [38] study suffered from neck disorders if they had a lower level of education, but inconsistent with the results of this study Haines et al. [39] were stated in their study that a higher level of education have no significant effect on neck disorders recovery.

5 Conclusion

As the results showed, If the workers perceive imbalance between effort and rewards, they would be more likely to experience neck disorder. If this Imbalance is between effort scale and respect at work as an occupational reward, the worker will be more prone to have neck disorder than Imbalance between Effort scale and monetary or job security as occupational rewards. According to the results, it can be concluded that If interventions for neck pain are requisite, managers would be taking into account the different components of the effort-reward imbalance model and especially focus on adjustment of the various components of occupational rewards in this model.

Acknowledgment. This research has been supported by Tehran University of medical sciences. Also Researchers are thankful for all participations in this study.

References

1. Tsutsumi A, Kawakami N (2004) A review of empirical studies on the model of effort-reward imbalance at work: reducing occupational stress by implementing a new theory. Soc Sci Med 59(11):2335–2359
2. Shimazu A, de Jonge J (2009) Reciprocal relations between effort-reward imbalance at work and adverse health: a three-wave panel survey. Soc Sci Med 68(1):60–68
3. Siegrist J (2005) Social reciprocity and health: new scientific evidence and policy implications. Psychoneuroendocrinology 30(10):1033–1038
4. ur Rehman S, Khan MA, Afzal H (2010) An investigative relationship between efforts-rewards model and job stress in private educational institutions: a validation study. Int J Bus Manag 5(3):42

5. Siegrist J (2010) Effort-reward imbalance at work and cardiovascular diseases. Int J Occup Med Environ Health 23(3):279–285
6. Rugulies R, Krause N (2008) Effort–reward imbalance and incidence of low back and neck injuries in San Francisco transit operators. Occup Environ Med 65(8):525–533
7. Loerbroks A, Herr RM, Li J, Bosch JA, Seegel M, Schneider M et al (2015) The association of effort–reward imbalance and asthma: findings from two cross-sectional studies. Int Arch Occup Environ Health 88(3):351–358
8. Siegrist J, Starke D, Chandola T, Godin I, Marmot M, Niedhammer I et al (2004) The measurement of effort–reward imbalance at work: European comparisons. Soc Sci Med 58 (8):1483–1499
9. Niedhammer I, Tek M-L, Starke D, Siegrist J (2004) Effort–reward imbalance model and self-reported health: cross-sectional and prospective findings from the GAZEL cohort. Soc Sci Med 58(8):1531–1541
10. Van Vegchel N, de Jonge J, Bakker A, Schaufeli W (2002) Testing global and specific indicators of rewards in the effort-reward imbalance model: does it make any difference? Eur J Work Organ Psychol 11(4):403–421
11. Babamiri M, Nisi A, Arshadi N, Mehrabizadeh HM, Bashlideh K (2015) Survey of occupational stressors and personality characteristics as a predictors of psychosomatic syndrome in workers. Psychol Achivements 22(1):187–208 [in Persian]
12. Fahlen* G, Peter R, Knutsson A (2004) The effort-reward imbalance model of psychosocial stress at the workplace—a comparison of ERI exposure assessment using two estimation methods. Work Stress 18(1):81–88
13. Babamiri M, Nisi A, Arshadi N, Shahroie SH (2015) Job stressors as predictors of psychosomatic symptoms. J Ilam Univ Med Sci 23(1):45–55
14. Preckel D, Meinel M, Kudielka BM, Haug HJ, Fischer JE (2007) Effort-reward-imbalance, overcommitment and self-reported health: is it the interaction that matters? J Occup Organ Psychol 80(1):91–107
15. Van der Heijden B, Kümmerling A, Van Dam K, Van der Schoot E, Estryn-Béhar M, Hasselhorn H (2010) The impact of social support upon intention to leave among female nurses in Europe: secondary analysis of data from the NEXT survey. Int J Nurs Stud 47 (4):434–445
16. Chandler CI, Chonya S, Mtei F, Reyburn H, Whitty CJ (2009) Motivation, money and respect: a mixed-method study of Tanzanian non-physician clinicians. Soc Sci Med 68 (11):2078–2088
17. Hampton GM, Hampton DL (2004) Relationship of professionalism, rewards, market orientation and job satisfaction among medical professionals: the case of certified Nurse-Midwives. J Bus Res 57(9):1042–1053
18. Faragher EB, Cass M, Cooper CL (2005) The relationship between job satisfaction and health: a meta-analysis. Occup Environ Med 62(2):105–112
19. Siegrist J (2000) Place, social exchange and health: proposed sociological framework. Soc Sci Med 51(9):1283–1293
20. Siegrist J, Marmot M (2004) Health inequalities and the psychosocial environment—two scientific challenges. Soc Sci Med 58(8):1463–1473
21. Marchand A, Demers A, Durand P (2005) Does work really cause distress? The contribution of occupational structure and work organization to the experience of psychological distress. Soc Sci Med 61(1):1–14
22. Dahl E (1994) Social inequalities in ill-health: the significance of occupational status, education and income-results from a Norwegian survey. Soc Health Illn 16(5):644–667

23. Li J, Galatsch M, Siegrist J, Müller BH, Hasselhorn HM, group ENS (2011) Reward frustration at work and intention to leave the nursing profession—prospective results from the European longitudinal NEXT study. Int J Nurs Stud 48(5):628-635
24. De Jonge J, Van Der Linden S, Schaufeli W, Peter R, Siegrist J (2008) Factorial invariance and stability of the effort-reward imbalance scales: a longitudinal analysis of two samples with different time lags. Int J Behav Med 15(1):62–72
25. James P, Mayhew C, Quinlan M (2006) Economic pressure, multi-tiered subcontracting and occupational health and safety in Australian long-haul trucking. Empl Relat 28(3):212–229
26. Yousefi H, Habibi E, Tanaka H (2017) Prevalence of work related musculoskeletal disorders among the iranian working population in different sectors of industries. In: Advances in social and occupational ergonomics, pp 271–281. Springer
27. Hildebrandt V, Bongers P, Van Dijk F, Kemper H, Dul J (2001) Dutch musculoskeletal questionnaire: description and basic qualities. Ergonomics 44(12):1038–1055
28. Eftekhar Sadat B, Babaei A, Amidfar N, Jedari Eslami MR (2013) Prevalence and risk factors for low back pain in nursing staffs of Tabriz hospitals in 1387. J Nurs Midwifery Urmia Univ Med Sci 11(9)
29. Simon M, Tackenberg P, Nienhaus A, Estryn-Behar M, Conway PM, Hasselhorn H-M (2008) Back or neck-pain-related disability of nursing staff in hospitals, nursing homes and home care in seven countries—results from the European NEXT-Study. Int J Nurs Stud 45 (1):24–34
30. Peter R, Geißler H, Siegrist J (1998) Associations of effort-reward imbalance at work and reported symptoms in different groups of male and female public transport workers. Stress Health 14(3):175–182
31. Yu S, Nakata A, Gu G, Swanson NG, He L, Zhou W et al (2013) Job strain, effort-reward imbalance and neck, shoulder and wrist symptoms among Chinese workers. Ind Health 51 (2):180–192
32. Maslow AH (1943) A theory of human motivation. Psychol Rev 50(4):370
33. Hooftman WE, van der Beek AJ, Bongers PM, van Mechelen W (2005) Gender differences in self-reported physical and psychosocial exposures in jobs with both female and male workers. J Occup Environ Med 47(3):244–252
34. Yue P, Liu F, Li L (2012) Neck/shoulder pain and low back pain among school teachers in China, prevalence and risk factors. BMC Publ Health 12(1):1
35. McLean SM, May S, Klaber-Moffett J, Sharp DM, Gardiner E (2010) Risk factors for the onset of non-specific neck pain: a systematic review. J Epidemiol Commun Health 64 (7):565–572
36. Trinkoff AM, Le R, Geiger-Brown J, Lipscomb J, Lang G (2006) Longitudinal relationship of work hours, mandatory overtime, and on-call to musculoskeletal problems in nurses. Am J Ind Med 49(11):964–971
37. Fredriksson K, Alfredsson L, Köster M, Thorbjörnsson CB, Toomingas A, Torgén M et al (1999) Risk factors for neck and upper limb disorders: results from 24 years of follow up. Occup Environ Med 56(1):59–66
38. Luo X, Edwards CL, Richardson W, Hey L (2004) Relationships of clinical, psychologic, and individual factors with the functional status of neck pain patients. Value Health 7(1):61–69
39. Haines T, Gross AR, Burnie S, Goldsmith CH, Perry L, Graham N et al (2009) A cochrane review of patient education for neck pain. Spine J 9(10):859–871

Ergonomics Risk Factors Prevailing in Kota Doria Loom Weavers of India

Nabila Rehman[✉]

Faculty of Home Science, Banasthali University, Vanasthali, India
nabila.rhmn@gmail.com

Abstract. Weaving of Kota Doria fabric has been a traditional activity of Ansari community of the Kota region. After abandoning this art by men in order to reach more stable and money yielding means of livelihood, women of the community have adopted this traditional activity in order to continue their culture and generate least minimum income. Weaving of Kota Doria fabric is carried on pit loom. 100 women weavers from Kaithoon area of Kota district participated in the study. In depth observation of the workplace followed by the subjective assessment of body discomfort and Workplace Ergonomic Risk Assessment (WERA) method was carried out.

On analysis of WERA, the leg score for WERA body part was >4 in 73% of weavers whereas discomfort in knees and legs were reported by 87% yielding a significant association between WERA body part score and self-reported pain (χ^2 = 20.51; p = 0.000). The body part score for shoulder region during Kota Doria operation yielded a score of >4 in 68% and cause pain in 72% association being significant (χ^2 = 23.86; p = 0.000). The neck region for WERA body part score was >4 in 80% of the weavers with 89% reporting pain or neck discomfort with significant association (χ^2 = 5.96; p = 0.014). the back region score for WERA body part was >4 in 79% of weavers whilst discomfort reported by 97% with a significant association (χ^2 = 4.20; p = 0.040).

Therefore it can be concluded that existing Kota Doria pit loom weavers depict significant prevalence of MSDs and need intervention for enhancing productivity and occupational wellbeing of the weavers.

Keywords: Kota doria pit loom · Musculoskeltal disorders
Ergonomic risk assessment · Observation method

1 Introduction

Weaving is acknowledged to be one of the oldest surviving crafts in the world. In present era of mechanization and standardization, the handloom sector has been able to endure the oldest craft through diverse manual skills representing various cultural and traditional art forms across the globe. In India, the handloom industry is one of the largest employment generating trade after agriculture It has been estimated that there are about 4.60 million handlooms in the world out of which about 3.9 million are in India [2] and 77.9% of the workforce in this sector is reported to be women [3].

One of the pit handloom fabric manufactured in the western region of India is known across the world for its square check patterns called Kota Doria (also spelled as

S. Bagnara et al. (Eds.): IEA 2018, AISC 820, pp. 326–337, 2019.
https://doi.org/10.1007/978-3-319-96083-8_44

Kota Dori). The name Kota Doria is taken from it place of origin, Kota in province of Rajasthan, India. Kota Doria industry plays a vital role in economic development of the rural masses of Hadoti region(comprises Baran, Bundi and Kota district). At present Kota Doria cluster reports to contribute the total business of around 850,000,000 INR (or 13,035,600 USD approx.) every year with 2500 looms engaging around 3000 weavers.

Weaving appears to be a very simple process but is one of the most tedious professions which entail long hours of static, highly repetitive and monotonous tasks to be performed in a sitting working posture with upper back curved and head bent over the loom without any kind of lumbar support. The work is visually demanding and requires a high degree of concentration and accuracy which can contribute to occupational hazard by developing musculoskeletal disorders through awkward posture, repetitive movement and contact stress [4]. Interestingly, weaving of Kota Doria is an inherent art among the Ansari community which traditionally was performed by both male and female. But due to low remuneration for their skills, males have engaged in post-production activities or have pursued some other jobs thus leaving this art entirely on the shoulders of the women of Ansari community. Each abode of the community possess handloom moreover the number of handloom is directly proportional to the number of females in the family. These women prefer working from as it gives them feasibility to work part time, full time and sometimes beyond full time.

Handlooms used by women are traditional which has been designed by local carpenters with little or no knowledge of female anthropometry and physiology and awareness of requirements of domestic weaving. This leads to various ergonomic issues at the handloom and the women weavers.

Various studies have identified glitches confronted by handloom weavers, the most widespread of them are musculoskeletal disorders (MSDs) such as back pain, shoulder pain, knee pain, palm pain, stiffness of hand joints along with diminished visual acuity, COPD (Chronic Obstructive Pulmonary Disease) which is associated with breathing problems due to inhalation of fiber dust during weaving followed by psychological stresses and headaches [9–12]. Ergonomic study in textile industry of India is very rare and particularly with reference to women workforce [1] however, few investigations carried out on female weavers divulges that women weavers were more prone to musculoskeletal disorders as compared to men counterpart [9, 12].

A study has even claimed that increase in number of years of weaving has found to be significantly associated with backache, diminished visual acuity, COPD and musculoskeletal pain.

A range of studies have been carried out on probing the ergonomic issues prevalent in Iran carpet industry, it was found that ergonomic intervention have resulted in enhanced quality, productivity and occupational health of the carpet menders [5–8].

However, several studies have been conducted on diverse type of handlooms, but there is lack of study on ergonomic issues on Kota Doria pit loom weavers. Therefore, the existing study have been designed to discover the ergonomic problems and relevant scopes of ergonomic intervention in order to improve the existing loom workstation with the aim of enhancement of operational easiness and occupational wellbeing. The observations attained through this paper can be regarded as an initiative for improving the occupational situations of weavers engaged in Kota Doria pit looms.

It can be expected that ergonomic approach of assessing the Kota Doria pit loom's present working status may help in finding core problems exist with the weavers and their working environment. Ergonomic interventions would be vital in reducing occupational hazards which increase the efficiency of the weavers, thereby enhancing production.

2 Materials and Methods

The present study was carried out in three phases.

2.1 Phase 1

From Kaithoon, Kota Women's Weaving Organizations was approached and after discussing with the President and Vice President of the organization, a visit has been organized to the Common Facility Center (provided by Rajasthan Government and now functioning as their office), whereby weavers of the association around 450 women weavers (belonging to various Self-help groups) were collected. In order to understand the workstation related design problems prevailed in the existing Kota Doria pit looms, 100 weavers were selected who had worked on pit loom for more than 8 h in previous month. (all of them have pit looms installed at their home).

Task assessment of Kota Doria work cycle and Observational study of Kota Doria pit loom workstation has been performed with the help of video recording. Video recording of weaving process was conducted in various pit looms and analyzed in slow motion. In order to study the Kota Doria weavers it is crucial to understand the tasks performed by them in order to conduct in-depth study with the help of task assessment whereas the purpose of observational study of pit loom is to demonstrate how man-machine interaction takes and identifying the underlying ergonomic problems which have/might lead to occupational hazards.

2.2 Phase II

Subjective assessment was carried out to quantify discomfort or pains in different body parts with the help of Body Map [13] Fig. 1 among the Kota Doria weavers followed by ergonomic workplace assessment with the help of Workplace Ergonomic Risk Assessment Method (WERA).

2.3 Phase III

To establish relationship between WERA assessment and self-reported musculoskeletal disorders in form of pains, aches and discomfort in the relevant body part. Chi square test (χ2-test) has been used to determine the association between physical risk factor score defined by WERA tool and reported pain, aches and discomfort of body part based on number of tasks. Statistical analysis was performed using SPSS for Windows (version 16.0).

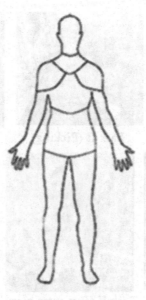

Fig. 1. Body map

3 Results

3.1 Phase I

3.1.1 Stage I: Task Assessment
Seven tasks were analyzed of the Kota Doria pit loom weaving operation. Firstly, the weavers are required to sit on the pit loom with legs placed inside the pit in a piece of brick or stone with a rope inserted between the lower limb's thumb and finger to maneuver the loom. The weavers then are required to insert weft yarn through fabric by shedding, a process where weavers are required to rise and lower the warp yarn, then inserting weft yarn with the help of shuttle (a device with the bobbin carrying thread of the weft) called picking, next beating i.e. packing the weft yarn into the cloth to make it more compact. Subsequently inserting the extra weft yarn for designing motifs as per the requirement. Next, Kota Doria weavers are required to treadle/peddle in between to raise the threads as per the requirement. Thenceforward taking up where the woven fabric is winded on the cloth beam followed by checking and finishing of weaved fabric whereby any minor flaws or loose threads etc. are identified and rectified (Fig. 2).

3.1.2 Stage II: Observational Study
Stage II comprises direct observation method in order to investigate the problems prevailing in Kota Doria handloom. In-depth observational study of video recording was conducted in order to analyze the weaving task. It helped in having better insight of the situation, important for identifying the possible areas of ergonomic intervention to solve occupational health problems. Finally, eight major problem areas associated with seating, treadling, shuttling, beating, taking up, load, work-rest regime and working environment were identified.

Task I (Shredding)   Task II (Picking)   Task III (Beating)

Task IV (Pedalling)   Task V (Inst.extra warp,   Task VI (Taking up)

Task VII (Checking and finishing)

Fig. 2. Tasks involved in Kota Doria weaving

Table 1. Video event analysis of critical points and associated risk factors

Critical point	Associated risk factors
(1) Seating	In order to carry out weaving operation, Kota Doria weavers are required to sit on floor with their feet inserted inside the pit. The pit is square in shape with 16 in. dimension and 19 in. depth (Fig. 3e). Also no back support has been provided to relieve fatigue when not working (Fig. 3c)
	Insufficient thigh clearance due to narrow passage between thigh and cloth beam, it also hits thigh muscles while upward movement and while getting out of the pit loom (Fig. 3c)
	Constant sitting on floor for more than 8 h create pressure on hips, posterior thigh muscles and under knees
	No change in posture due to restricted space leads to fatigue and swelling in feet

(continued)

Table 1. (*continued*)

Critical point	Associated risk factors
(2) Shuttling	Repetition 33.9 ± 4.5 rpm and contributes to fatigue in trapezius muscles
(3) Beating	Pulling and pushing force required for beating operation of reed frame involves repetition 41.1 ± 3.31
(4) Treadling	Treadle are made up of small piece of wood which are rough, unfinished and provide minimum support, resulting pressure development in ankle and toe with frequent operational force (Fig. 3d)
	Weavers maneuver treadle with nylon rope inserted between big toe and index toe of the foot, which contribute to rashes, cuts and sores in feet (Fig 3e)
	Weavers are required to move warps which tied up with 22 kg of weight with high repetition (61.9 ± 12.66 rpm) leading to development of fatigue in leg
	Also treadling makes the thigh repetitively hit against the floor which digs into posterior thigh muscles and creates pressure on buttocks as well
(5) Taking up	The iron rod does not have any tool handle or any protective equipment while taking up the fabric (Fig. 3f)
	During the process 22 kg of load need to be lifted by weaver while sitting on the pit loom by bending on the right hand excessively
(6) Load	In order to provide tension to warp on the horizontal loom, the threads are tied up with load of 22 kg in cloth bag s with small stones in combination of 4 bags * 2.5 kg = 10 Kgs, 4 bags * 2 kg = 8 Kgs, 4 bags * 0.5 kg = 2 kg, 2 bags * 1 kg = 2 kg (Fig. 3g)
	Loads are being held on iron rod attached to the ceiling with the hooks, however rusting of iron leads to fall of load bags and cause accident (Fig. 3g)
(7) Working environment	
(i) Poor lighting	Weaving of Kota Doria is regarded as precision work due to fineness and color recognition
	Since the weaving of Kota Doria was carried out by 100% of weavers at home, there was insufficient lighting and often worn out paint and greasy walls with light level of 31.6 lx ± 7.29 lx
	Inadequate illumination contributes to more awkward postures as weavers are required to incline their head, neck and back to be able to look closer to the work along with headache and eye strain

(*continued*)

Table 1. (*continued*)

Critical point	Associated risk factors
(ii) Poor ambient condition	Since weaving is carried out by cent percent weavers, often cooling and heating system at weaving workplace is not satisfactory as the single room serves as kitchen, bedroom, workplace, dining area as well as play area for children. The average temperature recorded in April for Doria pit looms was 39.5 °C ± 5.5 °C Lack of ventilation in room/house of weavers often contributes to poor quality with fragment of fibers which serves as risk to various respiratory diseases
(8) Lack of work rest regime	Motivation towards earning more money causes weavers to work continuously for long duration without or very less rest periods

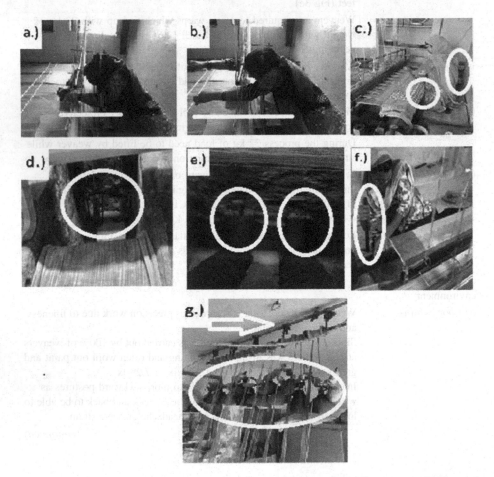

Fig. 3. Certain observations of video analysis method. (a) Forward bending posture, (b) Forward stretching posture, (c) Seat offering no back support and lack of clearance between thigh and cloth beam along with improper plane width for sitting, (d) Treadle offering least support to feet, (e) Legs in pit loom, (f) No protective covering for iron rods used while taking up, (g) 22 kg load bags hanged with iron rod on ceiling.

3.2   Phase II

3.2.1   Stage 1

Region wise mapping of pain illustrates that 97% weavers reported pain in back followed by pain in neck with 89% of responses. Of the upper limbs, 72% reported pain in shoulder, and 91% in the wrist area. In case of lower limb, 87% of weavers reported pain in knees as shown in Fig. 4.

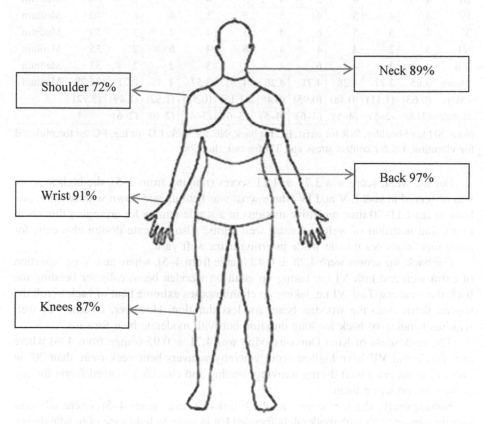

Fig. 4. Body discomfort survey in different body parts after weaving operation

From the reporting of pain frequency, it was found that neck, back and foot were the most common affected area where 80% and more Doria weavers reported having pain –"always" followed by hand and wrist where more than 70% weavers reported facing pain– "always".

3.2.2   Stage 2

From the WERA Assessment in seven tasks in Kota Doria weaving operation, the shoulder score was 3.85 ± 0.69 (range from 3–5). The highest score for shoulder was in task I which is Shedding where the weavers are required to lift hands in the vertical plane above the chest level to arrange the threads of the loom (Table 2).

Table 2.

Tasks	Score of WERA assessment									Final score	Action level
	SC	WR	BC	NC	LG	FC	VB	CS	TD		
I	5	5	4	4	5	4	4	6	2	39	Medium
II	3	3	4	4	4	3	3	3	2	30	Medium
III	4	4	4	5	4	3	4	3	2	38	Medium
IV	4	4	5	4	5	5	3	4	6	40	Medium
V	4	5	5	6	4	3	4	4	3	39	Medium
VI	3	2	4	4	4	6	4	6	2	35	Medium
VII	4	3	4	6	4	4	3	2	2	33	Medium
Mean	3.85	3.71	4.28	4.71	4.28	4	3.57	4	2.71	36.28	Medium
(SD)	(0.63)	(1.11)	(0.48)	(0.95)	(0.48)	(1.15)	(0.53)	(1.52)	(1.49)	(3.72)	
Range	(3–5)	(2–5)	(4–5)	(4–6)	(4–5)	(3–6)	(3–4)	(2–6)	(2–6)		

Note: SH for Shoulder, WR for wrist, BC for back, NC for neck, LG for leg, FC for forceful, VB for vibration, CS for contact stress and TD for task duration.

For the wrist, score was 3.71 ± 1.11 scores (ranging from 2–5), the highest score was observed in task I, V and IV where wrist was bent up and down with twisting and have at least 11–20 time repetitive motions in a single minute for arranging threads in loom, and insertion of weft and extra weft yarns. Also intricate design also calls for more repetitions per minute while inserting extra weft yarn.

For back, the scores were 4.28 ± 0.48 (range from 4–5), where task V i.e. insertion of extra weft and task VI i.e. taking up cloth on wooden beam calls for bending the back the weavers. Task VI i.e. taking up cloth imposes extreme bent of back to roll the woven fabric onto the wooden beam for less duration. However, insertion of weft requires bending of back for long duration but with moderate bent forward.

The neck score of Kota Doria weaving was 4.71 ± 0.95 (range from 4–6) where task III, V and VII have highest score whereby weavers bent neck more than 20° in order to insert extra weft during weaving, beating and checking weaved fabric for any flaws and correcting them.

Subsequently, the leg score was 4.28 ± 0.48 (range from 4–5) where all tasks require weaver to sit with moderately forward leg in order to hold rope of treadle during task II, III, V, VI and VII between the thumb and finger of lower limbs and moving the treadle/peddle with foot in task I and IV. Forceful scores was 4 ± 1.15 (ranged 3–6) whereby load maneuver is 22 kg for task IV and VI in form of hanged stones in order to give tension to warp threads to be woven. The scores for vibration and contact stress were 3.57 ± 0.53 (range from 2–6) and 4 ± 1.52 (range from 2–6) respectively. However weavers never used any vibrational tools and in task VI while taking up fabric on cloth beam an iron rod is used to revolve the beam as indicated in Fig. 3(f) which is devoid of any kind of handle and no gloves were used while handling the iron rod. The score of the task duration was 2.71 ± 1.49 (ranging from 2–6) for various tasks (during 8 h of job duration). Task II have WERA score of 30 which is lowest among all the tasks. Tasks IV, VI and VII have final score 35, 34 and 33, followed by tasks I, V, III with final scores 39, 39 and 38 (reaching overhead, bending neck and stretching hands).

The entire final score eight tasks of Doria weaving indicated medium risk in action level. Therefore, the total final score for VII tasks in Kota Doria weaving was 36.28 ± 3.72 score (range from 30–39) in medium risk level. The above specified results illustrates that Kota Doria weaving tasks need further investigation and require ergonomic changes at Kota Doria Pitloom. Table 3 elucidates the final score and action level for seven tasks of Kota Doria weaving process.

3.3 Phase III

χ^2 analysis of WERA body-part score and number of workers reporting pains, aches and discomfort in the body area. The presence of pains, aches and discomfort was recorded as "pain", their absence as "no pain". For the WERA score, all the body part were scores either in 1–3 (Low), 4 (Medium) or 5–6 (High) for the risk level.

In Kota Doria weaving operation (n = 100), the relationship of individual WERA body part scores to the presence of pain and discomfort is statistically significant for shoulder, neck, back and leg region.

The shoulder score for WERA body part was >4 in 83% of weavers while shoulder pain was reported by 72% of Doria weavers. Yielding a significant association between WERA body part score and self-reported pain ($\chi^2 = 23.86$; p = 0.000). The WERA body part score for neck region during weaving operation yielded a score >4 in 93% of the weavers and caused neck pain/discomfort among 89%, the association being significant ($\chi^2 = 5.96$; p = 0.014). The back region WERA body part score was >4 in 80% of workers with 97% reported pains and discomfort in the upper and lower back area with significant association ($\chi^2 = 4.20$; p = 0.04). The leg region WERA body part scoring was >4 in 78% of the weavers, with 85% complaining for pains and aches in legs, with the association being significant ($\chi^2 = 20.51$; p = 0.000). the wrist scores for WERA body part was 1–3 in 48% and >4 in 52% of weavers, however no association was found in wrist scores for WERA tool and reported pains and discomfort in the wrists.

Table 3. Chi-square statistical analysis (χ^2-test) of the WERA body part scores (1–3 = Low and >4 = Medium) and reported pains and discomfort in Kota Doria pit loom weaving operation.

Body part	Pain	WERA score		χ^2	p < 0.05
		1–3	>4		
Shoulder	No	12	15	23.86	0.000
	Yes	4	68		
Neck	No	4	7	5.96	0.014
	Yes	9	80		
Back	No	2	1	4.20	0.040
	Yes	18	79		
Wrist	No	5	4	0.22	0.634
	Yes	43	48		
Leg	No	10	5	20.51	0.000
	Yes	12	73		

Table 3 shows the chi square statistical analysis (χ^2-test) of WERA body part score and number of weavers reporting pains, aches and discomfort in the Kota Doria weaving operation carried out at pit loom.

4 Discussion

The results revealed that Kota Doria weaving operation reported discomfort in their back (97%), wrist (91%), neck (89%) and knee (87%) was reported to be higher than the previous studies conducted on handloom weavers in India [12, 14, 15]. The entire population of Kota Doria weavers in the study is females. One reason for higher reported MSDs could be unawareness of balancing work rest pattern during weaving operation, which turns out to be strenuous for various muscles and ligaments. Another reason could be high physiological demand to perform duties at home front such as cleaning, cooking, washing clothes etc. without any aid of mechanization. Even when weavers are not performing their job, they are looking towards finishing up of certain household activities which gives very menial time for physical recovery throughout the day. However, present study does not hold information on non-work related social factors for MSDs, but similar observations revealed that women weavers were more affected with MSDs in handloom sector because of juggling both household and work activities [9, 16].

On correlating the individual WERA body part scores to development of pains and discomfort to body regions such as shoulder, neck, back and leg were found to be statistically significant. The WERA score of Kota Doria pit loom workstation calls for investigate and suggest further change for all the seven tasks performed during the operation.

On in-depth observation of all the tasks, it has been found that Kota Doria pitlooms have been constructed by local carpenters and blacksmith with no knowledge of ergonomics and women anthropometry, have been designing these pitlooms through prior knowledge, feedback and special requirement of weavers, thus various features of the pitloom serves at critical hazardous points which often lead to occupational hazards (refer Table 1). Various studies affirm that unergonomically or poorly designed workstation lead to array of occupational hazards ranging from physical and emotional stress leading to low productivity and poor quality of work [17, 18]. The content of the work and design of job is also believed to have a direct impact on workers' comfort and possibly the development of work related injuries [19]. Therefore shunning ergonomic principles while designing workstation brings inefficiency and pain to the workforce.

5 Conclusion

Thus it can be concluded that poor ergonomic approach adopted while designing and operating Kota Doria pit loom expose weavers to numerous ergonomic risks and lead to various occupational hazards. Kota Doria weaving operation calls for ergonomic intervention in design of the existing workstation in order to improve the wellbeing and enhance productivity of the weavers.

Acknowledgments. Whole hearted cooperation of Kota Women's Weaving Organizations, weavers and loom owners of Aman Self Help Group, Shehnaaz and Aamna are gratefully acknowledged.

References

1. Metgud DC, Khatri S, Mokashi MG, Saha PN (2008) An ergonomic study of women workers in a woolen textile factory for identification of health-related problems. Int J Occup Environ Med 12(1):14–19
2. Burdorf A, Naakrgeboren B, Post W (1998) Prognostic factor for musculoskeletal sickness absence and return to work among welders and metal workers. Occ Env Med 55:490–495
3. Choobinch AR, Shahnavaz H, Lahmi MA (2004) Major health risk factors in Iranian hand – woven carpet industry. Int J Occ Safety Ergon 10(1):65–78
4. Cole DC, Hudak PL (1996) Prognosis of non – specific work – related musculoskeletal disorders of the neck and upper extremity. Am J Ind Med 29:657–668
5. Motamedzade M, Choobineh A, Mououdi MA, Arghami S (2007) Ergonomic design of carpet weaving hand tools. Int J Ind Ergon 37:581–587
6. Choobineh A, Hosseini M, Lahmi M, Jazani RK, Shahnavaz H (2007) Musculoskeletal problems in Iranian Hand-woven carpet industry: guidelines for workstation design. Appl Ergon 38:617–624
7. Choobineh AR, Shahnavaz H, Lahmi MA (2004) Major health risk factors in Iranian hand-woven carpet industry. Int J Occup Safety Ergon 10(1):65–78
8. Choobineh AR, Tosian R, Alhamdi Z, Davarzanie M (2004) Ergonomic intervention in carpet mending operation. Appl Ergon 35:493–496
9. Mukopadhyay PB (2008) Morbidity profile in handloom weavers of Hoogly district, West Bengal, India (Unpublished Master's thesis) National Institute of epidemiology (ICMR), India
10. Goel A, Tyagi I (2012) Occupational health hazards related to weaving. Int J Appl Math Stat Sci 1(1):22–28
11. Pandit S, Kumar P, Chakrabarti DC (2013) Ergonomic problems prevalent in handloom units of North East India. Int J Sci Res Publ 3(1):1–7
12. Nag A, Vyas H, Nag PK (2010) Gender differences, work stressors and musculoskeletal disorders in weaving industries. Ind Health 48:339–348
13. Corlett EN, Bishop RP (1976) A technique for assessing postural discomfort. Ergonomics 19 (2):175–182
14. Tiwari RR, Pathak MC, Zodpey SP (2003) Low back pain among textile workers. Indian J Occup Environ Med 7(1):27–29
15. Pandit S, Kumar P, Chakrabarti DC (2015) Ergonomic risk assessment on women's handloom weavers in assam with the introduction of Jacquard. In: ICoRD 2015 - Research in to design across boundaries, vol 1, pp 431–441. Springer, India. https://doi.org/10.1007/978-81-322-2232-3_38
16. Motamedzade M, Moghimbeigi A (2012) Musculoskeletal disorders among female carpet weavers. Ergonomics 55(2):229–236
17. Ayoub MA (1990) Ergonomic deficiencies: I. Pain at work. J Occ Med 32(1):52–57
18. Ayoub MA (1990) Ergonomic deficiencies: II. Probable causes. J Occ Med 32(2):131–136
19. US Department of Labor (OSHA) (2004) Ergonomics for the Prevention of Musculoskeletal Disorders: Guidelines for Retail Grocery stores. OSHA Publications, 3192

Effect of a Passive Exoskeleton on Muscle Activity and Posture During Order Picking

R. Motmans[(⊠)], T. Debaets, and S. Chrispeels

Department Health and Safety at Work, Colruyt Group, 1500 Halle, Belgium
roeland_motmans@yahoo.com

Abstract. Order picking is a standard task in logistics that is difficult to fully automate due to the variety of products. A passive exoskeleton can therefore be a strategy to support the order picker. This field study investigated the effect on muscle activity and posture of the back and shoulders.

Ten operators performed the task of order picking cheese in real life conditions 1,5 h without and 1,5 h with exoskeleton. The electromyography (EMG) of m. Erector Spinae and m. Trapezius pars descendens was measured. The posture of the back and upper arms were recorded by motion sensors. The subjective experience was questioned on a 5 point scale.

During order picking the back muscle activity was 9 and 12% lower when wearing the exoskeleton, respectively for the left and right side. The back was bent more than 30° for 26% of the time with exoskeleton and 23% without. The difference was not significant. The muscle activity and postures in the right and left shoulder also didn't show meaningful differences. Subjectively the physical load was experienced positively with the exoskeleton. However, fatigue and safety require some attention.

The practical implication of an exoskeleton in the prevention of low back pain is discussed. Due to the rather small benefit this will probably not be the only solution to reduce the physical load of a job.

Keywords: Exoskeleton · Muscle activity · EMG · Posture

1 Introduction

Order picking in logistics consists of repetitive lifting and awkward positions in limited spaces. In the retail sector fresh food is often stored in cold environments. To relieve the load on the low back several developments are ongoing. Automation would avoid manual lifting. This is possible when the products are stored in standardized boxes. The robot takes the box out of the warehouse and puts the box on a pallet. In reality however there is a high mix of products. The standard carrier is not always a pallet, but also a closed and cooled chart. This makes full automation impossible.

A goods to man system is an automated warehouse. The box is delivered by a conveyor belt to the man. The operator places the boxes in the cooled chart. The productivity can theoretically be increased by 500%. The human however remains the limiting factor because the frequency of manual order picking is already high. Job

© Springer Nature Switzerland AG 2019
S. Bagnara et al. (Eds.): IEA 2018, AISC 820, pp. 338–346, 2019.
https://doi.org/10.1007/978-3-319-96083-8_45

rotation is a strategy to reduce the lifting frequency on a daily basis, but a goods to man system will increase the overall physical load of the task.

Another evolution is to support the order picker himself by an exoskeleton. An exoskeleton is an external structure on the body that supports or augments the human force. For the industry there exist active and passive exoskeletons. Exoskeletons can also be distinguished by the supported body part: lower limbs, back, upper limbs or a full-body exoskeleton [1]. Passive exoskeletons for the back store energy during the downward movement or bending the back. This energy may support the person to keep that position or to move upwards while lifting an object. The activity of the back muscles reduces with 10 to 40% [2–8] when wearing a passive exoskeleton. During static holding the endurance time was three times longer wearing the external support [8]. However post trial endurance showed mixed results [7, 9, 10]. The compression forces at L4/L5 are estimated to be 23 to 29% lower [11]. However there is an increase in leg muscle activity [6]. Subjectively discomfort at the shoulders and the knees were also mentioned. Local discomfort is an important factor in the social acceptability of an exoskeleton on the work floor. It may not disturb the normal work [3].

Active exoskeletons contain actuators who deliver extra force in the joints. The can support and strengthen the human force during lifting the upper arms, bending in the elbow and bending the back. A positive effect is found during dynamic lifting and static holding. Muscle activity reductions were reported in the range of 20–55% for the shoulders in dynamic lifting [12]. While holding a weight above the head, the active exoskeleton resulted in a decrease in muscle activity of 30–70% for the upper arms and shoulders [13]. Commercially available active exoskeletons are still rare and expensive.

In this study the effect of a passive exoskeleton was evaluated. The posture and muscle activity of the back and shoulders were evaluated both subjectively and objectively during order picking of cheese in a 6° environment. Postural analysis and electromyography (EMG) were used to quantify the physical load.

2 Methodology

2.1 Participants

Ten order pickers volunteered to participate in this study. They were all men with an average age of 37 years (SD 8) and mean height of 1,73 m (SD 0,08). None of the participants reported low back pain at the time of the study. Half of the order pickers mentioned low back pain during the previous 12 months. The subjects signed an informed consent after being informed about the procedure of the study.

2.2 Exoskeleton

A passive exoskeleton of Laevo 2.5 was used in this study. It contains a support on the chest and the upper legs. Both supports are connected by a spring. In that way the forces on the back are transferred to the chest and legs. There are four different sizes of the springs to accommodate small and tall operators. The angle at which degree of bending the back the exoskeleton supports the most can be adjusted. For order picking the support was chosen to be the most pronounced at 35° (Fig. 1).

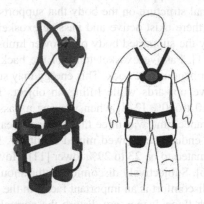

Fig. 1. Laevo exoskeleton 2.5 used in this study.

2.3 Task

The order pickers performed their normal job in real life. Two situations were evaluated: 1,5 h picking without exoskeleton and 1,5 h picking with passive exoskeleton. The week before the test the participants could already use the exoskeleton to get used to it. The load of the task was controlled because they only collected boxes of cheese. For the experiment they used a manual picking list due to the sensitivity of the measurement apparatus for radiation. Normally voice picking is used to give the commands while the hands are free. This means that the working pace during the study was lower than the real working pace.

2.4 Electromyography (EMG)

The muscle activity of the back and shoulders was measured. Surface electrodes were positioned to m. Erector Spinae at the level of L3–L4 and 2 cm from the spinal cord. The EMG activity of the shoulders was measured at m. Trapezius pars descendens. Before the electrodes were applied the skin was shaved, scrubbed and cleaned with alcohol. Data were sampled at 1000 Hz, amplified, band-pass filtered (15–500 Hz) and relayed to a computer-based data acquisition and analysis system (Captiv). The raw data are normalized to the values of working without exoskeleton.

2.5 Postural Analysis

The posture was captured by 4 T Motion Sens (Captiv) wireless sensors. The sensors were placed at the back (lumbar and thoracic) and the upper arms (right and left). They were attached to the body by straps or a harness. The sensors register the movements of the person according to the three axes (x, y, z). All data are wireless transmitted to the computer system. To calibrate the zero position of the sensors, the test persons stood in a resting position. The postures and movements are registered and visualized by a 3D computer model. The postures are summarized into four parameters: % time shoulder flexion >60°, % time shoulder flexion >90°, % time back bending >30°and %time back bending >45°.

2.6   Questionnaire

After the measurements the participants filled out a questionnaire on a 5 points rating scale. The subjectively evaluated items during the work were: putting the exoskeleton on and off, taking a box from the pallet, stepping on and off the electric pallet jack, putting the box in the cooled chart and walking. The questions about the comfort of the suit handled about the support at the chest, support on the upper legs and the hip belt. Concerning the effectivity the following items were questioned: physical load on the back, fatigue, posture of the back.

2.7   Statistical Analysis

The raw EMG data were normalized to the muscle activity without wearing an exoskeleton. The difference in the mean normalized EMG amplitude was analyzed by a paired sample t-test. The results of the postural analysis were expressed as a percentage of time that the back or shoulders were positioned in an awkward posture. These averages were also analyzed by a paired sample t-test. The subjective evaluation on a 5 point scale were only measured after wearing the exoskeleton. The averages are qualitatively described together with the subjective comments.

3   Results

3.1   Muscle Activity (EMG)

Order picking with an exoskeleton reduced the muscle activity in the back, but not in the shoulders. The EMG activity was 12% lower at the right side of m. Erector Spinae and 9% lower at the left side when wearing a passive exoskeleton for the back (see Fig. 2). For m. Trapezius there was a minimal and non-significant improvement of 1% (right side) and 2% (left side).

3.2   Posture

There was no meaningful difference in the position of the back when the order pickers were wearing an exoskeleton or not. During 1,5 h order picking the back of the participants was 26% of the time flexed more than 30° when wearing the exoskeleton. Without exoskeleton this time was 23%. The difference of 3% was not significant. The time working with the back bent more than 60° was equally with or without support (3% of the time).

Comparable results were found for the position of the shoulders. At the right side the upper arm was 11% of the time raised more than 60° when wearing the exoskeleton. Without exoskeleton this time was 10%. The percentage of time that the

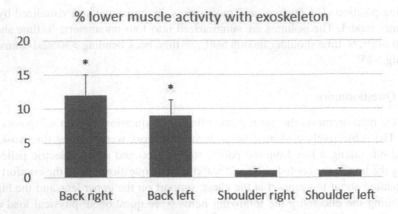

Fig. 2. Percentage of improvement in muscle activity when wearing an exoskeleton for m. Erector Spinae (right and left) and m. Trapezius pars descendens (right and left). Significant results are marked with an * (p < 0,05).

left shoulder was flexed more than 30° was 5% without and 9% with exoskeleton (see Fig. 3). The upper arms did not work above shoulder height in any condition. The angle between trunk and upper arm was never more than 90°.

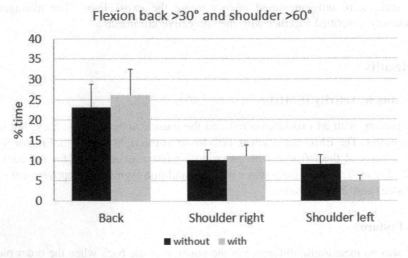

Fig. 3. Percentage of time that the trunk is flexed more than 30° and the shoulders are flexed more than 60°. The two conditions with and without exoskeleton are compared. Significant results are marked with an * (p < 0,05).

3.3   Questionnaire

On a five point scale the physical load on the back was evaluated the most positive (4,5), stepping on and off the electrical pallet jack the most negative (1,75). Taking and placing the boxes, walking, support at the chest and the posture of the back were all perceived well with an average of 3,75 points. The support at the legs and the hip belt were two determinants of the comfort of the exoskeleton. With 3,5 points this was in general suitable. Putting the exoskeleton on and off scored still positive (3,25). A negative evaluation however was noted for fatigue (2,25). The downward movement was experienced as fatiguing (Table 1).

Table 1. Subjective evaluation of the exoskeleton.

Parameter	Points on 5
Work	
Putting on and off exoskeleton	3,25
Taking a box from the pallet	3,75
Stepping on and off the electrical pallet jack	1,75
Putting the box in the cooled chart	3,75
Walking	3,75
Comfort	
Support at the chest	3,75
Support on the upper legs	3,5
Hip belt	3,5
Effectivity	
Physical load on the back	4,5
Fatigue	2,25
Posture of the back	3,75

4 Discussion

4.1 Lower Back Muscle Activity

The muscle activity of the low back decreased with 9 and 12% when wearing an exoskeleton. The EMG values at the shoulder didn't change. The passive exoskeleton in this study only supports the back. In lab situations a positive benefit of 37–44% less back muscle activity with the same exoskeleton was found when the operators were working in a static posture with the back bent 40°. During a dynamic assembly task this reduction was 35–38% [8].

This result was not achieved in this study of dynamic lifting up and down. A possible explanation can be that this was a field study where the conditions are less controlled compared to a lab set up. In a review study the benefit of passive exoskeletons was summarized as 10 to 40% for the low back [1]. However only one field study was included. During automotive assembly tasks on the work floor the EMG activity of m. Erector Spinae at L3 was 15% lower [5]. This result is more in accordance to the values measured during order picking in this study.

The posture of the back and shoulders remained the same with or without the support. The benefit of the lower back muscle activity in this study can be explained by the exoskeleton itself. Subjectively the operator had the feeling that they had to keep to back more straight because of the support on the chest. Objectively this could not be confirmed. Other studies however found that test persons adjusted their lifting technique towards a more squatting-like pattern [12]. During stooped lifting the passive exoskeleton was also found to reduce trunk flexion [6]. This study shows that even without changing the working posture an exoskeleton can have a beneficial impact on the lower back muscles.

4.2 Less Physical Load Experienced

Subjectively the physical load was perceived the most positive (4,5/5). The order pickers mention that less energy was needed to lift a box, the weights felt lighter. The support of the exoskeleton is clear. In the downward movement they have to deliver more power, but there is an even bigger gain during upward lift movement. This is in agreement with other studies. All subjects reported the feeling of an exoskeleton assisting them in the up phase of a lift [9]. However in one lab study with asymmetric lifting the benefit was also experienced in the down phase [3]. This is in contrast with the order pickers on the field who report that they have to deliver more energy in the downward movement. Moreover this is the reason that they experienced slightly more fatigue when working 1,5 h with the exoskeleton (2,25).

The downward movement to store energy in the springs costs an extra effort of the order picker. They had the feeling that the abdominal muscles had to work harder. Subjectively putting the boxes down in the cooled chart required more energy. This could not be confirmed. The rate of perceived exertion (RPE) during an assembly task on the field was 16% lower while the abdominal muscle activity didn't change [5]. However the results of perceived exertion are mixed. During prolonged lifting/lowering in a lab the RPE decreased 25% [13] while the same set up showed no difference [14]. Objectively endurance also showed mixed after 45 min lowering and lifting objects [7, 8, 14]. Oxygen consumption was not affected [7]. Research over a longer working time is necessary.

4.3 Subjective Remarks

The support at the chest was accepted well (3,75). In this study only men were selected because of the support at the chest. For women the shape of the support is not always comfortable at the breasts. However the new version of the exoskeleton had already an adjusted shape to accommodate half of the women's body figure. The fitting of the suit was generally evaluated positive. The support at the legs didn't always remain at the right place. Straps around the leg were proposed by the order pickers. Fitting was an advantage of this exoskeleton. With one suit of four sizes of spring every operator can wear the same exoskeleton. However, the right adjustment of the suit required some experience.

Getting on and off the electrical pallet jack was evaluated very negative (1,75). Because of the springs at the side of the body, the operator becomes a little bit broader. This made that they couldn't turn anymore on the pallet jack. They had to move sideways to get off the pallet jack. This is also less safe because you can't see the other operators approaching. The springs hit often the pallet jack, which is not good for the sustainability.

During the work the exoskeleton was well accepted. Social acceptation was a point of remark at the beginning of this study. All supervisors and operators got a short information session to explain what a passive exoskeleton does and why what they can expect.

4.4 Long Term Effects Are not Known

During order picking the operators experience more fatigue. They say it's not possible to wear the exoskeleton during the whole day. It becomes also hot. That's why they propose to alternate periods of order picking with and without exoskeleton. Another strategy can be to implement the external support only for specific tasks. There are no studies yet that investigated the use of an exoskeleton during a whole day.

Another argument for not wearing the exoskeleton during the whole day is the weakening of the back muscles. Based on the results of this on the field study this fear is not really reasonable because the back muscle activity was only 9 and 12% reduced. The passive exoskeleton supports the back, but the muscles remain doing most of the work.

4.5 Practical Implications

An exoskeleton can be used for three purposes: preventing low back pain, supporting operators with low back pain and accelerating the return to work after a long period of sick leave because of the back. With a benefit of only 10% this last track is not much reasonable. However in combination with a half time schedule, job rotation, less expected productivity,... this extra 10% benefit can be an additional measure to reintegrate an order picker.

Another strategy can be to use an exoskeleton to prevent low back complaints. In this way every operator should wear the external support. That makes it more costly. When it is necessary that every operator should wear an exoskeleton, there is something wrong with the basic ergonomics of the task. An exoskeleton can't be the first solution for a physical load that is too high. During order picking the trunk is bent more than 30° for around 25% of the time. To attack this physical load the first measurement should be to avoid the risk or decrease the exposure time by job rotation. A second strategy can be a collective measure by increasing the height of the pallets in the racks or by providing height adjustable pallet jacks. All order pickers will experience a benefit. A personal protective like an exoskeleton is only a third line strategy to reduce the physical load on the back.

References

1. de Looze MP, Bosch T, Krause F, Stadler KS, O'Sullivan LW (2016) Exoskeletons for industrial application and their potential effects on physical work load. Ergonomics 59 (5):671–681
2. Abdoli-Eramaki M, Agnew MJ, Stevenson JM (2006) An on-body personal lift augmentation device (PLAD) reduces EMG amplitude of Erector Spinae during lifting tasks. Clin Biomech 21(5):456–465
3. Abdoli-Eramaki M, Stevenson JM (2008) The effect of on-body lift assistive device on the lumbar 3D dynamic moments and EMG during asymmetric freestyle lifting. Clin Biomech 23:372–380
4. Frost DM, Abbodli-Eramaki M, Stevenson JM (2009) PLAD (personal lift assistive device) Stiffness affects the lumbar flexion/extension moment and the posterior chain EMG during symmetrical lifting tasks. J Electromyogr Kinesiol 19(6):403–412
5. Graham RB, Agnew MJ, Stevenson JM (2009) Effectiveness of an on-body lifting aid at reducing low back physical demands during an automotive assembly task: assessment of EMG response and user acceptability. Appl Ergon 40(5):936–942
6. Ulrey BL, Fathallah FA (2013) Subject-specific whole-body models of the stooped posture with a personal weight transfer device. J Electromyogr Kinesiol 23(1):206–215
7. Whitfield BH, Costigan PA, Stevenson JM, Smallman CL (2014) Effect of an on-body ergonomic aid on oxygen consumption during a repetitive lifting task. Int J Ind Erg 44 (1):39–44
8. Bosch T, van Eck J, Knitel K, de Looze M (2016) The effects of a passive exoskeleton on muscle activity, discomfort and endurance time in forward bending. Apll Ergon 54:212–217
9. Lotz CA, Agnew MJ, Godwin AA, Stevenson JM (2009) The effect of an on-body personal lift assist device (PLAD) on fatigue during a repetitive lifting task. J Electromyogr Kinesiol 19(2):331–340
10. Godwin A, Stevenson JM, Agnew MJ, Twiddy AL, Abdoli-Eramaki M, Lotz CA (2009) Testing the efficacy of an ergonomic lifting aid at diminishing muscular fatigue in women over a prolonged period of lifting. Int J Ind Erg 39:121–126
11. Abdoli-Eramaki M, Stevenson JM, Reid A, Bryant TJ (2007) Mathematical and empirical proof of principle for an on-body personal lift augmentation device (PLAD). J Biomech 40 (8):1694–1700
12. Muramatsu Y, Kobayashi H, Sato Y, Jiaou H, Hashimoto T, Kobayashi H (2011) Quantitative performance analysis of exoskeleton augmenting devices – Muscle Suit – for manual worker. Int J Autom Tech 5(4):559–567
13. Kobayashi H, Nozaki H (2007) Development of muscle suit for supporting manual worker. In: IEEE/RSJ international conference on intelligent robots and systems, San Diego, pp 1769–1774
14. Sadler EM, Graham RB, Stevenson JM (2011) The personal lift-assist device and lifting technique: a principal component analysis. Ergonomics 54(4):392–402

Simulation Study on the Effects of Adaptive Time for Assist Considering Release of Isometric Force During Elbow Flexion

Jeewon Choi[1(✉)], Ping Yeap Loh[2], and Satoshi Muraki[2]

[1] Graduate School of Design, Kyushu University, Fukuoka, Japan
Jeewon.choi@outlook.com
[2] Faculty of Design, Kyushu University, Fukuoka, Japan

Abstract. The increasing trend of development of assistive technology allows for the use of assistive robots such as power assist devices to be prevalent in various social domains. Such power assist devices usually provide incidental power to their users, requiring human-machine force interaction. If the power assist device requires users to release their muscular force without considering adaptive time, users might be confused to control the level of their manual performance in response to the external force. This study investigated adaptive time with varying release rates of isometric force during one-arm elbow flexion, focusing on muscle activity and force control. Eight participants conducted graphical force-tracking tasks designed to simulate power-assist condition. Electromyography signals and the tension forces of the biceps brachii and triceps brachii were measured. The results implied that sufficient adaptive time for muscular force release induced better performance level with a smaller difference between the target force and the actual force. However, higher subjective exertion was also accompanied during the longer time for muscular force release. This study suggests that in designing power assist devices, the duration for muscular force release and consequent characteristics should be considered to maintain the precise level of force control.

Keywords: Power-assist simulation · Adaptive time · Muscular force release

1 Introduction

Assistive technology focuses on developing any device or system that is usually attached to the human body and helps humans physically. Power assist devices, for instance, have been designed to generate assistive movement to users to facilitate easier human action and relieve physical stress. However, since assistive devices are usually attached directly to the human body and involves physical interaction, users inevitably experience various issues of practical use, such as adaptability to the assistive force [1, 2]. In the interaction stage, during which users receive and adapt to the assistive force from the devices, muscular force is released down to a certain level, allowing muscular ability to control the task performance. However, if such a force release is induced by the power assist device without considering the appropriate time to adapt, users might be confused to control the level of their task performance in a moment.

© Springer Nature Switzerland AG 2019
S. Bagnara et al. (Eds.): IEA 2018, AISC 820, pp. 347–350, 2019.
https://doi.org/10.1007/978-3-319-96083-8_46

In this study, the power assist condition was simulated with a series of graphical force-tracking tasks performed in isometric elbow flexion. We aimed to elucidate the effect of adaptive time with varying release rates of isometric force toward certain levels of assistive force, mainly exploring force control performance.

2 Methods

Eight healthy male participants without prior or current functional disorders were recruited (age, 23.3 ± 1.3 years). The hand dominance of the participants was determined using the Edinburgh Handedness Inventory.

The participants were positioned in an upright sitting posture with a display in front of them. The upper body was straightened, and the right forearm was supinated and fixed at 90° elbow flexion. A wrist strap was equipped at the styloid process of the right radius, and a tensile sensor (T.K.K. 1269f, Takei Scientific Instruments Co., Ltd, Niigata, Japan) was connected with chain between the strap and the floor.

On the basis of the environment for isometric elbow flexion, the force tracking task was simulated [3]. This simulation facilitated the modeling and observation of muscular force release for varying adaptive time, before taking account of practical assistive devices. In this study, three sequential phases were defined as follows: (1) before assist (40% MVC, 7 s), (2) adaptive time for muscular release, and (3) constant assist (7 s). The adaptive time was set at 1 and 5 s, and each force was sloped and linearly guided to be released to the phase of constant assist, at a level of either 33% or 67% assist for the 40% MVC. The corresponding graphics of three segments were sequentially presented on a computer screen as a target force. The actual force output of each participant, which was measured using a tensile sensor in real time, was also graphically demonstrated. The graphical visualization was implemented using data analysis software (LabChart 7, ADInstruments, Dunedin, New Zealand).

Before the simulation, a maximal isometric voluntary contraction (MVC) and the corresponding maximum exerted force (N) for the elbow flexion of each participant were obtained to construct a graphical guide for the force tracking task. In the main experiment, the participants were asked to track the force lines as accurately as they could by exerting isometric contraction of elbow flexion. The learning effect was minimized by considering two additional conditions of fake slope and complete counterbalancing. Each trial was repeated twice.

A multi-channel data acquisition system (Powerlab 16/30, ADInstruments) was used to obtain the force signals from the tensile sensor. Samples were calculated using the relative percentage difference (RPD) of the individual force output against the target force from the graphics, which is the percentage of (target force − output force)/target force (Fig. 1).

The data for the two phases of adaptive time and constant assist were considered in this study. Two-by-two repeated-measures analysis of variance (ANOVA) was used to analyze the two factors of adaptive time (1 and 5 s) and the simulated level of assist (assist level; 33% and 67%). All statistical analyses were conducted using SPSS Statistics 23.0 (IBM, NC, USA).

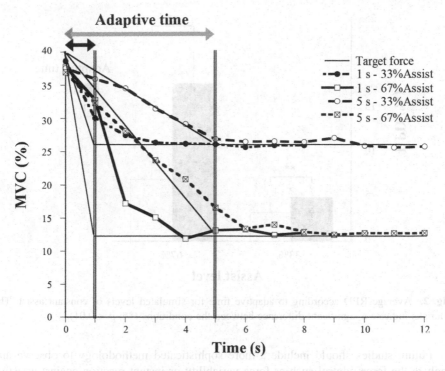

Fig. 1. A representative result for the two latter phases of adaptive time and constant assist during one trial. The difference between the target force and the individual force output for each condition was calculated for the RPD.

3 Results and Discussion

The data normality was verified using the Shapiro-Wilk test (all p values > 0.185). The ANOVA revealed the significant main effect of adaptive time ($F(1,15) = 100.31$, $p < 0.01$), indicating a higher RPD in the adaptive time of 1 s ($M = 11.37$, $SE = 0.70$) than that of 5 s ($M = 5.13$, $SE = 0.31$). The main effect of the assist level was also significant ($F(1,15) = 112.75$, $p < 0.01$), indicating that a 67% assist demonstrated a higher RPD ($M = 12.39$, $SE = 0.81$) than a 33% assist ($M = 4.11$, $SE = 0.21$). The interaction between the adaptive time and the assist level was significant (F $(1,15) = 49.80$, $p < 0.01$), and the follow-up t tests revealed that the longer adaptive time for muscular force release showed a significant decrease in RPD throughout the assist level, especially for the 67% assist (Fig. 2).

As was expected, the longer adaptive time lessened the gap between the target force and the actual force, deriving less percentage difference in force control. The interaction showed, however, that the gap was dramatically reduced at the 67% assist, demonstrating that the sufficient adaptive time for muscular force release should be considered if a higher assist level needs to be delivered with more accuracy.

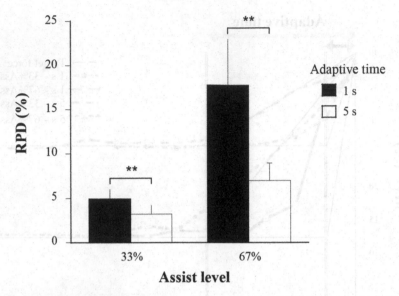

Fig. 2. Average RPD according to adaptive time for simulated levels of constant assist. The asterisks indicate a significant difference between the conditions (**: $p < 0.01$).

Future studies should include a more sophisticated methodology to observe and analyze the force control such as force variability or instant reaction against assistive force, taking into account the related muscle activity as well.

References

1. Anttila H, Samuelsson K, Salminen AL, Brandt Å (2012) Quality of evidence of assistive technology interventions for people with disability: an overview of systematic reviews. Technol Disabil 24(1):9–48
2. Nasir N, Hayashi K, Loh PY, Muraki S (2017) The effect of assistive force on the agonist and antagonist muscles in elbow flexion. Mov Health Exerc 6(2):35–52
3. Naik SK, Patten C, Lodha N, Coombes SA, Cauraugh JH (2011) Force control deficits in chronic stroke: grip formation and release phases. Exp Brain Res 211(1):1

Reducing Musculoskeletal Pains of Operating Theatre Nurses

Peter Hoppe[1(✉)], Karl Schableger[4], Brigitte König[2],
Brigitte Eichinger[2,3], Anna Gabriel[1], Tanja Holzmann[1],
and Iris Frenner[1]

[1] Bachelor Program Physiotherapy, University of Applied Science for Health
Professions Upper Austria, Paula-Scherleitner-Weg 3, 4020 Linz, Austria
peter.hoppe@fhgooe.ac.at
[2] Kepler University Hospital Med Campus III, Krankenhausstraße 9, 4020 Linz,
Austria
[3] Department Sport and Health, City Government, Hauptstr. 1-5, 4020 Linz,
Austria
[4] Department Health Economics, The Upper Austrian Regional Health
Insurance, Gruberstr. 77, 4020 Linz, Austria

Abstract. Musculoskeletal pains top the list of the most common work-related
ailments in Austria. In 2016, they were the second-leading cause for taking sick
leave. The Austrian Employee Protection Strategy for 2013–2020 aims to reduce
work-related health risks. The first goal of the presented monocentric, randomised, con-trolled study was to survey the intensity and regional distribution
of musculoskeletal pain of operating theatre nurses. Secondly, it was to investigate the effectiveness of a combined programme consisting of awareness
training and an exercise programme for reducing pain. The intervention group
received a two-stage intervention, and the control group received no intervention. The first stage of the intervention was an awareness training. The second
stage was a specific corrective exercise. The musculoskeletal pain was recorded
with a modified version of the Nordic questionnaire at three different times.
Fifty-six individuals took part in the programme. The two most common pain
areas reported by all respondents were the shoulder and neck region with an
average pain level of 5.62 (SD = 3.03) and the lumbar region of the spine with
an average pain level of 5.54 (SD = 3.14). The awareness training led to a 4%
(1.1 points) increase in the reported pain level across all pain areas compared
with the control group. The pain relief programme led to a 16% (4.9 points)
decrease in pain levels across all pain areas. In conclusion, a programme based
on awareness training and corrective exercises can increase awareness of pain
during the occupational activities of operating theatre nurses.

Keywords: Awareness training · Exercise · Operating theatre nurses
Musculoskeletal pain

© Springer Nature Switzerland AG 2019
S. Bagnara et al. (Eds.): IEA 2018, AISC 820, pp. 351–359, 2019.
https://doi.org/10.1007/978-3-319-96083-8_47

1 Introduction

Musculoskeletal pains top the list of the most common work-related ailments in Austria [1]. In 2016, they were the second-leading cause for taking sick leave [2]. Amongst other things, the Austrian Employee Protection Strategy for 2013–2020 aims to reduce work-related health risks, in particular, the stresses placed on the muscular and skeletal systems [3]. Operating theatre nurses work in sterile zones under the supervision of surgeons and are responsible for preparing surgical instruments before procedures, keeping count of surgical instruments, monitoring all members of the team to maintain a sterile zone, transporting patients to and from the operating room (OR) and passing instruments to surgeons and surgical assistants. A sterile zone, which is a clearly defined zone, is the area closely surrounding the OR table or instrument tray [4]. The results of a survey in 2009 on the self-reported occupational risk factors indicated that 84% of the subjects perceived that 50% or more of their time was spent in a standing position, and only 22% reported that they could take breaks [4]. Standing for long hours (up to 10 h for some surgical procedures, such as lumbar fusion) is a significant contributor to lower back pain and pain in the ankle/foot [4]. These musculoskeletal symptoms, along with fatigue and pain in the neck and the shoulders, were accelerated when the nurses were required to wear lead gowns (weight of approx. 5 lb) during surgical procedures that required frequent x-rays [4]. Other professions, e.g. in the industry, show that working in a standing position for an extended period of time decreases the workers' performance. This de-crease may be a result of occupational injuries, a decrease in productivity, and an increase in treatment and medical costs [5]. Prolonged standing can lead to subjective discomfort in workers. When workers per-form jobs in a prolonged standing position, static contraction occurs, particularly in their back and legs, resulting in discomfort and muscle fatigue [5]. In a previous study, Halim et al. identified four risk factors which significantly contribute to discomfort and muscle fatigue for employees associated with jobs that require prolonged standing. They are: working posture, muscle activity, standing duration and holding time [6]. In another study by Halim et al. [7], the relationship between psychological fatigue and measurable muscular fatigue was determined using a questionnaire and an EMG measurement. The evaluation showed that the subjectively perceived fatigue of the left M. erector spinae correlates with the EMG measurement ($p < 0.05$). Messing et al. [8] examined the working population in Québec, more than 50% of whom spend most of their time standing at work. There was a strong correlation between standing in a fixed or relatively fixed position and pain in the foot and ankle, lower leg, and calf muscles ($p < 0.001$). Strength training can demonstrably reduce this musculoskeletal pain. Several studies describe a large effect of training on pain in the foot or ankle [9–15]. Comparable effects are de-scribed in several studies for pain reduction in the lumbar spine [16–20] and upper extremity [21–25]. In order to perceive one's own posture during the activity, for example, a training with basic body awareness therapy (BBAT) can increase the awareness of one's own musculoskeletal symptoms [26]. On one hand, the goal of the present study was to survey the intensity and regional distribution of musculoskeletal pain of operating theatre nurses at the Kepler University Hospital in Linz. On the other hand, it was to investigate the effectiveness of a two-stage intervention programme consisting of training for awareness of the physical stresses encountered during occupational activities as well as an exercise programme for reducing pain during occupational activity.

2 Methods/Design

The presented monocentric, randomised study was a two-arm randomized control trial with one intervention (IG) and one control group (CG) to survey the intensity and regional distribution of musculoskeletal pain of operating theatre nurses at the Kepler University Hospital in Linz (see Fig. 1).

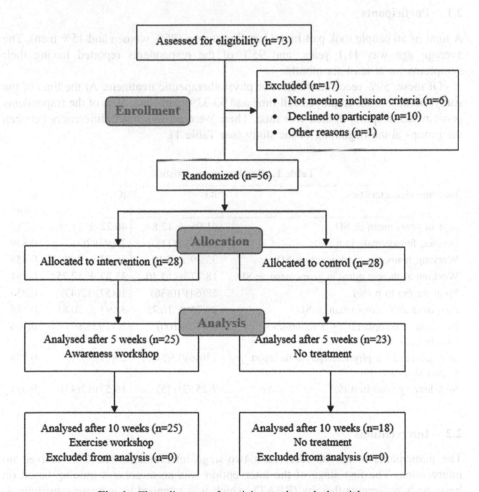

Fig. 1. Flow diagram of participants through the trial.

After the baseline assessment was completed, the participants were randomly allocated to either the intervention or the control arm. The experimental group subjects received a two-stage intervention, and the control group received neither intervention nor encountered any change in their regular work environments. The first stage of the intervention was an awareness training, which was intended to increase sensitivity to musculoskeletal pain during occupational activities. After this was completed, specific corrective exercises were taught. These were also intended to reduce pain in the areas subject to stress during occupational activities. Subjective pain levels were recorded at the beginning of the investigation, after the end of the awareness training (after 5 weeks), and after the instruction in corrective exercises (after 10 weeks). The Ethics Committee of the State of Upper Austria granted approval for the implementation of the research (C-99-15).

2.1   Participants

A total of 56 people took part in the baseline survey (85% women and 15% men). The average age was 41.1 years, and 95% of the respondents reported having their symptoms for at least six months.

Of these, 59% received medical or physiotherapeutic treatment. At the time of the study, 66.66% were employed full-time and 33.33% part-time. 60% of the respondents participated in sports in their free time. There were no significant differences between the groups at the beginning of the study (see Table 1).

Table 1. Baseline characteristics.

Baseline characteristics	IG	IC	p value
Age in years mean ± SD	41.95 ± 12.84	40.22 ± 11.56	0.786
Gender, female/male in n (%)	25(89)/3(11)	26(93)/2(7)	0.839
Working hours per week mean ± SD	33.09 ± 9.19	35.33 ± 8.86	0.438
Working as theatre nurse in years mean ± SD	18.77 ± 13.40	35.33 ± 12.25	0.781
Sport yes/no in n (%)	18(64)/10(36)	16(57)/12(43)	0.834
Pain over all regions mean ± SD	29.75 ± 16.75	32.93 ± 20.81	0.588
Duration of complaints (≤ 3 months/>3 months) in n (%)	0(0)/28(100)	3(11)/25(89)	0.104
In treatment by a physician/physiotherapist yes/no in n (%)	13(46)/15(54)	20(71)/8(29)	0.084
Sick leave yes/no in n (%)	7(25)/21(75)	16(57)/12(43)	0.091

2.2   Interventions

The participants in the IG received a two-stage intervention. The CG received no intervention. The first stage of the intervention was an awareness training based on basic body awareness therapy (BBAT), which was intended to increase sensitivity to musculoskeletal pain during occupational activities. After this was completed, specific corrective exercises were taught. These were also intended to reduce pain in the areas subject to stress during occupational activities. The duration of this workshop was

45 min for every participant. Basic body awareness therapy (BBAT) is a mind/body approach that aims to normalize posture, balance, and muscular tension, thereby increasing the body awareness and realistic body image [27, 28]. Within BBAT philosophy, gaining awareness is described as the gateway to movement learning. The training process includes the gradual awareness of how to relate to the ground, the vertical axis, centering, breathing, and flow [29]. Mainly the training focused on the awareness of the vertical posture with the focus on ankle/foot, knee, hip, spine, shoulder and neck position in space. This exercise workshop was 45 min long, as well. In the sessions, the training focused on the individually affected body regions without adding weight. For example, for the complainants with physical pain in the knees were trained in squats with a small range of motion. All gymnastic exercises were based on recommendation by Fiona Wilson et al. [30] with some special conditions. The exercises must be able to be done in the OR during occupational activities of operating theatre nurses and be able to be done with a small range of motion.

2.3   Outcome Measures

The subjective musculoskeletal pain was recorded with a modified version of the Nordic questionnaire [31] at three different times. Subjective pain levels were record-ed at the beginning of the investigation, after the awareness training, and after the instruction in corrective exercises. The nine body areas from the Nordic questionnaire were supplemented with visual analogue scales (VAS) for recording the intensity [32, 33]. On a 11-point (0–10) numerical rating scale pain intensity can be measured individually [32]. Each of the other commonly used methods of rating pain intensity, including VAS, numerical rating scales (NRS), and verbal rating scales (VRS) are reliable and valid, and no one scale consistently demonstrates greater responsiveness in detecting improvements associated with pain treatment [34].

2.4   Analysis

In general, the collected Data's were analyzed with descriptive statistics methods. The mean was used as the measure of location and for the measure of variation was applied the standard deviation. Baseline characteristics in the Intervention and Control groups were compared with $\chi2$ tests, Fisher exact tests, or Mann–Whitney tests when appropriate. The data were evaluated using IBM SPSS Statistics 24 and Microsoft Excel 2016.

3   Results

As the two most common regions, 94% of all respondents indicated the shoulder-neck region with an average pain of 5.62 (SD $\pm$ 3.03) and 89% the lumbar spine with an average pain of 5.54 (SD $\pm$ 3.14). The third most common was the thoracic spine with (3.20; SD $\pm$ 3.01) followed by the knees (2.89; SD $\pm$ 3.08), the hips (2.77; SD $\pm$ 3.45), the hands (2.58; SD $\pm$ 3.11), and the ankle/foot (2.24; SD $\pm$ 2.91). The elbow region came last with 1.36 (SD $\pm$ 2.61). Awareness training led to a 4% (1.1 points) increase in pain across all pain regions compared with the CG. In the exercise pro-gramme, pain was reduced by 16% (4.9 points) across all pain regions compared with the CG (see Fig. 2). In summary, a small Cohen effect (d = −0.18) results in the awareness training. For the balancing exercises, there was a medium Cohen effect (d = −0.45).

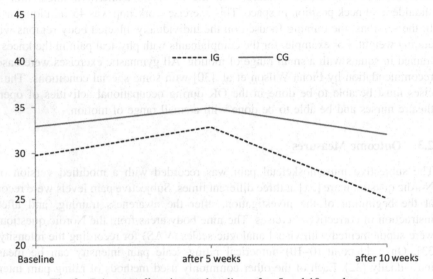

Fig. 2. Pain over all regions at baseline, after 5 and 10 weeks.

4 Discussion

Our study showed that a programme of individual balancing exercises can reduce pain by an average of 21% in the intervention group. Depending on the region of the body, the literature confirms this effect on pain reduction [9–25]. Awareness training showed a 10% increase in pain (3.0) in the intervention group and a 6% increase in pain (1.9) in the control group. This result was surprising because the results in the literature showed effects for a reduction of pain [27–29]. In the context of BBAT, several studies have reported that a more positive self-image and a lower level of pain have a negative effect on the outcome [28, 35]. Because the average pain level at the beginning of the study was 31.34 pain points (out of a maximum of 90 possible) in all subjects, a lower pain level can be assumed. The fact that all participants completed training in nursing may also indicate a better self-image. This would have to be taken into account in a further study. The authors of the study are aware that the methodology has some weaknesses. For example, it would make sense to develop a suitable sham treatment during a new examination in order to motivate the participants of the control group even more. This could reduce the failure by a good third of the subjects in the control group. Finally, from the authors' point of view, the large variance and group size also led to the fact that this study could not show any significant group differences.

5 Conclusion

This study shows that a combination of perception training and strength training can be clinically relevant to reduce the pain of instrumental caregivers. In the sense of increasing health-related self-knowledge, all professions involving long periods of standing could benefit from such a programme. The skills can either be taught in workshops by physiotherapists or integrated into vocational training.

References

1. Arbeitsinspektorat: Arbeitsbedingte Erkrankungen. https://www.arbeitsinspektion.gv.at/inspektorat/Gesundheit_im_Betrieb/arbeitsbedingte_Erkrankungen/. Accessed 8 Dec 2017
2. Leoni T, Uhl A (2016) Fehlzeitenreport. Krankheits- und unfallbedingte Fehlzeiten in Österreich.   http://www.hauptverband.at/cdscontent/load?contentid=10008.637786&version=1487075768. Accessed 8 Dec 2017
3. Federal Ministry of Labour, Social Affairs and Consumer Protection: Occupational Safety and Health Strategy. Structures, processes and tasks. https://www.arbeitsinspektion.gv.at/inspektorat/Uebergreifende_Themen/ArbeitnehmerInnenschutzstrategie/. Accessed 8 Dec 2017
4. Baird CL, Sands L (2004) A pilot study of the effectiveness of guided imagery with progressive muscle relaxation to reduce chronic pain and mobility difficulties of osteoarthritis. Pain Manag Nurs Official J Am Soc Pain Manag Nurs 5(3):97–104
5. Halim I, Arep H, Kamat SR, Abdullah R, Omar AR, Ismail AR (2014) Development of a decision support system for analysis and solutions of prolonged standing in the workplace. Saf Health Work 5(2):97–105
6. Halim I, Omar AR (2012) Development of prolonged standing strain index to quantify risk levels of standing jobs. Int J Occup Saf Ergon JOSE 18(1):85–96
7. Halim I, Omar AR, Saman AM, Othman I (2012) Assessment of muscle fatigue associated with prolonged standing in the workplace. Saf Health work 3(1):31–42
8. Messing K, Tissot F, Stock S (2008) Distal lower-extremity pain and work postures in the Quebec population. Am J Public Health 98(4):705–713
9. Beyer R, Kongsgaard M, Hougs Kjær B, Øhlenschlæger T, Kjær M, Magnusson SP (2015) Heavy slow resistance versus eccentric training as treatment for achilles tendinopathy: a randomized controlled trial. Am J Sports Med 43(7):1704–1711
10. Rompe JD, Nafe B, Furia JP, Maffulli N (2007) Eccentric loading, shock-wave treatment, or a wait-and-see policy for tendinopathy of the main body of tendo Achillis: a randomized controlled trial. Am J Sports Med 35(3):374–383
11. Rompe JD, Furia J, Maffulli N (2008) Eccentric loading compared with shock wave treatment for chronic insertional achilles tendinopathy. A randomized, controlled trial. J Bone Joint Surg Am 90(1):52–61
12. Rompe JD, Furia J, Maffulli N (2009) Eccentric loading versus eccentric loading plus shock-wave treatment for midportion achilles tendinopathy: a randomized controlled trial. Am J Sports Med 37(3):463–470
13. Silbernagel KG, Thomeé R, Eriksson BI, Karlsson J (2007) Continued sports activity, using a pain-monitoring model, during rehabilitation in patients with Achilles tendinopathy: a randomized controlled study. Am J Sports Med 35(6):897–906

14. Stevens M, Tan C-W (2014) Effectiveness of the Alfredson protocol compared with a lower repetition-volume protocol for midportion Achilles tendinopathy: a randomized controlled trial. J Orthop Sports Phys Ther 44(2):59–67
15. Yu J, Park D, Lee G (2013) Effect of eccentric strengthening on pain, muscle strength, endurance, and functional fitness factors in male patients with achilles tendinopathy. Am J Phys Med Rehabil 92(1):68–76
16. Hagen EM, Eriksen HR, Ursin H (2000) Does early intervention with a light mobilization program reduce long-term sick leave for low back pain? Spine 25(15):1973–1976
17. Hlobil H, Staal JB, Twisk J, Köke A, Ariëns G, Smid T, van Mechelen W (2005) The effects of a graded activity intervention for low back pain in occupational health on sick leave, functional status and pain: 12-month results of a randomized controlled trial. J Occup Rehabil 15(4):569–580
18. Lindström I, Öhlund C, Eek C, Wallin L, Peterson L-E, Fordyce WE, Nachemson AL (1992) The effect of graded activity on patients with subacute low back pain: a randomized prospective clinical study with an operant-conditioning behavioral approach. Phys Ther 72 (4):279–290
19. Staal JB, Hlobil H, Twisk JWR, Smid T, Köke AJA, van Mechelen W (2004) Graded activity for low back pain in occupational health care. Ann Intern Med 140(2):77
20. Verbeek JH, van der Weide WE, van Dijk FJ (2002) Early occupational health management of patients with back pain: a randomized controlled trial. Spine 27(17):1844–1851 discussion 1851
21. Peterson M, Butler S, Eriksson M, Svärdsudd K (2014) A randomized controlled trial of eccentric vs. concentric graded exercise in chronic tennis elbow (lateral elbow tendinopathy). Clin Rehabil 28(9):862–872
22. Stasinopoulos D, Stasinopoulos I, Pantelis M, Stasinopoulou K (2010) Comparison of effects of a home exercise programme and a supervised exercise programme for the management of lateral elbow tendinopathy. Br J Sports Med 44(8):579–583
23. Peterson M, Butler S, Eriksson M, Svärdsudd K (2011) A randomized controlled trial of exercise versus wait-list in chronic tennis elbow (lateral epicondylosis). Upsala J Med Sci 116(4):269–279
24. Pedersen MT, Andersen CH, Zebis MK, Sjøgaard G, Andersen LL (2013) Implementation of specific strength training among industrial laboratory technicians: long-term effects on back, neck and upper extremity pain. BMC Musculoskelet Disord 14:287
25. Martinez-Silvestrini JA, Newcomer KL, Gay RE, Schaefer MP, Kortebein P, Arendt KW (2005) Chronic lateral epicondylitis: comparative effectiveness of a home exercise program including stretching alone versus stretching supplemented with eccentric or concentric strengthening. J Hand Therapy Official J Am Soc Hand Therapists 18(4):411–419 quiz 420
26. Yagci G, Yakut Y, Simsek E (2018) The effects of exercise on perception of verticality in adolescent idiopathic scoliosis. Physiotherapy Theory Pract 34(8):579–588
27. Gard G (2005) Body awareness therapy for patients with fibromyalgia and chronic pain Disabil Rehabil 27(12):725–728
28. Malmgren-Olsson E-B, Armelius B-A, Armelius K (2001) A comparative outcome study of body awareness therapy, feldenkrais, and conventional physiotherapy for patients with nonspecific musculoskeletal disorders: changes in psychological symptoms, pain, and self-image. Physiotherapy Theory Pract 17(2):77–95
29. Skjaerven LH, Kristoffersen K, Gard G (2010) How can movement quality be promoted in clinical practice? A phenomenological study of physical therapist experts. Phys Ther 90 (10):1479–1492

30. Wilson F, Gormley J, Hussey J (2011) Exercise therapy in the management of musculoskeletal disorders. Wiley, West Sussex
31. Kuorinka I, Jonsson B, Kilbom A, Vinterberg H, Biering-Sørensen F, Andersson G, Jørgensen K (1987) Standardised Nordic questionnaires for the analysis of musculoskeletal symptoms. Appl Ergon 18(3):233–237
32. Carlsson AM (1983) Assessment of chronic pain. I. Aspects of the reliability and validity of the visual analogue scale. Pain 16(1):87–101
33. Breivik H, Borchgrevink PC, Allen SM, Rosseland LA, Romundstad L, Hals EKB, Kvarstein G, Stubhaug A (2008) Assessment of pain. Br J Anaesth 101(1):17–24
34. Jensen MP (2011) Measuring pain behavior. In: Jensen MP (ed) The pain stethoscope: a clinician's guide to measuring pain. Springer Healthcare, Tarporley, pp 19–21
35. Mohr DC, Beutler LE, Engle D, Shoham-Salomon V, Bergan J, Kaszniak AW, Yost EB (1990) Identification of patients at risk for nonresponse and negative outcome in psychotherapy. J Consult Clin Psychol 58(5):622–628

Is the Work Safe? Do I Feel Safe? A 'Choose Your Own' Psychosocial Adventure

Alison Gembarovski[✉], Ian Sutcliffe, and Lachlan Hislop

WorkSafe Victoria, Melbourne, VIC, Australia
alison_gembarovski@worksafe.vic.gov.au

Abstract. Physical and psychosocial demands are known to affect the development of musculoskeletal disorders (MSD). Psychosocial factors are also known to affect the length of time for employees to return back to work after a work-related injury. Internationally, despite the vast body of research and multitudes of different approaches in this area over the last 2–3 decades, many employers report that they are unsure what psychosocial factors are or what to do prevent injury or to enable easier and quicker return to work. In light of this challenge, WorkSafe Victoria, Victoria's occupational health and safety regulator, has developed a simple and interactive visual engagement concept to support the publication of work-related stress guidance material and manual handling guidance material. It is designed to show employers what psychosocial factors are, and the role they play in the development of injury and how they can affect an employee returning to work after injury. The concept includes how the Victorian occupational health and safety legislation applies and how discharging of health and safety duties can impact both positively and negatively on injury, wellbeing, workplace safety culture, business productivity and life at home and in the greater community. More importantly, the visual concept can provide a talking point for what good interventions can look like for organisational psychosocial factors. Be part of the 'choose your own' adventure during the poster presentation "Is the work safe? Do I feel safe?" and see creativity in practice from WorkSafe Victoria.

Keywords: Musculoskeletal disorders · Psychosocial · Prevention
Return to work

"Any intelligent fool can make things bigger and more complex... It takes a touch of genius - and a lot of courage to move in the opposite direction." EF Schumacher

1 Background

The Victorian WorkCover Authority, as WorkSafe Victoria, is the occupational health and safety regulator in Victoria, Australia. It administers and enforces legislation including the *Occupational Health and Safety Act 2004* (Vic) [1] (OHS) and the *Workplace Injury Rehabilitation and Compensation Act 2013* (Vic) [2] (WIRC). Broadly, WorkSafe Victoria's responsibilities are to help prevent workplace injuries occurring, enforce Victoria's OHS laws, provide reasonably priced workplace injury insurance for employers, help injured workers back into the workforce and manage the

© Springer Nature Switzerland AG 2019
S. Bagnara et al. (Eds.): IEA 2018, AISC 820, pp. 360–368, 2019.
https://doi.org/10.1007/978-3-319-96083-8_48

workers' compensation scheme by ensuring the prompt delivery of appropriate services and adopting prudent financial practices.

As such, the Authority has many functions, including the dissemination of information about duties, and obligations under the OHS and WIRC laws and the formuation of guidance to assist duty holders to comply with their duties and obligations, and to promote public awareness and discussion. Another function of the Authority is to monitor and enforce compliance with those laws.

Work-related musculoskeletal disorders (MSD) account for significant proportion over half of workers' compensation claims in Victoria, Australia. Whilst claim numbers have been reducing over time, there is still much work to do in the prevention of musculoskeletal disorders and in assisting workers back to healthy and safe work following injury.

2 The Challenge for the Regulator

2.1 Research Evidence

One of the challenges for any Regulator is translating evidence into policy and practice. Over the last few decades there has been an increasing body of research outlining the significance of psychosocial factors.

In the development of MSD, both physical and psychosocial demands are known to contribute [3, 4] and whilst there is mixed evidence as to their relative contributions, it is widely acknowledged that there is contribution from both and therefore both must be addressed for successful MSD risk prevention [5, 6]. Psychosocial factors are also known to impact on return to work [7, 8]. There is a vast array of conceptual models in the return to work sphere, including biomedical, psychosocial, ergonomics, biopsychosocial and integrated models. All include elements of psychosocial factors and their effect on the length of time for employees to return back to work after work-related injury. Research evidence [9] also supports that the earlier someone with a health problem, such as MSD, can be helped to return to work, the better for the person, the employer and society.

2.2 Legislation and Guidance Material

One of the objects of the Victorian OHS legislation is to secure the health, safety and welfare of employees and other persons at work. The definition of 'health' under the OHS Act includes physical as well as 'psychological health' - therefore any reference to OHS obligations in relation to the health of employees extends to their psychological health.

The objects of the WIRC legislation includes provisions for workers' compensation payments, the effective occupational rehabilitation of injured workers and their early return to work. This includes increasing the provision of suitable employment to workers who are injured to enable their early return to work and assisting injured workers who cannot return to their usual duties to find suitable alternative employment. In addition, the

WIRC Act seeks to improve the health and safety of people at work and reduce the social and economic cost to the Victorian community of accident compensation.

OHS and WIRC laws contain provisions for prevention of MSD arising from hazardous manual handling, and outline employer duties to ensure a safe return to work. Many of these laws encapsulate psychosocial factors without specifically naming them as such.

The Victorian OHS Act
The Victorian OHS Act outlines duties for Victorian employers in order to provide for a working environment that is safe and without risks to health. Elements of psychosocial factors align with the OHS Act, for example, consultation with employees and health and safety representatives, if present, is required under the OHS Act. This aligns broadly with psychosocial factors relating to control over the way work is done, hazards and risks identified, and risk control measures implemented.

In this way, where there are failures in the discharging of duties articulated in the OHS legislation, workers may not only continue to be exposed to risks to health and safety, but psychosocial factors may also come into play [10]. Consider the following; if there are failures in the general duties relating to health and safety (e.g.: failure to eliminate risk when reasonably practicable) this may impact on the perception of support and control over the way work is done; if there is a failure to consult over risk of MSD or risk control measures; this may impact on the perception of support from management and supervisors or with regard to resources. These examples suggest that the degree of compliance by duty holders with their obligations under the OHS legislation could impact indirectly on the development of MSD.

The Victorian WIRC Act
Elements of psychosocial factors are also encapsulated in the Victorian WIRC Act. The WIRC Act outlines employer obligations if a worker is injured including that employers must, to the extent that it is reasonable to do so, provide employment for workers when they have a capacity to return to suitable duties.

Consider the following; if there are failures in the general duties relating to return to work (e.g.: failure to assess and propose suitable duties in return to work, or to consider reasonable workplace support, aids or modifications to assist a workers return to work) this may impact on the perception of support and control; if there is a failure to consult around return to work this may impact on the perception of support from management and supervisors. These examples suggest that the degree of compliance by duty holders with their obligations under the WIRC legislation could impact further on return to work following an MSD.

Occupational Health and Safety Regulations 2017 (Vic) (OHS Regulations)
The Victorian OHS Regulations contain specific regulations for hazardous manual handling. Duties owed under the OHS Regulations sit under the overarching duties in the OHS Act. Risk of MSD arising from hazardous manual handling is currently based largely on physical demands however some psychosocial factors can be addressed under the risk control and review and revision of risk control measures provisions. The OHS Regulations require an employer to eliminate the risk of MSD, or if it is not reasonably practicable to eliminate the risk then to reduce it, in accordance with a

hierarchy of controls. This may involve altering the workplace layout, the workplace environment, the systems of work used to undertake the work, changing the objects used in the work, using mechanical aids or any combination of the above. There is also a requirement to review and, if necessary revise, manual handling risk control measures following the report of an MSD.

Statutory and Non-Statutory Guidance Material

WorkSafe Victoria produces much statutory and non-statutory guidance material to assist duty holders to comply with their duties including the prevention and management of MSD and return to work following MSD. It is clear that failure to comply with duties under the OHS or WIRC Acts may affect both development of injury and return to work. Resources include the recently published statutory guidance the Hazardous Manual Handling Compliance Code [11], industry guides on manual handling risk controls and guidance on reviewing and revising risk control measures [12, 13]. Guidance material outlining organisational psychosocial factors includes 'Preventing and Managing Work-Related Stress' [14] where the factors are outlined as work demands, level of control, level of support, relationships, change management and role clarity and conflict, similar to other jurisdictional guidance material e.g.: WorkCover Queensland and the Health and Safety Executive (HSE) Management Standards in the United Kingdom.

Resources also exist for employees and employers following an injury. Several Compliance Codes exist, such as 'Providing employment, planning and consultation about return to work' [15]. There is also other information such as guides for employers outlining what to do if a worker is injured [16], templates for employers regarding proposed suitable or pre-injury employment [17] and step by step guides to assessing suitable employment options [18].

2.3   Evidence into Practice

WorkSafe is aware that, despite the legislation, and the guidance material, many employers report that they are unsure what to do in order to prevent MSD and are unclear what psychosocial factors are or what do to enable easier and quicker return to work. In the absence of knowing what to do, employers are sometimes doing nothing.

Internationally, there is also an enormous array of websites, documents, tools, toolkits and jurisdictional guidance that can assist employers. Simply typing some key words into search engines on the internet can give half a million to more than 750 million results. A theme stressed in the knowledge management literature is the paradoxical situation that, although there is an abundance of information available, it is often difficult to obtain useful, relevant information when it is needed [19].

If employers still lack understanding and are unsure on the appropriate action to take, WorkSafe must review the way we are presenting the messages. Have we overwhelmed workers and employers by making things too big and too complex? Should WorkSafe move in the opposite direction?

3 What Workers and Employers Are Telling Us

WorkSafe has conducted extensive consultation and engagement with Victorians over the last two years in order to successfully plan for our future and address the challenges we collectively face to prevent injury and illness in workplace.

WorkSafe's insight into workers and employers demonstrates that, for some, there can be a significant difference between what they expect to gain from interacting with us, compared to what they actually gain. Many feel that the compensation system in Victoria is too complex and difficult to understand. Whilst employers want helpful and effective guidance and support, some have commented that our 'one-size fits all' approach does not provide the best experience or outcomes [20]. Workers and employers want WorkSafe to understand their needs and make things simpler and easier to understand. They want their interactions with WorkSafe and our agents to be simplified and to integrate the platforms with a self-service capability. They expect to be informed and educated about occupational health and safety and return to work and want us to provide consultative support and industry specific advice.

4 Our Approach to Date

WorkSafe uses a Constructive Compliance Strategy in its compliance and enforcement activities, which balances a combination of positive motivators and deterrents to improve workplace health and safety. In order to support workplace health and safety, WorkSafe provides practical and constructive advice and information to duty holders on how to comply, engages and communicates with stakeholders, fosters consultative and cooperative relationships with employers and employees, and supports and involves stakeholders in the provision and promotion of OHS education and training [21, 22].

WorkSafe also leverages its broader media and advertising campaigns to engage with the Victorian community to encourage and promote greater awareness around preventing injury in the workplace and the importance of returning to work. It is about acknowledging the supporting role that all workers, employers and the wider community play in maintaining workplace health and safety and preventing incidents and injuries from occurring. Better safety outcomes have a broader impact than just in a workplace. It creates social and economic benefits for all Victorians.

By using an integrated approach to enforce Victoria's laws, WorkSafe combines the provision of advice and guidance material with the use of one or more enforcement measures (for example, issuing an improvement or prohibition notice). This approach recognises that the right tools need to be used in the right circumstances.

Integral to WorkSafe's Constructive Compliance Strategy is the recognition that real and sustainable improvement in workplace health and safety is achieved primarily by the active involvement of employers and employees in hazard identification, management and elimination.

In practice, in relation to manual handling and MSD, this has meant a mix of compliance and enforcement activity largely based around the manual handling regulations and physical work demands. In relation to organisational psychosocial factors, this activity has focused on education and awareness largely based around work-related stress guidance material. Most of the resources, including the recent Hazardous Manual Handling Compliance Code are text based in an electronic pdf format on the website. The more comprehensive Codes and guidance material can be 30–50 pages in length, whilst shorter information sheets may be 1–6 pages long. The links between manual handling organisational psychosocial factors and development of MSD, and on return to work outcomes are not easy to see in our discrete, hazard based guidance approach which means there are opportunities for improvement.

5 Developing a New Approach

5.1 "Any Intelligent Fool Can Make Things Bigger and More Complex… It Takes a Touch of Genius – and a Lot of Courage to Move in the Opposite Direction" EF Schumacher

In 2016, WorkSafe Victoria commenced on a journey to simplify and link the messages. A simple 'choose your own adventure' stick figure presentation was developed, firstly encapsulating the OHS Act and Regulations in relation to hazardous manual handling and organisational psychosocial factors. It was initially developed and tested with Victorian stakeholders via formal stakeholder working groups involving peak bodies from employer groups and unions representing workers. The concept then added the return to work process and was tested and refined repeatedly via external presentations with employers, employees, health and safety representatives, OHS consultants, universities and stakeholders. The concept had five objectives designed to demonstrate:

1. The links between organisational psychosocial factors and development of musculoskeletal disorders,
2. How the Victorian legislation encompasses many organisational psychosocial factors,
3. What interventions can look like,
4. How discharging of OHS duties can impact positively on:
   a. Reducing hazards and risks associated with organisational factors
   b. Organisational health, occupational health, safety and wellbeing
   c. Positive workplace climate and culture
   d. Business productivity
   e. Life at home and the greater community
5. How lack of discharging of OHS duties can impact negatively on:
   a. Risk and development of injury (MSD and poor psychological health outcomes),
   b. Wellbeing and return to work
   c. Business productivity

The Stickperson message is an alternative way of communicating and engaging with employers and employees in Victorian workplaces. It aims to educate employers in a way which articulates and visualises legislation and guidance material by simplifying messages and the language we use. It will allow employers to more clearly understand their obligations and in turn increase their knowledge in the design and management of systems of work by applying adequate controls to organisational psychosocial factors that may give rise to risk of injury and protracted return to work. It also allows employers to understand, if an MSD does develop, how both Acts still apply (a) to review and revise the risk control measures, and (b) to best prevent further injury and to enable the safest return to work.

5.2   Feedback to Date

Feedback collected at sessions over the last two years suggests that this simplification of the messages resonates with Victorian duty holders. Rating by attendees has consistently showed agreement and strong agreement with following statements:

1. The presentation showed how lack of OHS can impact on development of an injury
2. The presentation showed how lack of OHS can impact on return to work
3. The presentation showed the links between work-related stress factors and development of MSD
4. After seeing this presentation I understand how my organization could do things differently in the workplace.

The 'choose your own adventure' style allows the duty holder to understand the consequences of particular decisions made. Feedback from attendees has included feedback such as "That really happens, you know", "OMG. We do that", "Can I have a copy of the presentation? I need to show it to my senior management.", "A serious message with humour – it really hit home", and "That was really close to the bone…"

6   Next Steps

On the basis of the testing to date and worker, employer and stakeholder feedback, this innovative way to engage differently with employers will be used to promote education and awareness through digital and interactive channels.

It represents a new way of identifying and addressing OHS risks in the workplace for most employers and is a simplification of the way we communicate, in plain English with simplified visuals. The proposal is to represent an online journey for employers and workers, introducing new concepts relevant to each stage of the risk management framework to identify and address OHS risks in the design and management of work, providing practical examples of higher order risk controls to be considered by the employer and information to assist in safer and healthier return to work.

The educational tool will also provide comparative scenarios and outcomes, both good, bad and options in between whilst also demonstrating ways to improve the journey and hence outcomes for both the worker and employer. Links to relevant information, for both the worker and employer will be shown along the journey, enabling easier access to the right information at the time it is needed. This will help to address the existing gap between what employers and workers want from WorkSafe and what we are currently providing.

References

1. Victorian Occupational Health and Safety Act 2004 Act No. 107/2004
2. Authorised Version Victorian Workplace Injury Rehabilitation and Compensation Act 2013 No. 67 of 2013
3. Macdonald W, Evans O (2006) Research on the prevention of work-related musculoskeletal disorders stage 1 – literature review. Australian Safety and Compensation Council. https://www.safeworkaustralia.gov.au/system/files/documents/1702/research_prevention_workrelated_musculoskeletal_disorders_stage_1_literature_review.pdf Accessed 13 Dec 2018
4. Macdonald W, Oakman J (2015) Requirements for more effective prevention of work-related musculoskeletal disorders. BMC Musculoskelet Disord 16:293 https://doi.org/10.1186/s12891-015-0750-8
5. Da Costa BR, Vieira ER (2010) Risk factors for work-related musculoskeletal disorders: a systematic review of recent longitudinal studies. Am J Indust Med 53:285–323
6. Hauke A et al (2011) The impact of work related psychosocial stressors on the onset of musculoskeletal disorders in specific body regions. A review and meta-analysis of 54 longitudinal studies. Work Stress 25(3):243–256
7. Kilgour E et al (2015) Interactions between injured workers and insurers in workers' compensation systems: a systematic review of qualitative research literature. J Occup Rehabil 25(1):160–181
8. Schultz IG, Gatchel RJ (eds) (2018) Handbook of Return to Work: From Research to Practice. https://doi.org/10.1007/978-1-4899-7627-7
9. Waddell G, Burton KA (2008) Is work good for your health and well-being? Department of Work and Pensions, UK Government. https://cardinal-management.co.uk/wp-content/uploads/2016/04/Burton-Waddell-is-work-good-for-you.pdf. Accessed 13 Dec 2018
10. Gembarovski A (2015) Psychosocial factors and musculoskeletal disorders - the challenges for the Victorian Regulator. International Ergonomics Association Congress. http://www.iea.cc/congress/2015/1159.pdf
11. Compliance Code – Hazardous Manual Handling. WorkSafe Victoria (2018). https://www.worksafe.vic.gov.au/__data/assets/pdf_file/0005/218084/ISBN-hazardous-manual-handling-compliance-code-2018-03.pdf
12. Manual Handling (2016) Review and revision of risk control measures. WorkSafe Victoria. https://www.worksafe.vic.gov.au/__data/assets/pdf_file/0020/211196/ISBN-Hazardous-manual-handling-review-revision-of-risk-control-measures-2017-06.pdf
13. Manual Handling (2016) Improving review and revision of risk control measures. WorkSafe Victoria. http://www.worksafe.vic.gov.au/__data/print-to-pdf/api.php?url=http://www.worksafe.vic.gov.au/pages/forms-and-publications/forms-and-publications/improving-the-review-and-revision-of-manual-handling-risk-control-measures%3FSQ_DESIGN_NAME=blank%26SQ_PAINT_LAYOUT_NAME=pdf
14. Preventing and managing work-related stress. WorkSafe Victoria (2017). https://www.worksafe.vic.gov.au/__data/assets/pdf_file/0006/211299/ISBN-Preventing-and-managing-work-related-stress-guidebook-2017-06.pdf
15. Compliance Code 1 of 4 (2014) Providing employment, planning and consulting about return to work. WorkSafe Victoria. https://www.worksafe.vic.gov.au/__data/assets/pdf_file/0003/212268/ISBN-Compliance-code-1-of-4-providing-employment-planning-consultation-rtw-2014-07.pdf

16. A guide for employers - What to do if a worker is injured. WorkSafe Victoria (2013). https://www.worksafe.vic.gov.au/__data/assets/pdf_file/0011/207749/ISBN-What-to-do-if-a-worker-is-injured-web-a-guide-for-employers-2013-07.pdf

17. Return to Work Arrangements Template: Return to work toolkit. WorkSafe Victoria (2011)

18. Suitable employment for injured workers, a step by step guide to assessing suitable employment options. WorkSafe Victoria. FOR726/04/03.13

19. Edmunds A, Morris A (2000) The problem of information overload in business organisations: a review of the literature. Int J Inf Manag 20:17–28

20. WorkSafe Strategy 2030 Discussion paper (2017) Creating safer and healthier workplaces for all Victorians. https://www.worksafe.vic.gov.au/__data/assets/pdf_file/0003/208794/ISBN-strategy-2030-discussion-paper-2017-06.pdf

21. WorkSafe 2017 (2017) WorkSafe Victoria. https://www.worksafe.vic.gov.au/__data/assets/pdf_file/0004/209488/ISBN-Worksafe-strategy-summary-2017-01.pdf

22. Victorian State Government (2016) Independent Review of Occupational Health and Safety Compliance and Enforcement in Victoria. https://www.dtf.vic.gov.au/sites/default/files/2018-03/Independent-Review-of-OHS-Compliance-and-Enforcement-in-Victoria-Report%20%281%29.pdf

ERIN: A Practical Tool for Assessing Exposure to Risks Factors for Work-Related Musculoskeletal Disorders

Yordán Rodríguez Ruíz$^{(\boxtimes)}$ (iD)

National School of Public Health, Universidad de Antioquia, Cl. 62, #52-59
Medellín, Colombia
yordan.rodriguez@udea.edu.co

Abstract. The prevention of work-related musculoskeletal disorders (WMSDs) has become one of the greatest of daily challenges that faces ergonomics professionals. One of the actions carried out as part of prevention is the assessment of exposure to WMSDs factors. To date, many different ergonomic assessment methods have been developed. However, in many cases, the correct use of these methods is conditioned by having personnel with high training and experience in ergonomics, which is scarce, mainly in Latin American countries, where there is a presumption of a greater presence of WMSDs risk factors. For this reason, a necessity of having practical methods to evaluate a wide variety of tasks and generate improvement actions has been identified and stated in these countries. In this paper, Evaluación del Riesgo Individual (Individual Risk Assessment) (ERIN) method is presented, which can be used by non-expert personnel in the ergonomic assessment of tasks. ERIN has proven to be easy to learn, easy to use, requires little training time and is a useful tool in identifying opportunities for ergonomic improvement. Finally, a case study showing how to evaluate a task with the ERIN method is included and some results of studies carried out to evaluate the reliability, validity and usability of the method are mentioned.

Keywords: Musculoskeletal disorders · Ergonomic assessment
Postural load

1 Introduction

For several years, it has been recognized and accepted that the emergence of WMSDs is largely due to the existence of jobs/tasks that were designed without taking into account ergonomic principles, exposing workers to risk factors related to these diseases [1]. This has led many professionals and researchers to increase their interest in refining their assessment of exposure to WMSDs risk factors, resulting in the creation of new methods and tools and the updating of some existing ones [2–4].

On the other hand, the statistics on WMSDs continue to be alarming, which encourages increased efforts to prevent this type of disease, which today, we could say, has become an occupational epidemic. An important stage in the prevention process is the ergonomic assessment, which allows us to identify which aspects of the task to be evaluated should be intervened in order to reduce exposure to risk. In this regard, Latin

© Springer Nature Switzerland AG 2019
S. Bagnara et al. (Eds.): IEA 2018, AISC 820, pp. 369–379, 2019.
https://doi.org/10.1007/978-3-319-96083-8_49

American countries have been issued an increasingly number of regulations, that require and encourage ergonomic assessments. However, the number of experts in the use of these methods is very limited. This is one of the reasons that may lead to a considerable number of errors in the ergonomic assessments currently being carried out [5] or not being carried out. In this way, the effectiveness of actions to prevent and improve working conditions is hindered, and therefore the incidence and prevalence rates of WMSDs do not decrease.

In an effort to mitigate the panorama described above, the ERIN method was developed, which took into account the needs, limitations and capabilities of non-expert personnel, trying to respond to the demands of ergonomic evaluation of workplace/tasks in real contexts, while at the same time maintaining acceptable levels of reliability and validity [6–8].

This paper presents the ERIN method, an example of how it can be used in the ergonomic evaluation of tasks and a summary of the results of reliability, validity and usability studies of the method.

2 Development of ERIN

The ERIN method was developed in three stages: (1) Prototype construction, (2) Prototype improvement and (3) conformation of ERIN method (see Fig. 1).

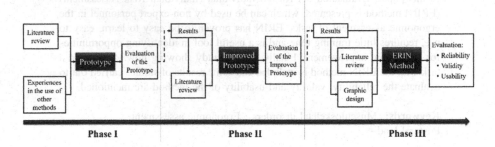

Fig. 1. Development process scheme of ERIN method. Source: from [6]

On the first stage, a prototype of the method was constructed, tested and the aspects to be changed and improved were identified from the results obtained from the designed experiments and the feedback of the assessor. The second stage was aimed at improving the aspects identified in the prototype in terms of variables, scoring system, worksheet design and procedure. This improved prototype was field tested and analyzed to determine which elements should be modified. In the third and final stage, the ERIN method was developed as a result of the changes made to the improved prototype and the incorporation of graphic design aspects. The main results of the reliability, validity and usability assessment are shown in Sect. 5.

3 ERIN: Individual Risk Assessment

ERIN is an easy-to-use observational method developed for used by non-experts with minimal training and few resources, to enable them to carry out large-scale assessments of individuals exposed to WMSDs. It was designed taking into account existing methods, WMSDs epidemiological evidence and the practical needs of non-expert users in assessing exposure to WMSDs risk factors [7, 9, 10].

ERIN evaluates the posture of the trunk, arm, wrist, neck and their frequency of movement; the rhythm, given by the speed of work and the effective duration of the task; the intensity of the effort, the result of the effort perceived by the evaluator and its frequency, and the self-perception of the stress referred by the subject on the task being performed. The method recommends levels of risk and action according to the total risk (see Table 1), which is calculated by adding the risk of the seven variables evaluated. The additive model used, allows to easily identify the influence of each factor and to locate which elements must be intervened to reduce the level of risk.

Table 1. Risk levels and recommended action based on total risk in ERIN.

Total risk	Risk level	Recommended action
6–14	Low	No changes are required
15–24	Medium	Further investigations are needed and changes may be required
25–34	High	Investigations and changes are required soon
≥ 35	Very high	Investigations and changes are required immediately

The short time spent on the evaluation process with ERIN allows for the study of a large number of jobs/tasks in different occupational settings, at relatively low costs and with minimal work interruptions. For its application only, a pencil and sheet is required, although its effectiveness will be influenced by the training and previous knowledge of the observer. ERIN can be used to assess, design and redesign static and dynamic tasks. Its scoring system allows criteria to be established to evaluate the impact of changes made (before and after), assuming that low scores correspond to more favorable conditions.

The results with ERIN can guide occupational risk management and prevention staff on what changes need to be made and in what direction. This same approach can be used to quickly demonstrate the extent to which risks have been minimized and working conditions improved once changes have been made.

A detailed guide on how the ERIN method can be used in the evaluation and jobs intervention, as well as some notes to increase the effectiveness of its application and to perform multi-tasking risk calculation, can be found in Rodríguez [7].

4 ERIN Method Example Application

The following is an example of how the ERIN method can be used to evaluate a task, which consists of closing sacks made of fique fibers by the sides using an industrial sewing machine and scissors to cut the sewing thread. Empty sacks weigh between 290 and 1000 grams and are used to pack agricultural products such as coffee, cocoa, among others. The production standard established for this task varies according to the type of sacks. The worker performs the task in a seated position throughout the 8-h workday, taking only a 30-min break from the workday for lunch and personal needs. The example shows the conformation of the coffee sacks, which is the heaviest (1000 g) and largest (two pieces of 70 × 95 cm). The approximate cycle time for closing a coffee sack is 7 s.

Evaluation of the Closing Sack Task Using the ERIN Worksheet

Figure 2 shows the evaluation of the sack closing task using the ERIN worksheet. As proposed in the ERIN method, the critical posture of the trunk, arm, wrist and neck, adopted by the worker during the work cycle and the frequency of movement of each of these body parts is selected. In the case of the wrist and arm, the observer should choose the side with the highest postural risk, left or right. In case of doubt, both sides are evaluated and the one with the highest postural load according to ERIN is chosen.

Figure 3 shows that the critical posture of the trunk occurs when the worker takes the sacks behind the chair to build it. The worker is sitting without lumbar support, with a score of (2) and an adjustment of (+1) for having the trunk turned. Therefore, the score of the postural load of the trunk is: 2 + 1 = 3. Trunk movement during the task is "very frequent", more than 10 times/min. Obtaining a score of 6 for the Trunk.

The critical position of the shoulder/arm, in this case, coincides with the instant when the critical position of the trunk occurs (see Fig. 4). The arm selected for the evaluation is the left arm, as it is the one with the greatest postural load. In this case the left arm is in severe extension (greater than 20°), with a score of (2) and an adjustment of (+1) for having the arm abducted. Therefore, the score of the postural load of the arm is: 2 + 1 = 3. The movement of the arm during the task is "very frequent" (almost a continuous movement). Getting a score of 8 for the Arm.

Figure 5 shows the critical posture taken by the left wrist, which occurs when the worker turns the sack to sew it. The wrist is in severe flexion (greater than 45°) corresponding to a score of (2) and is deviated (+1) and the hand holds the sacks for more than 50% of the total cycle time (+1). It should be noted that there are other moments during the execution of the task where the wrist performs flexo/extensions greater than 45°. Therefore, the wrist postural load score is: 2 + 1 + 1 = 4. The movement of the left wrist is "very frequent", more than 20 times per minute. Getting a score of 6 for the Wrist.

Figure 6 shows the critical neck posture, which occurs when the worker takes the sacks behind her. The neck is slightly flexed (0–20°), which corresponds to a score of (1) and an adjustment of (+1) for being rotated. Therefore, the neck postural load score is: 1 + 1 = 2. The movement of the neck during the execution of the task is "continuously". Obtaining a score of 6 for the Neck.

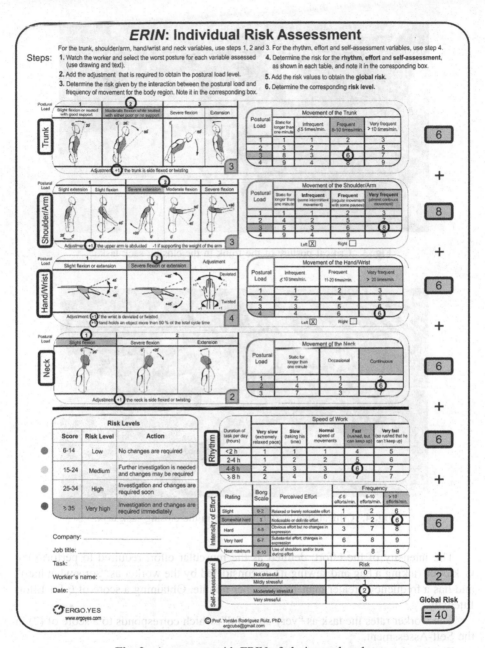

Fig. 2. Assessment with ERIN of closing sack task.

To meet the assigned workload during the working day, the worker carries out the activities at a "fast" work speed (able to support) and the effective work time dedicated to the task during the working day is around 7 h. Getting a score of 6 for the Rhythm.

Fig. 3. Critical posture of the trunk in the closing sack task.

Fig. 4. Critical shoulder/arm posture in the closing sack task.

The intensity of the effort, defined as the muscular effort required to perform the task once (taking a bag and sewing it), is considered by the worker as "somewhat hard" and has a frequency greater than 10 times per minute. Obtaining a score of 6 for Effort Intensity.

The worker rates the task as "very stressful", which corresponds to a score of (2) in the Self-Assessment.

Finally, the risk scores per variable are added and a total risk of 40 is obtained, corresponding to the risk level of "very high". This indicates that changes are required immediately.

Fig. 5. Critical wrist posture in the closing sack task.

Fig. 6. Critical neck posture in the closing sack task.

5 Reliability, Validity and Usability

Following are summarized the results of some studies conducted to evaluate the reliability, validity and usability of the ERIN method [6–8].

5.1 Reliability Studies

Inter-observer reliability: 42 non-expert subjects evaluated with ERIN seven varied industry tasks from videos. The projection of the videos was random and the observers could not stop the video. To analyze the influence of training time, half of the evaluators received 3 h of training, while the other half received 6 h.

Results: In general, the inter-observer reliability was moderate to good, from 18 variables studied, in seven an excellent agreement was obtained (adjust of trunk, arm and neck; wrist, trunk and neck posture; and dichotomized risk level), in five of moderate to good (trunk frequency of movement, arm posture, effort, global risk and risk level) and in six poor (adjust of wrist; arm, wrist and neck frequency of movement; speed of work and frequency of efforts). Agreement index were usually higher for the group of evaluators who received the longest training time, although the differences were not large enough to provide definitive criteria on the influence of this factor.

Intra-observer reliability: 17 non-experts evaluated twice with the ERIN method, a total of seven selected tasks from industry. The time between the first and second evaluation was three weeks.

Results: In general, the intra-observer reliability was moderate to good, since of the 16 variables studied, ten had moderate to good agreement (adjust of trunk, arm and neck; wrist, neck posture; speed of work; arm, trunk frequency of movement; frequency of efforts, and risk levels), three excellent (trunk, arm posture, and effort) and three poor (adjust of wrist; wrist, neck frequency of movement).

5.2   Validity

Study 1: Evaluations of seven video-based tasks by 42 observers were compared to a pattern defined by a panel of three experts. These experts had the possibility to stop the videos, freeze the images to observe the postures in detail, exchange criteria and slow down the projection speed of the filming, to minimize the error.

Results: Of the variables studied, 12 present-ed agreement percentages higher than 60%. Of these, five had values greater than 70%. Five variables had agreement per-centage values of less than 60%, of which four are related to the estimation of the frequency of movement of body parts (trunk, arm, wrist and neck). Showing that the determination of the frequencies of movement of body parts in real time is an aspect that is difficult for non-expert evaluators.

Study 2: Evaluations of two video-based tasks by 8 health professionals (physicians and specialists in traumatology, physiology and rehabilitation) were com-pared with a pattern defined by a group of three experts. These professionals did not have experience in the use of ergonomic assessment methods, but in related subjects. The evaluators received one hour of training in the use of ERIN.

Results: All variables, except wrist and neck movement with percentages of 63%, presented percentages of total agreement above 70%. Of the 17 variables studied, 12 presented values above 80% and of these, four above 90%. No variables differed in more than one category. This suggests that staff with a similar profile, receiving about one hour of training, are able to deliver valid and consistent results when using ERIN.

Study 3: The risk levels obtained by evaluating 32 workstations were compared with the ERIN, RULA and REBA methods. The RULA and REBA methods were chosen from among the existing methods because of the overlaps between the body regions assessed and the risk factors for ERIN, which propose a final risk level resulting from the combination and weighting of the risk factors assessed and which have been used for several years in the evaluation of jobs. Two observation strategies were used for the evaluation, the critical and the most common posture.

Results: The agreement between the ERIN and RULA methods was for the critical posture of an ICC (2.1) = 0.403 and a percentage of agreement of 56%, while for the most common posture it was of an ICC (2.1) = 0.457, and a percentage of agreement of 56%. The agreement between the ERIN and REBA methods was for the critical posture of an ICC (2.1) = 0.554 and a percentage of agreement of 66%, while for the most common posture it was of an ICC (2.1) = 0.556, and a percentage of agreement of 59%. It was suggested that regardless of the observation strategy used, by using ERIN instead of RULA or REBA, the agreement as to the level of final risk is moderate to good. However, the characteristics, strengths and weaknesses of each of these methods should be analyzed when selecting them to evaluate a given situation (e.g., personnel who will use it, dynamic or static task, body parts involved, among others).

5.3   Usability

Study 1: A group of 42 subjects, classified as non-experts, were provided with a tool to assess the usability of ERIN after evaluating 7 ERIN-projected video tasks. This instrument is a modification of one developed to measure the usability of software. The instrument consists of 10 questions focused on exploring the following aspects: easy learning, short training time, time needed to perform the assessment and applicability to various tasks. The usability index obtained varies between 0 and 100 points. The time it took each subject to evaluate each task was also recorded.

Results: The usability index varied between 60 and 100, with an average of 80 and a deviation of 10. The results show that ERIN's usability index is high, according to users' criteria. In addition, the agreement between the evaluators was analyzed when answering the questions related to usability. To analyze the agreement between the subjects when answering the questions asked, the ICC (2.1) was calculated with index of 0.642 individual and 0.987 average. Indicating an agreement to regulate good for each individual and excellent for the group. The average assessment time per task was 6.5 min, with a tendency to decrease as more assessments were conducted. This was supposed to be because the evaluators became increasingly familiar with the method.

Study 2: A group of 8 health professionals were provided with the tool designed to assess the usability of ERIN (mentioned in study 1), after evaluating two tasks. The evaluators received one hour of training prior to using the ERIN method.

Results: The usability index ranged from 62.5 to 92.5 with an average of 84 and a deviation of 12. The results show that ERIN's usability index is high, according to users' criteria. To analyze the agreement among the professionals in answering the questions asked, the ICC (2.1) was calculated with indices of 0.747 individual and 0.947 average. Indicating an agreement very close to excellent for each individual and excellent for the group.

6 ERIN Software

To promote the use of the ERIN method, an open-access online application (available at www.ergoyes.com) has been made available to the public in both English and Spanish. Also, the software ERIN 2.0 has been developed, which allows for structured storage of evaluations, generation of task-specific and global reports by company, multi-tasking risk calculation, and generation of recommendations to reduce risk based on the results of each evaluation (see Fig. 7).

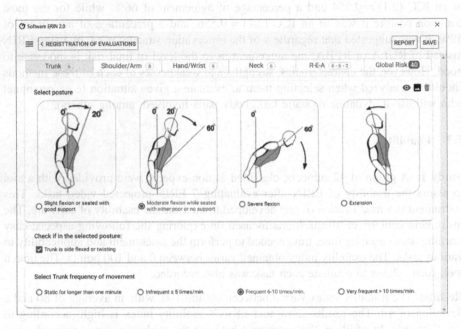

Fig. 7. Software ERIN 2.0. Source: www.ergoyes.com

7 Conclusions

In recent times, reports on the use of the ERIN method in ergonomic task assessment have been gradually increasing. In general, professionals, workers, students and other stakeholders who have used it have found it a practical, simple and useful method in evaluating and improving tasks/jobs. The fundamental value of the ERIN method is that, maintaining acceptable levels of reliability and validity, with little time, resources and training, non-experts people in the field of ergonomics, can perform large-scale evaluations of the exposure to WMSDs risk factors in tasks/workplaces and take corrective/preventive actions quickly, thus favoring the evaluation and intervention process, in accordance with the needs of the people responsible for the prevention of WMSDs in our Latin American countries. It should be highlighted that ERIN is only one method in a comprehensive effort to prevent WMSDs.

Lastly, I would like to mention that, although the improvement of the ergonomic evaluation process must not be stopped and will be influenced by constant technological developments and future research, a balance must be found between what is "theoretically recommended" and what is required in practice, since in many cases, a considerable effort is devoted to evaluation compared to the effort devoted to intervention. Just by evaluating, no matter how well we do it, we will not be able to reverse the current situation of WMSDs, we must intervene in the jobs and work systems from a systemic perspective, if we want to show better results.

References

1. NIOSH, Bernard BP, Putz-Anderson V (1997) Musculoskeletal disorders and workplace factors. A critical review of epidemiologic evidence for work-related musculoskeletal disorders of the neck, upper extremity, and low back. U.S. Department of Health and Human Services, Public Health Service, Centers for Disease Control and Prevention, National Institute for Occupational Safety and Health
2. Takala E-P, Pehkonen I, Forsman M, Hansson G-A, Mathiassen SE, Neumann WP, Sjogaard G, Veiersted KB, Westgaard RH, Winkel J (2010) Systematic evaluation of observational methods assessing biomechanical exposures at work. Scand. J. Work Environ. Health 36:3–24. https://doi.org/10.5271/sjweh.2876
3. David GC (2005) Ergonomic methods for assessing exposure to risk factors for work-related musculoskeletal disorders. Occup Med (Lond) 55:190–199. https://doi.org/10.1093/occmed/kqi082
4. Li G, Buckle P (1999) Current techniques for assessing physical exposure to work-related musculoskeletal risks, with emphasis on posture-based methods. Ergonomics 42:674–695. https://doi.org/10.1080/001401399185388
5. Diego-Mas J-A, Alcaide-Marzal J, Poveda-Bautista R (2017) Errors using observational methods for ergonomics assessment in real practice. Hum. Factors 59:1173–1187. https://doi.org/10.1177/0018720817723496
6. Rodríguez Y (2011) ERIN: método práctico para evaluar la exposición a factores de riesgo de desórdenes músculo-esqueléticos. Departamento Ingeniería Industrial, Instituto Superior Politécnico José Antonio Echeverría
7. Rodríguez Y (2018) Individual risk assessment (ERIN): method for the assessment of workplace risks for work-related musculoskeletal disorders. In: Handbook of research on ergonomics and product design, pp 1–27. IGI Global
8. Rodríguez Y, Heredia JJ (2013) Confiabilidad ínter-observador del método de Evaluación de Riesgo Individual. Revista Hacia la Promoción de la Salud 18
9. Rodriguez Y, Vina S, Montero R (2013) A method for non-experts in assessing exposure to risk factors for work-related musculoskeletal disorders-ERIN. Ind. Health 51:622–626. https://doi.org/10.2486/indhealth.2013-0008
10. Rodríguez Y, Viña S, Montero R (2013) ERIN: a practical tool for assessing work-related musculoskeletal disorders. Occupat Ergon 11:59–73. https://doi.org/10.3233/OER-130210

A Study on Posture Analysis of Assembly Line Workers in a Manufacturing Industry

Jingyun Li[1($\boxtimes$)], Yabo Lu[2], Yajun Nan[2], Lihua He[3], Xin Wang[4], and Dongsheng Niu[5]

[1] National Center of Occupational Safety and Health, SAWS, Beijing, China
15810439672@126.com
[2] North China University of Science and Technology, Tangshan Hebei Province, China
[3] School of Public Health, Peking University, Beijing, China
[4] Smith & Nephew Orthopedics (Beijing) Co., LTD, Beijing, China
[5] Beijing Prevention and Treatment Hospital of Occupational Disease for Chemical Industry, Beijing, China

Abstract. Objective: In order to reveal and analyze the feature of surface electromyography (sEMG) signal and assembler's posture of repetitive work, assembler's posture and electrical manifestations of localized muscles was collected by using the inertial motion capture technology and sEMG during the work. **Methods:** Six male workers were chosen in laser lettering workshop of a medical equipment manufacturing industry. They performed repetitive pull-and-push operation, meanwhile, sEMG and posture signal in 10 min period was collected at the beginning and end of work. Results: The MFs of LST, LBB and LFCR were significantly lower at the beginning than at the end of work, while the RMSs of LST, LBB and LFCR were significantly higher at the beginning than at the end of work. And the differences of LST between the beginning and the end of work was most distinct. Keeping fixed joint angle for a long time had a greater effect on muscle fatigue, and the result was statistically significant ($P < 0.05$). **Conclusions:** The fatigue strength of muscle increased significantly with repetitive push-and-pull operation of workers for a long time, especially of the shoulder; And fatigue was easily caused by keeping fixed joint angle for a long time.

Keywords: Repetitive work · sEMG · Inertia motion capture Fatigue

1 Introduction

In the course of labor work activities, long-term repetitive stress and maintaining a certain working posture, could be a kind of workload and cause musculoskeletal injuries- common diseases related to work methods. Its main features are pain and activity limitation such as work-related musculoskeletal disorders (WMSDs).

European countries and the United States have already listed WMSDs as an occupational disease category. Studies have shown that WMSDs are negatively correlated with work ability, affecting the work and production efficiency of operators

S. Bagnara et al. (Eds.): IEA 2018, AISC 820, pp. 380–386, 2019.
https://doi.org/10.1007/978-3-319-96083-8_50

seriously, and causing serious economic losses to society. Therefore, wide attention has been paid to WMSDs. In recent years, researches on WMSDs have focused on their causes and analyzed to find effective preventive measures.

With the development of science and technology, there have emerged some new high-tech methods for evaluating workload, such as surface electromyography and simulation technology. Surface electromyography may be an objective method for assessing muscle fatigue during field work, but the current research in related areas mostly focuses on the effects of static load and/or intensity on muscle function status. There were few researches on dynamic load, and real-time motion analysis of work. We collect the workers' electromyographic signals and postures in real time, and measure and analyze the related indicators to analyze the characteristics of the working posture, which provided a reference for the study of dynamic load analysis.

2 Subjects and Methods

2.1 Subjects

In a medical device manufacturing company, male workers who performed repetitive push-pull operations in a laser-engraving workshop were chosen as the subjects of the study. The normal operation of the laser lettering machine was: first, open the sliding door of the device, place the workpiece in the device, and close the sliding door; then, start the lettering program; at last, open the sliding door and replace the workpiece after lettering. The above operation was repeated hundreds of times every day.

2.2 Instruments and Materials

Inertial motion capture device: Wearable motion capture system and three-dimensional gait analyzer (Technaid S.L, Spain), 8-channel wireless Bluetooth, sampling frequency 100 Hz, with head as a reference.

sEMG measuring instrument: wireless portable surface electromyography (OT Bioelettronica, Italy), 8-channel wireless Bluetooth, sampling frequency of 1000 Hz, band-pass filtering of 10–500 Hz, input impedance >109 Ω, using OT BioLab software for sEMG signal analysis. Muscular surface electrode: Ag-AgCl sensor electrode, diameter 1 cm, with conductive paste on the surface.

2.3 Methods

Working Posture. Based on the anatomical surface, the left upper extremity was higher than the shoulder level and the elbow flexion approaches 180° during laser lettering machine operation.

Sensor Position. According to the requirements of the inertial motion capture device, the sensor is fixed on the outer edge of the left and right upper and lower arms and the upper edge of the scapula. The head is used as a reference for movement and the sensor is attached to the forehead to capture motion of the left arm. Balance action test was performed before data collection.

Electrode Paste Position. Shoulder and elbow pain was the most obvious symptoms according the complaint of workers. Therefore, three muscles were chosen as test target: the left shoulder trapezius muscle (LST), the left biceps brachii (LBB) and the left temporal flexor tendon (LFCR). The electrodes were stuck in pairs in the corresponding direction along the direction parallel to the muscle fibers. The electrode paste positions were 3 cm below the shoulder peak, the middle of the biceps brachii and radial wrist muscle; the reference electrode was attached to the olecranon at the elbow joint.

Sampling Hours. Employees had a short break of 15 min per 2 h and a long break of 1 h at noon in the entire 8-h workday. Inertial motion and surface EMG signals were collected for 10 min at the beginning of work in the morning and 10 min before the end of the work in the afternoon.

Analysis methods. The analysis of surface electromyogram included time domain analysis and frequency domain analysis. The main index in time domain analysis was root mean square (RMS), and the main index in frequency domain analysis was median frequency (MF). The RMS and MF values of the surface myoelectric signals were used to analyze muscle load conditions. The inertial motion capture technology was used to collect motions in real time, and the angles of the upper extremity were decomposed in three-dimensional space, and the change of the work posture was analyzed. SPSS20.0 statistics software was used for data analysis. The paired t-test was used to analyze the difference between RMS and MF myoelectric signals of the same worker during 10 min at the beginning of work and 10 min before the end of the work; the definitive fisher's exact test was used to analyze the effect of muscle fatigue after a long-time operation in a certain posture. P value less than 0.05 was considered statistically significant.

3   Results

3.1   General Conditions

Six male workers were selected for this study, with an average age of (34.0 ± 2.0) years old, an average working age of (6.8 ± 1.2) years, an average height of (172.7 ± 5.3) cm, and an average body weight of (66.4 ± 13.6) kg. Time of operating repetitive pull and push were more than 7 h per day.

Subjects were in good health, had no history of musculoskeletal disorders and upper extremity and shoulder trauma.

3.2   Changes in MF and RMS of 10 Min at the Beginning of Work and 10 Min Before the End of Work

The effect of dynamic load on MF and RMS can be seen in Tables 1 and 2. Compared with the 10 min at the beginning of work, the MF value decreased and the RMS value increased of 10 min before the end of work. There was a statistically significant difference ($P < 0.05$) between them, and the most obvious change occurred in LST. It indicated that muscle fatigue increased.

Table 1. Changes in MF of 10 min at the beginning of work and 10 min before the end of work (Hz, $\bar{x} \pm s$)

Subject	Time	LFCR	LBB	LST
1	Beginning	0.085 ± 0.001	0.051 ± 0.001	0.071 ± 0.001
	End	0.056 ± 0.001	0.050 ± 0.001*	0.040 ± 0.001*
2	Beginning	0.054 ± 0.001	0.052 ± 0.001	0.072 ± 0.001
	End	0.045 ± 0.001*	0.042 ± 0.001*	0.051 ± 0.001*
3	Beginning	0.078 ± 0.001	0.070 ± 0.001	0.063 ± 0.001
	End	0.067 ± 0.002*	0.053 ± 0.001*	0.050 ± 0.001*
4	Beginning	0.054 ± 0.001	0.107 ± 0.002	0.101 ± 0.001
	End	0.050 ± 0.001*	0.900 ± 0.001*	0.084 ± 0.001*
5	Beginning	0.138 ± 0.005	0.128 ± 0.001	0.130 ± 0.001
	End	0.108 ± 0.003*	0.106 ± 0.002*	0.103 ± 0.001*
6	Beginning	0.108 ± 0.002	0.101 ± 0.001	0.100 ± 0.001
	End	0.069 ± 0.001*	0.050 ± 0.001*	0.050 ± 0.001*

Note: * $P < 0.05$

Table 2. Changes in MF of 10 min at the beginning of work and 10 min before the end of work (μV, $\bar{x} \pm s$)

Subject	Time	LFCR	LBB	LST
1	Beginning	182.76 ± 2.86	322.67 ± 1.91	413.96 ± 4.74
	End	183.04 ± 1.04*	325.76 ± 1.64*	447.18 ± 4.35*
2	Beginning	191.27 ± 1.61	316.54 ± 0.77	329.88 ± 3.26
	End	196.10 ± 4.03*	331.07 ± 7.72*	445.49 ± 3.19*
3	Beginning	224.18 ± 4.62	337.47 ± 6.95	435.66 ± 1.54
	End	240.14 ± 5.66*	344.95 ± 8.93*	445.34 ± 3.55*
4	Beginning	189.66 ± 1.77	339.68 ± 8.13	431.29 ± 4.61
	End	202.79 ± 4.98*	344.68 ± 7.98*	445.81 ± 7.94*
5	Beginning	224.39 ± 5.42	328.99 ± 9.25	445.68 ± 6.21
	End	242.82 ± 4.72*	336.93 ± 7.52*	458.11 ± 7.55*
6	Beginning	191.56 ± 4.05	318.32 ± 2.31	424.75 ± 1.25
	End	195.23 ± 5.09*	319.52 ± 3.14*	447.03 ± 3.21*

Note: * $P < 0.05$

3.3   Activity Analysis of Repetitive Stress

Change in motion Frequency. The frequency of workers' repetitive activities was showed in Table 3 for details. The frequency of movement of employees before and after their work mostly reduced.

Table 3. The frequency of workers' repetitive activities (times/min)

Subject	At the beginning of work	Before the end of work
1	7	4
2	8	6
3	6	5
4	7	5
5	5	5
6	6	4

Change of Joint Position. Figure 1 showed the change of joint position during repetitive activities, and there were two conditions: repetitive activities with maintaining a certain joint angle, and with regularly change of the joint angle, as detail was showed in Table 4.

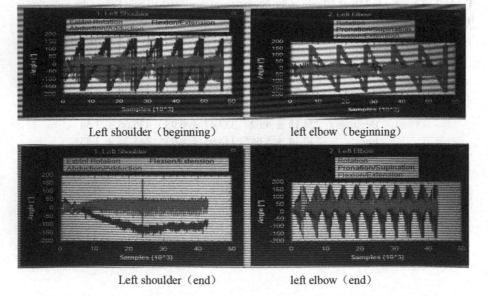

| Left shoulder（beginning） | left elbow（beginning） |
| Left shoulder（end） | left elbow（end） |

Fig. 1. Change of joint position of a subject

The Relation Between Muscule Load and Joint Position It was indicated that dynamic load level, and fatigue increased with increasing RMS and decreasing MF. According to the changes in MF, RMS and motion frequency analysis, fatigue degree of subject 1, 2, 4 and 6 was the highest. Fisher's exact test result showed that the change of joint position had significant influence on the degree of fatigue ($P = 0.01 < 0.05$).

Table 4. Changes of joint position in 3D space

Subject	Joint	Directions	Angles
1, 2, 4, 6	Shoulder	X	0°–60°
		Y	70°–120°
		Z	20°–50°
	Elbow	X	0°–70°
		Y	0°–40°
		Z	0°–180°
3, 5	Shoulder	X	0°–60°
		Y	0°–180°
		Z	0°–110°
	Elbow	X	0°–70°
		Y	0°–40°
		Z	0°–180°

4 Discussion

Fatigue is an acute effect of the muscles during the work. Skeletal system damage could be caused if long-term repetitive muscle load can't be completely recovered, and eventually leading to WMSDs.

The time-domain index RMS and frequency-domain index MF of sEMG signal are often used to study the myoelectric load. RMS mainly reflects the number of units participating in exercise during muscle activity, the types of participating in exercise units, and the degree of synchronization. MF mainly reflects the conduction speed of action potentials in muscle motor units, and types of the unit. The RMS increases and the MF decreases with muscle fatigue. In this study, for repetitive activities, the combination of RMS and MF indicators is used to analyze the changes in the load, and it can be accurately determined that the change of the sEMG signal was caused by fatigue.

Inertial motion capture technology was used to collect three-dimensional joint position changes of working postures in real time to explore the effect of work posture changes on the degree of fatigue during the repetitive operation. Repetitive activities with maintaining a certain joint angle, and with regularly change of the joint angle were more likely to cause fatigue, and the results were statistically significant. Therefore, in the actual practice, above work posture should be avoided and the working time should be reduced.

During the repetitive work of the employees in this study, the RMS of the sEMG signal increased, and MF decreased, meant that the workload and the fatigue increased with work. Meanwhile, the trend of sEMG signal detected at the shoulder was more obvious during the operation, showed that shoulders were most prone to fatigue damage during work procedure. The main reason for this was the fixed work postures and the repetitive stress of arms. Related studies have shown that arm extension and abduction are one of the important causes of shoulder fatigue. Therefore, how to take appropriate ergonomic design to prevent and control the WMSDs in the shoulder should be paid attention to.

At present, the research on workload is mainly focused on the use of sEMG technique to analyze and evaluate static load. Evaluation of dynamic load in actual factories field is a problem. In this study, dynamic load was analyzed by using motion capture instrument and surface electromyography technology, and results would be statistically significant, but the number of samples was less, and results could be contingency. Further study should be performed to approve the reliability and repeatability of research method.

5   Conclusion

The RMS and MF changes of the sEMG in this study showed that repetitive activities could cause fatigue, and the shoulder is the mostly injured in the repetitive push-pull operation.

It is feasible to analyze dynamic load by using motion capture instrument and surface electromyography technology. And this research results could provide reference for the study of dynamic load.

Introduction and Testing of a Passive Exoskeleton in an Industrial Working Environment

Steven Amandels[1,2(✉)], Hans Op het Eyndt[2], Liesbeth Daenen[3,4], and Veerle Hermans[5,6]

[1] University College Odisee, Brussels, Belgium
steven.amandels@idewe.be
[2] Department of Ergonomics, Group IDEWE (External Service for Prevention and Protection at Work), Belgium Sciences Vrije Universiteit Brussel, Brussels, Belgium
[3] Knowledge, Information and Research Center, Group IDEWE (External Service for Prevention and Protection at Work), Louvain, Belgium
[4] Department of Rehabilitation Sciences and Physiotherapy, Human Physiology and Anatomy (KIMA), Faculty of Physical Education and Physiotherapy, Vrije Universiteit Brussel, Brussels, Belgium
[5] Department Head of Ergonomics, Group IDEWE, Louvain, Belgium
[6] Department of Experimental and Applied Psychology, Work and Organisational Psychology (WOPS), Faculty of Psychology and Education, Vrije Universiteit Brussel, Brussels, Belgium

Abstract. *Background:* Commercial availability of passive exoskeletons, based on the use of springs and durable in an industrial setting, made the these devices much more accessible. However, there is limited evidence of the effect of these devices on physiological aspects, factors outside the single-task regulated laboratory environment and, discomfort and acceptance by the employee. *Objective:* This study aimed at evaluating the effect of a passive exoskeleton on muscle activity, kinematics and acceptance by employees on multi-task workstations on the shop floor. *Methodology:* Starting with an individual instruction course, nine participants were encouraged to test the device (Laevo, V2.4) during three weeks. After this period, motion and surface EMG were measured during two consecutive periods of executing daily tasks for at least 30 min (with exoskeleton and without exoskeleton). *Results:* Results show that RMS values were significantly higher for the Trapezius muscle with the exoskeleton (Mdn = 44.02) compared to the measuring period without the device (Mdn = 34.83, T = 0, $p < 0.05$, r = −.73); No differences were found for Erector Spinae and Biceps Femoris muscle activity. Participants reported significantly higher discomfort scores for the upper back/chest and thigh region with the exoskeleton (both $p < 0.05$, r = −.68). *Conclusion:* Results show high discomfort scores and no difference in lower back muscle activity possibly as a result of an inadequate amount of resistance given by the exoskeleton while bending over.

Keywords: Passive exoskeleton · Muscle activity · Discomfort Acceptance · Industry

© Springer Nature Switzerland AG 2019
S. Bagnara et al. (Eds.): IEA 2018, AISC 820, pp. 387–392, 2019.
https://doi.org/10.1007/978-3-319-96083-8_51

1   Background and Objective

An exoskeleton can be described as a wearable, external mechanical structure that can augment the strength of the wearer (active exoskeleton) or lower the impact of a movement on the human body (passive exoskeleton). New types emerge frequently and their commercial availability made the use of these devices much more accessible for employers. Employers are seeking for opportunities to support employees obtain a job with acceptable and feasible working conditions, in order to work 'longer' and reach their retirement in good physical condition and wellbeing. For industrial employees, different types of passive exoskeletons (from support of the upper limb or the lower back during overhead or heavy work to full support that makes manipulating heavy tools almost effortless) are available.

Based on results of previous studies, exoskeletons are promised to be applicable in a large variety of physically demanding tasks with both dynamic or static muscle activity. A review of de Looze et al. (2016) found 40 papers describing 26 different exoskeletons, of which only 13 studies evaluate the effect of exoskeletons on physical load (e.g. muscle activity). Beneficial results in muscle activity of the Erector Spinae muscle ranging from 10%–40% (depending on the task being executed and the dynamic or static characteristics of the task) are shown. In addition, Bosch et al. (2016) found for a dynamic and static task an effect of 35% and 38% on the Erector Spinae muscle activity, with a minimal degree of discomfort experienced by the participants. These results indicate a positive effect of a passive exoskeleton to reduce physical load.

Exoskeletons should apply to three important milestones. First, a mechanical working device is necessary. Secondly, its effectiveness has to be evaluated in a controlled environment. Thirdly, measurement of end-user's perception and usability of the device during physically demanding tasks has to be done. However, only a limited number of existing exoskeletons meet these criteria. Tasks conducted in controlled lab settings differ dramatically from those in a more complex, multitasking workstation on the field. This makes transferring of lab results to the shop floor quite difficult.

To our knowledge, only one study evaluating the effect of a passive exoskeleton on the physical load and its usability and feasibility on the workfloor is available (Graham et al. 2009). A 15% lower Erector Spinae activity during a car assembly task was found for the PLAD exoskeleton, with very high comfort scores. It could however be questioned whether an effect of 15% can be perceived by employees and is enough to motivate them to wear an exoskeleton. Furthermore, despite efforts in preserving freedom of movement, it is doubtful whether employees are motivated to wear a device when they cannot feel its effect.

This study aimed at evaluating the effect of a passive exoskeleton on muscle activity, kinematics and acceptance by employees in a multitasking workstation on the shop floor. It is hypothesized that activity in the muscles of lower back and knee joint movement decreases due to the use of a passive exoskeleton. Secondly, it is hypothesized that usability and acceptability are high after testing the device on the shop floor.

2 Methodology

2.1 Study Population

Ten employees from a press and shear department of a multinational manufacturer volunteered to the study. The management and union representatives of the participating organisation pursued an introduction session about the working mechanism, evidence and possible implications of the passive exoskeleton and the aim of the study. The shop floor teamleaders participated in an individual instruction session in order to feel and test the exoskeleton and choose two workstations that are most applicable. Participants were informed about the aim of the study and both workstations by their teamleaders. Informed consent was obtained from all participants.

2.2 Study Procedure and Materials

After a medical history, participants received an individual instruction session on the use of the exoskeleton Laëvo V2.4. Participants were motivated to test the passive exoskeleton during three weeks at 2 different workstations. The tasks executed on these workstations consisted of frequent far reaching with bending over to place or collect items. During the testing phase, periodical comments of the participants and their team leaders were collected.

After three weeks of using the exoskeleton at a self-appointed frequency, motion and surface EMG were measured during two consecutive periods of at least 30 min, one with and one without the exoskeleton, using the CAPTIV software (TEA). Surface EMG electrodes were placed and fixed on the dominating side on m. Erector Spinae, m. Biceps Femoris and m. Trapezius pars descendens, according Seniam Guidelines. After each period of at least 30 min with and without the exoskeleton, a body map with 9 locations and a 10-point Likert-scale ranging from 0 (no discomfort) to 10 (extreme discomfort) was shown and participants were asked to score their discomfort. At the end of the second period, participants were asked to fill in a structured user-experience questionnaire (Laugwitz et al. 2008), evaluating user's perception on the attractiveness, perspicuity, efficiency, dependability, stimulation and novelty of the device. Participants had to score 26 questions on a 7-point scale ranging from −3 to +3. Perception was rated as negative (below −0,8), neutral (between −0,8 and +0,8) or positive (above +0,8).

Wilcoxon signed rank test by means of the SPSS 23.0 software package (SPSS Inc; Chicago, IL) were used to test the study hypotheses.

3 Results

3.1 Participants' Characteristics

Ten employees were invited to participate in the study, and nine men agreed to do so. One participant did not participate because of sick leave at the test moment. The mean age and BMI of the participants was respectively 45.6 (SD 11,64) and 26.9 (SD 2,78). All participants had a history of low back pain, with 6 out of 9 having complaints during the last 12 months and 3 out of 9 during the last 7 days. Eight participants related their low back pain to work-related tasks.

3.2 Muscle Activity of M. Erector Spinae, M. Biceps Femoris and M. Trapezius Pars Descendens

A statistically non-significant reduction of m. Erector Spinae muscle activity can be seen as a result of wearing the exoskeleton ($-12,02\% \pm 0,21$, $p = 0,68$), presented in Table 1. Large differences between participants are found. No difference was observed for m. biceps femoris. ($p = 0,26$) A significant higher muscle activity was observed in m. trapezius pars descendens using the exoskeleton ($T = 0$, $p < 0.05$, $r = -.73$).

Table 1. Erector Spinae muscle activity as a result of wearing the passive exoskeleton

Participant number	Workstation number (1/2)	Difference in RMS* value (%)
1	1	−27,92
2	2	+2,94
3	2	−1,76
4	1	−56,02
5	1	−10,48
6	2	+8,20
7	1	−8,39
8	1	−2,73
9	1	/**

* RMS: Root Mean Square
** No measurement was obtained because of loose electrode

3.3 Low Back and Knee Motion

Results show higher relative duration of detrimental back rotation angles ($p < 0.05$, $r = -.74$) and knee flexion angles ($p < 0.05$, $r = -.67$) during tasks when the exoskeleton is used.

3.4 Discomfort

After each measuring period (with and without exoskeleton) participants were again asked to indicate locations of discomfort and again rate between 0 and 10 for each location. Five participants experienced discomfort at the chest and thighs while wearing the exoskeleton (with an average score of 6,6/10 for the chest and 4,6/10 for the thigh), compared to 0 participants for the chest and 1 participant for the thighs (average score of 4/10) without the exoskeleton. Higher significant discomfort scores wearing the exoskeleton were found ($p < 0.05$, $r = -.68$ and $p < 0.05$, $r = -.68$) for both locations (chest and thigh), compared to not wearing the exoskeleton.

3.5 Effect on Users' Experience

Participants rated their perception of the exoskeleton as 'neutral' on aspects of attractiveness, perspicuity, efficiency, dependability and stimulation (average score -0.17, $+0.17$, $+0.28$, -0.14, $+0.00$), but were 'positive' about the aspect of novelty (average $+0,92$).

4 Discussion and Conclusions

Although the passive exoskeletons have evolved rapidly over the past few years and are already commercial available, introducing them to the workfloor is still not evident. Despite the intensive sensibilisation of employees and their team leaders, the individual instructions for each participant and the close guidance during the test phase of 3 weeks, results on reduction in muscle activity and awkward postures are mostly not significant. The perception of discomfort as a result of wearing the exoskeleton was high and overall user experience was neutral.

Results in this study showed high variations in relative difference of surface EMG of the m. Erector Spinae pars longissimus between a period of using the exoskeleton compared with not using it. Previous studies demonstrated a reduced muscle activity during the execution of controlled dynamic or static tasks while wearing the exoskeleton in a controlled lab environment (de Looze et al. 2016) up to 38% (Bosch et al. 2016). The job responsibilities of a modern factory floor worker are, however, more versatile implicating a great variety of postures and movements. Our results revealed a (not-significant) average decline in Erector Spinae muscle activity of 12% while using the exoskeleton, which is in line with the study results of Graham et al. (2009). During a car assembly task, they found a decline of 15% in Erector Spinae muscle activity. The exoskeleton tested in our study gives a support of 16 kg taking into account an upper body weight of 35 kg. As our male study population had a higher overall body weight than participants in the study of Bosch et al. (2016), a lower effect could have been predicted.

Strangely, a significant increase in activity for the Trapezius pars descendens muscle compared with the situation without exoskeleton was found. This suggests a prior inclination towards reaching with the shoulder instead of immediately bending over in the lower back to reach for an item. This was the case in 8 out of 9 tested participants. In a recent study analyzing the effect of an exoskeleton for the upper limbs regarding reduction of physical load, also a transfer of load was found to other body regions (Weston et al. 2018). It is important to analyze such compensation mechanisms in future research.

Although comfort is a primary aim for the manufacturer, participants still pointed out high rates of discomfort located at the chest and thighs being the physical contact points of the device. These discomfort scores were much higher as experienced by study participants in a lab situation in the study of Bosch et al. (2016) but also as on the shop floor by Graham et al. (2009). First of all, discomfort can be quite different between a lab environment and a factory floor where noise, heat and/or production-driven stress causes participants to experience frustration more quickly. Secondly and in contrast to car assembly, there is no immediate feedback given to the employees about the status of their work related to the goals that were set, resulting in a high working pace.

The overall user experience indicates that participants are only distinctly positive about the originality of the device appreciating the aspect of the factory having an open mindset for new aids.

In interpreting the results, we should consider the effect of a limited frequency in testing before the surface EMG measurements. As the participants tested the exoskeleton during 3 weeks at a self-appointed frequency, limited testing could have resulted in a limited predictability of the exoskeleton where muscle activation patterns were possibly 'overshooting' the need to exert force.

It is important to notice that the company found it very important to test and introduce new helping devices to reduce physical load, with the contribution of the workers However, the tested passive exoskeleton didn't perform as expected regarding its effect on reducing the lower back muscle and its comfort towards the employee. Although the manufacturer aims to increase user comfort with every new version of the exoskeleton, there is still a large gap between the effects in a controlled lab environment and the factory floor.

References

Bosch T, van Eck J, Knitel K, de Looze M (2016) The effects of a passive exoskeleton on muscle activity, discomfort and endurance time in forward bending work. Appl Ergon 54:212–217

de Looze MP, Bosch T, Krause F, Stadler KS, O'Sullivan LW (2016) Exoskeletons for industrial application and their potential effects on physical work load. Ergonomics 59(5):671–681

Graham RB, Agnew MJ, Stevenson JM (2009) Effectiveness of an on-body lifting aid at reducing low back physical demands during an automotive assembly task: assesment of EMG response and user acceptability. Appl Ergon 40(5):936–942

Laugwitz B, Schrepp M, Held T (2008) Construction and evaluation of a user experience questionnaire. In: Holzinger A (ed) HCI and usability for education and work. Springer, Berlin, pp 63–76

Weston EB, Alizadeh M, Knapik GG, Wang X, Marras WS (2018) Biomechanical evaluation of exoskeleton using on loading of the lumbar spine. Appl Ergon 68:101

Ergonomics Introduction and Management of Risk to Biomechanical Overload in a Mechanical Engineering Factory Production Chain Saws and Trimmers

Marco Placci[1,2(✉)]

[1] EPM IES - Ergonomics of Posture and Movement International Ergonomics School, Milan, Italy
marco.placci@libero.it
[2] Studio Eur. Erg. Ing. Marco Placci, via Donati 4, 48018 Faenza, Italy

Abstract. The biomechanical overload risk, especially in the upper limbs and the spine, resulting in Italy one of the most significant security issues for manufacturing companies due to possible occupational diseases to which it may be related.

Already for several years the most careful companies consider necessary proper management of this risk which involves all corporate entities and which, following an adequate risk assessment, addresses the main variables that determine a concrete exposure by the employees. As the risk is multifactorial, only with the involvement of all the factory entities, a gradual reduction in exposure conditions is possible.

The management process started in 2014 by EMAK SPA, the leading manufacturer of industrial tools for greening and gardening, initially saw the involvement of the Factory Production and Process Engineering in a thorough training and operational process.

This produced in 2015 a specific procedure defined "of policy and guidelines" for the design of workstations with ergonomic features, suitable for the entire working population, even in view of its gradual aging.

In the following years, training was given to workers on specific biomechanical overload risk that included behavioral routes to avoid extreme exposure conditions.

In 2016, the Company's Management strongly wanted department of industrial design to be part of the project, by also proposing changes in the design of the models sold, that would improve the working conditions of the assembly lines.

Finally, in 2017, the involvement of the Medical Worker was concretized, which together with the company's workgroup identifies the critical conditions still present in every position of the company to include not only workers with normal characteristics, but above all, workers with reduced capacity working or advanced age.

Over the years, EMAK's "Ergonomics Project" has allowed to improve workplaces and reintroduce about 50% of workers with restrictions.

© Springer Nature Switzerland AG 2019
S. Bagnara et al. (Eds.): IEA 2018, AISC 820, pp. 393–403, 2019.
https://doi.org/10.1007/978-3-319-96083-8_52

Keywords: Biomechanical overload risk · Risk assessment in the company
Improvement activities in the metal mechanics factory
Management of biomechanical overload pathologies

1   Emak Project "Ergonomy in the Manufacture"

1.1   Business Presentation

The Biomechanical Overload risk, is responsible or co-responsible for a significant number of musculoskeletal disorders and diseases of professional origin. It can be re-dimensioned through a correct risk assessment and a subsequent planning or redesign of the work stations. The study of the ergonomics of workplaces in fact has as its main objective that of "contributing to the design of objects, services, living and working environments, which respect the limits of man and strengthen their operational capabilities". The Emak Group operates on the world market, directly managing distribution in Italy and in ten other foreign markets - United States of America, France, Germany, Great Britain, Spain, Poland, Ukraine, China, South Africa, Chile and Brazil – through offering a wide range of products with recognized brands and addressing a highly diversified customer target.

Outdoor Power Equipment includes the development, manufacture and marketing of gardening products, forestry and small agricultural machinery, such as brushcutters, lawnmowers, tractors, chainsaws, rotary tillers and rotary cultivators. The Group distributes its products with the main brands Oleo-Mac, Efco, Bertolini, Nibbi and Staub (Fig. 1).

Fig. 1. Production units, commercial and commercial branches in the world.

1.2   The Introduction of Ergonomics in the Production Lines

The management of the Bagnolo in Piano production plant - Reggio Emilia - Italy, which has long been sensitive to issues related to the management of safety and health of workers, noted the need to face, in a preventive manner and following the indications of the Occupational Physician, the issues related to the Ergonomic conditions of work activities and those resulting from the gradual aging of the operational staff, has decided to set up in 2014 the "ERGONOMIA IN LINE" PROJECT.

The main purpose of the project was to create an interdisciplinary company group able to face the problems related to the Biomechanical Overload of the various body districts through a specific training of its technicians and managers, resulting from an inadequately designed working methodology. Initially the working group was composed of Technicians of Production Engineering and of Production Staff figures. Following a specific propaedeutic training, the group, assisted by the Ergonomist specializing in issues of Biomechanical Overload, ing. Marco Placci had the task of using the knowledge acquired to propose "guide" solutions for the design and construction of workstations and assembly lines.

The trade union part, in the persons of the Workers' Safety Representatives was directly involved in the project, both in the training process and in the operational phase of determining the Ergonomic rules for designing workstations. The process started in 2014 has seen a differentiated development in recent years:

(A) **First phase 2014 - training objective** of the Technical Working Group: within the company the Management set up a "technical work group" designating 11 figures belonging to different Plant Entities such as: Production, Technologies (Engineering of Process), the Prevention Service, the Warehouse to follow three basic modules of Ergonomics, organized according to the guidelines of the International School of Ergonomics of EPM - Research Unit Ergonomics of Posture and the Milan Movement with the following contents:

A. **Risk by repetitive movements** - check list ocra evaluation method - 16 h
B. **Risk of manual load handling** - niosh assessment method - 16 h
C. **Ergonomic re-placement of workplaces** - international orthodontal and biomechanical standards - 16 h

The three modules, attended by all the members of the working group, were planned and carried out in June, July and September 2014 and were given to carry out the second phase of the project.

(B) **Second phase 2014 - operational objective**: following the preventive training, and through the Evaluation of Risks by Repetitive Movements and Manual Handling of Loads (OCRA and NIOSH Method), the working group was involved in determining the unsuitable working conditions in workstations at assembly lines and to define intervention priorities mainly from a purely ergonomic point of view.

The resolutive approach to reduce or, if necessary, eliminate the negative conditions detected was subsequently carried out taking into account the multidisciplinary nature of the problems considering three different approaches:

(1) Structural/design: involving the geometric and physical factors that determined the constitution of the biomechanical overload risk in the assembly lines;
(2) Organizational: regarding the lack of recovery times or rotations on several positions, in particular for the upper limbs and the correct quantity and placement of the exposure pauses from the risk in question;

(3) Training for workers: since the assumed improvements must also be considered as those carried out by the direct manufacturing activity. It was necessary to transmit both the main risk concepts and the correct behavioral modalities proposed and introduced in the improvement actions at all workers present in the plant (about 200 units overall);

The proposals and suggestions made by the working group were an active part of the study of improvements and changes to the line stations and contributed to the design of new equipment to allow workers a better postural/productive condition. For the dissemination of optimal working conditions and to allow the execution of activities even for a prolonged time.

2   Individuation of the Pilot Line for Evaluation and Re-Design

The production of the plant subdivided into brush cutters, chainsaws, motorized tractors, blowers/atomizers, motor pumps etc. has been analyzed to determine the intervention priorities on the most recurring models using the production forecast for 2014 (about 155,000 pieces) (Table 1):

Table 1. The lines present are summarized in the table below

Chainsaws	n. 4
Brush Cutters	n. 3
Motor pumps/motorized drills	n. 1
Motorized truncators	n. 1
Motorized hoes "Bartolini"	n. 2
Shredders	n. 1
Big cultivators	n. 1
Transmission line - subgroup	n. 1
Preassembly - cultivators	n. 1

Each line is divided into stations (from 6 to 10) numbered in sequence.
In the organization of the plant there are also pre-assembly areas:
- Valves; - Envelopes supplied; - Spare parts kit; - Tank pre-assembly.

To define the priorities of intervention to ensure that the placement of participants in the training was useful in practice as well as in the formation, was performed and analyzed the risk mapping from Repetitive Movements which showed that the highest values of risk were relevant to the product brush cutter and in particular for the DCS43 model. Both for this obvious reason and because it represented 41% of the production of the plant and could be found in production on three different lines, the brushcutter and, in particular, the DCS model 43 were chosen for the analysis and redesign evaluations. The members of the working group were subdivided into 5 subgroups to which the line stations provided for the assembly of the DCS 43 were assigned.

The objectives of the five groups related to the assigned positions were:

(1) Verification of the existing evaluation through the execution of new films and comparison with the existing evaluation. Calculation of the value of Check list and Lifting Index.
(2) Determinations of critical variables that raise risk values. Identification of unsuitable working conditions.
(3) Determination of possible structural improvement actions for workstations and equipment.
(4) Control of the activities performed and of the hypothesised improvements.
(5) At the end of the structural interventions proposed of the possible activities of improvement of organizational typology for the overall reduction of the risk indexes.

2.1 General Analysis of Problems Detected

The study of the assembly of the DCS 43 brushcutter in the stations 10, 20, 30, 40 and 60 through the verification of the films performed by the working groups, did not reveal substantial differences to the guidelines on the respect of the Ergonomic Principles, but only some recurring conditions that could be correlated with long-established and unaddressed operating procedures as considered non-determinant.

For the redesign of activities and lines, we are therefore directed to the presence or absence of the determinants of risk such as the high frequency of technical actions, the presence of force, the postures incongruous due to improper equipment or lay outs. Although planned, the improvement activities did not include the organization of the days and breaks as they focused mainly on the technical variables.

The improvement activities determined for the workstations were divided between specific location conditions and common cross-cutting problems. In order to tackle the topics dealt with in a multidisciplinary manner, the Research and Development Agency was also involved in presenting the main themes on which it is supposed to continue with the study:

(1) The high number of screw types, both for the thread and the head. Consequently this forces the use of different screwdrivers with different inserts and tightening torques at the line stations;
(2) The use in all the stations of locking product that increases considerably in pinch of the hand and frequency of technical actions;
(3) Common to all the trimmer models is the difficulty of mounting the polyurethane tube for the fuel;
(4) A problem has been detected on almost all the stations determined by the positioning of the screwdrivers, placed too far behind the working point defined by components and templates on the bench.

With the support of the Plant Management the improvement activities were then divided by competence: those concerning the work stations were followed by the Production Engineering (Technology) and the Assembly Department, while for all the problems concerning the product the Management considered the involvement of the Research and Product Development Agency (Product Design) to be necessary.

In this advanced phase of the project the Occupational Doctor was also involved, to design, on his advice, workstations suitable for the reintegration or relocation of workers with limited work skills, and to allow the production of adequate management of these workers.

Following the Kaizen methodologies, four technical families analyzed were identified: Means/working environment; Method; Facility; Materials.

From these observations and considering the fact that the postural problems, but also those that identified variables specific to the Risk from Repeated Movements, were mainly related to the workplace the company first studied a prototype workstation that would be adaptable to the measures anthropometric of each worker and subsequently produced an internal procedure that would be mandatory for the construction of new lines or work stations.

It is however necessary to clarify that the main objective of the procedure and of the technical guidelines has been directed to an adequate planning of the workplace to guarantee the health of workers in a mid-term and long-term vision. In fact, it is undeniable that the ergonomic approach and the choices of adaptability of the positions can be advantageous choices for the work force that gradually will be found in old age and that will require specific aids to guarantee, however, an adequate productivity.

Finally, also considering the inclusion of new staff, the presence from the beginning of the employment relationship, of equipment and enslavements that ensure the maintenance of a low level of risk of the biomechanical overload can be, together with an adequate training in the field that its use and consequent control are explicit, an optimal prevention condition to avoid the onset of directly related diseases.

2.2 Drafting the Safety Procedure

The internal security procedure called "ERGONOMIC DESIGN OF THE STATIONS AND WORKING LINES" was formally presented to the management on 03/08/2015.

The procedure defined the conditions for attacking the individual risk-setting variables and were effective in the overall reduction of problems due to biomechanical overload considering the indications of the Standards EN 1005-2, EN 1005-3, EN 1005-4, EN 1005-5.

For the specific design indications, the ISO 14738 standard concerning the design of industrial workstations was used, while the ISO 11226 standard was used to comply with the postural conditions.

Using all the indications collected, the working group has defined the project for a prototype of a modular work station that can be considered the main block for the block realization of a production line in conformity with the International Ergonomic Standards.

The workstation has been implemented in practice by the company's technical staff with the support of the Certified Ergonomist and for particular processing through the contribution of an external supplier.

The table, has been assembled following the proposals made of adaptability with respect to the anthropometric differences of the single workers.

In particular, the adaptability refers to the height adjustment, which according to the indications of the ISO 14738 Standard, for a table to be used by a standing worker, with operations that do not require excessive precision, must have an adjustable height between 103 and 129 cm (standard excursion of 265 mm).

The built table has an area of 160 × 38 cm. The minimum height from the ground is 76 cm and the maximum height of 103 cm. The height adjustment provides a sliding device to allow different heights of the worktop. The adaptability of the work table has been tested by having several workers try to complete the whole percentile spectrum for anthropometric measurements.

It should be emphasized that the table height values have been further implemented by a workpiece template of the assembly to be assembled, which allows 360° rotation of the semi-finished product.

The inserted support, about 25 cm high, brings the height of the adjustable table according to Norma's indications to adequate values.

To verify the correctness of the table heights, a high worker 185 was used, placed on wedges 14 cm high (total 194 cm equal to the 95th percentile), and a worker with a height of 155 cm (to be considered 5th percentile).

The tests considered the width, depth and height (adjusted) of the table to be adequate. The heights of the two collection shelves of the warehouse have also been studied, considering and revaluing the withdrawal depths so that the working areas of the workers are within the appropriate values of the Technical Standard for the postures of the upper limbs, both for the priority components and both for those of second order of importance. The same height measurements of the shelves have also been defined according to the boxes that will be placed in the workstation serving table containing the components necessary for assembly (rectangular blue plastic boxes, dimensions 145 × 90 × H70 mm, 210 × 148 × H122 mm, 310 × 210 × H145 mm). The first shelf of the warehouse was placed so that the edge of the boxes was not to be over-ridden by hand because it was higher than the work table.

2.3   Final Activities and Drafting of the Procedure

At the end of the operational test of the prototype realized, all the technical indications were subsequently collected within a company safety procedure entitled "PS1 W - DESIGN OF WORK STATIONS AND PRODUCTION LINES: GUIDELINE" divided into four paragraphs thus defined and exploded:

"Physical, methodological and organizational indications for the design of stations and lines for the assembly of products".

A. **Conception (design) of workplaces**

- Work table (dimensions, settings); - Frontal support structure; o Shoulder - incongruous conditions; o Elbow - incongruous conditions; o Wrist - incongruous conditions; - Platforms;

B. **Definition of work and production methods**

- Development of new products

C. **Choice of equipment**

- Piece holder dime; - Screwing tools; - Containers and loading units;

D. **Materials/products/components**

- Design choices;

The procedure was drafted and included in the Safety Management System Manual and was presented officially to the entire company management in a specific meeting held on 03 August 2015.

Subsequently, in a further meeting held at the beginning of September 2015, the procedure was presented to the union representatives and to the coordinators of all the production lines for its adoption and their applicative involvement.

2.4 Implementation of the Procedure and Training of Workers

Once the first work stations were created, according to the guidelines of the procedure, all the workers dedicated to assembly were instructed in a specific training session, in the classroom, on the issues of biomechanical overload and on the use of new equipment and new tables present in the assembly lines.

At the end of this session about 200 production operators trained for two hours were involved.

2.5 Product Issues - Involvement of the Planning Body

The assembly problems of some components were found to be detrimental to the biomechanical overload of the workers at the assembly lines. In some cases they determined high frequencies, use of force, incongruous postures.

The plant management at the beginning of 2016 then implemented the project requiring the involvement of the Product Design Organization to eliminate the existing obsolete components and to define new construction methods for the products to be placed on the market. In the first quarter of the year, therefore, all plant departments (16 members involved) underwent basic training on the Biomechanical Overload and in particular on how incorrect product concepts could characterize large problems for workers during assembly. Following the training, the design management requested to tackle the problems detected during the assembly phase analysis according to an improvement plan with the end of December 2016. During the three-year period 2017/2019, all the new models will be designed according to the ergonomics guidelines of the assembly. A specific meeting has been set up and organized that will involve the Design, Technology and Production that will monitor the progress of the projects and periodically verify the assembly conditions of the new models.

3 Health Surveillance and Reintegration of Workers with Problems

The Occupational Doctor (according to Italian legislation) has constantly followed the path taken by the company, intervening with advice and suggestions.

To allow the Production Organization to manage all the workers and to complete the project, from the beginning of 2017, periodic visits to the Prevention Service Manager, the Occupational Doctor, the Ergonomic Consultant, a Production representative and one of the Technologies/Process Engineering.

Particular attention was given to all workers with particular health problems that have been subdivided, by the Doctor, into families of homogeneous limitations. The positions were then classified according to the physician's guidelines and a "Mapping of non-suitability" was produced in order to adequately place the workers without aggravating their state of health but also with particular attention to the advancement of their age (Table 2).

Table 2. Table of exclusions per job position

OK T	suitable location for all operators even with limitations
A	Limited for handling weights - check weights present
B	Not suitable for operators with limitation to the district hand presence of pinch
C	Not adequate for the presence of incongruous posture of the shoulder and use of force
D	Not suitable for operators with pinch - palmar - wrist - force limitation

During the inspection it is expected that each workstation is analyzed with a critical eye to identify the correct classification and the families of workers with restrictions that can still be accessed. For the positions that are not adequate, always during the inspection stage, the technical department has the task of recording the criticalities detected and subsequently proposing improvement and modification actions to the existing workstations designed to eliminate the non-conformities detected.

4 Final Account and Dedicated Resources

The EMAK project officially started on 01/22/2014 and is still continuing. The striking result obtained today is that of the possible reintegration of personnel with limitations of difficult use, and in particular to have prepared assembly lines that adapting manually can contribute to the worker's well-being even considering the gradual aging that the company's staff will undergo in the future.

The financial disbursement addressed by the company is related to the costs of training and design and implementation of the prototypes of the stations.

Approximately summarize the commitments in dedicated hours relating to the activities carried out and to the dedicated figures.

TRAINING - They were given in total
Training for the Technicians and Production group (11 members)   528 h
Basic training for workers (200 operators)   280 h
Training for Research and Development staff (16 members)   384 h

EVALUATION AND IMPROVEMENT ACTIVITIES
Evaluation and verification visits   110 h
Research activities improvement actions   528 h
Prototype realization and modifications   72 h
Research and development design improvements   320 h
Drafting of the procedure and disclosure   480 h

SANITARY SURVEILLANCE MANAGEMENT ACTIVITIES (2017 only)
Surveys for verification of production stations   192 h

Over the years, EMAK's "Ergonomics Project" has allowed to improve workplaces and reintroduce about 50% of workers with restrictions.

MSDs: Recommendations for Prevention, Rehabilitation and Occupational Reinsertion – Results from a Survey by the Ergonomics Working Group of the ISSA Health Services Section

Jean-Pierre Zana[1]([✉]), Sigfried Sandner[2], Barbara Beate Beck[3], Martine Bloch[4], Stefan Kuhn[5], and Irène Kunz-Vondracek[6]

[1] ZConcept, Paris, France
jeanpierrezana@gmail.com
[2] BGW, Hamburg, Germany
[3] Forum fBB, Hamburg, Germany
[4] INRS, Paris, France
[5] BGW, Mainz, Germany
[6] Suva, Lucerne, Switzerland

Abstract. For most countries, musculoskeletal disorders (MSDs) are still a major occupational health concern. It is therefore in everyone's interest to have effective prevention and rehabilitation practices.

In order to compare the approaches of different countries for the prevention of work-related musculoskeletal disorders as well as the rehabilitation and reinsertion of workers with MSDs, the Ergonomics Working Group of the Health Services Section of the International Social Security Association has decided to carry out a questionnaire-based survey.

This questionnaire has been addressed to stakeholders that, directly or indirectly, are faced with the issue of work-related MSDs: accident insurance bodies, rehabilitation centres and their therapists, occupational physicians, human resource departments, professional organisations, trade unions, companies, etc.

The results have been analysed, they show a great disparity of modes of action both in prevention and rehabilitation. The Ergonomics Working Group of the ISSA Health Services Section will provide its conclusions and recommendations to advance the return to employment and job retention for these pathologies, which predominate in all countries.

Keywords: MSD · Prevention · Rehabilitation
Recommendation of prevention · International survey

© Springer Nature Switzerland AG 2019
S. Bagnara et al. (Eds.): IEA 2018, AISC 820, pp. 404–409, 2019.
https://doi.org/10.1007/978-3-319-96083-8_53

1 Introduction

In the context of its work, the Ergonomics group – "Manual handling of Patients" – wished to address a questionnaire to the "general public" to understand the methods used in different countries to prevent and recover from Musculo Skeletal Disorders (MSDs). The questionnaire developed was broadly communicated to healthcare professionals and relevant company representatives known to our group, and to some professional journals, in particular in physiotherapy (FMT Mag).

The study coordinated by Eurogip (1): "Musculo Skeletal Disorders - What recognition as occupational diseases? A study on ten European countries" (2016) indicated that even though the lists of occupational diseases (OD) are currently quite similar between countries, in terms of MSD, the data relating to cases of recognised MSDs show significant differences between European countries. The report indicated that these differences can be explained by how cases are considered by the social insurance organisations, and are not related to the MSDs likely to be handled. Indeed, numerous countries have opted for a system based on defining a definite causal link between disease and work; in this case, the list is only indicative. Scientific expertise and the opinion of the case manager are decisive. In other countries, the regulatory bodies use a list of OD which expresses a social consensus surrounding diseases which must be covered by the insurer. This system leaves little room for manoeuvre to determine the true link between the disease and work conditions, since the list includes diagnoses and presumed occupational exposure.

The study report underlines that applications of one or other model has little incidence in terms of volumes of recognition of diseases for which a link to occupational exposure has often been confirmed. However, it does affect recognition of multifactorial disorders such as MSDs. In the ten countries covered by the study, MSDs make up a very significant proportion of all recognised occupational diseases:

France 88%
Spain 75%
Belgium and Italy 69%
Germany, Austria, Denmark, Finland and Switzerland <20%

In light of these observations, it was important to extend the analysis by examining prevention and rehabilitation practices in the different countries to propose recommendations through which cases could be handled in such a way as to promote progress and improve workers' occupational quality of life.

Although the rate of response to the questionnaire developed by ISSA was low relative to the number of connections to the site hosting it, the respondents – through the quality of their answers and their professions – demonstrated good knowledge of their subject. Their professional practice leads them to prevent or engage in therapy for MSDs. Numerous professionals involved in rehabilitation responded, as did a large number of company practitioners.

Because of the diversity of respondents and the various origins of the responses, the data was complex to treat. Nevertheless, it was possible to conclude on a few directions for recommendations to help prevent and promote rehabilitation from MSDs.

2 Methodology

The questionnaire was prepared and published on a specifically-created site, the address of which was communicated to the various stakeholders to allow them to respond directly on-line. A cover letter presented the objectives and expected outcomes for the survey and was sent by e-mail to a panel of professional addresses communicated by the participants to the workgroup. The same e-mail was addressed as a press release to various French professional journals in the field of physiotherapy.

The questionnaire was available in three languages: French, German and English. It was split into two parts: the first consisted of 13 questions and related to MSD prevention; the second focused on rehabilitation through 19 questions.

Responses were analysed by the computer scientist who created the site, and the overall results were analysed by the workgroup.

3 Results on MSD Prevention

The First Part of the Questionnaire Related to MSD Prevention. After a few questions to gather information on the origin and quality of the respondents, the questions focused on MSD-prevention practices applied in each country, and if possible, specific to different activity sectors. The respondents were from 24 countries; 51 worked in the healthcare sector, 13 were employed by insurance organisations and 88 worked in a range of industrial and service sectors.

Risk Assessment
For nearly all European respondents, the European directives, transcribed into national law, were used to make risk assessment in companies obligatory (87%). Beyond the employer, who is named as responsible for risk assessment, there is apparently no person specifically named as responsible for this task.

Preventive Measures
Due to the large disparity of responses it appears difficult to identify a common methodology. Preventive measures, when implemented, were communicated by respondents in a very theoretical manner. Technical measures were most often mentioned, but adaptations of workstations or design of work zones appear to be catching up with equipment purchases.

In organisational terms, a large range of responses remained very theoretical.

Training
The titles of training courses or the brief descriptions received were not significant. There is apparently no consensus on the training methods used, probably due to the differences between the respondents' professions, but similar conclusions can be drawn from practices and the literature. Indeed, each country has its "own" methods, and no overall consensus is applied. A CEN/ISO standardisation workgroup did attempt to establish a consensus, through an ISO technical report on manual handling of patients (ISO TR 12296). This report resulted in a presentation of methods that are not linked to each other.

No consensus on the funding sources and durations of training was found either.

The Second Part of the Survey Related to Rehabilitation and Occupational Reinsertion.
It seems there is no single country that provides a specific national approach to rehabilitation of workers following a diagnosis of MSD. However, local initiatives by branch are being validated or developed in Germany and Switzerland, for example.

The prescriber remains the physician but it appears that the insurer or coverage system for each branch can request rehabilitation.

With regard to the institutions ensuring this rehabilitation, when offered, all the existing models have been cited, from handling by a medical practitioner in their own practice to outpatient or inpatient treatment in public or private rehabilitation centres. Given the low number of responses, it appears that rehabilitation approaches in the workplace are only used incidentally.

It is difficult to find a consensus on durations, funding and the programme for rehabilitation as the responses provided are very diverse.

4 Discussion

The number of responses for each individual country, except France (101 respondents), was insufficient to allow comparison of practices. However, not even the French responses reveal a consensus, and even less concordance is observed between countries.

The prevention-related questions provided a greater number of responses; they do not indicate specific approaches or major methodological principles. Although MSD-prevention campaigns have been launched in Germany and France, they were never mentioned in any response.

In the rehabilitation and occupational reinsertion questions, the differences between countries were even more striking. Rehabilitation is not considered in the same way, perhaps due to a lack of European harmonisation of initial and in-service training.

Given the professions of the respondents, from ergonomists to occupational physicians, Social Insurance technicians to occupational health and safety trainers, we would have expected more precise or pragmatic responses in terms of prevention. Does the lack of European and international consensus on the subject of MSD-prevention and the lack of training on occupational diseases in the courses for those consulted as specialists in re-education explain the different answers?

It would be therefore of great importance to communicate on the driving principles developed by ISSA - which brings together the Social Insurance organisations from more than 150 countries.

5 Conclusion

To attempt to implement an integrated approach to prevention, ISSA has developed internationally-recognised Guidelines relating to three fields: prevention of occupational risks, health-promotion in the workplace, return to work and occupational reinsertion.

The ISSA's Guidelines provide Social Insurance institutions with access to international standards on specific aspects of prevention, creating a basis of comparison when implementing administrative improvements. Each series of Guidelines is associated with complementary resources - references and links to examples of good practice - which can help extend knowledge and facilitate the application of the Guidelines.

The ISSA Guidelines in terms of return to work and occupational re-insertion relate to how the social insurance institutions can collaborate with other interested parties with a view to supporting people who are off work for medical reasons but retain a link with a specific employer. The return to work occupies a central place within a group of processes aimed at facilitating occupational reinsertion of those whose work capacity has been decreased as a result of an occupational or non-occupational accident or disease. Because they consider individual needs, the work environment, and the legal needs and obligations of companies, the process of return to work can be considered as a strategy aiming to promote continued employment, and as a first step to avoiding premature exclusion from active life of those with a reduced work capacity.

It is therefore appropriate to direct most efforts towards reducing occupational risks through preventive approaches and the promotion of occupational health. If necessary, emphasis should be placed on continued employment of those who are victims of a disease or accident and as a result are at risk of losing their positions, and suffering all the ensuing economic, social and psychological consequences.

This new orientation constitutes a paradigm shift, which as a result causes modern social insurance institutions to move from being "payers" to assuming a role of "active partner".

If all these aspects, which are mentioned in the ISSA Guidelines, are systematically taken into consideration, we can expect continuous improvement in terms of prevention. As a general rule, these aspects are integrated into a prevention strategy which sets objectives to reduce the number of occupational accidents and diseases within a precise timeframe, and defines a context of collaboration with others, including social partners and the authorities responsible for health, safety, rehabilitation and occupational reinsertion.

ISSA also carried out an international study to determine the potential return on investment in prevention for companies, entitled "Calculating the International Return on Prevention for Companies: costs and benefits of investments in occupational safety and health". Based on a scientific methodological approach, this study revealed that for every 1 euro (EUR) an employer invests in preventive measures in the workplace, they can generate a return of up to 2.2 EUR. Prevention thus has an exceptionally high return on investment of 1/2.2.

This result provides the systems insuring against occupational accidents with a powerful argument to convince workers and employers covered by their regimes to invest in prevention. The results of these studies also demonstrate the need for insurance regimes which only cover the cost of accidents to review their strategy and develop prevention programmes which reward companies obtaining excellent results in terms of occupational safety and health.

Like prevention of occupational risks, the potential advantages of return to work programmes are also considerable, not only for employers and companies but also for social insurance systems. The potential return on investments in return to work programmes would be around 1/2.4.

Reference

Musculo Skeletal Disorders - What recognition as occupational diseases? A study on ten European countries   EUROGIP - Ref. Eurogip-120/F - 2016 - ISBN: 979-10-91290-76-0 Paris

The Biomechanical Overload of the Rachis in Push and Pull Activities: Historical Revision, State of the Art and Future Prospects in the Light of the New High-Sampling Digital Dynamometers and the Multitask Features of Work in the Workplace

Marco Placci[1,2(✉)], Marco Cerbai[1], and Leonardo Bonci[1]

[1] EPM IES - Ergonomics of Posture and Movement International Ergonomics School, Milan, Italy
marco.placci@libero.it, cerbaimarco@safetywork.it
[2] Studio Eur. Erg. Ing. Marco Placci, via Donati 4, 48018 Faenza, Italy

Abstract. The risk of biomechanical overload for the rachis is characterized by two different exposure situations: lifting of loads and push and pull of carts or objects. If in the first case the risk is determined by the continuous and repeated compression of the intervertebral disks, in the second case the damage is generated differently by lateral-lateral and anteroposterior cutting forces acting on the disk.

The study of the risk caused by Towing and Driving activities was based on three main areas: epidemiological studies, psychophysical studies and physiological studies developed mainly since the 1970s and continued until today.

The most concrete reference to these investigations is, however, the famous study that uses the psychophysical approach conducted in the Liberty Mutual Research labs for more than 25 years, synthesized in 1991 by Snook and Ciriello and integrated by Mital et al. (1993).

Substantially, the psycho-physical study of Snook and Ciriello that generated the resulting tables is performed to determine how risk factors can influence the maximum acceptable forces during a working day. The problem with these studies is the correct identification by the operator of its real tolerance limit, and the absence of evidence of direct correlation with health disorders during work.

Keeping these references today, through sophisticated instruments, but with good price, is possible to obtain a significant amount of relevant information from the measurements that describes in a much deeper way the working conditions.

Last but not least, it is necessary to expand the calculation of the risk index in push and pull activities which must adequately describe the different working conditions that occur on the working day.

A proposal consistent with the formula N.I.O.S.H. of multiple tasks defines a computational scenario similar to the multitask, proposed by Waters Thomas, Putz-Anderson v. Garg A., End LJ, in the ergonomics publication "Revised niosh equation for the design and evaluation for manual lifting tasks", 37, 7, 1993 pp. 749–776.

© Springer Nature Switzerland AG 2019
S. Bagnara et al. (Eds.): IEA 2018, AISC 820, pp. 410–422, 2019.
https://doi.org/10.1007/978-3-319-96083-8_54

Calculation proposals should, however, be able to reflect actual worker exposure in push and pull multitask activities but may not yield results different from the values used so far in the Snook and Ciriello tables. For this reason, hundreds of work-related measures have been performed to compare historical values with those derived from new measurement modes

Keywords: Biomechanical overload risk for the rachis in push and pull activities
Push and pull conditions determined by professional activities
Operational indications for determining traction and thrust characterization variables · The dynamometer as a tool for measuring forces in push and pull
New possibilities for measuring with high sampling digital dynamometers and multitask calculations

1 The Activities of Pulling and Pushing in the Working Realities: Introductory Notes and Synthetic Bibliography

1.1 A Subsection Sample

The manual handling of objects or materials is present in practice in all work contexts.

The current legislation, also derived from the implementation of Directive 90/269/EC indicates that the Manual Movement refers to all those "operations of transport or support of a load by one or more workers, including the actions of lifting, depositing, push, pull, carry or move a load "(Legislative Decree 81/2008)".

Already in the mid-90s, different studies have said that during the working day, towing and pushing activities represent about half of the situations related to Material Handling. Moreover, towing and pushing activities are common in many industrial contexts, and have been the subject of various analyzes and studies:

- in warehouses,
- in distribution centers
- in urban waste collection
- in agriculture and construction
- in the cleaning sector (Marras et al. 2009), (Sogaard 1994; Sogaard et al. 1996)
- in the maintenance of gardens (Kumar et al. 1990; Kumar 1995)
- in the transport sector (Van der Beek et al. 1994),
- in firefighting (Gledhill et al. 1992; Nuwayhid et al. 1993),
- in healthcare (Estryn-Behar et al. 1990; Garg et al. 1991; Stubbs et al. 1983).

Approximately half of the manual material handling activities have been estimated to consist of towing and pushing activities (Baril-Gringas and Lortie 1995; Kumar, 1995), which usually find themselves replacing lifting and manual transport of objects, operations to which historically greater attention has been paid (NIOSH, 1981; Chaffin and Anderson 1991; Marras et al. 2009).

The studies that have had as an object of analysis the towing and thrusting activities say that they can increase the onset of musculoskeletal disorders, in particular those whose target organ is the lumbar spine, their physiological influence is however very different. with respect to lifting activities and this has led to an increasing need for an in-depth study of these issues in the last period. (Marras et al. 2009) (Theado et al. 2007).

In the literature there are studies that show how pulling and pushing activities can contribute to the risk of skeletal muscle disorders at the rachis (Hoozemans et al. 1998; Hoozemans et al. 2002a; Hoozemans et al. 2002b; Damkot et al. 1984; Frymore et al. 1980; Kelsey et al. 1990; NIOSH 1981; Snook 1978).

Furthermore, epidemiological studies have focused their attention mainly on the population of industrialized countries, such as the USA and Europe. In developing countries, the lack of automation generates a predictable increase in manual handling operations. It is therefore plausible that, in these countries, the prevalence of muscu-loskeletal disorders and consequently costs both for industry and society may be higher (Todd 2005).

However, uncontrolled rolling of heavy loads and objects can introduce new risks of injury. For example, such inadequately controlled actions could cause slips and falls.

Other additional risks are then hypothesized in the execution of activities with towing and pushing, in particular the onset of problems is accentuated by the non-physical and physical suitability of the workers. Despite this, the results of the studies indicate that almost half of the activities with manual handling include towing and pushing actions (Baril-Gringas and Lortie 1995).

2 Pushing and Pulling Actions: Definitions

2.1 The Definitions of the Bibliography

The aim of this text is to resume the indications of today's literature on the study of the application of human force during towing and pushing activities, as well as the identifi-cation of musculoskeletal risk factors associated with the use of this type of manual force.

There are few definitions that describe the application of human fatigue involved in pushing and pulling. This could derive from the notable complexity and variety of the bodily actions underway during the execution of this type of use of force.

Hoozemans et al. (1998), in a review of musculoskeletal risk factors associated with thrusts and tractions, he chose to use the definitions provided by Martin and Chaffin (1972) and Baril-Gingras and Lortie (1995):

"Pushing and pulling could be defined as the application of the force (of the hand), whose direction of the main component of the resulting force is horizontal, from someone on another object or person. In pushing the force (of the hand) it is direct far from the body and in pulling the force it is directed towards the body. ".. the same authors also specified that":

"To be called pushing or pulling force, the application of force does not have to be always directed horizontally, for example, when pulling a rope to start a mower's engine (Garg et al. 1988)"

The current definition of towing and pushing can certainly be obtained from the paragraph of terms and definitions of the ISO 11228-2 Standard:

Pulling: Human physical exertion in which the driving force is exerted in front of the body and directed towards the body itself, while the body is standing upright or moving backwards.

Pushing: Human physical exertion in which the driving force is exerted in front of the body and in the opposite direction to it the body is standing in an upright position or moving forward.

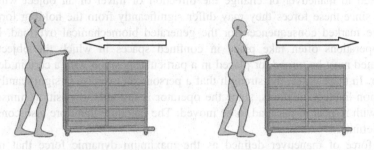

2.2   The Physical Quantities Involved

In addition to the classical definitions it is certainly necessary to also involve the study of physical quantities that come into play during towing and pushing activities, the main ones being:

- Applied force (initial and maintenance)
- Mass
- Acceleration
- Speed
- Static friction
- Dynamic friction
- Amount of motion

The main greatness present in the towing and pushing activities is strength. Strength is a physical quantity linked to the mass and acceleration of an object. The most known force is the weight force, it is in effect a function of the mass of an object and of the acceleration of gravity.

Gravity acceleration is the acceleration that a body undergoes, when not constrained, in a gravity field (for example, the gravitational field of the Earth). This acceleration is directed towards the center of the earth.

$$\vec{G} = m \times \vec{g}$$

In the same way a body that varies its velocity in space is necessarily subjected to a field of external forces, where, in place of the acceleration of gravity, the acceleration of the body itself will always be present.

In the area of towing and pushing, the interest is directly linked to determining the force to be supplied to a trolley to move it or keep it in motion.

$$F = m \times a$$

Likewise, for a complete analysis of the forces in place during the handling of trolleys and equipment, it would also be necessary to distinguish the thrust and traction forces used to maneuver or change the direction of travel of an object while it is moving, since these forces they may differ significantly from the holding forces and may have marked consequences for the generated biomechanical overload. Maneuvering operations often take place in confined spaces in which the object to be manipulated must be turned, or placed in a particular position with a certain degree of precision. In these cases, the strength that a person can exercise is significantly lower than in non-limited situations, since the operator is not able to position himself adequately with respect to the load to be moved. The literature therefore also considers a further definition:

The force of maneuver defined as the maximum dynamic force that must be expressed to change the direction or movement of an object (maneuvering force) Rodgers et al. (1986)

In most cases, towing and pushing activities will include a combination of each of these forces. For example, when moving a cart loaded with components, an initial force will be needed to put the carriage in motion and a prolonged force will be needed to keep the carriage moving. During the operation of the trolley, it may be necessary to maneuver around·machines or plants or to place the trolley inside a work area; at the end of the movement the trolley must be stopped.

Returning to the forces evaluated by the standard and the example just produced, the external force applied is correlated:

- to the mass of the cart
- forces opposed to motion: friction, soil and wheel conditions, these last conditions greatly affect the performance of the applied force.

Basically we can say that the Initial Force represents the force necessary to disturb the state of quietness of the trolley and to bring it in motion. It represents the external force level that the operator must apply directly on the handles or on the handle of the trolley to obtain the initial displacement.

The Maintenance Force, on the other hand, describes the force necessary to maintain, once achieved, the desired speed. In succession to the first phase of the movement, performed to overcome the initial quiet state of a trolley, a second phase intervenes which, theoretically, allows us to keep the speed reached constant.

Basically, when you want to move a cart, you initially move its state of quiet by imposing an initial force and then continue to keep the truck moving with the contribution of the holding force.

The initial force and the holding force therefore depend on:

- from the object to be moved, in particular from his MASS,
- the conditions of the working environment, the flooring, the wheels and their relative friction
- from the modalities with which the movement that characterizes the: speed.

If we look at the graph of speed variation with respect to time, we will notice a vertical line with an inclination proportional to the acceleration impressed on the carriage at the start. This curve leaves more or less gradually a horizontal line that instead describes the limited speed variation that takes over for the maintenance force.

3 The Risk of Pushing and Pulling Actions - Exposure-Damage Relationships

3.1 Considerations on the Bibliography

From what emerges from the epidemiological studies in the literature it is possible to conclude that the pulling and pushing activities are associated with rachis and shoulder disorders. There is no evidence for disorders at other body districts. (Garg et al. 2012)

None of the studies mentioned was able to provide exposure-damage relationships also because of the transversal modality used in the design of the experiments. In the future, to deepen this causal relationship, longitudinal epidemiological studies focusing on towing and pushing activities appear necessary. It may also be interesting to understand more precisely what the variables of the towing and pushing activities are that may be able to influence the onset of musculoskeletal disorders and what their contribution is in quantitative terms.

The results of psychophysical studies do not provide information on some potentially at risk body districts. (Resnick et al. 1995) show how, through perceived stress measures, the most stressed parts of the body during towing and pushing are the arms and legs rather than the rachis. The conclusion to which they arrive is that perceived effort measures may not be reliable in identifying possible risks to back health. The authors conclude that with a psychophysical approach, many aspects of towing and pushing operations that influence the maximum acceptable forces can be identified. These aspects can be considered as risk factors for musculoskeletal disorders when these forces approach or exceed the values of the maximum acceptable forces. Furthermore, the shoulders represent the bodily district seat of perceived higher subjective effort in the pulling and pushing operations.

The basic assumption of a psychophysical approach in determining the maximum acceptable forces is that an individual is able to define his maximum values and such as not to generate negative effects for his health. (Snook 1978; Harrin et al. 1986) showed how the reference value of the population to be protected was determined (75% for (Snook 1978), 90% for (Harrin et al. 1986)) and as a consequence of it the ways of carrying out the manual handling operations of the materials have been designed, the musculoskeletal disorders of the rachis can be concretely reduced.

There are no studies in the literature that have deepened the association between the maximum acceptable pulling forces and thrust and the risk of musculoskeletal disorders. Similarly, the definition of a relationship between the pulling and pushing forces exerted and the risk of musculoskeletal disorders in the spine and shoulders has still to be studied in a limited way. There is evidence that a certain combination of pulling forces and thrust, frequency, distance, height, etc., may be able to generate backside rachis disorders. What is not clear is what combinations of these factors can be. In the literature there is awareness of how much these studies are necessary, however, the objective difficulty of their realization in industrial contexts that are characterized by the simultaneous presence of lifting and deposition and pulling and pushing actions is also well known. In fact, it is very difficult to separate the causes and the respective weights. A careful design of the experiments in contexts where the lifting and deposition activities are not carried out or are carried out in a very limited way, while instead there are reference populations with both low and high exposure to pulling and pushing operations could be able to define the relationships between maximum acceptable forces defined with psychophysical criteria and risk of musculoskeletal disorders for the spine and shoulders.

Although it has been stated that the maximum acceptable workload is based on a subjective interpretation of biomechanical and physiological efforts (Ayoub 1992; Waters et al. 1993), the maximum acceptable forces determined with a psychophysical approach are overestimated at high levels. frequencies due to physiological stress that exceeds the acceptability values (Ciriello and Snook 1983; Ciriello et al. 1990; Snook et al. 1991). The result is that the maximum acceptable pulling and pushing forces must be corrected in such a way that the physiological criteria (33% VO2 max) are respected (Snook et al. 1991; Mital et al. 1993).

At present, the maximum acceptable forces determined with a psychophysical approach provide recommendations for practical application for the design and assessment of risk that appear valid until more consolidated data from biomechanical studies will not be able to bring out different indications (Garg et al. 2012).

The maximum allowable holding forces for certain combinations of frequency, distances and heights can result in excessive physical fatigue. (Snook et al. 1991) identified these combinations. However, as the extent of the reduction necessary to fit within acceptable physiological limits is not clear, attention must be paid to using this type of combination.

The methodologies adopted in determining the initial maximum acceptable forces for low frequency tasks (some times a day) appear to be inadequate. For the determination of the same, a biomechanical approach is suggested (Garg et al. 2012).

Studies that have investigated the physiological effects of towing and pushing activities are scarce and only a few risk factors can be studied by the influence with this approach. The influence of distance, the height of the hands from the ground, the weight of the trolley, the strength of the hands and the kind were studied occasionally as well as the difference between towing and pushing. The relationship between tow and thrust and local muscle activity and corresponding fatigue remains unclear.

There are a limited number of studies that have evaluated the oxygen demand (VO2) and the heartbeat (HR) for maximum acceptable forces in pulling and pushing operations determined with a psychophysical approach (Ciriello and Snook 1983; Snook et al. 1991; Ciriello et al. 1993; Dempsey et al. 2008). These studies have shown that the values of HR and VO2 may be too high for certain combinations of towing and pushing activities at certain frequencies and distances traveled. (Snook et al. 1991) identified the combinations of distance, frequency and height of hands from the ground that exceed the physiological criteria on 8 h (0.7 l/min for women, 1.0 l/min for men). In general the physiological criteria are exceeded at relatively high frequencies for a given distance. During the pulling and pushing of trolleys, the oxygen demand is influenced by, among other variables, towing and pushing force, body posture, towing frequency and thrust, speed, gender (van der Beek et al. 2000; Dempsey et al. 2008). At present the relationship between oxygen and force demand, speed, frequency, distance and body posture in towing and pushing operations is not sufficiently clear. Therefore, the extent of the reduction in the value of the maximum acceptable forces of driving and pushing to meet the physiological criteria is not clear. The validity of an indication of caution remains when the reference values of the maximum acceptable forces for the situations mentioned above are adopted (Garg et al. 2012).

A model of the human body with connected segments was used to calculate the net moments around the joints (de Looze et al. 1992). The net moment around the joints is the result of all the moments around the joints caused by the different anatomical structures. The net moments do not provide information on the mechanical stresses on the structures (e.g. compressive forces on the intervertebral discs). The studies conducted in the biomechanical field have highlighted the need for more detailed models able to distribute the net momentum on the different structures. Mono-muscular models were used to define the compressive forces on the intervertebral discs (Ayoub and McDaniel 1974; Lee et al. 1991; Gagnon et al 1992; de Looze et al 1995; Resnick and Chaffin 1995) but the validity of these models proved to be low for the modeling of towing and pushing activity. More detailed models are needed to determine the stresses at the shoulders, which also take into account the mechanisms of coactivation of the antagonist muscles (Van der Helm 1994 a, b).

Large pulling and pushing forces can generate great strain on both the spine and shoulders. There are few examples in the literature of a joint study of stresses from pulling and pushing operations both for the spine and for the shoulders (de Looze et al. 2000; Schibye et al. 2001; Hoozemans et al. 2004).

Possible design indications such as the height of the sockets or the maximum permissible forces should aim at minimizing stress both for the back and for the shoulders. From the biomechanics studies it emerges how much the combination of the different variables determine the onset or less of risk, what is not clear is what are the combinations of these variables that entail an increase in risk. For example, (Lett et al. 2006) recommend a gripping height above shoulder height to minimize compression and shear forces that urge the spine; (Hoozemans et al. 1998; Hoozemans et al. 2004) recommend positioning the hands at shoulder height to minimize the moments acting on the shoulders while keeping the wrists, elbows and shoulders as close to the line of action of the forces exerted.

Examination of risk factors related to the development of musculoskeletal disorders associated with towing and pushing activities according to the conceptual model of literature presentation based on Westgaard and Winkel (1996), Van Dijk et al. (1990) through the epidemiological, psychophysical, physiological and biomechanical perspective lead to the conclusion that scientific research focused mainly on the forces exerted, and therefore on the part related to external exposure. The relationship between pulling and pushing forces exerted and musculoskeletal disorders has not been sufficiently deepened although an increase in the forces exerted is plausibly associated with an increased risk of musculoskeletal disorders. Many factors that characterize the driving and pushing activities seem to have influence on the forces exerted and therefore their influence on health risks is probable (Hoozemans et al. 1998).

The internal exposure and the physiological response during the pulling and pushing operations have not been studied as extensively as the influence of the forces exerted. Greater attention should be paid to the mechanical load at the back and shoulders. The action limit proposed by NIOSH for compression forces on intervertebral discs at L5–S1 is based on in vitro studies and is assumed to be tolerated by the young and healthy male population (NIOSH 1981; Genaidy et al. 1993; Waters et al. 1993). Compression forces above the maximum allowed limit, as proposed by NIOSH, are associated with a significant increase in musculoskeletal disorders. The same considerations apply to the cutting forces at the back. Although (Garg et al. 1991) report higher values of shear forces in lifting tasks than in thrust operations, shear forces can contribute to the development of back disorders. However, little is known about the maximum values of shear forces that can be tolerated (Lamy et al. 1975).

In addition, these limits consider the effect of force peaks on the development of damage to the intervertebral discs. The effect of cyclic loads is still insufficiently understood (Brinckmann et al. 1987; Hansson et al. 1987; Kumar et al. 1990; de Looze et al. 1996).

4   Conclusion

To date, epidemiological studies have considered towing and pushing activities as a broad concept. Cross-sectional studies have shown association between towing and pushing operations and musculoskeletal disorders. It would be interesting to understand which aspects of towing and pushing activities can be considered potentially dangerous. From the research carried out it emerges how many factors can influence the risk of musculoskeletal disorders. However, if these aspects can be influential risk factors it should be explored with longitudinal epidemiological studies designed specifically for towing and pushing activities. These studies should include internal and external exposure measures in order to achieve dose-response relationships (Winkel et al. 1992).

A concrete result of this path could be to provide guidelines based on solid scientific bases for the prevention of musculoskeletal disorders able to provide answers to the typical operational contexts in which the towing and pushing activities are present in non-negligible terms.

The maximum acceptable forces, determined with a psychophysical approach, in the thrust operations are slightly higher than those of towing, this implies that the thrust actions are preferable to those of towing. These conclusions are not supported by evidence in biomechanical studies. At present it is unclear whether operators should be encouraged to favor push actions rather than towing.

The psychophysical approach is based on the assumption that the subjective perception of the maximum admissible effort on the part of the operator corresponds to a tolerance of the stress on the body of the subject itself (Snook 1978). However, the literature has shown that individuals are not, in general, able to correctly estimate the loads handled and that there is generally a tendency for them to overestimate their lifting capacity (Karwowski et al. 1992).

Another aspect of practical relevance concerns the definition of the optimal grip height. The differences in anthropometric characteristics between men and women and the non-concordance in studies of a biomechanical nature make it difficult to formulate reliable indications, especially if both the stresses at the spine and those at the shoulders are considered. At present the indications coming from psychophysical studies appear to be of practical importance and such as to be the expression of an integrated response by the operators. A contribution from biomechanics appears necessary. (Garg et al. 2012)

Many "simple" biomechanical models are static and their adoption makes it difficult to model dynamic and complex tasks such as towing and pushing. Also as a consequence of this the psychophysical tables were often used to determine the percentage of individuals able to carry out specific operations of pulling and pushing in safety (Le et al. 2012).

References

Abel EW, Frank TG (1991) The design of attended propelle wheelchairs. Prsthet Orthot Int 42:38–45

Al-Eisawi KW et al (1999a) Factors affecting minimum push and pull forces of manual carts. Ergonomics 30(3):235–245

Al-Eisawi KW et al (1999b) The effect of handle height and cart load on the initial hand forces in cart pushing and pulling. Ergonomics 42(8):1099–1113

Andres RO, Chaffin DB (1991) Validation of a biodynamic model of pushing and pulling. J Biomech 24:1033–1045

Astrand P, Rodahl K (1986) Textbook of work physiology: physiological bases of exercise. McGraw-Hill, New York

Ayoub MM, MaDaniel JW (1974) Effect of operator stance on pushing and pulling tasks. Am Inst Ind Eng Trans 6:185–195

Ayoub MM (1992) Problems and solutions in manual material handling: the state of art. Ergonomics 35:713–728

Baril-Gringas G, Lortie M (1995) The handling objects other than boxes:univariate analysis of handling techniques in a large transport company. Ergonomics 38:905–925

Bernard TE, Joseph BS (1994) Estimation of metabolic rate using qualitative job descriptors. Am Ind Hyg Assoc J 55:1021–1029

Berndsen MB (1990) Appliances for paviours: an evaluation of purchase and use. Ergonomics 33 (3):361–366

Boocock MG et al (2006) Initial force and postural adaptations when pushing and pulling on floor surfaces with good and reduced resistence to slipping. Ergonomics 49(9):801–821

Brinckmann P et al (1987) Fatigue fracture of human lumar vertebrae. Clin Biomech 2:94–96

Chaffin DB, Anderson GBJ (1991) Occupational biomechanics. Wiley, New York

Chaffin DB, Andres RO, Garg A (1983) Volitional postures during maximal push/pull exertions in the sagittal plane. Hum Factors 25(5):541–550

Chaffin DB, Anderson GBJ, Martin BJ (1999) Occupational biomechanics, 3rd edn. Wiley, New York

Cheng TS, Lee TH (2004) Human pulling strenghts in different conditions of exertion. Percept Mot Skills 98:542–550

Ciriello VM (2001) The effects of box size, vertical distance, and height on lowering tasks. Int J Ind Ergon 28(2):61–67

Ciriello VM (1999) Maximum acceptable forces of dynamic pushing: comparison of two techniques. Ergonomics 42:32–39

Ciriello VM (2007) Revisited: comparison of two techniques to establish maximum accptable forces of dynamic pushing for male indutrial workers. Int J Ind Ergon 37:877–882

Ciriello VM et al (2008) Secular changes in psycophysically determined maximum acceptable weights and forces over 20 years for male industrial workers. Ergonomics 51(5):593–601

Ciriello VM, Snook S (1983) A study of size, distance, height and frequency effects in manual handling tasks. Hum Factors 25:473–483

Ciriello VM (2005) Psycophisically determined horizontal and vertical forces of dynamic pushing on high and low coefficient of friction floors for female industrial workers. J Occup Environ Hyg 2:136–142

Ciriello VM, Snook SH, Hughes GJ (1993) Further studies of psycophysically determined maximum acceptable weights and forces. Hum Factors 35:175–186

Ciriello VM et al (1990) The effects of task duration on maximum acceptable weights and forces. Ergonomics 33:187–200

Daams BJ (1993) Static force exertion in postures with different degrees of freedom. Ergonomics 36:397–406

Damkot DK et al (1984) The relationship between work history, work environmnet and low-back pain in men. Spine 9:395–399

Das B, Wang Y (1995) Determination of isometric push and pull strength profiles in workplace reach envelopes. In: Advances in industrial ergonomics and safety VII: proceedings of the tenth annual international industrial ergonomics and safety conference, Seattle, WA, USA, pp 13–16

Das B, Wimpee J (2002) Ergonomics evaluation and redesign of a hospital meal cart. Appl Ergon 33(4):309–318

Datta SR, Chatterjee BB, Roy BN (1983) The energy cost of pulling handcarts. Ergonomics 26:461–464

Datta SR, Chatterjee BB, Roy BN (1978) The energy cost of rickshaw pulling. Ergonomics 21:879–886

David GC, Nicholson AS (1985) Aids to lifting and handling. The safety practitioner/Institution of Ocuupational Safety and Health, Leicester, UK, pp 4–7

Davis PR, Stubbs DA (1977a) Safe levels of manual forces for young males I. Appl Ergon 8:141–150

Davis PR, Stubbs DA (1977b) Safe levels of manual forces for young males II. Appl Ergon 8:218–219

Davis PR, Stubbs DA (1978) Safe levels of manual forces for young males III. Appl Ergon 9:33–37

Davis PR (1981) The use of intra-abdominal pressure in evaluating stresses on the lumbar spine. Spine 6:90–92

de Looze MP et al (2000) Force directionand physical load in dynamic pushing and pulling. Ergonomics 43(3):377–390

de Looze MP et al (1995) Mechanical loading on the low back in three models of refuse collecting. Ergonomics 38:1993–2006

de Looze MP et al (1992) Validation of a dynamic linked segment model to calculate joint moments in lifting. Clin Biomech 7:161–169

de Looze MP et al (1996) Weight and frequency effect on spinal loading in a bricklaying task. J Biomech 29:1425–1433

Dempsey PG et al (2008) Oxygen consumption prediction models for individual and combination manual material handling tasks. Ergonomics 51:1776–1789

Dempster WT (1958) Analysis of two-handed pulls using free body diagrams. J Appl Phys 13:469–480

Drury CG, Barnes RE, Daniels EB (1975) Pedestrian operated vehicles in hospitals. In: Proceedings of the 26th spring annual conference and world productivity congress/ed. Engineers American Institute of Industrial, Norcross, GA

Eastman Kodak Co (1986) Ergonomic design for people at work, vol 2. Van Nostrand Reinhold, New York

Estryn-Behar M et al (1990) Strenous working conditions and musculoskeletal disorders among female hospital workers. Int Arch Occup Environ Health 62:47–57

Fothergill DM, Grieve DE, Pheasant ST (1991) Human strength capabilities during on-handed maximum voluntary exertions in the fore and aft plane. Ergonomics 2(35):203–212

Frings-Dresen MHW et al (1995a) Guidelines for energetic load in three methods of reuse collecting. Ergonomics 38:2056–2064

Frings-Dresen MHW et al (1995b) The daily work load of refuse collectors working with three different collecting methods: a field study. Ergonomics 38:2045–2055

Frymore JW et al (1980) Epidemiologoc studies of low-back pain. Spine 5:419–423

Gagnon M, Beaugrand S, Authier M (1992) The dynamics of pushing loads onto shelves of different heights. Int J Ind Ergon 9:1–13

Garcin M et al. Physiological strains while pushing or hauling. Eur J Appl Physiol 72:278–482

Garg A et al (1991) A biomechanical and ergonomic evaluation of patient transferring tasks: bed to wheelchair and wheelchair to bed. Ergonomics 34:289–312

Garg A et al. (2012) Psycophysical basis for maximum pushing and pulling forces: a review and raccomandations. Int J Indust Ergon, 1–11

Garg A, Beller D (1990) One-handed dynamic pulling strenght with special reference to speed, handle height and angles of pulling. Int J Ind Ergon 6:231–240

Garg A, Moore J (1992) Epidemiology of low back pain in industry. Occup Med 7:593–608

Garg A, Chaffin DB, Herrin GD (1978) Prediction of metabolic rates for manual material handling jobs. Am Ind Hyg Assoc J 39:661–674

Genaidy AM et al (1993) Spinal compression tolerance limits for the design of manual material handling operations in the workplace. Ergonomics 36:415–434

Genaidy AM, Houshyar A, Asfour SS (1990) Physiological and psychophysical responses to static, dynamic and combined tasks. Appl Ergon 21:62–67

Gledhill N, Jamnik VH (1992) Characterization of the physical demands of firefighting. Can J Sports Sci 17:207–213

Glitsch U et al (2007) Physical workload of flight attendants when pushing and pulling trolleys aboard aircraft. Int J Ind Ergon 37(11–12):845–854

422   M. Placci et al.

Granata KP, Bennet BC (2005) Low-back biomechanics and static stability during isometric pushing. Hum Factors 47(3):536–549

Grieve DW (1979a) Environmental constraints on the static exertion of forces: PSD analysis in task-design. Ergonomics 22:1165–1175

Grieve DW (1983) Slipping due to manual exertion. Ergonomics 26:1155–1164

Grieve DW (1979b) The postural stability diagram (PSD): personal constraints on the static exertion of force. Ergonomics 22:1155–1164

Hansson JE (1968) Work physiology as a tool in ergonomics and production engineering. National Institute of Occupational Health. Stockholm, Sweden

Hansson TH, Keller TS (1987) Spengler D.M., mechanical behaviour of the huamn lumbar spine II. Fatigue strength during dynamic compressive loading. J Orthop Res 5:479–487

Harkness EF et al (2003) Mechanical and physiological factors predict new onset shoulder pain; a prospective cohort study of newly employed workers. Occup Environ Med 6:850–857

Recent Changes to the Manual Handling Law and Implementation in Chile

Paulina Hernández Albrecht(✉)

Latin American Unión of Ergonomics, Valparaíso, Chile
pauhernandez@ergonomiachile.cl

Abstract. The Law No. 20.944 that modifies the Chilean Work Code in regulating the maximum human load weight was published in 2016.

This Law replaced the No. 20.001 Law (2005) and reduced from 50 kilos to 25 kilos the maximum limit of manual handling of load for adult male population and maintained the maximum limits of 20 kg for both men and women under 18 years. It prohibits these tasks for pregnant women and indicates that the maximum limit will be modified to the extent that other aggravating factors exist, in which case, the manipulation must be carried out in accordance with the guidelines established in the Technical Guide. This guide, developed for the first time according to Law No. 20.001, had to be reviewed and updated was carried out by a team of Chilean ergonomists and it was published on March 1, 2018. This article is a brief analysis and a reflection of the components involved in the process of creating a technical guide to be used in the country, which not only includes technical aspects regarding the tools for its application, also involves considering other aspects such as political and context management, to ensure that the regulation is effective and widely applied.

Keywords: Manual handling law · Updating

1 Introduction

The National Policy on Health and Safety at Work in Chile (2016) [1], provides, in its principles, that the prevention of occupational risks will be a priority, from the design of production systems and jobs, focused on the elimination or control of the risks at the source. It also indicates, that the danger of productive processes, heavy work, high-risk work, the gender factor and the specific risks of economic sectors must be considered.

The Chilean Work Code [2], indicates is mandatory to the employer, to take all necessary measures to effectively protect the life and health of workers and must adopt those appropriate measures to ensure effective protection, preventing any risk to which employees may be exposed in the work execution.

The Law No. 20.001 [3], was published on 5 February 2005, which incorporated the regulation of the maximum weight of human loads into the Chilean Work Code,

---

P. H. Albrecht—Chilean ergonomist. Consultant. President of Ergonomics Latin American Unión. Leader of the team for the elaboration of the technical guide of manual handling 2017.

© Springer Nature Switzerland AG 2019
S. Bagnara et al. (Eds.): IEA 2018, AISC 820, pp. 423–431, 2019.
https://doi.org/10.1007/978-3-319-96083-8_55

established preventive mechanisms to manage the risks of manual handling of loads. The Supreme Decree No. 63 [4], as a regulatory framework, established that the procedure for assessing health risks or the conditions of workers derived from these risks, shall be governed by the Technical Guide for the Evaluation and Control of Risks Associated with Manual Load Handling.

In 2016 and at the request of a group of senators of the Republic, the Law No. 20,949 [5], was published that modified the Chilean Work Code, reducing from 50 kg (2005) to 25 kilos the maximum limit of manual handling of load (MMH) for male population adult, maintaining the maximum limits of 20 kg for both men and women under 18 years, established by the previous law. The law prohibits these tasks for the pregnant woman and it indicates that the maximum limit will be modified to the extent that other aggravating factors exist, in which case, the manipulation must be carried out in accordance with the guidelines established in the Technical Guide for the Evaluation and Control of the Risks Associated with Handling or Manual Load Handling [6]. This guide, developed for the first time according to Law No. 20.001, had to be reviewed and updated by a team of six Chilean ergonomists and its was published on 1 March 2018.

2 Purpose

The purpose of this article is to reflect on the components involved in a process of creating regulations, which not only includes technical aspects regarding the tools for their application, but also involves considering other aspects such as political and context management, to ensure that the regulations are effective and widely applied.

3 Updating Process of the Technical Guide for the Evaluation and Control of the Risks Associated with Manual Handling or Handling of Load in Chile

Within the characteristics of the process for the development of this technical guide, aspects of the governmental political context, its adherence to OSH policies, and some of scientific consideration and experiential learning in addressing these risks are highlighted.

3.1 Governmental Political Context Aspects

The requirement for updating this guide was addressed by the Undersecretary of Social Welfare of the Ministry of Labor of Chile, through a public tender process. This same entity led the enactment of laws regarding the maximum weight of human load (20.001 and 20.949), its decrees and the previous guide (2008).

It is important to consider that, in general, the processes of revision and updating of the regulations in Chile haven't been frequent and occur after several years of its application. In this case, eight years have passed for the modification of the Law and almost ten years for updating the technical guide. Changes in regulations isn't a matter that depends only from technical aspects, but rather on political aspects or can also be influenced by global governing bodies, for example, ILO and WHO.

For the elaboration of the bidding rules and their requirements, a commission was formed composed of representatives of several government agencies related to occupational health of the Ministries of Labour and Health (agencies, standards, qualifiers and auditors) and it was their representatives who established the requirements of the new technical guide and monitored its progress in all its stages. This integration made it possible to respond to the different realities of the State agencies have to face in this matter and tends to ensure that the different State actors in OSH matters assume the responsibility for compliance and effective application of the technical guide.

The Ministry of Labour established a limit of eighty days for the development of the new guide, with milestones of compliance of stages during this period. It deduces that this high temporary demand was due the fact that one month after the delivery of the guide, there was a change of government in the country and the government authorities requesting the guide wanted to finish the entire process before it happened.

3.2   Adherence to National Policies on Matters of Health and Safety at Work

Chile, like many countries, has had for fifty years, regulations on health and safety at work, whose main body is the Law No. 16.744 (1968) [7], that establishes rules on compulsory insurance for work accidents and occupational diseases. Since the creation of this Law, a series of decrees and technical guides have emerged to address the country in the prevention and control of risks for the safety and health of its workers. The Occupational Safety Institute (ISL) and the Employers' Mutual Societies are responsible for the administration of insurance against risks of occupational accidents and diseases in Chile. To the above, it must be added to the companies with delegated administration, which administer certain insurance benefits. These bodies are called "Law Administration Bodies" (OAL) [8]. The normative framework that inspired the elaboration of the MMH technical guide and the National Policy on Safety and Health at Work (2016), which complies with ILO Convention No. 187 [9], where the State of Chile committed to carry out a process of construction and development of a culture of prevention of work accidents and occupational diseases. This is the regulatory framework that also supports Law 20.949, which modifies the Chilean Work Code to reduce the weight of manually handled loads. From this modification and its regulation, the updating of the technical guide for manual handling of loads arose.

The requirement of the Ministry of Labor included that, associated with the technical guide MMH, an analysis of the Chilean economic sectors be incorporated in which a greater impact of accidents and musculoskeletal injuries associated with MMH is observed, to focus preventive measures in specific to these economic sectors. The Health sector is one of the most affected in Chile by these risks, for that the Ministry explicitly requested to be addressed in this guide. This is how the management process integrated manual handling of loads (MMH) and manual handling of people (MMP).

426   P. H. Albrecht

3.3   Scientific and Experiential Learning in Addressing the Risks Associated with MMC Considerations

The demand indicated that the executing team of the new MMH technical guide should be made up of a doctor in ergonomics, an occupational health doctor, an engineer in risk prevention and other ergonomics professionals with a minimum master's degree. These professionals had to demonstrate, in turn, a vast experience of field work.

The updating process was technically based on the search for updated references in researching associated with the management of these risks, both in the identification, evaluation and intervention stages with control measures. Around a hundred and forty bibliographical references and sixty technical aid websites were consulted.

Considering that the first evaluation guide and control of these risks began to be applied in Chile in 2008, the executing team defined it was important and necessary to rescue the experience of application of the previous guide by ergonomic professionals of the country and to know its proposals for improvement. Interviews and focus groups were carried out with thirty-seven ergonomists from the OAL, universities and the Chilean Society of Ergonomics, collecting positive and negative aspects of the previous guide and their opinions on aspects to be improved. Also, some of these professionals collaborated in the final revision of the new technical guide.

The 2008 Technical Guide was very extensive, difficult to use and did not orient professionals or companies in its application. It didn't provide clear guidance on how to manage risks and requires the use of certain evaluation instruments, of low sensitivity. Furthermore, the Technical Guide did not give answers for different work situations or their complexities. Neither our country hasn't researched the usability or impact the old guide. A mistake made by some companies was to believe that workers could handle loads close to 50 kg without any risk and, therefore, employers did not take any preventive action. On the other hand, it was observed that the control of companies in its application of the previous technical guide was very limited and that, in general, companies only applied an evaluation with the same methodology for all work situations and did not implement control measures to eliminate the risks.

4   Results of the Updating of the Technical Guide for the Evaluation and Control of the Risks Associated with Handling or Manual Load Handling

4.1   Sectors with the Greatest Impact of Manual Handling Loads in Chile

A descriptive analysis was made of 25.811 cases of reported complaints of lumbar pain related to work during the period 2014–2016, of a population covered by two million workers (Mutual de Seguridad C.Ch.C, 2017). The most frequent diagnosis was "lumbago" in 93% of the cases. The complaints corresponded, mainly, to young workers for both sexes. Men workers were affected workers in approximately 75% of the cases, who work mainly in "elementary occupations" (CIUO-2008), that is, workers with a lower level of qualification. For example, pawns of the construction, pawns of load, personnel of cleaning, laborers of agricultural exploitations.

The five highest reporting rates correspond to Industry (average rate, 681 × 100,000), Transport (average rate, 630 × 100,000), Agriculture and Fisheries (average rate, 532 × 100,000), Trade (average rate, 524 × 100,000) and Construction (average rate, 456 × 100,000 Tb). It was observed women reported the most complaints of lumbar spine who work in the areas of Personal Services, Trade and Financial and Professional Services. Men reported more frequent complaints in Construction, Commerce, Industry and Transportation. The Personal Services category is the only one in which the complaints of women frequently exceeded to men. This would be explained since this item includes health services activities, which is widely recognized internationally for the specific risk of manual manipulation of people [10].

4.2 Strengthening Management in the Risks Associated with Manual Handling of Loads

One of the points most demanded by both government agencies and ergonomists interviewed, that it was very necessary that the risk management of MMH/MMP be structured as a system and integrated into the risk management of the companies.

The technical guide MMH/MMP established the following basic and mandatory aspects in management:

- The responsibility in the management of these risks corresponds to the employer and includes prevention action, risk control, risk reduction and risk protection of the workers who carry out these tasks.
- The company can perform this management, through the Risk Prevention Department of the company, with technical assistance from the corresponding OAL, with advice from competent professionals and with the support of the Joint Committee on Hygiene and Safety.
- The OAL responsibilities are, to carry out permanent activities to prevent these risks, according to the nature and magnitude of the risk associated with the productive activity, formally prescribing the necessary measures aimed at the maintenance of safe and healthy work environments.
- The workers responsibilities are, compliance with instructions, regulations and working procedures for the MMH/MMP; collaborate with the employer in regulatory compliance and preventive management; inform about risk conditions; participate and promote participation in occupational safety and health activities; report to regulatory bodies the non-compliance with regulations; disseminate and promote preventive actions.
- The supervision of compliance with the rules of MMH/MMP will correspond to the Labor Directorate and other supervising entities, according to their scope of competence, including the Regional Ministerial Health Secretariats.
- Establishes deadlines for compliance with the different stages of management and which are the backups that companies will be asked to verify.

4.3   Stages of Management

Companies must have a risk management program of MMH/MMP that contains, at least, job identification, number of workers involved differentiated by sex, risk identification and risk evaluation, the result of those evaluations, the corresponding preventive and corrective measures and, the corrections to the work situation according to the level of risk.

Stage I: Identification of the risks associated with the MMH/MMP

Its objective is to determine the absence of risk or the existence of a critical condition in the tasks of MMH/MMP. This stage should be applied to all the jobs of the companies and leads to take actions to reduce or eliminate risk.

Unlike the previous technical guide (2008), this new guide strengthens the early adoption of corrective measures from the identification stage. The procedure established for this stage is based on ISO/TR 12295: 2014 [11], and it consists of two activities: Initial identification and Advanced identification. It does not involve a specific evaluation.

Its application leads to determine the presence of two opposite conditions: acceptable condition (absence of risk) or critical risk condition. If a critical condition is identified, the risk should be reduced immediately through corrective actions.

This stage allows the application for users not specialized in ergonomics, which are all those people who, within the companies, are responsible for executing or supporting the management of these risks such as, risk prevention professionals, CPHS, owners, among others; who do not have specific training in the discipline.

Stage II: Evaluation of the risks associated with the MMH/MMP

Defining the MMH/MMP risk assessment tools was a complex process to solve given that the counterpart of the ministries insisted on opting for a single evaluation method as defined in the previous technical guide. It was necessary to work with them in the understanding that having a variety of tools allows us to account for different characteristics and diverse conditions of work situations, since at present no method can assess all the situations, especially complex ones. For this, criteria for the selection of MMH/MMP evaluation methods were established, using decision trees based on the different realities of the jobs and their tasks. Two types of evaluation were proposed according to complexity: Initial Evaluation and Advanced Evaluation:

- Initial Evaluation: corresponds to methods without action level in the overall result and easier in their application. It can be applied by trained professionals, as risk prevention professionals and other like a physical therapist, occupational therapists, industrial designers, nurses, among others who have approved a training in the methods of initial evaluation of MMH included in the technical guide. The training can be given a competent national organization (OAL, universities or other bodies authorized to simulate nature) and will have a minimum of eight hours, 50% of which must be practical, including a final evaluation.
- Advanced Evaluation: are methods with action level in the overall results and more difficult in their application. They can only be applied by ergonomist who have been trained in the methods of specialized evaluation of MMH/MMP (minimum of forty hours).

Stage III: Control of the risk factors associated with MMC/MMP

The aim of controls is to prevent accidents and health damage that are a consequence of work, that are related to work or occur during work, eliminating and minimizing, to the extent that is reasonable and feasible, the causes of the risks inherent in the work environment. (ILO Convention No.155, article 4) [12]. This technical guide is strengthened in its orientation to users, establishing the following criteria:

- Prioritization of the application: in situations identified as "unacceptable", "critical" or "risky" that can cause occupational accidents or disease; in jobs where an occupational disease or accident is caused by the MMH/MMP has been generated; in cases of symptoms or injuries of workers associated with MMH/MMP, especially if they are frequent; in jobs with several tasks in critical condition.
- Priority of control measures: the preventive approach emphasizes the prevention of job risks over the protection of these, therefore, the criteria to avoid the MMH/MMP as much as possible should be incorporated from the design of the production systems and jobs. As a first step, the elimination of risk is promoted over mitigations or administrative measures.
- Considerations for management: It includes, the promotion of a preventive culture as a basis for the transformation towards healthy and safe jobs; the type of measures to be implemented and how to combine them; the route to follow in the intervention processes and the importance of the participatory approach.
- Guidance on aspects related to training: it is oriented towards the minimum contents and recipients of training, depth and its modalities.

Stage IV: Ensure continuity of the action plans

Stage of important value, so that the changes or improvements implemented remain in time and comply with its preventive function. For this, this technical guide indicates that verifications must be carried out by the OALs and the companies themselves; permanently observe the safe and efficient use of new measures or solutions; check the efficiency and effectiveness of the control measures implemented.

5   Conclusions

a. The creation of regulations associated with OSH is a decision of the country that has to do with its public policies on these issues, so it is essential that there is alignment and coordination among the various government actors towards a common objective. The executing professionals of the regulation, in general, specialists in their subjects and not very political, should not forget that they are responding to a requirement at a macro level that will have a high impact and, therefore, it is necessary to deal with political aspects and the context in what happens to ensure that the development of the regulation is "technically correct".

b. In the regulations associated with ergonomics, the role of the team of professionals who elaborate it is fundamental. Although this seems obvious, this is more remarkable for ergonomists than for other professionals, due to the holistic training of the discipline, its multidisciplinary nature and its knowledge of the work activity. Ergonomics could be defined as a bounding science, a discipline that can encompass a series of other basic disciplines for the creation of an artefact and, also for the creation of a normative. Team work in this case was decisive to meet the temporary demand and the quality of the product requested. In this sense, the different competences of the members, leadership and permanent coordination are valued.

c. A novelty in this regulation was to generate a differentiated tool for users with different competencies and training in OSH matters, since in general, most of these tools require specialized professionals for their application. In the advance flow in the detection of risks, this technical guide allows its application to train non-specialist users, reserving only the situations of greater complexity to trained ergonomists. This feature makes it easier for companies to move quickly towards the correction of work conditions that, in themselves, generate risk and prevent companies from being paralyzed because they do not have specialized professionals or cannot finance them.

d. As a learning to be disseminated, OSH regulations should be reviewed and updated periodically, at least every two or three years, evaluating their real impact on the objectives for which they were conceived, evaluating their application and incorporating new scientific advances in the matter.

References

1. Ministry of Labor and Social Security (2016) National Policy on Health and Safety at Work in Chile. https://www.leychile.cl/Navegar?idNorma=1094869. Accessed 16 May 2018
2. Ministry of Labor and Social Security, Sub Secretariat of the Work. https://www.leychile.cl/Navegar?idNorma=207436. Accessed 16 May 2018
3. Ministry of Labor and Social Security, Law No. 20.001 (2005) Regulates the Maximum Weight of Human Load. https://www.leychile.cl/Navegar?idNorma=235279. Accessed 26 Apr 2018
4. Ministry of Labor and Social Security (2005) Sub Secretariat of Social Protection. Supreme Decree No. 63. https://www.leychile.cl/Navegar?idNorma=241855. Accessed 26 Apr 2018
5. Ministry of Labor and Social Security, Law No. 20.949 (2016) Modify the Work Code to Reduce the Weight of Manual Handling Loads. https://www.leychile.cl/Navegar?idNorma=1094899. Accessed 20 Mar 2018
6. Ministry of Labor and Social Security (2018) Guide for the Evaluation and Control of the Risks Associated with Handling or Manual Load Handling. https://www.previsionsocial.gob.cl/sps/guia-tecnica-la-evaluacion-control-riesgos-asociados-al-manejo-manipulacion-manual-carga/. Accessed 25 May 2018
7. Ministry of Labor and Social Security, Law No. 16.744 (1968) Establishes Standards on Work-related Accidents and Occupational Diseases. https://www.leychile.cl/Navegar?idNorma=28650. Accessed 20 Mar 2018
8. Ministry of Labor and Social Security (1968) Regulations for the Application of Law No. 16.744, Supreme Decree No. 101. https://www.leychile.cl/Navegar?idNorma=9231. Accessed 20 Mar 2018
9. ILO, Chile ratifies ILO Convention No.187 on the Promotional Framework for Occupational Safety and Health. http://www.ilo.org/global/standards/information-resources-and-publications/news/WCMS_154812/lang--es/index.htm. Accessed 25 May 2018

10. June KJ, Cho S-H (2011) Low back pain and work-related factors among nurses in intensive care units. J Clin Nurs 20(3–4):479–487
11. ISO (2014) ISO/TR 12295, Application document for International Standards on manual handling (ISO 11228-1, ISO 11228-2 and ISO 11228-3) and evaluation of static working postures (ISO 11226)
12. ILO (1981) C-155, Occupational Safety and Health Convention. http://www.ilo.org/dyn/normlex/es/f?p=NORMLEXPUB:12100:0:NO:p12100_instrument_id:312300. Accessed 20 Mar 2018

Update on the Musculoskeletal Health of Office Employees in Hong Kong

M. Y. Chim Justine[✉]

Chim's Ergonomics and Safety Limited, 17/F, no. 80 Gloucester Road,
Wan Chai, Hong Kong, China
jchim@my-ergonomics.com

Abstract. Employers have legal responsibilities to protect their employees' health and safety at work. The Occupational Safety and Health (Display Screen Equipment) Regulation (Cap 509B) was enacted in Hong Kong in 2003 which aims at protecting the health and safety of employees who use computer for prolonged period of time. The study result showed that from 2015 to 2017, 96% of 1,618 employees spent at least six hours a day in computing work. The number of employees spending prolonged hours on computers was the highest compared to previous studies. With regard to the musculoskeletal health for the period in 2015 to 2017, 74% of sample employees reported musculoskeletal symptoms for at least one body region. The top three most commonly reported body region with discomfort, in 2015 to 2017, was Shoulder (69%), Neck (49%) and Lower Back (39%). Shoulder, Neck and Lower Back have been the highest reported regions experiencing discomfort since 2011 in Hong Kong. For the seeking of medical treatment, from 2015 to 2017, 39% of sample employees reported that medical treatment was needed for treating their musculoskeletal symptoms. A similar percentage of 42% of sample employees reported requiring medical treatment from 2013 to 2014. The high number of reported cases requiring medical treatment represents the high medical cost borne by employers. In conclusion, the reported rate of musculoskeletal discomfort among office employees has been high in the past 10 years. Regular analysis and updating of the musculoskeletal health of office employees in Hong Kong is needed.

Keywords: Occupational health · Musculoskeletal disorders
Musculoskeletal health · Health and safety · Computer users
Display screen equipment · Hong kong

1 Introduction

The Occupational Safety and Health (Display Screen Equipment) (DSE) Regulation (Cap 509B) was enacted in Hong Kong in 2003. The DSE Regulation prescribes six provisions: (1) Workstation Risk Assessment; (2) Record of Risk Assessment; (3) Reduction of Risks; (4) Requirements for Workstations; (5) Health and Safety Training; and (6) Users' Compliance with the Safe System of Work. It aims at

---

M. Y. Chim Justine—CPE & RSO, Director and Principal Consultant.

S. Bagnara et al. (Eds.): IEA 2018, AISC 820, pp. 432–437, 2019.
https://doi.org/10.1007/978-3-319-96083-8_56

protecting the health and safety of employees who use DSE at work for prolonged periods [1]. Chim (2014) reported that among three Asian countries, only Hong Kong has enacted legislation to specify the legal requirements to protect prolonged computer users, whereas Singapore and Japan have published codes of practice and guidelines on office ergonomics without implementing legislation [2].

Past research studies investigated the implementation of individual ergonomics interventions to improve musculoskeletal health. The interventions include office ergonomics education, wellness and ergonomics promotion, rest break software and alternative keyboards [3–10]. To promote office ergonomics programs in a systematic way, Chim (2013) proposed a FITS model office ergonomics program to eliminate or minimize the risk of musculoskeletal disorders among computer users and promote employees' wellness. The FITS Model office ergonomics program was developed according to the practical industrial practices in ergonomics, occupational health and safety management, and human resources management in Hong Kong and overseas with consideration of the legal requirements regarding DSE Regulation. The model includes (1) Furniture Evaluation and Selection; (2) Individual Workstation Assessment; (3) Training and Education; (4) Stretching Exercises and Rest Breaks as elements of an effective program. [2] A case study of implementation of the FITS Model office ergonomics program generated a positive result for promoting workstation ergonomics and avoiding musculoskeletal disorders for computer users [11].

2 Aims

A number of previous studies have reported the musculoskeletal health of office employees in Hong Kong for 1997 to 2014. This paper reports the update of musculoskeletal health statistics in Hong Kong in 2015 to 2017. It will summarize and compare the prevalence rate of musculoskeletal symptoms of office ergonomics in Hong Kong in the most recent years.

3 Methods

Chim's Ergonomics and Safety Limited conducted 1,618 individual ergonomics workstation assessments between 2015 and 2017 in Hong Kong. During the face-to-face individual ergonomics workstation assessment, self-reported musculoskeletal symptoms, received treatment, and average working hours of computer operation were examined during the workstation assessment.

4 Summary of Previous Studies on the Musculoskeletal Health in Hong Kong

A survey by the Occupational Safety and Health Council, Hong Kong in 2002 concluded that office employees reported multiple sources of physical discomfort in relation to computing work and the neck (63%), shoulders (60%) and back (40%) were top three reported regions of discomfort [12].

A study of occupational health condition of government office employees in 2010 concluded that 87.9% of 217 surveyed employees reported musculoskeletal symptoms in the past 12 months and 21.8% of employees who reported musculoskeletal symptoms were occasionally absent from work due to musculoskeletal health problem [13].

A study conducted in Hong Kong between 2011 and 2012, reported that over 80% of survey employees who reported musculoskeletal symptoms for at least one body region. Furthermore, over 50% of office employees received treatment for musculoskeletal symptoms in the past [14]. In the same study, it retrieved that over 73% of 618 office employees in Hong Kong spent at least six hours a day in DSE operation whereas the study by Occupational Safety and Health Council in Hong Kong ("OSHC") in 2002, 48% of 368 office employees spent at least four hours a day with using computer. The results indicated that office employees spent much longer hours with using computer at work compared to 10 years ago [12, 15].

The result showed that in 2011 and 2012, 73% of 618 employees who spent at least six hours a day in computing works, whereas in 2013 to 2014, the higher percentage of employees (87% of 245 employees) spent at least six hours a day with using computer. For consolidated period of 2010 to 2014, 84% of Hong Kong office employees whose reported musculoskeletal symptoms for at least one body region. If consider the more recent data for 2013 and 2014, 91% of employees who reported musculoskeletal symptoms for at least one body region. 23.7% and 28.6% of employees who reported two body regions and three body regions suffered from musculoskeletal discomfort or injuries [15].

The most common report discomfort body region was Shoulder – both sides (31%) and second highest reported body region musculoskeletal symptoms was Lower Back (26%). The third highest report body region with musculoskeletal symptoms was Neck (14%).

5   Update of Musculoskeletal Health of Office Employees from 2015 to 2017

5.1   Time Spent at Least Six Hours a Day in Computing Work

The result showed that from 2015 to 2017, 96% of 1,618 employees spent at least six hours a day in computing work. The number of employees spending prolonged hours on computers was the highest compared to previous studies. The study of prolonged working hours in computing works of office employees for 2011 and 2012 reported 73% and for 2013 to 2014 the figure was 87% (Table 1).

Table 1. Time spent at least six hours a day in computing work (2011 to 2017)

Items	2011–2012	2013–2014	2015–2017
Sample size (number of office employees)	618	245	1618
Time Spent at least six hours a day in computing work (%)	73%	87%	96%

5.2   Self-reported Musculoskeletal Symptoms

With regard to the musculoskeletal health for the period in 2015 to 2017, 74% of sample employees reported musculoskeletal symptoms for at least one body region. When comparing the previous study with the statistics for 2013 to 2014, the reporting percentage of musculoskeletal symptoms was much higher at 91%.

The top three most commonly reported body region with discomfort, in 2015 to 2017, was Shoulder (69%), Neck (49%) and Lower Back (39%). In the study for 2013 to 2014, Shoulder (31%), Lower Back (26%) and Neck (14%) were also the top three most common body discomfort regions which showed that Shoulder, Neck and Lower Back have been the highest reported regions experiencing discomfort since 2011. When looking at the rate of reported discomfort, all three body regions were much higher in 2015 to 2017. Although the reported rate of musculoskeletal symptoms decreased from 2015 to 2017, there were a significantly higher percentage of reported cases with discomfort in these three body regions compared to 2013 to 2014. This shows the severity of musculoskeletal discomfort in the most recent statistics (Table 2).

Table 2. Self-reported musculoskeletal health (2011 to 2017)

Items	2011–2012	2013–2014	2015–2017
Sample size (number of office employees)	618	245	1618
Reported musculoskeletal symptoms (%)	81%	91%	74%
Commonly reported body region with discomfort	Neck, Shoulder, Lower Back	Shoulder (31%)	Shoulder (69%)
		Lower Back (26%)	Neck (49%)
		Neck (14%)	Lower Back (39%)

5.3   Seeking of Medical Treatment

For the seeking of medical treatment, from 2015 to 2017, 39% of sample employees reported that medical treatment was needed for treating their musculoskeletal symptoms. A similar percentage of 42% of sample employees reported requiring medical treatment from 2013 to 2014. The high number of reported cases requiring medical treatment represents the high medical cost borne by employers (Table 3).

Table 3. Seeking of medical treatment (2011 to 2017)

Items	2011–2012	2013–2014	2015–2017
Sample size (number of office employees)	618	245	191
Seeking of medical treatment	52%	42%	39%

6 Conclusion

In conclusion, the reported rate of musculoskeletal discomfort among office employees has been high in the past 10 years. Although the Occupational Safety and Health (Display Screen Equipment) Regulation was enacted in 2003 in Hong Kong, further promotion of healthy computing should be enhanced. Regular analysis and updating of the musculoskeletal health of office employees in Hong Kong is needed.

Acknowledgement. This work was supported by Chim's Ergonomics and Safety Limited. The paper was prepared by considering the real case experience from working with clients in workplace design and applies the ergonomics principles in the whole design process.

References

1. HKSAR Government (2013) Labour Department. Code of Practice for Working with Display Screen Equipment. Hong Kong
2. Chim JMY (2014) Ergonomics for the prevention of musculoskeletal disorders of computer users in Hong Kong, Singapore and Japan. J Ergon S4:004. https://doi.org/10.4172/2165-7556.s4-004
3. Chim JMY (2013) The FITS model office ergonomics program: a model for best practice. Work J Prev Assess Rehabil https://doi.org/10.3233/wor-131806
4. Chim JMY, Ng P, Tai S (2009) Winning telebet centre, design: apply participatory ergonomics to promote work health & safety, employee wellness and operational efficiency. Paper presented at: 17th world congress on ergonomics, international ergonomics association. Beijing, China
5. Robertson M, Amick BC, DeRango K, Rooney T, Bazzani L, Harrist R et al (2009) The effects of an office ergonomics training and chair intervention on worker knowledge, behavior and musculoskeletal risk. Appl Ergon 40:124–125
6. Robertson MM, O'Neill MJ (1999) Effects of environmental control on performance, group effectiveness and stress. In: Proceedings of the 43rd annual meeting of the human factors and ergonomics society. Human Factors and Ergonomics Society, Santa Monica, pp 552–556
7. Bohr PC (2000) Efficacy of office ergonomics. J Occupat Rehabil 10:243–255
8. Marcoux BC, Krause V, Nieuwenhuijsen ER (2000) Effectiveness of an educational intervention to increase knowledge and reduce use of risky behaviors associated with cumulative trauma in office workers. Work 14:127–135
9. Szeto GP, Ng JK (2000) A comparison of wrist posture and forearm muscle activities while using an alternative keyboard and a standard keyboard. J Occupat Rehabil 10:189–197
10. Trujillo L, Zeng X (2006) Data entry workers perceptions and satisfaction response to the "Stop and Stretch" software pro-gram. Work 27:111–121
11. Justine MYC (2015) Effective ergonomics project on implementation of FITS TM office ergonomics program. In: Proceedings 19th Triennial Congress of the IEA, Melbourne 9–14 August 2015
12. Occupational Safety and Health Council of Hong Kong (2002) A follow-up survey on the occupational health issues of users of the display screen equipment. Occupational Safety and Health Council of Hong Kong, Hong Kong

13. Hong Kong Workers' Health Centre (2011) Hong Kong government civil service employees: action research on occupational health for the use of display screen equipment. 香港政府文書職系公務人員使用顯示屏幕設備的職業健康行動研究報告. http://www.hkwhc.org.hk/wp-content/uploads/2018/01/research_report_12.pdf. Accessed 28 May 2018

14. Chim JMY (2013) Musculoskeletal disorders among office employees in Hong Kong and best practice office ergonomics solutions. In: Proceedings of the eighth international conference on prevention of work-related musculoskeletal disorders, Busan, Korea, pp 330–331

15. Justine MYC (2015) Healthy computing and ergonomics: review of musculoskeletal health problems and workplace setting. In: Proceedings 19th triennial congress of the IEA, Melbourne 9–14 August 2015

Risk Assessment in an Industrial Hospital Laundry

Giulio Arcangeli[1(✉)], Manfredi Montalti[1], Francesco Sderci[1],
Gabriele Giorgi[2], and Nicola Mucci[1]

[1] Department of Experimental and Clinical Medicine,
University of Florence, Florence, Italy
giulio.arcangeli@unifi.it
[2] Department of Human Sciences, European University, Rome, Italy

Abstract. Background. The laundries present in hospitals are responsible for washing, disinfection and distribution of working clothes. In healthcare services, workers from laundry services are exposed to various occupational hazards, including the ergonomic ones, related in particular to repeated movements of lumbar flexion, raising the arms above shoulder level and transportation of loads. These conditions can provoke musculoskeletal disorders (MSD), one of the most common occupational disease present in developed countries.

Methods. Step one: evaluation of the baseline condition by two different approaches. (1) Risk assessment with the study of working conditions using a specific integrated tool; (2) analysis of referred symptoms from workers using a specific validated questionnaire for MSD, the Nordic Musculoskeletal Questionnaire. 167 workers are included, 51 males (age $42,8 \pm 10,3$ years), 118 females (age $45,1 \pm 8,8$ years). Step two: changes of working organization on the basis of the results of step one.

Results. The analysis of questionnaires showed higher prevalence of symptoms in the female groups, in particular in the iron area. Symptoms increase with age and BMI but not with years of employment. On these basis a reorganization of the job processes were carried out, with changing in the machinery and in job organization (e.g., time of work at the workstation, including more frequent rotations of workers), and a specific individual practical training of workers was realized.

Discussion. The incidence of MSD symptoms appear very high. The inherent characteristics of the job associated with individual factors and unfavorable ergonomic conditions can contribute in the occurrence of MSD, injuries and aggravation of existing diseases.

Keywords: Musculoskeletal disorders · Industrial laundry · Ergonomic risks

1 Introduction

The laundries present in hospitals are responsible for washing, disinfection and distribution of working clothes. In healthcare services, workers from laundry services are exposed to various occupational hazards, including the ergonomic ones, related in particular to repeated movements of lumbar flexion, raising the arms above shoulder level

© Springer Nature Switzerland AG 2019
S. Bagnara et al. (Eds.): IEA 2018, AISC 820, pp. 438–445, 2019.
https://doi.org/10.1007/978-3-319-96083-8_57

and transportation of loads. These conditions can provoke musculoskeletal disorders (MSD), one of the most common occupational disease present in developed countries. These injuries can be result of ergonomic and psychosocial factors by organization and management of work, such as the type of equipment used, poor fitness machines and furniture, adoption of incorrect postures and positions, and unhygienic local conditions that generate discomfort, shift work, monotony or excessive pace of work. On the other hand, also gender and age may influence the onset and develop of symptoms.

MSD is therefore a priority for the European Union in the context of the "Community strategy on health and safety at work".

The causes of work-related MSDs are multifactorial (https://oshwiki.eu/wiki/Pathophysiological_mechanisms_of_musculoskeletal_disorders) and there are numerous work-related risk factors for the various types of MSDs. Several risk factors including physical and mechanical factors, organizational and psychosocial factors, and individual and personal factors may contribute to the genesis of MSDs. Workers are generally exposed to several factors at the same time and the interaction of these effects are often unknown [2, 3].

Work-related MSDs refer to injuries developed over time that are caused by a combination of risk factors that act simultaneously on a joint or body region, in a synergistic effect. Until now the biological pathogenesis associated with the development of the majority of the work-related MSDs is unknown. Several models have been proposed to explain the biological mechanisms. Usually three sets of factors are considered [1]:

- Physical factors (e.g., sustained or awkward postures, repetition of the same movements, forceful exertions, hand-arm vibration, all-body vibration, mechanical compression, and cold);
- Psychosocial factors (e.g., work pace, autonomy, monotony, work/rest cycle, task demands, social support from colleagues and management and job uncertainty); and
- Individual factors (e.g., age, gender, professional activities, sport activities, domestic activities, recreational activities, alcohol/tobacco consumption and, previous work-related MSDs).

Most of the factors can occur both at work and in leisure time activities. Thus, it is important to include all the relevant activities performed both at work and outside work when of a specific employee developing MSDs in a particular working environment.

As referred before, risk factors act simultaneously on a worker joint or body region in a synergistic effect. To manage the risk factors it is advisable and important to take into account this interaction rather than focus on a single risk factor.

1.1 Types of Work Related MSDs

As mentioned before, most of the recorded work related MSD affect the lower back, neck, shoulders and upper limbs. MSD affect less often the lower limbs. It is important to recognize however that not all MSDs are caused by work, although work may provoke symptoms and the problem may prevent a person from working, or make it more difficult. For example, a recent study found that age, gender and BMI made a bigger proportional contribution to developing carpal tunnel syndrom (CTS) than work-related factors [4].

Work-Related Upper Limb Musculoskeletal Disorders
Work-related Upper Limb Disorders (WRULDs) can affect any region of the neck, shoulders, arms, forearms, wrists and hand. Some of WRULDs, such as tendonitis, carpal tunnel syndrome, osteoarthritis, vibration white finger and thoracic outlet syndrome have well-defined signs and symptoms, while others are less well-defined, involving only pain, discomfort, numbness and tingling. EU-OSHA has produced a series of reports about upper limb and neck work-related MSDs, see for instance [5]. Also very useful information about MSDs prevention can be found in the two following reports [6, 7].

However the designation of WRULDs in international literature is not consensual. In addition to MSDs, other terms are sometimes used referring to similar symptoms and health problems. Examples are: - cervico-brachial syndrome, occupational cervico-brachial disorders; - occupational overuse syndrome; - repetitive strain injury, repetitive stress injury, repetitive motion injuries; - cumulative trauma disorders; - upper limb disorders, upper extremity musculoskeletal disorders, upper limb pain syndromes [8, 9].

Despite all the available knowledge some uncertainty remains about the level of exposure to risk factors that triggers MSDs. In addition there is significant variability of individual response to the risk factors exposure.

The most common WRULDs are:

- Neck: Tension Neck Syndrome, Cervical Spine Syndrome;
- Shoulder: Shoulder Tendonitis, Shoulder Bursitis, Thoracic Outlet Syndrome;
- Elbow: Epicondylitis, Olecranon Bursitis, Radial Tunnel Syndrome, Cubital Tunnel Syndrome;
- Wrist/Hand: De Quervain Disease, Tenosynovitis Wrist/Hand, Synovial Cyst, Trigger Finger, Carpal Tunnel Syndrome, Guyon's Canal Syndrome, Hand-Arm Syndrome, Hypothenar Hammer Syndrome.

Low Back Work-Related Musculoskeletal Disorders
Low back work-related MSDs include spinal disc problems, muscle and soft tissue injuries. These disorders are mainly associated with physical work, manual handling (https://oshwiki.eu/wiki/Risk_factors_for_musculoskeletal_disorders_in_manual_handling_of_loads) and vehicle driving activities (https://oshwiki.eu/wiki/Driving_for_work), where lifting, twisting, bending, static postures, and whole body vibration are present.

Work-Related Lower Limb Musculoskeletal Disorders
Until now little attention has been given to the epidemiology of work-related lower limb MSDs. However, lower limb MSDs is a problem in many workplaces and they tend to be related with conditions in other areas of the body. Lower Limb Disorders affect the hips, knees and legs and usually happen because of overuse. Acute injury caused by a violent impact or extreme force is less common. Workers working over a long period in a standing or kneeling position are most at risk. The most common risk factors at work are: - repetitive kneeling and/or squatting; - fixed postures such as standing for more than two hours without a break; - frequent jumping from a height [3, 10].

Despite the lack of attention given to this type of work-related MSDs they deserve significant concern, since they often give up high degrees of immobility and thereby can substantially degrade the quality of life. Lower limb work-related MSDs that have been reported in occupational populations are:

- Hip/thigh conditions: Osteoarthritis (most frequent), Piriformis Syndrome, Trochanteritis, Hamstring strains, Sacroiliac Joint Pain;
- Knee/lower leg: Osteoarthritis, Bursitis, Beat Knee/Hyperkeratosis, Meniscal Lesions, Patellofemoral Pain Syndrome, Pre-patellar Tendonitis, Shin Splints, Infra-patellar Tendonitis, Stress Fractures;
- Ankle/foot: Achilles Tendonitis, Blisters, Foot Corns, Halux Valgus (Bunions), Hammer Toes, Pes Traverse Planus, Plantar Fasciitis, Sprained Ankle, Stress fractures, Varicose veins, Venous disorders [11].

However, although these may occur in specific occupational groups (for example Piriformis Syndrome and Trochanteritis have been reported amongst dental personnel; and hamstring strains amongst athletes) the extent to which these have been generally shown to be caused by work is unclear and there are many non-work related factors that can contribute, possibly making the major contribution.

Non-specific Work Related Musculoskeletal Disorders

The non-specific work related MSD are musculoskeletal disorders that have less well-defined symptoms, i.e. the symptoms tend to be diffuse and non-anatomical, spread over many areas: nerves, tendons and other anatomical structures [12]. The symptoms involve pain (which becomes worse with activity), discomfort, numbness and tingling without evidence of any discrete pathological condition.

2 Methods

This study want to evaluate the impact of ergonomic risks among workers in a clothes central distribution service of a large hospital, to understand the causes and to suggest the measures to minimize this risk.

167 workers are included, 51 males (age 42,8 ± 10,3 years), 118 females (age 45,1 ± 8,8 years), divided for area of working production (collecting, washing, ironing, wardrobe/distribution area).

2.1 First Step: Evaluation of the Baseline Condition

Using two different approaches:

(1) Risk assessment with the study of working conditions (postures, loads, etc.) using a specific integrated tool (inertial movement analysis MVN Biomech, ErgoCert, Udine, Italy); Results ars shown in Table 1.
(2) Analysis of referred symptoms from workers using a specific validated questionnaire for MSD, the Nordic Musculoskeletal Questionnaire which is composed by 2 items:

Table 1. Results from first risk assessment.

Workstation	Ocra Index 6h 15' 3 Breaks		Variable Lifting Index (VLI)		
	Right Arm	Left Arm	Men 18-45 y.o.	Men > 45 & Women 18-45 y.o.	Women > 45 y.o.
Load LC1	2,6	2,6	1,15	1,44	/
Load LC2 + LC3	3,1	3,1	1,20	1,50	/
Load LC4	6,4	6,4	2,82	3,52	/
Sorting LC1	3,8	3,6		Not Applicable	
Sorting LC2	4,3	4,1		Not Applicable	
Sorting LC3	4,3	4,1		Not Applicable	
Sorting LC4	5,8	5,2		Not Applicable	
Mangle 1	4,3	4,3		Not Applicable	
Mangle 2	4,6	4,6		Not Applicable	
Mangle 3				Not Working	
Mangle 4	4,3	4,3		Not Applicable	
Mangle 5	3,0	3,0		Not Applicable	
Tunnel Hanging Station	2,7	2,4		Not Applicable	
Press	3,4	3,4		Not Applicable	
Stir Pants	4,0	4,0		Not Applicable	
Dummy	4,1	4,1		Not Applicable	
Tunnel Bending	2,5	0,8		Not Applicable	
Press Towels	1,1	2,2		Not Applicable	
Mangle Outputs 1-2	/	/	1,21	1,52	2,02
Mangle Outputs 4-5	/	/	0,91	1,14	1,52
Bagging Machine	/	/	1,93	2,42	3,22
Bagging Machine	/	/	0,45	0,56	0,75

a. Section 1: a general questionnaire of 40 forced-choice items identifying areas of the body causing musculoskeletal problems. Completion is aided by a body map (Fig. 1) to indicate nine symptom sites being neck, shoulders, upper back, elbows, low back, wrist/hands, hips/thighs, knees and ankles/feet. Workers interviewed are asked if they have had any musculoskeletal trouble in the last 12 months and last 7 days which has prevented normal activity.

b. Section 2: additional questions relating to the neck, the shoulders and the lower back further detail relevant issues. Twenty-five forced-choice questions elicit any accidents affecting each area, functional impact at home and work (change of job or duties), duration of the problem, assessment by a health professional and musculoskeletal problems in the last 7 days.

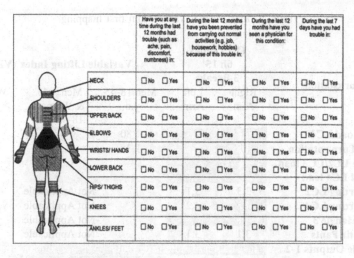

Fig. 1. Nordic musculoskeletal questionnaire body map.

2.2 Second Step: Changes of Working Organization

The second step changes of working organization on the basis of the results of step one: the analysis of questionnaires showed higher prevalence of symptoms in the female groups, in particular in the iron area. Symptoms increase with age and BMI but not with years of employment. The joints mostly affected are in order: lumbar spine, shoulders, cervical spine and wrists.

On these basis a reorganization of the job processes were carried out, with changing in the machinery and in job organization (e.g., time of work at the workstation, including more frequent rotations of workers), and a specific individual practical training of workers was realized.

3   Results

Table 2 shows the summarizes the risk index changed with respect to the first mapping following the interventions implemented.

Table 2. Risk indexes varied from first mapping.

Workstation	Ocra Index 6h 15' 3 Breaks		Variable Lifting Index (VLI)		
	Right Arm	Left Arm	Men 18-45 y.o.	Men > 45 & Women 18-45 y.o.	Women > 45 y.o.
Load LC4	5,3	5,3	1,80	2,25	/
Half-Turn rotation between LC2/LC3 load and LC4 load	4,8	4,8	1,38	1,72	/
Sorting LC2	3,3	3,1	Not Applicable		
Sorting LC4	4,8	4,2	Not Applicable		
Mangle 3	4,8	4,8	Not Applicable		
Stir Pants	3,1	3,1	Not Applicable		
Mangle Outputs 1-2	Installation of Automatic Exits – Activity no longer performed				
Mangle Outputs 4-5	/	/	0,55	0,68	0,91
Half-Turn Rotation between the bagging machines	/	/	1,15	1,44	1,92

4 Discussion

The technical improvements made led to a significant reduction in risk indices. The main technical adjustments made by the company following the first mapping are:

A. Installation of a support bench for the bags to the loading of the washing machine 4 which, if correctly used, reduces the over-shoulder elevations, avoids the double movement of the bags and reduces the maintenance/shaking times of the loads.
B. Installation of 2 routers to sorting 2 and 4.
C. Adjustment of the quotas of 2 sorting belts.
D. Lowering of the upper insertion point of the stretch-trouser machine, which completely avoids over-speeding for all operators.
E. Installation of a connecting belt ("automatic exit") between the stackers of the ironers and in the envelopes of the envelopes (complete elimination of manual movements of loads).
F. Modification of the ribbons and roller conveyors of the flat ironing inserters, which caused bending of the rachis in the unloading of the packaged packs.
G. Test and installation of anti-fatigue mats to reduce the effects of maintained static upright posture.

The on-the-job training activities have allowed us to deepen the specific operating procedures (providing the Company with targeted technical indications) and to share specific and applicable instructions with operators on the technical gestures that minimize the risk, compatibly with the specific nature of the different activities. The maintenance over time of the shared operating procedures requires the participation of activities of the different levels of the company.

References

1. Nunes IL (2009) FAST ERGO_X – a tool for ergonomic auditing and work-related musculoskeletal disorders prevention. Work J Prev Assess Rehabil 34:133–148
2. PEROSH – Sustainable workplaces of the future – European research challenges for occupational safety and health. Multifactorial genesis of work-related Musculoskeletal Disorders (MSDs) 2012
3. INRS – Dossier Troubles Musculosquelettiqes (TMS). Accessed 20 May 2015
4. Eurostat – Health and safety at work in Europe (1999–2007) – A statistical portrait, Inna Šteinbuka, Anne Clemenceau, Bart De Norre, August 2010
5. EU-OSHA – European agency for safety and health at work, Work-related musculoskeletal disorders: Back to work report 2007
6. EU-OSHA – European survey of enterprises on new and emerging risks (2010) Managing safety and health at work. Accessed 27 Feb 2015
7. Putz-Anderson V (1988) Cumulative trauma disorders: a manual for musculoskeletal diseases of the upper limbs. Taylor & Francis, Philadelphia
8. EUROFOUND – European foundation for the improvement of living and working conditions (2012) Fifth European working conditions survey 2010
9. EUROFOUND – Health and well-being at work: a report based on the fifth European working conditions survey 2012
10. HSE – Health and safety executive. Lower limb disorders. Accessed 22 May 2015
11. HSE – Health and safety executive (2002) Upper limb disorders in the workplace
12. PEROSH – Sustainable workplaces of the future – European research challenges for occupational safety and health (2012). Multifactorial genesis of work-related Musculoskeletal Disorders (MSDs)

Job Seniority and Time of Daily Exposure to Biomechanical Risk Factors in Claims of Work-Related Upper Limb Musculoskeletal Disorders in Chile

Marta Martínez[1]($\boxtimes$) and Paulina Hernández[2]

[1] Mutual de Seguridad, Cámara Chilena de la Construcción,
Chilean Society of Ergonomics (SOCHERGO), Teatinos 258, Santiago, Chile
mamartinez@mutual.cl
[2] Latin American Union of Ergonomics,
Predio El Médano S/N, Maitencillo, Chile
pauhernandez@ergonomiachile.cl

Abstract. Since March 2016 in Chile, the Workplace Assessment Tool (WAT) has been used in order to provide the professional disease qualification process with elements that ensure greater objectivity. Six WAT formats were developed according to upper limb segment or pathology: shoulder, elbow, wrist, carpal tunnel syndrome (CTS), thumb and fingers. This tool collects information about the worker, seniority and time of exposure to risks during the workday. At the international level, no consensus has been reached regarding the development times of work-related musculoskeletal disorders. The objective of this study was to estimate the relationship between seniority, daily exposure during working hours and biomechanical risk factors. Daily exposure presented an average of 435 min (SD = 84), without significant differences when analyzing according to the segment evaluated. Seniority presented significant values for the Wrist and CTS segment (X^2 = 59.25, df = 25, P = 0.00). In addition, for the Wrist segment, a positive difference was generated in the range of less than 6 months old, in the presence of two risk factors (X^2 = 36.314, df = 15, P = 0.002). It is concluded that there is a relationship between seniority and the pathologies that affect the wrist segment, however this relationship with the rest of the segments is not fulfilled, with the exposure time in the day nor with the biomechanical risk factors.

Keywords: Job seniority · Time of daily exposure
Workplace assessment tool · Work-related musculoskeletal disorders

1 Introduction

In Chile, Law 16.744 establishes Social Insurance for accidents and occupational diseases. This insurance is administered by three private, non-profit entities, denominated "Mutuals" (a Workplace Health and Safety Organization), in addition to a public entity. Companies must become a member of one of these entities and pay the established insurance for each worker. Mutual societies provide their adherent companies with

© Springer Nature Switzerland AG 2019
S. Bagnara et al. (Eds.): IEA 2018, AISC 820, pp. 446–453, 2019.
https://doi.org/10.1007/978-3-319-96083-8_58

advice on the prevention of accidents and occupational diseases, and in the event that these events occur, workers have the right to all corresponding medical attention until their rehabilitation and re-employment, in addition to the payment of the corresponding financial subsidies. Under this modality, the insurance coverage reached 71% of the workers in 2016 (5, 736, 416) according to the information provided by the Superintendence of Social Security (SUSESO), a government entity that supervises mutuals. Since this insurance was created, the accident rates in Chile have fallen from 20 per 100 workers in the early 70's [1], to 3.6 accidents per 100 workers in 2016 [2]. The statistics of occupational diseases are known since 2000, and practically the rate has not experienced variations corresponding in 2016 to 0.15 per 100 workers, according to the numbers reported by SUSESO. The most reported pathologies are musculoskeletal (52% of total complaints of diseases to 2016), followed by mental health, however, these are the ones that are classified in a lesser proportion as occupational diseases, since only 11% of complaints are accepted by the system [2]. Until 2015, each mutual entity developed the process of qualifying complaints of diseases according to their own guidelines, there being important differences between one entity and another. Considering the above, during 2015, SUSESO issued Circular No. 3167 to instruct mutuals on the "Protocol for minimum evaluation standards that must be met in the process of qualifying the origin of diseases reported as occupational." The objective of this Circular was "to provide the qualification process of occupational diseases with elements that ensure greater transparency, uniformity, specificity and objectivity". The document indicates the general protocol applicable to the qualification of pathologies, whatever their nature, reported as of presumably occupational origin and describes the specific protocols for upper limbs musculoskeletal pathologies and mental health pathologies. Main guidelines regarding musculoskeletal pathologies contained in the document are:

- Qualification of pathologies by a committee composed of at least 3 professionals (a physician specialized in work medicine and a traumatologist or physiatrist), recommending the participation of an occupational therapist or kinesiologist or other professional with training in ergonomics (diploma or master).
- For the qualification, the committee must have available: clinical evaluation, corresponding examinations, Workplace Evaluation (WAT) and background of surveillance of work-related musculoskeletal disorders (WMSDs).
- It presents WAT formats by pathology that must be performed in the workplace by an occupational therapist, kinesiologist or other professional, the latter with training in ergonomics, trained in the specific use of these formats.
- Training in WAT formats of at least 32 h, 16 of which correspond to supervised application of the instrument.
- Instructs the obligation of the employer to change the worker from the job position or readjust it in case of qualification of work origin, with a compliance period of 90 days.

This new qualification process took effect on March 1, 2016. The upper limb musculoskeletal pathologies for which six WAT formats were developed are: Trigger Finger, Flexor and Wrist Extender Tendinitis, De Quervain's Tendonitis, Carpus Tunnel Syndrome, Epicondylitis and Epitrocleitis, Biceps Tendonitis, Rotator Cuff Tendinopathy and Sub-Acromial Bursitis.

448   M. Martínez and P. Hernández

The WAT instrument records the seniority and laterality of the worker, as well as personal data. From the information on the organizational conditions, it includes: name of the position, total hours of work, performance of shifts and overtime, existence of breaks (programmed, unscheduled and inherent to the process and its duration) and if rotation of tasks during the day is being conducted. In the identification of the biomechanical risk factors, the evaluator must verify the presence of risk posture (s) (static or dynamic) for the studied pathology, frequency of movements per minute, and estimated force according the Borg scale [3], analyzing a maximum of three tasks, which should be representative of the working day. To determine exposure, the effective exposure time considering the duration of the working day (including overtime, if any) and the existing breaks should be indicated. It also reports the presence of additional factors that have been associated with WMSDs, such as the presence of vibrations, exposure to cold, use of gloves, among others. Circular 3167 does not indicate criteria for qualification, but these remain under the judgment of the evaluation committee.

Risk factors for WMSDs are known, various personal conditions, work organization, biomechanical and psychosocial factors that affect its development [4–13] have been described, and the magnitude, especially of the biomechanical factors that would cause damage, has also been studied. [3, 9, 10, 14, 15]. However, there is no consensus about the exposure times, both during the]working day, and historical. International manuals have been developed to guide diagnosis [16–18] however, the criteria differ on the sufficient time for the development of a certain pathology, and some refer to months of exposure (considering the historical exposure in the workplace) or exposure during the working day. But in very few cases both times are described together. Clearly, given the multiple factors that influence the development of these conditions, this is a very complex task, since not only the magnitude of each factor, but also its interrelation during the development of the job, as well as the exposure time, during the working day and the historic data should be considered. Despite the difficulty it presents, in order to contribute to an objective qualification, it is necessary to generate criteria that help physicians in the decision regarding the relationship of the job with the pathology developed.

Considering the above, this paper intends to estimate the relationship between seniority, daily exposure during working hours and biomechanical risk factors in WMSDs complaints in Chile.

2 Methodology

A total of 3344 WATs corresponding to workers who denounced a WMSD during 2016 in one of the private mutuals were analyzed descriptively and by non-parametric hypothesis testing.[1]

---

[1] The Total Exposure Time variable was not adjusted to normal distribution.

It is relevant for the interpretation of data that in 2016 this mutual covered 97,428 companies and a total of 1,979,390 workers, corresponding to 34% of workers covered nationwide. In terms of economic activities, these correspond in a greater percentage to Construction, Commerce, Transport and Public Administration. Another relevant fact is that 97% of the adherent companies correspond to companies with 100 or less workers.

For the analysis, for each type of WAT specific data was considered, both about the consulting population (gender and age), as well as the seniority at the job, the hours of the working day and the biomechanical risks present in them.

Seniority was measured in months and years in the evaluated job position, generating 6 ranges. Total time of exposure to risks (TTE) during the working day was measured in minutes, to calculate it a formula was used in which the total time of the working day and overtime was added, and the scheduled breaks were subtracted (e.g. lunch), unscheduled breaks (e.g. use of the restroom, drinking a coffee), and frequent process breaks (e.g. lack of material), in addition, cases without data and outliers were eliminated. For the biomechanical factors, the presence of risk positions was considered according to the format of each WAT, the repetitiveness for each segment was considered as risk in the case of being equal to or greater than the criteria defined by Kilbom. (4), and the force was considered as a risk if being equal to or greater than 3 in the Borg Scale (3). The presence of risks for each task (3217 tasks) was added and a score of 0, 1, 2 or 3 was assigned, according to the amount of risks present.

3 Results

As seen in Table 1, the highest frequency of WATs corresponds to the segment Elbow (32%), followed by Shoulder (28%) and Wrist (24%). The evaluations for both CTS, Thumb and Fingers, together account for only 16%. Women presented a statistically significant higher frequency for Wrist segments, including CTS, and men presented a statistically significant higher frequency for the Elbow segment (X^2 = 88.220, df = 5, p = 0.00). Regarding age, a lower mean was observed in the evaluations of Wrist and Thumb (37 and 38 years respectively), being the highest corresponding to the finger

Table 1. Frequency type of workplace assessment (WAT), gender, age and total time of exposure during the working day (TTE).

Type of WAT			Gender		Age	TTE
	Frequency	Percentage	Male	Female	Mean (SD)	Mean (SD)
Shoulder	926	27.7	432	494	43.55 (11.34)	433.55 (88.94)
Elbow	1084	32.4	583*	501	43.42 (9.96)	438.63 (84.75)
Wrist	797	23.8	278	519*	37.44 (11.01)	432.13 (82.12)
CTS	235	7	74	161*	40.4 (11.28)	444.88 (73.22)
Thumb	209	6.3	83	126	38.68 (11.64)	436.20 (83.63)
Fingers	93	2.8	46	47	44.11 (11.91)	442.07 (75.26)
Total	3344	100	1496	1848	41.54 (11.17)	436.04 (84.27)

* Significant difference at 0.05

segment (44 years). The TTE presents an average of 435 min, equivalent to just over 7 h a day, with a standard deviation of 84 min (Fig. 1). No significant differences were observed for this variable according to the segment evaluated (Table 1), nor when related to seniority (F = 1,049, df = 25, p = 0.39).

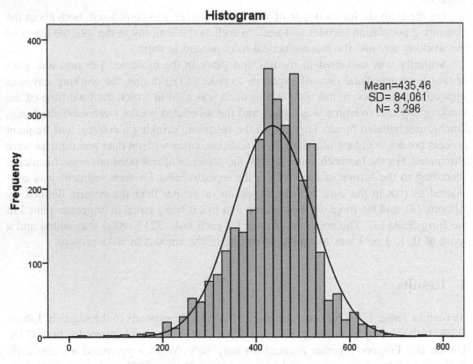

Fig. 1. Histogram of the total time of exposure (TTE) during the working day

Using the Chi-square test, significant differences were found between seniority range and the type of WAT (X^2 = 59.25, df = 25, p = 0.00). Based on the normalized standardized residuals, more cases are observed for the Wrist segment, in the "Less than 6 months" and "1 to 3 years" ranges, and fewer cases than expected for the range "10 years or more". For the WAT corresponding to CTS, a greater number of cases is observed for the "Less than 6 months" range. In the case of the Elbow segment, there is a lower number of cases for the "1 to 3 years" seniority range. The Shoulder segment also presents fewer cases for the "Less than 6 months" range, and more cases than expected for the "10 years or more" range. The Thumb and Fingers segments do not present significant differences in terms of seniority (Table 2).

Table 2. Crosstabs for seniority range according to WAT.

Type of WAT	Seniority range						
	<6 months	6 months to 1 year	1 to 3 years	3 to 5 years	5 to 10 years	10 years or more	Total
Shoulder	72*	94	227	181	169	183*	926
Elbow	105	131	246*	192	212	198	1084
Wrist	110*	87	223*	137	132	108*	797
CTS	40*	23	59	32	46	35	235
Thumb	31	21	61	35	27	34	209
Fingers	15	9	22	17	14	16	93
	373	365	838	594	600	574	3344

* Significant difference at 0.05

Table 3 shows the analysis of the relationship between seniority range in the workplace and the level of risk of the tasks according to type of WAT, finding significant differences for Elbow ($X^2 = 25,652$, df = 15, p = 0,042), and for Wrist ($X^2 = 36,314$, df = 15, p = 0,002). For the Elbow segment, there are fewer cases than expected for the exposure range of less than 6 months, in the presence of a risk factor, and in the presence of 3 risk factors for seniority between 5 and 10 years. On the other hand, for the Wrist segment, a smaller number of cases is observed in the presence of a risk factor with less than 6 months old, and more cases than expected for the age range of 3 to 5 years. In addition, for the same segment, in the presence of two risk factors, more cases than expected were observed in the range of less than 6 months old. It is necessary to emphasize that the risk factor that presented the highest frequency in the totality of the evaluations was posture, and the most frequent combination was posture and force. On the other hand, the least frequent risk factor was repetitiveness.

Table 3. Crosstabs for seniority range according to number of biomechanical risk (Only for Elbow and Wrist segments)

WAT type	No. of risk factors	<of 6 months	6 months to 1 year	1 to 3 years	3 to 5 years	5 to 10 years	10 years or more	Total
Elbow	0	9	3	14	20	17	16	79
	1	21*	40	76	55	65	54	311
	2	62	71	126	90	121	106	576
	3	12	16	26	24	8*	20	106
Wrist	0	6	8	11	18*	8	9	60
	1	24*	38	104*	49	59	49	323
	2	68*	34	93	57	53	41	346
	3	10	6	11	9	12	9	57

* Significant difference at 0.05

4   Discussion

The limitations of this study refer mainly to the fact that only cases of complaints made in a mutual society are analyzed and that they do not necessarily reflect the variability of mutuals in general. On the other hand, it is necessary to mention that values for exposure times were very outside of normal values, considering that in Chile there are regulations regarding the duration of the working day, establishing a maximum of 8 h for ordinary working days (45 h weekly), in addition to a limitation of the extra weekly hours of 12 h maximum. This may be due to errors in the verification of these data during the evaluation, or typing errors. In addition, the variable of seniority in the workplace does not consider previous similar exposures, therefore it could underestimate the total exposure time of the workers.

Findings are mainly related to the differences in gender and age of the workers whose jobs were evaluated with the WAT instrument. Significant differences were observed especially for the wrist segment, including the evaluation of CTS, where women report a greater number of complaints. On the other hand, men are more frequent in the evaluations of the Elbow segment.

When relating seniority with risk factors, significant findings were obtained, especially in the evaluations of Wrist and CTS, where the presence of one or two risk factors increased the number of cases higher than expected. It should be remembered that in both segments women significantly outnumber men in frequency. The literature refers as specific risk factors for this segment, extreme postures, repetition and direct compression of the palmar area, however in this study the most relevant factors were posture and force.

Despite the above, it was possible to observe that the majority of complaints are generated after the year of seniority, regardless of the number of risk factors present, contradicting what is expressed in the literature and what was expected by the research team. Therefore, this parameter could be influenced by other variables, beyond the risk factors present in the workplace.

The data obtained from the WAT instrument could shed light on the relationship between the exposure times and the magnitude of the risk factors, if a more accurate analysis of the historical exposure and its relationship with the onset of symptoms were included, so as to specify the time elapsed until the beginning of the damage.

We convey our appreciation to:
Leonardo Aguirre Aranibar
Head of Applied Research Projects
Mutual of Safety C.Ch.C

References

1. Comisión Asesora Presidencial para la, Seguridad en el Trabajo. Seguridad en el Trabajo Informe Final (2010). http://datos.gob.cl/uploads/recursos/informeComisiónSeguridaddel Trabajo.pdf. Accessed 6 May 2018
2. Superintendencia de Seguridad Social. Informe Anual Estadísticas de Seguridad Social (2016). https://www.suseso.cl/607/articles-40371_archivo_01.pdf. Accessed 6 May 2018

3. Borg G (1982) Psychophysical bases of perceived exertion. Med sci Sport Exerc. http:// fcesoftware.com/images/15_Perceived_Exertion.pdf. Accessed 6 May 2017
4. Kilbom Å (2000) Repetitive work of the upper extremity: Part I-Guidelines for the practitioner. Elsevier Ergon B Ser 1(C):145–150
5. Taylor P, Grieco A, Molteni G, Vito GDE, Sias N, De Vito G et al (2013) Epidemiology of musculoskeletal disorders due to biomechanical overload. Ergonomics 1998(41):1253–1260
6. Kuijer PPFM, Van Der Molen HF, Frings-Dresen MHW (2012) Evidence-based exposure criteria for workrelated musculoskeletal disorders as a tool to assess physical job demands. Work 41(SUPPL.1):3795–3797
7. Muggleton JM, Allen R, Chappell PH (1999) Hand and arm injuries associated with repetitive manual work in industry: a review of disorders, risk factors and preventive measures. Ergonomics 42(5):714–739. http://www.tandfonline.com/doi/abs/10.1080/0014 01399185405. Accessed 16 May 2018
8. Grieco A, Molteni G, Vito G de, Sias N (1998) Epidemiology of musculoskeletal disorders due to biomechanical overload. Ergonomics 41(9):1253–1260. http://www.tandfonline.com/ doi/abs/10.1080/001401398186298. Accessed 16 May 2018
9. Putz-Anderson V, Bernard B, Burt S (1997) Musculoskeletal disorders and workplace factors. ... Related Musculoskelet, 97–141 m, 1-1-7–11. http://scholar.google.com/scholar? hl=en&btnG=Search&q=intitle:Musculoskeletal+disorders+and+workplace+factors#1% 5Cnhttp://www.cdc.gov/niosh/docs/97-141/pdfs/97-141.pdf
10. Punnett L (2014) Musculoskeletal disorders and occupational exposures: how should we judge the evidence concerning the causal association? Scand J Public Health 42(13 Suppl):49–58. http://sjp.sagepub.com/cgi/doi/10.1177/1403494813517/324%5Cnpapers3:// publication/doi/10.1177/1403494813517324
11. Andersen JH, Haahr JP, Frost P (2007) Risk factors for more severe regional musculoskeletal symptoms: a two-year prospective study of a general working population. Arthritis Rheum 56(4):1355–1364
12. Silverstein BA, Stetson DS, Keyserling WM, Fine LJ (1997) Work-related musculoskeletal disorders: comparison of data sources for surveillance 31:600–608. https://deepblue.lib. umich.edu/bitstream/handle/2027.42/34815/15_ftp.pdf?sequence=1&isAllowed=y. Accessed 26 April 2017
13. Punnett L, Wegman DH (2004) Work-related musculoskeletal disorders: the epidemiologic evidence and the debate. J Electromyogr Kinesiol. 4(1):13–23. http://www.ncbi.nlm.nih.gov/ pubmed/14759746. Accessed 15 July 2012
14. Kilbom Å (1994) Repetitive work of the upper extremity: Part II - the scientific basis (knowledge base) for the guide. Int J Ind Ergon 14(1–2):59–86
15. Nordander C, Ohlsson K, Åkesson I, Arvidsson I, Balogh I, Hansson G-Å, et al (2013) Exposure–response relationships in work-related musculoskeletal disorders in elbows and hands – a synthesis of group-level data on exposure and response obtained using uniform methods of data collection. Appl Ergon 44(2):241–253. http://linkinghub.elsevier.com/ retrieve/pii/S0003687012001160. Accessed 26 April 2017
16. de Sanidad Consejería (2016) Comunidad de Madrid. Enfermedad Profesional Valoración de la Sospecha, Documento de Consenso en la Comunidad de Madrid
17. European Commission (2009) European Commission Information notices on occupational diseases: a guide to diagnosis. European Communities, Luxembourg, 282 p
18. Sluiter J, Rest K, Frings-Dresen M (2001) Criteria document evaluating for the work-relatedness of upper-extremity musculoskeletal disorder. Scand J Work Environ Health 27 (Suplement 1):1–102

The SIN-DME Questionnaire
(Symptoms of INcomfort Associated
with Muscle Skeletal Disorders)

Juan A. Castillo-M$^{(\boxtimes)}$ and María C. Trillos Ch

Health and Medicine School, ErgoMotion-Lab,
Universidad del Rosario, Bogota, Colombia
juan.castillom@urosario.edu.co

Abstract. This research was carried out between 2010 and 2017; the objective was to develop the Intervention Protocol for the Prevention of Musculoskeletal Disorders, the research followed three phases as follows: the first one dealt with the design of the protocol and the analysis tools, the second one focused on the adjustment of the intervention and modeling processes of the tools and the third one on the validation of the protocol and its operation in the companies.

In total, the applications of the tools and the validation of the development were carried out with the results obtained from 97 companies from 11 economic sectors that participated; from this universe of companies a sample of 2648 workers was constituted. This is a self-assessment questionnaire of WMSD's-related symptoms, designed in the category of questionnaires that focus on identification through the perception of health by individuals.

In this regard, it should be noted that previous studies carried out by the scientific team of the ergonomics laboratory of the University of Rosario, started from the hypothesis that groups of workers with some pathology associated with WMSD's would present a low perception of health, however, it was found that most of the workers perceived their own state of health as excellent. This is explained because when the symptomatology of the pathology presented by the worker does not prevent or condition the activities of daily life, people will perceive that they have an adequate state of health [1]. The questionnaire uses validated information, scales and the development of visual identification items to bring the representations of individuals closer to the estimates made. The conceptual objective for the development of the questionnaire is to record and measure the prevalence of symptoms in workers who may be associated with a musculoskeletal disorder, with the purpose of verifying the presence of symptoms later through a more exhaustive evaluation of health conditions.

Keywords: Ergonomics · Prevention · Musculoskeletal disorders
Manufacturing system

© Springer Nature Switzerland AG 2019
S. Bagnara et al. (Eds.): IEA 2018, AISC 820, pp. 454–462, 2019.
https://doi.org/10.1007/978-3-319-96083-8_59

1 The Research Findings Framework

1.1 The Development Process

This research was carried out between 2010 and 2014; the objective was to develop the Intervention Protocol for the Prevention of Musculoskeletal Disorders, the research followed three phases as follows: the first one dealt with the design of the protocol and the analysis tools, the second one focused on the adjustment of the intervention and modeling processes of the tools and the third one on the validation of the protocol and its operation in the companies [2].

In total, the applications of the tools and the validation of the development were carried out with the results obtained from 97 Colombian companies from 11 economic sectors that participated, from this universe of companies a sample of 2648 workers was constituted, 412 workers were the object of an in-depth psychosocial and clinical evaluation. According to the results obtained, more than 50% of the population in the sample was between the ages of 30 and 49, of whom 60% worked 8 h a day and 30% worked 10- or 12-h shifts. Regarding gender, there was a 70% male; 30% female distribution.

In the third phase, analysis of a group of 509 workers, 44% of the men and 40% of the women presented problems of overweight. Fifty-four per cent of them did not engage in any physical activity and 45 per cent did so on weekends. In this regard, it was recorded that 55% of the workers reported pain, discomfort or nuisances; when evaluating the intensity of pain with the EVA scale, it was found that 26.7% rated pain between 0 and 3/10, 53.4% between 4 and 7/10 and 19.9% rated it between 8 and 10/10.

In the clinical evaluation of this group it was found that with respect to the work posture 63% declared that they worked sitting and 46.7% with their back flexed, of which 90.2% presented this posture frequently or permanently, in this same evaluation it was found that the most declared symptomatology was lumbago, followed by rotator cuff syndrome. From the population evaluated, it was evident that workers who declared symptoms and presence of pain had not consulted their doctor for a diagnosis and to receive treatment according to their needs.

In this respect, of the total number of workers who were evaluated in the third phase, a sample of 274 workers was taken for in-depth clinical evaluation; in this phase, it was found that the sample consisted of 53% men and 47% women, of whom 69.7% of the sample was in the 36 to 56 year-old age range, with an average of 16.4 years of experience in the position.

In relation to the history of the sample, it was found that nearly half of the workers had a previously diagnosed disease, of which 36% were skeletal muscle type. On the other hand, 31 cases of qualified occupational illness were found.

In this sub-sample, it is important to note that 100% of the women indicated the presence of pain, and 83.2% of them described the pain as chronic, and it was found that in 43% of the cases the duration was greater than twelve months. Regarding the segments, the following symptoms were found in order of importance: 43% lumbar spine; 37.4% shoulder; 26.7 wrist; 16.7 cervical spine; 15.9% hand and 13.7% dorsal spine; the percentages do not add up to 100% given that they are not mutually exclusive categories, that is, a worker may have had more than one segment involved.

In this regard, 248 of them were found to have between 1 and 4 segments involved, representing 90.9%, 14 workers had five or more segments involved, representing 5.1% and only 11 male workers reported having no segments involved.

In relation to the distribution by sex of the number of segments involved, it is more frequent that the population of working women in the sample had more than one segment involved, in relation to men, this difference was statistically significant. The median of this variable indicates that at least 50% of the sample of workers had two segments involved and 25% of the sample more than three segments involved. When evaluating the behavior of the segments involved according to age category, it was found that the category in which symptomatic workers with the greatest number of segments involved were found was between 47 and 56 years old, followed by workers between 36 and 46 years old.

When reviewing the behavior of the presence of symptoms according to quadrants, it was found that there was a differential distribution in relation to gender, since in the upper, left and right quadrants, the frequency of affectation in the women in the sample was greater in relation to the men and this difference is statistically significant. In the case of the lower segments, the frequency of symptoms was higher in men than in the female workers in the sample, and this difference is also statistically significant. Some type of posture disturbance was also found in 70% of cases, with similar behavior in relation to sex. Likewise, 80% of the population of workers in the sample had some type of neurological evaluation problem.

1.2   The Protocol Design

The development of the Intervention protocol for the prevention of WMSD's focused on the instrumentalization of a comprehensive approach to the analysis and intervention of work situations. The development of data collection methods was proposed that could help to make an integral reading of the different events that are part of a work situation [3].

To this end, an approach based on a multi-method approach was launched, i.e., approaches based on action research were combined with elements of quantitative research [4]. The objective is to include a double point of view in the analysis: the worker's point of view and the point of view of the expert analyst of the work situation.

The selection of this approach is intended to overcome the "expert view" in the analysis of occupational health problems, it is well known that experts in occupational health and safety use checklists based on standardized parameters to judge situations and determine diagnoses, this has been found to be an insufficient intervention model and is also poorly adapted to the complexity of the WMSD's problem in companies [5, 6].

In the expert view, one-dimensional scales are used, which are frequently used in checklists (such as OWAS, OCRA, etc.). These tools are based on the following principle; the expert observes and qualifies a property in relation to a reference value (for example: range of motion), which can vary over time over a continuous execution of a task. The result is obtained by taking the significant value from the variation in the expert's opinion and then judging it against the normal value that the expert has assigned to the selected property.

This type of procedure significantly and importantly reduces the inclusion of the set of variables that are specific to work situations and that also condition the production of productive action of workers. In order to achieve a global and systematic understanding of the WMSD's problem in companies, it is necessary to include in the analysis, perception and point of view of all prevention actors, since this is the only way to develop sustainable prevention in the company.

1.3 The Questionnaire Design

This is a self-assessment questionnaire of WMSD's -related symptoms, designed in the category of questionnaires that focus on identification through the perception of health by individuals.

In this regard, it should be noted that previous studies carried out by the scientific team of the ergonomics laboratory of the University of Rosario, started from the hypothesis that groups of workers with some pathology associated with WMSD's would present a low perception of health, however, it was found that most of the workers perceived their own state of health as excellent. This is explained because when the symptomatology of the pathology presented by the worker does not prevent or condition the activities of daily life, people will perceive that they have an adequate state of health [7, 8].

It must also be recognized that the experience of a symptom becomes a complex phenomenon where biological, psychological, social and cultural aspects intersect. These may reflect a complex variety of sensations, which may not necessarily be linked or related to an illness or injury. In addition, it should be considered that, in general, people manage symptoms in a private manner, without making expression of it with their relatives or in their context. Therefore, the design and development of a questionnaire that explores the problem of the symptom is essential to understand "the experience of the symptom, the interpretation of the symptom and its management in daily life" [9].

In the study of WMSD's, because pain is the most frequently evoked symptom, and because there are no concepts of general application in this regard, it is necessary to gain knowledge about it, therefore the objective of this questionnaire is to comprehensively address the symptom from the perspective of the worker's experience.

The questionnaire uses validated information, scales and the development of visual identification items to bring the representations of individuals closer to the estimates made. In this analysis framework, it should be remembered that pain is influenced by two cognitive components: catastrophic view, which is defined as the set of cognitive and emotional processes that contribute to increasing pain levels. The second component is the acceptance of pain, which definitely influences the perception of pain; according to the evidence gathered, it is said that people with greater acceptance of pain consistently report less pain, less anxiety, depression, less disability and better work status.

The conceptual objective for the development of the questionnaire is to record and measure the prevalence of symptoms in workers who may be associated with a musculoskeletal disorder, with the purpose of verifying the presence of symptoms later through a more exhaustive evaluation of health conditions. Prevalence is defined as "the proportion of individuals in a population who present the event at a given time, or period of time,". As prevalence studies are strongly influenced by the speed of occurrence of the event (symptom) and its duration, the results of this questionnaire must necessarily be verified with the use of a more precise analysis instrument, which counts the signs that allow the health conditions of the workers to be determined. The development method used included the execution of the following steps:

a. Definition of the conceptual framework of reference, through the exploration of the scientific literature on the study of the symptoms associated with WMSD's. The purpose of the literature review was to establish the domains that should be registered and measured in order to establish the proportion of individuals whose working and health status should be studied more deeply.

b. Similar questionnaires were identified, criteria used were established and commonly applied scales were verified.

c. A pilot application was conducted in the first phase of the study to verify the acceptability, consistency and relevance of the information collected through the questionnaire.

d. In the second and third phases of the study, the workers of the participating companies were applied, taking as an application criterion the absence of information regarding the health conditions of the musculoskeletal workers health.

e. A sample of workers was defined following the sampling model developed for the research, in cases where no information was available, the production units were selected and applied to 100% of the workers.

According to what Rasmussen et al. (2014) [10] established, it is necessary to differentiate in the conceptual framework of analysis the concept of "Symptom (subjective evidence)" and "experience of the Symptom (multidimensional construction)"; this allows us to approach an understanding of how the worker experiences the symptom within the framework of his work activities and within the framework of private activities.

1.4   The Questionnaire Domains

According to the conceptual framework defined, the questionnaire was developed including four domains identified as relevant to identify the workers who should be the object of a detailed study that will allow specifying signs that should finally be delimited to functional structures. The four domains [D], considered achieving this objective were (Table 1):

D1:   This corresponds to personal information that allows us to specify certain characteristics of the worker. A component helps to pinpoint the influence of this as an activator of WMSD's.

D2:   Includes two components considered relevant by the literature regarding the availability of the worker's physical reserve and considered as protective elements to the exposure of events associated with WMSD's.

D3:   Explores the nature of the work activity from the perspective of the frequency of exposure in terms of temporality and the influence of various actions performed by the worker

D4:   Explores the experience of the symptom through four components that seek to characterize this experience from a subjective point of view.

Table 1. Example of the domains and subdomains considered in the questionnaire design.

Domains	Sub-domain	Component	Note	Specific weight	Global weight
D2 habits and lifestyle	Tobacco consumption	1–5 cigarettes	3	7, 5	15
		6 to 15 cigarettes	4		
		16 and more cigarettes	5		
	Years of Tobacco consumption	Less than 1 year	1		
		1 a 2 years	2		
		3 a 4 years	3		
		5 a 9 years	4		
		10 years and older	5		
	Frequency of physical activity	Never	5	7, 5	
		Weekend	4		
		Three times a week	3		
		Twice a week	2		
		Everyday	1		
	Duration of physical activity	15 min	5		
		30 min	4		
		60 min	3		
		More than 60 min	2		

For the sub domains considered, we have identified in the literature the reduction of respiratory capacity in the case of tobacco consumption, which implies overexertion when performing physically demanding tasks, and also imply longer recovery times, in addition to the known health implications.

On the other hand, physical activity is considered a protective factor as it improves aerobic capacity by allowing workers to better tolerate exposure to demanding physical activities, and it also helps to improve joint functionality when working on activities that involve holding fixed work positions for long periods of time.

It should also be considered that sporadic physical activity, including short exercise routines without a physical background, can cause discomfort and even injury. Therefore, even though physical activity is considered protective if it is not carried out under the appropriate frequencies and conditions, it can be a precursor to discomfort.

In the domain of work characteristics, it was taken into account that the physical and cognitive complexity of the analyzed tasks should be integrated in the analysis, since the duration of the tasks will have a specificity related to this complexity.

It has also been considered that the actual execution times of the tasks must be subtracted from the nominal times, i.e. the times referred to breaks (power supply) or operational breaks (component changes, maintenance, etc.) in order to obtain the actual times of exposure to the activities and demands of the work.

On the other hand, it has also been considered that the use of rotations is implemented, specifically in manufacturing and operational tasks, however, these rotations have more of a vocation for controlling repeatability. These generally do not consider physical involvement at the muscular or cognitive level, so the increase in rotations can be considered as a factor that makes activity more complex and can increase levels of demand.

On the other hand, if there is no rotation and tasks must be performed in closed cycles where the same set of gestures and positions are systematically reproduced, then the effect at the level of the WMSD's can be seen over time, i.e. it is installed progressively, however it must be considered that workers are capable of developing protective operational resources, however these only delay the time of manifestation of the signs associated with WMSD's.

In the domain of health status, it is taken into account that an analysis is made at a specific moment in the execution of the task, for which reason aspects such as the moment of productivity, the instantaneous state of the worker and the nature of the productive demands must be integrated.

The sub domains then focus on the location of the discomforts, the characterization of these, the presence of pain, the moment of perception of the same, the permanence resented by the worker and how this disappears or is integrated into the perception of the worker.

In the same way, using a scale of intensity we obtain the instantaneous perception of the same with respect to the location, in many cases this reveals the difficulties of carrying out actions and should be analyzed considering sub domains of the previous sub domains related to habits, work and BMI.

2   The Final Properties of the Questionnaire

A questionnaire was developed, designed to respond to the need to guide the analysis and development of WMSD's intervention strategies. This questionnaire facilitates the identification of workers with discomfort or symptoms, and also provides guidance on the management and addressing of specific analyses that allow the signs to be specified and thus determine the therapeutic and operational management at the work level. In the same way, this questionnaire allows the analysis to establish a well characterized line of action according to the domain and the qualification obtained.

The questionnaire was evaluated by a panel of experts in terms of linguistic coherence and continuity, this allowed the formulation of the questions to be clarified, as well as the pilot test to verify the importance of having visual resources to facilitate the location of areas with discomfort and the coding of discomfort, this because a worker may manifest more than one of them and may find that they manifest differently in different areas of the body. That's why the graphic code and visual help were added.

The questionnaire was applied to a sample of workers from the following regions in Colombia: Bogotá, east and west, in the departments of Cundinamarca, Boyacá, Antioquia, Santander, Norte de Santander, for analysis, data were taken from seven cities and municipalities; Bogotá, Tausa, Sutatausa, Ubaté and Segovia, corresponding to two departments; Cundinamarca and Antioquia.

Eighteen companies were included, representing four economic sectors. Since one of the companies has partial data, the analyses reported below are the results for 17 companies. In this study, the statistical analysis was carried out after consensus had been reached by the research group, based on an initial analysis proposed. Given the non-probabilistic type of sampling, a uni-variate analysis of the data collected was considered.

The results indicated that the domains allowed the delimitation of workers with symptoms, in order to determine if the symptoms were associated with signs and thus determine the quality of the information collected, a sample of 84 workers (30%) was taken, of which 78 were evaluated, from this evaluation of health conditions with ISO-WMSD's, from this application it was concluded that in 23 cases, the worker required a specific evaluation of a segment (Process I) and in 59 cases it was concluded that the worker required a thorough clinical evaluation of the quadrant. This means that the questionnaire made it possible to locate workers who were in the process of developing a WMSD's, and that by carrying out an in-depth evaluation not only of their health conditions, but also of their work situation, the lines of action in terms of prevention could be determined.

According to Selltiz (1977), [11] validity is established as "the degree to which differences in test scores reflect actual differences in people's scores with respect to the measured characteristic, rather than systematic or random errors. The questionnaire gives an account of these differences, and also allows the analyst to understand that the cross-analysis of the domains will not only allow him/her to better explain the differences between the workers, but will also give him/her the possibility of directing the efforts of a deep and precise evaluation of the health conditions in workers who actually present signs that must be evaluated and analyzed.

In short, a questionnaire has been developed structured into four domains and eighteen sub domains, which cover the experience of the symptom and make it possible to construct a global image of the condition of the work and the worker, where this perception takes place.

Similarly, the application of the questionnaire allows us to appreciate its consistency and structure, since it was completed by workers of different origins and training, and also facilitated the addressing processes for the evaluation of workers' health conditions. At the same time, it offers the analyst a first global approach through the sub domains to understand the social, organizational and cultural implications of the reported symptoms, in such a way that it facilitates the selection of resources for the development of interviews and for the application of tools with greater focus on specific aspects such as comfort, quality of life, physical activity, working conditions and health conditions.

References

1. Work-related musculoskeletal disorders_facts and figures National report_Spain.pdf, EU-OSHA –European Agency for Safety and Health at Work (2010)
2. Castillo (2011) Protocolo de Intervención para la prevención de los DME Preparado por Juan A. Castillo M. Ph.D.
3. Abdullah MZ, Othman AK, Ahmad MF, Justine M (2010) The mediating role of work-related musculoskeletal disorders on the link between psychosocial factors and absenteeism among administrative workers
4. Hauke A, Flintrop J, Brun E, Rugulies R (2011) The impact of work related psychosocial stressors on the onset of musculoskeletal disorders in specific body regions. a review and meta-analysis of 54 longitudinal studies 25(3):243–256
5. World Health Organization (2012) National Profile of Occupational Health System in Finland: World Health Organization, Geneva, p 12
6. Perdomo-Hernández M (2014) Grado de pérdida de capacidad laboral asociada a la comorbilidad de los desórdenes músculo esqueléticos en la Junta de Calificación de Invalidez, Huila, 2009–2012. Rev Univ Ind Santander Salud 46(3):249–258
7. Ministerio de trabajo (2013) II Encuesta Nacional de Condiciones de Seguridad y Salud en el Trabajo en el Sistema General de Riesgos Laborales. Bogotá. D.C, p 49–50
8. Ministerio de la Protección Social (2007) Primera encuesta Nacional de condiciones de salud y trabajo en el sistema general de riesgos profesionales. Bogotá. D.C: Ministerio de la Protección Social, p 88
9. Nübling M, Stößel U, Hasselhorn H-M, Michaelis M, Hofmann F (2006) Measuring psychological stress and strain at work - evaluation of the COPSOQ questionnaire in Germany. Psychosoc Med 3:Doc 05. http://www.ncbi.nlm.nih.gov/pubmed/19742072%5Cn
10. Rasmussen A, Katoni B, Keller AS, Wilkinson J (2011) Posttraumatic idioms of distress among darfur refugees: hozun and majnun. Transcult Psychiatry 48(4):392–415
11. Selltiz C, Wrightsman LS, Cook SW (1977) Les méthodes de recherche en sciences sociale. Les Editions HRW, Montréal

Issues with the Implementation of Material Handling Regulations in Switzerland

Maggie Graf(⊠)

Working Conditions Research Division, Department of Labour, State Secretariat
of Economic Affairs, Bern, Switzerland
maggie.graf@seco.admin.ch

Abstract. In Switzerland, there are specific legal requirements for companies to
provide lifting tools and organisational measures to avoid the necessity to lift.
From international surveys on risks and at work and company prevention
measures it does seem that these requirements are having an effect, however, in
order for laws to be implemented, companies must be aware of the requirements
and labour inspectors must insist on compliance. In both of these areas, there is
still work to be done. The survey show that physical risks at work are generally
increasing, although they still remain within the European average. Companies
seem to have introduced the technical measures more frequently than the
organizational measures prescribed by the law.

Keywords: Manual handling · Laws · Compliance

1 Relevant Swiss Framework

1.1 Relevant Laws

There are two ordinances in Switzerland that deal with manual handling. Ordinances
are legally binding subsidiary documents to laws. Laws in Switzerland, as in many
countries, state general provisions such as that an employer has a duty to prevent injury
and illness to employees while they are at work. An ordinance gives detailed technical
specifications about the prevention of specific known risks. Ordinances may refer to
Standards, but Standards are not legally binding in the sense that they do not need to be
followed to the letter. The Swiss law sees Standards as guidelines, which can be
adapted according to the specific situation in a company. Ordinances, on the other
hand, must be followed as a minimum standard in all cases.

The Labour Law of Switzerland [1] prescribes general duties of employers
regarding working hours, health protection, building requirements for workplaces and
social provisions such access to drinking water, cloakrooms and rest areas. Ordinance 3
to the Labour Law (Occupational Heath) [2] specifically addresses health protection
measures. Along with other provisions concerning ergonomic aspects, it contains an
article with four clauses that specifically require

- the provision of appropriate tools and organisational measures to avoid the
  necessity to lift loads. It further specifies that

S. Bagnara et al. (Eds.): IEA 2018, AISC 820, pp. 463–467, 2019.
https://doi.org/10.1007/978-3-319-96083-8_60

- when organisational and technical measures cannot remove the need to lift, then lifting aids must be provided.

  And...

- Employees must be informed about lifting risks and must be instructed how they are to do it.
- They must be informed of the weight of loads that they are required to lift and the weight distribution.

In comparison with many countries, this ordinance is quite detailed and specific about manual handling. Most developed countries require adequate measures to reduce manual handling risks but do not stipulate the means to achieve it, or mention the types of measures that should be considered, or in what order.

1.2 Law Enforcement

The State Secretariat for Economic Affairs SECO is responsible for overseeing compliance with the Labour Law, its Ordinances, and the Ordinance to the Accident Insurance Law on the Prevention of Accidents and Occupational Diseases, which is very similar in content. Inspections are carried out by cantonal (state) labour inspectors who have access to all companies and are trained and supported by SECO. To assist with this, SECO has produced Guidelines (administrative regulations) to the application of the Ordinances. These have a similar legal status to Standards, in that companies have some freedom in how they implement them. However, if a case is taken to court, the court will use the Guidelines to understand the necessary level of protection and decide if the measures taken by the company are equivalent to it. The administrative Guidelines contain weight lifting limits that vary according to age and gender. Pregnant women have the lowest limits.

If a labour inspector is not satisfied that a legal requirement has been met, according to the Ordinance and Guidelines, he or she will advise the company and request, in writing, an improvement within a specific period of time, according to the seriousness of the infringement. He or she will then check at the end of this period and assess whether the situation has been rectified. If there is a dispute, the inspector will refer the matter to the courts. If the courts agree with the labour inspector's assessment, a fine will be given. Only in the case of imminent life-threatening risks are labour inspectors allowed to stop the work in a company. It is not the view of the lawmakers and enforcers that manual handling risks fall into this category.

2   Real World Application

2.1   The Use of Assessment Tools

In developing and regularly reviewing the Guidelines, the authorities rely on evidence-based scientific studies, whenever they are available. However, the studies are often confined to specific situations that do not reflect the diversity of the real world very well. For example, it is not always possible to lift a load close to the body, or to avoid overhead lifting and carrying. Labour inspectors are frequently faced with decisions in practice that require special consideration and possibly a detailed study. As we know, there are various assessment tools available that they, or company personnel, may use to assess the risk in such situations and the Guidelines developed by the authorities refer them to these. The validity of an assessment tool must be scientifically accepted, before it receives such official endorsement. The issue here is that neither the labour inspectors nor the employees of most companies, most of which are very small, really understand how to use these tools or even where to find them. The inspectors may attend training courses but they are not mandatory.

2.2   The Competence of the Available People

There is an additional article in the Occupational Health Ordinance to the Labour Law that allows a labour inspector to require an employer to consult an appropriate professional to obtain a professional assessment, if there is grounds to doubt that the health protection measures are adequate. I know of a case where a labour inspector did require the consultation of a competent ergonomist to solve some lifting and carrying issues in a textile manufacturing plant. Another labour inspector required such an expertise to assess the situation for order pickers in an online marketing warehouse, but accepted the assessment of a physiotherapist untrained in ergonomics. However, most labour inspectors are very reluctant to oblige companies to consult external experts in relation to manual handling issues. This has to do with the status of ergonomists in Switzerland. The profession is not officially recognised as one of the "occupational safety and health specialists", which include safety specialists, occupational hygienists and occupational physicians. The reason for this, is that the ergonomics society did not exist at the time that the law was drafted and up until this year the society had no system for ensuring the quality of their members (national certification system). In the case of a legal dispute, labour inspectors must be able to rely on the acceptance by the courts of the experts that they consult.

2.3   Law Enforcement in Practice

In practice, there are very few sanctions on firms. Courts are limited by the law to specific levels of fine. Sometimes the fines are not adequate to motivate a company to make a change and they will happily pay the fine, rather than the cost of improvement. However, this very rarely occurs, as companies are often more afraid of damage to their reputation than the cost of the fine. Where public sympathy may fall with the workers, they are very unwilling to dispute the authority of the labour inspectors and almost all orders from labour inspectors are accepted by companies.

Most firms are concerned to "do the right thing" and are willing to comply with instructions from authorities and professional experts. On the other hand, the chances of inspection, particularly in the service sector, is quite low and companies usually only consult an expert if they have a serious issue to resolve. Labour inspectors try to concentrate their efforts on areas where they have most effect. This means that they tend to inspect medium to larger sized enterprises with known risks. Small enterprises, particularly in the service sector, may not see a labour inspector within a decade. It is often these types of enterprises where the manual handling risks are high.

Companies may be aware of the risks but most companies have little competence to manage appropriate prevention. Many ergonomists who work in the field can confirm that even rudimentary measures taught in lifting classes are "new" to many people. The availability of lifting aids is also not well known, however a comparison of data for Switzerland with the average of all member countries of the European Union (Switzerland is not a member) [3], shows that the specific articles in the Occupational Health Ordinance may have a positive effect, although it is confined to the provision of technical measures (equipment) rather than organisational measures. Figure 1 shows that the use of lifting devices is slightly higher in Switzerland, whereas other prevention measures to do with physical risks are less frequent.

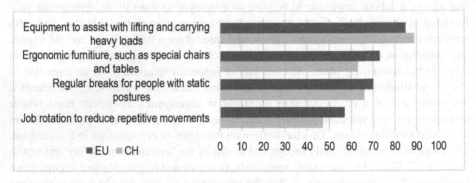

Fig. 1. Percent of Swiss and European companies who report the provision of various manual handling prevention measures (ESENER, EU-OSHA [3]).

The prevalence of several physical risks are increasing in Switzerland, although they have become equivalent to the average in the European countries. Repetitive movements were comparatively rare a decade ago, and painful or tiring working postures were also less frequent than in other European countries (see Fig. 2). These results also indicate that the legal requirement to provide lifting equipment may be helping to reduce the number of people who need to lift, as lifting and carrying remains slightly less frequent in Switzerland than in the other European countries.

3   What Does Work?

The most effective method of enforcing the law is increasing the awareness of risks through measures aimed at the public and directly at enterprises. Laws are only as good as the awareness of them by the people who are required to implement them. This awareness can be done by adequately trained occupational health and safety specialists working with and in companies. In Switzerland, as in all other European countries, there are laws that require companies to consult occupational health and safety specialists, although the Swiss laws are "lighter" than the European laws on this matter,

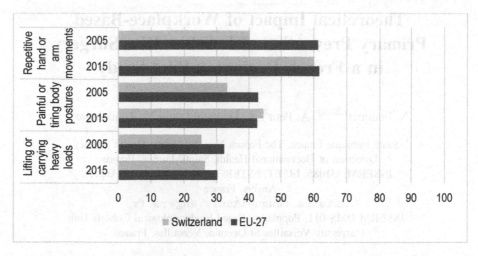

Fig. 2. Percent of people who report being exposed to various manual handling risks for at least ¼ of their working time (European Working Conditions Survey, Eurofound, [4]).

requiring consultation only by companies with specific risks. Lifting of loads over the limits recommended in the administrative Guidelines are included in the list of risks.

In Switzerland, although ergonomists are not recognised as safety and health specialists under the law, a lot of effort has been put into including the prevention of manual handling risks into the training of safety specialists, occupational hygienists and occupational physicians. This is done by qualified ergonomists.

References

1. Labour Law of Switzerland (only available in German, French and Italian). Arbeitsgesetz ArG. http://www.gesetze.ch/inh/inhsub822.11.htm. Accessed 28 May 2015
2. Ordinance 3 to the Labour Law (only available in German, French and Italian). Verordnung 3. https://www.admin.ch/ch/d/sr/c822_113.html. Accessed 28 May 2015
3. 2nd European Survey of Enterprises on New and Emerging Risks (2014) European Agency for Health and Safety at Work
4. 5th and 6th European Working Conditions Survey (2005, 2015) Eurofound

Theoretical Impact of Workplace-Based Primary Prevention of Lumbar Disc Surgery in a French Region: A Pilot Study

N. Fouquet[1,2(✉)], A. Petit[2], A. Descatha[3], and Y. Roquelaure[2]

[1] Santé Publique France, The French National Public Health Agency,
Direction of Occupational Health, Saint-Maurice, France
[2] INSERM, U1085, IRSET, ESTER Team, University of Angers,
Angers, France
natacha.fouquet@univ-angers.fr
[3] INSERM UMS 011, Population Based Epidemiological Cohorts Unit,
University Versailles St-Quentin, Versailles, France

Abstract. *Aim:* To assess the theoretical impact of workplace-based primary interventions designed to reduce exposure to personal and/or work-related risk factors for LDS. *Methods:* Cases of LDS were assessed using hospital discharge records for persons aged 20–64 in 2007–8 in the French Pays de la Loire region. We estimated the number of work-related cases of LDS (WR-LDS) in high-risk industry sectors. Three theoretical scenarios of workplace-based primary prevention for sectors at risk have been simulated: a mono-component work-centered intervention reducing the incidence of WR-LDS by 10% (10%-WI), and two multicomponent global interventions reducing the incidence of all cases of LDS by 5% (5%-GI) and 10% (10%-GI) by targeting personal and occupational risk factors. *Results:* Four industry sectors were at high risk of LDS in the region, amounting to 435 [351–532] LDS cases, of which 152 [68–253] were WR-LDS: construction and information & communication for men; wholesale & retail trade and accommodation & food service activities for women. AFE was limited for each industry sector, 30%, 50%, 33% and 55%, respectively. The 10%-WI, 5%-GI and 10%-GI scenarios hypothetically prevented 15 [7–25], 22 [18–27] and 44 [35–53] LDS cases among sectors at risk, respectively. *Discussion:* Prevention scenarios combining actions on personal and occupational risk factors would be the most effective, compared to prevention focused only on occupational risk factors. This suggests that, to reduce the incidence of LDS, implementing actions of promotion of health at work would be necessary in addition of actions on occupational risks.

Keywords: Lumbar disc-related disorders · Simulation · Preventive efficiency

1 Introduction

Low back pain (LBP), defined as spinal pain below the thoracic-lumbar junction (T12-L1) to the inferior gluteal fold [1], is a major public health issue in many industrialized countries [2] that extends worldwide [3], with high prevalence in the general population.

S. Bagnara et al. (Eds.): IEA 2018, AISC 820, pp. 468–477, 2019.
https://doi.org/10.1007/978-3-319-96083-8_61

LBP is among the first six health problems in terms of costs and among the first three sources of disabling pain in developed countries [4], especially in working age groups [5]. Indeed, LBP is responsible for pain, disability and handicap. It causes considerable human and social costs in terms of pain and discomfort in the workplace and everyday life [1, 6] and of health-related quality of life [7]. It is a source of sometimes severe functional sequelae, impairment of working capacity and career disruption for people and considerable costs for the society [8]. It generates substantial direct costs associated with seeking medical and paramedical care and diagnostic procedures [9–11] and more particularly indirect costs (compensation, job loss, etc.) which are 5–6 times higher. In addition to the intensity of pain, the severity of LBP is mainly due to the disability it causes [12, 13].

LBP is a complex symptom, and the type of contributors to both the pain and associated disability are multiple: lifestyle, psychological, social, biophysical, occupational... [5].

To prevent LBP and associated disability, guidelines recommend mainly self-management, physical and psychological therapies. According to Foster *et al.* [14], promising solutions to prevent LBP include focused *"implementation of best practice, the redesign of clinical pathways, integrated health and occupational care, changes to payment systems and legislation, and public health and prevention strategies"*.

The multifactorial origin of LBP makes it difficult to distinguish the relative contribution of personal and work-related factors at the individual level. However, at the population level a substantial number of LBP are mainly related to workers' personal characteristics and medical conditions. Cases that occur regardless of work exposures will be called 'personal-related' (PR) LBP in the remaining part of the text. Other cases occurring in excess in workers employed in jobs at high risk for LBP can be considered as mainly work-related or 'attributable to work'. The proportion of 'work-related' (WR) LBP can be estimated by the work-related attributable fraction of risk (AFE).

The reduction of LBP and associated disability in the workforce is a priority for policy makers due to the human, social and economic costs. Work-centered ergonomic interventions (WI) include ergonomic and organizational adaptation of the workplace. Some multifaceted global interventions (GI) add to the WI component various components of personal interventions (PI), such as worksite behavioral programs (e.g., social health promotion, exercise), education programs (e.g., education and training on risk-reducing working techniques) and diet programs to manage overweight [14, 15].

Multi-component global interventions (GI) including both personal behavioral interventions (PI components) and collective technical, ergonomic and organizational interventions (WI components) are considered the most promising preventive approach for LBP among workers [14, 16]. However, It's accepted in the literature that primary prevention of LBP is inadequate whereas secondary prevention programs on associated disability are more appropriate [14].

From a theoretical point of view, WIs focusing on work-related risk factors are expected to reduce mainly WR-LBP; their impact will depend on the proportion of cases attributable to work. Higher values can be expected for interventions focusing on some jobs at particularly high risks (high AFEs). Primary multi-component global interventions (GI) are expected to be the most efficient in targeting both PR-LBP and

WR-LBP, regardless of the AFE value [17, 18]. However, we still lack of information on the joined effects of reducing occupational and non-occupational risk factors.

The French national public health agency, Santé publique France, implemented a multilevel epidemiological surveillance system for work-related musculoskeletal disorders (MSDs) in the Pays de la Loire region (5% of French working population) [19] to identify occupations and sectors at high risk. The diseases analyzed in this program comprised those clearly identified as having a significant occupational component and risk factors whose effects are sufficiently established, and identified determinants and misunderstood phenomena [20, 21]. Although it has been shown that beliefs, expectations and preferences of patients and health-care practitioners may influence health care-seeking behavior for LBP, [22, 23] lumbar disc surgery (LDS), which can be identified in the medical databases from public and private hospitals [20], was chosen as the sentinel event for disc-related sciatica (DRS) and generally for LBP. A study was conducted in 2007–2008 in the Pays de la Loire region [20] that assessed the proportion of LDS attributable to work according to industry sectors.

On the same way of a recent article about carpal tunnel syndrome [24], the aim of this study was to assess the theoretical impact of workplace-based primary interventions designed to reduce exposure to personal and/or work-related risk factors for LDS.

2   Methods

Population: The population included in this surveillance program was made up of residents of the Pays de la Loire region (Loire Valley area, west central France) in the 20–64 age group (999,396 women and 995,883 men), whether they were professionally active or not, in 2009. According to the 2009 census data, the region has 5.7% of the French population, with a diversified socioeconomic structure [26].

Outcomes: The hospital discharge database of the French National Medical Information Systems Program (PMSI) was analyzed to include all patients aged 20 to 64 years residing in the region and having undergone LDS (list of the codes for surgical acts selected from PMSI database is detailed in a previous article [20]) in 2009.

Occupational History: Due to a lack of information on employment status in the PMSI database, we used data collected by a pilot study conducted among 3,150 persons having undergone LDS in the participating surgical centers (10 of the 14 regional centers for spinal surgery, representing 93% of LDS in the region in 2007) [20]. Each eligible patient was informed of the study by the surgeon and a consent form and a self-administered questionnaire were sent to collect medical and surgical history and employment history (industry, occupation and description of tasks throughout employment). For each occupation, industry sector was coded using the 2-digit codes of the French version of the statistical classification of economic activities in the European Community. The analysis was performed on the longest occupation of each entire working life before LDS, only for those employed at the time of surgery.

Scenarios of Prevention: As in a previous study of carpal tunnel syndrome [24], in absence of precise data in the literature, we arbitrarily hypothesized that interventions could reduce the incidence of LDS by 10% in high-risk jobs, and simulated three scenarios of workplace-based primary prevention, differing by their main target:

- *10% WI:* mono-component work-centered intervention targeting only work-related risk factors for LDS (e.g., ergonomic intervention: workstation redesign, establishment of an ergonomics task force, job rotation, ergonomics training, etc.) expected to reduce WR-LDS by 10%;
- *5%-GI* and *10% GIs:* multi-component global interventions targeting both personal and work-related risk factors for LDS and expected to reduce both PR-LDS and WR-LDS by 5% or 10%, respectively.

Statistical Analysis: Incidence rates of LDS in the whole population were computed separately for each gender. Using the information from the 2007–2008 pilot study, three indicators were computed: the age-adjusted standardized incidence ratios of LDS calculated for each sector with all other sectors as reference ($SIR_{sector-LDS}$), the age-adjusted relative risks of LDS according to sector computed using the Mantel-Haenszel method ($aRR_{sector-LDS}$), with the whole sample of subjects included in the study as reference, and the age-adjusted attributable fraction of risk in exposed individuals (AFE) which estimates the proportion of LDS attributable to work in the sectors at high risk of LDS [20, 27]: $AFE_{sector-LDS} = (aRR_{sector-LDS} - 1)/aRR_{sector-LDS}$.

These indicators were computed for each sector when (i) more than 10 men or women were employed and (ii) $aRR_{sector-LDS}$ was significantly higher than 1. Sectors at high risk of LDS in comparison with the whole population were thus detected and called "high-risk sectors". Then, specific incidence rates ($I_{sector-LDS}$) were computed according to high-risk sectors. The total number of LDS ($N_{sector-LDS}$) in the sector considered was computed by multiplying the number of workers employed in this sector ($N_{e-sector}$) by the incidence rate in this sector ($I_{sector-LDS}$). The number of WR-LDS ($N_{sector-WR-LDS}$) was calculated by multiplying the total number of LDS ($N_{sector-LDS}$) by the AFE in the sector considered ($AFE_{sector-LDS}$) [27].

The preventive efficiency (PE) was estimated as the ratios of LDS hypothetically avoided/total number of LDS (%) in the sector considered. A 95% confidence interval was computed only for $SIR_{sector-LDS}$ and for $aRR_{sector-LDS}$. For other indicators, a range was calculated using the lower and upper limits of the considered indicator in the calculation formula.

3   Results

Four industry sectors were at high risk of LDS in the region, amounting to 435 [351–532] LDS cases, of which 152 [68–253] were WR-LDS (Table 1): construction and information & communication for men; wholesale & retail trade and accommodation & food service activities for women. AFE was limited for each industry sector, varying between 30% for men construction and 55% for women of accommodation & food service activities.

Table 1. Standardized incident ratios (SIR) and incidence of lumbar disc surgery (LDS), attributable risk fractions among the exposed (AFE) and estimated number of cases of LDS in the high-risk sectors in men or women

	N Workers #	% Workers a	SIR_sector b	95% CI c	I_LDS d	Range e	AFE f	Range g	N_sector-LDS h	Range i	N_WR-LDS j	Range k
Construction												
Men*	95,372	12.5	1.3	1.0 1.5	1.4	1.1 1.7	30.3	14.6 43.1	134	110 161	41	16 69
Women	12,595	1.8	0.6	0.2 1.3	0.4	0.1 1.0						
Total	107,966	7.4										
Wholesale & retail trade												
Men	98,417	12.9	0.7	0.6 0.9	0.8	0.6 1.0						
Women*	88,213	12.8	1.4	1.1 1.7	1.1	1.0 1.3	32.7	15.5 46.4	117	76	31	12 54
Total	186,630	12.9										
Accommodation & food service activities												
Men	19,969	2.6	1.0	0.5 1.6	1.1	0.6 1.7						
Women*	23,305	3.4	1.9	1.3 2.7	1.5	1.0 2.1	54.9	35.1 68.6	34	23 49	19	8 34
Total	43,274	3.0										
Information & communication												
Men*	19,358	2.5	0.7	0.3 1.2	0.7	0.4 1.3	49.6	10.5 71.6	14	7 25	7	1 18
Women	9,390	1.4	0.9	0.3 2.0	0.7	0.3 1.5						
Total	28,748	2.0										
All high-risk sectors												
Men	114,730	15.1							148	117 186	48	17 87
Women	111,518	16.1							129	99 166	50	20 88
Total	366,619	25.2							435	351 532	152	68 253
Active population												
Men	760,849	100			1.1				827			
Women	691,273	100			0.8				614			
Total	1,452,122	100			0.9				1441			

* sectors at high risk of LDS; # 2009 INSEE Census data; a. % of the active regional population; b. Standardized incidence ratios of LDS (SIR_sector-LDS); c. 95% confidence interval (CI); d. Incidence of LDS per 1000 persons-years (I_LDS); e. Range computed using the lower and upper limits of the 95% CI of SIR_sector-LDS; f. Attributable fractions of LDS (AFE_sector-LDS(%)); g. Range computed using the lower and upper limits of the 95% CI of aRR_sector-LDS (data not shown); h. Total number of LDS (N_sector-LDS) in the sector considered; i. Range computed using the number of workers of each sector and gender (N) and the lower and upper limits of the range of I_sector-LDS; j. Number of WR-LDS cases (N_sector-wr-LDS); k. Range computed using the lower and upper limits of the range of AFE_sector-LDS(%).

Table 2. Estimated number of preventable cases of lumbar disc surgery (LDS) and preventive efficiency according the 10-WI, 5%-GI and 10%-GI preventive intervention scenarios in the high-risk sectors in men or women

	Preventable LDS according to preventive scenarios[a]						Preventive efficiency[b] (%)					
	10%-WI		5%-GI		10%GI		10%-WI		5%-GI		10%-GI	
	N	Range[c]	N	Range[d]	N	Range[d]	Mean	Range[e]	Mean	Range[f]	Mean	Range[f]
Construction												
Men*	4	2 7	7	6 8	13	11 16	3.0	1.2 6.4	5.2	3.7 7.3	9.7	6.8 14.5
Women	-	-	-	-	-	-	-	-	-	-	-	-
Total												
Wholesale & retail trade												
Men	3	1 5	5	4 6	10	8 12	3.2	0.9 6.6	5.3	3.4 7.9	10.5	6.8 15.8
Women*												
Total												
Accommodation & food service activities												
Men	2	1 3	2	1 2	3	2 5	5.9	2.0 13.0	5.9	2.0 8.7	8.8	4.1 21.7
Women*												
Total												
Information & communication												
Men*	1	0 2	1	0 1	1	1 3	7.1	0.0 28.6	7.1	0.0 14.3	7.1	4.0 42.9
Women												
Total												
All high-risk sectors												
Men	5	2 9	7	6 9	15	12 19	3.4	1.1 7.7	4.7	3.2 7.7	10.1	6.5 16.2
Women	5	2 9	6	5 8	13	10 17	3.9	1.2 9.1	4.7	3.0 8.1	10.1	6.0 17.2
Total	15	7 25	22	18 27	44	35 53	3.4	1.3 7.1	5.1	3.4 7.7	10.1	6.6 15.1

* sectors at high risk of LDS; # 2009 INSEE Census data; a. 10% WI: mono-component work-centered interventions targeting only work-related risk factors and expected to reduce WR-LDS cases by 10%; 5%-GI and 10% GI: multi-component global interventions targeting personal and work-related risk factors for LDS and expected to reduce both PR-LDS and WR-LDS by 5% or 10%, respectively; b. ratios of LDS hypothetically avoided / total number of LDS (%) in the sector considered, computed only if one of the inferior range is superior than 0; c. Range calculated using the lower and upper limits of the range of the number of WR-LDS cases ($N_{sector-wr-LDS}$); d. Range computed using the lower and upper limits of the range of WR-LDS and Upper limit of range of PE = lower limit of range of WR-LDS hypothetically avoided / Upper limit of range of total number of WR-LDS; e. Lower limit of range of PE = lower limit of range of WR-LDS hypothetically avoided / Upper limit of range of total number of WR-LDS and Upper limit of range of PE = lower limit of range of LDS hypothetically avoided / Upper limit of range of total number of WR-LDS; f. Lower limit of range of LDS hypothetically avoided / Upper limit of range of total number of LDS and Upper limit of range of PE = Upper limit of range of LDS hypothetically avoided / Lower limit of range of total number of LDS

As shown in Table 2, the number of avoidable LDS varied between the different preventive scenarios and sectors. The 10%-WI, 5%-GI and 10%-GI scenarios hypothetically prevented 15 [7–25], 22 [18–27] and 44 [35–53] LDS cases among sectors at risk, respectively. For each sector at risk, the hypothetical preventive efficiency was lower for the 10%-WI scenario compared to the 10%-GI and even the 5%-GI scenarios. Thus, for accommodation & food service activities for women (the highest AFE) the preventive efficiency was 5.9% [2.0–13.0] for the 10%-WI scenario, 8.8% [4.1–21.7] for the 10%-GI scenario and 5.9% [2.0–8.7] for the 5%-GI scenario. For construction for men (the lowest AFE) the preventive efficiency was 3.0% [1.2–6.4] for the 10%-WI scenario, 9.7% [6.8–14.5] for the 10%-GI scenario and 5.2% [3.7–7.3] for the 5%-GI scenario.

4 Discussion

Among the four sectors at high risk, AFEs were limited, varying between 30% for men in construction and 55% for women in accommodation & food service activities. This study found that a limited proportion of LDS in a French general working-age population were work-related, and that work-related LDS were concentrated in several high-risk industries.

Surveillance data used for the computation of potentially preventable LDS included data from one of the largest and most complete surveillance programs for LDS, covering an entire region of France [20, 28]. The French PMSI database registering only LDS underestimated cases potentially preventable since disc-related sciatica (DRS) requiring only medical treatment and more generally LBP were not counted. The proportion of DRS requiring surgery is unknown in France. Nevertheless, the rates of disc surgery computed in this region were close to those in France (data not shown), suggesting that no specific regional features of healthcare use or medical practice could explain our results. Given that the PMSI database lacked information relating to occupation and no more recent data were available, we used information on employment of patients undergoing LDS collected in 2007–2008 from all region's hand surgery centers to estimate AFEs of LDS [20]. No exhaustive job exposure data of the working population was available in the Pays de la Loire region, except the job titles collected by the 2009 Census, almost contemporary to our study data.

Certain very high-risk jobs involving few workers may not have been identified in the present study due to the lack of statistical power, and this might have led to underestimating the impact of work-centered prevention. The computation of the preventable cases of LDS supposed several hypotheses [27], namely (i) causal relationships between the occurrence of LDS and work exposure and (ii) substantial impact of interventions reducing exposure to risk factors at the workplace [27].

While there are many treatment options to chronic LBP, none are universally endorsed [29]. Since few years, the literature shows that the role of surgery is limited and recommendations in clinical guidelines vary [14]. The idea that chronic LBP is a condition best understood with reference to an interaction of physical, psychological and social influences, the 'biopsychosocial model', has received increasing acceptance.

Biopsychosocial framework, including psychological, social and occupational (organizational, biomechanical...) components, combined to psychological programs for patient with chronic symptoms, is recommended to guide management [29]. A recent paper underlines the necessity to integrate health and occupational interventions to improve in function and return to work and to reduce the economic and societal burden of work disability pensions due to LBP [14].

We did not evaluate the hypothetical preventive efficiency of interventions that focus only on personal risk factors, expecting that changes in "personal risk factors" would be an essential component of multifaceted workplace interventions (10%-GI scenarios) [24]. Combining interventions on personal and work-related factors was assumed to have a higher impact than interventions targeting only on personal or work-related factors [14]. To the best of our knowledge, we still lack data on the impact of multiple global interventions to estimate their joint effects. We have therefore adopted a simplistic additive model. We focused prevention only at the workplace level, although interventions to prevent LDS at the population level might worth investigating. Indeed, mass-media campaigns about back pain have namely proved to have some success in four countries [14].

This study suggests that prevention efforts to reduce exposure to work-related risk factors should focus on high-risk jobs. Simulated workplace-based mono-component work-centered interventions and multi-component global interventions showed that preventive efficiency varied depending on the intervention design, the number of workers in different jobs and the proportion of work-related LDS. Given that personal risk factors such as obesity are also risk factors for LDS, reducing rates of LDS in the general working-age population will also require strategies to reduce personal risk factors, particularly in jobs with low levels of work-related risk for LDS [18].

In conclusion, prevention scenarios combining actions on personal and occupational risk factors would be more effective than prevention scenarios focused only on occupational risk factors, even with higher incidence reduction targets. Thus, to reduce the incidence of LDS, implementing actions of promotion of health at work would be necessary in addition of actions on occupational risks.

References

1. Burton AK, Balagué F, Cardon G, Eriksen HR, Henrotin Y, Lahad A, Leclerc A, Müller G, van der Beek AJ (2006) European guidelines for prevention in low back pain. Eur Spine J Off Publ Eur Spine Soc Eur Spinal Deform Soc Eur Sect Cerv Spine Res Soc 15(Suppl 2): S136–S168
2. Balagué F, Mannion AF, Pellisé F, Cedraschi C (2012) Non-specific low back pain. Lancet 379:482–491
3. Louw QA, Morris LD, Grimmer-Somers K (2007) The prevalence of low back pain in Africa: a systematic review. BMC Musculoskelet Disord 8:105
4. Lamb SE, Hansen Z, Lall R, Castelnuovo E, Withers EJ, Nichols V, Potter R, Underwood MR (2010) Group cognitive behavioural treatment for low-back pain in primary care: a randomised controlled trial and cost-effectiveness analysis. Lancet 375:916–923

5. Hartvigsen J, Hancock MJ, Kongsted A, Louw Q, Ferreira ML, Genevay S, Hoy D, Karppinen J, Pransky G, Sieper J, Smeets RJ, Underwood M (2018) Lancet low back pain series working group: what low back pain is and why we need to pay attention. Lancet London, England

6. Punnett L, Prüss-Utün A, Nelson DI, Fingerhut MA, Leigh J, Tak S, Phillips S (2005) Estimating the global burden of low back pain attributable to combined occupational exposures. Am J Ind Med 48:459–469

7. Yamada K, Matsudaira K, Takeshita K, Oka H, Hara N, Takagi Y (2014) Prevalence of low back pain as the primary pain site and factors associated with low health-related quality of life in a large Japanese population: a pain-associated cross-sectional epidemiological survey. Mod Rheumatol 24:343–348

8. Dagenais S, Caro J, Haldeman S (2008) A systematic review of low back pain cost of illness studies in the United States and internationally. Spine J Off J North Am Spine Soc 8:8–20

9. Becker A, Held H, Redaelli M, Strauch K, Chenot JF, Leonhardt C, Keller S, Baum E, Pfingsten M, Hildebrandt J, Basler H-D, Kochen MM, Donner-Banzhoff N (2010) Low back pain in primary care: costs of care and prediction of future health care utilization. Spine 35:1714–1720

10. Ritzwoller DP, Crounse L, Shetterly S, Rublee D (2006) The association of comorbidities, utilization and costs for patients identified with low back pain. BMC Musculoskelet Disord 7:72

11. Walker BF, Muller R, Grant WD (2003) Low back pain in Australian adults: the economic burden. Asia-Pac J Public Health Asia-Pac Acad Consort Public Health 15:79–87

12. Loisel P, Durand P, Abenhaim L, Gosselin L, Simard R, Turcotte J, Esdaile JM (1994) Management of occupational back pain: the Sherbrooke model. Results of a pilot and feasibility study. Occup Environ Med 51:597–602

13. Loisel P, Lemaire J, Poitras S, Durand M-J, Champagne F, Stock S, Diallo B, Tremblay C (2002) Cost-benefit and cost-effectiveness analysis of a disability prevention model for back pain management: a six year follow up study. Occup Environ Med 59:807–815

14. Foster NE, Anema JR, Cherkin D, Chou R, Cohen SP, Gross DP, Ferreira PH, Fritz JM, Koes BW, Peul W, Turner JA, Maher CG (2018) Lancet low back pain series working group: prevention and treatment of low back pain: evidence, challenges, and promising directions. Lancet London, England

15. Société française de médecine du travail (2013) Recommandations de Bonne Pratique - Surveillance médico-professionnelle du risque lombaire pour les travailleurs exposés à des manipulations de charges - Argumentaire scientifique. Société française de médecine du travail, Paris

16. Petit A, Mairiaux P, Desarmenien A, Meyer J-P, Roquelaure Y (2016) French good practice guidelines for management of the risk of low back pain among workers exposed to manual material handling: hierarchical strategy of risk assessment of work situations. Work Read Mass 53:845–850

17. Punnett L, Cherniack M, Henning R, Morse T, Faghri P (2009) CPH-NEW research team: a conceptual framework for integrating workplace health promotion and occupational ergonomics programs. Public Health Rep Wash DC 1974(124 Suppl 1):16–25

18. Sorensen G, McLellan DL, Sabbath EL, Dennerlein JT, Nagler EM, Hurtado DA, Pronk NP, Wagner GR (2016) Integrating worksite health protection and health promotion: a conceptual model for intervention and research. Prev Med 91:188–196

19. Roquelaure Y, Fouquet N, Ha C, Bord E, Arnault N, Petit Le Manac'h A, Leclerc A, Lombrail P, Goldberg M, Imbernon E (2011) Epidemiological surveillance of lumbar disc surgery in the general population: a pilot study in a French region. Jt Bone Spine Rev Rhum 78:298–302

20. Fouquet N, Descatha A, Ha C, Petit A, Roquelaure Y (2016) An epidemiological surveillance network of lumbar disc surgery to help prevention of and compensation for low back pain. Eur J Public Health 26:543–548

21. Rutstein DD, Mullan RJ, Frazier TM, Halperin WE, Melius JM, Sestito JP (1983) Sentinel health events (occupational): a basis for physician recognition and public health surveillance. Am J Public Health 73:1054–1062

22. Mannion AF, Wieser S, Elfering A (2013) Association between beliefs and care-seeking behavior for low back pain. Spine 38:1016–1025

23. Main CJ, Foster N, Buchbinder R (2010) How important are back pain beliefs and expectations for satisfactory recovery from back pain? Best Pract Res Clin Rheumatol 24:205–217

24. Roquelaure Y, Fouquet N, Chazelle E, Descatha A, Evanoff B, Bodin J, Petit A (2018) Theoretical impact of simulated workplace-based primary prevention of carpal tunnel syndrome in a French region. BMC Public Health 18:426

25. Ha C, Roquelaure Y, Leclerc A, Touranchet A, Goldberg M, Imbernon E (2009) The French musculoskeletal disorders surveillance program: pays de la Loire network. Occup Environ Med 66:471–479

26. INSEE: Population active, emploi et chômage en 2009 - Région des Pays de la Loire (52). https://www.insee.fr/fr/statistiques/2044042?geo=REG-52

27. Rockhill B, Newman B, Weinberg C (1998) Use and misuse of population attributable fractions. Am J Public Health 88:15–19

28. Fouquet N, Bodin J, Chazelle E, Descatha A, Roquelaure Y (2018) Use of multiple data sources for surveillance of work-related chronic low-back pain and disc-related sciatica in a French region. Ann Work Expo Health

29. Kamper SJ, Apeldoorn AT, Chiarotto A, Smeets RJEM, Ostelo RWJG, Guzman J, van Tulder MW (2014) Multidisciplinary biopsychosocial rehabilitation for chronic low back pain. Cochrane Database Syst Rev CD000963

Matching New Ergonomics Regulations to Stakeholder Competence in South Africa

Andrew Ivan Todd(✉)

Human Kinetics and Ergonomics, Rhodes University, Grahamstown,
South Africa
a.todd@ru.ac.za

Abstract. In 2013 the South African department of labor convened a new technical committee to develop ergonomics regulations for the country and in 2016 a draft of the regulation was released for public comment. Although the existing occupational health and safety act refers to ergonomics, the new regulations would be the first to emphasize the requirement for South African employers to have an established ergonomics program. In order to further contextualize these regulations, it is important to understand several key factors relating to each of the four identified stakeholder groups (business, government, labor and specialists) identified in the regulations. Firstly, the nature of business ownership in the country. The small and medium size enterprise sector contributes approximately 36% of the gross domestic product and constitutes up to 90% of formal businesses, while the informal economy contributes a further 8% of gross domestic product. These sectors typically function under several constraints and whether these sectors have the finances and resources required to effectively implement the regulations is questionable. Secondly, it is necessary to ensure that the department of labor and their inspectors are in a position to not only implement the regulations but also to assist the various sectors of the economy in the development of their programs. Currently, only a small proportion of the inspectorate have any formal training and the necessary skills in ergonomics. Thirdly, from a labor perspective the level of knowledge and training of the workers and those likely to be responsible for the actual implementation and running of the ergonomics program needs very careful consideration and better understanding. Lastly, due to the fact that the number of ergonomics practitioners (at all levels) and training programs in South Africa are small there is a very real risk that they are inadequately resourced to deal with the increased demand associated the regulations being introduced. It is clear that within each group of stakeholders there are legitimate barriers to the effective implementation of the regulations. In order to overcome these barriers, it is imperative that the level of competence of each stakeholder is matched to the demands placed on them by the new regulations. This presentation will highlight the ways in which the stakeholders have gone about trying to address the issues of ensuring their competence to deal with the introduction of the new ergonomics regulations. It will further explore some of the challenges that remain to be overcome and question whether or not the system (and the various stakeholders) are currently mature enough to cope with the introduction of broad and wide ranging ergonomics regulations.

Keywords: Regulations · Stakeholders · Constraints

© Springer Nature Switzerland AG 2019
S. Bagnara et al. (Eds.): IEA 2018, AISC 820, pp. 478–483, 2019.
https://doi.org/10.1007/978-3-319-96083-8_62

1 Background

The South African department of labor hosted a conference in 2012 with a focus on occupational health and safety: Road to zero injuries and diseases in Johannesburg, during which the Ergonomics Society of South Africa was requested to present on ergonomics and its role within a South African context [1]. An outcome of this conference was the formation of a technical committee for ergonomics by the South African department of labor in 2013. The mandate of the technical committee was to develop the first potential regulations for ergonomics to compliment the existing occupational health and safety act. The technical committee is comprised of the three tripartite members of the South African economy, namely, labor, government and business and voting members of the committee. There are additional non voting members (considered as specialists) in the ergonomics technical committee made up of human factors and ergonomics academics and practitioners. The technical committee released a draft of the regulations for public comment in 2016 [2]. In order for the regulations to be successful and to have the desired impact of improving both worker well-being and productivity it is vital that are relevant stakeholders are in a position to respond in an appropriate manner. Dul and colleagues [3] in the 2012 white paper commissioned by the IEA identified four key stakeholder groups; system actors, system decision makers, system influencers and system experts. Within the context of the draft regulations in South Africa these could be viewed as labor, business, government and specialists. The purpose of the current paper is to explore the ability of each of these stakeholder groups to respond to the impending ergonomics regulations.

2 Characteristics of Stakeholder Groups

In order to understand the challenges associated with developing ergonomics regulations that are contextual specific and tailored to the local needs it is necessary to further unpack the characteristics of the various stakeholders of interest within the South African context.

2.1 South African Labour

A key characteristic of the South African labour force is it diverse nature. According to Statistics South Africa [4] in 2014 only 25% of the workforce in the country were considered to be skilled workers, while a further 46% were semi-skilled and the remaining 29% classified as low-skilled workers. Furthermore, from 1994–2014 there has been uneven distribution of progress in terms of moving the workforce from low-skilled to skilled work. Due to the historical injustices of the apartheid regime in South Africa Black South Africans have traditionally been the majority of the low-skilled labor. Unfortunately, within the Black African employment the growth in skills, as a proportion, was much lower than in the other population groups in the last 20 years [4].

Another important characteristic of the labor force in South Africa is high unemployment rates. Figure 1 provides useful insights into not only the very high levels (around 25%) of unemployment but also the inequitable distribution of unemployment across different race groups. These characteristics are important because provide important contextual insights into the diverse nature of workers and likely constraints for the implementation of ergonomics regulations.

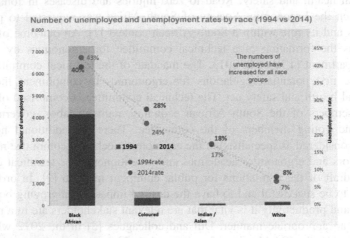

Fig. 1. Unemployment rates and unemployment by race within South Africa (Figure from Statistics South Africa [4])

2.2   South African Business

The recent shift in terminology used by the World bank from industrially developing countries to more specific income brackets [5] is a useful distinction within the context of this paper. This is due to fact that South Africa is characterized by diversity in industrial development from highly industrial multinational companies to informal traders. This is best depicted in the income distribution of the country as shown in Fig. 2, where no single income bracket represents the South African economy. Rather there is a broad continue of incomes from below the poverty line to very high income individuals. The nature of business ownership in South Africa follows a very similar trend; small and medium size enterprise sector contributes approximately 36% of the gross domestic product and constitutes up to 90% of formal businesses [6], while the informal economy contributes a further 8% of gross domestic product.

The economic resources of the small and medium size enterprises are typically very limited meaning that they function under several constraints and whether these sectors have the finances and resources required to effectively implement the regulations is questionable.

2.3   South African Government

The South African department of labour is responsible for the implementation of the occupational health and safety act and would be for the proposed ergonomics regulations as well. The department of labour has been proactive in attempting to be prepared for the implementation of ergonomics regulations in a variety of ways. Firstly,

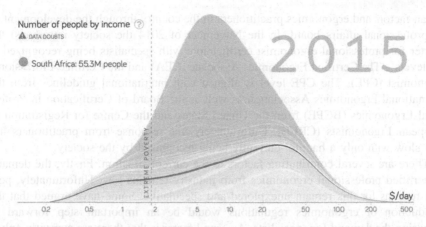

Number of people by income ⑦
△ DATA DOUBTS

● South Africa: 55.3M people

2015

EXTREME POVERTY

$/day

0.2   0.5   1   2   5   10   20   50   100   200   500

Fig. 2. South African income distribution (shown in blue) in 2015 (Drawn using the Gapminder tools) (Color figure online)

they have hosted numerous workshops that have highlighted the basic premise of human factors and ergonomics and the purpose of introducing the new regulations. These have primary focused on two groups of stakeholders; firstly business and secondly groups of interested practitioners from ergonomics and other cognate disciplines such as occupational hygiene and occupational therapy. The second proactive approach undertaken by the department of labour has been the training of inspectors in ergonomics where several groups of inspectors have undertaken approximately 360 h of ergonomics training.

These proactive steps of the department of labor are to be applauded and could go a long way to ensuring the success of the potential regulations. However, there are still only a limited number of inspectors who have been trained; and those that have been trained have very limited experience. Furthermore, the training that has been obtained is focused on gaining skills at the level of a certified ergonomics associate rather than at the level of a certified professional ergonomist. A likely consequence of this is that the inspectorate may not have the necessary understanding of the systemic nature of macro ergonomic issues within the workplace. The emergent characteristics of this dynamic are yet to be seen; for example, how effective is formal training in helping to implement ergonomics regulations within resource scarce small, micro and medium sized enterprises. These are just a few of the potential issues and the emergent characteristics of the system once the regulations are implemented remain unknown.

2.4   South African Ergonomics Association

The Ergonomics Society of South Africa (ESSA) was officially inaugurated in 1985, but was only recognized as a federated society of the International Ergonomics Association in 1994 [7]. Since then the growth of the society has been slow, with current membership fluctuating between 50 and 100 members. ESSA has also had a proactive approach to the potential introduction of the regulations. In particular, the society has developed a strong professional basis for the assurance of competence of human factors and ergonomics practitioners in the country through the development of the professional affairs board. In the November of 2014 the society introduced the Charter for professional ergonomist certification; with specialists being recognized at two levels – The Certified Ergonomics Associate (CEA) and the Certified Professional Ergonomist (CPE). The CPE level is aligned with international guidelines from the International Ergonomics Association as well as the Board of Certification in Professional Ergonomics (BCPE) from the United States and the Centre for Registration of European Ergonomists (CREE). Unfortunately, the response from practitioners has been slow with only a handful currently being recognized by the society.

There are several contributing factors to the current situation. Firstly, the demand for certified professional ergonomics from industry remains low. Unfortunately, possible reasons for this remain unexplored and speculative. Some have argued that the introduction of ergonomics regulations would be an important step forward in increasing the demand for specialists. A second factor is that there are currently only a few tertiary institutions providing formal recognized courses in human factors and ergonomics. Furthermore, there has been a traditional emphasis on laboratory research with little attempt to demonstrate the benefits of ergonomics to other stakeholders. This has started to change in the last decade but there remains a large amount of work to be done.

3   Regulations and Stakeholders

In order to ensure that ergonomics regulations can be successfully implemented it is important that they are matched to the needs and abilities of the various stakeholders. From a South African perspective several key considerations are clear:

- There is a large proportion of the labor force that is characterized as low-skilled workers.
- There is a large proportion of business that is characterized as small, medium or micro enterprises, many with significant resource scarcity
- Government that only has a small number of inspectors with any formal ergonomics training and little practical experience.
- An ergonomics society that only has a handful of certified professional ergonomists.

How these contextual characteristics are managed by the various stakeholders will play an important role in determining the likely success of the impending regulations. Furthermore, as ergonomics places an increasing emphasis on the systemic nature of the discipline it is important for the universities to respond appropriately. Wilson [8] laboratory research has an important role but not a primary one in systems ergonomics and a closer relationship between research and practice needs to be fostered. In fact, the notions that are advocated by Wilson; systems, interactions, context, holism, emergence and embedding would be a sound starting point for the successful implementation of regulations. This will allow for the recognition that there is no such thing as global solutions, but rather that solutions need to be highly contextualized to the context that they will be applied in. From a South African perspective this means ensuring that the ergonomics community is trained to be adaptive to the emergent characteristics of the rapidly changing economic and work environment landscape.

4   Conclusions

It is clear that several of the stakeholder groups have actively attempted to ensure that they are in a good position for the implementation of potential ergonomics regulations. However, the complexity associated with the South African labor market, the large diversity in business characteristics and small number of competent practitioners make this a daunting challenge. However, a systems focus that elucidates the affordances and constraints of each of the groups of stakeholders with an emphasis on a participatory approach may be able to overcome these challenges.

References

1. Todd AI (2012) Ergonomics and its role in developing a sustainable future for South Africa. South African department of labour international conference on occupational health and safety: Road to zero injuries and diseases. 7–8 March, Birchwood Conference Centre, Johannesburg
2. South African Department of Labour (2016) Occupational Health and Safety Act 1993 – Draft Ergonomics Regulations. Department of Labour, No. R. 64
3. Dul J, Bruder R, Buckle P, Carayon P, Falzon P, Marras WS, Wilson JR, van der Doelen B (2012) A strategy for human factors/ergonomic: developing the discipline and profession. Ergonomics 55(4):377–395
4. Statistics South Africa (2016) Employment, unemployment, skills and economic growth. An exploration of household survey evidence on skills development and unemployment between 1994 and 2014
5. World Bank: World Bank country and lending groups. https://datahelpdesk.worldbank.org/knowledgebase/articles/906519-world-bank-country-and-lending-groups. Accessed 24 May 2018
6. Global Entrepreneurship Monitor (2017) South Africa 2016–2017 report – can small business survive in South Africa
7. James J, Scott PA (2009) Ergonomics in South Africa, and beyond the borders. In: Scott PA (ed) Ergonomics in developing regions: needs and applications. CRC Press, Boca Raton, pp 343–348
8. Wilson J (2014) Fundamentals of systems ergonomics/human factors. Appl Ergonomics 45 (1):5–13

Utility Analysis of the Application of the Variable Lifting Index (VLI)

E. Alvarez-Casado[✉]

Centro de Ergonomia Aplicada (CENEA), Barcelona, Spain
enrique.alvarez@cenea.eu

Abstract. Introduction: The Variable Lifting Index (VLI) is one of the proposal extensions for the application of the Revised NIOSH lifting equation (RNLE) in highly variable work conditions. But is it necessary to apply these complex mathematical models to identify the main risk factor to be improved?

Methods: 10 warehouse clerk workplaces from different companies of logistics sectors have been assessed using VLI. 118 OHS practitioners have been asked about the strategy usually used to assess the manual lifting risk. After applying these strategies to the analyzed workplaces, the potential to identify efficient risk reduction interventions is compared.

Result: The most frequently used strategy to assess the manual lifting risk in highly variable conditions is to analyze the most frequent posture with the average weight. This strategy has identified the most significant risk factor, as VLI identified it, only in 2 out of 10 cases analyzed.

Discussion: VLI permits to identify the most efficient intervention to reduce the exposure level in a complex and variable working conditions. The use of other strategies will identify insufficient or misleading interventions. From a practical point of view, VLI will contribute to manage occupational risk related to manual lifting loads in a more efficient manner.

Keywords: Manual lifting · Variable lifting index · Risk assessment

1 Introduction

The relevance and utility of the Revised National Institute for Occupational Safety and Health Lifting Equation (RNLE) to assess the risk of manual lifting of loads is widely known [6]. But this equation also has mathematical limitations, mainly to analyze labor conditions of high variability [1].

In recent years, the Variable Lifting Index (VLI) has been presented as an extension of the RNLE to analyze and evaluate the risk related to manually lifting loads in workplaces where the weights of the handled loads and the postures and movements are very variable [4, 7, 8]. The VLI has already been validated in an epidemiological study as a risk assessment method; it can be said that this index is predictive of the acute low back pain due to the manual lifting of loads [2].

But the great utility of an analytical method of evaluating an occupational risk should be the potential to plan and design the work to avoid the risk of injury.

© Springer Nature Switzerland AG 2019
S. Bagnara et al. (Eds.): IEA 2018, AISC 820, pp. 484–494, 2019.
https://doi.org/10.1007/978-3-319-96083-8_63

Although, obviously, the design of an intervention to improve working conditions must be participatory and have a comprehensive view of the organizational, productive and technical factors, the effectiveness of interventions to reduce the probability of injury is important.

This study aims to show the utility of the VLI to identify the most priority risk factors in which to intervene. In addition, a comparative analysis of the use of VLI and the strategies most used by Occupational Health and Safety (OHS) practitioners to design interventions has been carried out.

Given that the logistics sector is a business activity that involves manual handling of loads under highly variable conditions, the study has been carried out in real cases of this type of company.

2 Methods

2.1 Selection of Workplaces

The workplaces that are the cases of this study have been selected from different CENEA client companies in the logistics sector.

The inclusion requirements have been the following:

- Exposure to manual handling of loads is continuous throughout the shift
- The duration of the work shift is 8 h.

The cases where tasks of pushing and/or pulling loads are present have been excluded.

For each company with workplaces that meet these requirements, the workplace with the largest number of exposed workers has been selected. A total of 10 jobs have been obtained.

2.2 Selection of Practitioners

To analyze the strategies currently used in companies to assess the risk due to manually lifting loads and to define improvement interventions, 150 occupational health and safety practitioners have been selected.

The inclusion requirements have been the following:

- More than 5 years of experience in occupational risk management
- Their main function is occupational risk management
- Experience in the logistics sector.

2.3 Workplace Analysis

The author has collected the organizational and productive data of each workplace and filmed the work activity for 15 min. The filming was made of the most senior worker in the workplace.

Dimensional measurements have been taken (heights, depths and asymmetries). The frequency of lifts has been obtained by dividing the number of objects handled in a

486   E. Alvarez-Casado

shift (data recorded in production) by the net time of work. Table 1 describes the data obtained from each workplace, where V means vertical distance and H means horizontal distance. The horizontal distance and the vertical distance are described following the criteria established in [3].

Table 1. Description of cases: measured data.

Case	1	2	3	4	5	6	7	8	9	10
Frequency	3,92	0,37	0,15	3,25	1,37	0,63	1,37	0,19	2,98	0,77
Assimmetry	45	45	0	45	0	0	0	30	30	45
V (cm)										
0 <= V <= 50	21,5%	29%	22%	22%	22%	22%	36%	46%	39%	
50 < V <= 125	57,0%	57%	67%	67%	67%	67%	46%	31%	42%	100%
125 < V <= 175	21,5%	14%	11%	11%	11%	11%	18%	23%	19%	
H (cm)										
25 <= H <= 40		50%	67%	100%		100%	46%	100%	100%	13%
40 < H <= 50	100%	50%	33%		100%		54%			63%
50 < H <= 63										25%

The weight of the loads has been obtained from the 1 month record. Table 2 describes the weights of the loads manually lifted in each case.

Table 2. Distribution of the weights of the lifted loads

Weight (kg)	Case 1	Case 2	Case 3	Case 4	Case 5	Case 6	Case 7	Case 8	Case 9	Case 10
W < 3	2,0%	28,7%	27,3%	28,7%	31,9%	35,8%	0,0%	45,0%	7,7%	58,6%
3 <=W < 5	16,0%	17,4%	26,8%	17,4%	23,6%	9,5%			9,6%	25,1%
5 <=W < 7	44,0%	16,2%	19,2%	16,2%	19,1%	10,5%	16,7%	20,0%	13,5%	10,6%
7 <=W < 9	16,7%	13,2%	10,5%	13,2%	12,6%	4,2%	11,1%		5,8%	1,1%
9 <=W < 11	6,7%	12,5%	8,9%	12,5%	7,4%		33,3%		5,8%	1,0%
11 <=W < 15	14,7%	12,1%	7,2%	12,1%	5,4%	10,5%	33,3%	10,0%	36,5%	3,6%
15 <=W < 20						21,1%	5,6%	5,0%	9,6%	
20 <=W < 25						7,4%		15,0%	7,7%	
25 <=W < 30						1,1%		5,0%	3,8%	
Average (kg)	7,3	5,8	5,2	5,6	4,9	8,9	10,3	8,0	11,5	3,6
Min (kg)	3	3	3	3	3	3	5	5	3	3
Max (kg)	14	12	12,5	13	12	26	15	25	28	13

2.4 Participation of the Practitioners

After presenting the scope of this study, the data of all cases were sent to the experts in digital format. The data sent included the filming of the work activity.

There were 3 videoconferences to explain the cases and resolve the doubts of the participants.

Answers to the following questions were collected during a month:

1. What strategy do you use to analyze this type of case?
2. For each case, what is the priority risk factor to improve?

The objective of the first question is to understand which is the most used procedure to evaluate the risk due to the manual lifting of loads in situations of high variability.

The objective of the second question is to identify which risk factor is considered by practitioners as the most priority to improve. The technical or economic viability of the intervention was not considered. The criterion is to identify which intervention will result in a greater risk reduction due to the manual lifting of loads (greater efficiency).

This second question was asked twice to each expert. The first time was done after receiving the files. In this way, the criterion is based solely on observation. The second time, the experts had 1 month to respond, taking time to do the necessary calculations.

Finally, the responses of 118 practitioners were collected.

2.5 Definition of the Risk Factor to Intervene

For the definition of interventions, the RNLE model has been followed. According to this equation, the factors that influence the risk are:

- Organizational factors: Frequency and duration
- Technical factors: Vertical distance, horizontal distance and vertical displacement
- Habit factors: Asymmetry
- Characteristics of the objects: Weight and quality of grasp.

To homogenize the responses, the following interventions to reduce risk were defined and provided to all participants:

- Intervention on horizontal distance: Reduce the horizontal distance to a value equal to or less than 40 cm ($25 <= H <= 40$).
- Intervention on vertical distance: Reduce the vertical distance to a value between 50 and 125 cm ($50 < V <= 125$).
- Intervention on the frequency: Reduce the frequency to half of the current frequency (Freq/2).
- Intervention on duration: Reduce the duration of exposure to short-duration (D = Short).
- Intervention on the asymmetry: To completely avoid the asymmetry ($A = 0°$).
- Intervention on the weight: Avoid the manual lifting of objects with weight equal to or greater than 11 kg ($W < 11$ kg).

Interventions on quality of grasp and vertical displacement have not been proposed. Improving the quality of grasp requires modifying the handled object, its container or handling the object with a manual tool. These types of interventions are often not viable in companies and difficult to imagine. The vertical displacement will be automatically improved by improving the vertical distances.

3 Results

3.1 Risk Assessment of Workplaces

The VLI of each workplace has been calculated (see Table 3). 25 kg of reference mass has been used [5].

Table 3. VLI of each case (m_{ref} = 25 kg)

Case	1	2	3	4	5	6	7	8	9	10
VLI	2,68	1,66	1,28	1,80	1,50	2,18	1,91	2,52	3,61	1,88

In all cases, the obtained risk level is significant [2]

3.2 Identification of the Most Effective Intervention

To identify which intervention of the proposals is the most effective in each case, the following procedure has been followed:

1. Select a risk factor to improve;
2. Modify the value of this risk factor as if the intervention had already been done.
3. Calculate the VLI of the new situation.

This procedure has been applied for each of the factors defined above.

Table 4 shows the VLI values obtained for each case when intervening in each factor and the expected reduction in the current risk index.

Table 4. VLI values and expected reduction in the VLI (%) after improving every risk factor

Case	1	2	3	4	5	6	7	8	9	10
VLI (m_{ref} = 25 kg)										
25 <=H <=40	2,12	1,36	1,01	1,80	1,19	2,18	1,52	2,52	3,61	1,09
50 < V<= 125	2,18	1,32	1,00	1,46	1,23	1,71	1,51	1,97	2,87	1,88
Freq/2	2,22	1,64	1,21	1,50	1,40	2,08	1,72	2,27	3,15	1,80
A = 0	2,17	1,34	1,28	1,46	1,50	2,18	1,91	2,04	2,93	1,53
W < 11 kg	2,08	1,41	1,07	1,41	1,26	0,79	1,37	0,62	1,48	1,55
D = Short	1,83	1,50	1,21	1,28	1,26	1,90	1,54	2,27	2,72	1,71
VLI reduction (%)										
25 <=H <=40	21%	18%	21%	0%	21%	0%	20%	0%	0%	42%
50 < V<= 125	19%	20%	22%	19%	18%	22%	21%	22%	20%	0%
Freq/2	17%	1%	5%	17%	7%	5%	10%	10%	13%	4%
A = 0	19%	19%	0%	19%	0%	0%	0%	19%	19%	19%
W < 11 kg	22%	15%	16%	22%	16%	64%	28%	75%	59%	18%
D = Short	32%	10%	5%	29%	16%	13%	19%	10%	25%	9%

In general, the least effective intervention to reduce the risk index is to reduce the frequency by half. The most effective intervention is different in each case: Short duration for cases 1 and 4; improve the vertical distance for cases 2 and 3; improve the horizontal distance for cases 5 and 10; avoid manual lifting of heavier objects for cases 6, 7, 8 and 9.

3.3   Strategies Used by Practitioners

The strategy used by all the practitioners to evaluate these cases is to reduce the variability to a simple case.

The difference between the experts lies in the strategy to reduce the variability (see Table 5).

Table 5. Strategy used by every practitioner to reduce variability

Strategy	% Practitioners
The most frequent case with the average weight	57,6%
The worst case with the average weight	31,4%
The worst case with the maximum weight	11,0%

Strategy 1: Most experts (57.6%) apply the RNLE to the most frequent posture (movement) lifting the load with a weight equal to the average of the weights of the loads lifted. Table 6 shows the values of the risk factors and the result of applying the RNLE with a reference mass equal to 25 kg. With this strategy, only 3 cases have significant risk. There is a clear underestimation of the risk with respect to the VLI.

Table 6. Lifting indexes for the strategy 1 (the most frequent posture with the average weight)

Case	1	2	3	4	5	6	7	8	9	10
Weight	7,3	5,8	5,2	5,6	4,9	8,9	10,3	8,0	11,5	3,6
Frequency	3,92	0,37	0,15	3,25	1,37	0,63	1,37	0,19	2,98	0,77
Assimmetry	45	45	0	45	0	0	0	30	30	45
V (cm)	75	75	75	75	75	75	75	0	75	75
H (cm)	45	45	35	35	45	35	45	35	35	45
LI (mref = 25 kg)	1,49	0,66	0,32	0,78	0,55	0,7	1,16	0,81	1,43	0,43

Strategy 2: The second most used strategy (31.4%) to reduce variability is to consider the worst posture (movement) and the weight equal to the average of the weights of the loads lifted (see Table 7). With this strategy, 4 cases have a totally acceptable level of risk.

Strategy 3: The least used strategy (11%) is to consider the worst posture (move-

Table 7. Lifting indexes for the strategy 2 (the worst posture with the average weight)

Case	1	2	3	4	5	6	7	8	9	10
Weight	7,3	5,8	5,2	5,6	4,9	8,9	10,3	8,0	11,5	3,6
Frequency	3,92	0,37	0,15	3,25	1,37	0,63	1,37	0,19	2,98	0,77
Assimmetry	45	45	0	45	0	0	0	0	30	45
V (cm)	175	175	175	175	175	175	175	175	175	75
H (cm)	45	45	45	35	45	35	45	35	35	63
LI (mref = 25 kg)	2,52	1,11	0,7	1,31	0,93	1,18	1,95	0,93	2,42	0,61

ment) and maximum weight (see Table 8). With this strategy, all cases have a significant level of risk, just as when applying VLI.

To the question of identifying the most priority risk factor (most influential in the

Table 8. Lifting indexes for the strategy 3 (the worst posture with the maximum weight)

Case	1	2	3	4	5	6	7	8	9	10
Weight	14,0	12,0	12,5	13,0	12,0	26,0	15,0	25,0	28,0	13,0
Frequency	3,92	0,37	0,15	3,25	1,37	0,63	1,37	0,19	2,98	0,77
Assimmetry	45	45	0	45	0	0	0	30	30	45
V (cm)	175	175	175	175	175	175	175	175	175	75
H (cm)	45	45	45	35	45	35	45	35	35	63
LI (mref = 25 kg)	4,83	2,29	1,69	3,04	2,27	3,44	2,84	2,91	5,9	2,19

risk level) based on observation, most practitioners point out the vertical distance for all cases (see Table 9). The extreme postures to lift objects in very low or very high locations is the factor that attracts the most attention of practitioners as a priority to improve. There are also practitioners who point out avoiding asymmetry as the most effective intervention. The rest of the risk factors are indicated as priorities by very few experts in some cases. The duration is ignored by all the practitioners.

Table 9. Distribution of practitioners indicating each risk factor as a priority for each case

Case	1	2	3	4	5	6	7	8	9	10
25 <= H <= 40	0%	0%	5,9%	0%	9,3%	0%	9,3%	0%	0%	10,2%
50 < V <= 125	81,4%	73,7%	94,1%	83,1%	90,7%	94,9%	90,7%	89,0%	87,3%	54,2%
Freq/2	5,9%	0%	0%	0%	0%	0%	0%	0%	0%	10,2%
A = 0	12,7%	26,3%	0%	16,9%	0%	0%	0%	5,1%	7,6%	25,4%
W < 11 kg	0%	0%	0%	0%	0%	5,1%	0%	5,9%	5,1%	0%
D = Short	0%	0%	0%	0%	0%	0%	0%	0%	0%	0%

3.4   Effectiveness of the Strategies Used

When the most determinant factor obtained by applying the VLI (Table 4) is compared with that indicated by the practitioners based on observation (Table 5), the result is that, in most cases, all the practitioners would improve a risk factor that is not the most effective by reducing the risk (see Table 10).

Table 10. Distribution of practitioners indicating the highest priority factor correctly based on observation

Case	1	2	3	4	5	6	7	8	9	10
Right priority	0,0%	73,7%	94,1%	0,0%	9,3%	5,1%	0,0%	5,9%	5,1%	10,2%

Only in cases where vertical distance was the most significant risk factor (cases 2 and 3), most practitioners would perform an effective intervention.

In addition, to evaluate the effectiveness of the analytical strategies applied by the practitioners, the following procedure has been applied: Select a risk factor, assign it the improved value; and recalculate the RNLE [9]. The result obtained for strategies 1, 2 and 3 mentioned above is shown in Tables 11, 12 and 13.

Table 11. Strategy 1 (the most frequent posture with the average weight): LI values after improving every risk factor and expected reduction on LI

Case	1	2	3	4	5	6	7	8	9	10
LI (mref = 25 kg)										
25 <= H <= 40	1,16	0,51	0,32	0,78	0,43	0,70	0,90	0,81	1,43	0,34
50 < V <= 125	1,49	0,66	0,32	0,78	0,55	0,70	1,16	0,55	1,43	0,43
Freq/2	1,04	0,54	0,30	0,59	0,50	0,66	1,05	0,80	1,13	0,41
A = 0	1,28	0,56	0,32	0,66	0,55	0,70	1,16	0,73	1,30	0,37
W < 11 kg	1,49	0,66	0,32	0,78	0,55	0,70	1,16	0,81	1,36	0,43
D = Short	0,81	0,55	0,28	0,47	0,42	0,58	0,89	0,78	0,90	0,35
LI reduction (%)										
25 <= H <= 40	22%	23%	0%	0%	22%	0%	22%	0%	0%	21%
50 < V <= 125	0%	0%	0%	0%	0%	0%	0%	32%	0%	0%
Freq/2	30%	18%	6%	24%	9%	6%	9%	1%	21%	5%
A = 0	14%	15%	0%	15%	0%	0%	0%	10%	9%	14%
W < 11 kg	0%	0%	0%	0%	0%	0%	0%	0%	5%	0%
D = Short	46%	17%	13%	40%	24%	17%	23%	4%	37%	19%

When applying strategy 1, duration is identified as the most significant risk factor in all cases, except in case 2 (horizontal distance) and case 8 (vertical distance).

When applying strategy 2, vertical distance is the most significant risk factor in all cases, except for case 1 (duration) and case 10 (horizontal distance).

Table 12. Strategy 2 (the worst posture with the average weight): LI values after improving every risk factor and expected reduction on LI

Case	1	2	3	4	5	6	7	8	9	10
LI (mref = 25 kg)										
25 <= H <= 40	1,96	0,86	0,55	1,31	0,72	1,18	1,52	0,93	2,42	0,34
50 < V <= 125	1,49	0,66	0,42	0,78	0,55	0,70	1,16	0,55	1,43	0,61
Freq/2	1,76	0,92	0,68	1,00	0,84	1,12	1,77	0,91	1,91	0,57
A = 0	2,15	0,95	0,70	1,12	0,93	1,18	1,95	0,84	2,19	0,52
W < 11 kg	2,52	1,11	0,70	1,31	0,93	1,18	1,95	0,93	2,30	0,61
D = Short	1,37	0,93	0,65	0,91	0,71	0,97	1,50	0,89	1,52	0,49
LI reduction (%)										
25 <=H <=40	22%	23%	21%	0%	23%	0%	22%	0%	0%	**44%**
50 < V<= 125	41%	**41%**	**40%**	**40%**	**41%**	**41%**	**41%**	**41%**	**41%**	0%
Freq/2	30%	17%	3%	24%	10%	5%	9%	2%	21%	7%
A = 0	15%	14%	0%	15%	0%	0%	0%	10%	10%	15%
W < 11 kg	0%	0%	0%	0%	0%	0%	0%	0%	5%	0%
D = Short	**46%**	16%	7%	31%	24%	18%	23%	4%	37%	20%

Table 13. Strategy 3 (the worst posture with the maximum weight): LI values after improving every risk factor and expected reduction on LI

Case	1	2	3	4	5	6	7	8	9	10
LI (mref = 25 kg)										
25 <= H <= 40	3,75	1,78	1,31	3,04	1,77	3,44	2,21	2,91	5,90	1,22
50 < V <= 125	2,86	1,36	1,00	1,80	1,35	2,04	1,68	1,72	3,49	2,19
Freq/2	3,38	1,89	1,68	2,32	2,06	3,28	2,57	2,89	4,64	2,06
A = 0	4,13	1,96	1,69	2,60	2,27	3,44	2,84	2,63	5,33	1,87
W < 11 kg	3,76	2,08	1,47	2,55	2,07	1,44	2,07	1,27	2,30	1,83
D = Short	2,62	1,93	1,60	1,83	1,75	2,84	2,18	2,83	3,70	1,78
LI reduction (%)										
25 <= H <= 40	22%	22%	22%	0%	22%	0%	22%	0%	0%	**44%**
50 < V <= 125	41%	**41%**	**41%**	**41%**	**41%**	41%	**41%**	41%	41%	0%
Freq/2	30%	17%	1%	24%	9%	5%	10%	1%	21%	6%
A = 0	14%	14%	0%	14%	0%	0%	0%	10%	10%	15%
W < 11 kg	22%	9%	13%	16%	9%	**58%**	27%	**56%**	**61%**	16%
D = Short	**46%**	16%	5%	40%	23%	17%	23%	3%	37%	19%

Strategy 3 only differs from strategy 2 in 3 out of 10 cases: cases 6, 8 and 9, in which the predominant risk factor obtained is the weight.

To evaluate the potential of identifying the intervention priorities of these strategies, a comparative analysis with the results of applying the VLI has been carried out (see Table 14). Considering the percentage of experts that applies each strategy, the percentage of experts who would have been successful in identifying the highest

Table 14. Priority factors obtained with each strategy (D: Duration; H: Horizontal distance; V: Vertical distance; W: Weight)

Case	1	2	3	4	5	6	7	8	9	10
VLI	D	V	V	D	H	W	W	W	W	H
Strategy 1	D	H	D	D	D	D	D	V	D	H
Strategy 2	D	V	V	V	V	V	V	V	V	H
Strategy 3	D	V	V	V	V	W	V	W	W	H
Effective intervention (% of practitioners)	100%	42,4%	42,4%	57,6%	0%	11,0%	0%	11,0%	11,0%	100%

priority factor has been identified. As seen in this table, only in 3 out of 10 cases (cases 1, 4 and 10), most experts would have identified the factor correctly.

4   Discussion

My hypothesis that VLI is a great help in designing effective interventions has been supported. In this article, a procedure of using the VLI to identify the most priority risk factors before designing an intervention in the workplace has been proposed.

Comparatively, the strategies applied by the practitioners, based on the RNLE instead of the VLI, can negatively affect the effectiveness of the improvement interventions generating an additional cost to the companies.

In my experience, many practitioners identify intervention priorities based on observation. As shown in this study, observation tends to give greater relevance to vertical distance (very high or very low locations of loads). Although improving vertical distance is always an improvement intervention, it is not always the most effective way to reduce the probability of developing a work-related musculoskeletal disorder.

The strategies currently used by practitioners to quantitatively assess the manual lifting of loads risk in warehouse are based on reducing the variability of postures and weights to obtain a simple lifting task and then, to apply the RNLE. To reduce the variability, the most successful procedure is to consider the worst movement with the maximum weight (this strategy has been successful in 7 out of 10 case studies). Even so, this procedure overestimates risk levels and fails in some cases to correctly identify the most priority risk factor.

In warehouses, the weight of the manually handled objects is, in general, a problem. This factor is underestimated by practitioners, except when using the strategy of considering only the maximum weight. It is worth mentioning that in recent years in Europe, an important effort has been made to reduce the weight of the loads in the logistics sector (some companies have limited the maximum weight of the packages to 15 kg).

Although the scope of this study is quite limited (small number of case studies and practitioners), it seems reasonable to conclude that the VLI is a recommendable method, both to assess the manual lifting of loads risk, and to design the improvement intervention.

Acknowledgments. I thank the OHS practitioners and companies who participated in this study. I also want to thank Sonia Tello for her help in formatting and analyzing the data.

References

1. Alvarez-Casado E (2013) Análisis de la exposición al riesgo por levantamiento manual de cargas en condiciones de alta variabilidad (Doctoral dissertation). Retrieved from TDX http://hdl.handle.net/10803/117066
2. Battevi N, Pandolfi M, Cortinovis I (2016) Variable lifting index for risk assessment of manual lifting: a preliminary validation study. Hum Factors 58:712–725. https://doi.org/10.1177/0018720816637538
3. Colombini D, Occhipinti E, Alvarez-Casado E, Hernandez A, Waters T (2009) Procedures for collecting and organizing data useful for the analysis of variable lifting tasks and for computing the VLI. In: Proceedings of the 2009 IEA Conference, China
4. Colombini D, Occhipinti E, Alvarez-Casado E, Waters T (2013) Manual lifting: a guide to the study of simple and complex lifting tasks. CRC Presss, Boca Raton
5. ISO (2014) Ergonomics – Application document for International Standards on manual handling (ISO 11228-1, ISO 11228-2 and ISO 11228-3) and evaluation of static working postures (ISO 11226). TR 12295
6. Lu-M L, Putz-Anderson V, Garg A, Davis KG (2016) Evaluation of the impact of the revised National Institute for Occupational Safety and Health lifting equation. Hum Factors 58(5):667–682
7. Waters T, Occhipinti E, Colombini D, Alvarez-Casado E, Fox R (2016) Variable lifting index (VLI): a new method for evaluating variable lifting tasks. Hum Factors 58:695–711. https://doi.org/10.1177/0018720815612256
8. Waters T, Occhipinti E, Colombini D, Alvarez-Casado E, Hernandez A (2009) The variable lifting index (VLI): a new method for evaluating variable lifting tasks using the Revised NIOSH lifting equation. In: Proceedings of the 2009 IEA Conference, China
9. Waters TR, Putz-Anderson V, Garg A (1994) Application manual for the revised NIOSH lifting equation [DHHS (NIOSH) Publication 94–110]. National Institute for Occupational Safety and Health, Centers for Disease Control and Prevention, Cincinnati, Ohio

Postural Deviation Gestures Distinguish Perceived Pain and Fatigue Particularly in Frontal Plane

Nancy Black[1]([⊠]) [iD], Andrew Hamilton-Wright[2], Joshua Lange[2], Clément Bouet[1], Mariah Martin Shein[2], Marthe Samson[1], and Maxime Lecanelier[1]

[1] Department of Mechanical Engineering, Université de Moncton, Moncton, NB, Canada
nancy.black@umoncton.ca
[2] School of Computer Science, University of Guelph, Guelph, ON, Canada

Abstract. Office workers frequently report discomfort in the neck, back and shoulders. Extreme and prolonged static postures are known risk factors. Sit-stand workstations encourage regular postural changes and tend to reduce discomfort. Head and upper back deviations were recorded for ten adults entering data using a seated desk and two sit-stand desks following three twenty-minute cycles with 6 or 9 min standing. Participants reported their perceived back pain, neck pain and fatigue following each recording.

Postural deviation was discretized into levels in sagittal and frontal planes. Series of four consecutive postural levels, termed 'Postural gestures', were tracked for each plane, workstation, and participant. Averages by quintile for perception values of each feature revealed inflection between low ("absent" or "No") and high ("present" or "Yes") slopes. Postural gestures were 'important' when associated with perceived variations. Significance occurred with contingency calculations pairing different postural gesture occurrence relative to (perception) "absent" and "present" quintiles.

Some postural gestures occurred more when perception was "absent" than "present", and vice versa. These differences may indicate postural behavior that increases risk or protects from pain or fatigue. All of the 81 postural gestures that occurred for all ten participants only occurred when using sit-stand workstations. Eighty-three percent (83%) of those occurred in the frontal plane with similar numbers for neck pain, back pain and fatigue. More patterns occurred for absent than present perceptions.

Matching postural deviation patterns with present and absent perceived back and neck pain, and fatigue distinguish between groups, complementing known Musculoskeletal Disorder (MSD) contributors.

Keywords: Postural gesture · Perceived pain and fatigue Sit-stand workstation

© Springer Nature Switzerland AG 2019
S. Bagnara et al. (Eds.): IEA 2018, AISC 820, pp. 495–501, 2019.
https://doi.org/10.1007/978-3-319-96083-8_64

1 Introduction

Musculoskeletal disorders (MSDs) are the leading cause of disability in developed countries [1], accounting for nearly 42% of the worker's compensation claims in Ontario [2] and 7.2 billion dollars in direct costs in Canada [3]. Approximately 75% of office workers report significant discomfort in one or more body regions [4] with these discomforts most frequently occurring in the neck, back and shoulders [5]. Discomfort is known to be a precursor of MSDs.

Office work tends to be sedentary and computer-focused. Research consistently links sedentary behaviors to health degradation including MSDs [6–9]. A recent expert statement on the sedentary office recommends at least two hours daily standing and light activity initially, progressing to four hours daily [10]. These should be evenly spaced over the day. Microbreaks are recommended as frequently as every 20 min without adversely affecting productivity while reducing discomfort [11]. What actually occurs over the day when incorporating such breaks or macro-postural changes is largely unknown, but likely includes variations in the amplitude and direction of postural adjustments.

This study considered repeated hour-long recordings during simulated data entry tasks using three different workstations: one seated and two varying regularly between seated and standing heights. Continuous postural measurement and variations in perceived neck pain, back pain and fatigue from start to end of were studied together. This article describes how postural deviation patterns match with high and low perceived back and neck pain, and fatigue, thus distinguishing between groups potentially complementing known MSD contributors.

2 Method

2.1 Data Source

Data used for this study was collected following ethical approval at the Université de Moncton. Each of the 10 working aged participants gave written informed consent prior to being recorded for one-hour simulated data entry tasks while wearing inclinometers on the head and upper back in a laboratory environment. At the end of each period, participants reported their level of back pain, neck pain and fatigue on a 10 cm Visual Analogue Scale. Each participant completed this activity in three workstation environments: a static, seated desk and two variable height desks that moved automatically to appropriate standing height for 6 min or 9 min during a 20-min cycle. These workstation conditions are identified as "seated," "AD30" and "AD45". In all cases, work surface height was located to respect ergonomics recommendations, placing the surface within 5 cm below relaxed elbow-height when standing erect.

The planes of measurement of the inclinometers were "head frontal," "head sagittal," "back frontal," and "back sagittal." Inclinometers were sampled across all channels at 15 Hz.

2.2   Postural Deviation

Postural deviation was grouped into deviation states defined in Table 1. Four levels for the head and five for the back in the sagittal plane were based on the Rapid Upper Limb Assessment (RULA) [12]. Five levels of deviation were defined for both head and back in the frontal plane based on Keyserling's 1986 data [13].

Table 1. Postural State Definitions with shaded zone indicating most frequently observed levels

State ID	Head		Back	
	Sagittal (HS)	Frontal (HF)	Sagittal (BS)	Frontal (BF)
−2		−∞ to −10		−∞ to −10
−1	−∞ to −5	−10 to −2	−∞ to −5	−10 to −2
0	−5 to 10	−2 to 2	−5 to 5	−2 to 2
+1	10 to 20	2 to 10	5 to 20	2 to 10
+2	20 to ∞	10 to ∞	20 to 60	10 to ∞
+3			60 to ∞	

Sequences of four postural states were identified by noting the state changes for each inclinometer channel, for each workstation, and each participant. As an example, an identified postural sequence of length 4 for the "back/frontal plane" channel was represented as "BF + 1 → BF + 2 → BF + 1 → BF00" meaning a change from between 2° and 10°, followed by a period with frontal deviation over 10°, and then deviation again between 2° and 10° followed by a neutral position between −2° and 2°. This study considers only postural state ordering, and not duration. Four states series were chosen since they are long enough to characterize clearly oscillatory variations (for example "BF + 1 → BF00 → BF + 1 → BF00") which are more repetitive and thus potentially higher risk.

2.3   Perceptual Features

The importance of each postural sequence was identified in terms of its association with neck pain, back pain and fatigue perceptual features indicated by each participant at the completion of the one-hour study for a given workstation. Features were scored measuring the distance along the line of each mark with a ruler and transforming into a [0–10] score indicating the degree of agreement of each value for each participant for each perceptual feature.

These feature values were then discretized into five levels, each containing 1/5 of the observed points for a given postural feature and together exactly covering the range of values observed for that feature.

The presence or absence of a given perception was defined by noting where the point of inflection occurred when plotting the means of the quintiles (Fig. 1). The regions below and above this point of inflection were identified as "No" (absent) or "Yes" (present), respectively, for each feature.

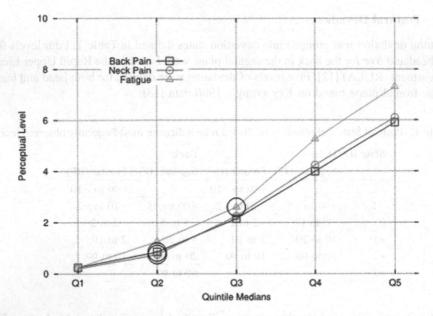

Fig. 1. Perceptual Quintile Medians with value separating 'No' and 'Yes' categories indicated by black circle

2.4 Combining Posture and Perception

The next step was to identify postural gestures that are significantly associated with either the presence ("Yes") or the absence ("No") of pain or fatigue. "Yes" and "No" perceptual feature response data quintiles were associated with each four-state pattern. If the pattern occurred significantly differently when the perceptual feature was present relative to its general occurrence, as measured by a χ^2 contingency, then it was deemed associated significantly positively with that perceptual feature. This was done for each of neck pain, back pain and fatigue. Similarly, any patterns associated with the absence of fatigue, or neck or back pain were identified as significant negative associations. This comparison was calculated using the χ^2 contingency comparison implemented using the chi2_contingency() function in Python SciPy library, version 0.15.1.

3 Results

Using these significant postural gestures, the relationship across participants was examined by considering the number of separate participants whose data included each significant postural gesture. The significant postural gestures that involve 9 or 10 of 10 participants are presented since this suggests these are representative in the general population (Table 2 and 3).

Notably, while all three stations included significant postural gestures for nine of 10 participants (Table 2), only those alternating between sitting and standing were present across all 10 participants (Table 3). Furthermore, significant postural gestures occurred

Table 2. Number of significant 4-state postural gestures observed across at least nine participants for each inclinometer channel, workstation style and perceptual feature

Workstation	Channel	Present	Absent	Present	Absent	Present	Absent
		Neck pain		Back pain		Fatigue	
AD30	HF	9	10	9	8	9	8
	HS	5	0	3	0	5	3
	BF	2	4	4	6	3	2
	BS	0	0	0	0	0	0
AD45	HF	8	6	11	6	9	6
	HS	5	4	6	6	5	6
	BF	0	2	2	6	4	4
	BS	3	0	3	1	2	0
Seated	HF	7	0	6	2	0	0
	HS	2	0	2	2	0	0
	BF	2	6	2	4	0	0
	BS	0	0	0	0	0	0

Table 3. Number of significant 4-state postural gestures observed across 10 participants for each inclinometer channel, workstation style and perceptual feature

Workstation	Channel	Present	Absent	Present	Absent	Present	Absent
		Neck pain		Back pain		Fatigue	
AD30	HF	6	2	2	5	1	5
	HS	2	0	0	0	2	0
	BF	2	2	2	6	1	2
	BS	0	0	0	0	0	0
AD45	HF	0	5	0	4	2	2
	HS	2	0	2	2	2	2
	BF	0	2	2	6	4	4
	BS	0	0	0	0	0	0

more frequently in the frontal plane than in the sagittal plane for both head and trunk (73% for 9 participants and 83% for 10 participants). Significant back-sagittal postural gestures were fewer than head- and back-frontal gestures, and occurred only for AD45 (eight observed with present perception and only one with absence). Somewhat more gestures were associated with increased risk (presence) than absence (56%) when considering nine of 10 participants, however this tendency reversed (40% risk) when considering all 10 participants.

Considering individual significant postural gestures, all those included in Table 3 included only +1, 0 and −1 levels, indicating that all such gestures were of limited amplitude. Furthermore, all postural gestures include the neutral (0) level. In contrast, when considering as few as nine participants, significant deviations includes ±2 and gestures excluding the neutral level. Further analysis will concentrate on the significant gestures only observed for all 10 participants.

The gestures resulting in perceptual feature absence were more likely to be associated with an oscillation than those observed with perception present (67% versus 44%). Indeed, oscillatory gestures occurred for all significant cases of the back frontal plane and absent neck pain and fatigue (AD30 and AD45) and head-frontal plane for absent neck pain (AD30 only), and fatigue (AD45). Absent back pain had more non-oscillatory results than either fatigue or neck pain. In contrast, when considering gestures associated with perception presence, only back pain was uniformly oscillatory, and even then only for AD30 (a total of 4 patterns: two for each of head and back frontal deviations).

For present postural features, the head-sagittal deviations were significant for one pair of patterns for AD30 neck pain (oscillatory) and for AD45 for each of fatigue, neck and back pain (all non-oscillatory).

The number of significant postural gestures varied by body segment region, most, but not all occurring in pairs, for example (BF − 1 → BF00 → BF − 1 → BF00 and BF00 → BF − 1 → BF00 → BF − 1).

4 Conclusions

The combination of postural and perception measurements with pattern recognition methods were used to explore the links between certain postural gesture patterns with increased (or decreased) pain risk. Significant differences in frequencies of postural gestures were noted with some patterns occurring more when perception was present ("Yes") than absent ("No"), and vice versa. These differences may indicate postural behavior that increases risk or protects from (avoids) pain or fatigue. All of the 81 postural gestures that occurred for all participants only occurred when using sit-stand workstations. Furthermore, 83% of those occurred in the frontal plane with similar distributions for neck pain, back pain and fatigue.

This combination of perceived and objective measurements with pattern recognition methods allows exploration of links between postural gesture patterns and perceived MSD risk indicators.

The complexity of the relationships between one pattern and several workstation modalities and with multiple perceptions, as well as the relationship between one pattern and another, indicate a problem of sufficient complexity that we next expect to turn to machine learning to discover complex patterns of interest. By using association mining techniques, such as those described in [14], we can represent each such pattern in terms of their information content.

Acknowledgements. The contributions of Martin Shein and Lange were supported by the Natural Sciences and Engineering Research Council of Canada (NSERC) discovery grant number DDG-2015-00007 through Hamilton-Wright. The contributions of Lecanelier were supported by the Research Assistant Initiative of the New Brunswick Innovation Fund through Black.

References

1. Mustard CA, Chambers A, Ibrahim S, Etches J, Smith P (2015) Time trends in musculoskeletal disorders attributed to work exposures in Ontario using three independent data sources, 2004–2011. Occup Environ Med 72:252–257
2. Government of Ontario, Ministry of Labour, Why are Pains and Strains/Musculoskeletal Disorders (MSDs) a Problem? Ontario Ministry of Labour. https://www.labour.gov.on.ca/english/hs/pubs/pains/problem.php
3. Economic Burden of Illness in Canada 2005–2008 (2014). http://www.phac-aspc.gc.ca/publicat/ebic-femc/2005-2008/assets/pdf/ebic-femc-2005-2008-eng.pdf
4. Bhanderi D, Choudhary S, Parmar L, Doshi V (2008) A study of occurrence of musculoskeletal discomfort in computer operators. Indian J Community Med Off Publ Indian Assoc Prev Soc Med 33:65–66. https://doi.org/10.4103/0970-0218.39252
5. Black N, Scoliège J (2016) Ergonomic study of workstations and employee's posture to minimise the injuries and improve quality. WSP Company, Moncton, Dartmouth
6. Chau J, Grunseui A, Chey T, Stamatakis E, Brown WJ, Matthews CE, Bauman AE, van der Ploeg HP (2013) Daily sitting time and all-cause mortality: a meta-analysis. PLOS One 8
7. Davis KG, Kotowski SE (2015) stand up and move; your musculoskeletal health depends on it. Ergon Des 23:9–13. https://doi.org/10.1177/1064804615588853
8. Dempsey PC, Owen N, Biddle SJH, Dunstan DW (2015) Managing sedentary behavior to reduce the risk of diabetes and cardiovascular disease. Curr Diab Rep 14:1–11
9. Owen N, Sparling PB, Healy GN, Dunstan DW, Matthews CE (2010) Sedentary behavior: emerging evidence for a new health risk. Mayo Clin Proc 85:1138–1141
10. Buckley JP, Hedge A, Yates T, Copeland RJ, Loosemore M, Hamer M, Bradley G, Dunstan DW (2015) The sedentary office: an expert statement on the growing case for change towards better health and productivity. Br J Sports Med 49:1357–1362. https://doi.org/10.1136/bjsports-2015-094618
11. McLean L, Tingley M, Scott RN, Rickards J (2001) Computer terminal work and the benefit of microbreaks. Appl Ergon 32:225–237. https://doi.org/10.1016/S0003-6870(00)00071-5
12. McAtamney L, Corlett EN (1993) RULA: a survey method for the investigation of work-related upper limb disorders. Appl Ergon 24:91–99
13. Keyserling WM (1986) Postural analysis of the trunk and shoulders in simulated real time. Ergonomics 29:569–583
14. Hamilton-Wright A, Stashuk DW (2008) Statistically based pattern discovery techniques for biological data analysis. In: Smolinski TG, Milanova MG, Hassanien A-E (eds) Applications of computational intelligence in biology: current trends and open problems. Springer, Heidelberg, pp 3–31

Identifying Situational Operational Leeway for Subcontract Supervisors so as to Progress in MSD Prevention

A. Cuny-Guerrier[1($\boxtimes$)], S. Caroly[2], F. Coutarel[3], and A. Aublet-Cuvelier[1]

[1] INRS, Vandœuvre, France
aude.cuny@inrs.fr
[2] PACTE, Communauté Université de Grenoble Alpes, Grenoble, France
[3] ACTé, EA 4281, Université Clermont Auvergne, Chamalières, France

Abstract. Supervisors seek to regulate critical situations influencing operators' capacity to deal with risks of musculoskeletal disorders (MSD). The regulation adopted depends on the supervisors' situational operational leeway (SOL), which is influenced by their individual characteristics and by characteristics related to their working environment (2–3). The aim of this study was to identify the components of SOL for first-line supervisors working through subcontracting, in order to allow them to contribute to the design of a working environment favourable to prevention of MSDs in the supervised operatives. Two qualitative case studies were conducted on meat-cutting supervisors employed by an external company (EC) and working in a user company (UC). From interviews with the supervisors and observation of their work (for 2 to 6 h on 4 days each), 15 critical situations were identified. These situations were then examined through self-confrontation interviews. The components of the SOL for the two supervisors were then constructed from the analysis of those situations. The SOL of the subcontracted supervisors breaks down as 16 components: 4 individual ones (i.e., the supervisors' cutting skills), 4 organizational ones related to the EC (i.e., procedural autonomy), 4 organizational ones related to the UC (i.e., flexibility of the production modes), and 4 collective components related to opportunities to cope collectively by interacting with EC-supervised operatives and/or with UC employees (i.e., with support from UC supervisors). Individual components were mobilized most often, whereas organizational components were not systematically identified. The collective components were mobilized particularly in situations lacking organizational components.

Keywords: Musculoskeletal disorders · Supervisor
Subcontracting meat-cutting sector

1 Context and Theoretical Framework

This study focuses on the prevention of musculoskeletal disorders (MSD) for employees of external companies (EC) working in meat-cutting in user companies (UC) in a subcontracting relationship [1]. A number of studies across multiple countries

© Springer Nature Switzerland AG 2019
S. Bagnara et al. (Eds.): IEA 2018, AISC 820, pp. 502–510, 2019.
https://doi.org/10.1007/978-3-319-96083-8_65

have linked the cutting work associated with meat-cutting with a high prevalence of MSD [2–5]. The effects of some of the characteristics of the working environment and the difficulties meat-cutters face to achieve their work targets have been highlighted. The influence of workstation design – temporal organization [6, 7], organization of sharpening [4, 8] or learning conditions [9] – has been studied to identify mechanisms potentially leading to the occurrence of MSD (biomechanical hypersollicitation, psychosocial exposure, etc.).

First-line supervisors contribute in part to defining the working conditions of meat-cutters. They deal with the "critical situations" which arise during the work shift, such as contradictions between the different activity goals [10]. How local supervision responds to these critical situations will influence the working conditions of the meat-cutters supervised [6, 7]. For example, Slappendel, Moore and Tappin [8] described the effects of work rate management by supervisors, which consists of placing experienced operators at the beginning of the chain and imposing their pace on the slower operators downstream. Dealing with critical situations requires supervisors to consider their situational operational leeway (SOL). The SOL is defined as the "space" necessary, in a given situation, to complete procedures while respecting the target objectives and overall health (mental, physical, emotional, etc.) [9]. The SOL emerges at the interface between the supervisors' individual characteristics and the characteristics of their working environment.

Some studies have focused on the individual characteristics of supervisors as part of MSD prevention. For example, by according trust to their subordinates or adopting a participatory style [10]. Other authors have mentioned difficulties related to supervisors' working conditions (workload, difficulty reconciling the objectives of production and prevention) [6, 11, 12]. However, few studies have attempted to describe how subcontractors can take action within this context. Subcontracting poses specific difficulties, for example leading to contradictions between prescriptions or the multiplication of dependency relationships between workers [13].

2 Aims

The aim of this study was to identify the components of SOL for first-line supervisors working through subcontracting, in order to participate in designing a working environment favourable to preventing MSDs in the supervised operatives.

3 Methods

3.1 A 2-Case Study

A double-case design was used to analyze how subcontracting supervisors deal with critical situations and the SOL available in this context. The aim was to identify similarities while also contributing to the analysis of the specificities of each case, so as to gradually formalize a theory [14]. This approach is used to understand processes in real-life situations when they are too complex to be analyzed experimentally [14, 15].

The case analysis unit was defined as the work activity of a subcontracted supervisor. The activities supervised were "secondary cutting activities", such as trimming, deboning and packaging, of pork and beef. Two cases were studied to take into account the diversity of the criteria useful for the analysis of local supervisory activities in an outsourcing situation, with constraints relating to site access and implementation.

Selection of Participants

Two supervisors were selected from two subcontracting companies who had consulted us for help relating to MSD prevention. Both supervisors were from the industry, having received initial training followed by internal promotion to the position of proximity supervisor. Supervisor A was older and had more years of experience in the meat-cutting industry and on the site analyzed than supervisor B. Neither supervisor had received training or awareness-raising on the issue of MSDs.

3.2   Data-Collection and Analysis

Identification of the components of the supervisors' SOL was based on the following steps.

Identification of Critical Situations

From video-recordings of their work (for 2 to 6 h, on 4 days each) and observation thereof, 15 critical situations were identified based on the following criteria: emergence of hazards, difficulties, contradictions between the requirements and the means available due to organization of the operators' work as instigated by the supervisor. Seven critical situations were thus selected for supervisor A and eight for supervisor B,

Table 1. The 15 critical situations analyzed for supervisors A and B

No.	Category	Definition	Number of situations analyzed	
			A	B
1	Lack or unavailability of human resources	Late arrival or absence, reduced number of staff present	2	1
2	Inappropriate cutting techniques	Inappropriate meat-cutting gesture or strategy, from the point of view of productivity, quality or health criteria	3	1
3	Quality defects	Problems related to the quality of finished products	0	1
		Problems related to the quality of raw materials	1	0
4	Requisition of subcontracted meat-cutters	Requisition of meat-cutters by the UC	1	0
5	Bottleneck in deboning	Difficulty maintaining the pace, resulting in a bottleneck (accumulation of pieces on tables, slowing of the pace, operators express difficulties)	0	2
6	Fall of meat	Problems related to meat falling on the floor	0	1
7	Organizational dependence of the employees of the EC	Rhythm problems related to organizational dependence of employees of the EC	0	1
8	Non-compliance with sharpening rules	Failure of the UC to apply the knife maintenance rule centrally	0	1

resulting in a total of fifteen situations. These situations were grouped into eight categories, of which three were common to both supervisors (category 1 to 3) and five were specific to one or the other of the supervisors (See. Table 1).

Analysis of the Regulation Process
Supervisors were then confronted with extracts of video-recordings presenting each critical situation, in line with the self-confrontation interview method. A series of open-ended questions was used to explore the processes used to regulate critical situations in relation to the supervisors' SOL, for example: What is the problem in this situation? How did you deal with it? What were your goals? What were the difficulties and resources available to you in this situation? What were the effects of this regulation? Self-confrontations were later transcribed in full and thematically analyzed by a researcher using NVivo software to allow a thematic approach to analysis.

Reconstitution of the Regulation Process
The data produced allowed us to reconstitute the regulation process, based on observation of activity combined with cognitive and subjective elements. By confronting the discourse and the observations, we sought to limit reconstruction or deformation of the situations by the supervisors.

Identification of SOL Components
The components of the supervisors' SOL were thus identified after the fact by reconstituting the regulation process for each critical situation. The previously defined theoretical concepts guided, but did not limit, the analysis process and data interpretation [16]. The components identified were first grouped into the following deductive categories: individual components related to the supervisor, environmental components related to the EC and the UC. An additional category relating to the collective dimension of the activity also emerged during analysis.

4   Results

Six out of the eight categories of critical situation were favourably regulated for meat-cutters (categories 1, 2, 4, 5, 7, and 8). Two categories were unfavourably regulated (categories 3 and 6) (see Table 2).

Individual components included the supervisor's cutting skills, their experience of supervision (e.g. task distribution), representation of performance and their ability to encourage autonomy.

The organizational components related to the EC were: whether the supervisor could participate in meat-cutting activities, his procedural autonomy when assigning resources and prescribing tasks to be performed by operators, the provision of appropriate rules or means to deal with the critical situation arising (e.g. the possibility to ask for help), and a recruitment strategy whereby meat-cutters had polyvalent skill-sets.

The organizational components related to the UC were: workstation design (al-

Table 2. Effects on working conditions for supervised operators for each critical situation analyzed for supervisors A and B

No.	Critical situation	Effects on operators' environment	
		A	B
1	Lack or unavailability of human resources	(+)No increase in the pace of work (+)Mutual aid from supervisor for meat-cutting activity	(+)Objectives adjusted to the available human resources (+)No increase in the pace of work (+)Mutual aid from supervisor for meat-cutting activity
2	Inappropriate cutting techniques	(+)Support for the development of gestural skills integrating intra and inter-individual diversity	(+)Support for the development of gestural skills
3	Quality defects	(-)Increased requirements, (-)Physical exposure	(-)Increased requirements, (-)Physical exposure
4	Requisition of subcontracted meat-cutters	(+)No increase in the pace of work	Not observed
5	Bottleneck occurrence in deboning	Not observed	(+)No increase in the pace of work
6	Fall of meat	Not observed	(-)Physical exposure
7	Organizational dependence of the employees of the EC	Not observed	(+)Increased physical space for operators (+) No increase in the pace of work
8	Non-compliance with sharpening rules	Not observed	(+)Tolerance of individual regulation for an operator to promote the prevention of MSDs

lowing mentoring), flexibility of production modes (compatible with temporary storage and a fluctuating work rhythm), flexibility of production requirements (adjustable to real working conditions), and tolerance of procedural decisions made by the supervisor.

The individual and organizational components were inter-connected. For example, the supervisor's cutting skills were more frequently mobilized by supervisor A through participation in production. Supervisor B was more likely to mobilize operators' autonomy to deal with weaker components provided by the EC.

Finally, the collective components were related to the opportunity for collective coping through interaction between EC-supervised operatives and/or with UC employees (i.e., with support from UC supervisors). These components therefore related to the polyvalent skills of the meat-cutters (more evident for supervisor A), the absence of competition between individual meat-cutters from the outside team and/or towards meat-cutters and supervisors employed by the user company, opportunities for exchange and interaction with the UC or support from the UC for the reorganization of resources (relates more to supervisor B). These collective components within the team were generally more frequently observed for supervisor A. Among others, these components supported the possibility of handling the difficulties within the team by redistributing resources or sharing experience on cutting strategies. These collective components were less frequently used in the case of supervisor B who tended to deal with critical situations not observed for supervisor A, involving components related to employees of the UC.

Situations in categories 3 and 6, which were observed to be unfavourably regulated for meat-cutters, correspond to situations that are particularly devoid of characteristic components, thus limiting the supervisor's SOL. The adaptations implemented to deal with these critical situations consisted, for example, in "prescribing" additional tasks for operators (to correct defects, for example). This resulted in a work overload for the meat-cutters.

Collective components were particularly mobilized in situations lacking organizational components.

SOL components for supervisors "A" and "B"	Critical situation category							
	1	2	3	4	5	6	7	8
Individual								
▪ supervisor's cutting skills	a	ab	/	/	b	/	/	/
▪ experience of supervision	a	/	/	/	/	/	/	/
▪ representation of performance	a	/	/	/	/	/	/	/
▪ promotion of autonomy	/	/	a	/	b	/	b	b
Organizational, imposed by the EC								
▪ Procedural autonomy in assigning resources and specific tasks to operators	ab	ab	ab	/	/	b	/	/
▪ Supervisor permitted to participate in meat-cutting	a	a	/	/	/	/	/	/
▪ Recruitment strategy promoting polyvalence of meat-cutters	a	a	/	/	/	/	/	/
▪ Possibility to request help	/	a	/	/	/	/	/	/
Organizational, imposed by the UC								
▪ Tolerance of procedural decisions made by the supervisor	a	/	/	/	/	/	/	/
▪ Workstation design allowing monitoring between meat-cutters	a	a	/	/	b	/	/	/
▪ Flexibility of production modes (temporary storage on the table authorized)	a	a	/	/	/	/	/	/
▪ Flexibility of production requirements	b	/	/	a	/	/	/	/
Collective								
▪ Polyvalent meat-cutters	a	a	/	a	b	/	/	/
▪ Absence of competition between meat-cutters	a	a	/	/	/	/	/	/
▪ Possibility of exchange with the UC	b	/	/	/	/	/	/	/
▪ Mutual aid from UC workers to reorganize work conditions	/	/	a	/	b	/	b	/

Legend:
1. Lack or unavailability of human resources
2. Inappropriate cutting technique
3. Quality defects
4. Requisition of subcontracted meat-cutters by UC
5. Bottleneck in deboning
6. Fall of meat
7. Organizational dependence of employees of the EC
8. Non-compliance with sharpening rules

a: Components for supervisor "A"
b: Components for supervisor "B"

Critical situations analyzed for supervisor "a"

Critical situations analyzed for supervisor "b"

5 Discussion

5.1 Limitations

Additional observations would have allowed a more holistic vision and a more detailed comparison of work situations, but would be difficult to organize. The results presented are thus limited to the two cases observed and the supervisors analyzed. The two cases studied do not represent the full range of supervisory functions in terms of profiles, initial training, conditions of access to the position and conditions in which the activity is performed. The results produced feed into an intermediate theory which can only be generalized by replication in other case studies [15]. To this end, the context in which the data were produced has been specified elsewhere [16] to allow comparison of the results obtained with those resulting from future studies on the same topic.

5.2 Discussion of Results

As other authors have shown, supervisors' SOL depends in particular on how "the organization allows or tolerates procedural autonomy of the supervisor" and the resources it provides [17]. In this context of subcontracting, the supervisors are working in a system constrained by both the EC and the UC. Autonomy and resources are determined by the UC, which influences the work objectives, the means available and the social environment. The emergence of collective components in our findings appears to contrast with studies highlighting the individualization of work in a sub-contracting context. The possibility to develop collective resources observed here can be explained by the conditions in which activities were performed. These conditions were quite different from those more commonly described in the literature relating to outsourcing (more permanent contracts, lower geographical dispersion, etc.). Beyond the context of outsourcing, but with similarities in terms of the working conditions, Gotteland-Agostini, Pueyo and Béguin [16] also described the development of horizontal networks with external peers to cope with difficulties (lack of time, etc.). From this point of view, the components are "expanded" in this context where activities are performed on the premises of a UC. This development of team work by supervisor B, extended to the employees of the UC, could be explained by the mutualization of peer resources.

5.3 Perspectives for MSD Prevention and for Future Research

The development of an overt MSD-prevention approach, including the use of resources available through local supervisors, appears to be relevant for the prevention of MSDs. The ability of the supervisor to create, negotiate, and maintain relationships at different levels as well as to demonstrate tasks, support the collective work of the operators or the characteristics of the team, and ensure consistent application of the rules between the UC and the EC, are all elements likely to enrich the diagnosis of the situation by prevention specialists interested in MSD prevention.

The collective dimension of the supervisor's activity as generator of resources for meat-cutters' activities should be a focus of particular attention. The critical situations empirically identified in this study may help and guide the data-collection process for future studies, to allow comparison across a greater number of regulations, components and effects.

6 Conclusion

The subcontracting context changes the organizational and collective resources available to supervisors in their management role. Supervisors can seek to influence the determinants of the cutting work, which is largely determined by the UC, e.g. by developing collective resources with employees of the UC. By identifying the components of supervisors' SOL on a larger scale and in various activity sectors it will become possible to enrich the recommendations on the characteristics of the organizational systems to be preferred or avoided for efficient organization of work by supervisors and to help prevent injury of operators.

References

1. Conseil économique et social, J officiel, avis et rapports: p 305 (1973)
2. Tappin D et al (2006) Musculoskeletal disorders in meat processing: a review of the literature for the New Zealand meat processing industry, Massey university, New Zealand. https://www.acc.co.nz/assets/injury-prevention/meat-strains-sprains-review.pdf
3. InVS and MSA (2007) État de santé des salariés de la filière viande du régime agricole en Bretagne. Relations avec leurs contraintes de travail physiques, organisationnelles et psychosociales. Rapport-Enquête épidémiologique. http://www.invs.sante.fr/publications/2007/salaries_filiere_viande/rapport_salaries_filiere_viande.pdf
4. Karltun J, Vogel K, Bergstrand M, Eklund J (2016) Maintaining knife sharpness in industrial meat cutting: a matter of knife or meat cutter ability. Appl Ergon 56:92–100
5. Botti L, Mora C, Regattieri A (2015) Improving ergonomics in the meat industry: a case study of an italian ham processing company. Proc IFAC-PapersOnLine 48(3):598–603
6. Dixon S, Theberge N, Cole D (2009) Sustaining management commitment to workplace health programs: the case of participatory ergonomics. Relat industrielles 64(1):50–74
7. Gotteland C, Pueyo V (2015) An analysis of supervisor's framework design in the horticole sector. In: Proceedings 19th triennal congress of the IEA, Melbourne, Australia
8. Slappendel C, Moore D, Tappin D (1996) Meat industry injury prevention project. case study two: preventing injuries by reducing work compression. New Zealand Council of Trade Unions, Wellington, New Zealand
9. Coutarel F, Caroly S, Vézina N (2015) Marge de manoeuvre situationnelle et pouvoir d'agir: des concepts à l'intervention ergonomique. Le travail humain 78(1): 9–29
10. Goldenhar L (2016) Making a positive difference in construction safety and health by improving safety culture, safety climate and safety leadership. In: PREMUS, Toronto, Canada
11. Coutarel F, Daniellou F, Dugué B (2003) Interroger l'organisation du travail au regard des marges de manœuvre en conception et en fonctionnement. La rotation est-elle une solution aux TMS? Pistes 5(2). http://pistes.revues.org/3328

12. Brown O (2005) Participatory ergonomics. In: Stanton N, Hedge A, Brookhuis A, Salas E, Hendrick HW Handbook of human factors and ergonomics methods. CRC Press, Boca Raton
13. Mayhew C, Quinlan M, Bennet L (1996) The effects of subcontracting/outsourcing on occupational Health and safety 25(1–3): 163–178 (1996). Industrial Relations Research Centre, The University of New South Wales, Sydney
14. Stake RE (1995) The art of case study research. Sage publications, Thousand Oaks
15. Yin RK (2009) Case study research: design and methods. applied social research methods series, 3rd edn. Sage publications, Thousand Oaks
16. Huberman M, Miles MB (2013) Analyse des données qualitatives: recueil de nouvelles méthodes. De Boeck Université, Belgique
17. Bolduc F, Baril-Gingras G (2010) Les conditions d'exercice du travail des cadres de premier niveau : une étude de cas. Pistes 12(3). http://pistes.revues.org/2777

Work-Related Upper Extremity Musculoskeletal Disorders and Low Back Pain in Japan

Hiroyuki Izumi(✉) and Seichi Horie

University of Occupational and Environmental Health,
Japan, 1-1, Iseigaoka, Yahatanishi-ku, Kitakyushu, Fukuoka 807-8555, Japan
izumi-h@med.uoeh-u.ac.jp

Abstract. In Japan, compensable occupational diseases are officially listed in Appended Tables 1 and 2 of Ordinance for Enforcement of the Labor Standards Act. The category No. 3 (diseases caused by work with extreme physical tension) includes subcategories of (b): Low back pain due to work to handle heavy objects, those done in unnatural postures or others which involve excessive tension to low back, and (d): musculoskeletal disorders of the back of the head, neck, shoulder girdle, upper arm, forearm, or fingers due to work which require repeated input into a computer or other operation involving excessive tension on the upper limbs. Local labour standard bureau is in charge of judging the work-relatedness.

The detailed criterion for upper extremity disorders was first published as a notice from the Labour Standard Bureau, Ministry of Labour on Feb 5, 1975, as "Judgment criterion for work-relatedness of upper extremity disorders from keypunching work" (Notification No. 59, 1975) and revised on Feb 3, 1997, as "Judgment criterion for work-relatedness of diseases from upper extremity work" (Notification No. 65, 1997).

The detailed criterion for low back pain was published as "Certification criteria of occupational low back pain" (Notification No. 750, 1976).

In this paper, minimum exposure criteria for upper extremity musculoskeletal disorders and low back pain is described from the view point of the certification criteria for compensable occupational diseases in Japan.

Keywords: Minimum exposure criteria · Upper extremity disorders and LBP Compensable occupational diseases in Japan

1 Introduction

In Japan, compensable occupational diseases are officially listed in Appended Tables 1 and 2 of Ordinance for Enforcement of the Labor Standards Act [1]. The diseases caused by work with extreme physical tension are listed in the category No. 3 (Table 1). It includes subcategories of (b): Low back pain due to work to handle heavy objects, those done in unnatural postures or others which involve excessive tension to low back, and (d): musculoskeletal disorders of the back of the head, neck, shoulder girdle, upper arm, forearm, or fingers due to work which require repeated

S. Bagnara et al. (Eds.): IEA 2018, AISC 820, pp. 511–514, 2019.
https://doi.org/10.1007/978-3-319-96083-8_66

Table 1. The disease caused by a form of work which involve extreme physical tension

(a)	Muscle, tendon, bone, or joint disease or prolapse of internal organs due to strenuous work
(b)	Low back pain due to work to handle heavy objects, those done in unnatural postures or others which involve excessive tension to low back
(c)	Peripheral circulatory disorder, peripheral nerve disorder, or motive organ disorder of fingers or forearm etc. due to work which vibrate the body due to use of equipment or machinery such as rock drill riveter, or chain saw
(d)	Motive organ disorder of the back of the head, neck, shoulder girdle, upper arm, forearm, or fingers due to work which require repeated input into a computer or other operation involving excessive tension on the upper limbs
(e)	In addition to the illness listed in (a) to (d) inclusive their annexed disease and other that are clearly caused by work executed in ways which involve excessive tension to the body

input into a computer or other operation involving excessive tension on the upper limbs.

Local labour standard bureau is in charge of judging the work-relatedness.

2 The Detailed Criterion for the Disease Caused by a Form of Work Which Involve Extreme Physical Tension

2.1 Upper Extremity Disorders

The detailed criterion for upper extremity disorders was first published as a notice from the Labour Standards Bureau, Ministry of Labour on Feb 5, 1975, as "Judgment criterion for work-relatedness of upper extremity disorders from keypunching work" (Notification No. 59, 1975) and revised on Feb 3, 1997, as "Judgment criterion for work-relatedness of diseases from upper extremity work" (Notification No. 65, 1997) [2]. Target diseases of the criterion include humeral epicondylitis, cubital tunnel syndrome, supinator syndrome, wrist tendonitis, carpal tunnel syndrome and cervico-omo-brachial syndrome (COBS; nonspecific symptoms of neck, shoulder and upper extremity). Three indispensable conditions of the required exposure for determining work-relatedness are listed below.

1. The symptoms develop after the engaging the work which put burden on upper limbs for long period (more than 6 months in principle). The work which put burden on upper limbs are listed below.
   a. The worker was involved in the task which had workload by 10% or more for about 3 months compared with the similar task in which the same-sex and similar-age worker is involved.
   b. There was workload per day by 20% or more than usual and the worker had such days about 10 days a month and such circumstance continued 3 months (If the total workload a month is not different from the usual workload, it is included).

   c. During about 1/3 working hours a day, the workload was over by 20% or more than usual, and the worker had such days about 10 days a month and such circumstance continued about 3 months (If the average workload a days is not different from the workload, it is included).

2. In the case that a worker was involved in the task which put burden on upper limbs for 3 months before the onset of symptoms in the following circumstances.

   a. The worker was involved in the task which had workload by 10% or more for about 3 months compared with the similar task in which the same-sex and similar-age worker is involved.

   b. There was workload per day by 20% or more than usual and the worker had such days about 10 days a month and such circumstance continued 3 months (If the total workload a month is not different from the usual workload, it is included).

   c. During about 1/3 working hours a day, the workload was over by 20% or more than usual, and the worker had such days about 10 days a month and such circumstance continued about 3 months (If the average workload a days is not different from the workload, it is included).

3. Engaging excessive workload and the course of the onset of symptoms are approved as medically reasonable ones.

The criterion requires careful and comprehensive judgment of the work-relatedness of individual cases referring to the working environment, heteronomous and restrictive nature of the work, their age, physical strength, life style at home, etc. It recommends avoiding diagnosing as COBS; however, it still allows use of COBS when the specified diagnoses are difficult.

2.2   Low Back Pain

The detailed criterion for low back pain was published as "Certification criteria of occupational low back pain" (Notification No. 750, 1976) [3]. There are 2 types of low back pain (low back pain resulting from accident and not resulting from accident) which Industrial Accident Compensation covers and medical treatment is necessary. Approval requirement is set for each type. Low back pain not resulting from accident is defined as a low back pain due to jobs to handle heavy objects, those done in unnatural postures or others which involve excessive tension to low back. Low back pain not resulting from accident is divided into 2 types according to the causes as muscle fatigue and bone deformation. Two indispensable conditions determining work-related chronic low back pain are listed below.

1. Backache caused by muscle fatigue after being involved in the task in relatively short period (about 3 month or more) is covered by Industrial Accident Compensation.

   a. Task with handling heavy goods about 20 kg or more handling different in weight heavy goods in a half-crouching position repeatedly.

   b. Task handling heavy goods of about 30 kg or more for 1/3 working hours or more.

2. Backache caused by bone deformation is approved to be covered by Industrial Accident Compensation only when the deformation "obviously exceeds the normal change by aging"
   a. Task handling heavy goods of about 30 kg or more for 1/3 working hours or more.
   b. Task handling heavy goods of about 20 kg or more for 1/2 working hours or more.

2.3   The Workers' Compensation Statistics

According to the workers' compensation statistics in 2016, there were 153 cases of upper extremity diseases and 29 cases of chronic low back pain out of 7,361 cases of occupational diseases in total with 4 or more lost work-days.

3   Medical Grounds of the Certification Criteria

The certification criteria for compensable occupational diseases were discussed by the expert committee under the initiative of Ministry of Health, Labour and Welfare in Japan. It is very important to make clear the medical grounds of these criteria. However, it is very difficult to explain clearly the medical grounds of the certification criteria for compensable occupational diseases in Japan. The minutes of meeting is not usually open to the public at this moment. We are trying to find information about medical grounds of the certification criteria but we couldn't get it at this moment.

References

1. Ordinance for Enforcement of the Labor Standards Act (1947) Ministry of Health, Labour and Welfare
2. Judgment criterion for work-relatedness of diseases from upper extremity work (Notification No. 65) (1997) Ministry of Health, Labour and Welfare
3. Certification criteria of occupational low back pain (Notification No. 750) (1976) Ministry of Health, Labour and Welfare

Smart Work Clothes Give Better Health - Through Improved Work Technique, Work Organization and Production Technology

Jörgen Eklund[1]([⊠]) [iD] and Mikael Forsman[2] [iD]

[1] Division of Ergonomics, KTH Royal Institute of Technology,
141 57 Huddinge, Sweden
jorekl@kth.se
[2] Institute of Environmental Medicine, Karolinska Institutet,
171 77 Stockholm, Sweden

Abstract. Musculoskeletal disorders (MSDs) constitute a major health problem for employees, and the economic consequences are substantial for the individuals, companies and the society. The ageing population creates a need for jobs to be sustainable so that employees can stay healthy and work longer. Prevention of MSD risks therefore needs to become more efficient, and more effective tools are thus needed for risk management. The use of smart work clothes is a way to automate data collection instead of manual observation.

The aim of this paper is to describe a new smart work clothes system that is under development, and to discuss future opportunities using new and smart technology for prevention of work injuries.

The system consists of a garment with textile sensors woven into the fabric for sensing heart rate and breathing. Tight and elastic first layer work wear is the basis for these sensors, and there are also pockets for inertial measurement units in order to measure movements and postures. The measurement data are sent wireless to a tablet or a mobile telephone for analysis. Several employees can be followed for a representative time period in order to assess a particular job and its workplace. Secondly, the system may be used for individuals to practice their work technique. The system also gives relevant information to a coach who can give feedback to the employees of how to improve their work technique. Thirdly, the data analysis may also give information to production engineers and managers regarding the risks. The information will support decisions on the type of actions needed, the body parts that are critical and the emergency of taking action.

Keywords: Prevention · Observation methods · Wearables

1 Introduction

1.1 Musculoskeletal Disorders and Prevention

Musculoskeletal disorders (MSDs) constitute a major health problem for employees in industrial countries as well as in developing countries. As a result, in addition to the individual suffering, individuals, organizations and societies are affected by the

© Springer Nature Switzerland AG 2019
S. Bagnara et al. (Eds.): IEA 2018, AISC 820, pp. 515–519, 2019.
https://doi.org/10.1007/978-3-319-96083-8_67

enormous economic consequences. In has been estimated that the costs might be between 0.5 to 2% of GNP (EU-OSHA 2017).

The ageing population is another challenge, due to difficulties to finance the pension system. From a political point of view, people get healthier and live longer, which means that retirement age needs to be increased. But still today, many employees need a pre-mature retirement due to disorders or too demanding work. An important societal challenge is to improve jobs and work conditions towards better sustainability so that people can stay healthy and work more years before retirement.

Prevention of MSD risks therefore needs to become more efficient, and more effective tools are thus needed for prevention and risk management. One new observation tool for this is RAMP (Risk Assessment and Management Tool for Manual Handling Proactively) (Rose 2017; Lind et al. 2017). It is a freely available tool, based on research and developed for company use. It includes risk assessment, visualization of the risks and support for making action plans. However, it requires experts and training to be implemented in practice. In addition, there are recommendations of limits for physical work load regarding oxygen consumption during work. One example of a recommendation is that the relative aerobic strain (RAS, average oxygen consumption during an 8 h work day in relation to the maximal oxygen uptake) should not exceed 33% (Smolander and Louhevaara 2011).

Observational methods have their strength in facilitating many different aspects, they need no particular specialized equipment and there results are usually given as risk traffic light colours, which are very easy to understand and visualize. However, in a project the inter- and intra-observer reliability of six observational methods used by 12 experienced ergonomists, via 3–5 min video-recordings of ten different jobs were evaluated (Forsman 2017). The results showed low reliabilities for the risk assessments, especially for the items concerning repetition, movements and postures. The project also revealed that the inter-method reliability was often low, i.e. when the same work is assessed with different methods different risk estimates are often obtained.

1.2   Aim

The aim of this paper is to describe a new smart work clothes system that is under development, and to discuss future opportunities using new and smart technology for prevention of work injuries.

2   The Vision

This paper describes an ongoing development of a smart work clothes system. As a starting point, a vision was created to use smart and wearable technology for data collection and analyses of postures, movements and physical loads as a basis for assessment of risks for musculoskeletal injury. The assessment criteria should be based on research and for issues where there is a lack of research on experience from experts and practitioners. By using wearable technology, risk assessments could be automated, not only relying on manual observations. This kind of automation has a potential to decrease the costs for the risk assessments in the future due to lower need of manual time resources. Further, reliability is also expected to increase due to the automation, since manual observations have low reliability (Takala et al. 2010).

The vision is that the system in the future should be capable of assessing and visualizing risks for MSDs and cardiovascular disorders at workplaces. Especially risks connected to movements and postures that are, as stated above, difficult to assess visually (Forsman 2017). These risks should be possible to assess via measurements, and to show the results in clear ways, with inspiration from observational methods. The system should time-efficiently cover whole workdays, while observational methods rarely are used for periods longer than ten minutes, which are hard to transform to a whole work day exposure. Visualization of risks imposed from the work and workplace is an important first step towards prevention.

The system under development consists of work clothes worn closest to the body. Knitted or woven textile sensors are used to measure body functions such as heart rate (ECG) and breathing through the thoracic electrical bio-impedance (TEB). Miniaturized inertial measurement units (IMUs), including accelerometers, magnetometers and gyroscopes are attached to the work clothes on body segments such as upper arms, back, thighs and neck for real time measurements of postures and movements. The clothes or wearables that might be used for the fully developed system include vests, T-shirts, shorts, gloves, socks, insoles, a cap or a helmet. All measurement data are transferred wireless to a computer, a mobile telephone or a tablet and recorded. The data are analyzed using algorithms that relate to the different risk assessment criteria.

In that way, the assessment criteria from the RAMP tool could be integrated in a smart work clothes system, and an automated risk assessment could be performed. The results are then visualized or converted to audio messages according to the needs of the different stakeholders. Tactile feedback, e.g. using vibratory feedback is another alternative. Examples of outcomes are real time or summarized feedback to the worker regarding work technique, summarized feedback to an ergonomist, coach, supervisor or manager, or as a summary assessment of the risk imposed by the work and workplace.

This system does not only allow assessment of risks from over-exertion and too high loads but also risks from lack of physical activity such as prolonged sitting or standing, as well as risks due to lack of recovery and rest pauses.

3   The Measurement and Analysis System

Different types of textile electrodes have been tested, and improvements are continuously carried out. For example, a strap has been tested in order to exert higher pressure on the electrodes for better accuracy of the measurements and fewer movement artifacts. Also, different ways of preventing electrodes from drying out have been tested. At present, the following equipment is used.

A vest with four textile electrodes is used to record the ECG and TEB signals. The sampling frequency is 250 Hz and 100 Hz, respectively.

Data from the ECG, TEB and IMUs are transmitted through Bluetooth to an Android tablet (Samsung, SM-T713). The data are stored and processed in this tablet, and partly visualized. Heart rate is calculated from the R-R intervals.

The IMU (LPMS-B2, LP Research, Tokyo, Japan, size $39 \times 39 \times 8$ mm) is used to measure postures and movements. The sampling frequency is set to 30 Hz for the IMU data (accelerometer, gyroscope, magnetometer).

An Android-compatible application has been developed which integrates both the graphical user interface and the communication system that is connected to the wearable system via Bluetooth. The use of the Android application enables real-time analysis of the measurements in parallel with the data collection. The application works for phones or tablets with the operative system Android Marshmallow 6.0 or higher.

4   Results and Discussion

The system has been well received by the users who have tested it so far. It has evoked a lot of curiosity and many people are eager to test the system. Since the smart work clothes can be worn under the normal clothes used at work, it is not possible for others to see the measurement equipment. This also enables risk assessment to be performed in environments where observers are not allowed, for example health care professionals in interaction with patients. The measurement system has not been found disturbing in tests with occupational users, but rather easy to use.

Measurements of postures, movements and heart rate function well, but more detailed studies are needed in order to evaluate the validity of these measurements. Further development of pressure and force measurements for hands and feet are needed. In the future there are opportunities to develop measurements of stress, including heart rate variability, skin impedance, blood pressure and possibly stress hormone sensors. Not only prevention of musculoskeletal disorders but also cardiovascular diseases is possible to develop, based on work load and its distribution over time, physical activities and recovery, as well as stress.

This type of measurements and assessments necessitate ethical considerations, for example who owns the data, how can the data be used, how can the integrity of the individual be protected? One solution is that Occupational Health Services can be given the task to manage these measurements and processes around, since they are used to handle patient data and integrity.

5   Conclusions

The ongoing research is approaching the vision of systems for automated risk assessments that are easy to use and comfortable to wear. The here described new system, has a potential of becoming an inexpensive and reliable tool for posture, movement and force measurements, since it demands less time resources than manual observations. Further, the system can provide feedback, on an individual training level, and on a group level for workplace and organization development. However, more development regarding technical solutions and the user interface, and detailed validity studies are still needed.

Acknowledgements. The research includes a handful number of projects with different participants in each. It is also a collaboration between KTH Royal Institute of Technology, Karolinska Institutet, University of Borås and other partners. The authors would like to acknowledge the following financiers, AFA Insurance, Vinnova and EIT Health, participating test persons and research colleagues including Kaj Lindecrantz, Fernando Seoane, Farhad Abtahi, Liyun Yang, Ke Lu, Carl Lind and Jose Diaz-Olivares.

References

European Agency for Safety and Health at Work (EU-OSHA) (2017) OSH in figures: work-related musculoskeletal disorders in the EU - Facts and figures

Rose L (2017) https://www.ramp.proj.kth.se/publications/publications-and-presentations-on-the-ramp-tool-1.731781

Lind CM, Forsman M, Rose LM (2017) Development and evaluation of RAMP I. A practitioner tool for screening of musculoskeletal disorder risk factors in manual handling. Int J Occup Saf Ergon 10:1–56

Smolander J, Louhevaara V (2011) Muscular Work. In: Encyclopedia of occupational health and safety. International Labor Organization. http://www.iloencyclopaedia.org/component/k2/item/487-muscular-work

Takala EP, Pehkonen I, Forsman M, Hansson GÅ, Mathiassen SE, Neumann WP, Sjøgaard G, Veiersted KB, Westgaard RH, Winkel J (2010) Systematic evaluation of observational methods assessing biomechanical exposures at work. Scand J Work Environ Health 36(1):3–24

Forsman M (2017) The search for practical reliable risk assessment methods – a key for successful interventions against work-related musculoskeletal disorders. Agron Res 15 (3):680–686

AUVAfit

AUVA Strategy for the Prevention of Musculoskeletal Disorders

Julia Lebersorg-Likar[✉]

Austrian Workers' Compensation Board,
Head Office, Adalbert-Stifter-Strasse 65, 1200 Vienna, Austria
Julia.lebersorg-likar@auva.at

Abstract. "Work should not make us sick – work can make us fit", to bring this slogan to life, it is important to consider some principles and to constantly optimize work and workplaces. Presently, musculoskeletal disorders (MSDs) cause nearly a quarter of the overall days of sick leave in Austria. According to EU-OSHA (European Agency for Safety and Health), MSDs are one of the most common work-related diseases. Throughout Europe, they affect millions of workers and cost employers billions of euros. Tackling MSDs helps to improve the health related quality of workers life, reduces cost for the employers thus making sense also in an economic way. To minimize work-related risk factors the Austrian Workers' Compensation Board (AUVA) designed a program called "AUVAfit". The aim of this program is to improve the workplace environment by reducing physical and mental risk factors. AUVAfit is available for the modules "Ergonomics" and/or "Occupational Psychology". The AUVAfit ergonomics team – in cooperation with the company – works out measurements to improve the work places according to ergonomic guidelines.

Keywords: Musculoskeletal disorders · Ergonomics
Occupational psychology

1 Introduction

According to the "Fehlzeitenreport 2017" (absence report of employees), in Austria, 21.4% of sick leave is caused by MSD. Therefore, MSD are the most common cause of illness alongside with the diseases of the respiratory system. Beyond the country's frontiers, MSDs affect millions of workers across Europe and result in billions of euros in costs for employers. Especially persons between 50 and 64 years of age are affected by this problem. Depending on the age, the causes of sick leave are varying. For example, in accordance with the "Fehlzeitenreport 2017" health problems in the group of under-30 s age group caused by MSDs make out only 10% of sick leave. If we take a closer look at the older employees, MSDs cause one-third of sick leave. On the other hand, the injury rate declines significantly with age.

© Springer Nature Switzerland AG 2019
S. Bagnara et al. (Eds.): IEA 2018, AISC 820, pp. 520–525, 2019.
https://doi.org/10.1007/978-3-319-96083-8_68

The notion musculoskeletal disorders has to be seen as an umbrella term. It summarizes disorders of the musculoskeletal system, including ligaments, blood vessels, cartilage, tendons, bones, etc. The spectrum ranges from slight, temporary health problems to serious illness with severe chronical, irreversible damage. The mostly affected body areas are the back and the upper extremities.

MSDs are not induced by single causes - several factors produce their occurrence. Physical work-related risk factors include:

- Handling loads, especially when bending and twisting
- Repetitive or forceful movements
- Awkward and static postures
- Vibration, poor lighting or cold working environments
- Fast-paced work
- Prolonged sitting or standing in the same position.

As a precaution measure against MSDs, it is important to take seriously the occupational health and safety of the employees and to set preventive measures.

AUVA (Allgemeine Unfallversicherungsanstalt - the Austrian Workers' Compensation Board) is the social insurance for occupational risks for more than 4 million employees, pupils and students.

Its legal duties are occupational medical care, first aid for occupational accidents, posttraumatic treatment, rehabilitation, financial compensation and research. Nevertheless, the most important duty is the prevention of occupational accidents and diseases.

As part of its prevention framework, AUVA is offering a programme, called AUVAfit, to obviate musculoskeletal disorders. Under the slogan "Analyze – Optimize – and receive the benefits", AUVAfit supports companies in different lines of business for reducing physical and mental risk factors the employees are exposed to.

AUVAfit consists of the modules "Ergonomics" and "Occupational Psychology". Both modules are available separately or in combination. Companies, where the majority of the employees is insured by AUVA, and which fulfil the basic rules of employment protection (OSH) can take advantage of this programme.

2   Material and Method

2.1   Key Points of AUVAfit

We want to and have to go through working life for a long and hopefully healthy time. To make this possible, it is important to minimize physical and mental risk factors at workplace.

The AUVAfit team analyses and evaluates up to 8 different types of workplaces. As a next step, together with the company, appropriate improvements are developed to reduce possible strain factors. These improvements are implemented according to the so-called "TOP principle" – technical before organizational before personal measures. In the field of personal improvements AUVA also supports companies with various services, such as workshops, training courses, visualization, consulting, etc.

The AUVAfit program is free of cost. Nevertheless investments that the company wants to implement on the basis of advice must be paid by the company.

Any company whose management is standing behind this project, where the majority of the employees is insured by AUVA and which fulfills the basic rules of employee protection (OSH) can take advantage of the AUVAfit program. The scope of the analysed workplaces is determined after an initial interview by the AUVA.

Below you will find a closer look at the ergonomic part of AUVAfit. Subsequently, examples of ergonomic improvements, which have already been implemented in cooperation with the companies, are described.

3   The Process of AUVAfit Ergonomics

1. Informative preliminary talk
   In order to clarify the mutual expectations and the ideas to the project, as a first step an informative preliminary discussion takes place.
2. Project assignment
   AUVAfit is assigned by the management, so a smooth project flow is guaranteed.
3. Definition of workplaces
   Together with the company, workplaces or work tasks to be improved are filtered out. Especially those workplaces need to be analysed where the company needs ergonomic advice.
4. Workplace Analysis
   The previously determined types of workplaces and work tasks are analysed and evaluated by the AUVAfit team. Depending on the type of workplace or work task, a suitable assessment tool is selected.
5. Development and joint implementation of appropriate improvements
   Following this survey and rating, joint development of improvements according to ergonomic guidelines has to be found. The effect is a possible reduction of physical risk factors that could result in musculoskeletal disorders. Each company gets its own tailor-made concept to optimize the quality of its workplaces. All the suggestions are according to the "TOP" principle (Technical before Organizational before Personal measurements).
6. Review of the improvements
   In a final step, the implemented measures are inspected again at a second visit and the result is discussed with the company.

4   Results – Practical Examples of Ergonomic Improvements

AUVAfit responds individually to the needs of each company, therefore the program can be carried out in different branches of industry. To illustrate this, in the following implemented improvements from different companies are presented.

4.1   Example 1: Gas Cutting Machine

Initial Situation: The first example is from a steel processing plant. In the gas cutting machine different sized steel parts are cut and then removed. The parts have to be lifted manually or with the crane into a container. The parts shown are approx. 15 × 15 cm in size and weigh approx. 15 kg each. The parts have to be pulled out by force-locking holding with a magnetic gripper. This means a very high hand-arm load.

Technical Improvement: By attaching a handle, the need for force-locking holding when pulling out the steel element is reduced. Because of the reduced fist closure, the hand-arm load is considerably reduced.

Before   Afterwards

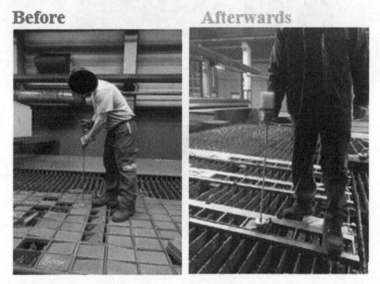

4.2   Example 2: Laundry Cart

Initial Situation: In a nursing home, a laundry cart was used in order to carry dirty laundry. The laundry cart was loud and rough-running because of old and dirty wheels.

Technical Improvement: In order to reduce the volume and improve the handling of the cart, the laundry trough was fixed on the cart. In addition, the wheels were renewed, having new rubber on the wheels and a wheel fixation.

Before

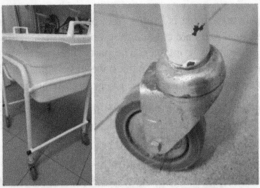

Afterwards

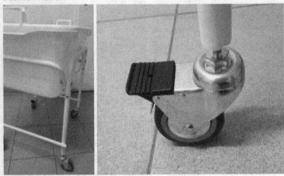

4.3 Example 3: Sorting the Laundry

Initial Situation: In a nursing home, laundry troughs are placed on the floor to sort the laundry by color and texture. Afterwards the laundry has to be washed, so the troughs have to be lifted up from the floor to the height of the washing machine. The washing machine is one meter away from the ground.

Technical Improvement: As an immediate improvement, tables were placed in the washing room and the troughs were placed on the tables. Because of this improvement, the troughs can be lifted in an upright posture.

5 Message

AUVAfit is an actual project which benefits from practical and professional exchange of experience. We want to offer an overview of the prevention programme and to present the ergonomic module of AUVAfit. According to the already implemented improvements and practical experiences, we would like to show how companies benefit from this program.

References

1. Caffier G, Steinberg U, Liebers F (1999) Praxisorientiertes Methodeninventar zur Belastungs- und Beanspruchungsbeurteilung im Zusammenhang mit arbeitsbedingten Muskel-Skelett-Erkrankungen. Schriftreihe der Bundesanstalt für Arbeitsschutz und Arbeitsmedizin
2. Leoni T, Schwinger J (2017) Fehlzeitenreport 2017 – Krankheits- und unfallbedingte Fehlzeiten in Österreich – Die alter(n)sgerechte Arbeitswelt. Österreichisches Institut für Wirtschaftsforschung
3. Sandrock S (2009) In: angewandte Arbeitswissenschaft, No 202
4. https://osha.europa.eu/de. Europäische Agentur für Sicherheit und Gesundheitsschutz am Arbeitsplatz
5. http://www1.arbeitsinspektion.gv.at/ew07/artikel/artikel01-05.htm
6. https://www.auva.at/portal27/auvaportal/content?contentid=10007.670939&viewmode=content

Musculoskeletal Disorders Among Occupational Drivers Caused by Whole Body Vibration and Awkward Posture

Nastaran Raffler[1,2(✉)], Jörg Rissler[1], Rolf Ellegast[1], Thomas Kraus[2], and Elke Ochsmann[2,3]

[1] Institute for Occupational Health and Safety of the German Social Accident Insurance (IFA), Alte Heerstr. 111, 53757 Sankt Augustin, Germany
nastaran.raffler@dguv.de
[2] Institute and Outpatient Clinic for Occupational Medicine, University Hospital, Aachen University of Technology, Pauwelsstr. 30, 52074 Aachen, Germany
[3] Institute of Occupational Health, Lübeck University, Ratzeburger Allee 160, 23563 Lübeck, Germany

Abstract. Musculoskeletal disorders (MSD) can be caused by multifactorial workloads such as whole-body vibration (WBV), awkward posture and heavy lifting. Due to the complexity of field measurements, there is so far no epidemiological study investigating posture quantitatively. Therefore, we investigate the association between MSD outcomes and these exposures among 102 professional drivers by field measurements.

At different workplaces the combined exposures of WBV and posture were measured for 58 professional drivers. These measured data were extrapolated for subjects with the same workplaces and job tasks. The CUELA measuring system was used to capture and analyse the exposure of the posture. Further, the percentage of time spent in a non-neutral angular range was used to describe the upper body posture. Health and personal data as well as information about psychosocial factors and lifting tasks were collected by a questionnaire.

While an index for non-neutral posture shows significant association with the most MSD outcomes, daily vibration exposure value only effects lumbar spine disorder significantly. Also in case of sick leave, lifting and awkward posture appear to be more strongly associating than WBV exposure.

Keywords: Whole body vibration · Posture · Musculoskeletal disorders

1 Introduction

According to a survey in Germany from 2012 [1], Musculoskeletal disorders such as Low back pain (LBP) and local pain at the cervical spine and the shoulders are the most reported disorders during or immediately after the professional activities. Besides, more than 60% of the persons who reported pain are undergoing medical treatment. A large number of employees reported manual materials handling (MMH), exposition to awkward posture and high vibration and shocks. These exposures are often associated with musculoskeletal symptoms [2, 3]. Therefore other cofactors such as MMH,

© Springer Nature Switzerland AG 2019
S. Bagnara et al. (Eds.): IEA 2018, AISC 820, pp. 526–536, 2019.
https://doi.org/10.1007/978-3-319-96083-8_69

psychosocial stress, and especially awkward posture need to be considered [3, 4] while investigating the adverse health effect of WBV.

Observational analysis, self administered questionnaires and biomechanical calculations have been common methods to investigate the combined exposures. For instance, the exposures "trunk bent at work" and "lifting with bending/twisting" have shown a significant effect in terms of increasing the risk of LBP while exposed to WBV [5, 6] by using observational methods and questionnaires. In biomechanical research Fritz et al. [7] assess the forces within a rigid-mass model of the lumbar spine and compare the effects of different postures during exposure to WBV. The bent-forward postures result in an increase of the compressive and shear forces in the dorsoventral direction compared to the upright sitting posture. However, due to the complexity of measuring posture quantitatively, no epidemiological analysis has so far investigated the combination of these exposures by means of quantitative data and their relationship to MSD.

The CUELA posture measuring system ("computer-assisted recording and long-term analysis of musculoskeletal loads") has been introduced for field measurements of combined exposure to WBV and posture [8]. This system permits the quantitative analysis of posture as body angles during exposure to WBV. Thus, the measurements provide an objective and quantitative description of such exposures [9]. In another study the relationship between combined exposure (WBV and posture) and LBP are introduced [10]. However, its relationship to adverse health effect concerning other regions of the spine is still unknown.

Therefore, the aim of this study is to investigate measurement data of WBV exposure and postural stress in terms of their relationship to MSD.

2 Subjects and Methods

2.1 Study Population

The study population has been selected based on the following criteria: The measured exposures should reflect the previous lifetime exposure of the subjects (WBV exposure for over 10 years, at least one year of WBV exposure in the current company). The subjects should have a similar age (40 to 50 years). Also the subjects should not have had any musculoskeletal disease or disorder before beginning their occupational training.

Based on job activities subjects have been divided in four vehicle groups: bus and locomotive drivers, crane operators, earth moving machine operators and forklift truck drivers.

All drivers have been in good health and have not been suffering from noteworthy physical complaints at the study time. The Ethics Committee of the Medical Faculty, RWTH Aachen University has approved the study.

2.2   Measurement and Analysis of WBV

In accordance with [11], WBV measurements are conducted in 3 orthogonal axes $l = \{x, y, z\}$ at the seat surface and at the seat mounting point. Vibration signals are detected at 480 Hz and weighted according to [12] to yield frequency-weighted vibration signals $a_{wl}(t)$, which are averaged by using the root-mean-square (RMS) method:

$$a_{wl} = \left(\frac{1}{T_M} \int_0^{T_M} a_{wl}^2(t) dt \right)^{\frac{1}{2}} \tag{1}$$

According to EU-Directive 2002/44/EC 2002 [13], the daily vibration exposure A (8) is defined as the largest resulting value, after correcting the RMS value a_{wl} in x- and y-direction by a constant and normalising the duration dependence to $T_0 = 8$ h:

$$A(8) = \max\left\{ 1.4a_{wx}\sqrt{\frac{T}{T_0}}; 1.4a_{wy}\sqrt{\frac{T}{T_0}}; a_{wz}\sqrt{\frac{T}{T_0}} \right\} \tag{2}$$

2.3   Measurement and Analysis of Posture

Drivers' posture has been detected at 50 Hz by using the CUELA system [8]. Making use of inertial/kinematic sensor technology, this system records the detected posture continuously as an angular measurement. It can be attached to the subject's clothing, without hindering the subjects during their work. Table 1 shows the sensor arrangement, the regions of the body, the locations of sensor attachment, and the respective body angles or degrees of freedom (DOFs) with definition for neutral range of movement (based on ISO 11226 [14]). In addition to the measurements, video recording is used for investigating the tasks and activities of the drivers during a shift and also for monitoring the alignment of the sensors.

The percentage of working time spent in non-neutral category can thus be shown for each DOF. In the previous published study [8], an index R_{DOF} has been introduced to summarise non-neutral posture. If the observed duration of the ith DOF in the non-neutral category (t_a, i) is greater than 30% of the daily exposure duration (T), the DOF in question is regarded as an "awkward" DOF. The index R_{DOF} is quantified as follows:

$$R_{DOF} = \sum_{i=1}^{i=11} c_i; \; c_i = \begin{cases} 0 \text{ if } \frac{t_{a,i}}{T} \cdot 100\% < 30\% \\ 1 \text{ if } \frac{t_{a,i}}{T} \cdot 100\% \geq 30\% \end{cases} \tag{3}$$

where c_i counts if a DOF exceeds 30% of the exposure duration in the non-neutral category, and in this study $0 \leq R_{DOF} \leq 11$. The index R_{DOF} is a straightforward way to combine the measured postural exposure data.

Table 1. CUELA posture measuring system. DOF with definition for neutral category.

Body region for sensor attachment	Degree of freedom derived from sensor data	Neutral category
Head Thoracic spine Lumbar spine	Head inclination (sagittal)	0° to 25° or <0° Full head support
	Neck flexion (lateral/sagittal)	-10° to 10°or 0° to 25°
	Neck torsion	-45° to 45°
	Thoracic inclination (lateral/sagittal)	0° to 10° or 0° to 20° or <0° Full head support
	Trunk inclination (lateral/sagittal)	0° to 10° or 0° - 20° or <0° Full back support
	Back flexion (lateral/sagittal)	-10° to 10° or 0° to 20°
	Back torsion	-10° to 10°

2.4 Questionnaire

A self-administered questionnaire has been sent to each driver one week before the medical investigation. Medical doctors have collected and checked the filled-in questionnaires during the medical examination.

The questionnaire requests information on the driver's occupational and medical history as well as further load factors such as MMH and psychosocial factors. MMH has been divided into lifting, carrying, pulling/pushing. Also the weight and the percentage of daily exposure duration have been included in the question. However, often the questions have not been answered in such detail, so that MMH and other items had to be treated as dichotomous questions (yes/no).

The outcome variables are assessed in the last section of the questionnaire, which records health complaints. It uses a modified version of the Nordic questionnaire on musculoskeletal symptoms [15]. The drivers are asked about the occurrence of pain in the neck, shoulder, upper and lower back region in the last 12 months or ever in their occupational lives. Drivers who report musculoskeletal symptoms are requested to answer additional items concerning duration, frequency, pain radiation, pain intensity and disability, symptom-related health care use, treatment (e.g. medication or physical therapy) and sick leave due to symptoms in the previous 12 months.

Based on the items of medical section of the questionnaire, following outcomes are defined:

(1) Cervical spine pain: pain or discomfort in the cervical spine area in the previous 12 months.
(2) Shoulder-arm pain: pain or discomfort in the shoulder arm area in the previous 12 months.
(3) Thoracic spine pain: pain or discomfort in the thoracic spine area.

(4) Lumbar spine pain: pain or discomfort in the low-back area between the twelfth rib and the gluteal folds, with or without radiating pain in one or both legs, lasting one day or longer in the previous twelve months.

(5) Scaitic pain: pain going down the leg from the lower back in one or both sides.

(6) Sick leave: sick leave due to MSD in the previous 12 months.

3 Results and Discussion – Description of Factors

3.1 Results – Study Population

Altogether 58 persons have been visited for WBV and posture measurements: 10 Bus and locomotive drivers, 19 crane operators, 20 earth moving machine operators and 9 forklift drivers. They were on average 46.1 ± 8.4 years old. They are on average 22.9 ± 9.7 years exposed to WBV, with 21.6 ± 9.9 years in the current company.

3.2 Results – WBV

The results of the frequency-weighted RMS accelerations measured at the driver-seat interfaces on the machines and vehicles used by professional drivers are presented in Table 2.

Table 2. Frequency-weighted RMS acceleration magnitude (a_{wl}) of vibration measured on the three orthogonal axes $l = \{x, y, z\}$ on the seat surface and the vibration total value, as mean values (standard deviation).

Machine group	Vehicle (number)	Duration of measurement [minutes]	Frequency weighted acceleration magnitude [ms^{-2}]				
			$a_{v1.4}$	a_{wx}	a_{wy}	a_{wz}	$A(8)$
1	Bus/locomotive (10)	94.1 (21.2)	0.30 (0.03)	0.11 (0.03)	0.14 (0.03)	0.19 (0.03)	0.21 (0.02)
2	Crane (19)	81.4 (18.7)	0.31 (0.12)	0.14 (0.06)	0.11 (0.05)	0.17 (0.09)	0.18 (0.04)
3	Earth moving mashine (20)	80.7 (15.3)	0.85 (0.27)	0.36 (0.12)	0.36 (0.15)	0.41 (0.19)	0.57 (0.15)
4	Forklift truck (9)	79.3 (25.1)	0.56 (0.06)	0.24 (0.03)	0.23 (0.04)	0.27 (0.13)	0.36 (0.03)

The duration of the measurements ranges from 79.3 min for the forklift trucks to 94.1 min for bus and locomotives, which is sufficient to capture the representative working conditions. The z-axis (vertical) weighted acceleration is the dominant component in all vehicle groups. Earth moving mashines show the highest vibration exposures. The total vibration value ($a_{v1.4}$) of the weighted RMS accelerations ranged on average from 0.30 to 0.85 ms^{-2} for bus/locomotive and earth moving mashines, respectively.

3.3   Results – Posture

Figure 1 contains the measured percentages of time spent in non-neutral posture for each DOF. The average time spent in non-neutral postures is largest for sagittal body angles for crane operators. Sagittal thoracic inclination and head inclination show the highest percentages, 76% and 64%, respectively for crane operators.

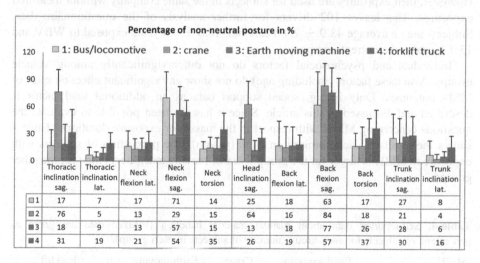

	Thoracic inclination sag.	Thoracic inclination lat.	Neck flexion lat.	Neck flexion sag.	Neck torsion	Head inclination sag.	Back flexion lat.	Back flexion sag.	Back torsion	Trunk inclination sag.	Trunk inclination lat.
1	17	7	17	71	14	25	18	63	17	27	8
2	76	5	13	29	15	64	16	84	18	21	4
3	18	9	13	57	15	13	18	77	26	28	6
4	31	19	21	54	35	26	19	57	37	30	16

Fig. 1. Measures of exposure to non-neutral posture as percentage of measurement duration among machine groups for each DOF. Data are given as mean values.

Concerning sagittal neck flexion, all the three groups except crane operators show on average high percentages in the non-neutral range of movement (54–71%). By means of the time spent in the non-neutral range of movement the R_{DOF} value was calculated (Eq. 4, Table 3). The maximum R_{DOF} is 4 for the crane operators, which is caused by awkward sagittal body angles; 16% of the subjects have reached this value. For the other groups the R_{DOF} value did not differ remarkably.

Table 3. The percentage of subjects who showed certain amount for R_{DOF}

Percentage of subjects with	Machine group				Percentage of subjects with	Machine group			
	Bus/locomotive	Crane	Earth-moving machine	Forklift truck		Bus/locomotive	Crane	Earth-moving machine	Forklift truck
$R_{DOF} = 0$	20	10	35	45	$R_{DOF} = 3$	30	26	5	0
$R_{DOF} = 1$	30	11	30	33	$R_{DOF} = 4$	0	16	0	0
$R_{DOF} = 2$	20	37	30	22					

3.4   Results – Questionnaire

Altogether 129 persons participated, out of which 58 persons have been visited for WBV and posture measurements. There have been 27 drivers who participated in the measurements, but have been unable to participate in the medical examination. However, their exposures are used for subjects in the same company without measured exposures. This leaves 102 drivers for further analysis of the questionnaire data. Subjects are on average 43.9 ± 7.9 Years old, 18.6 ± 9.8 years exposed to WBV and 15.7 years in the current company.

Individual and psychosocial factors do not differ significantly among vehicle groups. Also these factors (including age) do not show any significant effect on most of MSD outcomes. Only lifting, social support data as an additional load factor is described and discussed in this article. Since it has not been possible to evaluate the questions concerning MMH with respect to the masses or exposure durations, Table 4 shows the results of dichotomised answers (yes/no). The proportion of subjects with exposure to MMH is larger for the group 3 (earth-moving machinery) than for the other groups.

Table 4. Self-reported data about manual material handling (MMH): data are given as percentage of subjects for each group (number of subjects in each group).

MMH	Bus/locomotive (12)	Crane (39)	Earth-moving machinery (26)	Forklift truck (25)
Lifting [%]	8	26	50	28
Carrying [%]	8	26	35	24
Pulling/pushing [%]	8	15	42	24

Table 5 reports on the proposed outcome variables with respect to the exposure groups. Crane operators reported the highest number for sick leave because of MSD significantly (48%, chi-square p = 0.007). In general more MSD outcomes were reported by crane and earth moving mashine operators than the other groups.

Table 5. Information on prevalence of musculoskeletal disorders (MSD) in the last 12 months and sick leave due to MSD. Data are given as percentages of subjects for each group.

Musculoskeletal disorders	Group				chi^2 (Pearson)
	Bus/locomotive	Crane	Earth-moving machinery	Forklift truck	
Cervical spine pain	41,7	64,1	42,3	36,0	0,113
Shoulder arm pain	33,3	48,7	50,0	36,0	0,586
Thoracic spine pain	16,7	38,5	30,8	12,0	0,102
Lumbar spine pain	33,3	53,8	69,2	36,0	0,062
Sciatic pain	16,7	15,4	7,7	24,0	0,462
Sick leave	0,0	48,6	26,9	20,8	**0,007**

3.5   Linear Dependencies and Regression Analysis for a Final Model for MSD

Bivariate analyses have been used to identify factors that depend (linearly) on each other. Since non-neutral body angles are highly dependent on each other, the R_{DOF}-value has been selected to represent awkward posture. WBV is described by the tenfold daily vibration exposure A(8)x10. Thus, the range of the data from WBV and other factors has been equalised for the interpretation of regression analysis.

Finally, a logistic regression analysis is started with this selection of factors. The factor vibration is entered in the model regardless of its significance. The selected independent variables are then added to the respective model. The regressions are calculated stepwise backwards (exclusion criteria: p >= 0.05).

In Table 6 the outcomes for MSD are described as final model with the tenfold daily vibration exposure A(8)x10, posture with the R_{DOF} value and other factors in case of significancy. While posture seemed to have significant effect on cervical, thoracic, lumbar spine disorders and sick leave, the daily vibration value only showed a significant effect on lumbar spine disorder. Other factors as age and social support showed tendencies for shoulder arm disorder, while lifting showed a very significant effect on sick leave beside posture. No significant correlation was found in terms of sciatic pain.

4   Discussion

Measurement data of WBV exposure and postural stress for the study population are detected. The magnitude of RMS values are similar to those published in other reports [5, 6, 8]. Also for the postural workload a reasonable variation is discernible (Fig. 2).

Other studies have found a correlation between age [16] or lifetime dose values [5] with outcome variables that describe LBP. In this study age was only describing shoulder arm outcome (p = 0.055). However, a significant effect of age in an univariate analysis for MSD in this study did not appear. This could follow from the fact that the age distribution among the drivers in this study has been homogenous enough to suppress the effect in most of MSD outcomes.

Table 6. Regression models for the 12 months prevalence of MSD and sick leave

Final model	Variables	OR (95% CI)	P-value	Other variables in the model	OR (95% CI)	P-value
Cervical spine pain	A(8)x10	1,17 (0,87–1,57)	0,304			
	R_{DOF}	**1,69 (1,12–2,55)**	0,012			
Shoulder arm pain	A(8)x10	1,27 (0,81–1,98)	0,294	Age	**1,08 (1–1,17)**	0,055
	R_{DOF}	1,62 (0,84–3,11)	0,149	Social support	**1,67 (1,02–2,75)**	0,041
Thoracic spine pain	A(8)x10	1,37 (0,97–1,94)	0,070			
	R_{DOF}	**1,69 (1,05–2,71)**	0,029			
Lumbar spine pain	A(8)x10	**1,72 (1,21–2,46)**	0,003			
	R_{DOF}	**1,84 (1,18–2,84)**	0,007			
Sciatic pain	A(8)x10	0,97 (0,65–1,46)	0,901			
	R_{DOF}	1,09 (0,65–1,84)	0,735			
Sick leave	A(8)x10	1,07 (0,71–1,61)	0,758	Lifting	**6,26 (2,17–18,07)**	0,001
	R_{DOF}	**1,94 (1,13–3,32)**	0,016			

The body posture index R_{DOF} in Eq. (3) describing the adopted upper posture has led to a significant correlation with most of the MSD outcomes and sick leave in combination with WBV exposure. Also in another study the interaction test did not show over-additive effects from the product of these two variables A(8) and R_{DOF} [10]. Other variables only seemed to have correlation between MMH and sick leave, age and social support with shoulder arm pain outcome. This indicates that for this small study population no other effects have been overlooked. At the same time it cannot be excluded that those other factors have a correlation with MSD outcomes: the small size of the study population and the fact that the factors other than WBV and posture have not been assessed with the same accuracy may have led to the absence of a statistically significant correlation. It has to be mentioned in this respect that other studies also failed to see a correlation with psychosocial factors, while MMH has been attributed with LBP in the literature [5, 10, 17].

Finally, the correlation of the posture index R_{DOF} with MSD outcome variables and the A(8) for LBP show that they are suitable quantities describing the combined exposure of WBV and awkward posture in further studies.

5 Conclusion

In the multifactorial context of MSD it is advantageous if not necessary to assess the relevant exposures on the same level of accuracy, and this study successfully used measurements to describe combined exposures to WBV and awkward postures. It has been shown, in addition, that quantities based on a quasi-static assessment of posture, especially R_{DOF}, are significantly associated with MSD prevalences and sick leave in this context.

References

1. Brennscheidt F, Nöllenheidt C, Siefer A (2012) Zahlen - Daten - Fakten, Federal Institute for Occupational Safety and Health (BAuA)
2. Bernard, B., Musculoskeletal Disorders and Workplace Factors, ed. N.I.f.O.S.a. Health (1997) 4676 Columbia Parkway, Cincinnati, OH 45226–1998
3. Bovenzi M (2015) A prospective cohort study of neck and shoulder pain in professional drivers. Ergonomics 58(7):1103–1116
4. Lötters F et al (2003) Model for the work-relatedness of low-back pain. Scand J Work Environ Health 29(6):431–440
5. Bovenzi M et al (2006) An epidemiological study of low back pain in professional drivers. J Sound Vib 298:514–539
6. Tiemessen I, Hulshof C, Frings-Dresen M (2008) Low back pain in drivers exposed to whole body vibration: analysis of a dose-response pattern. Occup Environ Med 65(10):667–675
7. Fritz M, Schäfer K (2010) Berücksichtigung der Haltung des Oberkörpers bei der Beurteilung von Ganzkörper-Schwingungen. Zeitschrift für Arbeitswissenschaft 64:293–304
8. Raffler N et al (2010) Assessing combined exposures of whole-body vibration and awkward posture–further results from application of a simultaneous field measurement methodology. Ind Health 48(5):638–644
9. Raffler N et al (2016) Factors affecting the perception of whole-body vibration of occupational drivers: an analysis of posture and manual materials handling and musculoskeletal disorders. Ergonomics 59(1):48–60
10. Raffler N et al (2017) Combined exposures of whole-body vibration and awkward posture: a cross sectional investigation among occupational drivers by means of simultaneous field measurements. Ergonomics 60(11):1564–1575
11. ISO 8041 (2005) Human response to vibration - Measuring instrumentation, in DIN
12. ISO 2631-1 (1997) Mechanical vibration and shock - Evaluation of human exposure to whole-body vibration - Part 1: General requirements, pp 1–31
13. EU-Directive 2002/44/EC (2002) On the minimum health and safety requirements regarding the exposure of workers to the risks arising from physical agents (vibration): Official J Eur Communities. 6 July 2002, L 117/13-19
14. ISO 11226 (2000) Ergonomics - Evaluation of static working postures

15. Kuorinka I et al (1987) Standardised Nordic questionnaires for the analysis of musculoskeletal symptoms. Appl Ergon 18(3):233–237
16. Notbohm G, Schwarze S, Albers M (2009) Ganzkörperschwingungen und das Risiko bandscheibenbedingter Erkrankungen. Arbeitsmedizin Sozialmedizin Umweltmedizin 44:327–335
17. Lotters F et al (2003) Model for the work-relatedness of low-back pain. Scand J Work Environ Health 29(6):431–440

RAMP – A Comprehensive MSD Risk Management Tool

Linda M. Rose[1]([⊠]), Jörgen Eklund[1], and Lena Nord Nilsson[1,2]

[1] Division of Ergonomics, Department of Biomedical Engineering
and Health Systems, KTH School of Engineering Sciences in Chemistry,
Biotechnology and Health, Hälsovägen 11C, 141 57 Huddinge, Sweden
lrose@kth.se
[2] Unit of Safety and Health, Scania CV AB, Södertälje, Sweden

Abstract. The objective of this paper is to describe the development, dissemination and preliminary effects of the use of a new musculoskeletal disorder (MSD) risk management tool for manual handling, RAMP (Risk Assessment and Management tool for manual handling Proactively). RAMP is research based and developed in close collaboration between researchers and practitioners with a participative iterative methodology. A broad strategy is used for the dissemination, including the use of professional networks, conferences, a specially developed homepage, and Massive Open Online Courses which also provide training on the tool use. The tool has been spread widely to about 45 countries since the release 2017. E.g. Scania CV uses RAMP as its global standard method for managing MSD risks at logistics and machining departments. Among the preliminary effects results show that at one department risk reduction measures had been taken for more than 2/3 of the work stations with assessments signalling elevated risk levels after 1.5 years. Further studies on RAMP are discussed. It is concluded that the development and the dissemination of RAMP can be seen as successful. Preliminary reports on the tool use effects indicate that the RAMP tool supports the MSD risk management process in the work to reduce MDS risks at workplaces.

Keywords: Manual handling · Risk management · Observation method

1 Introduction

1.1 Background

In line with the theme of the IEA2018 congress 'Creativity in Practice', the RAMP can be seen as a result of the challenge and aim to transform research results into a concrete applicable tool for actions to improve the quality of peoples work and life.

Musculoskeletal disorders (MSDs) still lead to a large amount of negative consequences for those affected by them, as for companies and the society. In the European Union MSDs are the most frequent occupational diseases [1]. About 20% of the approved occupational diseases in the Swedish construction industry are carpal tunnel syndromes [2]. In the US the costs for overexertion injuries are estimated to more than 13 billion US Dollars solely as compensation to employees [3] and at societal level the

© Springer Nature Switzerland AG 2019
S. Bagnara et al. (Eds.): IEA 2018, AISC 820, pp. 537–546, 2019.
https://doi.org/10.1007/978-3-319-96083-8_70

costs of MSDs within the European Union are estimated to be around 0.5–2% of the Gross National Product, GNP [1].

A number of risk assessment tools for assessing MSD risks have been developed over the past decades, e.g. the NIOSH Lifting Equation, [4], RULA [5], QEC [6], and HARM [7]. The existing tools have strengths, but also limitations. These include that methods solely assess certain body parts or certain types of work (e.g. RULA, NIOSH Lifting equation) or only support part of the systematic risk management process (most methods). Many of these are observation based tools and an overview of some of these tools is given by Takala et al. [8].

Two large global companies identified the need of a more comprehensive risk assessment and risk management tool, which would include a larger amount of relevant risk factors than the existing tools do, especially regarding manual handling work, be research based and support the whole MSD risk management process [9]. To meet industrial needs of a comprehensive MSD risk management tool for manual handling jobs, that supports the whole systematic MSD risk management process and risk communication in organizations, a new risk assessment and risk management tool, RAMP (Risk Assessment and Management tool for Manual Handling Proactively) was developed.

The objective of this paper is to describe the development, dissemination and preliminary effects of the use of the RAMP tool.

2   Methods

The development project was led by researchers at KTH and carried out in close collaboration with other researchers and practitioners at companies. After a needs analysis and development of requirements on the tool, and literature studies, which also included studying several existing tools, RAMP prototypes were developed. This was done iteratively with tests among and feedback from users-to-be. The tool is based on over 250 research publications and developed iteratively with a participative methodology using feedback from over 80 practitioners. During the development the process several evaluations on the tool usability and reliability were carried out [10, 11].

With the objective to provide suitable training for tool users-to-be, a Professional Certificate Program on the RAMP tool use is under development. It consists of three Massive Open Online courses (MOOCs) supported by the edx.org platform. In the development of the courses a multi-disciplinary collaboration-model for the design is used, including content experts, media producers and educationalists. The course design includes lectures, applied authentic examples from industry, interviews with users in different contexts and has continuous assessment and feedback throughout the courses.

A broad dissemination strategy is being used to spread knowledge about the tool, where it can be downloaded, as well as on the how to register and follow the three RAMP MOOCs. This includes the use of professional networks, presentations at conferences, the development of an especially dedicated RAMP homepage, and Massive Open Online Courses which also provide training on the tool use.

The digitalized RAMP tool (Version 1.02) was evaluated using a survey at the end of two separate 90-min workshops in Canada in the summer 2017 as well as at the end of a one day workshop on the RAMP tool in Estonia in the autumn 2017. The first two were held for ergonomists in a city, and at an international ergonomics conference respectively, while the latter was held for ergonomics specialists, managers and university faculty and students. Among the 62 participants 27 stated being ergonomists or ergonomic consultants, 6 managers or directors, 10 senior researchers, and 11 students (PhD and university students), while the others did not state profession. At the workshops an introduction of the RAMP tool was given, followed by a walk-through of the tool and applied use of the tool based on an example from industry. At the end of the workshops the participants were invited to participate in a survey. This participation was voluntary.

3   Results

3.1   The RAMP Tool

The RAMP tool consists of four modules: RAMP I, RAMP II, (see Fig. 1 for an illustration of pdf- version of these) the Results module and the Action module. Together they support the whole MSD risk management process from identifying and assessing risk factors, supporting presentation and communication of results and the development, implementation and follow-up of risk reduction measures.

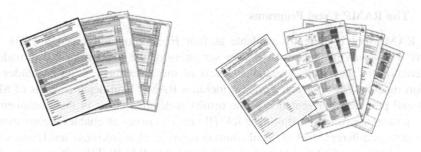

Fig. 1. Layout of the pdf-versions of (a) RAMP I checklist and (b) RAMP II In depth analysis.

RAMP I is a checklist based assessment tool for scanning of MSD risks. Assessment items are assessed choosing between 'Yes' and 'No' alternatives and the assessment results are presented with a three-colour coding regarding the risk-and-priority levels. In RAMP I green signals low risk for most employees, but for some employees risk reduction measures may be needed; red signals high risk for most employees and risk reduction actions should be prioritized; grey signals that this has to be investigated further before the risk and priority level can be determined.

RAMP II enables a more in depth analysis. It consists of more detailed and complex alternatives in the assessment. In RAMP II the results are presented with a similar three-colour coding system: green and red, as in RAMP I, but instead of grey as in RAMP I, in RAMP II the result at the intermediate level; yellow, signals an elevated risk for certain employees and that risk reduction measures should be taken. In addition, there is also a Risk score in RAMP II.

In RAMP I and RAMP II seven risk categories are investigated: Postures, Work movements and repetitive work, Lifting work, Pushing and pulling work, Influencing factors, Reports at the company on strenuous work and finally Perceived physical discomfort among the employees carrying out the work task.

The Results Module presents the results at different level of detail and scope, ranging from detailed results, results at risk category level to results at overview level. Also results from a single work station, to a department, site or a whole company can be presented in one results table. This design was chosen to meet different stakeholders and professional role's needs within an organization: those who work with risk reduction measures need to know what the risks are to be able to develop adequate risk reduction measures, while at management level less detail but better overview is more useful as a base for informed decision making.

The Action module is developed to support the risk management process by providing the Action model, which can be used for systematic support in finding risk reduction measures; automatically generated Action suggestions for those Assessment items/risk factors which have been assessed as red in RAMP I, or red or yellow in RAMP II. This module also includes an Action plan template with the assessment results presented and provides a structure to plan for risk reduction measures, when they shall be implemented, who is responsible as well as a check when this is achieved and when a follow-up is planned to be carried out.

3.2   The RAMP Excel Programs

The RAMP tool is currently available as four Excel programs (version 1.03). *The RAMP I program* includes RAMP I for screening of the MSD risks and provides a presentation of the results at detailed levels of the assessment. It also includes the Action module. *The RAMP II program* includes RAMP II in depth analysis of MSD risks and provides a presentation of the results at detailed levels of the assessment. It also includes the Action module. The *RAMP I Results program* enables presentation of the results at different level of detail (from detailed level to overview level) and scope (ranging from a work station to a whole company). The RAMP II Results program does the same, but with the RAMP II results.

An example
In Fig. 2 a brief example of assessment of one of the assessment items is given. In this example the posture of the head is assessed, which in this case is bent forward or twisted more than 30° for approximately 1.5 h per work day. In RAMP I (Fig. 2a) this is assessed as that the head is 'clearly bent or twisted – forward or towards a side' for about 1 h per work day or more.

This is assessed as a 'Yes', which leads to a grey result, signaling that this has to be investigated further before a risk-and priority level assessment can be established. Figure 2b shows the assessment of the 'Posture of the head - forwards or to a side' with illustrations to support the assessment as well as a table with exposure time. Since the forward bending or twisting of the head is more than 30° for approximately 1.5 h per

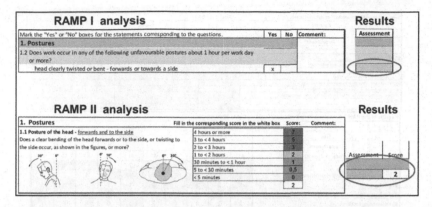

Fig. 2. Assessment of one assessment item in (a) RAMP I and (b) RAMP II, posture of the head – twisted or bent forwards, and the colour coded results of the assessment.

work day, this results in an assessment with the Risk score of '2' in this case. This results in a yellow assessment, signaling that there is an elevated risk for certain employees and that risk reduction measures should be taken.

Figure 3 illustrates how these results can be presented at detailed, risk category and at overview level using the RAMP II Results program. In Fig. 3a the row '1.1' shows the assessment of the posture of the head – forwards or to a side and 'A1' denotes the case in the example above, while results from other work stations are shown in other columns. This visualisation shows *where* risk-and-priority levels are increased, *what* causes the increased risks, and this presentation can also be used for planning work. For example, one should avoid rotating between work station A3 and A5, since they are assessed as red for the same assessment item.

In Fig. 3b the colour of a cell at the risk category level shows the highest risk-and-priority level in the risk category and the number in the cell how many assessments in that risk category are assessed to be at that risk-and-priority level. In the example used above this means that for 'A1', one of the assessment items under 'Postures' is assessed as red while the others are assessed as yellow or green.

Figure 4a shows part of the Action model, which can be used as a support when developing risk reduction actions at the organisation. The model suggests that solutions are sought in five different areas: Technology and design, Employees, Organisation, Vision and strategies, and Environment. In addition, automatically generated examples of Action suggestions are presented for those assessment items assessed as red or yellow. Figure 4b shows Actions suggestions for the assessment item used in the example (1.1 Posture of the head - forwards or to a side).

3.3   The Professional Certificate Program and the MOOCs

The Professional Certificate Program on the RAMP tool is freely available at edx.org. It and its three courses are developed by KTH in close collaboration with several industrial companies. The courses are taken online, are self-paced, and free of charge to audit. Participants can choose to study each course individually. To be eligible for

Results of the RAMP II analysis at detailed level   Date: 2017-03-27

	Country		Sweden								
	Site		Sthlm					Sthlm		Sthlm	
	Department		A					B		C	
	Work station ID	A1	A2	A3	A4	A5	B1	B2	B3	C1	C2
1.1. Postures											
1.1 Posture of the head - forwards and to the side											
1.2 Posture of the head - backwards											
1.3 Back posture - moderate bending											
1.4 Back posture - considerable bending and twisting											
1.5 Upper arm posture - hand in or above shoulder height*											
1.6 Upper arm posture - hand in or outside the outer work area*											
1.7 Wrist posture*											
1.8 Leg and foot space and surface											

Results of the RAMP II analysis at risk category level   Date: 2017-03-27

	Country			Sweden							
	Site				Stockholm						
	Department		A				B			C	
	Work station ID	A1	A2	A3	A4	A5	B1	B2	B3	C1	C2
1. Postures		1		1	1	3	2	2		1	
2. Work movements and repetitive work		1	1	2	1	1	1	2	1	2	
3. Lifting work		1	2	2	2	2	1	1		1	1
4. Pushing and pulling work							2	2	2		
5. Influencing factors		1	2	1	1	3	1	6	6	6	
6. Reports on physically strenuous work				1	1	1	1	1	1		
7. Perceived physical discomfort				1	1	1	1	1		1	
Results summary:											
Number of red assessments (high risk)		3	1	5	4	6	2	4	0	2	0
Number of yellow assessments (risk)		4	6	4	4	5	12	11	11	12	1
Number of green assessments (low risk)		28	28	26	27	24	21	20	24	21	34

Fig. 3. Results presentation using the RAMP II Results program for presentation of results at (a) detailed level for one the risk category 'Postures', and (b) for all seven risk categories at risk category level and overview level. (Color figure online)

(a) The Action model:

ORGANISATION — TECHNOLOGY & DESIGN — CHANGE — EMPLOYEES — ENVIRONMENT — VISION & STRATEGIES

(b)

1.1 Posture of the head – forwards and to the side

Type of action	Examples of suggestions for solutions
T&D	Investigate the visual conditions and secure that the lighting is appropriate for the work that is carried out (e.g. illuminance, glare, and contrast) and that the work area is arranged in an appropriate way to the light. See visual ergonomics guidelines. Maybe the employees visions need to be checked and visual aids obtained.
T&D	Redesign the work/work area, also considering the visual design, so that the unfavourable postures are eliminated or reduced. For example, adjustable surfaces may be needed. Altered shelf heights or tilted surfaces to improve vison and access may be appropriate solutions, or secure that it is easy to visually inspect or physically feel that the work is performed correctly.
ORG	Consider work organisational changes, e.g. job enrichment, job enlargement, and job rotation. Review the work content regarding the amount/magnitude and frequency of the exposure.
EMPL	Inform, educate and train the employees and secure knowledge.
V&S	Work with aims, visions and strategies for decreasing the MSD risks.
ENV	Aim at smooth logistics access, a layout that enables easy movements and good flow and also consider physical (e.g. noise), thermal (cold/heat) and chemical factors.

(c) Action plan

Date of assessment: 2016-04-20			Work task/ Employee load: Work station 3		Department: A2		
Work/Work task: Packaging at WS3			Site: Stockholm		Country: Sweden		
Orderd by: Jens Andersson	Formed by: Julia Riviera		Date of action plan: 2016-04-29		Note: Urgent to solve!		
	Assessment	Score	Comments	Planned actions	When	By whom	Ready (date)	Follow-up
1. Arbetsställningar								
1.1 Posture of the head - forwards and to the side		2		Redesign work surface	May 2016	OSH/CP	2016-05-27	2016-06-03
1.2 Posture of the head - backwards		1.5		Lower shelves	May 2016	OSH/CP	2016-05-27	2016-06-03
1.3 Back posture - moderate bending		1						
1.4 Back posture - considerable bending and twisting		1						
1.5 Upper arm posture - hand in or above shoulder height*		7		Lower shelves, redesign worksp.	May 2016	OSH/CP	2016-05-27	2016-06-03
1.6 Upper arm posture - hand in or outside the outer work area*		2		Redesign work space	May 2016	OSH/CP	2016-05-27	2016-06-03
1.7 Wrist posture*		1						
1.8 Leg and foot space and surface		2		Redesign work space	May 2016	OSH/CP	2016-05-27	2016-06-03
2. Work movements and repetitive work								
2.1 Movements of the arm (upper and lower arm)*		0						
2.2 Movements of the wrist*		1		Rotate btwn jobs, redesign task	Sept 2016	Consultant	2016-11-25	2017-03-25
2.3 Type of grip - frequency*		1						
2.4 Shorter recovery/variation during work		0						
2.5 Longer recovery/variation during work		0						

Fig. 4. The Action module with (a) The Action model, (b) the Action suggestions for the assessment item 1.1 in RAMP II, and (c) part of the Action plan for assessment item 1.1 using RAMP II's Results program. (Color figure online)

course certificates the participants have to pass the course examination and the certificates also come with a fee. To be eligible for the Professional Certificate the participants have to pass all three courses. The first course 'Assessment of Work-Related Injury Risks using RAMP I' was launched in April 2018 and had participants from over 45 countries from all continents except Antarctica within the first week. Course II will start in September and course III is planned to start in October. No specific prerequisites are required. Each course requires approximately 15–25 h of work from the participants.

3.4   Dissemination of the RAMP Tool

The RAMP tool has been disseminated using a broad strategy. This includes using the researchers' and practitioners' professional networks, presentations at conferences, work-shops, seminars, short courses, publications in peer-reviewed scientific journals, a PhD Thesis, networking, and articles in professional magazines. An important platform for the dissemination is the especially developed RAMP homepage, *ramp.proj.kth.se*. Also the edx.org platform, hosting the Professional Certificate Program and the RAMP MOOCs are a dissemination node.

The RAMP tool has received attention from a number of different stakeholders, ranging from interested individuals, ergonomists, students, companies, organisations and institutes to research groups and authorities. Companies have started to implement RAMP and the tool has also gained attention from researchers around the world. This has led to 10 presentations after special invitations at conferences (in addition to 9 'normal' conference presentations), six presentations at different universities and authorities after specific invitations and more than a dozen presentations at companies. In addition, about a dozen organisations and authorities have explicitly outspoken their interest in the tool and to collaborate in studies on it. The RAMP project leader has also been contacted with inquiries about the tool in over 60 emails. Approximately 1700 unique individuals visited the RAMP homepage within the first year and the digitalized RAMP tool was downloaded by over 600 users from over 40 countries from six continents during that period. The first RAMP MOOC had over 180 participants from over 45 countries and six continents within the first week and around 400 learners from 80 countries during the first month.

3.5   Evaluation of the RAMP Tool

Out of the 62 participants participating in the survey, 95% agreed fully or partly on that the RAMP tool is usable for risk assessment and for communication of risks, results, and risk reduction actions within the organisation, whereas 3% were neutral and 2% did partly not agree or not agree at all. 97% agreed fully or partly that on the RAMP tool is usable as a decision base for prioritizing where actions should be taken, while the other 3% were neutral.

Regarding how the users perceive the RAMP tool regarding the reliability of the assessments, 68% stated very good or good, 29% were neutral and 3% as poor or very poor. On how well the RAMP tool assesses the risk of developing MSDs, 92% stated very well or well, 7% neutral and 2% poorly or very poorly. On the question on how

well the RAMP tool supports the systematic risk management work (from risk identification and risk assessment to risk communication and risk management) within the
organisation, 95% stated very well or well, while the other 5% were neutral. During the
development process early versions of RAMP have been evaluated with good results
regarding usability and reliability [10, 11].

3.6   Effects Using the RAMP Tool

So far information about effects of the RAMP use has been gained by personal contacts
mainly with representatives from organisations which started applying early versions of
the tool, before it was made officially available from the homepage.

At Scania, who also was one of the participating companies in the RAMP development project, RAMP was tested at 8 different departments. All of them were positive
to continue using the method and do so. The experience at Scania is that using RAMP
it is possible to assess a larger number of relevant risk factors than with other methods,
such as KIM, NIOSH, and SES, which the company had used previously. The company also has evaluated the RAMP tool positively on a number of features, e.g. that it
provides an overview of the current state regarding MSD risks and improvement needs,
as well as presentation of assessments at different levels of detail and scope. Further the
company has identified that the method supports goal setting and can be used as a basis
for priorities as well as follow-up of improvements.

In addition, preliminary results from a logistics department at Scania, where the
RAMP tools was implemented in 2016, show that at the end of a 1.5 year period risk-
reduction measures had been taken for more than 2/3 of the assessments signalling
elevated risk levels, i.e. those assessed as red (signalling increased risk for many
employees and that risk reduction measures should be prioritized) and yellow (signalling increased risk for certain employees and that risk reduction measures should be
taken). The wellbeing among the employees had improved and work attendance
increased, mainly considered to be attributed to the systematic risk reduction initiated
by the use of RAMP. In addition, also positive quality effects were perceived. The
results from this evaluation show that RAMP contributes to improved work environment and increased health attendance [12].

Since then Scania has decided to use and RAMP as the global standard method for
managing MSD risks at its logistics and machining departments. It has been implemented at Scania in Sweden, France, Netherlands, Poland, and India.

Also GKN Driveline in Köping, Sweden, implemented RAMP in 2016, and also
report positive effects of the tool usage. Both companies use RAMP proactively, and
GKN Driveline also uses it as a criteria method when ordering new production equipment.

One of the largest Occupational Health Services (OHS) in Sweden, Avonova, has
decided to implement RAMP as their standard MSD risk assessment and risk management tool. They intend to let all ergonomists who perform risk assessments follow
the first, already available RAMP MOOC. The company recommends their customer
companies to take the first RAMP MOOC and to use RAMP I themselves. The strategy
is that the OSH company thereafter offers support the customer companies with the
RAMP II in depth analysis and how to proceed in the risk management process, if the
customer companies do not have that competence themselves.

4 Discussion

RAMP and the first RAMP MOOC have been disseminated to a rather large amount of countries and to different types of users ranging from individuals, companies and organisations to research groups and authorities. The RAMP tool developers have received feedback from companies (e.g. Avonova and Scania) that RAMP is considered as the most comprehensive and current MSD risk assessment and risk management method and the only tool that supports the entire MSD risk management process [8] in manual handling work, and not only results in risk assessments.

However, longer use of the tool, as well as studies on the tool use should be studied. The current digitalized version of the RAMP tool (1.03) consists of four Excel-programs. There is also a call from industry for a databased RAMP tool, for even more improved usability.

There is also a call from several industrial branches to enhance the application range of the RAMP tool by including more risk factors and action suggestions, e.g. on hand-intensive work. By including such risk factors and risk reduction suggestions, the applicability of the RAMP tool would widen from manual handling to including e.g. assembly work and dental work.

The criteria for risks that were developed for the RAMP tool are also used in projects in which postures are measured instead of observed. One example of this is the development of "smart work clothes" for automated assessments of risks for MSDs. This development for the future, after it has been commercialized, has a potential to further disseminate the use of risk assessment and risk handling in working life.

We suggest and encourage future studies to evaluate the effects of the RAMP tool, and studies on its usability, validity, and reliability, by others than the tool developers. Also, the results from such studies are important in the revision of the criteria limits used in the RAMP tool.

5 Conclusion

It is concluded that the development and the dissemination of RAMP can be seen as successful measured in terms of a large number of downloads, users and requests for information. A Professional Certificate Program on the RAMP tool use is under development and the first of three MOOCs that has been launched engaged learners from six continents within the first week. A survey on the perception of RAMP among ergonomists and other professionals in the ergonomics field shows good results.

Preliminary reports on the tool use effects indicate that the RAMP tool supports the MSD risk management process in the work to reduce MDS risks at workplaces. These results also show that the RAMP tool has contributed to improved work environment and increased health attendance.

References

1. European Agency for Safety and Health at Work (EU-OSHA) (2017) OSH in figures: Work-related musculoskeletal disorders in the EU - Facts and figures
2. Försäkring AFA (2017) Arbetsolyckor och sjukskrivningar i byggbranschen. Report (In Swedish)
3. Liberty Mutual: 2017 Liberty Mutual Workplace Safety Index. http://image.email-libertymutual.com/lib/fe541570726d02757312/m/1/2017+WSI.pdf. Accessed 03 Mar 2018
4. Waters TR, Putz-Anderson V, Garg A et al (1993) Revised NIOSH equation for the design and evaluation of manual lifting tasks. Ergonomics 36(7):749–776
5. McAtamney L, Corlett EN (1993) RULA: a survey method for the investigation of work-related upper limb disorders. Appl Ergon 24(2):91–99
6. David G, Woods V, Li G et al (2008) The development of the Quick Exposure Check (QEC) for assessing exposure to risk factors for work-related musculoskeletal disorders. Appl Ergon 39(1):57–69
7. Douwes M, Boocock M, Coenen P et al (2014) Predictive validity of the Hand Arm Risk assessment Method (HARM). Int J Ind Ergon 44(2):328–334
8. Takala EP, Pehkonen I, Forsman M, Hansson GA, Mathiassen SE, Neumann WP, Sjogaard G, Veiersted KB, Westgaard RH, Winkel J (2010) Systematic evaluation of observational methods assessing biomechanical exposures at work. Scand J Work Environ Health 36:3–24
9. International Organization for Standardization (ISO) (2009) 31000:2009 Risk management – Principles and guidelines, Geneva, Switzerland
10. Lind C, Rose L (2016) Shifting to proactive risk management: risk communication using the RAMP tool. Agron Res 14(2):513–524
11. Lind CM, Forsman M, Rose LM (2017) Development and evaluation of RAMP I - A practitioner's tool for screening of musculoskeletal disorder risk factors in manual handling. Int J Occup Saf Ergon, 17 October 2017
12. Scania (2018) Krafttag för bättre ergonomi inom logistikhanteringen. (In Swedish) In Scania Inside, Nr 1/2018

The Speed Calculated Hand Activity Level (HAL) Matches Observer Estimates Better Than the Frequency Calculated HAL

Oguz Akkas[1], Stephen Bao[2], Carisa Harris-Adamson[3], Jia-Hua Lin[2], Alysha Meyers[4], David Rempel[3], and Robert G. Radwin[1(✉)]

[1] University of Wisconsin-Madison, Madison, WI 53706, USA
rradwin@wisc.edu
[2] Washington State Department of Labor and Industries, Olympia, WA 44000, USA
[3] University of California-San Francisco, San Francisco, CA 94143, USA
[4] National Institute for Occupational Safety and Health, Cincinnati, OH 45226, USA

Abstract. Hand Activity Level (HAL) can be estimated from observations (HAL_O), calculated from exertion frequency, F and duty cycle D (HAL_F) or from speed, S and D (HAL_S). Data collected by prospective cohort studies were used to compare these methods. There was 75% agreement between HAL_O and HAL_S ($HAL_S = 1.02 \times HAL_O -0.2$, $R^2 = 0.78$, $F(1,1003) = 43665$, $p < .001$), but only 30% agreement between HAL_O and HAL_F ($HAL_F = 0.21 \times HAL_O + 2.2$, $R^2 = 0.04$, $F(1,1003) = 71.71$, $p < .001$). HAL_S was more consistent with HAL_O since both are dependent on speed, and because HAL_S can be automated, it is more objective than HAL_F or HAL_O in this sample.

Keywords: Repetitive motion · Threshold limit value · Musculoskeletal disorders

1 Introduction

The Hand Activity Level (HAL) of the ACGIH TLV is estimated on a 10-point visual-analog scale by trained observers rating exertion speed and hand pauses. Only exertions greater than 10% of posture-specific strength are counted. It can also be estimated from measurements of exertion frequency (F) and percent duty cycle (D) using a lookup table or an equation [1]. HAL can automatically be estimated using computer vision [2] from videos by tracking hand speed (S) and D [3, 4].

Bao et al. [5] observed poor correlation between HAL observations, self-reports and direct-measurements for several reasons, including using different definitions of repetitive exertions. Wurzelbacher, et al. [6] compared more than 700 tasks for 484 workers and found that while correlated, the agreement (within ± 1point) between observations and calculated HAL was 61%. The current study compares observer estimated HAL (HAL_O), frequency F and D calculated HAL (HAL_F) estimated from

S. Bagnara et al. (Eds.): IEA 2018, AISC 820, pp. 547–549, 2019.
https://doi.org/10.1007/978-3-319-96083-8_71

frame-by-frame analysis of videos, and speed S and D calculated HAL (HAL_S) using computer vision. The effect of exertion definition was also considered.

2 Methods

Video and exposure data from prospective cohort studies [7–11] were used. A total of 1004 videos containing HAL_O values were selected in the order of subject ID, containing at least five contiguous cycles, no breaks, limited camera motion, and no visual obstructions. Hand motion, including S, was measured from the videos using computer vision. The estimates of total exertion F and D from the data were used to calculate HAL_F, and S and D were used to estimate HAL_S.

3 Results

There was 75% agreement (within $\pm$ 1point) between HAL_O and HAL_S ($HAL_S = 1.02$ HAL_O -0.2, $R^2 = 0.78$, $F(1,1003) = 43665$, $p < .001$), however, there was only 30% agreement (within $\pm$ 1point) between HAL_O and HAL_F ($HAL_F = 0.21 \times HAL_O + 2.2$, $R^2 = 0.04, F(1,1003) = 71.71, p < .001$). We selected a subset of 111 videos and checked the exertion annotation. After modifying F to count every forceful finger, hand or wrist exertion, as recommended in Bao et al. [5], agreement was 57% between HAL_F and HAL_O ($HAL_F = 0.71 \times HAL_O + 1.21$, $R^2 = 0.54$, $F(1,110) = 128.4, p < .001$).

4 Discussion

HAL_S was more comparable to HAL_O than HAL_F. HAL_F varied considerably depending on the exertion definition. For example, if a power hand tool was operated several times and the entire time holding the tool was counted as a single exertion, the resulting HAL_F was small. However, if every instance of exertion was counted, a larger F and thus a larger HAL_F was obtained. When recommended rules for F were applied, HAL_F and HAL_O were more closely aligned, and therefore these rules should be used for more reliability among measures. HAL_S was more consistent with HAL_O since both are dependent on speed, and because HAL_S can be automated, it was more objective than HAL_F or HAL_O in this sample.

References

1. Radwin RG, Azari DP, Lindstrom MJ, Ulin SS, Armstrong TJ, Rempel D (2015) A frequency-duty cycle equation for the ACGIH hand activity level. Ergonomics 58(2):173–183
2. Chen CH, Hu YH, Yen TY, Radwin RG (2013) Automated video exposure assessment of repetitive hand activity level for a load transfer task. Hum Fact J Hum Fact Ergon Soc 55(2):298–308

3. Akkas O, Azari DP, Chen C-H, Hu YH, Ulin SS, Armstrong TJ, Rempel D, Radwin RG (2015) A hand speed and duty cycle equation for estimating the ACGIH hand activity level rating. Ergonomics 58(2):184–194
4. Akkas O, Lee CH, Hu YH, Yen TY, Radwin RG (2016) Measuring elemental time and duty cycle using automated video processing. Ergonomics, 1–12
5. Bao S, Howard N, Spielholz P, Silverstein B (2006) Quantifying repetitive hand activity for epidemiological research on musculoskeletal disorders–Part II: comparison of different methods of measuring force level and repetitiveness. Ergonomics 49(4):381–392
6. Wurzelbacher S, Burt S, Crombie K, Ramsey J, Luo L, Allee S, Jin Y (2010) A comparison of assessment methods of hand activity and force for use in calculating the ACGIH® Hand Activity Level (HAL) TLV®. J Occup Environ Hyg 7(7):407–416
7. Bao SS, Kapellusch JM, Garg A, Silverstein BA, Harris-Adamson C, Burt SE, Dale AM, Evanoff BA, Gerr FE, Hegmann KT, Merlino LA, Thiese MS, Rempel DM (2015) Developing a pooled job physical exposure data set from multiple independent studies: an example of a consortium study of carpal tunnel syndrome. Occup. Environ. Med. 72(2):130–137
8. Harris-Adamson C, Eisen EA, Dale AM, Evanoff B, Hegmann KT, Thiese MS, Kapellusch JM, Garg A, Burt S, Bao S, Silverstein B, Gerr F, Merlino L, Rempel D (2013) Personal and workplace psychosocial risk factors for carpal tunnel syndrome: a pooled study cohort. Occup. Environ. Med. 70(8):529–537
9. Harris-Adamson C, Eisen EA, Kapellusch J, Garg A, Hegmann KT, Thiese, MS, Dale AM, Evanoff B, Burt S, Bao S, Silverstein B, Merlino L, Gerr F, Rempel D (2014) Biomechanical risk factors for carpal tunnel syndrome: a pooled study of 2474 workers. Occup. Environ Med. oemed-2014
10. Kapellusch JM, Garg A, Bao SS, Silverstein BA, Burt SE, Dale AM, Evanoff BA, Gerr FE, Harris-Adamson C, Hegmann KT, Merlino LA, Rempel DM (2013) Pooling job physical exposure data from multiple independent studies in a consortium study of carpal tunnel syndrome. Ergonomics 56(6):1021–1037
11. Kapellusch JM, Garg A, Hegmann KT, Thiese MS, Malloy EJ (2014) The strain index and ACGIH TLV for HAL risk of trigger digit in the WISTAH prospective cohort. Hum Fact J Hum Fact Ergon Soc 56(1):98–111

Automated Video Lifting Posture Classification Using Bounding Box Dimensions

Runyu Greene[1], Yu Hen Hu[1], Nicholas Difranco[1], Xuan Wang[1],
Ming-Lun Lu[2], Stephen Bao[3], Jia-Hua Lin[3],
and Robert G. Radwin[1(✉)]

[1] University of Wisconsin-Madison, Madison, WI, USA
rradwin@wisc.edu
[2] National Institute for Occupational Safety and Health, Cincinnati, OH, USA
[3] Washington Department of Labor and Industries, Olympia, WA, USA

Abstract. A method is introduced for automatically classifying lifting postures using simple features obtained through drawing a rectangular bounding box tightly around the body on the sagittal plane in video recordings. Mannequin postures were generated using the University of Michigan 3DSSPP software encompassing a variety of hand locations and were classified into squatting, stooping, and standing. For each mannequin posture a rectangular bounding box was drawn tightly around the mannequin for views in the sagittal plane and rotated by 30 ° horizontally. The bounding box dimensions were measured and normalized based on the standing height of the corresponding mannequin. A classification and regression tree algorithm was trained using the height and width of the bounding box to classify the postures. The resulting algorithm misclassified 0.36% of the training-set cases. The algorithm was tested on 30 lifting postures collected from video recordings a variety of industrial lifting tasks, misclassifying 3.33% of test-set cases. The sensitivity and specificity, respectively were 100.0% and 100.0% for squatting, 90.0% and 100.0% for stooping, and 100.0% and 95.0% for standing. The algorithm was capable of classifying lifting postures based only on dimensions of bounding boxes which are simple features that can be measured automatically and continuously. We have developed computer vision software that continuously tracks the subject's body and automatically applies the described bounding box.

Keywords: Computer vision · Musculoskeletal disorders
Exposure assessment

1 Introduction

Computer vision was previously used for evaluating hand activity level [1], exertion frequency and duty cycle [2], and visualizing repetitive motion task factors [3]. The current study investigates automatic classification of lifting postures using simple features extracted from video. Rather than measuring joint angles, we take a practical approach that is insensitive to challenging workplace conditions, such as poor illumination, poor vantage points, and obstructions, by relaxing the need for high precision. We explore if features of a simple sagittal plane "elastic" rectangular bounding

© Springer Nature Switzerland AG 2019
S. Bagnara et al. (Eds.): IEA 2018, AISC 820, pp. 550–552, 2019.
https://doi.org/10.1007/978-3-319-96083-8_72

box encompassing the entire body while continuously tracking the subject in a video, can classify standing, stooping and squatting while lifting.

2 Methods

Mannequin postures were systematically generated using the University of Michigan 3DSSPP software to encompass the range of hand locations (20 ACGIH TLV lifting zone boundary points [4]) and anthropometries (5th, 50th, and 95th percentile height for males and females). After excluding locations that smaller mannequins cannot reach, 105 cases (training-set) were generated for analysis. Based on torso (40°) and knee angles (130°), there were 43 squats, 13 stoops, and 49 standing postures.

A bounding box was drawn tightly around the subject for each training-set case for views in the sagittal plane as well as rotated by 30° with respect to the sagittal plane. The stature normalized height and width were measured. After randomly ordering the data, a classification and regression tree (CART) algorithm was trained [5] to classify the postures [6]. Decision trees for splits ranging between 1 to 10 were generated to determine the optimal thresholds based on cross validation error. To test the classifier, ten industrial video clips [7–12] in each class, totaling 30 cases (test-set), were randomly selected if the full body during lifting was visible in the sagittal plane.

3 Results

The resulting tree had four levels and four splits, misclassifying 0.36% training-set cases. For the test-set, the algorithm correctly classified 10 of 10 squats, 9 of 10 stoops, and 10 of 10 stands, misclassifying 3.3% cases. The sensitivity and specificity, respectively was 90.0% and 100.0% for squat, 90.0% and 100.0% for stoop, and 100.0% and 95.0% for standing.

4 Discussion

These posture classifications, which are related to hip and knee angles, were identified without direct angle measurements, instrumentation, markers, or fitting the image to a skeletal model. Although this study was limited to the sagittal plane, future work will investigate additional views. We have developed software to track a worker using this bounding box approach which may be used for continuously quantifying the frequency and duration postures are assumed during work. It is anticipated that this simple algorithm can be implemented on a hand-held device such as a smart phone, making it readily accessible to practitioners.

References

1. Chen CH, Hu YH, Yen TY, Radwin RG (2013) Automated video exposure assessment of repetitive hand activity level for a load transfer task. Hum Fact 55(2):298–308
2. Akkas O, Lee CH, Hu YH, Yen TY, Radwin RG (2016) Measuring elemental time and duty cycle using automated video processing. Ergonomics 59(11):1514–1525
3. Greene RL, Azari DP, Hu YH, Radwin RG (2017) Visualizing stressful aspects of repetitive motion tasks and opportunities for ergonomic improvements using computer vision. Appl Ergon
4. ACGIH (American Conference of Governmental Industrial Hygienists). TLV®/BEI® Introduction (2017) http://www.acgih.org/tlv-bei-guidelines/tlv-bei-introduction
5. Mathworks (2017) Decision Trees. https://www.mathworks.com/help/stats/classification-trees-and-regression-trees.html
6. Breiman L, Friedman J, Stone CJ, Olshen RA (1984) Classification and regression trees. CRC Press
7. Bao S, Howard N, Spielholz P, Silverstein B (2006) Quantifying repetitive hand activity for epidemiological research on musculoskeletal disorders–Part II: comparison of different methods of measuring force level and repetitiveness. Ergonomics 49(4):381–392
8. Lu M, Waters T, Krieg E, Werren D (2014) Efficacy of the revised NIOSH lifting equation for predicting risk of low back pain associated with manual lifting: a one-year prospective study. Hum Fact 56(1):73–85
9. Safetyvideopreviews (2012) Manual Material Handling/Safe Lifting. https://www.youtube.com/watch?v=rrl2n8qehrY&t=8s
10. University of Michigan Center for Ergonomics. 3DSSPP: Background Information (2017) https://c4e.engin.umich.edu/tools-services/3dsspp-software/3dsspp-background-information/
11. University of Michigan Center for Ergonomics (2014) Paper Flopping - Job Modification. https://www.youtube.com/watch?v=61cu5qvH0kM&index=54&list=PLn5IJRj74S88rnFFV6ObxS6nFdDXUFiGW
12. University of Michigan Center for Ergonomics (2017) Stacking, Facing Line2 CE. https://www.youtube.com/watch?v=MxTgvuhVAJA&t=55s

Can the Revised NIOSH Lifting Equation Be Improved by Incorporating Personal Characteristics?

Menekse Salar Barim[1]([⊠]), Richard F. Sesek[2], M. Fehmi Capanoglu[2],
Sean Gallagher[2], Mark C. Schall Jr.[2], and Gerard A. Davis[2]

[1] Oak Ridge Institute for Science and Education (ORISE) Research Fellow,
Cincinnati, OH 45202, USA
mzs0053@auburn.edu
[2] Auburn University, Auburn, AL 36830, USA

Abstract. The impact of manual material handling such as lifting, lowering, pushing and pulling have been extensively studied. Many models using these external demands to predict injury have been proposed and employed by safety and health professionals. However, ergonomic models incorporating personal characteristics into a comprehensive model are lacking. This study explores the utility of adding personal characteristics such as the estimated L5/S1 Intervertebral Disc (IVD) cross sectional area, height, age, gender and Body Mass Index (BMI) to the Revised NIOSH Lifting Equation (RNLE) with the goal to improve injury prediction. A dataset with known RNLE Cumulative Lifting Indices (CLIs) and related health outcomes was used to evaluate the impact of personal characteristics on RNLE performance. The dataset included 29 cases and 101 controls selected from a cohort of 1,022 subjects performing 667 jobs. RNLE performance was significantly improved by incorporation of personal characteristics. Adding gender and intervertebral disc size multipliers to the RNLE raised the odds ratio for a CLI of 3.0 from 6.71 (CI: 2.2–20.9, PPV: 0.60, NPV: 0.82) to 24.75 (CI: 2.8–215.4, PPV: 0.86, NPV: 0.80). The most promising RNLE change involved incorporation of the multiplier based on the estimated IVD cross-sectional area (CSA). This multiplier was developed by normalizing against the IVD CSA for a 50th percentile woman. This multiplier could assume values greater than one (for subjects with larger IVD CSA than a 50th percentile woman). Thus, CLI could both decrease and increase as a result of this multiplier. Increases in RNLE performance were achieved primarily by decreasing the number of RNLE false positives (e.g., some CLIs for uninjured subjects were reduced below 3.0). Results are promising, but confidence intervals are broad and additional, prospective research is warranted to validate findings.

Keywords: Revised NIOSH Lifting Equation (RNLE)
Personal characteristics · BMI · Age · Gender · Low back pain
Intervertebral disc cross sectional area · L5/S1

© Springer Nature Switzerland AG 2019
S. Bagnara et al. (Eds.): IEA 2018, AISC 820, pp. 553–560, 2019.
https://doi.org/10.1007/978-3-319-96083-8_73

1   Introduction

Musculoskeletal disorders (MSDs) are a major burden on individuals, health systems and social care systems, with indirect costs being predominant, and their impact is pervasive [1]. MSDs affect hundreds of millions of people around the World and the most common MSD is low back pain (LBP) which is the leading cause of activity limitation and work absence [2]. It has been recognized that low back pain (LBP) risk is associated with a combination of personal factors, psychological or psychosocial factors, as well as physical exposures [3]. Several case-control studies have revealed that high BMI (overweight) has a significant association with low back pain [4, 5]. According to a systematic review, heavy physical work, awkward postures, lifting, psychosocial factors, BMI and age all have a strong impact on low back pain [6].

Several risk assessment tools have been developed to evaluate LBP risk resulting from manual material lifting tasks. The most well-known and widely–used tool among the ergonomics community is the Revised NIOSH Lifting Equation (RNLE) [7–14]. However, most ergonomic assessments do not consider personal characteristics directly, rather, they focus on physical factors associated with the job demands. Suggestions have been made on how to modify the equation or multipliers used in the equation to improve its reliability, better estimate stressors faced by varying populations, expand the functionality, or simplify the RNLE [15–17].

This research explores the potential impact of these factors and proposes several ways to incorporate such characteristics into the RNLE. Specifically, multipliers were created to explore age, gender, BMI, and a scaling factor based upon intervertebral disc diameter.

2   Methodology

This paper modified the RNLE by considering additional multipliers: including age, gender, Body Mass Index (BMI), intervertebral disc (IVD) cross-sectional area (CSA) and a new coupling multiplier with lower coefficients for non-optimal couplings. A retrospective, case-control methodology was employed to determine the predictive ability of the RNLE and modified RNLE measures. A database was modified to allow multipliers to be "switched on or off" so that various combinations could be explored. First, multipliers were added individually followed by various combinations to determine their impact on model performance. All combinations were evaluated based on odds ratios, sensitivity, specificity, positive predictive value (PPV), and negative predictive value (NPV) compared to baseline ("normal") RNLE performance with all six original multipliers in place. All outputs were recorded in tables comparing new models to baseline RNLE data. A database from an epidemiological study [15] involving a large automotive manufacturer was used to explore modifications to the RNLE.

2.1   An Automotive Manufacturing Ergonomic Field Study

The data were collected from six different automotive plants and consist of 667 manufacturing jobs with 1,022 participants as well as job-specific, historical injury data. Well-defined lifting activities meeting the RNLE criteria for analysis (e.g., two-handed, symmetric lifts) were selected for this study. Personal characteristic variables investigated for this study included height, weight, age and gender and self-reported ratings of perceived discomfort. Subjects were asked to report their LBP discomfort on the day they were interviewed as well as to report any LBP symptoms for the previous year. Cases were defined as subjects who had both LBP symptoms in the previous year and whose job had one or more LBP-related medical visits in the previous year. There were 130 subjects meeting all inclusion criteria: 29 cases and 101 controls. The subject population was composed of 101 males and 29 females aged 23–65 (mean 42 ± 11.2 years), heights from 59-76 inches (mean 69.5 ± 3.6), weights from 115–350 lb (mean 191 ± 45.1), and Body Mass Index from 17.0 to 54.8 kg/m² (mean 27.6 ± 5.6). The prevalence of low back pain for this population was 22% (29/130).

2.2 A Morphometric Study of Low Back Geometry Using MRI Technology

Previous research has yielded a regression equation to predict the size of an individual's IVD cross-sectional area [17, 18]. That study used subjects without current or chronic episodes of LBP and examined them using a whole body 3T Magnetic Resonance Imaging machine (Siemens Verio open-bore). The IVD cross-sectional area used for this study [18] was the L5/S1 IVD measured at its center (see Line "B" in Fig. 1 below).

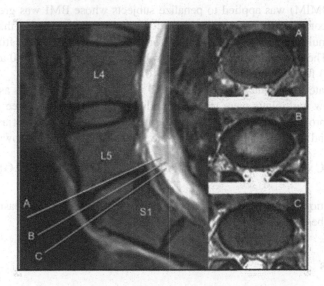

Fig. 1. Sample of MRI scan in sagittal and transverse planes [18]

IVD area was used to scale risk up or down for smaller and larger subjects, respectively. A 50th percentile female IVD area was used to normalize risk. Subjects with smaller estimated IVD areas were considered at higher risk and those with larger

IVD areas were considered to be at lower risk. The IVD multiplier could have been normalized to any size disc but was targeted to a smaller than average size to account for false positives common with the RNLE.

$$L5/S1\ IVD\ CSAs = [-16.959 + 0.179 * Height * 2.54 + 1.7 * Gender]\ cm^2$$
$$(Gender\ (G) = 0\ for\ females\ and\ 1\ for\ males)$$

3   Experimental Design

Modifications to the RNLE were proposed and several novel multipliers were selected for evaluation. These multipliers are gender (GM), body mass index (BMIM), age (AGEM), an approximation of the low back intervertebral disc (IVD) size (IVDM).

$$IVDM = Subject\ L5/S1\ IVD\ Area\ /\ 50th\ percentile\ female\ L5/S1\ IVD\ Area$$

A new, more conservative CM was also proposed and tested. The RNLE uses the following multipliers "good coupling" = 1.0, "fair coupling" = 0.95, and "poor coupling" = 0.90. The proposed new coupling multiplier (NCM) uses 1.0, 0.80, and 0.70 for good, fair, and poor couplings, respectively. A gender multiplier (GM) of 2/3 was applied to female subjects as proposed in the *Applications Manual for the Revised NIOSH Lifting Equation* [10]. Males were assigned 1.0 for GM. A BMI multiplier (BMIM) was applied to penalize subjects whose BMI was greater than 30. The BMIM consisted of 30/BMI for BMIs > 30 and 1.0 for BMIs less than or equal to 30. An age multiplier (AGEM) to account for strength losses expected from aging was also tested. The age multiplier was 1.0 for subjects under the age of 40 and decreased by 1% (0.01) for each year of age beyond 40.

To evaluate RNLE multipliers, a LI of 3.0 was used to classify jobs as more or less risky. All new multipliers were tested individually and in groups to see if predictions could be improved for the RNLE CLI. The new multipliers work just as the original multipliers and can be easily included in the RWL calculations as shown in below:

$$RWL = LC \times HM \times VM \times DM \times AM \times FM \times CM \times GM \times BMIM \times AGEM \times IVDM$$

Modifications to the RNLE were proposed to account for an increasingly diverse, aging, and obese population of workers.

4   Results

Table 1 shows the impact of adding personal multipliers to the RNLE compared to the baseline (original) RNLE CLI. Each column represents the addition of a single multiplier to the baseline RNLE.

Table 1. Addition of new personal multipliers

	CLI (Baseline)	+BMIM	+AGEM	+GM	+IVDM
Odds Ratio	6.71	6.71	6.71	7.83	19.8
(95% CI)	(2.2–20.9)	(2.2–20.9)	(2.2–20.9)	(2.6–24.0)	(2.2–177.2)
p-value	0.0057	0.0057	0.0057	0.0015	0.0073
Sensitivity	0.30	0.30	0.30	0.33	0.17
Specificity	0.94	0.94	0.94	0.94	0.99
PPV	0.60	0.60	0.60	0.63	0.83
NPV	0.82	0.82	0.82	0.82	0.80

The IVD multiplier (IVDM) had the greatest impact on the RNLE, significantly improving overall odds ratio. However, the sensitivity dropped substantially (from .30 to .17). The GM modestly improved the odds ratio and overall model performance. The addition of an AGEM or BMIM did not alter the RNLE. Next, multipliers were added in combinations to see if RNLE performance could be further increased. Table 2 and 3 illustrate the impact of various combinations of proposed multipliers, ranging from lower to higher performance.

Table 2. Combinations of new multipliers

	CLI (Baseline)	+BMIM +AGEM	+BMIM +AGEM +GM	+GM +AGEM/ +BMIM +GM	+IVDM +AGEM
Odds Ratio	6.71	4.33	4.5	6.64	9.8
(95% CI)	(2.2–20.9)	(1.5–12.2)	(1.7–12.3)	(2.3–19.6)	(1.8–53.5)
P-value	0.0010	0.0057	0.0032	0.0006	0.0084
Sensitivity	0.30	0.30	0.33	0.33	0.17
Specificity	0.94	0.91	0.90	0.93	0.98
PPV	0.60	0.50	0.50	0.59	0.71
NPV	0.82	0.81	0.82	0.82	0.80

RNLE performance was maximized by adding both the IVDM and GM. The IVDM had the greatest impact and combinations that included it performed the best. An odds ratio near 25 was achieved by combining the IVDM and the GM. It should be noted, however, that sensitivity remained significantly lower than baseline with performance (odds ratio; improvements coming from the reclassification of false positives as true negatives. Also, the confidence interval, while significant, is very large (2.8–215.4). Unlike all of the other multipliers, the IVDM can actually reduce estimated risk; reducing some over-estimation of risk.

Table 3. Combinations of new multipliers

	CLI (Baseline)	+GM +AGEM +BMIM +IVDM	+BMIM +IVDM +GM or +BMIM +IVD +AGEM or +IVDM +AGEM +GM	+BMIM +IVDM	+IVDM +GM
Odds Ratio	6.71	9.84	12.25	19.80	24.75
(95% CI)	(2.2–20.9)	(2.4–41.0)	(2.3–64.5)	(2.2–177.2)	(2.8–215.4)
P-value	0.0010	0.0017	0.0031	0.0076	0.0036
Sensitivity	0.30	0.23	0.20	0.17	0.20
Specificity	0.94	0.97	0.98	0.99	0.99
PPV	0.60	0.70	0.75	0.83	0.86
NPV	0.82	0.81	0.80	0.80	0.80

5 Discussion

This research indicates that personal characteristics can be successfully and simply factored into ergonomic assessment tools such as the RNLE to improve their performance. Further, some factors may be removed from tools without a decrement in performance. In the case of the RNLE, personal characteristics may even be integrated after job level data collection to improve risk estimation for individuals.

This study demonstrates that model performance cannot solely be assessed by univariate analyses. Various combinations of multipliers should be explored to determine the best performing models. This is particularly true for the traditional multipliers, all of which can hold maximum values of 1.0. In other words, risk estimates increase (or stay the same) when these multipliers are employed. The IVDM, on the other hand, can increase or decrease risk since it can have values both less than and greater than 1.0. Future work should consider other multipliers that can hold values greater than 1.0 and/or consider modifying existing multipliers to allow values above 1.0. Multipliers exceeding 1.0 may especially help to minimize false positive classifications.

While model performance was significantly enhanced by incorporating personal characteristics, model sensitivity (detecting cases) was relatively low and was, in fact, lower than the baseline sensitivity in the best performing models (sensitivity reduced from .3 to .2). While positive predictive value (PPV) was relatively high, with 86% of subjects with CLIs over 3.0 properly identified as cases, only 1 in 5 cases; however (.2), were identified using the new RNLE model (+IVDM, +GM). More research is needed to produce effective models. However, ergonomists can also alter decision points to impact sensitivity. For example, Table 4 below shows the impact of reducing the CLI decision cut-point from 3.0 to 2.5.

Table 4. RNLE model with new multipliers and reduced CLI Cut-point of 2.5

	CLI (Baseline)	+IVDM +GM
Odds Ratio	6.71	10.29
(95% CI)	(2.2–20.9)	(2.9–36.6)
P-value	0.0010	0.0003
Sensitivity	0.30	0.30
Specificity	0.94	0.96
PPV	0.60	0.69
NPV	0.82	0.82

Sensitivity returned to 0.30 (baseline) along with modest improvements to specificity and PPV.

6 Conclusion

Personal characteristics appear to drive a significant proportion of manual material handling (MMH) risk and should be considered when assessing MMH risk. Models incorporating a subject's estimated intervertebral disc size were the most promising and should be explored further. This study demonstrated the potential value of including these personal characteristics on diverse set of subjects and lifting tasks from 6 different automotive manufacturing sites. The subjects included a wide range of ages, BMIs, and were comprised of 22% female workers. Likewise, future research should also include subject populations that are as diverse as possible, particularly since the workforce is aging and increasingly obese. Identifying the contributions of obesity to MMH risk may further demonstrate the value of wellness programs aimed at assisting workers in maintaining healthy lifestyles and physical conditions.

Acknowledgement. This publication was partially supported by Grant # 2T420H008436 from NIOSH. The findings and conclusions in this report are those of the authors and do not necessarily represent the official position of the Centers for Disease Control and Prevention. Use of trade names is for identification only and does not constitute endorsement by the Public Health Service or by the U.S. Department of Health and Human Service.

References

1. Woolf AD, Pfleger B (2003) Burden of major musculoskeletal conditions. Bull. World Health Organ. 81(9):646–656
2. Lidgren L (2003) The bone and joint decade 2000–2010. Bull. World Health Organ. 81 (9):629
3. Institute of Medicine and National Research Council (2001) Musculoskeletal disorders and the workplace. The National Academies Press, Washington, DC

4. Shiri R, Karppinen J, Leino-Arjas P, Solovieva S, Varonen H et al (2007) Cardiovascular and lifestyle risk factors in lumbar radicular pain or clinically defined sciatica: a systematic review. Eur. Spine J. 16:2043–2054
5. Schumann B, Bolm-Audorff U, Bergmann A, Ellegast R, Elsner G et al (2010) Lifestyle factors and lumbar disc disease: results of a German multi-center case-control study (EPILIFFT). Arthritis Res Ther 12:193
6. Da Costa BR, Vieira ER (2010) Risk factors for work-related musculoskeletal disorders: a systematic review of recent longitudinal studies. Am. J. Ind. Med. 53:285–323
7. Dempsey PG (2002) Usability of the revised NIOSH lifting equation. Ergonomics 45 (12):817–828 Liberty Mutual Research Center for Safety & Health, 71 Frankland Road, Hopkinton, Massachusetts 01748, USA
8. Dempsey PG, McGorry RW, Maynard WS (2005) A survey of tools and methods used by certified professional ergonomists. Appl Ergon 36:489–503
9. Waters TR, Putz-Anderson V, Garg A, Fine L (1993) Revised NIOSH equation for design and evaluation of manual lifting tasks. Ergonomics 36(7):446–749
10. Waters TR, Putz-Anderson V, Garg A (1993) Applications manual for the revised NIOSH lifting equation. US Department of Health and Human Services, Cincinnati
11. Waters TR, Putz-Anderson V, Garg A, Fine LJ (1993) Revised NIOSH equation for the design and evaluation of manual lifting tasks. Ergonomics 36:749–776
12. Waters TR, Putz-Anderson V, Garg A (1994) 'Applications manual for the revised NIOSH lifting equation' DHHS (NIOSH) Publication No. 94–110, U. S. Department of Health and Human Services, National Institute for Occupational Safety and Health, Cincinnati, OH
13. Gallagher S, Sesek RF, Schall M Jr, Huangfu R (2017) Development and validation of an easy-to-use risk assessment tool for cumulative low back loading: the lifting fatigue failure tool (LiFFT). Appl Ergon 63(142–150):13
14. Garg A (1995) Revised NIOSH equation for manual lifting: a method for job evaluation. AAOHN J 43(4):211–216
15. Sesek RF (1999) Evaluation and refinement of ergonomic survey tools to evaluate worker risk of cumulative trauma disorders. Doctoral Dissertation. University of Utah, Salt Lake City, UT
16. Sesek R, Gilkey D, Drinkaus P, Bloswick DS, Herron R (2003) Evaluation and qualification of manual materials handling risk factors. Int J Occup Saf Ergon 9(3):271–287
17. Sesek R, Tang R, Gungor C, Gallagher S, Davis GA, Foreman KB (2014) Using MRI-derived spinal geometry to compute back compressive stress (BCS): a new measure of low back pain risk. In: Duffy V (ed) Proceedings of the 5th AHFE Conference, pp 13–18
18. Tang R (2013) Morphometric analysis of the human lower lumbar intervertebral discs and vertebral endplates: experimental approach and regression models. Doctoral Dissertation. Auburn University, Auburn, AL

Difference of Actual Handled Weight and the Recommended Limit for Dynamic Asymmetrical Manual Handling Tasks in Chilean Construction Workers

Olivares Giovanni, Villalobos Victoria, Rodríguez Carolina, and Cerda Eduardo[✉]

Ergonomics and Biomechanical Laboratory, Department of Physical Therapy, Faculty of Medicine, Universidad de Chile, Santiago, Chile
encerda@med.uchile.cl

Abstract. Manual handling is a risk factor with high attributable fraction in lumbar pain, frequent injury among construction workers due to dynamic and complex tasks. The aim of this study was to establish the differences between the actual handled weight and the recommended weight limit (RWL) according to EC2 (Ergo Carga Construcción) evaluation method.

The study was an analytic and non-experimental cross-section carried out in 32 construction sites with a sample of 186 workers. The positions assessed were bricklayer, scaffolding assembler, carpenters, hand laborers, construction laborers (excavation laborer/safety carpenter) and ironworkers during weight handling tasks. The EC2 method is designed to assess Dynamic Asymmetrical Manual Handling Tasks (DAMHT) and as result, estimates the Recommended Weight Limit (RWL) per task.

Within the group 179 DAMHT were assessed. In the sample the actual handled weight was between 3 kg and 80 kg, with a median of 20 kg (ICR = 20 kg). Meanwhile the values of the RWL were between 1.8 kg y 28.7 kg, with a median of 7.31 kg (ICR = 3.37 kg). Generally, all positions handle weights above the RWL established by EC2. The analyzed sample has a difference of 10.98 kg (ICR 23.4), between the actual handled weight and the RWL. The exception are the construction laborers (excavation laborer/safety carpenter), who present a negative difference, while hand laborers (15.69 kg/ICR 25.84), bricklayer (15.17 kg/ICR 23.28) and ironworkers (10,7 kg/ICR 24,62 kg) presents the highest difference among the group.

Contrasting research data with limits allowed by Chilean Law (Law 20.949 – maximum limit of 25 kg), a 33% of the sample performs DAMHT above lawful limits. 5% handle weights above 50 kg with a maximum of 80 kg. Regarding the RWL 83.2% of the manual handling observed is above this limit, therefore they imply high physical workload, thus the intervention must be not only with a technical approach but with administrative and engineering actions.

Keywords: Manual material handling · Construction · Ergonomics

© Springer Nature Switzerland AG 2019
S. Bagnara et al. (Eds.): IEA 2018, AISC 820, pp. 561–569, 2019.
https://doi.org/10.1007/978-3-319-96083-8_74

1   Introduction

Based on the epidemiological evidence, the risk factor for Manual Material Handling (MMH) has a high attributable fraction in the thoracolumbar disorders. Particularly in the construction sector, there is a high presence of tasks with MMH in the construction processes and their various phases, such as the earth movement phase, structure phase, closing phase and finishing phase. Due to the above, this productive sector becomes a vulnerable sector for the development of musculoskeletal disorders (Buchholz et al. 1996; Punnet and Wegman 2004; Umer et al. 2018). In this context, it has been observed that some work-related tasks in the construction sector represent a high physical workload which increases the risk of suffering musculoskeletal disorders in the lower back. Therefore, ergonomic measures should be implemented to decrease the appearance of these musculoskeletal disorders (Villumsen et al. 2016).

Based on national studies, it is described that a high percentage of workers are exposed to ergonomic factors represented by 32% of companies, where MMH is a relevant risk factor. In the last National Survey of Labor Conditions (Encuesta Nacional de Condiciones de Trabajo - ENCLA) of the Government of Chile in the construction sector, 26.7% of companies reported to have ergonomic risks (Dirección del trabajo 2014). That is why, in the construction sector, the identification and evaluation of risk factors associated with the physical workload (with an emphasis on MMH) are relevant, to improve preventive actions and strategies in the different tasks. Moreover, in the specific offices that act in the building construction process.

Musculoskeletal disorder (MSD) is one of the main problems faced by construction workers, who, due to the nature of their physical work, show a significantly higher prevalence of MSD in different regions of the body. In many types of occupational groups, MSDs are the leading causes of work-related disability and time lost due to illness (Chang et al. 2009).

Among workers in the construction sector, there are high rates of injuries that outnumber workers in other areas of work, so this sector continues one of the most risky sectors due to its association with a high incidence of deaths and injuries (Leung et al. 2012; Yi and Chan 2016).

1.1   Dynamic-Asymmetric Manual Material Handling

The most common operations carried out in the construction sector in specific housing and commercial building construction, involves MMH, especially in tasks related to scaffolding, formwork, steel structures, masonry, construction fabrics, plumbing, suspension of ceilings and pavements (Albers et al. 2005; Albers and Estill 2007).

In critical sectors such as Construction, processes are presented characteristically with variable work cycles and multiple incident variables, and which, as a whole, will condition the process of evaluating the "ergonomic" risk to determine the Physical workload, as well as the risk of developing MSD.

In this productive sector the MMH tasks are often from a "Dynamic-Asymmetric" nature. The common denominator is the execution of MMH with lifting, transport and deposit in continuous form, as well as, executed in perimeters greater than two meters from where the activity originates.

This presents a difficulty for the evaluation because these instruments do not manage to objectively determine the risk. An evaluation process must be structured that considers specific variables, with an evaluation approach oriented to the study of tasks with dynamic asymmetric manual material handling (Cerda 2006, 2013).

1.2 Weight Handled

In September 2017 in Chile, the Law 20.949 came into force. Modifying the Labor Code and the maximum weight limit allowed at the national level, currently up to 25 kg, for the male adult population. This legislative change is in accordance with the provisions of ISO/TR 12295: 2014, which establishes the 25-kg limit as a critical condition in tasks that involve lifting and transporting materials for male subjects between 18 and 45 years old (ISO 2014).

In the literature, there is a vast number of investigations related to the different risk factors that influence the handling of materials and the potential development of MSD. Among them are: lifting frequency, grip, asymmetric posture, vertical and horizontal lifting distance, transfer distance and, of course, the weight of the handled material. In construction activities involving MMH are also carried out in extreme environmental conditions, with postural limitations, high repetition and high weight of materials, tools and equipment (Ray et al. 2015). Generating high stress on the lumbar spine at segments L5/S1 level based on biomechanical studies performed on construction workers (Ray et al. 2015), in addition to a more significant increase in energy expenditure (Villagra 2000).

In Construction sites the handled materials have different sizes and weights. In Brazil, for example, the loading and unloading of 50 kg bags of cement and loaded trolleys with 49.7 kg on average have been studied to determine risk (Debiase et al. 2015). Construction workers have pointed out in studies utilizing surveys, that they handled weights around 15 kg on average per day (Fang et al. 2015). Likewise, it has been established in investigations that bricklayers handle weight between 2.5 kg and 10 kg of weight in stonework masonry (Villagra 2000). Scaffolding assemblers, while manipulating 17.3 kg or more, when building conventional brick walls (Hess et al. 2012). Given the risk represented by the weight handled, it is essential to accurately determine the handled weight limit to avoid overexertion (Lee 2012).

Based on what has been described, the development of this study is aimed at establishing differences between the actual weight handled and the Recommended Weight Limit (RWL) according to the EC2 evaluation method, which in its evaluation strategy considers the study of tasks with manual manipulation of Dynamic Asymmetric materials in specific trades of the building construction process.

2 Participants and Methods

The study was an analytical and non-experimental cross-section. The sample was selected for convenience and in a stratified form in main trade-tasks, in a two-stage manner. Companies were selected and then obtained through the selection of trade-workers who agree to participate in the study voluntarily and who meet the inclusion criteria.

Building Construction is considered, (large companies of more than 100 workers in the Metropolitan Region and Valparaíso Region of Chile) affiliated to an Organism Administrator of Law, the Instituto de Seguridad del Trabajo (IST). Of the companies defined to capture the sample, the participation of 32 different construction works, 17 works in the Metropolitan Region and 15 works in the Valparaíso Region which are specified.

The trades evaluated were: Bricklayer, scaffolding assembler, carpenters, hand laborers, construction laborers (excavation laborer/safety carpenter) and ironworkers. A total of 186 workers were evaluated. There were six evaluations which due to lack of data in the land evaluation instruments, were not counted in the summing up of the final sample.

For the execution of work in the field, a protocol of field study, development of evaluation material is designed, as well as informed consent presented to each worker. Land test planning and training of the evaluation team was made exclusively by Ergonomists of the Universidad de Chile. The EC2 method was used, which is designed to evaluate Dynamic Asymmetric Manual Handling Tasks (DAMHT) and as a result estimates the recommended weight limit (RWL) for the task (Cerda et al. 2014).

The ethics committee approved this study in human research of the Faculty of Medicine of the Universidad de Chile.

3 Results

Within the group of trades, a total of 179 DAMHT were evaluated, with distribution by trades as shown in Table 1. The group of "Hand laborers" has the highest representation in the sample with 41.34%, this group, which develops all the tasks of hauling and distributing materials within the works. The group with the least representation is the "Ironworker" with a representation of 8.37%.

Table 1. Distribution of the sample evaluated by trades in the 32 construction works between the Metropolitan Region and the Valparaíso Region, Chile.

Trade	Evaluation with EC2	
	Absolute frequency	Relative frequency
Bricklayer	27	15%
Scaffolding assembler	21	11.73%
Carpenter	24	13.40%
Hand laborers	74	41.34%
Construction laborers (excavation laborer/safety carpenter)	18	10.05%
Ironwork	15	8.37%

In the total sample, the handled weight was between 3 kg and 80 kg, with a median of 20 kg (ICR = 20 kg). Meanwhile, the RWL values were between 1.8 kg and 28.7 kg, with a median of 7.31 kg (ICR = 3.37 kg) (See Graph 1).

The median for both groups, actual weight handled and recommended, are under the limit of 25 kg established by Chilean Law 20.949. It has been observed outlier corresponding to handled weights of 70 kg in two cases, 74 kg in one case and two cases with 80 kg, while, for the Recommended Weight Limit (RWL) three extreme outlier were observed. 18.05 kg, 18.78 kg and 28.69 kg, the latter exceeding the weight limit established by Chilean law.

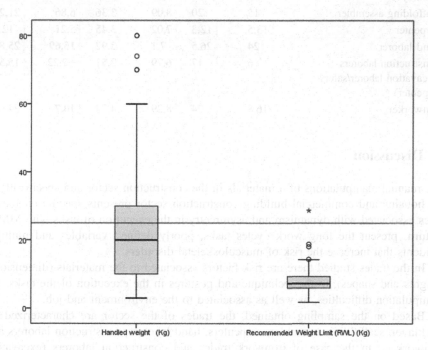

Graph 1. Handled weight distribution and Recommended Weight Limit (RWL) established by EC2 method.

Concerning the behavior of the actual weight handled and the recommended weight limit, as well as the difference between these two indices by trades evaluated, this is detailed in Table 2.

In general, all trades handle weights above the established RWL according to EC2. The analyzed sample has a difference of 10.98 kg (ICR 23.4), between the handled weight and the RWL. The exception is Construction Laborers (excavation laborer/safety carpenter) who present a negative difference, while Hand laborers (15.69 kg, ICR 25.84), Bricklayer (15.17 kg, ICR 23.28) and Ironworker (10.7 kg, ICR 24.62 kg) present the most significant difference between the group.

Table 2. Description of actual weight handled based on a total sample, weight limit recommended by EC2 method and difference of actual weight handled and recommended weight limit calculated by EC2.

Trade	Actual weight handled		Weight limit recommended		Difference of actual weight handled and recommended weight limit	
	Median	IQR	Median	IQR	Median	IQR
Bricklayer	25	20	7,65	2,75	**15,17**	23,28
Scaffolding assembler	15	20	8,09	3,36	6,86	21,28
Carpenter	13,5	12,3	7,02	3,45	5,21	12,6
Hand laborer	24	26,5	7,1	2,92	**15,69**	25,84
Construction laborers (excavation laborer/safety carpenter)	6	17	6,79	2,51	−2,22	15,52
Ironworker	16,8	24	8,29	4,74	**10,7**	24,62

4   Discussion

The manual manipulations of a materials in the construction sector and specifically in the housing and commercial building construction sector presents specific characteristics associated with dynamism and asymmetry in the execution of tasks with MMH, in turn, present the long work cycles tasks, poorly defined variables and multiple incidents that increase the risk of musculoskeletal disorders.

In the trades studied there are risk factors associated to the materials (dimensions, weights and shapes), to the technique and postures in the execution of the tasks, the manipulation difficulties, as well as associated to the environment and job.

Based on the sampling obtained, the trades of the sector are characterized by bricklayer, scaffolding assembler, carpenters, hand laborers, construction laborers and ironworkers. In the case of ironwork trades and construction laborers (excavation laborers/safety carpenter), there is a low frequency of presentation due to the subcontracting schemes existing at the sites of the companies participating in the study; work contract strategy that responds to a model that is carried out in the world in the construction sector (Bryan et al. 2017). In this study, the trades and tasks sampled characterize the dynamic system associated with the time of permanence in the work and linkage with a major company or subcontracted companies.

In relation to the results, they express the assessment of the handled weight. In general, they are below the maximum allowed limit. However, emphasis is placed on the point of analysis, that being under the weight limit established by the Law (in this case under 25 kg) does not imply that there is no risk since the context and conditions of how the task is performed must be considered based on the worker-environment relationship being able to, therefore, be an acceptable or unacceptable level of risk according to each case and with the same material for example.

Nevertheless, multiple analysis models calculate the risk based on different variables and their weighing factors for calculating the recommended weight limit. The most traditional being weight, vertical distances, horizontal distances, frequency, manual handling and grip. In this case, it is worth noting that the recommended weight limit (RWL) considers specific variables in the interpretation of risk for tasks with manual manipulation of dynamic-asymmetric loading. Some of these variables being relevant to weight, handling techniques (arm segment), manipulation posture (trunk segment), frequency, difficulty of manipulation (infrastructure, distances and organization), combined control and perceived effort (Cerda et al. 2014).

Concerning the weights handled, the trades, which have the most difference between the actual weight handled and that recommended by the method, are hand laborers (15.69 kg), bricklayer (15.17 kg) also ironworkers (10.7 kg). The only trade which presents a handled weight close to the recommended limit is the construction laborers (excavation laborer/safety carpenter). However, all are under the limit established by Chilean and international regulations for cargo handling (ISO 2014; MINTRAB 2018).

In this context, it is relevant to consider in the prevention of musculoskeletal disorders a dynamic model oriented to the understanding of the dynamic-asymmetric tasks of manual handling of cargo.

5 Conclusion

The approach in the identification and evaluation of risk in tasks with DAHM Tasks must consider the differentiation between this type of tasks and "more defined" tasks of manipulation that can be approached with traditional methodologies. In this context, considering that this work emphasis was placed on fulfilling the recommended weight limit with the actual manipulation and contrasting the research data with the limits allowed by Chilean Law (Law 20.949 - updated 2017, maximum limit of 25 kg), 33% of the sample performed DAMHT over the legal limits and 5% handled weights of over 50 kg with a maximum of 80 kg.

On the other hand, with respect to the RWL, 83.2% of the manual handling observed is above this limit, which implies a tremendous physical workload, so the intervention must not only be with a focus on the technique of manipulation, but rather a systemic approach where aspects of difficult manipulation such as; distance, obstacles, environmental characteristics, the perception of effort, the combined grip, considered with greater emphasis a prevention and correction approach oriented to administrative measures, work organization, to the incorporation of specific technical aids for the sector and as well as engineering measures.

Finally, based on the results obtained, it can be concluded that both in the normative instruments of the different countries, as well as in the supporting technical documents for the evaluation of the risk in manual handling tasks, an analysis must be considered. Differentiated in the different tasks with manual load handling, differentiating the analysis in those cases with Dynamic Asymmetrical Manual Handling Tasks through models adapted for the specific study of said condition.

Acknowledgments. In memory of Professor Jorge Rodríguez, Rest in Peace.

Funding. This work was selected in the "Convocatoria de Proyectos de Investigación e Innovación en Prevención de Accidentes y Enfermedades Profesionales año 2016 de la Superintendencia de Seguridad Social" (Chile) and was funded with the resources of Social Security Law No. 16,744 of Industrial accidents and occupational diseases. This project was executed with funds provided by the Instituto de Seguridad del Trabajo - IST.

References

Albers J, Estill C, MacDonald L (2005) Identification of ergonomics interventions used to reduce musculoskeletal loading for building installation tasks. Appl Ergon 36:4

Albers J, Estill CF (2007) Simple solutions: Ergonomics for construction workers, Department of Health and Human Services (DHHS), Centers for Disease Control and Prevention, National Institute for Occupational Safety and Health (NIOSH). NIOSH-Publications Dissemination, Cincinnati

Buchholz B, Paquet V, Punnet L et al (1996) PATH: a work sampling based approach to ergonomic job analysis for construction work path: a work sampling-based approach to ergonomic job analysis for construction and other non-repetitive work. Appl Ergon 27:177–187

Bryan D, Rafferty M, Toner P, Wright S (2017) Financialisation and labour in the Australian commercial construction industry. Econ. Labour Relat. Rev. 28(4):500–518. https://doi.org/10.1177/1035304617739504

Cerda E (2006) Ergonomics in the construction sector: the EC2 method. Maastrich, Netherland

Cerda E (2013) Modelo Conceptual de Proceso de Evaluación de Factores Ergonómicos en Tareas con Manipulación Manual de Carga Dinámico Asimétrica en el Sector Construcción. Tesis Doctoral. http://tdx.cat/handle/10803/183/discover

Cerda E, Rodríguez C, Olivares G et al (2014) Revisión de proceso de evaluación y fórmula de cálculo de límite de peso recomendado en Método EC2 para la Evaluación en Tareas con Manipulación Manual de Carga Dinámico Asimetrica. ORP J 1:19–39

Chang FL, Sun YM, Chuang KH, Hsu DJ (2009) Work fatigue and physiological symptoms in different occupations of high-elevation construction workers. Appl Ergon 40:591–596

Debiase D, de Farias J, Madeira K, Longen W (2015) Análise do risco ergonômico lombar de trabalhadores da construção civil através do método NIOSH/Analysis of lumbar ergonomic risk of construction workers using the NIOSH method. Revista Produção Online 15(3):914–924. https://doi.org/10.14488/1676-1901.v15i3.1888

Dirección del trabajo (2014) Octava Encuesta Nacional de Condiciones Laborales. Gobierno de Chile. http://www.dt.gob.cl/documentacion/1612/articles-108317_recurso_1.pdf

Fang D, Jiang Z, Zhang M, Wang H (2015) An experimental method to study the effect of fatigue on construction workers' safety performance. Saf Sci 73(C):80–91. https://doi.org/10.1016/j.ssci.2014.11.019

Hess JA, Mizner RL, Kincl L, Anton D (2012) Alternatives to lifting concrete masonry blocks onto rebar: biomechanical and perceptual evaluations. Ergonomics 55(10):1229–1242. https://doi.org/10.1080/00140139.2012.703694

Lee T-H (2012) Effects of range and mode on lifting capability and lifting time. Int J Occup Saf Ergon 18(3):387–391. https://doi.org/10.1080/10803548.2012.11076941

Leung MY, Chan IYS, Yu J (2012) Preventing construction worker injury incidents through the management of personal stress and organizational stressors. Accid Anal Prev 48:156–166

MINTRAB (2018) Guía técnica para la evaluación y control de riesgos asociados al manejo o manipulación manual de carga. Ministerio del Trabajo y Previsión Social. Gobierno de Chile. https://www.previsionsocial.gob.cl/sps/download/biblioteca/seguridad-y-salud-en-el-trabajo/ guia-manejo-cargas/guia-tecnica-manejo-manual-de-carga.pdf

ISO (2014) PD ISO/TR 12295:2014 - Ergonomics. Application document for International Standards on manual handling (ISO 11228-1, ISO 11228-2 and ISO 11228-3) and evaluation of static working postures (ISO 11226)

Punnet L, Wegman D (2004) Work-related musculoskeletal disorders: the epidemiologic evidence and the debate. J Electromyogr Kinesiol 14(1):13–23

Ray PK, Parida R, Sarkar S (2015) Ergonomic analysis of construction jobs in India: a biomechanical modelling approach. Procedia Manuf 3:4606–4612. https://doi.org/10.1016/j. promfg.2015.07.542

Umer W, Antwi-Afari MF, Li H, Szeto GPY, Wong AYL (2018) The prevalence of musculoskeletal symptoms in the construction industry: a systematic re-view and meta-analysis. Int Arch Occup Environ Health 91(2):125–144. https://doi.org/10.1007/s00420-017-1273-4

Villagra M (2000) Evaluation of the physical workload of bricklayers during the execution of block walls. Revista de Ingeniería de Construcción 15(2):91–99. https://repositorio.uc.cl/ handle/11534/10083

Villumsen M, Holtermann A, Samani A, Madeleine P, Birk Jørgensen M (2016) Social support modifies association between forward bending of the trunk and low-back pain: Cross-sectional field study of blue-collar workers. Scand J Work Environ Health 42(2):125–134. https://doi.org/10.5271/sjweh.3549

Yi W, Chan A (2016) Health profile of construction workers in Hong Kong. Int J Environ Res Public Health 13(12). Número de artículo: 1232. https://doi.org/10.3390/ijerph13121232

A Presentation of the Ergonomic Analysis of Risk Factors in Productive Sectors of Chile and Their Relation with Upper Limb Musculoskeletal Symptomatology

Cerda Leonidas[1], Cerda Eduardo[1], Olivares Giovanni[1],
Villalobos Victoria[1], Antúnez Marcela[2], and Rodríguez Carolina[1(✉)]

[1] Ergonomics and Biomechanical Laboratory, Department of Physical Therapy,
Faculty of Medicine, University of Chile, Santiago, Chile
carorodriguez@uchile.cl
[2] Department of Education in Health Sciences, Faculty of Medicine,
University of Chile, Santiago, Chile

Abstract. This study aims to identify musculoskeletal health conditions present in a population that works in several different productive sectors in Chile in order to determine the physical risk factors in that type of labor and relation between these risk factors and upper limb musculoskeletal symptomatology. An analytic, nonexperimental, transversal association study was carried out with a sample of 390 worker's tasks, confidence level of 95%, standard deviation of 5% and maximal variability. Representative task per trade-workers of six different sectors was evaluated. Results describe most representative physical risk factors. In industrial sector were: posture and strength; in service sector: repetitive movement and force factors; in mining sector: repetitive movement and posture and strength; in agriculture sector: repetitive movement, posture and recovery time; in construction sector: posture, repetitive movement and strength; in aquaculture sector: repetitive movement and recovery time. There was a high prevalence of musculoskeletal complaints, of 69,74% in the sample evaluated with Nordic Musculoskeletal Questionnaire. There was no association between musculoskeletal health condition and identification of risk factors (Fisher p-value 0,587). Statistically significant association was found between physical risk factors identified and specific productive sectors chosen (p-value = 0.0001). There is a high prevalence of musculoskeletal symptoms in all productive sectors studied. Productive sectors that present most specific risk factors are Aquaculture, Agriculture, and Industry Sector and this translates into a high prevalence of physical risk factors. The model of the applied tools is efficient to perform the surveillance of the ergonomic risk factors and the musculoskeletal health condition.

Keywords: Ergonomic risk factors · Upper limb
Work-related musculoskeletal disorders

© Springer Nature Switzerland AG 2019
S. Bagnara et al. (Eds.): IEA 2018, AISC 820, pp. 570–578, 2019.
https://doi.org/10.1007/978-3-319-96083-8_75

1 Introduction

In Chile, the risk prevention regarding work-related musculoskeletal disorders is regulated by the Supreme Decree N° 594 [1] and Law N° 16.744, on accidents and profession-related illnesses [2]. In September of 2012, the Ministry of Health in Chile published the Technical Norm for the Identification and Evaluation of Risk Factors in Work-Related Musculoskeletal Disorders in Upper Limbs, denominated the TMERT-EESS Norm.

The Technical Norm for the Identification and Evaluation of Risk Factors in Work-Related Musculoskeletal Disorders in Upper Limbs relies on an instrument for "Risk Identification and evaluation" that was developed based on criteria given by the ISO Norm 11228-3 for the evaluation of task with a low load and high frequency [3, 4].

Based on the last National Labor Condition Survey in Chile [5], it is observed that among the main risk factors identified in the companies surveyed. The ergonomic risks are found in a 32% of the companies. In the Agricultural Sector, a 24,3% of the companies presented ergonomic factors; in Services, there was a 37%; in Industry, a 27,3% and in the Construction Sector, there was a 26,7% of ergonomic factors. Also, there is a great number of workers exposed to ergonomic risk factors depending on the productive activity.

In the Agricultural Sector there are 824,79 thousand persons; in Services, which considers several branches, there are over 950 thousand affected persons; in the Industrial Sector, there are 886,89 thousand persons, and in Construction, there are 712,95 thousand persons exposed to risk factors. In relation to those persons occupied by an economic activity in the selected sectors, as a whole, represent approximately a 35% of those employed on a national level according to the data given by the National Institute of Statistics for the first trimester of 2017 [6].

The health surveillance of the workers that depend directly on the objectivity of the information obtained from the evaluation of the risk levels to which the workers are exposed. For this, it is also necessary to know more about the relationship that exists between the exposure to risk factors and the development of symptomatology, and, subsequently, the musculoskeletal pathology.

The public institutions in Chile, related to the regulation of Occupational Health matters, note in their statistics that there is a high prevalence of upper limb musculoskeletal illnesses related to work conditions. This is why it is relevant to know the specific manner in which risk factors that are present as well as the specific health condition that is musculoskeletal. This will allow for the development of specific strategies and actions for the correction and prevention of musculoskeletal disorders.

2 Methodology

For this present study, we are using a database built by the Ergonomy Laboratorio of the Univerisdad de Chile, which is based on a study of work conditions carried out during a period of 18 months. The data obtained from the database takes into account an analytical, nonexperimental, transversal association study base. All the data obtained by the workers, their trade and main tasks in the Industry, Aquaculture, Agriculture, Service, Mining and Construction Sectors will be taken into account.

The data obtained considers the study of 390 tasks (Unit of Analysis) and workers (Unit of Sampling), the tasks that are sampled considered workers and main tasks according to the productive sector and company. The sampling strategies from which the data is obtained in order to characterize the exposure to factors of ergonomic risk during the execution of their tasks. For this, the criteria of inclusion that are considered are tasks that involve the use of upper limbs that take place in the selected labor Sectors.

The data obtained from a stratified sampling according to region and area. The assignation of the social stratum were appointed freely with a minimum of 30 workers per stratum. The sampling size is of 390 and it is calculated by considering a trust of 95%, an error of estimation of 5% and a maximum variability. In the sample, the tasks considered are the most representative in each field, by the workers, in which there is an exposure of upper limbs to ergonomic factors such as repetition, posture, force and recovery periods.

In order to carry out this ground work, a fieldwork protocol was designed, as well as the necessary materials, the acceptance letters for the participation of the selected companies and the consent forms to be signed by the worker. We carried out the test planning on site. The evaluations were effected by specialist of the Universidad de Chile and engineers in prevention from the studied companies.

The obtainment of the data comes from a study of exposure of the worker to his task, studied with an observational method, which is a model developed for this purpose. The field protocol from which this information is obtained included a brief interview to the worker, task analysis, graphic records and the preliminary identification and evaluation of the risk. This preliminary evaluation of risk is carried out through the on site application of the evaluation instrument, according to the current norm in Chile (Check list TMERT) and the musculoskeletal health condition that was evaluated by the Nordic Kuorinka questionaire, currently in force in Chile.

3 Results

3.1 Sample Description

The treatment and result analysis consider the systematization of the database information, process done by the SPSS System, based on the description of the results through descriptive statistics and the application of statistic tests such as the Fisher Test and Chi-Case for the analysis of the association of study variables (TMERT Risk factors and the Musculoskeletal Health Conditions).

The results describe that of the sample of 390 tasks studied, 59,74% correspond to tasks carried out by men and 40,26% correspond to tasks carried out by women. 92,05% of the workers present a right handed laterality. The average age of the sample studied is 38 years. In the sample studied, the average work period described is of 48 months at a given work post.

The sampling distribution considers Agriculture (12%), Service (40,2%), Construction (12%), Manufacturing Industry (20,2%), Mining (7,7%) and Aquaculture (7,7%).

3.2 Identification of Risk Factors in Studied Tasks

Exposure is described as the frequency in which risk given by repetition, posture associated to movement, use of Strength and recovery times in the studied tasks are present.

The risk factor of movement is present in 349 (85%) of the tasks, posture is present in 367 (94%) of the tasks and Strength is present in 282 tasks, which represents a 72, 3%, the same as the recovery period risk factor (Table 1).

Table 1. Distribution of absolute and relative frequency of risk factors considering all studied tasks

	Repetitive movement		Posture		Strength		Recovery time	
	N°	%	N°	%	N°	%	N°	%
Present	349	89,5	367	94,1	282	72,3	282	72,3
Absent	41	10,5	23	5,9	108	27,7	108	27,7
Total	390	100	390	100	390	390	390	100

In the description of the risk factors identified, the preliminary most frequent risk level for each are analyzed. It is shown that, regarding the Repetitive Movement Risk Factor, the level of preliminary risk is on a red level; regarding the Strength Risk Factor, the most frequent preliminary level of risk is mixed (Green, yellow and red level), and finally, for the Recovery Time Risk Factor, the preliminary level of most frequent risk in level green (Table 2).

Table 2. Distribution of absolute and relative frequency of preliminary categorization of risk according to risk factors considering all studied tasks

	Repetitive movement		Posture		Strength		Recovery time	
Level risk	N°	%	N°	%	N°	%	N°	%
Green	86	25,74	109	30,8	94	33,2	112	39,8
Yellow	95	28,44	101	28,5	94	33,2	75	26,7
Red	153	45,8	144	40,7	95	33,7	94	33,4
Total	334	100	354	100	283	100	281	100

4 Level of Risk by Evaluated Area

The level of risk determined by risk factor and by area, considering the total of evaluations in which a risk factor is identified. For this, the analysis is carried out by grouping the information according to risk factor and area. It is highlighted that,

Table 3. Distribution of relative frequency according to the risk factor, level of risk and area

	Repetitive movement grouped by area			Total
	Green	Yellow	Red	
Industrial	33,9%	5,4%	60,7%	100%
Service	26,2%	36,6%	37,2%	100%
Mining	26,7%	36,7%	36,7%	100%
Agriculture	10,6%	29,8%	59,6%	100%
Construction	34,1%	24,4%	41,5%	100%
Aquaculture	6,7%	13,3%	80%	100%
Total (%)	24,6%	27,2%	48,1%	100%
	Posture grouped by area			Total
	Green	Yellow	Red	
Industrial	37,7%	24,6%	37,7%	100%
Service	34,2%	35,6%	30,1%	100%
Mining	24,1%	48,3%	27,6%	100%
Agriculture	10,6%	27,7%	61,7%	100%
Construction	34%	27,7%	38,3%	100%
Aquaculture	17,9%	14,3%	67,9%	100%
Total (%)	29,8%	30,9%	39,3%	100%
	Strength grouped by area			Total
	Green	Yellow	Red	
Industrial	47,5%	24,6%	27,9%	100%
Service	35,6%	40,6%	23,8%	100%
Mining	20,7%	44,8%	34,5%	100%
Agriculture	29%	16,1%	54,8%	100%
Construction	30%	40%	30%	100%
Aquaculture	5%	15%	80%	100%
Total (%)	33%	33%	34%	100%
	Recovery period by area			Total
	Green	Yellow	Red	
Industrial	42,9%	16,3%	40,8%	100%
Service	46,9%	16,3%	40,8%	100%
Mining	31,6%	47,4%	21,1%	100%
Agriculture	32,5%	17,5%	50%	100%
Construction	38,u%	45,2%	16,1%	100%
Aquaculture	24,1%	20,7%	55,2%	100%
Total (%)	39,9%	26,7%	33,5%	100%

in relation to the repetitive movement, as a risk factor, is frequently present in the Manufacturing, Agricultural and Aquacultural Industry. The Posture Factor is notable as it is relevant in the sectors of Agriculture, Aquaculture and Manufacturing Industry. The risk factor of Strength is notable as it is relevant in the Aquaculture, Agriculture and Mining Sectors. Finally, the Period of Recovery Factor is relevant in the Aquaculture, Agriculture, and Industry Sectors (Table 3).

4.1 Description of Musculoskeletal Health

It is determined that the sample studied presents a high prevalence of musculoskeletal discomfort, manifesting musculoskeletal discomfort in at least one segment is 69,74%. Of this general information, 39,74% of the evaluated workers present pain in the right hand/wrist segment, 33% present pain in the right shoulder, 28,21% in the neck, 24,36% in the left hand/wrist, 20,51% in the right elbow and forearm, 19,23% in the left shoulder and 12,05% have pain in their left elbow and forearm (Table 4).

Table 4. Distribution of absolute and relative frequency of the presence of musculoskeletal discomfort such as pain, numbness or tingling sensation according to body segment in the last 12 months according to Nordic Questionnaire in the population sample of 390 workers

Segment	N°	%
Neck	110	28,21
Right shoulder	130	33,33
Left shoulder	75	19,23
Right elbow/forearm	80	20,51
Left elbow/forearm	47	12,05
Right hand/wrist	155	39,74
Left hand/wrist	95	24,36

When the Nordic Questionnaire evaluation instrument was applied, it also explored the presence of the perception of musculoskeletal discomfort in the last seven days. The results describe a result similar to that which was presented in relation to the consult of the last twelve months. A 23,59% of the sample is described as having presented discomfort in the right hand/wrist and a 14,62% of the sample presents the perception of musculoskeletal discomfort in the last seven days in the right shoulder.

The following table (Table 5) determines the presence of Incapacity Associated to pain, by segment and average of pain perceived. The pain is evaluated through a visual analog scale of 10 points.

Table 5. Distribution of absolute and relative frequency according to the presence of musculoskeletal discomfort during the last seven days en average of the assessment of pain in each segment of the whole of the subjects evaluated

Segment	N°	%	Average EVA pain
Neck	55	14,10	4,35
Right shoulder	57	14,62	4,12
Left shoulder	55	14,10	4,36
Right elbow/forearm	43	11,03	4,44
Left elbow/forearm	26	6,67	4,50
Right hand/wrist	92	23,59	4,45
Left hand/wrist	57	14,62	4,44

In relation to the association between the risk factors identified (by the checklist of the Technical Norm) and the musculoskeletal health condition represented by the Nordic Questionnaire is described by the following results:

Considering the results of the identification of the risk factors and the presence of the musculoskeletal discomfort in the last twelve months is described as not having an association between the identification of the risk factors of Repetitive Movement, Posture, Strength and Recovery Time, with the musculoskeletal health condition (Table 6). In general terms, there is no existent association between the presence of risk factors and the musculoskeletal discomfort, given a Fisher p-value 0,587.

Table 6. Results of the association between the identification of the risk factors and the presence of the musculoskeletal discomforts in the last twelve months

	Perception of musculoskeletal discomfort in the last 12 months	
	Chi/Fisher	P value
Repetitive movement	2,7	0,07
Posture	2,02	0,119
Strength	0.006	0,514
Recovery period	0,435	0,298

The results, considering the grouping of the information linked to the identification of the risk factor, level of the preliminary evaluation of risk and area, describe that there is an association between the preliminary level of evaluated risk of each risk factor by means of the TMERT checklist and the specific studied areas. In this sense, the association of the identification of the risk factor with the areas allows noting that the characterization of the different areas is achieved with the applied instrument (Table 7).

Table 7. Results of the association between risk factors, preliminary evaluation of risk and area

	Area	
	Chi	P value
Repetitive movement	43,366	0,0001
Posture	33,711	0,0001
Strength	40,367	0,0001
Recovery period	28,717 (Fisher)	0,0001

5 Conclusion and Discussion

There is a high prevalence of musculoskeletal symptomology in all the productive sectors studied, as well as there is a high prevalence of TMERT ergonomic risks [7].

The factors that stand out in relation to the prevalence of risk factors of TME are the risk factors of Repetitively and Posture. The Aquaculture, Agriculture and Industrial Manufacturing Sectors present high prevalence in the different factors studied.

We conclude that there is no existent association (possible relations) between the conditions of musculoskeletal health and the identification of risk factors. It is noteworthy that the second stage of progression of the disorder was also evaluated, which corresponds to the perception of musculoskeletal discomfort, diagnosed cases have not been evaluated [8, 9].

It is also notable that the epidemiological evidence establishes a high fraction that is attributed to these risk factors and the development of the musculoskeletal disorders on the level corresponding to the upper extremities [10, 11]. With this, it can be concluded that we may face the presence of many potential cases of future work-related musculoskeletal disorders in the sectors that were studied, which is why prevention strategies and actions must be reinforced.

The checklist of the TMERT Norm, as well as the Nordic Questionnaire, are efficient tools to develop the activity of surveillance of the risk factors, understanding that they are not specific tools of evaluation as much as they are surveillance tools that use the preliminary identification of risk, to subsequently give orientation for more specific actions of evaluation.

In this context, it can be concluded that it is relevant to have public policies of adequate records and protocols of surveillance of risk factors and medical surveillance to establish health policies in the future for appropriate records classifies by regions and productive areas, by establishing the relevance of the concept of sectorization of the instruments of evaluation, as well as the preventive actions.

References

1. Decreto Supremo N° 594, del Ministerio de Salud de Chile (2000) Diario Oficial de la República de Chile, Santiago, Chile, 29 abril del año 2000
2. Ley N° 16.744 (1968) Diario Oficial de la República de Chile, Santiago, Chile, 1 de febrero de 1968

3. Norma Técnica de Identificación y Evaluación de Factores de Riesgo de Trastornos Musculo-esqueléticos Relacionados con el Trabajo, Extremidades Superiores, Resolución Exenta N° 804 del 26 de septiembre de 2012. http://dipol.minsal.cl/departamentos-2/salud-ocupacional/trastornos-musculos-esqueleticos-de-extremidades-superiores-relacionados-con-el-trabajo/

4. ISO 11228-3:2007. Ergonomics – Manual Handling – Part 3: Handling of low loads at high frecuency

5. Espinosa M, Damianovic N (2000) Encuesta Laboral ENCLA. Informe de Resultados. Dirección del Trabajo de Chile, Departamento de Estudios

6. http://www.ine.cl/

7. Roquelaure Y, Ha C, Leclerc A, Touranchet A, Sauteron M, Melchior M, Goldberg M (2006) Epidemiologic surveillance of upper-extremity musculoskeletal disorders in the working population. Arthritis Care Res 55(5):765–778

8. Kuorinka I, Jonsson B, Kilbom A, Vinterberg H, Biering-Sørensen F, Andersson G, Jørgensen K (1987) Standardised nordic questionnaires for the analysis of musculoskeletal symptoms. Appl Ergon 18(3):233–237

9. Maldonado MM, Muñoz RA Validación del Cuestionario Nórdico Estandarizado de Síntomas Musculoesqueléticos para la población trabajadora chilena, adicionando una escala de dolor. Revista de Salud Pública 21(2):43–53

10. Punnett L, Wegman DH (2004) Work-related musculoskeletal disorders: the epidemiologic evidence and the debate. J Electromyogr Kinesiol 14(1):13–23

11. Andersen JH, Haahr JP, Frost P (2007) Risk factors for more severe regional musculoskeletal symptoms: a two-year prospective study of a general working population. Arthritis Rheumatol 56(4):1355–1364

The Relationship Between MRI Parameters and Spinal Compressive Loading

Jie Zhou[1], Fadi Fathallah[1(✉)], and Jeffery Walton[2]

[1] Department of Biological and Agricultural Engineering,
University of California Davis, Davis, CA 95616, USA
fathallah@ucdavis.edu
[2] NMR Facility, University of California Davis, Davis, CA 95616, USA

Abstract. Intervertebral disc (IVD) is a leading source of Low back pain (LBP). The health and functions of the IVD are determined by the inherent biomechanical properties of the IVD and its interaction with external mechanical loading. Quantitative Magnetic Resonance Imaging (MRI) parameters have the potential in detecting loading-induced changes in biomechanical properties of the IVD. $T_{1\rho}$, T_2 and Apparent Diffusion Coefficient (ADC) were obtained with a 7T MRI scanner from 20 functional spinal units (FSU) before and after receiving compressive loading of 263.27 N for 60 min. Compressive loading was found to significantly reduce $T_{1\rho}$ and T_2 but not ADC, indicating that $T_{1\rho}$ and T_2 had the potential to detect loading-induced changes in biomechanical properties of the IVD. These parameters may provide more sensitivity and specificity to understand the injury mechanism of the IVD and contribute to early diagnosis of IVD degeneration.

Keywords: MRI parameters · Spinal compressive loading · Intervertebral disc

1 Introduction

Low back pain (LBP) remains a major socioeconomic problem, and one of the leading sources is the IVD [1]. Injuries to the disc itself can cause discogenic pain [2]. Also, compromise in IVD biomechanics may impair neuromuscular control of postural stability and make other spinal tissues more vulnerable to injury, and thus may cause LBP [3–5].

The biomechanical properties of the IVD are largely determined by the concentration of the proteoglycan (PG) in the nucleus pulposus (NP) of the IVD [6]. When subjected to compressive loading, water is expelled out from the NP, which increases the concentration of PG and the associated osmotic pressure, such that the compressive load can be balanced and an equilibrium can be restored. When the load is removed, water re-imbibes into the NP, which reduces osmotic pressure corresponding to the magnitude of the load, hence, achieving a new equilibrium state [7] (Fig. 1).

The health and functions of the IVD are determined by both the inherent tissue properties (e.g. PG content, osmotic pressure and hydration) and their interaction with external mechanical loading. The inherent properties of the IVD are not directly determinable in-vivo; whereas, quantitative MRI parameters, such as spin–lattice

© Springer Nature Switzerland AG 2019
S. Bagnara et al. (Eds.): IEA 2018, AISC 820, pp. 579–585, 2019.
https://doi.org/10.1007/978-3-319-96083-8_76

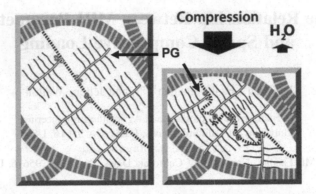

Fig. 1. The expulsion of water from the IVD

relaxation in the rotating frame ($T_{1\rho}$), spin–spin relaxation (T_2) and Apparent Diffusion Coefficient (ADC) may extract pertinent information about biomechanical properties of the IVD, which is useful for understanding the effects of loading on the health and functions of the IVD [7]. Therefore, the objective of the current study is to investigate how compressive loading changes $T_{1\rho}$, T_2 and ADC in the NP of the IVD.

2   Methods

2.1   Specimens

Porcine cervical spines were harvested from young pigs freshly-slaughtered and used as the specimens, due to the similarity in morphometry, geometry, and curvature to the human lumbar spine [8]. The cervical spines were then wrapped in phosphate buffered saline (PBS) soaked gauze, sealed in plastic bags, and frozen at −20 °C until testing. They were thawed for approximately 12 h before testing and each cervical spine was dissected to obtain 2 functional spinal units (FSU) (C3-C4 and C5-C6) [9]. Each FSU consists of 2 adjacent vertebral bodies, the IVD and all connecting ligamentous tissues between them. The posterior ligaments were removed to isolate loading to the IVD (Fig. 2). A total of 20 FSUs were harvested from 10 pigs, the average (standard deviation) of the weight and age were 10.92 kg (1.49 kg) and 45.4 days (3.58 days), respectively.

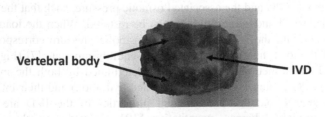

Fig. 2. The structure of Functional Spinal Unit (FSU)

2.2   Experimental Design

A repeated-measures experimental design was performed. $T_{1\rho}$, T_2 and ADC were obtained before and after receiving compressive load of 263.27 N for 60 min. This load magnitude corresponds to 1500 N on human lumbar IVD according to the scaling factor in IVD areas between the porcine cervical and human lumbar discs [10]. The 3D-SSPP software (University of Michigan, Ann Arbor, Michigan, USA) was used to determine such 1500 N compressive load when simulating a 50th percentile male worker holding a 23 kg load. The 23 kg has been suggested as the maximum recommended weight limit by NIOSH lifting equation [11].

2.3   Experimental Procedure

A computer controlled load-displacement device was built to apply compressive loading inside the MRI scanner (Fig. 3). FSUs were casted and drilled, then secured into the sample holder with pins. PBS soaked gauze was used to wrap the FSU to maintain the hydration of the IVD (Fig. 4a), sample holder was assembled into the tube (Fig. 4b), and the whole apparatus was installed onto the opening of the MRI bore with an aluminum flange plate (Fig. 4c). And the electrical components (e.g. pressure sensors and DAQ), pneumatic system (e.g. pneumatic actuator, proportional pneumatic control valve and compressed air supply) and LabView software were set up to achieve force-displacement control and monitor (Fig. 4d).

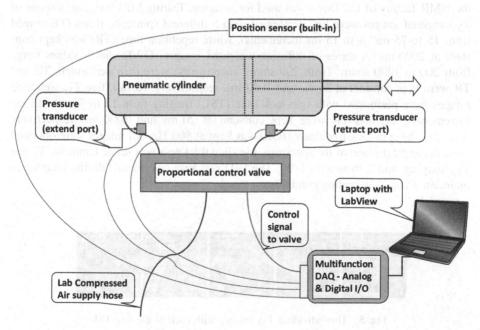

Fig. 3. The diagram of the apparatus

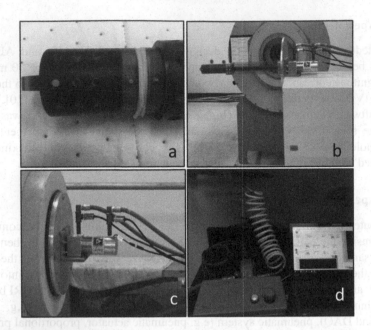

Fig. 4. The computer controlled pneumatic load-displacement apparatus

The 7T Bruker Biospec imaging instrument (Bruker Inc., Billerica, MA, USA) in the NMR facility of UC Davis was used for imaging. During MRI imaging, a series of T_2-weighted images were first performed with 5 different spin echo times (TE ranged from 15 to 75 ms, with 15 ms increments), while repetition time (TR) was kept constant at 2000 ms. A series of diffusion-weighted images (DWI) with b values range from 200 to 1000 s/mm^2 (with 200 s/mm^2 increments) were then performed, TE and TR were kept constant at 24 ms and 1000 ms, respectively. At last, five $T_{1\rho}$-weighted images were performed with spin lock time (TSL) ranging from 15 to 75 ms (15 ms increments), TE and TR were kept constant at 20 ms and 1000 ms, respectively (Fig. 5). The spin lock frequency (FSL) was kept at 500 Hz. All images were obtained from the sagittal plane of the specimen, the slice thickness was set at 1 mm for T_2 and $T_{1\rho}$ imaging, and 2.26 mm for DWI. The IVD was kept compressed during imaging to maintain a constant loading condition.

Fig. 5. The individual $T_{1\rho}$ images with each of the five TSL

The Bruker built-in ParaVision software was used to analyze the acquired images to obtain the MRI parameters. At first, the region of interest (ROI) was defined to cover only the NP of the IVD, and the signal intensity in the ROI of the series of images was calculated. The signal intensity of these images was then fitted with the exponential decay equation of each MRI parameter, thus the MRI parameters (i.e. $T_{1\rho}$, T_2 and ADC) can be calculated.

2.4 Statistical Analysis

Paired-t tests were performed to compare $T_{1\rho}$, T_2 and ADC obtained before and after the compressive loading. The α value of 0.05 was set as the significance criteria. Minitab software (Minitab Inc., State College, Pennsylvania, USA) was used to conduct all the statistical analyses.

3   Results

The effects of compressive loading on MRI parameters were demonstrated in Table 1. Significantly reduced $T_{1\rho}$ (from 112.43 to 93.47 ms) was observed in the NP of the IVD after receiving the compressive loading. T_2 was also significantly decreased from 50.78 ms in the baseline condition to 33.51 ms in the compressed condition. However, ADC was not significant affected (P = 0.353).

Table 1. Measured MRI parameters (and standard errors) with and without loading. Bold fonts indicate significant differences.

	$T_{1\rho}$ (ms)	T_2 (ms)	ADC (mm^2/s)
Baseline	**112.43 (6.22)**	**50.78 (4.70)**	0.002887 (0.000054)
Compressed	**93.47 (5.12)**	**33.51 (3.45)**	0.002790 (0.000084)

4   Discussion

The $T_{1\rho}$ relaxation time of the NP can be estimated as the weighted $T_{1\rho}$ relaxation times of all the chemical components that made up of the NP [12]. NP primarily constitutes of water (80%) and PG (15%) [13], change in either composition can affect the $T_{1\rho}$ relaxation time. In the current study, the applied compressive load is believed to expel water content out from the NP, which can be detected by $T_{1\rho}$ relaxation time. Also, the decrease in $T_{1\rho}$ relaxation time indicated reduced water content in NP, which is in line with the finding from previous studies [14, 15].

It is not surprising that T_2 relaxation time decreases with greater IVD degeneration grades [16], since the grading system was created according to the IVD morphology with T_2-weighted images [17]. Loss of water content is one of the most immediate and apparent changes, which occur at early stage of IVD degeneration [18]. Therefore, in the current study, T_2 relaxation time is also significantly reduced by the applied compressive loading, since such loading can expel water out from the NP of the IVD and induce early degenerative changes [7].

ADC was found to be affected by the IVD composition and matrix integrity [19]. In the current study, although water content was reduced in the NP of the IVD, ADC was not significantly affected by the applied compressive load, presumably due to the expelled water was not sufficient enough to alter ADC.

In summary, $T_{1\rho}$ and T_2 relaxation times were found sensitive to the applied compressive load, which indicates their potential to detect the biomechanical changes in the NP of the IVD induced by the external mechanical loading. Therefore, they may provide more sensitivity and specificity to understand the injury mechanism of the IVD and contribute to early diagnosis of IVD degeneration, thus help reduce the prevalence of LBP.

Acknowledgement. This project was partially funded by the Western Center for Agricultural Health and Safety (WCAHS) with NIOSH Grant No. 2U54OH007550.

References

1. Fourney DR, Andersson G, Arnold PM, Dettori J, Cahana A, Fehlings MG, Chapman JR (2011) Chronic low back pain: a heterogeneous condition with challenges for an evidence-based approach. Spine 36:S1–S9
2. Brisby H (2006) Pathology and possible mechanisms of nervous system response to disc degeneration. J Bone Joint Surg 88(suppl 2):68–71
3. Panjabi MM (2006) A hypothesis of chronic back pain: ligament subfailure injuries lead to muscle control dysfunction. Eur Spine J 15(5):668–676
4. Zhou J, Ning X, Fathallah F (2016) Differences in lumbopelvic rhythm between trunk flexion and extension. Clin Biomech 32:274–279
5. Ning X, Zhou J, Dai B, Jaridi M (2014) The assessment of material handling strategies in dealing with sudden loading: the effects of load handling position on trunk biomechanics. Appl Ergon 45(6):1399–1405
6. Adams MA, Roughley PJ (2006) What is intervertebral disc degeneration, and what causes it. Spine 31(18):2151–2161
7. Urban JP, Winlove CP (2007) Pathophysiology of the intervertebral disc and the challenges for MRI. J Magn Reson Imaging 25(2):419–432
8. Noguchi M, Gooyers CE, Karakolis T, Noguchi K, Callaghan JP (2016) Is intervertebral disc pressure linked to herniation?: an in-vitro study using a porcine model. J Biomech 49(9):1824–1830
9. Nikkhoo M, Kuo Y, Hsu Y, Khalaf K, Haghpanahi M, Parnianpour M, Wang J (2015) Time-dependent response of intact intervertebral disc–In Vitro and In-Silico study on the effect of loading mode and rate. Eng Solid Mech 3(1):51–58
10. Beckstein JC, Sen S, Schaer TP, Vresilovic EJ, Elliott DM (2008) Comparison of animal discs used in disc research to human lumbar disc: axial compression mechanics and glycosaminoglycan content. Spine 33(6):E166–E173
11. Waters TR, Putz-Anderson V, Garg A, Fine LJ (1993) Revised NIOSH equation for the design and evaluation of manual lifting tasks. Ergonomics 36(7):749–776
12. Levine, H., & Slade, L. (Eds.). (2013). Water Relationships in Foods: Advances in the 1980 s and Trends for the 1990 s (Vol. 302). Springer Science & Business Media

13. Raj PP (2008) Intervertebral disc: anatomy-physiology-pathophysiology-treatment. Pain Pract 8(1):18–44
14. Johannessen W, Auerbach JD, Wheaton AJ, Kurji A, Borthakur A, Reddy R, Elliott DM (2006) Assessment of human disc degeneration and proteoglycan content using T1ρ-weighted magnetic resonance imaging. Spine 31(11):1253
15. Souza RB, Kumar D, Calixto N, Singh J, Schooler J, Subburaj K, Majumdar S (2014) Response of knee cartilage T1rho and T2 relaxation times to in vivo mechanical loading in individuals with and without knee osteoarthritis. Osteoarthritis Cartilage 22(10):1367–1376
16. Takashima H, Takebayashi T, Yoshimoto M, Terashima Y, Tsuda H, Ida K, Yamashita T (2012) Correlation between T2 relaxation time and intervertebral disk degeneration. Skeletal Radiol 41(2):163–167
17. Griffith JF, Wang YXJ, Antonio GE, Choi KC, Yu A, Ahuja AT, Leung PC (2007) Modified Pfirrmann grading system for lumbar intervertebral disc degeneration. Spine 32(24): E708–E712
18. Stokes IA, Iatridis JC (2004) Mechanical conditions that accelerate intervertebral disc degeneration: overload versus immobilization. Spine 29(23):2724–2732
19. Antoniou J, Demers CN, Beaudoin G, Goswami T, Mwale F, Aebi M, Alini M (2004) Apparent diffusion coefficient of intervertebral discs related to matrix composition and integrity. Magn Reson Imaging 22(7):963–972

Analysis of the Activity: 12 Years of Experience in Using a Data-Acquisition Platform by a French Occupational Health Service Working in Various Companies

Regine Codron[1]([⊠]), Sonia Bahiri[1], Patrick Bruneteau[1],
Véronique Delalande[1], and Michel Dupery[2]

[1] ACMS, Suresnes, France
{regine.codron,Sonia.bahiri}@acms.asso.fr
[2] Saint-Maur, France

1 Introduction

In 2005, the ACMS (SIST) occupational health service introduced innovation, by using a data-acquisition platform in order to make the analysis of the situation of real work easier.
First of all, we'll consider:

1. How an occupational health service (SIST) works in France.
2. How the request for action reaches the health services, how to take it into account and how to treat it.
3. Then, after describing the methodology being used,
4. Through several examples, we will show the type of support offered to the companies and how, through this process, we have been able to help with the prevention of muscle and skeleton disorders, as well as help keep ageing or sick employees at work, deal with pregnant employees, or else give advice in the design or layout of workstations in various sectors of activity.

The experience of these cases has also allowed to set off the criteria of and the brakes on the success of this type of support.

2 How Does an Occupational Health Service (SIST) Work in France?

Our occupational health service takes care of over one million employees in Ile de France in 80,000 workplaces (3/4 of which employ fewer than 10 staff). All the sectors of activity are represented: the trade, business services, insurance, finance,

---

R. Codron and M. Dupery—Occupational health physician.
S. Bahiri—Ergonomist.

© Springer Nature Switzerland AG 2019
S. Bagnara et al. (Eds.): IEA 2018, AISC 820, pp. 586–595, 2019.
https://doi.org/10.1007/978-3-319-96083-8_77

building-trade, health and social services, the industry, nuclear industry, transport and logistics...

In order to follow up these employees, our service is made of about 40 teams led and coordinated by doctors specialized in health at work, who are the partners in charge of the companies. These teams are made of various professionals, such as ergonomists, occupational health nurses, engineers, technicians in hygiene, safety and environment, technical assistants, psychologists specialised in occupational health...

The 20-07-2011 French law reasserted the task of the occupational health services, that is, to avoid any impairment of workers' health due to their job.

This law is based on several principles:

- First and foremost, priority is given to actions work surroundings and to collective prevention: this is primary prevention.
- Setting off the counselling role of the multidisciplinary team, led and coordinated by an occupational health doctor and the occupational social services. The health professionals have free access to places of work and are subject to professional and industrial secrecy.
- Individual follow-up of the workers 'health.
- Contribution of the occupational health service to the traceability of professional exposures and to health monitoring.

3   How Does the Request for Action Reach the Health Services, How Is It Taken into Account and Treated?

Several types of cases may come up and lead to a request:

- A "TMS epidemic", numerous restrictions to working at a work-station, designing/implementing new methods...
- Action of the multidisciplinary team in some work-environment after spotting a risk context
- An issue mentioned by staff representatives, especially in a hygiene and safety meeting (chsct)
- Screening for medical problems requiring a study of the work-station
- A survey needed by an institutional organisation...

Exchanges between the occupational health team (the ergonomist and the doctor) and the company (employers, employees and staff representatives) allow to define the aims of the survey, the method, and the relevance of using the platform and of giving the explanations required to obtain better cooperation. After carrying the survey and analysing the collected data, the conclusions are sent back to the company. Then the ergonomist takes part in setting up a plan of action and in its following-up.

How Do We Implement the (CAPTIV) Data Acquisition Platform?
Since 2005, a work-group of expertise, made of doctors and ergonomists, has been innovating, using a data acquisition platform (CAPTIV). Coupled with a video image, it can simultaneously study up to 15 different parameters: heart rate, electromyogram,

"motion" captors, goniometer, "eye tracker", chemical risk, vibration, noise or temperature metrology…

The trained participants, ergonomists, doctors, choose the method of analysis and treat the collected data, which allows to assess the activity, method of work, operating strategies and possible dysfunctions, positions and gestures, the work environment, in order to spot the factors and situations of risk, to identify and assess positions and gestures undetected so far…

In each case observed, a protocol is set up with the participants: choosing one day typical of the activity, of the area, of the work-stations and work-methods to be observed, choosing the operator, the captors to be used…

4   From Some Examples of Actions, What Are Our Results?

Over twelve years, more than 300 short observations have been carried out in business of various sizes, in all sectors of activity.

4.1   In Large-Scale of Distribution

The initial request comes from both occupational health doctor and the business that wish to assess the use of a prototype cash-register being tested in a store, and to compare it with a traditional cash-register. The aim of this study is to identify the risk work-situations, to assess the physiological stress and the biomechanical pressures of this new cash-register. In order to carry out this study, we rely on the "CAPTIV" platform: putting together physiological data from the upper limb (recording muscle-activity) and videos of cash-register activity.

This study goes through some time, 3 observation periods are carried out over about 3 years. We have been following the evolution of the prototype, which has undergone a few charges, especially after our conclusions and our pieces of advice. Our methodology has also grown more sophisticated in the course of these studies. Indeed, on top of the observations and measurements, the second study was the subject of a questionnaire about taking into account the experience on the cashiers working with this prototype, and in the third study we also integrated the notion of customer satisfaction with this new type of check-out. Since then, this model was finally approved and has been set up all over the stores of this chain.

4.2   In the Building-Trade

An occupational doctor transmits the request of a building firm coming especially from members of the CHS-CT that wish to make a survey in order to prevent possible muscle-skeleton disorders, because of the sharp rise in the number of occupational diseases: peri-articular disorders of the elbow and shoulders in formworkers.

The CAPTIV platform is used to allow to synchronize videos representative of the activity with measurements of vibrations and electromyogram (EMG) data. We have used a vibration analyser (a HAVPro machine made by Quest Technologies) equipped with a triaxial accelerometer mounted on the handle of the tool for the Hand and Arm vibrations. The EMG captors are located on the back and upper limbs. This method is used to identify and objectivize the most exposure-prone work stages.

A few images will illustrate the results hereby collected:

– Vibrations

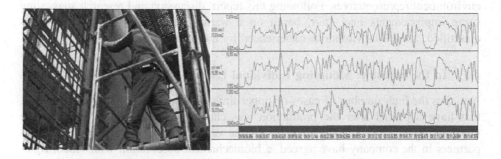

– Muscle strains, working gestures and positions

The cursor indicates the data from each captor corresponding to the image, and therefore the position and activity carried out by the operator.

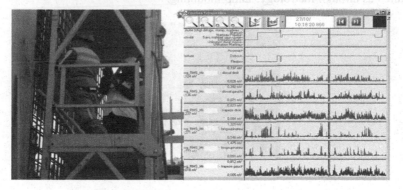

The muscle strains are analysed as the pneumatic drill is being used according to the work position:

Muscle zone	Standing working position	Bending working position	Squatting working position
Upper back	76%fmv	134%fmv	55%fmv
Shoulder	107%fmv	204%fmv	105%fmv
Forearm	226%fmv	156%fmv	164%fmv

The threshold recommended by specialists that must not be exceeded when straining while moving, is 30% of the Maximum Voluntary Strength (FMV). The average figure of the data of the highest recorded values has been compared to this threshold level.

Our results show the muscle strains are found to be more severe for the upper back and shoulders when in bending working position, whereas weaker when in squatting working position. Finally, the forearm and shoulder muscles are those that undergo the highest pressure, whatever the position.

The results of this study are then sent back to the company-management, the CHSCT members, and the method-services, building-site and quality, safety and environment representatives. Following this report, discussion and research suggesting some improvements may start, based on a warning diagnosis, in which effort is quantified, as well as vibrations experienced by the operator.

4.3 In the Industry: Mounting Individual Lamps

Noticing the high number of cases of upper limb tendinopathy, mostly in the elbow, that have appeared after the production was changed (new models, change in the working rhythm ...), the doctor asks for the ergonomist's help. After the various partners in the company have agreed, a biomechanical observation with video, right-wrist goniometer, and right upper-arm and shoulder muscle EMG is associated with a subjective assessment of the work through anonymous questionnaires. Several work-stations in the workshop – stocking, storing, assembling and packing- are surveyed, and one of them is pointed out as particularly exacting, as often through the questionnaires as through metrology: lamp-mounting.

– Muscle strains are particularly severe when passing the cable into the lampstand with uncomfortable wrist and elbow positions.
– Exposure to hand-arm vibration is confirmed, due to using screwer and belt-sander.

An accurate analysis of the videos has allowed to objectivise strains in very short stages of the various mounting tasks. Then the workgroup launched in the company have got to know these elements in order to look for solutions: widening the bole diameter, hanging tools, a specific support facilitating small part prehension...

4.4 In Catering

In order to suggest adapted health follow-up, the occupational doctor needed to assess the pressures in work-stations in a restaurant dining-room. After a first observation of real work, the ergonomist decides to use the CAPTIV platform, using both videos and recordings of cardiofrequencymetry captors, and EMG of back and shoulders.

Throughout the analysis, they identified:

- Loaded moving for 1/3 of their time, unloaded moving ¼ of the time
- About 4 tons were carried on trays by waiters
- Preparing orders for 1/3 of the time
- Heavy physical loading, worsened by moving around obstacles or up and down stairs
- Straining working gestures when laying and clearing tables, and cleaning chairs and lack of room adding to the problem.

At the end of this survey, various pieces of advice and suggestions of improvements are put forward, for the company to ponder on, in order to set up a plan of action. While working again in this company, it gave us great satisfaction to notice that most of our suggestions had been implemented.

Besides, the doctor considered that, in that environment, the workstation can't be allowed to a pregnant employee and should be adapted, for example by prioritizing reception or cash-desk.

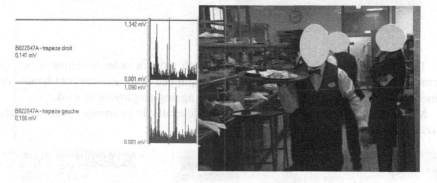

4.5   Caretakers

Let's consider the case of a business that builds and manages housing estates and blocks of flats in the Paris area, thus employing caretakers on various sites. Facing ageing staff, the company asked for occupational health service 'help in order to study how hard working on these sites was (linked to the task of taking wheelie-bins out) and to suggest any improvements with a view to maintaining those workers in their jobs.

A 2-step process has been implemented: first of all, an investigation was led on the sites about the containers 'way out; then an analysis of the activity in real work position was carried out on some sites with videos synchronized with measurements (heart rate, EMG). The results of the investigation and observation allowed to characterize hard work and implement a score of expected painfulness integrating various criteria (how bulky the volume that was handled, how far to go, how many obstacles...).

A few examples illustrate uneasy places - with a slope, or uneven surfacing, going through doorways - and a site equipped with handling assistance (tractor).

This survey helped endow the company with tools in order to define which sites were of easier work conditions and/or to be equipped with assistance to handle containers and thus to find solutions to keep their ageing employers at work.

Moreover, the SIST has developed and made available awareness – raising tools to caretakers.

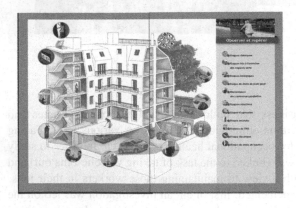

4.6   In a Medico-Educational Institute (MEI)

As requested by the institute consultant and the educational team in a Disabled Center, a survey was suggested, that would use the "CAPTIV" platform, that is, a video observation with EMG captors in the various working areas: a changing area, a classroom, an awakening room.

The analysis of these observations turned on the activity, the effort made, the positions, the educators' communication and the patients 'behaviour - conflicting, psychologically-induced attitude of a patient, or an energy-lacking patient -. In this case, it has been shown that it makes tasks more difficult, harms positions, and makes the educators 'efforts more painful.

Following of these observations, some pieces of advice have been suggested in order to improve:

- In all the areas of interventions, don't forget to put the seat and changing table in high position before handling the patient, and place him or her as near to the edge as possible.
- Let's add an important emotional involvement that occasionally requires support.
- Changing patients: the size of the room is not big enough for 2 staff to work with the patient.
- The classroom: helping hypotonic patients stand up is made easier thanks to good communication, which moreover allows more active contribution.
- The awakening room: several staff must take part in the preparation of the room, since the number of participants is always different.
- Advice about the choice of equipment, and about how to handle patients.

Here are a few examples of the seats being used:

4.7   Logistics in Thin Air

A new contract specifies that the employees of a trans-stocker maintenance company in a 40-meter-high logistics warehouse will have to work in an oxygen-depleted atmosphere: 13.5% instead of 21% of natural air.

The occupational health doctor warns the employer about the risks associated with this activity, but the latter encounters incomprehension from his client. Then, we propose an observation with a "CAPTIV" evaluation of the physical load of this post in a similar warehouse but rising only 27 m and in normal atmosphere. The results show a very high level of difficulty during the operations of crinoline climbing and climbing ladders with harnesses, the use of hoists and in uncomfortable postures. A second test is then proposed, the employee being equipped with a device with air supply. The values improve significantly but the use of the equipment proves dangerous because of the pipes limiting displacements. Finally, the relevance of this study convinced the client to organize the maintenance after temporarily reducing the oxygen level to 19%.

5 Discussion

Examples of actions and their results are described: potentially harmful work situations are avoided, help is provided to implement a plan of action, from the individual point-of-view we switch to the collective point-of-view, and psychosocial risk is taken in account through questionnaires.

Moreover, thus objectivized, observations are better perceived within the company, especially when the social climate is strained.

As the years go by, and we've been using the CAPTIV platform within the SIST, the process and methodology of our practice have changed: we switch from biomechanics to the analysis of activity and of the psychosocial environment.

The experience thus acquired allows to share knowledge and develop new skills.

At last, the effects of an alteration can be seen immediately.

However, we can see limits to our actions.

Interventions in SIST are generally carried out over short periods. Consequently, it is sometimes awkward to implement a survey, with difficulties in observing days representative of the activity, even if we are careful enough to choose them.

Moreover, interventions can be performed in a strained social background.

We've also had to face rather uncooperative operators, sometimes refusing being recorded on videos.

We've had to face technical problems, especially linked to the environment and in case of use of the wired version of the platform.

An example of the background in which the use of the CAPTIV platform in wired version requires a laptop on the site, near the operator with his equipment.

The analysis of the data requires expertise from trained professionals, who follow the technical improvements of equipment, and give incentives for suppliers to innovate. Besides, the SIST purchased the new version of CAPTIV (wireless captors), a model that is easier to handle on site.

Moreover, because of the scale of the analysing work, associated with the quantity of gathered and treated data, we have been forced to study limited work stages. The amount of time devoted to this analysis, to structuring the results of interviews and observations, do not necessarily always allow time to organise meetings with the operators, the manager, or colleague.

Eventually, technical improvements and our experience will contribute to still changing our practice, our measuring methodology, only aiming at helping businesses to implement primary prevention.

6 Conclusion

This innovating methodology uses the observation of real work associated to metrology. For 12 years it has proved how efficient it is in our practice and in prevention approach. It should be able to integrate data-innovations.

We wish to offer this service to a wider number of companies so as to meet their demands of prevention as well as possible as part of our new service project.

Acknowledgment. We thank all those who solicited us and who facilitated our interventions.

Glossary

ACMS

CAPTIV

SIST: occupational health service

References

1. Dupéry M, Alcouffe J, Delalande-Danet V (2012) Medico-ergonomic interventions in occupational health service to maintain employment: a record of four years of experience: Archives des maladies professionnelles et de l'environnement 73(6):868–876
2. Musculoskeletal Disorders (MSDs) in a laundry: contribution of the CAPTIV platform : Marchan MF, Muqa F, médecins du travail ACMS, Bruneteau P, Dupery M, ergonomes, CAMIP 2009 -3 - CAMIP.info - La revue de la santé au travail

The Influence of Psychosocial and Patient Handling Factors on the Musculoskeletal Health of Nurses

Mark G. Boocock[1]([⊠]), Fiona Trevelyan[1], Liz Ashby[1], Andy Ang[2], Nguyen Diep[3], Stephen Teo[3], and Felicity Lamm[2]

[1] Health and Rehabilitation Research Institute,
Auckland University of Technology, Auckland, New Zealand
mark.boocock@aut.ac.nz
[2] Centre for Occupational Health and Safety Research,
Auckland University of Technology, Auckland, New Zealand
[3] Edith Cowan University, Joondalup, WA, Australia

Abstract. Psychosocial work demands, the level of organisational support, and the provision and implementation of a patient handling programme have been identified as important determinants of musculoskeletal disorders (MSD) in nurses. The aim of this study was to describe the prevalence of MSD and explore the association of work-related psychosocial and patient handling factors on the musculoskeletal health of New Zealand (NZ) nurses. A sample (N = 201) of NZ nurses from the 2013 NZ Census completed an online survey in 2016–17 (45% of those eligible). MSD prevalence was measured using a modified version of the Nordic Musculoskeletal Questionnaire. Psychosocial work demands were measured using The Copenhagen Psychosocial Questionnaire (COPSOQ II). Components of "The Tool for Risk Outstanding in Patient Handling Interventions (TROPHI)" assessed patient handling practices. Binary logistic regression provided measures of association between psychosocial and patient handling factors, and MSD. The overall prevalence of MSD was 58% in the previous 12 months and 31% for the last 7 days. Low back (55%) and shoulder (54%) complaints were the most frequently reported MSD during previous 12 months, and co-morbidity of symptoms was high (59%). Higher work pace and emotional demands were significantly associated with MSD. Completing a patient handling task without equipment when equipment was prescribed, and perceived lack of suitable equipment, space, environment, skills or knowledge affecting patient care were also significantly associated with MSD. Interventions for the prevention and management of work-related MSD in nurses should take a multifaceted approach inclusive of physical and psychosocial components embedded within a comprehensive patient handling programme.

Keywords: Musculoskeletal disorders · Psychosocial risk factors
Patient handling · Nursing

1 Introduction

Psychosocial and organisational aspects of the work environment are important risk factors in the aetiology of work-related musculoskeletal disorders (MSD) [1]. In healthcare, nurses are at high risk of musculoskeletal injury, ill-health and early retirement [1]. Psychosocial work demands, the level of organisational support, and the provision and implementation of patient handling systems have been identified as important determinants of musculoskeletal injury reporting [2, 3]. High psychosocial demands and low job control have been linked to the prevalence and incidence of musculoskeletal pain, and more specifically low back pain in nurses [2]. Work-related injuries also appear to be associated with a poor psychosocial safety climate [4].

The aim of this study was to describe the prevalence of MSD and explore the association of work-related psychosocial and patient handling factors with these disorders in New Zealand (NZ) nurses. It was hypothesised that higher psychosocial workplace demands and a poor patient handling environment would be associated with increased risk of musculoskeletal conditions in nurses.

2 Methods

2.1 Study Design and Data Collection Methods

New Zealand nurses from the 2013 Census were invited to complete an online survey in 2016–17. The survey comprised 14 domains that included questions relating to organizational and personal factors affecting workplace behaviour and the health of nurses. This paper focuses on the questions concerned with psychosocial workload demands, patient handling factors, and self-reported musculoskeletal complaints. The study was approved by the University's research ethnics committee.

The prevalence of musculoskeletal conditions was measured during the previous 12 months and 7 days using a modified version of the Nordic Musculoskeletal Questionnaire, a self-report measure of pain or discomfort [5]. Respondents indicated aches, pains or discomfort across nine body regions, and the impact these may have had on functional activities, i.e. prevented doing normal activities (hobbies, housework).

Psychosocial work demands (Demands at Work) were determined using components of The Copenhagen Psychosocial Questionnaire (COPSOQ II) [6]. This comprised 15 items (questions) grouped into four scales: quantitative, cognitive and emotional demands (each with four questions) and work pace (three questions). Each item consisted of 5 response options, as defined by Pejtersen et al. (2010).

Perceived organizational commitment to a patient handling programme was evaluated using components of "The Tool for Risk Outstanding in Patient Handling Interventions (TROPHI)" [7]. Four questions relating to the reporting of patient handling accidents; four about the provision of suitable equipment, space, environment, skills or knowledge affecting patient care [8]; and a question on perceptions about the organisation's commitment (personnel, time and financial) to a patient handling programme (PHOQS Organisational Culture Tool) [9]. Sociodemographic information and work history were also collected.

2.2   Statistical Analysis

Descriptive statistics were used to analyse the sociodemographic of the sample population, the prevalence of self-reported musculoskeletal complaints (previous 12 months and last 7 days) and disability (that prevented normal activities), and patient handling accidents. For psychosocial workload demands, all questions in each item were scored from 0 to 100 (i.e. 0, 25, 50, 75 and 100 for the five response category items). A mean item score was calculated for each scale scores, with direction of the score appropriate to the scale name [6]. Questions relating to the lack of suitable equipment, space, environment, skills or knowledge affecting patient care were scored from 1 to 5, with a mean calculated as the patient handling score [7]. Binary logistic regression was used to determine the association of psychosocial and patient handling factors on musculoskeletal reporting. Those nurses who did not report a MSD were used as the reference category, and odds ratios with 95% confidence intervals (CI) provided a measure of the strength of association.

Statistical analysis was undertaken using IBM SPSS Statistics for Windows Version 24 (IBM Corp., USA). A statistical significance of $p = 0.05$ was applied throughout.

3   Results

3.1   Study Participants

Two hundred and one NZ nurses completed the online survey (45% of those eligible). The majority of respondents were female registered nurses (63%), aged between 26 and 40 years (45%), in full-time employment (71%) (Table 1). They worked primarily in the public hospital environment (52%) from the four largest populated regions of NZ: Auckland (36%); Canterbury (13%); Wellington (10%); and Waikato (9%). Approximately 39% of respondents had 5 or more years of experience in their current position.

3.2   Musculoskeletal Complaints

Fifty-eight percent of respondents reported musculoskeletal complaints in the last 12 months, and 31% in the last 7 days. Thirty percent of those who experienced pain in the last 12 months, reported that the pain had stopped them from doing normal activities (e.g., hobbies, housework). The distribution of MSD across the 9 body regions is shown in Fig. 1. The lower back and shoulders were the most frequently reported sites of MSD during the previous 12 months (55.2% and 53.7%, respectively) and last 7 days (17.4% and 15.4%, respectively).

Approximately 59% of respondents reported co-morbidity of symptoms during the previous 12 months, and approximately 55% experienced problems in four or more body regions during the previous 12 months.

Table 1. Sociodemographic and work history of respondents.

	Category	N = 201
Age (years)	18–25 yrs	15.9%
	26–30 yrs	24.4%
	31–40 yrs	20.9%
	41–50 yrs	14.9%
	51–60 yrs	15.4%
	>61 years	8.5%
Sex	Female	152 (75%)
	Male	49 (25%)
Health sector	Public hospital	104 (52%)
	Residential aged care	32 (16%)
	Private hospital	21 (10%)
	Public health	15 (8%)
	Home care	5 (2%)
	Other	24 (12%)
Tenure (current organization)	<12 months	12.4%
	1–<3 yr	31.8%
	3–<5 yr	16.9%
	5–<10 yr	21.9%
	10 years or more	16.9%

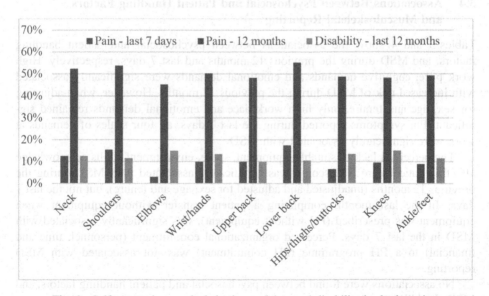

Fig. 1. Self-reported musculoskeletal complaints and disability by body region

3.3   Psychosocial and Patient Handling

Cognitive demands had the highest mean score for the four scales of psychosocial work demands (Table 2).

Table 2. Mean and 95% confidence intervals (CI) of work-related psychosocial factors.

	Number of items	Mean	95%CI
Quantitative demands	4	48.07	45.5–50.6
Work pace	3	63.23	60.4–66.0
Cognitive demands	4	66.32	63.6–69.1
Emotional demands	4	59.79	56.8–62.7

When asked about patient handling accidents and practices, approximately 27% of nurses said that they had reported a patient handling accident on the accident reporting system in the last 12 months. In the last month, approximately 22% reported using or seeing a patient handling method that they considered dangerous, and 13% reported undertaking a patient handling task without equipment when equipment was prescribed. When asked how often their organisation/department/area demonstrated a commitment (personnel, time and financial) to supporting a patient handling programme, approximately 29% of nurses considered it to be "hardly ever" or "never".

3.4   Associations Between Psychosocial and Patient Handling Factors and Musculoskeletal Reporting

Tables 3 and 4 show the associations between psychosocial and patient handling factors, and MSD during the previous 12 months and last 7 days, respectively. High work pace, cognitive demands and emotional demands were significantly associated with increased risk of MSD during the previous 12 months. However, when adjusting for sex, age and tenure, only high work pace and emotional demands remained significant. For symptoms reported during the last 7 days, all four scales of demands at work were significantly associated with MSD.

The perceived lack of suitable equipment, space, environment, skills or knowledge affecting patient care (PH score), was significantly associated with MSD during the previous 12 months (unadjusted and adjusted for sex, age and tenure), but not the last 7 days. In the last month, completing a patient transfer without equipment when equipment was prescribed (PH without equipment), was significantly associated with MSD in the last 7 days. Perceived organizational commitment (personnel, time and financial) to a PH programme (PH commitment) was not associated with MSD reporting.

No associations were found between psychosocial and patient handling factors, and MSD disability.

Table 3. Odds ratios (OR) and 95% confidence intervals (CI) for associations between work-related psychosocial and patient handling (PH) factors, and musculoskeletal disorders (MSD) reported during the previous 12 months.

	MSD past 12 months			MSD past 12 months		
	OR[a]	95%CI	P value	OR[b]	95%CI	P value
Demands at work						
Quantitative demands	1.009	0.994–1.025	NS[c]	1.009	0.993–1.025	NS
Work pace	1.026	1.010–1.041	0.001	1.025	1.009–1.041	0.002
Cognitive demands	1.016	1.001–1.031	0.037	1.014	0.999–1.030	NS
Emotional demands	1.026	1.010–1.042	0.002	1.024	1.008–1.041	0.003
Patient handling						
PH score	1.47	1.061–2.038	0.021	1.567	1.115–2.203	0.01
PH without equipment	2.268	0.912–5.641	NS	2.330	0.928–5.846	NS
PH commitment	1.159	0.916–1.467	NS	1.133	0.891–1.440	NS

[a]Unadjusted odds ratio (OR);
[b]Adjusted OR (sex, age, tenure);
[c]NS – non-significant

Table 4. Odds ratios (OR) and 95% confidence intervals (CI) for associations between work-related psychosocial and patient handling (PH) factors, and musculoskeletal disorders (MSD) reported during the last 7 days.

	MSD last 7 days			MSD last 7 days		
	OR[a]	95%CI	P value	OR[b]	95%CI	P value
Demands at work						
Quantitative demands	1.022	1.005–1.039	0.011	1.02	1.003–1.038	0.02
Work pace	1.029	1.012–1.046	0.001	1.028	1.01–1047	0.002
Cognitive demands	1.024	1.007–1.041	0.006	1.022	1.004–1.040	0.018
Emotional demands	1.026	1.010–1.042	0.002	1.024	1.008–1.041	0.003
Patient handling						
PH score	1.210	0.871–1.681	NS	1.31	0.931–1.844	NS
PH without equipment	2.303	1.011–5.246	0.047	2.537	1.092–5.895	0.031
PH commitment	0.904	0.705–1.160	NS	0.871	0.675–1.125	NS

[a]Unadjusted odds ratio (OR);
[b]Adjusted OR (sex, age, tenure);
[c]NS – non-significant

4   Discussion and Summary

As in many other countries, the prevalence of self-reported MSD was high in this sample of NZ nurses, although slightly lower than has been previously reported. In a comparative study of NZ nurses, Harcombe et al. [10] found a 91% prevalence of MSD (any region) for the previous 12 months, which is higher than the 58% prevalence rate identified in this study. For low back pain, they found a similar 12 month prevalence rate of 57% compared to 55% in the current study. Back pain in NZ nurses has been studied previously, with Coggan et al. [11] identifying an annual prevalence of 37% and point prevalence of 12%. Whilst this study also found low back pain to be the most frequently reported complaint, shoulder pain during the previous 12 months was the second most common complaint (54%). Other studies have suggested lower rates of shoulder complaints for nurses, 39% in NZ [10] and 31% in Estonia [12]. In the current study, neck complaints (52%) during the previous 12 months were high and comparable to other nursing studies, 52% in NZ [10] and 58% Estonia [12].

Higher work demands were found to be significantly associated with increased MSD, even after adjusting for sex, age and tenure. However, the strength of these associations could only be considered low. Across the four scales of work demands, work pace and emotional demands showed increased association with MSD reporting for the previous 12 months and last seven days. Freimann et al. [12] also showed higher work pace and emotional demands to be significantly associated with increased risk of musculoskeletal pain in the past month in Estonian hospital nurses. Odds ratios were comparable to those reported in the current study, 1.17 for both work pace and emotional demands.

Strong associations (OR range = 2.3–2.5) were found for completing a patient handling task without equipment when equipment was prescribed, and MSD in the last 7 days. The association between perceived lack of suitable equipment, space, environment, skills or knowledge affecting patient care and MSD during the previous 12 months had a modest, but significant association (OR range = 1.5–1.6). Previous studies have found a decreased injury rate and risk of MSD with the introduction of ergonomic devices and safe patient handling programmes [3, 13]. Perceptions about organizational commitment to patient handling was not found to be significantly associated with self-reported MSD.

This study is not without limitations. The cross-sectional design of the study does not allow a definitive causal relationship between dependent and independent measures to be established. The response rate was modest and therefore, potentially biased towards respondents with, or who had previously suffered a MSD. The self-reported measures also present difficulties when assessing the reliability of exposure measures.

In summary, this study found a high annual prevalence of self-reported MSD across most body regions, with low back, neck, shoulder and wrist/hand complaints being greater than 50%. Work-related psychosocial work demands (work pace and emotional demands) and patient handling factors appear to have an important impact on the occurrence of MSD. Interventions for the prevention and management of work-related MSD in nurses should take a multifaceted approach inclusive of physical and psychosocial components embedded within a comprehensive patient handling programme.

References

1. Solidaki E, Chatzi L, Bitsios P, Markatzi I, Plana E, Castro F, Palmer K, Coggon D, Kogevinas M (2010) Work related and psychological determinants of multi-site musculoskeletal pain. Scand J Work Environ Health 36(1):54
2. Bernal D, Campos-Serna J, Tobias A, Vargas-Prada S, Benavides FG, Serra C (2015) Review: work-related psychosocial risk factors and musculoskeletal disorders in hospital nurses and nursing aides: a systematic review and meta-analysis. Int J Nurs Stud 52:635–648
3. Humrickhouse R, Knibbe HJ (2016) The importance of safe patient handling to create a culture of safety: an evidential review. Ergon Open J 927–942
4. Zadow AJ, Dollard MF, Mclinton SS, Lawrence P, Tuckey MR (2017) Psychosocial safety climate, emotional exhaustion, and work injuries in healthcare workplaces. Stress Health 33(5):558–569
5. Kuorinka I, Jonsson B, Kilbom A, Vinterberg H, Biering-Sørensen F, Andersson G, Jørgensen K (1987) Standardised nordic questionnaires for the analysis of musculoskeletal symptoms. Appl Ergon 18(3):233–237
6. Pejtersen JH, Kristensen TS, Borg V, Bjorner JB (2010) The second version of the Copenhagen psychosocial questionnaire. Scand J Public Health 38(3_suppl):8–24
7. Fray M, Hignett S (2013) TROPHI: development of a tool to measure complex, multifactorial patient handling interventions. Ergonomics 56(8):1280–1294
8. Nelson A, Collins J, Siddharthan K, Matz M, Waters T (2008) Link between safe patient handling and patient outcomes in long-term care. Rehabil Nurs 33(1):33–43
9. Hignett S, Crumpton E (2007) Competency-based training for patient handling. Appl Ergon 38(1):7–17
10. Harcombe H, McBride D, Derrett S, Gray A (2009) Prevalence and impact of musculoskeletal disorders in New Zealand nurses, postal workers and office workers. Aust N Z J Public Health 33(5):437–441
11. Coggan C, Norton R, Roberts I, Hope V (1994) Prevalence of back pain among nurses. N Z Med J 107(983):306–308
12. Freimann T, Pääsuke M, Merisalu E (2016) Work-related psychosocial factors and mental health problems associated with musculoskeletal pain in nurses: a cross-sectional study. Pain Res Manag 2016
13. Yassi A, Lockhart K (2013) Work-relatedness of low back pain in nursing personnel: a systematic review. Int J Occup Environ Health 19(3):223–244

DUTCH: A New Tool for Practitioners for Risk Assessment of Push and Pull Activities

Marjolein Douwes[1(✉)], Reinier Könemann[1], Marco Hoozemans[2],
Paul Kuijer[3], and Hetty Vermeulen[4]

[1] TNO, PO Box 3005, 2301 DA Leiden, The Netherlands
marjolein.douwes@tno.nl
[2] Department of Human Movement Sciences,
Faculty of Behavioural and Movement Sciences, Vrije Universiteit,
Amsterdam Movement Sciences, Amsterdam, The Netherlands
[3] Netherlands Center for Occupational Diseases,
Coronel Institute of Occupational Health, Academic Medical Center (AMC),
University of Amsterdam, Amsterdam Public Health Research Institute,
Amsterdam, The Netherlands
[4] vhp human performance, The Hague, The Netherlands

Abstract. Pushing and pulling at work is an undervalued theme within occupational health policy. This is unjustified, because these activities are very common and potentially increase the risk of shoulder symptoms. Gaining insight into the possible health risks of specific push or pull activities at the workplace is a first step towards prevention of musculoskeletal symptoms. Existing instruments proved to be insufficiently suitable to give that insight in a simple way. This was the motivation for developing the Push and Pull Check (DUTCH). This method makes clear whether the push or pull activity is acceptable or not, which risk factors exist, and which measures can reduce the risk. This article describes the operation of the DUTCH, as well as the development of the tool.

Keywords: Physical workload · Pushing · Pulling · Risk assessment
Prevention · Shoulder · Low back

1 Introduction

Manual force exertion and pushing and/or pulling at work increase the risk of developing low back and shoulder pain [1, 2]. In particular, performing push or pull activities at work poses a high risk of shoulder symptoms. This was concluded from a systematic review of seven studies, covering a total of 8,279 employees, in which the risk of shoulder symptoms among workers with high exposure to pushing and pulling was between two and five times higher than in workers that did not perform pushing and pulling activities [2].

It is important that companies are aware of the health risks of pushing and pulling at work. Health and safety regulations do not offer specific health limits for pushing and pulling. It is also not possible to draw up these health limits based on available epidemiological literature [1]. The Dutch Health Council's recommendations stated

© Springer Nature Switzerland AG 2019
S. Bagnara et al. (Eds.): IEA 2018, AISC 820, pp. 604–614, 2019.
https://doi.org/10.1007/978-3-319-96083-8_79

that the 'Mital tables' [3] provide the best available information to assess and evaluate the physical work demands of pushing and pulling tasks. These tables provide data from psychophysical research on the - self reported – maximum manual force exertion under different conditions. However, the practical application of these tables requires force measurements at the workplace, which is often not feasible in practice. Existing practical instruments that do not require force measurements, such as the Key Indicator Method (KIM) [4, 5], the PushPullCalculator (PPC) [6, 7] and the pushing and pulling Operations Risk Assessment Tool [8] do not meet all criteria that apply to a good practical tool. For example, the KIM and the HSL tool are not sufficiently evidence-based, while the DTC does not take into account environmental factors of pushing and pulling tasks, such as characteristics of the floor surface and of the wheels of a trolley. These factors potentially have a large effect on the rolling resistance and thus on the hand force needed when pushing or pulling a container [9].

Thus, a new practical tool was developed, with the aim to give companies quick and simple insight into the presence of risk factors of pushing and pulling at work.

2 Pre-study on Inter-rater Reliability and Face Validity of Existing Tools

2.1 Objective

An important criterion for a risk assessment method is that it produces reliable and valid results. No information was available on the reliability and validity of existing tools for the evaluation of pushing and pulling at work. To get an impression of these qualities, we conducted a brief study on the inter-rater reliability and the 'face-validity' of the KIM and the DTC. In this paper, face-validity means: the level of agreement between the results of the instruments with the judgments of a group of experts in the field of physical workload on the evaluation of the potential health-risk associated with performing specific pushing and pulling activities. The final version of the HSL assessment tool and data presented by Weston et al. [10] were not yet available for this study.

2.2 Methods

For this study we selected ten push-pull tasks that had been studied before by one of the experts involved. These ten tasks involved various pushing and pulling activities with horizontally-oriented hand forces, with a variable level of force exertion, that had been measured according to a valid protocol. Characteristics of the tasks are listed in the first column of Table 1. The experts provided descriptions of these tasks in a standardized format, which enabled the KIM and DTC to be applied. Some of the tasks lacked information on the frequency of pushing or pulling, because they were experimental situations. In those cases, realistic estimates were used. Eight experts independently evaluated the push and pull tasks with the KIM and the DTC. They translated the task descriptions to the required input data of the two instruments. In addition, they answered some questions on the exact data and methods they used for the evaluation,

606   M. Douwes et al.

the difficulties they encountered, the advantages and disadvantages of the two instruments and the extent to which the outcome was in line with their expert judgement on the physical demands of the tasks (red, orange or green). Green means 'at least 90% of the population can maintain this task for eight hours', orange means 'between 25% and 90% can maintain this task' and red means '25% of the population or less can maintain this task for eight hours'.

2.3   Data Analysis

To determine the face validity of the KIM and DTC, their results were compared with that of the experts' judgments, which were used as the reference standard in this study. To reach a single final judgement from the three parts of the DTC (assessment of the hand force, low back load and shoulder load), the lowest limit value (strictest assessment) was used. For the inter-rater reliability, the percentage of absolute agreement between red-orange-green assessments of the evaluators was calculated. In addition, Kappa values were calculated for the individual scores of the DTC and the final score of the KIM. For the interpretation of Kappa results the following cut-off points were used: 'low' (0–0.20); 'moderate' (0.21–0.40); 'reasonable' (0.41–0.60); 'sufficient to good' (0.61–0.80) and 'almost perfect' (0.81–1.00) [11].

2.4   Results

Face-Validity

Table 1 shows the results of the KIM and DTC, and expert judgments for the ten push/pull tasks. When comparing the results from the KIM with the expert judgments, they are in agreement for four tasks, more strict (orange instead of green) for three tasks and less strict (orange instead of red) for three tasks. The ratings of the DTC are in agreement with the expert judgments for five tasks, more strict for four tasks (twice red instead of green and once red instead of orange) and less strict for one task (green instead of red). Overall, KIM reviews were orange for seven out of ten tasks; DTC reviews were red for seven out of ten tasks. Also notable is the fact that only one of the tasks (task 2) was judged the same by the KIM and DTC.

Inter-rater Reliability

In Table 2, the percentage of agreement in categorical scores (red, orange, green) between the experts is displayed for the DTC and the KIM. The overall percentage of agreement for the KIM is 81%. Since the DTC does not present an overall score, the level of agreement is presented for hand force, shoulder load and low back load separately. The lowest level of agreement for DTC (62%) was found for the shoulder load (sustained) and the highest level of agreement (99%) was found for the low back load. To give a final judgment for DTC the experts used a 'worst case' assessment; the agreement percentage of this judgment was 91%. In addition to the percentage of agreement, the kappa values and their classifications are presented for both tools. The kappa rating for the KIM risk score is 'sufficient to good' (0.705) and for the DTC it varies from 'reasonable' to 'almost perfect' (0.447–0.967), but is 'good' (0.833) for the DTC worst case assessment.

Table 1. Average reviews and corresponding red-orange-green judgements according to the KIM, most common judging according to the DTC and consensus judgments from experts

Tasks	KIM: mean final score KIM (sd)	DTC: most common score* (% agreement)	Consensus expert judgment
Pushing a small trolley with food in the train	44 (14)	Green (63%)	Green
Postal expedition: pushing/ pulling carts in distribution centre	54 (14)	Red (100%)	Red
Postal distribution: pushing/ pulling carts in distribution centre	43 (8)	Red (100%)	Green
Level out concrete floor with an electrical (vibrating) rei	85 (22)	Green (86%)**	Red
Move a hand pallet truck in a warehouse	35 (12)	Red (100%)**	Orange
Move trolleys to and in trucks	29 (8)	Red (100%)	Red
Move garbage containers along tiles with 1 or 2 persons	39 (13)	Red (100%)	Red
Move garbage containers along tiles with helping device	30 (13)	Red (100%)**	Green
Move garbage containers along tiles without helping device	38 (17)	Red (100%)**	Red
Pull money carts along carpet in casino	7 (1)	Orange (57%)**	Green

* The DTC does not present a 'overall score '. Therefore, the highest subscore per evaluator and most prevalent score among all evaluators was used; **N = 7.

2.5   Discussion

Face Validity

Both the DTC and KIM resulted in a large number of assessments that did not equal the expert assessment: six out of ten tasks for KIM and five out of ten for the DTC. Overall, the assessments of the KIM were more in line with the expert assessments because the

Table 2. Percentage of agreement and Kappa-score to indicate the agreement between the eight raters, for ten tasks, both for DTC and KIM categorical results

Evaluation	% agreement	Kappa	Kappa classification
DTC, initial hand force (to set in motion)	92%	0.853	Good
DTC, sustained hand force (to keep in motion)	79%	0.680	Sufficient to good
DTC, back strain at onset	97%	0.933	Almost perfect
DTC, back strain – sustained	99%	0.967	Almost perfect
DTC, shoulder strain at onset	78%	0.567	Reasonable
DTC, shoulder strain – sustained	62%	0.447	Reasonable
DTC, worst case	91%	0.833	Good
KIM, risk score	81%	0.705	Sufficient to good

number of overestimations was equal to the number of underestimations, whereas the DTC overestimated more severely and more frequently. However, some of the tasks that scored 'red' by the experts were not scored 'red' when using the KIM. In those cases, the KIM underestimated the physical load according to the experts. Moreover, both methods have less responsive character than the experts: seven out of ten reviews with the KIM are orange, nine out of ten reviews with the DTC are red.

The face-validity of an instrument also depends on its scientific background. The DTC is based on Hoozemans et al. [12] for the estimation of exerted hand forces, low back load and shoulder load, and based on Mital et al. [3] for the evaluation of the hand force; on Jäger [13] for the evaluation of the low back load and on Chaffin et al. [14] for the evaluation of the shoulder load. At the time of this study, the authors of the KIM could not provide scientific background for the KIM. Moreover, the KIM results do not appear to be associated with the Mital tables [3]. The KIM evaluates pushing and pulling with a low frequency as less hazardous and with a high frequency as (much) more hazardous than the Mital tables.

A possible explanation for the differences between assessments with both instruments on one hand and expert reviews on the other hand is that the KIM takes environmental factors into consideration while the DTC does not. Moreover, there is variation in the expert judgments that were used as a 'reference standard' in this study. This is probably due to different opinions of the experts about the risk of exposure for work-related musculoskeletal complaints.

Inter-rater Reliability. De kappa-classification for the DTC varied from 'reasonable' to 'almost perfect'(0.45–0.97), and was 'good' (0.83) for the 'worst case' assessment, which is the most important result of the DTC. Mutual differences in results can be explained by difference in interpretation of the information on the task provided for using KIM and DTC. Examples are differences in determining the male/female distribution in the population, working posture and working conditions; these factors are difficult to estimate and generalize per task.

2.6   Limitations and Conclusions

This research has limitations. Firstly, there was little variation between the experts' assessments of the tasks, which were 'red' relatively often. This may give a too one-sided view. Secondly, there was no need for an interpretation by the evaluator for some pre-provided data. However, in practice variation may arise for these factors, which would result in a lower reliability than we found in this study. Thirdly, this study was carried out with experts trained in the application of similar methods. Application of the KIM and DTC by users without prior knowledge is expected to provide lower reliability as they will have more difficulty in determining the requested input data. To apply a tool as we envision, with little or no prior expertise on physical workload assessment or ergonomics training is probably a major constraint.

Both the KIM and the DTC have a moderate face-validity and inter-rater reliability to assess the work-related risk of pushing and pulling. Therefore, and due to the lack of scientific evidence for the KIM, it was concluded that a new tool should be developed that is not based on either the KIM or the DTC but should incorporate the strengths of both methods. In addition, the new tool should be evidence-based, present a clear overall final judgement and provide insight in the main hazards, include both the average and peak load in the assessment to avoid 'means' of extreme situations, mention conditions for which the tool is not applicable (e.g. sliding of objects) and include recommendations for risk reduction and give insight into the effect of small improvements, to encourage workers and donors to take and use measures.

3   Development of DUTCH

3.1   Maximum Acceptable Push and Pull Forces

In theory, limits for an acceptable workload should be deduced from epidemiological literature, biomechanical models and/or psychophysical experiments. Through a concise literature review, which we do not discuss here, recent relevant epidemiological literature was studied. In the literature a strong relationship is described between pushing/pulling and the prevalence of shoulder symptoms [2]. However, the literature does not provide sufficient guidance to define clear limits above which the risk of shoulder symptoms strongly increases. Moreover, biomechanical shoulder models are insufficient for a scientific criterion. Available shoulder models, which include muscle load around the shoulder joint, indicate limited health limits. Therefore, it was decided to use the psychophysical Snook tables [15] for the determination of the maximum acceptable push and pull hand forces. These tables correspond to those of Mital et al. [3] but provide more extensive data on the frequency of pushing and pulling and population percentiles.

3.2   Snook Tables

Based on self-reporting, the Snook tables present the maximum acceptable horizontal hand force (Newton) for pushing or pulling rolling stock under different conditions, if that task would last all day. The acceptable hand force depends on the direction of force (pushing or pulling), the frequency (number of activities per day), the distance per activity, the hand height and the gender of the employees. There is also a distinction between the initial force (to start the motion) and the sustained force required (to maintain the motion of the load). To use these data in the DUTCH we made the following calculations and choices:

- based on the normal-distributed percentile values per push-pull situation, curves were assessed for maximal acceptable hand forces, as a percentage of the working population (see Fig. 1);
- in addition to separate values for men and women, limit values were calculated for an equal distribution of men and women (dotted line in Fig. 1);

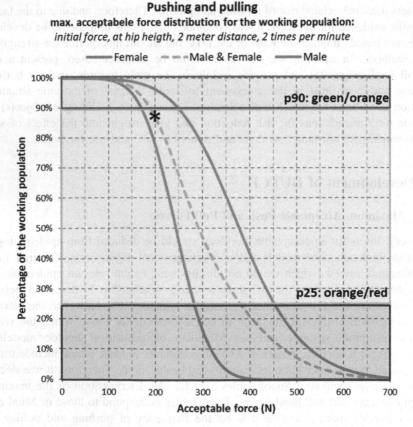

Pushing and pulling

max. acceptabele force distribution for the working population:

initial force, at hip heigth, 2 meter distance, 2 times per minute

━━━ Female   ━ ━ ━ Male & Female   ━━━ Male

Fig. 1. Maximal acceptable hand forces for different percentages of the working population, for a specific push/pull condition (initial force, hip height, 2 m, 2 times/min). The green dotted line shows values for a population of 50% men/50% women. The limits between the coloured planes show limits for p90 (green/orange) and p25 (orange/red). * For example: a force of 200 N is acceptable for 88% of the workforce (men and women), when pushing or pulling carts at waist level twice per minute. (Color figure online)

- a traffic light model was added (coloured planes in Fig. 1); The orange/red border represents the point where 90% of the workforce is protected, which is common in ergonomic guidelines and instruments. With the orange/red border 25% is protected, which is a boundary based on consensus by the expert group;
- because limit values for the initial motion of a load are lower than those for the sustained motion of the same load, limit values for the initial motion are used;
- because in practice there is almost always a combination of pushing and pulling, the average acceptable hand forces for pushing and pulling are used.

3.3   Convert Hand Force to Cart Weight

Because force measurements are difficult to perform in practice and often lead to errors, maximum hand forces from the Snook tables were converted to cart weights. For this purpose a formula was distracted from measurements of Hoozemans et al. [12] at 3 cart weights (85, 135 and 320 kg), 2 hand heights (hip and shoulder height), push and pull with 1 and 2 hand(s) and initial and sustained hand forces. The number of measurements and the different push and pull conditions in this study were limited. Therefore, the formula must be validated in future. To get an impression of the 'face-validity' of the formula, hand forces from the Snook table were converted to cart weights using the formula. These cart weights seemed to be realistic.

3.4   Influence of the Surface and Material (Qualitative Part of the Method)

The force required for pushing and pulling a cart also depends on factors such as the type of surface, wheel diameter, material of the wheels and state of maintenance. Because these factors are often difficult to determine for the user and because of insufficient scientific support for the impact of these features on the hand forces, it was decided to process them in a 'qualitative' way and not in a 'quantitative' way. This means that we indicate whether these characteristics are *favourable* or *unfavourable* to the required strength, without calculating the effect on an acceptable cart weight.

4   The Result: DUTCH - Description of the Tool

4.1   Structure of the Tool

Figure 2 presents an overview of (1) required input for using the tool (2) calculations that are being made using the input data, and (3) the results that are being presented to the user. The figure also shows the difference between the quantitative (upper part in blue) and qualitative part (lower part in green) of the method.

4.2   Results of an Assessment

Based on the quantitative input data, the DUTCH calculates the average and maximum cart weights (cart and load together) in a specific situation. For the evaluation of this score the traffic light model is used: 'green' means 'low work demands, with minimal

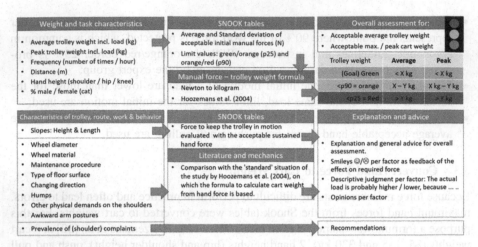

Fig. 2. Input, calculations and results of the DUTCH for the quantitative and qualitative assessment of pushing and pulling (Color figure online)

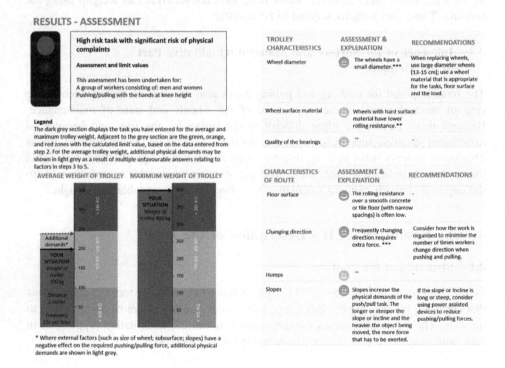

Fig. 3. Example of the quantitative assessment (left) and qualitative assessment (right) (Color figure online)

risk of physical symptoms; 'orange' means 'demanding task: there is a risk of physical symptoms and 'red' means 'high work symptoms, with high risk of physical symptoms' (Fig. 3, left side).

In addition, information is presented on the effect of the characteristics of the environment, material and behaviour. Emoticons and explanatory text show for each factor if they have a favourable or unfavourable effect on the evaluation. This is the result of the qualitative assessment (Fig. 3, right side). For example, a smooth and firm surface has a positive effect on the assessment (☺) because pushing a cart or trolley on a smooth firm surface requires less force than if the surface is rough (☹).

In case of an orange or red assessment, the DUTCH presents measures to reduce the physical work demands and it refers to a 5 steps risk reduction approach.

5 Conclusions

The DUTCH is an evidence and expert based, simple webtool for a quick evaluation of push and pull activities at work, providing insight into the presence of risk factors and potential measures to reduce the work demands. The English version of the tool is available at https://www.fysiekebelasting.tno.nl/en/ and is targeted at occupational health officers. The DUTCH has been tested on a small scale for usability by companies and experts. The reliability and validity of the evaluation method have not been studied extensively yet.

References

1. Dutch Health Council (Gezondheidsraad) (2012) Kracht zetten, duwen en trekken in werksituaties. Gezond-heidsraad, Den Haag. http://gr.nl/nl/adviezen/gezonde-arbeidsom standigheden/kracht-zetten-duwen-en-trekken-werksituaties. (in Dutch)
2. Hoozemans MJ, Knelange EB, Frings-Dresen MH, Veeger HE, Kuijer PP (2014) Are pushing and pulling work-related risk factors for upper extremity symptoms? A systematic review of observational studies. Occup Environ Med 71(11):788–795
3. Mital A, Nicholson AS, Ayoub MMA (1997) Guide to manual materials handling. Taylor & Francis, London
4. Jürgens WW, Mohr D, Pangert R, Pernack E, Schultz K, Steinberg U (2002) Handlungsanleitung zur Beurteilung der Arbeitsbedingungen beim Ziehen und Schieben von Lasten. LASI Ver-öffentlichung LV29. Hrsg. Länderausschuss für Arbeitsschutz und Sicherheitstechnik
5. Steinberg U, Caffier G, Liebers F (2006) Assessment of manual material handling based on key indicators: German guidelines. In: Karwowski W (ed) Handbook on standards and guidelines in ergonomics and human factors, Chap 18. Lawrence Erlbaum Associates, Mahwah, pp 317–335
6. Hoozemans MJM, Visser B, Van Dieën JH (2010) Evaluation of pushing and pulling at the work-place using an web-based PushPullCalculator. In: Seventh international scientific conference on prevention of work-related musculoskeletal disorders PREMUS 2010, Angers, France

7. Kuijer PPFM, Hoozemans MJM, Frings-Dresen MHWA (2007) A different approach for the ergonomic evaluation of pushing and pulling in practice. Int J Ind Ergon 37:855–862

8. HSL (2003) Pulling and pushing operations risk assessment tool, draft, 24 June 2013

9. Hoozemans MJ, van der Beek AJ, Frings-Dresen MH, van Dijk FJ, van der Woude LH (1998) Pushing and pulling in relation to musculoskeletal disorders: a review of risk factors. Ergonomics 41(6):757–781 Review

10. Weston EB, Aurand A, Dufour JS, Knapik GG, Marras WS (2018) Biomechanically determined hand force limits protecting the low back during occupational pushing and pulling tasks. Ergonomics 61(6):853–865. https://doi.org/10.1080/00140139.2017.1417643

11. Landis JR, Koch GG (1977) The measurement of observer agreement for categorical data. Biometrics 33(1):159–174

12. Hoozemans MJM, Kuijer PPFM, Kingma I, Van Dieën JH, De Vries WHK, Van der Woude LHV, Veeger HEJ, Van der Beek AJ, Frings-Dresen MHW (2004) Mechanical loading of the low back and shoulders during pushing and pulling activities. Ergonomics 47 (1):1–18

13. Jäger M (2001) Belastung und Belastbarkeit der Lendenwirbelsäule im Berufsalltag. Ein interdisziplinärer Ansatz für eine ergonomische Arbeitsgestaltung. Fortschr.-Ber. VDI Reihe 17 Nr. 208. VDI Verlag, Düsseldorf

14. Chaffin DB, Andersson GBJ, Martin BJ (1999) Occupational biomechanics, 3rd edn. Wiley, New York

15. Snook SH, Ciriello VM (1991) The design of manual handling tasks: revised tables of maximum acceptable weights and forces. Ergonomics 34:9

Ergonomic Intervention for Healthcare Workers and Patients: A Development of Patient Handling Device

Rex Aurelius C. Robielos[(✉)], Karla Coleen A. Sambua, and Joanna G. Fernandez

School of Industrial Engineering and Service Engineering Management, Mapúa University, Manila, Philippines

Abstract. Over the years, healthcare workers have suffered debilitating musculoskeletal disorders when lifting, transferring and repositioning patients manually. Hence this study focuses on risk-related issues experienced by healthcare workers as well as their patients due to manual patient handling. Issues regarding patient handling activities were analyzed by conducting a survey on healthcare workers and patients that includes ergonomic assessment tools like Rapid Upper Limb Assessment (RULA), NIOSH Lifting Equation, and Nordic Musculoskeletal Questionnaire (NMQ) etc. After which, one-way ANOVA and Tukey's HSD Test were conducted to determine a significant difference between the risk factors on patient handling activities. As a result, healthcare workers tend to experience discomfort mostly on their upper limbs due to equipment's dimensions and capacity to lift. To address the problem, this study suggested an intervention for the most critical patient handling activity by proposing a patient handling device. The design of the device was based on healthcare workers and patients' needs and requirements which were translated using Quality Function Deployment (QFD) and Detail Design. For validation of the product, Design Failure Mode and Effect Analysis (DFMEA) and survey regarding user's perspective of the product were conducted. To test the effectiveness of the proposed design, RULA and NIOSH Lifting Equation were also utilized and the results were compared to the ones obtained before the intervention. Through this, it was confirmed that the proposed design captures critical patient handling activities and offers more functions and features than the available lifting equipment with a lower cost.

Keywords: Patient handling device · Musculoskeletal disorder
Healthcare workers

1 Introduction

For decades, nurses and other healthcare workers have suffered debilitating musculoskeletal injuries when lifting, transferring and repositioning patients manually. These activities are called patient handling and movement tasks, which are often physically demanding, performed under unfavorable conditions, and are often unpredictable in nature. Patients offer multiple challenges including variations in size, physical disabilities,

© Springer Nature Switzerland AG 2019
S. Bagnara et al. (Eds.): IEA 2018, AISC 820, pp. 615–638, 2019.
https://doi.org/10.1007/978-3-319-96083-8_80

cognitive function, level of cooperation, and fluctuations in condition [1]. Surprisingly, in one typical 8-h shift, the cumulative weight lifted by a nurse is equivalent to 1.8 tons [2]. With this, it is more difficult for a healthcare worker to move since lowering and carrying distances have a strenuous effect in healthcare personnel as said in the Health and Safety Authority. According to the Bureau of Labor and Statistics, strains, sprains, and tears caused by overexertion, repetitive motion, and unexpected patient movements make up the largest proportion of patient-handling injuries among nursing, psychiatric, and other healthcare workers. These injuries and disorders can affect the human body's movement or musculoskeletal system and are also called Musculoskeletal Disorders (MSD).

Work related musculoskeletal disorders (WMSD) are important occupational health issue among all health care workers. WMSD is a collective and descriptive term for the symptoms caused or aggravated by work and characterized by discomfort, impairment, disability or persistent pain [3]. WMSD are also a major cause of the increased number of absenteeism and/or sick days every year and a study confirmed this by reporting that 56% of all sick days among professional nurses were due to WMSD [4]. As the rate of absenteeism increases, the number of healthcare personnel who acknowledges early retirement also increases. Nurses cite three key factors that trigger premature retirement; strenuous work activities, psychology demands and WMSD [5]. Along with these consequences, the U.S. Department of Labor, Occupational Safety and Health state that nurse injuries are also costly in terms of medical expenses, disability compensation and litigation.

The intensity within the healthcare profession tasks exposes the nurses to a variety of musculoskeletal disorder. Poor patient transfer technique is the primary culprit of low-back pain among nurses and it is claimed that low-back pain is the most common MSD inherited in the working environment [6, 7]. About 132 studies conducted over the past 30 years and provided a comprehensive report on prevalence of musculoskeletal disorders in nursing workers (e.g., the mean annual prevalence rates of 55% for low back pain, 44% for shoulder pain, 42% for neck pain, 26% for upper extremity pain, and 36% for lower extremity pain) [8]. Based on the facts enumerated, this study aims to address the most common musculoskeletal disorders a healthcare worker can acquire by manually handling patients.

Fortunately, MSD are preventable with the use of proper lifting, transfer equipment, and good body mechanics [9]. Some studies also confirmed that robotic bed movers and patient handling devices can lessen the manpower needed to manually transport the beds of the patient and manually carry the patients itself [10, 11]. It is also claimed that these interventions reduce physiological strains on the healthcare workers therefore lowering the risk of obtaining musculoskeletal injuries, especially low back pain. However, it is inferred that hospitals tend to underscore the importance of safety practices and rules regarding patient-handling tasks for injury prevention [12]. Also, there are no current studies which involve consideration of both healthcare workers and patients in developing patient handling devices. A study also suggested a need for a major paradigm shift towards evidence based practices that incorporate the extensive use of lifting devices, improved patient care ergonomic assessment tools, true no lift policies, effective and required training on the use of lifting devices [1].

To address the inadequacy of the available studies, this research aims to evaluate the current condition of healthcare workers and patients using ergonomic assessment tools such as RULA, NIOSH etc. which will then be utilize for designing a patient-handling device that could aid the most critical patient-handling activities using product design and development tools like FMEA and QFD, along with the equipment's safety precautions and rules.

2 Methodology

2.1 Design and Sample

This study employed a descriptive research which aims to determine the current condition of patients, workers and equipment concerning patient handling activities rendered for in-patient services. Observed data and surveys were utilized as the basis of the study and conclusions are drawn based on the assessments performed. In addition, a product will be developed to address the evaluated current conditions.

The participants were composed of healthcare workers directly involved in patient handling activities, specifically the nurses, nursing aides and assistants, as well as the attendants that had already been doing the job for at least a year. Also, this study also selected patients who undergo treatment that requires at least one-night stay in the hospital which requires minimal, moderate or completely dependent supervision as respondents for they are also affected in obtaining high-risk dilemma due to these activities. The research was conducted within the vicinity of Pasig City General Hospital (PCGH). Pasig City General Hospital possesses a complete hospital facility that provides general services for all kinds of illnesses, diseases and injuries. There are eight (8) main hospital units but five (5) units were only considered: emergency room, medicine ward, OB-GYNE ward, pediatrician ward and surgery ward. The other three units namely the Intensive Care Unit, operating room and delivery room were not included because patients in these units were undergoing special treatment due to their severe conditions which requires strict supervision. The sample size estimation was taken using stratified random sampling technique by integrating Slovin's Formula (1) where n is the total sample size, N is the total population size and e is the margin of error with standard value of 5%. After determining the sample size for the entire population, Proportionate Stratification (2) was utilized to determine the sample size per hospital unit where n_h and N_h . is the sample and population size of the stratum h respectively.

$$n = \frac{N}{1 + Ne^2} \tag{1}$$

$$n_h = \frac{N_H}{N} * n \tag{2}$$

2.2 Survey Development and Ergonomic Analysis

Survey forms were used to gather information from both healthcare workers and patients that were further used for analysis and evaluation of the study. One survey form was given to patients for the identification of their demographic profile (gender, age, height, weight etc.) and feedbacks or concerns regarding the existing patient handling equipment which can be translated using Body Part Discomfort Scale (BPDS). While three-part survey forms were given to healthcare workers: Part 1 is for Nordic Musculoskeletal questionnaire (NMQ) and Pairwise Comparison questionnaire, risk factor rating and importance of product features for Part 2 and product evaluation survey for Part 3. Nordic Musculoskeletal questionnaire focuses on musculoskeletal issues and work factors while PCQ was used as an input in the Analytical Hierarchy Process (AHP) to compare each hospital unit by pair for the identification of the most critical unit. However, Part 2 & 3 of the survey was only given to healthcare workers under the most critical hospital unit. Equation (3) shows how the number of pairwise comparisons is computed, with n being the number of criteria or alternatives to be compared.

$$Number\ of\ Comparisons = \frac{n(n-1)}{2} \tag{3}$$

To provide a quick and systematic assessment of the postural risks to a worker focusing on work-related upper limb disorders, Rapid Upper Limb Assessment (RULA) worksheet was utilized. Rapid Upper Limb Assessment also provides a method that could go into a more versatile ergonomic assessment, eliminating the need for assessment equipment (AIHA Ergonomic Committee, 2011). Also, the revised NIOSH Lifting Equation was used to assess the manual material handling risks and to calculate the recommended weight limit for these tasks. The recommended weight limit was computed by multiplying the load constant with the horizontal, vertical, distance, asymmetric, frequency and coupling multiplier.

2.3 Statistical Analysis

Differences between the risk factors experienced by healthcare workers were determined using the data gathered from the survey forms and were analyzed using One-Way Analysis of Variance (ANOVA). An additional test was created to further confirm which certain risk factor has a different effect from the others and was done by utilizing Tukey's HSD Test. Minitab Software is used for the calculation of these two analyses for an easier and faster calculation. Expert Choice Software is used on structuring the Analytical Hierarchy Process (AHP) for the simplification of the decision process on what is the most critical hospital unit because it provides a structured approach.

2.4 Product Development

Figure 1 presents the different processes needed to successfully develop a new product. Since the product development is just a part of the methodology, its processes will be simplified. The product planning, mission statement and concept development can be put together in one phase, as well as the system-level design and detail design and the testing and refinement, production ramp-up and product launch can be grouped at the last phase.

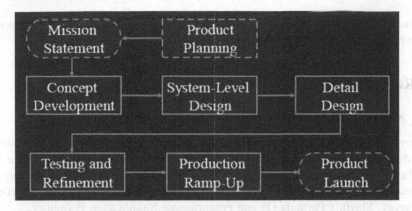

Fig. 1. Product development process

Phase 1:
To address the risk regarding the most critical hospital unit, this research aims to design a patient handling device. Past researches have conducted about the available patient lifting equipment which served as one of the considerations in designing the product and its functions and features. Quality Function Deployment (QFD) aims to translate user demands into design quality as well as the issues and suggestions regarding the existing equipment.

Phase 2:
The functions, features and anthropometric measurements that were developed in the first phase were translated and will be presented using 3D Illustration by SketchUp Software.

Phase 3:
Product validation is necessary for a newly designed product to identify its potential fault, which in turn allows the designer to make improvements. For the evaluation of the proposed design of the patient handling device, successive processes and analysis were done. Design Failure Mode and Effect Analysis (DFMEA) were utilized to ensure that the product was able to meet is requirements and specifications translated using QFD and is also used to identify all the possible failures and errors in the design as well as its respective consequences. Also, to determine how the product satisfies its potential users, an evaluation sheet was given to the respondents that represent the most critical hospital unit. The evaluation sheet demands to rate each product attribute from a scale of 1–5, being 1 as lowest and 5 as highest. Since a prototype will not be available, print-out illustrations were presented to the respondents with its functions, features and specifications as well as the procedures on how to use it.

Another process done to validate the product and measure its effectiveness is to once again perform an ergonomic analysis. It is important to confirm if the product design really helps to lessen the risks encountered by manual patient handling or by using such equipment. Rapid Upper Limb Assessment (RULA) and revised NIOSH Lifting Analysis were performed again, and its results were compared to the previous results. Lastly, product costing was done to discuss the total cost it will take to manufacture one unit of the proposed design, specifically for future use.

3 Results

3.1 Survey and Observation Results

Surveys were utilized to gather information like the demographic profile, feedbacks regarding their usage of the existing equipment and their comfortability from both healthcare workers and patients for further analysis and investigation. There are two formulas utilized to determine the number of patients and healthcare workers needed in the survey: Slovin's Formula (1) and Proportionate Stratification Formula (2).

Patients
Since there are currently 199 patients in the 5 hospital units, the computed sample size is 133. And by using (2), Table 1 shows the results of its computations. The 133 patients consist of 57 male and 76 female patients with an average data of 27, 122.87 lb and 4 feet and 8 in. for their age, weight and height respectively.

Table 1. PCGH patients: population & sampling size

Hospital unit	Population size (N)	Sampling size
Emergency room	14	9
Medicine ward	80	54
OB-GYNE ward	30	20
Pediatrician ward	40	27
Surgery ward	35	23
Total	199	133

Healthcare Workers
There are a total of 167 healthcare workers assigned in the 8 hospital units under study consisting of 115 registered nurses and 52 nursing attendants. Using Slovin's Formula, 118 healthcare workers were computed as the sample size and using this in (2), Table 2 shows the results of its computations. 47 of these 118 workers are male and 71 are female. The average data of workers in terms of age, weight and height are 28, 141.78 lb and 5 feet and 2 in. respectively (Table 3).

From the researcher's observations, twelve manual patient handling activities were being utilized in the hospital which are: Australian shoulder lift, orthodox lift, underarm drag lift, flip turn on bed, through-arm or top and tail lift, front assisted transfers with one worker, three-or-more patient lift, blanket lift, lift using spine/stretcher board, one worker repositioning the patient in a chair, assisted walking with one or two workers and independent walking with or without mobility aids. The first seven activities use upper body extremities, other activities were done using a handling device and the last four are assisting activities done by healthcare workers.

Table 2. PCGH healthcare workers: population & sampling size

Hospital unit	Total registered nurses	Total nursing attendant	Population size	Sampling size
Emergency room	19	9	28	20
Operating room	28	4	32	23
Deliver room	4	8	12	28
Intensive Care Unit	28	12	40	8
Medicine ward	16	8	24	17
OB-GYNE ward	4	4	8	6
Pediatrician ward	8	4	12	8
Surgery ward	8	3	11	8
Total	115	52	167	118

Table 3. Questionnaire results

Alternatives	Frequency	% Frequency
Years in service		
1–5 years	55	46.61%
6–10 years	30	33.90%
More than 10 years	23	19.49%
Daily frequency of lifting, transferring and repositioning		
Less than 10 times	54	45.76%
10–20 times	64	54.24%
Frequency usage of lifting equipment		
Sometimes	87	73.73%
Less frequent	31	30.51%
Workers who experienced pain during manual patient handling		
Yes	82	69.49%
No	36	30.51%
Workers that undergone any medical treatment		
Yes	19	23.17%
No	63	76.83%
Workers who have been absent from work due to patient handling activities		
Yes	27	32.93%
No	55	67.07%
Number of days workers are absent		
1–2 days	22	81.48%
3–4 days	4	14.81%
5 days and above	1	3.70%

3.2   Ergonomic Analysis Results

Body Part Discomfort Scale

The researcher utilized body part discomfort questionnaires to determine the respondent's body part along with the extent concerning their body pain and discomfort. Corlett and Bishop's body part discomfort scale (BPDS) was used to patients while the modified NMQ was used to workers. The patients and healthcare workers were asked to rate their comfortability level on each body part during patient handling operations with respect on how the workers lifts or assist patients along with the use of handling devices. There are 133 patients and 118 healthcare workers who answered the BPDS. They are given a rating of 1–5, 1 as extremely comfortable and 5 as extremely uncomfortable. Table 4 shows the BPDS results for patients and healthcare workers.

Table 4. BPDS results

Body part	Patients		Healthcare workers	
	Rank	Mean rate	Rank	Mean rate
Neck	9	2.04	7	2.31
Shoulder	3	2.77	2	2.9
Arms	5	2.66	3	2.56
Elbow	4	2.71	10	1.97
Wrist/Hand	8	2.07	4	2.43
Upper back	1	4.1	6	2.32
Lower back	2	3.07	1	4
Hips/Thigh	6	2.38	9	2.04
Knees	7	2.08	5	2.38
Ankles/Feet	10	2.02	8	2.28

Rapid Upper Limb Assessment

RULA was done to assess the biomechanical and postural loading on the whole body, concentrating on the neck, trunk and upper limbs. The researcher performed RULA on the 118 healthcare workers and was done on a random period. Since the researcher have time constraint, not all patient handling activities are assessed by RULA on each unit (Tables 5 and 6).

NIOSH Lifting Analysis

To assess the physical stress of two-handed manual lifting jobs, NIOSH lifting equation was utilized. The assessment was done only on lifting activities, particularly the activities that require standing and two hand lift. Among the 12 patient handling activities present in PCGH, there are only 8 activities that met the criteria. The expected value of this assessment determines if the lifting activity is safe or not. The lower the value of lifting index, the lower the risk a worker may experience doing that certain activity.

Table 5. RULA scores per patient handling activity

Activity	Average RULA score	Remarks
1. Australian shoulder lift	6.50	Investigate & change immediately
2. Orthodox lift	6.55	Investigate & change immediately
3. Underarm drag lift	6.67	Investigate & change immediately
4. Flip turn on bed	6.50	Investigate & change immediately
5. Through-arm or top and tail lift	6.50	Investigate & change immediately
6. Front assisted transfers with one worker	6.71	Investigate & change immediately
7. Three-or-more patient lift	6.56	Investigate & change immediately
8. Blanket Lift	6.50	Investigate & change immediately
9. Lift using spine board/stretcher board	6.40	Investigate further & change soon
10. One worker repositioning the patient in a chair	2.00	Acceptable
11. Assisted walking with one or two workers	1.90	Acceptable
12. Independent walking with or without mobility aids	2.00	Acceptable

Table 6. RULA scores per hospital unit

Ward/Unit	Average RULA score	Remarks
Emergency room	6.3	Investigate further & change soon
Intensive Care Unit	5.93	Investigate further & change soon
Operating room	5.87	Investigate further & change soon
Surgery ward	5.25	Investigate further & change soon
Medicine ward	4.94	Investigate further & change soon
OB-GYNE ward	4.88	Investigate further & change soon
Delivery room	4.38	Investigate further
Pedia ward	3.67	Investigate further

The summary of the results is clustered by activity and hospital unit. Table 7 indicates that all lifting activities are considered unsafe for the reason that they got higher percentages in the yellow and red category which according to NIOSH requires redesigning of lifting tasks. Table 8 shows that the most critical unit in the hospital is the Emergency room, followed by the Intensive Care Unit and Operating room.

Table 7. Lifting index percentage per lifting activity

Activity	Lifting index		
	Green category	Yellow category	Red category
Orthodox lift	0.00%	37.50%	62.50%
Underarm drag lift	0.00%	60.00%	40.00%
Flip turn on bed	0.00%	71.43%	28.57%
Through-arm or top and tail lift	11.54%	30.77%	57.69%
Front assisted transfers with one worker	0.00%	38.46%	61.54%
Three-or-more patient lift	0.00%	100.00%	0.00%
Blanket Lift	0.00%	0.00%	100.00%
Lift using spine/stretcher board	0.00%	9.52%	90.48%

Table 8. Lifting index percentage per hospital unit

Activity	Lifting index		
	Green category	Yellow category	Red category
Emergency room	0.00%	30.00%	70.00%
Intensive Care Unit	0.00%	32.14%	67.86%
Operating room	0.00%	43.48%	56.52%
Surgery ward	0.00%	50.00%	50.00%
Medicine ward	11.76%	41.18%	47.06%
OB-GYNE ward	0.00%	66.67%	33.33%
Delivery room	12.50%	62.50%	25.00%
Pedia ward	0.00%	75.00%	25.00%

3.3 Analytic Hierarchy Process

Using Expert Choice Software, Analytical Hierarchy Process was created to determine the most critical hospital unit, in terms of patient handling activities. Table 9 shows the possible risk factors on patient handling activities that were utilized as criteria for the AHP. These factors were based from observations and related literature conducted for the study. While alternatives are consisted of the eight (8) hospital units considered in the study.

Subsequently, pairwise comparison was conducted and answers in the pairwise comparison questionnaire served as inputs and priority weight values were computed as outputs. 118 healthcare workers are chosen to be the respondents in the decision-making process since they are assigned and experiencing the whole operation in handling activities. However, healthcare workers do not undergo job rotation, so this decision-making was derived from the worker's perspective and judgments drawn from their own observations to the other units.

Table 9. Risk factors on patient handling activities

Criteria	Brief description
Patient	This refers to the communication of the patient particularly the speech, vision and hearing, as well as the cognition (decision-making), coordination, physical status and diagnosis/illness
Worker	This refers to the physical status of the worker as well as the experience and training, posture and the frequency of lift/transfer
Equipment	This refers to the capacity to lift of the equipment as well as its dimensions
Environment	This refers to the workspace of the vicinity and the medical equipment installed to the patients (e.g. catheter, oxygen, chest tube, etc.)
Organization	This refers to the time pressure or how much time does the worker needs to do the handling activities (e.g. life-threatening illness, needs immediate operation)

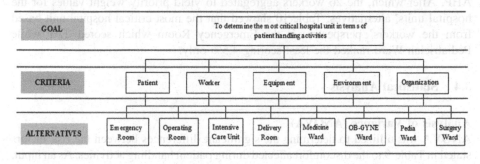

Fig. 2. AHP hierarchy model

The first type of pairwise comparison in this process is between pairs of criteria. Since there are five (5) criteria, ten (10) pairs of criteria were compared relative to its importance to the organizational goal. The second type of pairwise comparison is between the alternatives. Pairs of alternatives were compared against the criteria. Eight (8) hospital units were compared with respect to the given criteria, resulted to twenty-eight (28) pairwise of units. These pairwise comparisons between alternatives were used to obtain ranking of hospital unit relative to each objective. This leads to six (6) types of questions to be weighed by the workers within the survey form. However, this study aims to relate hospital units to the organizational goal. Thus, global priorities were derived by synthesizing the alternatives. After encoding the answers of the respondents to the pairwise comparisons, Fig. 2 shows the translated global priorities wherein hospital units are now related to the organizational goal of the study which is to determine the most critical unit. The figure represents the subjective judgment of Worker 1 and can be interpreted as follows:

Since there is an inconsistency ratio value of 0.17 and the acceptable value is 0.10 and below, a consistency check was applied. Using Expert Choice Software, it is seen that 26 out of 118 workers passed the consistency test therefore these 26 workers were used to develop a general priority weight values for hospital units. Geometric mean was applied to these priorities since it is more consistent with the meaning of priorities in

Table 10. Ranking of hospital unit (most critical to least critical)

Ranking	Hospital unit	Percentage score
1	Emergency room	27.1%
2	Intensive Care Unit	23.2%
3	Operating room	15.5%
4	Surgery ward	10.0%
5	Delivery room	8.4%
6	Medicine ward	6.0%
7	OB-GYNE ward	5.2%
8	Pediatrician ward	4.5%
Total		100%

AHP. After which, the 26 workers aggregated to yield priority weight values for the hospital units' alternatives. Table 10 showed that the most critical hospital unit based from the workers' perspective is the Emergency Room which scored 27% while Pediatrician Ward ranked the least scoring 4.5% only.

3.4   Statistical Analysis

Analysis of Variance (ANOVA)
ANOVA was utilized to determine the significant differences between the risk factors stated in Table 9 to the discomfort affected during patient handling activities. As an input, survey forms regarding how they felt pain or discomfortability per given risk factors and how these factors affect their comfort during patient handling activities. A 5-points scale was given, having 1 as significantly affecting and 5 as no affect. Using minitab software, results showed P-value (0.00) is less than the level of significance (0.05).

Tukey's HSD Test
Since results from ANOVA does not show whether what data specifically is significantly different from each other, Tukey's HSD Test was applied. Since there are 16 risk factors, 120 comparisons were then performed. Grouping information was shown to generalize the results and every risk factor is assigned on either Group A, B, C or D. As a result, equipment dimension, equipment's capacity to lift, experience & training of worker, and diagnosis/illness of patient are significantly different among all the risk factors. There is enough statistical evidence to conclude that there is a significant difference between the risk factors to the discomfort affected during patient handling activities. Also, Tukey's HSD Test indicates that equipment dimensions and capacity to lift has significantly affects higher discomfort among all the risks factors during patient handling, while diagnosis/illness of patient affects the least.

3.5   Equipment Evaluation

Present handling equipment in Emergency Room of PCGH is distinguished in two types, lifting and transferring equipment. Existing lifting equipment in emergency room are spine boards, stretcher boards and blankets while the transferring equipment are wheelchair and stretcher beds. However, not all lifting activities to patients are allowed to use lifting equipment. In some cases, workers would have to lift the patient manually by the use of their extremities. Some activities require the worker to lift the entire patient's body weight while others are just partial lift. Table 11 shows all the equipment present in the emergency room.

Table 11. Equipment evaluation summary

Lifting Equipment	Illustration	Description	Nature	Dimension	Capacity
Spine Board		A board made of hard plastic usually has holes surrounding the board that serves as the handle of the worker; Light-weight, portable and easy to carry-out	Non-Mechanic al Aid	184cm x 46cm x6.5cm (L x W x H)	Max Weight: 159 kg 1 person at a time
Stretcher		A foldable stretcher made up of leather and a two metal stick that serves as the handle of the worker; Light-weight, portable and easy to carry-out	Non-Mechanic al Aid	206cm x 52cm x14cm (L x W x H)	Max Weight: 159 kg 1 person at a time
Blanket		A plain piece of cloth usually in color white, used as a support to lift the patient	Non-Mechanic al Aid	90cm x 45cm (L x W)	Max Weight: N/A 1 person at a time

3.6   Product Design and Development

Quality Function Deployment

By executing QFD through the use of House of Quality, target designs of the proposed product have obtained. Assumptions to be made on the design are: Patients to be lift and transfer are already positioned and can perform instructions correctly and cooperative by the healthcare workers. Appendix B shows the complete HOQ which has the customer and technical requirements, inter-relationship matrix, trade-off triangle, technical priorities and targets. Competitive assessment and benchmarking were not done because there are no available modernized and mechanical lifting equipment available in the Philippines.

Table 12. Technical requirements

Technical requirements	Target design
1. Mechanical type	Battery operated
2. Metal material type	Aluminum, Steel
3. Load capacity	Can carry approximately 100 kg
4. Adjustable	Multi-purpose sling, Adjustable leg support
5. Ergonomically designed	Grip handle design, Soft padded, Anthropometric measurements
6. Light-weight	Not heavier than 32 kg
7. Simple operations needed	Easily Operated
8. Well body support	At least 4 point suspension
9. Mobile	Installed with wheels, Easy to maneuver

Recently QFD and the underlying principles of TQM have been applied to healthcare. One of the driving forces for this approach is the fact that healthcare is one of the fastest growing industries in the service sector [13]. Based from the HOQ, Table 12 discusses what the product may offer as well as the general attributes of the design, it shows the summary on what target design resulted in the HOQ diagram.

Anthropometric Measurements and Detailed Design
By using SketchUp software, the illustrated 3D design of the proposed patient handling device is presented above (Figs. 3, 4, 5 and 6).

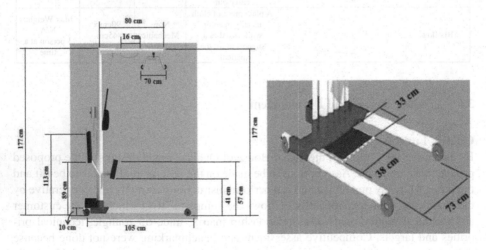

Fig. 3. Anthropometric measurements of the lifter

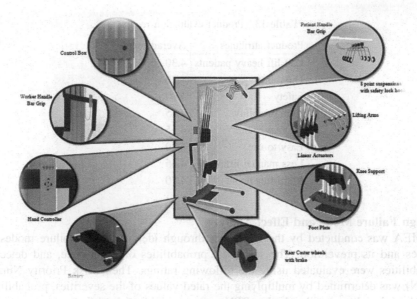

Fig. 4. Patient handling device and its parts

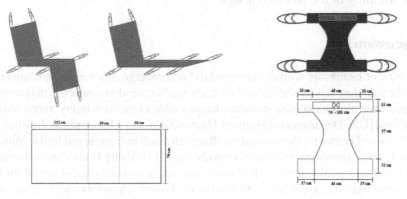

Fig. 5. Universal sling

Fig. 6. Stand aid sling

3.7 Product Validation

Evaluation of the Users

Since prototype testing is not applicable, print-out pictures of the design with brief description of the product are presented using a thorough survey to healthcare workers within the emergency room of PCGH using a sample size of twenty (20). Table 13 shows the evaluation of the potential users. Also, Appendix C to E discusses and illustrates the full specifications and designs of the proposed product.

Table 13. Product evaluation results

Product attributes	Average score
Can lift heavy patients	4.30
Durability	4.35
Safety	4.20
Comfortability	4.15
Maneuverability	4.15
Easy to use	4.20
Less manual lifting	4.45
Multi-functional	4.70

Design Failure Mode and Effect Analysis

DFMEA was conducted by the researcher through identifying the failure modes, its causes and its prevention. The severities, probabilities of occurrence, and detection capabilities were evaluated using the following ratings. Then, Risk Priority Number (RPN) was determined by multiplying the rated values of the severities, probabilities, and detection. Items with a higher RPN promote higher risk to the consumer, and so should be given a higher priority in the design review. These failure modes need the most attention and actions for future improvement of the design. Table 14 shows the complete DFMEA of the proposed design.

4 Discussion

Almost 80% of healthcare workers surveyed did not undergo any medical treatment to address the pain they felt for the reason that many workers tend to ignore the fatigue they experience since musculoskeletal disorders happen after a repetitive injury over a certain period of time (U.S. Department of Health & Human Services). This confirms the study by Op De Beeck and Hermans, that manual handling can result in fatigue, and lead to injuries of the back, neck, shoulders, arms or other body parts [14]. Using Body Part Discomfort Scale, it was found out that the body part which gives patients the most discomfort during handling activities is the upper back, followed by the lower back and shoulders. It was also determined that body part which experiences the most discomfort for healthcare workers are the lower back, followed by the shoulder and arms. The results confirmed a recent literature review which says that in the most recent year of work 55, 44, and 36 percent of nurses reported low back, shoulder, and lower extremity pain, respectively [7]. With these results, the patient handling device to be developed will give emphasis on adding comfort to the body parts (e.g. upper and lower back, shoulder) which promotes discomfort in patients during patient handling activities and will also address the issues of the healthcare workers regarding their lower back, shoulders and arms.

By utilizing Rapid Upper Limb Assessment, nine (9) were identified as activities that promotes risks in developing upper limbs postural disorders, requiring further investigation and immediate changes on current activities. This assessment agrees to two studies in the 1990's which states that those activities involving moving patients within

Table 14. DFMEA results

Item & function	Potential failure mode	Potential effect(s) of failure	Severity	Potential cause(s) of failure	Probability of occurrence	Controls	Detention capability	RPN	Recommended actions
Patient sling	Too large or too small for the patient	Patient may slip out or fall out	9	One size fits all design sling	6	Check patient's body measurements; Multi-colored loops having different lengths, use longer loops for big persons, use shorter loops for small persons	1	54	Assign a worker who looks after the patient during the whole operation
	Damaged, sling, shear, tears up, loose stitching	Patient fal, injury, trauma, facture	9	Excessive weight; Too much usage; During washing/laundry	5	Check the sling condition as well as patient's weight before using		45	Always check the sling before using; Replace the sling immediately if damaged; Warning labels on weight capacity as well as in washing instructions
Patient's handle bar grip	Too high, too low, too near or too far from the user	Uncomfortable lift, no adequate support	8	Equipment is not positioned properly to the patient	5	Check the distance of the equipment to the patient particularly the lifting arms	1	40	Double check patient's position before lifting; Assign a worker who looks after the patient during the whole operation
Control box	Equipment not operating properly or not working at all	Patient may stuck up on a position; may delay or add time for the lift	6	Insufficient battery power	5	Check the battery power indicator	1	30	Always monitor battery power; Charge to power outlet source if low battery; replace battery if outdated
				Short circuit; Power failure	4	Check patient's weight	2	48	Warning labels; Include emergency features
Lifting arms & suspension	Deforms or breaks	Patient falls, injury, trauma, facture	9	Excessive weight; Too much usage	5	Check the metal parts and patient's weight before using	1	45	Periodically check the metal parts; Warning labels; Replace metal parts if deforms/breaks

(continued)

Table 14. (*continued*)

Item & function	Potential failure mode	Potential effect (s) of failure	Severity	Potential cause (s) of failure	Probability of occurrence	Controls	Detention capability	RPN	Recommended actions
Knee support	Too high, too low, too near or too far from the user	Uncomfortable lift	4	Knees are not positioned properly	5	Check the patient's knee before lifting	1	20	Double check patient's position before lifting; Assign a worker who looks after the patient during the whole operation
Foot plate	Too high, too low, too near or too far from the user	Uncomfortable lift	4	Feet are not positioned properly	5	Check the patient's feet before lifting	1	20	Double check patient's position before lifting; Assign a worker who looks after the patient during the whole operation
	Damaged, broken down	Patient falls, injury, trauma, facture	8	Excessive weight; Too much usage	5	Check patient's weight	1	40	Warning labels
Caster wheels	Wheels are not functioning well, falls off or broken out	Difficulty in maneuvering; may delay or add time for the lift	6	Excessive weight; Too much usage; Accumulated dirt inside the wheel	5	Check patient's weight as well as the wheels; Avoid passing on uneven floors	1	30	Periodically clean the wheels; Replace caster wheels

bed and to/from the bed are most often reported as contributing to low back pain [15, 16]. The results of NIOSH Lifting Index also entail that all lifting activities are considered unsafe for the reason that they got higher percentages in the yellow and red category which according to NIOSH requires redesigning of lifting tasks. Since it is not possible to design a product that will address all types of manual handling activities, the researcher aims to design a product prioritizing the most crucial activities first, then as many activities as can be to equip on the product. The most critical patient handling activities which are: through-arm or top and tail lift, front assisted transfers with one worker, blanket lift and lift using spine/stretcher board. According to Ergonomics Center, these activities were considered the most critical since they got the highest percentage in the red category which entails that it exceeds the capabilities of safely performing the lift for nearly all workers so redesigning of the lifting task is recommended.

Moreover, after executing Analytical Hierarchy Process using expert choice software, the researcher was able to identify the most critical hospital unit in terms of patient handling activities. The result was primarily confirmed through ergonomic assessment performed like RULA and NIOSH Lifting Equation. Therefore, it is concluded that the emergency room is the most critical hospital unit.

Hence, the design should address the three most common set-ups for the most critical activities in the emergency room which are the following: sit to stand movements, lifting the patient from a sitting position and lifting the patient in a lying position. The current results are consistent with earlier evidence: McCoskey's study was designed to describe the type, frequency and physical demands of the patient-handling tasks in an acute care facility [17]. Repositioning transfers accounted for 47% of all transfers. It was further reported that over 50% of all lateral transfers, including repositioning transfers, required moderate or greater physical exertion. McCoskey's findings were also consistent with those of who found that nurses experience the greatest amount of physical stress when repositioning patients [18].

Although use of equipment can help decrease physical demands and lost workdays [19] it appears that in the majority of cases, nurses do not use equipment. The high prevalence of injuries may not be surprising considering how frequently paramedics are required to lift, carry, reposition, and transport patients using equipment including stretchers, stair chairs, back boards, transfer sheets, etc. [20]. These literatures confirm the state of the hospital because there are only three (3) lifting equipment present in the hospital are non-mechanical aid and are utilized by simple mechanics. Traditionally, stretchers have been manually operated by a team consisting of two paramedics. Stretchers are operated to: lower or raise the cot surface to facilitate patient transfer, transport, or repositioning; push or pull the stretcher while transporting a patient from the scene to the rear of the ambulance or from the ambulance into a hospital; and, to load and unload the stretcher into and out of the rear of the ambulance. The physical exposures to these activities, considering trunk strength and spine compression are high [21, 22, 23] in some cases described as exceeding well-established threshold limit values such as the NIOSH action limit for spine compression [24]. The usage of the blanket is almost the same with the spine and stretcher board; except for the placing of the device. Sometimes, blanket refers to the bed sheet within the patient's bed so the worker would not be needing to place the equipment under the patient's body but would just to adjust it on proper position. Thus, it will serve as their handle to move the patient.

Mechanical type is the topmost priority for the technical attributes of the proposed design. It is chosen since the product's objective is to lessen the human effort exerted to operate an equipment that could reduce the risks of body discomforts healthcare workers experiences. One study shows that the power load functionality significantly reduced peak low back compression compared to loading a manual stretcher [25] because several studies have shown that a large percentage of nurses report low back pain due to work-related activity, primarily from manual patient-handling [1, 26, 27]. The material to be used must be lightweight so that the device will be easy to maneuver but must be durable as well and can stand the target load capacity, which is 100 kg. Moreover, it should be comfortable with the primary and secondary users. Thus, the types of metals to be used are aluminum and steel; these two are most widely used metals for its versability according to Aalco on its introduction to aliminum. It is recyclable, lightweight, strong, non-toxic, and inexpensive, easily machined and accepts many types of finishes [28]. With these types of materials, the product design will be adjustable since a study concluded that adjustable furniture is currently being preferable due to its capability of increasing the comfortability while reducing the chance of MSDs in the long run [29]. It also has a multi-purpose sling to address different lifting positions which address the most critical patient handling activities. A study also verified that 96% of the nurses who had input on equipment selection and training prior to use, rated transfer equipment as extremely effective [11]. So, the design aims to add comfort and safety to the future users, with proper training and guidance. As much as possible, the device should be easily operated so that the workers would not find any difficulty in controlling the equipment. Moreover, it will be easier for the worker to control the device and will not consume a lot of time when using since the device is for the emergency department.

Anthropometric measurements whenever be considered for designing helps healthcare workers in achieving comfort level and reduces musculoskeletal disorders [30]. The patient handling device was aimed to address the needs of the Filipino healthcare workers therefore, all the necessary data needed for the anthropometric measurements of the product were based on the study of Jinky Leilanie Del Pradu-Lu which is entitled "Anthropometric Measurement of Filipino Manufacturing Workers". In this study, anthropometric data for standing, sitting, hand and foot dimensions, breadth and circumference of various body parts and grip strength were present. Another major consideration in designing the patient sling is the patient's safety and comfort. The patient must ensure to be in good sitting position, as well as lying position. Patient's body breadth and tallness must be fit on the sling. Furthermore, to add security and to distribute the weight evenly, the researcher decided to have an 8 loops sling with different sizes. The researcher came up on a design which is s stretcher sling that can be fold as ordinary sling. Since the sling would be up to the head of the patient, it will be a high back sling. Sling loops are basically the strings to be attached on the safety hook of the suspensions. In the proposed design, there will be eight (8) sling loops since it is the common number of loops present in a sling. As the target design of the stand aid sling, the researcher decided to integrate the shoulder strap and buttock strap in one in order to add security and safety for patients. Because present designs, buttock straps are only used for large or heavy patients (Table 15).

Table 15. Materials used

Materials used	Brief description
Electric line actuator	A type of motor that is responsible for moving or controlling a mechanism or system. Linear actuators provide safe, clean and quiet movement with accurate motion control
Aluminum and steel	Ideal structure types of metals to be used are round tubular since most parts are adjustable. Moreover, the ideal thickness for aluminum tubes and stainless steels are 8 mm and 4 mm, respectively. The mechanical properties of the materials are the most concern since it entails the hardness, elasticity, strength and toughness of the material
Polyster fabric	Polyester fabric is ideal as a material used for the slings because of its high tenacity and durability. Also, padded fabric is ideal to promote comfort for the patient
Battery	Lead Acid Rechargeable Battery, Rated Capacity: 80Ah, Charging Voltage: 12 V DC; Battery Charger; Power Cord

The proposed device, as evaluated by the primary users, shows versatility of functions of the product shows great satisfaction and helpful to workers, while comfort level and maneuverability as the least satisfaction. Through the evaluation, the overall rating of the product scores 4.31, having 5 as the highest. Henceforth, the product is a success since it scored above average rate. Furthermore, suggestions obtained from potential users were as follows: (a) much better if it is foldable equipment which can be used in wards with limited spaces and ambulatory services; and (b) higher weight capacity must be set, possibly 135 kg (300 lb) because it is approximately the average maximum weight of patients. After identifying the detail design of the proposed device, all the possible failures were listed for the researcher to address such failures. Presented below are the possible issues and its preventive measurements:

(a) Patient sling might be too large or too small for the patient on which patients may slip out or fall out since the sling design promotes a one size that could fit all. However, the design already has controls on which it offers sling loops having different lengths, longer loops for bigger persons while smaller loops for smaller persons. As further actions, assigning a specific worker who will look after the patient during the whole operation is necessary but since it is a newly developed worker, finding staff to fulfill these requirements may be difficult and even considered unfair [32].

(b) Equipment is not operating properly or not working at all which is caused by power failure and/or short circuits. This may delay the lift as well as the patient may be stuck on a particular lifting position. To avoid this, warning label regarding on the weight limit is recommended as well as including emergency features on the equipment

(c) Patient slings that are damaged, sheared, tear up, and has loose stitching which are caused by excessive weight, too much usage and during washing/laundry. As a recommendation, the sling must always be checked before using and must be replaced if damaged. Also, warning labels on weight capacity as well as in washing instructions must be noted.

(d) Lifting arms & suspension may deform or break which may result to patient falls, injury, trauma, facture. This may be caused by excessive weight and too much usage of the equipment. With these, maintenance check as well as placing warning label on safe working load capacity is necessary. Also, replacing of metal parts must be done if ever deform or break happen in the future.

5 Conclusion

The study analyzed the problems of healthcare workers and patients associated with manual patient handling and as a solution, a patient handling device was developed. The study results showed that manual patient handling have certain effects for both healthcare workers and patients because of the presence of unsafe lifting and incorrect position of patients. This research successfully identified the most critical hospital unit using analytic hierarchy process and the most critical patient handling activities using NIOSH, RULA and BDPS that was further used as a basis in developing a patient handling device.

It is thus inferred that manual patient handling can cause musculoskeletal disorders and discomfortness for both healthcare workers and patients and to address this problem, a mechanized patient handling device for both healthcare workers and patients was developed and was able to capture all identified critical patient handling activities as well as 90% of the total population of the patients. The effectiveness of the device were tested by performing ergonomic assessment and compared it to the results before the intervention, and it was identified that the device lessen the risks in performing patient handling activities. Moreover, it is found out through evaluation that the design offers more functions and features than the available lifting equipments, plus it offers a more cost affordable price. This study also addresses the lack of research in developing a patient handling device that considers both the needs of primary and secondary users which are thehealthcare workers and patients. However, the device is ready to use assuming that all factors are in stable condition and any intangible and/or uncontrollable factors may hinder the effectivity of the proposed design.

As a recommendation, future researchers should conduct ergonomic analysis and focus on the body parts not considered in this study, especially the lower parts of the body and activities that do not require motion of arms, hand and wrists. Also, a presentation of a more detailed design in terms of engineering and technical discussions of the proposed design, as well as mathematical computations which determines the safety of the patients. Lastly, an actual prototype of the product to provide more accurate product evaluation is much recommended for the testing to see how the device will help both users.

References

1. Nelson A (2007) Evidence-based practices for safe patient handling and movement
2. Lawler E (2008) Safe patient handling and movement during the perioperative continuum: mobile surgical platform use. Perioperative Nurs Clin 3:27–33. https://doi.org/10.1016/j.cpen.2007.11.002

3. Kee D, Rim S (2007) Musculoskeletal disorders among nursing personnel in Korea 37: 207–212. https://doi.org/10.1016/j.ergon.2006.10.020
4. Gabbe S, Melville J, Mandel L, Walker E (2002) Burnout in chairs of obstetrics and gynecology: diagnosis, treatment, and prevention. Am. J. Obstet. Gynecol. 186:601–612
5. Conrad KM, Reichelt PA, Lavender SA, Gacki-smith J, Hattle S (2008) Designing ergonomic interventions for EMS workers: concept generation of patient-handling devices. Appl Ergon 39:792–802. https://doi.org/10.1016/j.apergo.2007.12.001
6. Zhou J, Wiggermann N (2017) Ergonomic evaluation of brake pedal and push handle locations on hospital beds. Appl Ergon 60:305–312. https://doi.org/10.1016/j.apergo.2016. 12.012
7. Villarroya A, Arezes P, De Freijo SD, Fraga F (2017) Validity and reliability of the HEMPA method for patient handling assessment. Appl Ergon 65:209–222. https://doi.org/10.1016/j. apergo.2017.06.018
8. Elnitsky CA, Lind JD, Rugs D, Powell-cope G (2014) Implications for patient safety in the use of safe patient handling equipment: a national survey. Int. J. Nurs. Stud. 51:1624–1633. https://doi.org/10.1016/j.ijnurstu.2014.04.015
9. Guo Z, Bei R, Mun K, Yu H (2017) Experimental evaluation of a novel robotic hospital bed mover with omni-directional mobility. Appl Ergon 65:389–397. https://doi.org/10.1016/j. apergo.2017.04.010
10. Le Bon C, Forrester C (1997) An ergonomic evaluation of a patient handling device: the elevate and transfer vehicle. Appl Ergon 28(5–6):365–374
11. Lee S, Lee JH (2017) Safe patient handling behaviors and lift use among hospital nurses: a cross- sectional study. Int. J. Nurs. Stud. 74:53–60. https://doi.org/10.1016/j.ijnurstu.2017. 06.002
12. Lim PC, Tang NKH, Lim PC, Tang NKH (2005) Managing service quality emerald article: the development of a model for total quality healthcare techniques. The development of a model for total quality healthcare
13. Taylor P, Garg A, Owen B, Beller D, Banaag J, Beller D (n.d.) A biomechanical and ergonomic evaluation of patient transferring tasks: bed to wheelchair and wheelchair to bed, pp 37–41
14. De Beeck, RO, Hermans, V (2000) Research on work-related low back disorders. European agency for safety and health at work. ISBN 92 950007 02 06
15. Harber P, Billet E, Gutowski M, SooHoo K, Lew M, Roman A (1985) Occupational low back pain in hospital nurses. J Occup Med 27(7):518–524
16. Cato C, Olson D, Studer M (1989) Incidence, prevalence, and variables associated with low back pain in staff nurses. Am Assoc Occup Health Nurs J 37(8):321–327
17. McCoskey K (2007) Ergonomics and patient handling. AAOHN J 55(11):454–462
18. Sheth R, Margaret K, Rivera AJ (2017) Qualitative ergonomics/human factors research in health care: current state and future directions. Appl Ergon 62:43–71. https://doi.org/10. 1016/j.apergo.2017.01.016
19. Doss R (2017) Investigating the effectiveness of posture coaching and feedback during patient handling activities in a student nursing population
20. Coffey B, VanderGriendt C, Fischer SL (2016) Evaluating the ability of novices to identify and quantify physical demand elements following an introductory education session: a pilot study. Appl Ergon 54:33–40
21. Cooper G, Ghassemieh E (2007) Risk assessment of patient handling with ambulance stretcher using biomechanical failure criteria 29:775–787. https://doi.org/10.1016/j. medengphy.2006.08.008

22. Marras WS, Knapik GG, Ferguson S (2009) Clinical Biomechanics Loading along the lumbar spine as influence by speed, control, load magnitude, and handle height during pushing. Clin. Biomech. 24:155–163. https://doi.org/10.1016/j.clinbiomech.2008.10.007
23. Mawston GA (2012) The effect of lumbar posture on spinal loading and the function of the erector spinae: implications for exercise and vocational rehabilitation. N Z J Physiotherapy 40:135–140
24. Waters T, Putz-Anderson V, Garg A, Fine L (1993) Revised NIOSH equation for the design and evaluation of manual lifting tasks. Ergonomics 36(7):749–776
25. Hayashi S, Katsuhira J, Matsudaira K, Maruyama H (2016) Effect of pelvic forward tilt on low back compressive and shear forces during a manual lifting task. Phys Ther Sci 28(3):802–806
26. Wulff B, Dalgas S, Louis L, Sørensen J (2017) A multi-component patient-handling intervention improves attitudes and behaviors for safe patient handling and reduces aggression experienced by nursing staff: A controlled before-after study. Appl. Ergon. 60:74–82. https://doi.org/10.1016/j.apergo.2016.10.011
27. Van Niekerk S, Louw QA, Hillier S (2012) The effectiveness of a chair intervention in the workplace to reduce musculoskeletal symptoms. A systematic review
28. Wilson I, Desai DA (2016) Anthropometric measurements for ergonomic design of students' furniture in India. Eng Sci Technol Int J, 4–11. https://doi.org/10.1016/j.jestch.2016.08.004
29. McGill SM, Kavcic NS (2005) Transfer of the horizontal patient: the effect of a friction reducing assistive device on low back mechanics. Ergonomics 48(8):915–929
30. Rivilis I, Van Eerd D, Cullen K, Cole DC, Irvin E, Tyson J, Mahood Q (2008) Effectiveness of participatory ergonomic interventions on health outcomes: a systematic review. Appl Ergon 39:342–358
31. Caska BA, Patnode RE (2000) Reducing lower back injuries in VAMC nursing personnel. Research report#94 136. Veterans Health Administration

Up to Our Elbows in Ergonomics: Quantifying the Risks of Bovine Rectal Palpations

Robyn Reist[1]([✉]), Brenna Bath[1,3], Murray Jelinski[2],
and Catherine Trask[1]

[1] Canadian Centre for Health and Safety in Agriculture,
University of Saskatchewan, Saskatoon, Canada
robyn.reist@usask.ca
[2] Western College of Veterinary Medicine,
University of Saskatchewan, Saskatoon, Canada
[3] School of Rehabilitation Science,
University of Saskatchewan, Saskatoon, Canada

Abstract. Musculoskeletal disorders (MSD) are common in food animal veterinarians. A recent mail survey of Canadian bovine veterinarians found that discomfort was most common in the shoulder, and that bovine rectal palpations were considered one of the most physically demanding tasks. The present preliminary pilot further analyzed the potential association between MSD and rectal palpations using both survey data and observational methods. Statistical analysis of the mail survey results found a non-significant increased odds of shoulder trouble when performing high numbers of rectal palpations, significant protective effects due to years of experience, number of co-workers, and being female, and significant increased odds of shoulder trouble when devoting under 50% of practice time to cattle. A field ergonomic assessment of veterinary bovine rectal palpations using the Rapid Upper Limb Assessment (RULA) method identified several awkward postures associated with the task and consistent scores greater than 6. The combined results suggest bovine rectal palpations hold the potential for hazardous biomechanical exposure if performed at high frequencies, and warrants further study in order to develop prevention strategies.

Keywords: Musculoskeletal disorders · Veterinarian · Exposure assessment

1 Introduction

A number of studies have been performed investigating the occurrence of musculoskeletal disorders (MSD) in large animal veterinarians and other practitioners, such as veterinary assistants, who work with animals. The first documented cases of MSD in bovine practitioners were the subject of a short report in the Canadian Veterinary Journal by Ailsby in 1996 [1]. The issue has since been studied in greater depth and on a national scale in many countries [2–6]. Large animal (i.e. livestock) veterinarians in particular stand out in the literature as having an increased risk for MSD. Research performed in Germany and New Zealand found that large animal veterinarians reported higher numbers of MSD in nearly all body regions than small or mixed animal veterinarians [2, 5].

© Springer Nature Switzerland AG 2019
S. Bagnara et al. (Eds.): IEA 2018, AISC 820, pp. 639–649, 2019.
https://doi.org/10.1007/978-3-319-96083-8_81

640   R. Reist et al.

A recent study focused specifically on Western Canadian bovine veterinarians. Its objectives were to quantify the prevalence of MSD among Canadian bovine practitioners, describe their impact on veterinary work, and identify the most physically demanding tasks to be studied via ergonomic assessments in future phases of the project. The survey results indicated that rectal palpations and other obstetric tasks (e.g. calvings) were considered by the respondents to be the most physically demanding. Nearly 90% of respondents reported experiencing musculoskeletal symptoms in the past 12 months, with the most common body region affected being the shoulder (63.9%) [6].

The task of bovine rectal palpations is performed thousands of times per year by Western Canadian bovine practitioners [6]. Rectal palpations are performed primarily to confirm pregnancy status in beef and dairy cattle, but are also required for artificial insemination and bull breeding soundness evaluations. It requires the practitioner to insert their arm inside the rectum of the cow (Fig. 1). This may expose workers to awkward posture and high forces, potentially exacerbated by the repetitiveness of the task.

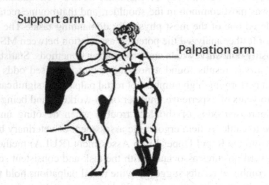

Fig. 1. Veterinarian performing a rectal palpation. Non-palpating arm ("support arm") often holds tail of animal for support during task.

New Zealand veterinarians surveyed by Scuffham et al. identified rectal palpations as a task that they hypothesized to lead to musculoskeletal discomfort, and cited the following reasons, in order: repetitive activity, position and activity of upper limbs, frequency of procedures, awkward posture, and poor facilities/slip hazards [7]. Repetition, frequency, and positioning of the upper limb were also suggested by Ailsby as musculoskeletal injury risk factors arising from rectal palpations [1]. In a survey of American bovine veterinarians, Cattell found increased odds of MSD symptoms occurring on the side of the body with the palpation arm, as well as highest prevalence of symptoms in the shoulder as compared to other body parts [3]. Kozak et al. found a significant association between number of rectal palpations performed per year by German veterinarians and severe elbow and hand/wrist MSD [5].

While there is a growing body of literature on MSD in veterinarians, the majority of articles are based on self-report surveys with low response rates, sometimes lower than 10% [4]. As well, no study ever appeared to lead to a "second phase" where results were used to identify specific MSD risk factors associated with veterinary-specific tasks and suggest prevention strategies. Thus, there is a clear need to fully investigate the relationship between veterinary tasks and MSD with an overarching goal to go beyond previous studies and use the results to identify risk factors, perform ergonomic assessments, and suggest prevention strategies.

This study is part of a larger project that aims to quantify the MSD risk factors associated with bovine rectal palpations so that adequate, tailored interventions can be developed. The objectives of this study were two-fold. The first objective was to identify predictors for musculoskeletal symptoms in the shoulder over a 12-month period among large animal veterinarians. The second was to perform a pilot ergonomic assessment of the task of bovine rectal palpations using the Rapid Upper Limb Assessment (RULA) [8] method in order to begin to quantify the specific ergonomic risk factors associated with this task, and determine if RULA is an appropriate method for assessing the task.

2 Methods

This study consisted of two distinct parts: regression analysis of survey results to identify risk factors for shoulder MSD, and a pilot observational ergonomic assessment to complement the findings from the survey analysis. A more detailed description of survey methods can be found in the descriptive analysis by Zeng et al. [6].

2.1 Target Population and Recruitment

The Western Canadian Association of Bovine Practitioners (WCABP) is an organization representing bovine veterinarians in the western provinces of Canada; 262 practicing and retired members of the organization were targeted to be participants in this study [6].

To create awareness among the target population, an oral presentation describing the upcoming study was presented at the annual WCABP conference in early 2017. Surveys were initially mailed to potential participants in spring 2017 via the WCABP quarterly newsletter [6]. Another presentation was given at the 2018 WCABP conference describing preliminary survey results and the next stages of the project, at which time several veterinarians volunteered to participate in ergonomic assessments.

2.2 Part 1 - Survey

Design

The cross-sectional mixed mode survey consisted of 25 questions regarding personal characteristics, work experience and work tasks, and general musculoskeletal health questions. An adapted Standardized Nordic Questionnaire [9] was used to determine the specific body region where musculoskeletal trouble (defined as "ache, pain, discomfort") was experienced over the past 12 months and its impact on work and quality of life. The survey was approved by the University of Saskatchewan Biomedical Ethics Board.

Data Collection

Participation in the survey was anonymous and voluntary. Participants had the option to return a paper copy of the survey or complete it online. In total, the survey was mailed three times to potential participants; every six weeks after a mailing, a follow-up mailing was conducted to those who had not responded. Email and social media reminders were periodically sent out to potential participants by the WCABP administration. The data was initially used for Zeng et al.'s descriptive analysis; no new data was collected for this secondary analysis. The overall survey sample size was n = 133 which corresponded to a response rate of 51% [6].

Statistical Analysis

Logistic regression was conducted using SPSS version 23.0 (IBM Corporation, 2015). The outcome (dependent) variable was experiencing shoulder trouble in the past 12 months (yes/no). The primary risk factor was considered to be number of palpations performed in a year. Because the range of this variable was extremely large and right-skewed [6], it was transformed into a binary variable either greater or less than the median number of palpations (8200). Other potential independent predictor variables were selected based on biological plausibility and known risk factors for shoulder MSD, such as age, sex, and anthropometry (e.g. height, body mass index) [10]. Bivariate analyses using simple logistic regression were performed for each independent variable; variables with a threshold p-value < 0.25 or considered to be biologically relevant were considered for inclusion in the final multiple logistic regression model. Variables were a mix of continuous and categorical; variables that were originally continuous but not significant were then categorized to check for improved significance. If no improvement emerged the continuous version of the variable was used. The level of significance for the final model was p < 0.05. Multicollinearity was assessed using Pearson correlation coefficients; when variable pairs had a coefficient $\rho > 0.5$ only one was offered to the final model. Significant variables were retained in the final model with the primary risk factor. Potential interactions between all variables in the final model were tested for using the likelihood ratio test. Independent variables chosen from the bivariate analysis were also checked for confounding status by comparing adjusted and crude odds ratios (OR) for the primary risk factor. OR and 95% confidence intervals (CI) were calculated for each variable in the final model to describe the strength of the association. The Hosmer-Lemeshow test was used to confirm goodness-of-fit for the final model.

2.3   Part 2 - Ergonomic Assessment

Data Collection

One bovine veterinarian was shadowed while performing rectal exams on 45 beef cattle in November 2017, over a period of approximately two hours. In total, 38 manual palpations and 10 ultrasound exams were performed by the veterinarian (see Fig. 2). The exams were performed in what the veterinarian described as near-optimal conditions, as there was access to a squeeze chute and multiple helpers, including a veterinary student to share the workload. The veterinarian was also taking blood samples from the cattle, and thus there were breaks of two or more minutes between palpations. Video recordings were collected by two observers. A follow-up assessment was conducted with the veterinarian at a later date to assess "inside the cow" wrist posture and conduct. Estimates of the palpation entry force by the veterinarian during the task were estimated in a laboratory setting using a microFET® 2 Digital Handheld Dynamometer (Hoggan Scientific, LLC, UT).

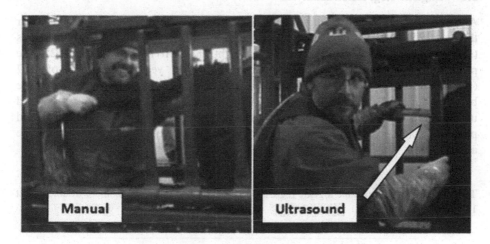

Fig. 2. Veterinarian demonstrating manual and ultrasound exams

RULA Analysis

The Rapid Upper Limb Assessment method (RULA) [8] was used to assess the potential exposure to UEMSD risk related to each rectal palpation. This method assesses the upper limb and the rest of the body as two separate regions, and assigns both posture and muscle use/force load scores to each region. The maximum posture score that can be attributed to each region of the body is 9, with a lower score corresponding to more neutral postures. The maximum combined muscle use and force/load score is 4, which may correspond to shocks or heavy forces combined with either long static or repetitive movements. These values dictate a final score which corresponds to the level of MSD risk associated with the task.

Video recordings and screenshots of the task were reviewed while completing Cornell University's RULA Employee Assessment Worksheet [11]. As different left and right arm/wrist postures occur during each palpation, RULA was used twice for each task with a separate analysis for each upper limb. Scores were assigned to the positions of the upper arm, lower arm, wrist, neck, trunk, and legs, as well as for muscle use and force/load encountered simultaneously by the arm/wrist and neck/trunk/legs. As the lower body was not consistently visible in the video playback, a typical lower body posture score was calculated for the lower body and applied to all palpations. Scores for "inside the cow" wrist posture and force/load were estimated via an interview with the worker and laboratory microFET® 2 measurements, and single estimates for these were applied to all palpations. Average, minimum, and maximum scores were calculated for the arm and wrist analyses (palpation and support arms) and the final score.

3   Results

3.1   Logistic Regression Analysis

As previously reported by Zeng et al., the overall prevalence for shoulder trouble over the past 12 months was 63.9% (85 out of 133 respondents) [6]. Physical characteristics (sex, dominant writing and palpation hand, height, and BMI) had a p-value > 0.25. The primary risk factor, average palpations performed per year, had an OR of 1.67 ($p = 0.17$) for respondents who performed more than 8200 palpations per year in the bivariate analysis. Years of experience, age, perception of overall health, ultrasound use, number of veterinary colleagues (i.e. other veterinarians in the practice) and percent of practice devoted primarily to bovine work had a p-value < 0.25 in bivariate analysis.

Table 1. Final multivariable logistic regression model for prevalence of shoulder trouble in last 12 months

Variable	Adjusted OR (95% CI)	p-value
Average palpations per year		
≤ 8200	1	
>8200	1.58 (0.65–3.82)	0.31
% of practice bovine work		
51–100	1	
0–50	2.83 (1.01–7.93)	<0.05
Sex		
Male	1	
Female	0.34 (0.12–0.98)	<0.05
Years of experience	0.93 (0.89–0.97)	<0.001
Number of veterinarians in practice	0.83 (0.70–0.98)	0.03

Variables meeting the p < 0.25 criteria from the bivariate analysis were candidates for inclusion in the final model. As sex and BMI are important known predictors of MSD [10], a decision was made to include them in the final model-building process despite not reaching the threshold of p < 0.25. The final model is described in Table 1.

Age was removed from the final model as it strongly correlated ($\rho > 0.9$) with years of experience. Ultrasound use, perception of overall health, and BMI were not retained as they did not significantly contribute to an improved fit of the final model. The OR for performing >8200 palpations per year was 1.58 (0.65–3.82), as compared to ≤ 8200; this was the only non-significant predictor in the model. The continuous variables "years of experience" and "number of veterinary colleagues" had slightly protective OR's of 0.93 (0.89–0.97) and 0.83 (0.70–0.98), respectively. Sex (female compared to male) had an OR of 0.34 (0.12–0.98). Devoting 50% or less of one's practice exclusively to bovine work had an OR of 2.83 (1.01–7.93) compared to those whose practice was 51% bovine work or more. No significant interactions were observed, but some degree of confounding on the primary risk factor (i.e. number of palpations per year) was observed for sex (13%) and percentage of practice devoted to bovine work (25%).

3.2   RULA Assessment

The average peak force on the palpation arm during manual palpations was estimated to be 97.9 N. The average maximum force on the palpation arm (arm holding the wand) during ultrasound palpations was estimated to be 46.7 N. Forces on the support arm for both manual and ultrasound palpations were considered negligible during most exams. "Inside the cow" wrist posture for manual palpations was scored as 3, corresponding to a flexion of 16° or more. Neck/trunk/leg posture was scored a 7 for all palpations, with a muscle/force score of 4 added for manual palpations and 0 for ultrasound palpations. These "typical" values obtained during force and posture matching assessments with the participant were added to the individual upper limb scores for the palpations available in video playback.

Table 2 presents upper limb posture and RULA assessment results for measured rectal exams. For manual palpations, more extreme postures were observed in the support arm than the palpation arm (5.1 vs 4.7), though the average muscle use and force/load scores increased the final scores for the palpation arm in comparison to the support arm (7.0 vs. 6.7). For ultrasound exams, more extreme postures were observed in the examination arm (4.1 vs. 3.6), however the upper limb posture scores were considerably lower for ultrasound exams compared to manual. The final RULA scores for both types of task were >6; the average score for manual palpations (averaging the full body scores for both arms) was 6.9 and the average score for ultrasound palpations was 6.6. The average RULA score when including the manually palpating arm only was 7, which is the maximum score RULA score possible, corresponding to the recommendation "investigate and change immediately" [11].

The average time to complete the task (i.e. duration of examining inside the rectum) was 15 s for manual palpations (standard deviation = 6.2 s) and 16 s (standard deviation = 11.9 s) for ultrasound exams. The average break time between palpations was 2 min; this time was typically spent obtaining a blood sample from the cow and changing the sanitary sleeve.

Table 2. Rapid Upper Limb Assessment (RULA) scores for types of bovine rectal exams

Arm (palpation or support) and palpation type (manual or U/S)	Number of observations	Posture score			Final RULA total score			Average RULA score
		Avg.	Max.	Min.	Avg.	Max.	Min.	
Palpation - Manual	37	4.7	7	4	7.0	7	7	6.9
Support - Manual	35	5.1	6	2	6.7	7	5	
Palpation - U/S	9	4.1	5	2	6.7	7	6	6.6
Support - U/S	9	3.6	5	3	6.4	7	6	

4 Discussion

This is the first published work attempting to quantify MSD risk factors specifically associated with both bovine rectal palpations and shoulder trouble in large animal veterinarians. While prevalence of shoulder trouble over the past 12 months and quantity of rectal palpations performed in a year were established in the previous report of WCABP members' MSD experience by Zeng et al. [6], the present study expanded the analysis to identify potential predictors of shoulder trouble, and began to investigate the mechanisms for this using a RULA analysis of bovine rectal palpations under optimal conditions.

Anthropometric characteristics of survey respondents were not found to be significantly associated with shoulder trouble. The majority of respondents were male, over 40, tall (>177 cm), self-rated their health as "very good" or "excellent", and had a high BMI (>25 kg/m^2). It is possible that selection bias occurred as veterinarians experiencing MSD self-selected more readily into the study. The small sample size in combination with these factors may have created a non-ideal sample for investigating significant associations between physical characteristics and shoulder trouble.

Sex was a significant predictor of shoulder trouble, though the protective OR for women is unusual and may be confounded by work-related factors. Existing literature generally shows trends of women experiencing a higher prevalence of work-related MSD than men, even when tasks performed are identical [12]. In this study, the crude rate of shoulder trouble was 60% for women and 65% for men. While sex was not found to be correlated with any of the other variables, female respondents did perform fewer palpations per year on average than the male respondents; only 32% of the female respondents performed more than 8200 palpations per year vs. 53% of male respondents. The variable "percentage of practice devoted to bovine work" was not observed to interact statistically with sex, but the differences between men and women in terms of this variable may also help explain the protective OR. More male respondents (73 men, 16 women) reported devoting 51% or more of practice time to bovine work; this corresponds to this category being composed of 18% women and 82% men. The distribution of practice for female veterinarians in this survey is similar to a 2015 survey of Western Canadian veterinarians by Jelinski and Barth, who reported that 53% of mixed animal veterinarians and 23% of food animal veterinarians were women. They also reported that female veterinarians were more likely to work part-time than their male counterparts [13]. Thus, it is likely that the protective OR for women in this study is due in part to indirectly reducing exposure by diversifying their practice and/or working part-time, in comparison to the male veterinarians. Unfortunately the survey did not ask whether participants worked on a full- or part-time basis, which may have been able to further explain this discrepancy.

Four work-related characteristics showed a higher point estimate for shoulder trouble: performing more than 8200 palpations per year, devoting 50% or less of practice to bovine work, more years of experience, and higher number of veterinarians in the practice. Though not significant, performing more than 8200 palpations per year showed higher odds of having shoulder pain, and was confounded by percent of practice devoted to bovine work and, to a lesser degree, sex. Increasing years of experience had a protective effect. This may be due to a form of healthy-worker effect; several survey respondents noted that the physical demands of large animal work are not conducive to performing this work full-time or for an extended career [6], thus the most experienced survey respondents may have been those who were better able to cope with MSD symptoms as they aged. An increasing number of veterinarians in a practice (i.e. co-workers) also showed a slight protective effect. Intuitively, having more colleagues may make it easier to take time off or share the workload during busy seasons. Devoting 50% or less of practice to bovine work had a nearly three-fold increase in odds of shoulder trouble. This could relate back to a healthy-worker effect in those who devote a high percentage of practice to bovine work; shoulder trouble may force veterinarians to reduce their bovine work as needing to scale back one's bovine workload was a noted impact of MSD by Zeng et al. [6]. On the other hand, over half (51%) of the veterinarians in this category still reported performing over 8200 palpations per year, so being involved in a mixed practice is not necessarily an accurate proxy for lower exposure.

The RULA assessments indicated there is risk to both shoulders during both manual and ultrasound-assisted rectal palpations. The palpation arm, while sustaining a less extreme posture than the support arm during manual palpations, experienced repetitive motions and forces around 10 kg. The support arm sustained an extreme posture (abducted, shoulder raised, elbow angle >100°, working across midline of body [11]) during manual palpations, but negligible forces. While the upper limb postures were less extreme during ultrasound use, the overall body posture still resulted in high RULA scores for this exam method (>6). The trunk and neck were generally bent or twisted, which also may contribute to awkward posture of the upper limb. The survey did not ask respondents to specify on which side of the body symptoms occurred, therefore it is not possible to attempt to connect the RULA results with the survey results, or to draw a conclusion regarding increased MSD symptoms in the palpation arm similar to Cattell [3]. Nonetheless, these results do suggest that overall this is a task that needs further investigation and development of appropriate targeted interventions.

There are several limitations in this study. The sample size (n = 133) for the survey is small and likely underpowered, and as described the sample was likely not diverse enough to determine statistically significant physical characteristics associated with shoulder trouble. It is possible that including more survey questions about work characteristics (e.g. part time vs. full time work) could have created a more complete understanding of some unexpected results, such as why being female had a protective effect. The ergonomic assessments were only performed on one practitioner under optimal conditions, and postural angles were visually observed rather than directly measured. Due to the task being performed in a cattle squeeze (and inside a cow), the camera angle on the video recordings was not optimal to view the veterinarian's entire body, so some posture assessments had to be recreated by the veterinarian after the fact.

5 Conclusions and Future Work

While the survey analysis was unable to identify significant associations between individual-level physical risk factors and shoulder MSD, it did find that several work-related characteristics were predictive of shoulder trouble. Specifically, the ability to share work with other veterinarians and/or working part time (which may affect frequency and volume of task performance) could reduce the risk of shoulder MSD. The RULA results support the postulations by Scuffham and Ailsby that repetition, frequency, awkward posture, and position of the upper limb(s) may be associated with musculoskeletal trouble of the shoulder in large animal veterinarians [1, 7]. Scuffham [7] proposed quality of facilities as an important risk factor, though in the current study this was not investigated as the ergonomic assessments were performed in a single setting of near-optimal conditions. However, as the study expands to include more assessments in the future, risks presented by the work environment will be able to be quantified.

The present study is part of a larger project to quantify the risk factors for MSD associated with bovine rectal palpations. Further statistical analyses will be performed on the data to assess other upper extremity data from the survey (neck and elbow trouble, and degrees of discomfort). The pilot ergonomic assessment will be expanded to measure more veterinarians, and direct measurement methods will be used to determine postural angles. As the final RULA scores for each version of the task in the present study were all greater than 6, studying the tasks with a method that investigates the postural risk factors in greater detail without combining them into a single score will be preferable. Results from the expanded study will be used to lay the groundwork for suggesting intervention or prevention strategies.

References

1. Ailsby RA (1996) Occupational arm, shoulder, and neck syndrome affecting large animal practitioners. Can Vet J 37(7):411
2. Scuffham AM, Legg SJ, Firth EC, Stevenson MA (2010) Prevalence and risk factors associated with musculoskeletal discomfort in New Zealand veterinarians. Appl Ergon 41 (3):444–453
3. Cattell MB (2000) Rectal palpation associated cumulative trauma disorders and acute traumatic injury affecting bovine practitioners. Bovine Practioner 34(1):1–4
4. O'Sullivan K, Curran N (2009) It shouldn't happen to a vet… Occupational injuries in veterinary practitioners working in Ireland. Ir Vet J 61(9):584–586

5. Kozak A, Schedlbauer G, Peters C, Nienhaus A (2014) Self-reported musculoskeletal disorders of the distal upper extremities and the neck in German veterinarians: a cross-sectional study. PLoS ONE 9(2):1–9
6. Zeng X et ala (2018, in press) Musculoskeletal discomfort among Canadian bovine practitioners: prevalence, impact on work, and perception of physically demanding tasks. Can Vet J 59(7)
7. Scuffham A, Firth E, Stevenson M, Legg S (2010) Tasks considered by veterinarians to cause them musculoskeletal discomfort, and suggested solutions. N Z Vet J 58(1):37–44
8. Mcatamney L, Corlett EN (1993) RULA: a survey method for the investigation of world-related upper limb disorders. Appl Ergon 24(2):91–99
9. Kuorinka I et al (1987) Standardised nordic questionnaires for the analysis of musculoskeletal symptoms. Appl Ergon 18(3):233–237
10. Bernard B et al (1997) Musculoskeletal disorders and workplace factors. U.S. Department of Health and Human Services (NIOSH) publication 97B141
11. Hedge A (2001) RULA employee assessment worksheet. Cornell University
12. Nordander C et al (2008) Gender differences in workers with identical repetitive industrial tasks: exposure and musculoskeletal disorders. Int Arch Occup Environ Health 81(8):939–947
13. Jelinski MD, Barth KK (2015) Survey of Western Canadian veterinary practices: a demographic profile. Can Vet J 56(12):1245–1251

Analyses of Musculoskeletal Disorders Among Aesthetic Students Applying the Methods: REBA, Nordic and FSS

Gabriela de Souza Raymundo(✉) and Ivana Salvagni Rotta(✉)

Centro Universitário Herminio Ometto, FHO, Araras, SP, Brazil
gabriolasr@hotmail.com, ivanarotta@gmail.com

Abstract. The number of professionals and the demand for aesthetic treatments has been increasing significantly. During a treatment, the worker could be in non-ergonomic positions and practice repetitive movements, that it may cause Work Related of Musculoskeletal Disorders (WRMDs). These injuries can be prevented if identified in through Ergonomic Analysis of Work (EAW), therefore the aim of this study consist in analyse the harm caused by the activities developed in the clinic during the procedures performed by forty four students of the Aesthetics course of a University. The research developed by applying ergonomics questionnaires REBA and Nordic that evaluate the movements, positions of work and identify pain/discomfort in the body during the performance. The fatigue severity scale also evaluate the fatigue at the beginning, middle and end of the beautician's session. The REBA was completed according to the observer's assessment, showed that the majority of the students presented medium-level and high risk, which means that they could develop musculoskeletal disorders. The fatigue severity scale measured the fatigue and verified that the students feel tired by performing repetitive movements. The results show that the majority of the students feel pain and/or discomfort in several parts of the body, the lower back was the region that obtained the highest number of complaints, followed by the upper back and neck. In addition to the analysis of muscle injuries, the research point out the importance of ergonomics studies to improve the quality of life of the aestheticians.

Keywords: Musculoskeletal disorders · Ergonomics · Aesthetic

1 Introduction

Brazilian people cares a lot about appearance, the culture focused most of the time on health and wellness care increases the demand for aesthetic treatments [1]. Professionals in the area are increasingly required to provide these treatments, as a result, they work more hours. When they are executing their tasks they usually put themselves in many non-ergonomic postures. Some of those positions combined with repetitive movements could cause a musculoskeletal disorder (MSDs).

According to Gowda et al. [9] Work-related Musculoskeletal Disorders increased in the past years among bodywork professionals such as aestheticians, beauticians and massage therapists.

© Springer Nature Switzerland AG 2019
S. Bagnara et al. (Eds.): IEA 2018, AISC 820, pp. 650–659, 2019.
https://doi.org/10.1007/978-3-319-96083-8_82

However, MSDs could be prevented using tools to identify and reduce the levels of risk factors in the workplace [14]. In order to prevent these muscular disorders and to improve the quality of life of the beauticians, ergonomic methods are suggested [13]. Ergonomics is a study that comprehend the man's interaction with principles and methods, and the application of these principles, in order to improve the well-being of this professional while performing the task. The word ergonomic comes from the Greek, and it means ergon (work) and nomos (norms, rule, and laws). The term of ergonomics could be described in many ways [12]. Wisner [25] considers scientific studies related to the human being and the equipment used during his work in a more comfortable and safe way, also seen as methodology that goes beyond providing improvement in the conditions of job, but also considers human anthropometry, associated with physiological and psychological factors. So the concept is used to prevent, possible damages that the labour can bring. Improving the work position is one of the measures taken in the view of ergonomics [10]. The importance of studying ergonomics is to improve the performance of this collaborator and his interaction with the environment in which he is working. Ergonomics prioritizes to provide effectiveness and efficiency, which is combined to well-being and health by adapting work to human being [5].

The concern about musculoskeletal disorders occurred after the Industrial Revolution, which was intensified through mechanization and the appearance of computers [21]. Repetitive tasks leads to Repetitive Strain Injuries or Work-Related musculoskeletal disorders (WMSDs), various symptoms are diagnosed, such as pain, feeling of heaviness, fatigue usually on the neck and upper limbs but may also affect lower limbs as well [2]. According to Gonçalves and Camarotto [8] analyses of the activity at the moment it is performed and it is known that these repetitive movements can be directly linked to the cause of pain and injury leading to cause these disorders [3].

Note that among the existing occupational diseases, WMSDs is the one with the highest number of medical diagnoses among professionals [18]. The relevance of ergonomic studies associated with the reduction of incidence of these disorders, since the workplace takes into account several aspects from the environment where the action takes place, the creation of more ergonomic equipment for the execution of these tasks and the organization of the job stations.

A study conducted by Devitt [7] and Surdival [24] in Podiatrists have shown that during practice there are risks of developing muscle injuries in the wrist and other parts of the body. The Rapid Entire Body Assessment (REBA) questionnaire proposed immediate changes to some activities developed by these professionals, due to the risk of developing such disorders. Also in a study of estheticians points out that during facial treatments aestheticians remain in a position that can often bring discomfort and/or pain on the neck, wrists, spine, shoulders, arms and legs and feet, the author compares beauticians to people who use microscopes [19]. By all those facts, notice the importance of improving the performance of this students that will be an Aesthetic professional, as well as prioritizing effectiveness and efficiency and looking through possible musculoskeletal disorders among them. The study was conducted on using ergonomics tools to find out the level of the risk and give them possible recommendations. The analysis of the possible musculoskeletal disorders are identified with the application of questionnaires that analyze the movements of the body during the performance of them.

In view of this context, this project is a study case that analyzes the activities performed by the students of the Aesthetics and Cosmetology course of the University during the patients' appointments in the clinic, verifying the non-ergonomic positions and the practice of repetitive movements. It is proposed the ergonomic analysis, with the application of questionnaires to verify and prevent such disorders in this way, guaranteeing the quality of life for future professionals.

2 Materials and Methods

The present work has a quantitative and qualitative approach. Initially, a bibliographical review was held regarding the concepts discussed. Next, a case study was elaborated. The analysis and evaluations of the tasks, studies the behavior of the professional, the researcher observes the worker during his activities thus proposing solutions, if it is necessary, and often improve the equipment used. There are many ergonomic methods used to check, and from these analyzes, risks are raised and improvements can be made [13]. The study was carried out with 44 female students of Aesthetic and Cosmetology Course in a University, which practiced the treatments at the clinic, doing facial, body treatments and therapies such as reflexology, shiatsu, massage and facial cleaning. A general information questionnaire was introduced to each student, followed by ergonomics evaluation methods as Nordic, REBA and Bipolar questionnaire that was evaluate using the Fatigue Severity Scale, the methods were conducted by the observer so notes, photos and videos were also recorded during the treatments, in order to compare the ergonomic procedures used. The students participating in the research signed and completed a consent form. The five main tasks analyzed on this research are facials and body treatments, shiatsu, reflexology and lymphatic drainage.

The Nordic questionnaire was developed for self-fulfillment, the evaluated professionals will be guided with a figure that divided the body into 9 parts and respond by indicating yes or no to the parts that feel some discomfort. The objective is to discover the occurrence of WRMDs [13]. This method aims to facilitate and standardize the reports of musculoskeletal disorders symptoms [17].

The Rapid Entire Body Assessment was developed by Sue Hignett and Lynn McAtmney [20] and uses the posture score that comes from the Rapid Upper Limb Assessment (RULA) another method of ergonomic evaluation. This tool provides a quick assessment that checks the posture of the neck, trunk and upper limbs with muscles function combined with the external loads experienced by the professional. The body it is divided in two groups in this assessment. Group A, which is trunk, neck and legs and group B is upper arms, lower arms and wrist. For each of these body parts a number it is applied. In addition, considerations could be made. A load/force will be marked. There are the table A and B where it is possible to find the rate for each section. For table A it is total average of load/force, while on table B the total of coupling. Combining results of A and B it will have a score C. At the end of the observed records, there will be a result that is the REBA Score, the evaluation of this value will be done with the tables of levels that presents the risks of the activity observed [11]. REBA was applied in the five tasks analyzed on this study case.

This method instantly assesses the practitioner's posture during the performance; the observer has the opportunity over the analysis to determine the most critical moment, in addition to allowing a more standardized data collection and taking into account risk factors that are directly linked to the musculoskeletal disorders [15]. The analysis of these muscle problems performed using this tool, applied in the process of salting meat in a factory, observed that this questionnaire is very efficient and easy to apply in the health area, presents reliable results and determines possible risks that the worker is could exist in the work environment [22, 23].

The fatigue severity scale and the bipolar questionnaire developed by Corlett used to measure tiredness during tasks [4, 16]. It is some questions that the observer asks the workers at the beginning, middle and end the task. The professional answers choosing a range between 1 to 7. After asking those questions, a relevant scale is obtained and then the task could cause mild, moderate or severe fatigue. The ways of it could cause stress during the performance and this ascertain the scale of it [6].

3   Results and Discussions

The figures and pictures present the results obtained in this study. The mean age of the 44 students' participants was $22 \pm 3,27$ years old and the mean height was $162 \pm 6,26$ cm. During the application of the general information questionnaire, the percentage of students who has any knowledge of the term ergonomics was verified (see Fig. 1).

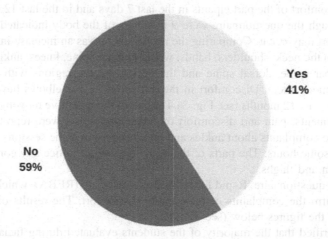

Fig. 1. Knowledge of the term Ergonomics among the students, about 59% of the participants do not know what is ergonomic, on the other hand, 41% of them said that they know or have heard about the term.

The general information questionnaire, associated with the Nordic analysis, which demonstrate the occurrence of pain and/or discomfort among the students. Figure 2 provides the percentage of students who feels muscular pain and those who do not it.

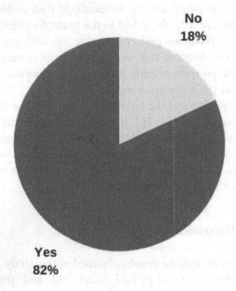

Fig. 2. All most of the students reported any pain or discomfort in their body, only 18% of them do not feel anything.

The Nordic Questionnaire was used to verify the possible problems associated with pain and discomfort of the participants in the last 7 days and in the last 12 months (see Fig. 3). Through the questionnaire were 9 members of the body indicated where pain and discomfort may occur. Comparing the results, there was an increase in pain and/or discomfort on the neck, shoulders, hands, wrists, lumbar spine, knees, ankles, and feet.

The lumbar spine, dorsal spine and the neck were the regions with the greatest complaints among them. Discomfort in the left shoulder and elbows have only been reported in the last 12 months (see Fig. 3). Despite of the repetitive movements adopted in some treatments, pain and discomfort on hands and wrists were reported by some students. Also complaints about ankles and feet, because in some sessions they remain standing for some hours. The parts of the body that do not notice any complaint is in the hip region and thighs.

The third questionnaire, Rapid Entire Body Assessment (REBA), which it also was used to confirm the complaints of pain and/or discomfort. The results of REBA are presented on the figures below (see Fig. 4).

It was verified that the majority of the students evaluated during facial treatments present medium risk level, about 17 (70.8%) and 7 (29.2%) presented high level.

It is noteworthy that most of the students assessed during body therapies such as shiatsu, reflexology, draining and modeler massage present medium-level risks, about 17 (85%) and 3 (15%) present low-level risks, being able to develop muscular

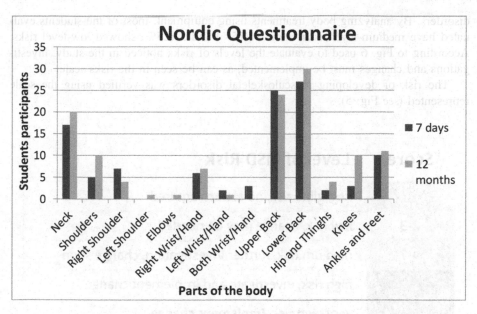

Fig. 3. The parts of the body presents on the Nordic questionnaire that the students reported pain or discomfort in the last 7 days and 12 months.

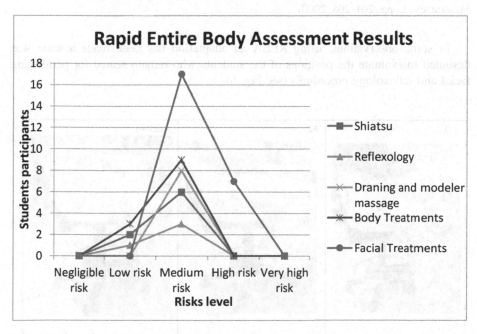

Fig. 4. Results from REBA after observing the treatments, most of the students present medium risk of developing musculoskeletal disorders.

disorders. By analyzing body treatments using equipment, most of the students evaluated have medium-level risks, about 9 (75%) and 3 (25%) showed low-level risks. According to Fig. 6 used to evaluate the levels of risks noticed in the study, investigations and changes must be implemented, as can be seen in the risks scale.

The risk of developing musculoskeletal disorders was verified using the scale represented (see Fig. 5).

Score	Level of MSD Risk
1	negligible risk, no action required
2-3	low risk, change may be needed
4-7	medium risk, further investigation, change soon
	high risk, investigate and implement change
	very high risk, implement change

Fig. 5. Score level of musculoskeletal disorders risk and suggestion needed (Source: Hignett, S.; Mcatamney, L. pp. 201–205, 2000)

In some observations, using REBA an adaptation has been made a scale was designed to evaluate the positions of the students who remain seated for performing facial and reflexology procedures (see Fig. 6).

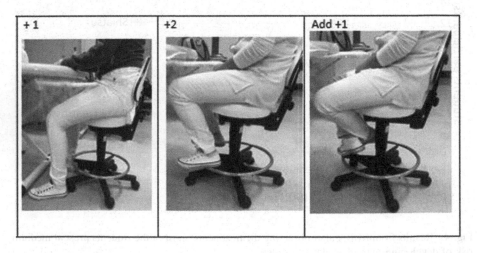

Fig. 6. Posture adopted by the students during sessions and adapted to use combined to REBA observation.

It was verified that the students of short stature could not support both feet on the ground, which make them to adopt the second or third position represented in figure below, while the higher trainees adopted the first position, which should be more comfortable to perform.

The bipolar questionnaire and fatigue severity scale measure student's fatigue, it was also applied during body therapies. Table 1 shows a scale determining the fatigue scale reported by them from the application of the questionnaire and the reviews made.

Table 1. Fatigue scale obtained by the observations during the procedures performed in this study

Fatigue scale	Level of fatigue	Treatments observed
1	*Lack of fatigue*	-Reflexology
2		-Body Treatments using equipment
3		
4	*Moderate fatigue*	Shiatsu
5		
6		
7	*Intense fatigue*	Draining and modeler massage

The fatigue scale is based on a previous research developed by COUTO [4].

In the present study noticed that the intense fatigue is present on executing the draining and modeler massage despite of the repetitive and intense movements performed by the professional.

During reflexology, they remain seated so the level of fatigue is low, also when they perform body treatments. Through shiatsu, the level of fatigue obtained was moderate because of the pressure they need to put against the body of the patient.

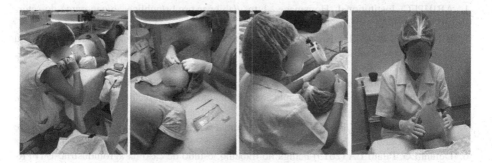

Fig. 7. Photos during facial treatments performed by the students.

Photos of facial treatments sessions performed by the trainees in the clinic were taken during the study (see Fig. 7). In the first and second photo the professional is performing the skin cleaning on the patient, this procedure requires a greater strength and precision of the wrists and hands, in the third and in the fourth is made the cleaning and moisture of the patient's skin note some repetitive movements with the hands on those tasks performed by them.

During the observations, the trainee's complain about the discomfort caused by the stool, due to some broken settings resulting in non-ergonomic positions, and a difficulty to adjust the height of some stretcher that was fixed, leading to non-suitability of the equipment for all individuals, causing discomforts during the performance.

4 Conclusion

Brazilian culture worried about appearance so much, despite of that the aesthetic area has been developed in the latest years. All the data obtained from the questionnaires used on this study, it was verified that the majority of the trainee students could develop musculoskeletal disorders. Due to the tasks performed during the sessions and the long journey of work, they are adopting. Interviewing the students that already work outside the university, most of them reported that they work more than 8 h on the weekend days without pausing between treatments. The results also reached between the medium and high-risk level of developing disorders. Therefore, it is important to propose alternatives for performing work and training, as well as new mechanisms to reduce the risks of muscle disorders such as MSDs/WMSDs. However, through the study exists some limitations, such as more observations were needed to develop a deep comparative, there is need to do further research in other to reduce the disorders and to propose new equipment that could be more ergonomic to the professional.

Electronic Supplementary Material: https://youtu.be/8egVwNUZH_s

References

1. ABIHPEC Institutional Homepage. https://abihpec.org.br/publicacao/panorama-do-setor-2017/. Accessed 19 Jan 2018
2. BRASIL, INSS (2003) Norma Técnica sobre Lesões por Esforços Repetitivos-LER ou Distúrbios Osteomusculares Relacionados ao Trabalho- DORT, Ministério da Saúde, Brasília 97p
3. Bridger, RS (2009) Introduction to Ergonomics. 3. ed. FL: CRC Press
4. de Couto HA (1995) Ergonomia aplicada ao trabalho: manual técnico da máquina humana, vol 01. Ergo Editora, Belo Horizonte
5. Cybis W, Betiol AH, Faust R (2007) Ergonomia e Usabilidade: Conhecimento Métodos e Aplicações, 2nd edn. Novatec, São Paulo
6. Defani LG, Pilatti LA (2015) Fadiga no trabalho: estudo de caso na agroindústria. UTFPR-PG. S/dv
7. Devitt K (2010) A comparative study of three observational techniques to establish if podiatrists are at risk of musculoskeletal ill-health. M.Sc. Occupational Health, Safety & Ergonomics. National University of Ireland, Galway (unpublished thesis)
8. Gonçalves JM, Camarotto JA (2015) Estratégias operatórias frente ao trabalho repetitivo. Production 25(1):190–200. https://doi.org/10.1590/S0103-65132013005000087

9. Gowda H, Mascarenhas SP, Patil DP, Pandit U (2016) Hand function assessment in beauticians. Int J Health Sci Res (IJHSR) 6(12):121–126
10. Guérin F, Kerguelen A, Laville A, Daniellou F, Duraffourg J (1997) Compreender o trabalho para transformá-lo: a prática da Ergonomia. Edgard Blücher Ltda, São Paulo
11. Hignett S, Mcatamney L (2000) Rapid entire body assessment (REBA). Appl Ergon 31(2):201–205
12. IEA. Associação Internacional de Ergonomia. http://www.iea.cc/whats/index.html. Accessed 26 Sept 2017
13. Iida I, de Guimarães LBM (2016) Ergonomia: projeto e produção. 3 edn. Blucher, São Paulo
14. Irishhealth.com, 'Musculoskeletal Disorders' (2009) http://www.irishhealth.com/article.html?con=444. Accessed 22 June 2016
15. Junior MMC (2006) Avaliação ergonômica: Revisão dos métodos para avaliação postural. Revista produção online, vol 6, no 3. https://doi.org/10.14488/1676-1901.v6i3.630
16. Krupp LB, LaRocca NG, Muir-Nash J, Steinberg AD (1989) The fatigue severity scale. Application to patients with multiple sclerosis and systemic lupus erythematosus. Arch Neuro, 46:1121–1123
17. Kuorinka I, Jonsson B, Kilbom A, Vinterberg H, Biering-Sørensen F, Andersson G, Jørgensen K (1987) Standardised nordic questionnaires for the analysis of musculoskeletal symptoms. Appl Ergon 18(3):233–237
18. Maeno M, Tavares DS, Lima CQB (2016) A precarização do trabalho, a desconstrução dos direitos, a desigualdade social e as LER/DORT. https://cut.org.br/system/uploads/ck/files/LER-DORT-28-fev-2016-final-2.pdf. Accessed 26 Sept 2017
19. Massambani EM (2011) Incidência de distúrbios músculo esqueléticos em profissionais de estética: suas repercussões sobre a qualidade de vida e de trabalho. Arq. Ciênc. Saúde UNIPAR, Umuarama, vol 15, no 1, pp 51–62, jan./abr. https://doi.org/10.25110/arqsaude.v15i1.2011.3692
20. Mcatamney L, Hignett S (1995) REBA: a rapid entire body assessment method for investigating work related musculoskeletal disorders. In: Proceedings of the 31st annual conference of the ergonomics society of Australia. The Society, Melbourne
21. Oliveira ER (2007) Prevalência de doenças Osteomusculares em cirurgiões dentistas da rede pública e privada de Porto Velho – Rondônia. Dissertação de Mestrado UNB, Brasília
22. Paludo HCM, Paludo V (2015) Analise Ergonomica do Trabalho (AET) Aplicado no Processo de Salga da Carne de uma fábrica de Charque. In: ENEGEP 35. Fortaleza, Anais
23. Pinto AMP (2009) Análise ergonômica dos postos de trabalho com equipamentos dotados de visor em centros de saúde da administração regional de saúde do centro. Dissertação de Mestrado da Faculdade de Medicina da Universidade de Coimbra
24. Surdival L (2010) Wrist posture and wrist repetition as possible risk factors during podiatry work. M.Sc. Occupational Health & Safety and Ergonomics National University of Ireland, Galway. (unpublished thesis)
25. Wisner A (1987) Por Dentro do Trabalho - Ergonomia: Métodos e Técnicas. FTD/Oboré, São Paulo

Push and Pull – Force Measurement Updates, Interpretation of Measurements and Modes, Peculiarities (Curves, Steps, Etc.). Multi-task Analysis

Marco Cerbai[1,2](✉) and Marco Placci[1]

[1] EPM IES - Ergonomics of Posture and Movement International
Ergonomics School, Milan, Italy
cerbaimarco@safetywork.it
[2] Safety Work Srl, Imola, Italy
http://www.epmresearch.org, http://www.safetywork.it

Abstract. Push and Pull measurement using conventional instruments such as mechanical dynamometers, does not allow to fully appreciate the peculiarities of a path with curves, gradients or variations in the terrain.

High-sampling digital dynamometers allow to accurately observe the performance of the force during push and pull activities.

It's possible to identify:

- The Initial Force by recording the peak in the initial phase;
- The Sustained Force by recording the entire phase after the peak: in this case, the value should be processed as a 50th, 75th or 90th percentile of the sample of data recorded, because the percentile concept approximates the classical interpretation of the Sustained Force, according to proposed criteria by Stover H. Snook and Vincent M. Ciriello, authors of the benchmarks for the push and pull activities of the International Standard ISO 11228-2.

In various measurement experiences carried out in working environments such as urban waste management, hospital departments, or manufacturing departments, it has been observed that sustained force is not constant, but also varies rapidly. These variations are intrinsic to any handling activity, but introduce a further difficulty in defining the value of this factor.

During the shift operators are exposed to multiple handling tasks: it's important to set up a multitask analysis push and pull activities.

Curves, steps and gradients introduce additional variables to the Multitask Analysis.

The analysis of the various situations allows to create a database of useful data to map the risk in advance.

Keywords: Push and pull · Risk assessment
High-sampling digital dynamometers · Ergonomics · Musculoskeletal disorders

© Springer Nature Switzerland AG 2019
S. Bagnara et al. (Eds.): IEA 2018, AISC 820, pp. 660–669, 2019.
https://doi.org/10.1007/978-3-319-96083-8_83

1 Introduction

Push and Pull measurement using conventional instruments such as mechanical dynamometers, does not allow to fully appreciate the peculiarities of a path with curves, gradients or variations in the terrain.

The digital high sampling rate dynamometers are measurement instruments which work as the mechanical ones, but they are different because they are able to test forces at high sampling rates (500, 1000 or more samples per second). These kinds of digital instruments usually have electronic devices that allow them to connect with a computer: all the measurement data can be saved and it is possible to watch at the moment all the measurement data and their evolution on the computer monitor. It is also possible to measure force for the whole path in all kind of situations (paths with twists, holes, steps, changes in slope or floor and so on). The measurement mode of digital high sampling rate dynamometers is more complex in the execution phase but these instruments provide accurate achievements.

High-sampling digital dynamometers allow to accurately observe the performance of the force during push and pull activities.

2 Method

The International Standard ISO 11228-2 requires the measurement of the following forces:

- The Initial Force by recording the peak in the initial phase of the movement;
- The Sustained Force by recording the entire phase after the peak.

In Fig. 1 it is possible to observe the trend of the force applied to an object in an ideal case of push and pull activities.

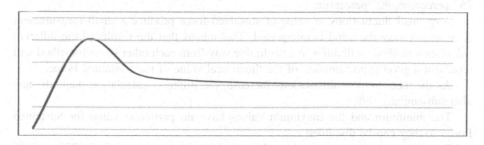

Fig. 1. Theoretical trend of the measured force

The Sustained Force is the hardest magnitude to be assessed and interpreted; it represents the magnitude of the necessary force to maintain in movement to constant speed the cart. This force can be very variable depending on floor/ground conditions (holes or depressions of the ground), the kind of ground (asphalt or sand), the kind of wheel (small or large diameter) and other variables.

Therefore the trend of the sustained force will not be like the one in Fig. 1, but it could show variable fluctuations.

In Fig. 2 it is possible to observe the trend of the force in a real case of push and pull activities with little variations of sustained force.

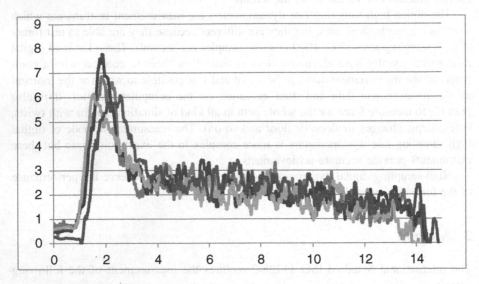

Fig. 2. Probable real trend of the measured force

Having a wide distribution of data available (1000 or more samples per second) it is possible to detect a method of calculating that best approximates the ideal theoretical data. Different calculation methods are assumed: the average value, the weighted average value, the maximum or minimum value, the 25° percentile, the 50° percentile, 75° percentile, 90° percentile.

Any small fluctuations in value of sustained force produce a small measurement error in any way the trend is interpreted. This means that the results of the different calculation methods will differ in a negligible way from each other so each method will represent a good approximation of the theoretical value of the Sustained Force.

As the measurement fluctuations increase, the different calculation methods may also substantially differ.

The minimum and the maximum values have no particular value for Sustained Force so they can be discarded.

For these variables measure the value should be processed as a 50th, 75th or 90th percentile of the sample of data recorded, because the percentile concept approximates the classical interpretation of the Sustained Force, according to proposed criteria by Snook and Ciriello [1].

- The Maximum Force recorded after the Initial Force (not required by ISO11228-2): in some cases there may be steps or imperfections in the floor that generate peaks in the initial phase and in the next stage of handling that are difficult to observe with the traditional methodologies of evaluation. Through high data sampling dynamometers (e.g. 1000 sampling per second or more) it's possible to measure these peaks and compare them with the limits proposed by the ISO11228-2 standard for initial forces.

• The Stopping Force (not required by ISO11228-2): it is the force necessary to stop or slow down the cart which in particular cases may be comparable or superior to the Initial Force.

3 Measurements with Fluctuations

In various measurement experiences carried out in working environments such as urban waste management, hospital departments, or manufacturing departments, it has been observed that Sustained Force is not constant, but also varies rapidly. These variations are intrinsic to any handling activity, but introduce a further difficulty in defining the value of this factor (Fig. 3).

Fig. 3. Push and pull activity

The reference remains the studies made by Snook and Ciriello [1], where the Authors measured the Initial Forces and the Sustained Forces through direct observation of the measurement during their tests in the Liberty Mutual laboratories: the subjects were on a specially constructed treadmill with a brake gear to simulate different cases; they moved with slow and regular movements, and forces were measured through the load cell.

With these modalities the values of the Initial Forces and the Sustained Forces are observed through the direct reading of the instrument and the annotation of the data.

The greatest difficulty is the estimate of the Sustained Force. It is important:

– POINT OF HOOKING: observe the task to identify the point of application of the force or the point of grip.
– HOW TO HOOK THE DYNAMOMETER: it is preferable to have an anchoring system equipped with two points of hooking to the handle.
– POSITIONING: the dynamometer must be in a position parallel to the walking surface (horizontal).

- EXECUTION: to avoid abnormal measurements, the test must be carried out with "fluid and controlled" movement.
- READING OF THE DATA: during the test it is possible to read the values of the Initial Force and the Sustained Force. In particular for the Sustained Force, first of all it is important to remove any data depending on any anomalies (these measured peaks during the maintenance phase will be considered separately). Then it will be possible to identify several values all around the theoretical measurements: the most recurrent value among the measured values will be chosen.

It is important to note that the measurements made in this way represent a subjective value that depends on the skills and the experience of the evaluator.

Through a direct observation it is possible to choose the most frequent value. This value is included in a range of values that can potentially all represent the Sustained Force.

- Choosing the maximum value or the minimum value of this range may be an overestimation or an underestimation of the risk assessment;
- Choosing the average value of the range may be an underestimation of the risk assessment, especially if the data have large fluctuations;
- The percentile calculation conceptually represents the measurement made by the direct observation of a discrete value represented on a display. A percentile greater than the 50th is considered in this study (for example 75th percentile, 90th percentile and 95th percentile).
- As a first analysis of the Authors, a value above the average could be appear to be correct even though there are not clinical data to be able to affirm it yet. In support of this hypothesis, also other highly reliable methodologies, for example OCRA Method [2, 3], Composite Lifting Index [4], Sequential Lifting Index [5] and Variable Lifting Index [6] proposed by NIOSH and EPM (Ergonomics of posture and movements International Ergonomics school), are strongly non-linear on some factors and they attribute greater weight to factors or tasks with higher overload (over the average value).

In Fig. 4 are represented some percentiles of a force measurement of a cart on a regular floor.

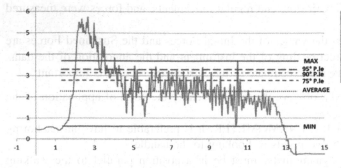

Percentile	
95° P.le	3,22
90° P.le	3,03
75° P.le	2,88
50° P.le	2,14
Average	2,16

Fig. 4. Calculation of percentiles for a measure with large fluctuations

In the Authors studies with more than 450 measurements made comparing for each test the values recorded digitally and the values defined through the observation, it is possible to define that the value of the 75th percentile is a good approximation of the value defined by direct observation.

It is possible to observe that with small fluctuations between the minimum value and maximum value the differences between the 75th, 90th and 95th percentile are very small, and they can be considered similar without generating a significant error in the risk assessment (see Fig. 5).

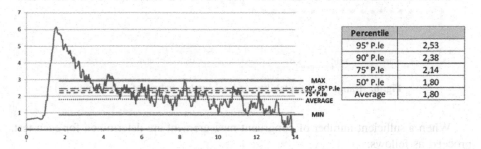

Percentile	
95° P.le	2,53
90° P.le	2,38
75° P.le	2,14
50° P.le	1,80
Average	1,80

Fig. 5. Calculation of percentiles for measurements with small fluctuations

This result is true as a consequence of the limits proposed in the tables of the standard that have a minimum variation of 1 kg between the different conditions.

4 Procedure for Push and Pull Test

First of all it is important to select the correct route, then it is possible to carry out the measurement test with slow and regular movements.

- With pivoting wheels it is necessary to test the wheels in line with the movement and perpendicular to the movement and consider the worst case;
- if there are blocks a test will be done to unlock the cart and another test will be necessary for the route.

Repeated measurements are necessary: in order to correctly evaluate the Initial Force and Sustained Force it's advisable to carry out three, five, seven or more measurements during the same test. In this way it is possible to obtain congruent measurement, with differences not exceeding 15%.

It is important to correctly evaluate the difference between the measurement. In particular according to the theory of measures there is only one conceptually correct value and each measure approximates this theoretical value with a variable error.

If several measurement approximate the theoretical value within an acceptable error range (previously determined) then a mathematical function of them (for example the average or the maximum value, etc.) can represent the theoretical value a with sufficient approximation (Fig. 6).

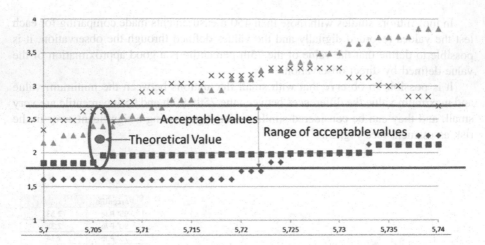

Fig. 6. Range of acceptability and acceptable values of a measure

When a sufficient number of congruent measurement are determined for each test, proceed as follows:

- For the Initial Force, obtained at least three congruent measurement, the highest value is taken;
- For the Sustained Force, obtained at least two congruent measurement, the average of the values is calculated.

The calculated values will be compared with the limits proposed by the tables of the standard and the risk assessment will be calculated as follows:

$$I.F.I. = \frac{Initial\ Force\ Measure}{Recommendend\ Initial\ Force}$$

$$S.F.I. = \frac{Sustained\ Force\ Measure}{Recommendend\ Sustained\ Force}$$

The Initial Force Index and the Sustained Force Index will be classified by the second level Method of ISO TR12295 relative to the Standard ISO 11228-2 Method 1.

5 Results of Application

During a study where over 250 measurements were collected between 2010 and 2015 and a further 200 measurements up to 2017, the values summarized in Fig. 7 were found.

In the Authors studies experience the collected data through the direct observation of the display during push and pull tests and the data of the same identical tests recorded with high sampling frequency (1000 samples/sec) were compared.

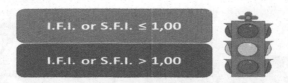

Fig. 7. Risk classification ISO TR 12295, ISO 11228-2 Method1

This data collection was then processed as the 50th percentile, 75th percentile, 90th percentile and 95th percentile and the results are summarized in the Table 1.

Table 1. Examples of measurements and calculations for different percentiles

Observed value	50° P.le	Δ% 50° P.le	75° P.le	Δ% 75° P.le	90° P.le	Δ% 90° P.le	95° P.le	Δ% 95° P.le
15,50	14,37	- 1,13	17,29	1,79	19,89	4,39	21,45	5,95
40,00	37,84	- 2,16	40,10	0,10	42,43	2,43	43,59	3,59
20,00	18,24	- 1,76	19,87	- 0,13	21,14	1,14	21,75	1,75
9,50	8,29	- 1,21	10,56	1,06	12,62	3,12	13,52	4,02
10,50	9,39	- 1,11	12,57	2,07	14,77	4,27	16,29	5,79
9,00	6,71	- 2,29	8,80	- 0,20	12,17	3,17	16,68	7,68
2,50	2,12	- 0,38	3,17	0,67	4,38	1,88	4,98	2,48
3,50	3,26	- 0,24	4,33	0,83	6,16	2,66	7,09	3,59
11,00	7,74	- 3,26	10,21	- 0,79	16,36	5,36	21,15	10,15
5,50	4,89	- 0,61	7,61	2,11	11,93	6,43	13,21	7,71
11,00	8,67	- 2,33	10,98	- 0,02	12,92	1,92	13,72	2,72
11,00	9,73	- 1,27	12,71	1,71	14,85	3,85	16,38	5,38
9,00	6,37	- 2,63	9,18	0,18	13,23	4,23	21,36	12,36
10,50	8,05	- 2,45	10,16	- 0,34	12,44	1,04	13,78	3,28

[...]

5,00	5,04	0,04	7,10	2,10	9,11	4,11	10,28	5,28
6,00	6,67	0,67	9,87	3,87	11,68	5,68	12,82	6,82
5,00	4,54	- 0,46	5,37	0,37	6,91	1,91	7,95	2,95
5,50	3,97	- 1,53	5,48	- 0,02	8,69	3,19	9,90	4,40
5,50	4,08	- 1,42	5,33	- 0,17	6,59	1,09	7,39	1,89
4,50	3,83	- 0,67	4,62	0,12	5,86	1,36	8,76	4,26
4,50	3,83	- 0,67	5,18	0,68	7,74	3,24	9,19	4,69
5,00	3,90	- 1,10	5,22	0,22	8,37	3,37	10,69	5,69
5,50	3,79	- 1,71	6,20	0,70	7,83	2,33	9,87	4,37

6 Multitask Analysis

There are many kinds of handling activities in workplaces. For example, in urban waste collection, two-wheeled and four-wheeled carts with different volumes and flow rates, on different types of terrain (Fig. 8).

During the work shift, the operators are exposed to several tasks: it is important to set up a Multitask Analysis of the push and pull activities.

The curves, the steps and the depressions of the ground introduce further variables to the Multitask Analysis.

The analysis of the different situations allows the creation of a database of data useful for mapping in advance the level of risk.

On a representative day of the year it is possible to define the type of area in which a homogeneous group of operators must work: the push and pull operations take place on routs with different characteristics (Fig. 9).

Even with the same type of flooring, differences in the results of the different tests are observed.

Fig. 8. Examples of push an pull tasks

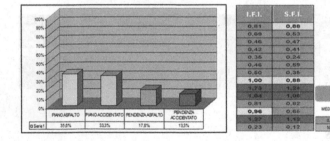

Fig. 9. Example of distribution of different tasks

The exposure level of the operators will be a mix of the various tasks that alternate in the work shift.

A multitask calculation can be used to identify acceptable values and critical values and a Final Index to estimate the operator exposure level.

7 Discussion

High-sampling digital dynamometers allow to accurately observe the performance of the force during push and pull activities.

The Sustained Force is the hardest magnitude to be assessed and interpreted; it represents the magnitude of the necessary force to maintain in movement to constant speed the cart. This force can be very variable depending on floor conditions (holes or depressions of the ground), the kind of ground (asphalt or sand), the kinds of wheel (small or large diameter) and other variables.

Therefore the trend of the sustained force will not be constant, but it could show variable fluctuations (see Fig. 2)

In various measurement experiences carried out in working environments such as urban waste management, hospital departments, or manufacturing departments, it has been observed that Sustained Force is not constant, but also varies rapidly. These variations are intrinsic to any handling activity, but introduce a further difficulty in defining the value of this factor.

The greatest difficulty is the estimate of the Sustained Force: through a direct observation it is possible to choose the most frequent value. This value is included in a range of values that can potentially all represents the Sustained Force (see Fig. 4).

In the Authors experience of over 450 measurements carried out between 2010 and 2017, the collected data through direct observation of the display during the Push and Pull tests and the data of the same identical tests recorded with high sampling frequency (1000 samples/sec) were compared.

It is important to note that the measurements made in this way represent a subjective value that depends on the skills and the experience of the evaluator.

As the Authors first analysis, a value above the average appears to be correct even though there are not clinical data to be able to affirm it yet. In support of this hypothesis, also other highly reliable methodologies, for example OCRA Method, Composite Lifting Index Sequential Lifting Index and Variable Lifting Index proposed by NIOSH, are strongly non-linear on some factors and they attribute greater weight to factors or tasks with higher overload.

It will be important for the future to collect the data from the Push and Pull tests and the clinical data of the exposed operators to these activities in order to accurately determine the best method to analyze the results.

References

1. Snook SH, Ciriello VM (1991) The design of manual handling tasks: revised tables of maximum acceptable weights and force. Ergonomics 34(9):1197–1213
2. Colombini D, Grieco A, Occhipinti E (1998) Ergonomics: special issue. occupational musculoskeletal disorders of the upper limbs due to mechanical overload. Ergonomics 41:9
3. Occhipinti E, Colombini D (2007) Updating reference values and predictive models of the OCRA method in the risk assessment of work-related musculoskeletal disorders of the upper limbs. Ergonomics 50(11):1727–1739
4. Waters TR, Putz-Anderson V, Garg A (1994) Application manual for the revised NIOSH Lifting Equation. NIOSH, Cincinnati
5. Waters TR, Lu ML, Occhipinti E (2007) New procedure for assessing sequential manual lifting jobs using the revised NIOSH lifting equation. Ergonomics 50(11):1761–1770
6. Colombini D, Occhipinti E, Alvarez-Casado E, Waters TR (2017) Manual lifting: a guide to the study of simple and complex lifting tasks. CRC Press, Boca Raton

The Characterization and Evaluation of an Intervention to Reduce Neonate Whole Body Vibration Exposures During Ambulance Transport

Dawn M. Ryan[1]([⊠])(iD), Adam Lokeh[2], David Hirschman[2], June Spector[1], Rob Parker[3], and Peter W. Johnson[1]

[1] University of Washington, School of Public Health, Seattle, USA
dmryan@uw.edu
[2] Children's Minnesota, Minneapolis, USA
[3] Bose Corporation, Framingham, USA

Abstract. Newborn infants delivered in a compromised health state often require transport between secondary and primary care hospitals. The objective of this study was to measure and characterize the WBV exposures during simulated newborn infant inter-hospital ground transport and determine how vehicle-based vibration is transmitted through the chain of equipment used to support newborn infants and whether there is a need and potential for mitigation of these exposures. A simulated newborn infant was transported over a 46-min, 32 km route between two hospitals to simulate a typical transport route. The route was completed with a standard transport system as well as a new, modified vibration dampening transport system. The average-weighted vibrations and the vibration dose values were calculated. Relative to the floor measured vibration (0.36 m/s^2), the standard transport system amplified the average weighted vibration through the chain of equipment nearly doubling the vibration at the interface where the simulated neonate rested (0.67 m/s^2). With the new system, the vibration at the point just above the suspension system was almost half (0.25 m/s^2) of the floor measured vibration (0.44 m/s^2), but then increased to a maximum of 0.48 m/s^2 at the interface where the simulated neonate rested. Results were similar for VDV exposures. When comparing the two systems, the standard stretcher amplified the floor measured vibration by 86% whereas the new stretcher system with the built-in suspension amplified the vibration by just 9%. Options for further investigation and mitigation of vibration in future studies is discussed.

Keywords: Action limit · Exposure limit · Average-weighted vibration VDV

© Springer Nature Switzerland AG 2019
S. Bagnara et al. (Eds.): IEA 2018, AISC 820, pp. 670–677, 2019.
https://doi.org/10.1007/978-3-319-96083-8_84

1 Introduction

1.1 Background

In adults, prolonged whole-body vibration (WBV) exposure has been linked to multiple adverse health outcomes. Potential health effects for children may be similar, or there may be other unknown and/or additional impacts due to their more fragile and developmental state. Newborn infants delivered in a compromised health state often require transport between secondary and primary care hospitals. During a period when they are already at increased vulnerability and risk for adverse health outcomes, the introduction of external stressors such as mechanical vibration during transport may cause increased morbidity and even increased mortality [1]. Neonates experience high levels of mechanical vibration and shocks during inter-hospital ground transport [2], but it is not clear how the transport equipment affects these WBV exposures. WBV exposures, when high, may impact the infants' near and longer-term health outcomes [3].

The current occupational WBV standard for adults [4] suggests two methods for evaluating WBV exposures: (1) the weighted root mean square (r.m.s.) acceleration (A_w) in m/s^2; and (2) the vibration dose value (VDV) in $m/s^{1.75}$ when the vehicle vibrations are expected to contain impulsive jarring and mechanical shocks. The A_w is averaged over time and was designed to measure the vehicle occupant's exposure to continuous, typically lower amplitude, cyclical vibration exposures. In contrast, the VDV is a cumulative measure and was designed to measure the cumulative impact on the vehicle occupant's body from the larger amplitude mechanical shocks and jolts. The ISO-2631-1 and EU Vibration Directive [5] 8-h daily vibration action values for A^w and VDV exposures for adult occupational exposures is 0.5 m/s^2 and 9.1 $m/s^{1.75}$, respectively.

1.2 Objectives

The primary objective of this study was to measure and characterize WBV exposures during simulated newborn infant inter-hospital ground transport and determine how the vehicle-based vibrations are transmitted through the stretcher and the chain of equipment used to support and protect the newborn infants. This study also characterized and compared WBV exposures between a standard stretcher system and a new prototype stretcher with a built-in suspension system designed to absorb and reduce vehicle-induced vibrations. Simulated neonate WBV exposures were measured and characterized using both the standard and new stretcher systems to characterize vibration exposure levels, determine whether the equipment in the stretcher systems altered vibration exposures and whether there were differences in vibration exposures between the stretcher systems.

2   Methods

WBV exposures were collected when a 1.3 kg simulated newborn infant was repeatedly transported by ambulance over a 46-min, 32 km route between two hospitals. Measurements were taken from two different stretcher systems: (1) a standard neonate ground transport stretcher system (two times), and (2) a custom-made stretcher system with a built-in suspension (four times). The standard stretcher system consisted of a standard power load cot (Power-PRO IT; Stryker, Kalamazoo, MI). The second stretcher system utilized the same power-load cot, but it also contained an air-assisted scissor suspension on the top of the stretcher for absorbing vertical vibration, and ball-bearing based translating actuators for absorbing fore/aft and side-to-side lateral vibrations. On top of each stretcher was a detachable aluminum transfer sled which contained life support equipment and a standard isolette (Airborne Voyager Transport Incubator; International Biomedical, Austin, TX) which housed and protected the simulated neonate.

Inside the isolette the simulated neonate rested on top of a fluidized positioner (Z-Flo; Molnlycke Health Care; Norcross, Georgia) which was supported by a standard gel mattress underlayer. To record vibration exposures, as shown in Fig. 1, six accelerometers (Model 356B41 and Model 352C33; PCB Piezotronics; Depew, NY)

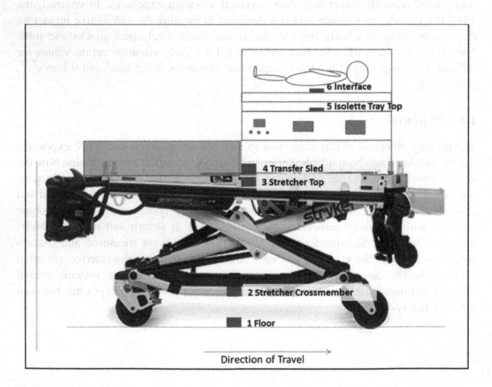

Fig. 1. Instrumentation diagram; red rectangles indicate position and location of accelerometers on the transport equipment.

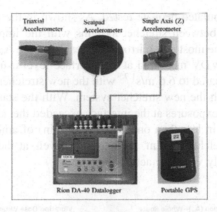

Fig. 2. Data collection equipment

were placed on the stretcher systems starting at the ambulance floor and ending at the interface between the fluidized positioner and the simulated, newborn infant [1]. Continuous vibration data were gathered and stored in two data loggers (DA-40; Rion Co. LTD; Tokyo, Japan) at a sample rate of 1280 Hz, and simultaneous global positioning system (GPS) data were collected at 1 Hz using a portable GPS unit (Model CR-Q1100P; Qstarz Co.; Taipei, Taiwan) (Fig. 2). After the data collection, the acceleration and GPS data were aligned and combined into one file so the multi-trip data could be segmented into single inter-hospital routes [6].

A LabVIEW program was used to apply vibration weighting as described in ISO 2631-1 [4] standards, and the predominant, vertical A_w WBV exposures and the VDV were calculated across the chain of transport equipment shown in Fig. 1. In order to better compare the stretcher systems, the vibration exposures of each stretcher system were normalized at the floor and compared up the chain of the transport equipment and between the stretcher systems. In addition, the relative time to reach vibration action limits was normalized and calculated relative to the vibration exposures at the floor of the ambulance.

3 Results

As shown in the left portion of Fig. 3, the average, vertical, WBV exposures (A_w) measured from the ambulance floor using both the standard and the new stretcher systems were below the adult, occupational-based, daily vibration action limits (0.36 m/s^2 and 0.44 m/s^2, respectively). However, the standard stretcher amplified the vibration exposures moving up the equipment chain from the floor to the interface supporting the neonate (solid lines, Fig. 3). Utilizing the standard stretcher system, the WBV exposures at the interface supporting the neonate nearly doubled (0.67 m/s^2) relative to the vibration measured at the floor (0.36 m/s^2). With the new stretcher system, which contained the built-in suspension, the vibration at the aluminum transfer sled, which was just above the built-in suspension, was lower (0.25 m/s^2) than the floor measured vibration (0.44 m/s^2), but then increased to a maximum of 0.48 m/s^2 at the interface where the simulated neonate rested. As shown in the right portion of Fig. 3, the trends within and between stretcher systems with the impulsive, vertical, WBV exposures (VDV) for the most part mirrored the results of the A_w. When comparing the stretcher systems, the VDV measured at the interface level using the standard system was 10.9 m/s$^{1.75}$, compared to 6.6 m/s$^{1.75}$ with the new stretcher system, almost a 40% exposure reduction with the new stretcher system. With the standard stretcher system, both the A_w and VDV exposures at the interface exceeded the adult, occupational daily vibration action limits in less than one hour ($\sim$46 min) of ambulance travel. In contrast, with the new stretcher system, the exposure levels at the interface were below adult, occupational daily vibration action limits.

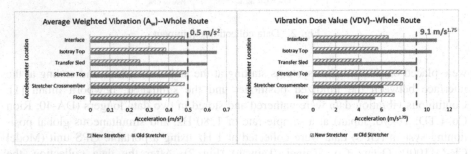

Fig. 3. Average weighted vibration exposures (A_w, left) and cumulative impulsive vibration exposures (VDV, right). Measurements averaged over the inter-hospital routes which were $\sim$46 min long. The vertical dashed lines indicate the daily (8-h) adult, occupational vibration action limits from the ISO 2631-1 standard.

The WBV exposures for both the standard and new stretcher systems were then compared to determine whether the stretcher systems were amplifying or attenuating vibrations relative to the input vibrations measured at the floor. As can be seen in Fig. 4, both systems transmitted vibrations up the equipment chain; the standard stretcher system amplified the average weighted vibration exposures and the cumulative, impulsive vibration exposures by 1.8- and 2-fold respectively (solid lines in left and right portions of Fig. 4). In contrast, with the new stretcher system (striped lines in Fig. 4), the average weighted vibration increased vibration exposures by just 6% relative to the floor, and the cumulative impulsive exposures were reduced by 8%.

The average weighted vibrations and impulsive vibrations experienced with both the standard and new stretcher systems were then used to determine the relative length of time, normalized and compared to the vibration inputs at the floor, that an infant could ride in the isolette before reaching vibration action limits. As can be seen in Fig. 5, average weighted vibration exposures and impulsive vibration dose values exposures experienced by the simulated neonate in the standard system (solid lines in Fig. 5) would reduce transportation times by 70% and 93% respectively due to the amplification of the vibrations through the stretcher system. With the new stretcher system (striped lines in Fig. 5), the transmission of the floor measured vibrations through the stretcher system was reduced. For the average weighted vibration

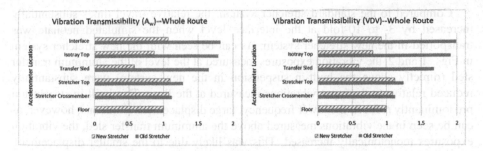

Fig. 4. Vibration transmissibility of the stretcher systems relative to the floor for the average weighed vibration exposures (A_w, left) and cumulative, impulsive vibration exposures (VDV, right). Measurements averaged over the inter-hospital routes which were ~46 min long.

exposures, the transportation times were only reduced by 12%, and for the impulsive vibration dose values exposures, the transportation times increased by 36%, due to the built-in suspension's ability to better absorb the bumps, jolts and jarring. Due to the reduced vibrations, when compared to the standard stretcher system, the simulated infant in the new stretcher system could ride 3- to 10-fold longer based on the A_w and cumulative, impulsive VDV exposures, respectively.

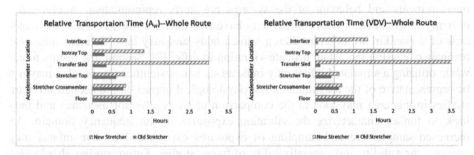

Fig. 5. Relative transportation times normalized relative to the input vibrations measured at the floor for the average weighted vibration exposures (A_w, left) and cumulative, impulsive vibration dose values exposures (VDV, right). Relative transportation times based on the intra-hospital routes which were ~46 min long.

4   Discussion/Conclusion

The exposures measured using the standard stretcher system may cause some concern, as the simulated ambulance ride lasted less than an hour yet the vibration exposures exceed the 8-h occupational daily vibration action levels set forth for adults. These adult-based, occupational exposure standards may not be ideal for comparison purposes; however, they are the only relevant comparisons that we can make at this time. The new stretcher system substantially reduced both the overall average weighted vibrations and the cumulative, impulsive vibrations.

Compared to the standard stretcher system, the relative travel times substantially increased by 3- to 10-fold at the interface level when the simulated neonate was transported in the new stretcher system. As can be seen with the new stretcher system in Figs. 1 and 2, the vibration exposures measured at the level of the aluminum transfer sled (directly above the built-in suspension in the new system) were substantially reduced relative to the input vibrations measured at the floor. This built-in suspension predominantly reduced the lower frequency, large displacement vibrations; however, as can be seen in the vibrations measured above the aluminum transfer sled, the vibration exposures monotonically increased. This was likely due to the smaller displacement, higher frequency vibrations passing through the built-in suspension and causing the components above the aluminum transfer sled to resonate and amplify the higher frequency vibration content. The end result was that the new stretcher considerably reduced the lower frequency, larger amplitude impulsive vibrations, but the equipment above the level of the aluminum transfer sled appeared to amplify the higher frequency vibration exposures. Further investigation may be merited to evaluate the structures at and above the aluminum transfer sled. Here, altering the weight and stiffness of the system components and/or adding vibration isolators between components may help. Finally, improvements may be made by altering the vibration properties of the mattress and/or fluidized positioner that support the baby in the isolette.

The limitations of this study include the small sample size. However, due to the standardized route, there was not that much variability between measurements, and the measurements and behavior of the systems are fairly representative. Additionally, utilizing a simulated newborn infant may have two different impacts. First, the materials of a manikin are different than a human body and may have different mass and resonance, causing the actual neonate vibration to differ [7]. Second, the results found while utilizing a simulated infant may indicate statistical significance, but this may not be representative of the magnitude of the physiological impact on the neonates. Future studies in this area may look at the comparison of tri-axial/vector measures and may look to further characterize the vibration exposures in the frequency domain. An increased sample size and sampling of exposures experienced by live infants may improve the validity and generalizability of future studies. Future studies should also look at additional and alternative mitigating strategies to further improve the attenuation of the experienced vibrations.

References

1. Shah S, Rothberger A, Caprio M, Mally P, Hendricks-Munoz K (2008) Quantification of impulse experienced by neonates during inter- and intra-hospital transport measured by biophysical accelerometery. J. Perinat. Med. 36:87–92. https://doi.org/10.1515/JPM.2008.009
2. Shenai JP, Johnson GE, Varney RV (1981) Mechanical vibration in neonatal transport. Pediatrics 68(1):55–57
3. Grosek S, Mlakar G, Vidmar I, Ihan A, Primozic J (2009) Heart rate and leukocytes after air and ground transportation in artificially ventilated neonates: a prospective observational study. Intensive Care Med. 35(1):161–165. https://doi.org/10.1007/s00134-008-1256-8

4. International Standards Organization ISO 2631-1 (1997) Mechanical vibration and shock-Evaluation of human exposure to whole-body vibration-Par1-General Requirements, ISO
5. European Union (2002) Directive 2002/44/EC of the European Parliament and of the Council of 25 June 2002 on the minimum health and safety requirements regarding the exposure of workers to the risks arising from physical agents (vibration) (sixteenth individual Directive within the meaning of Article 16(1) of Directive 89/391/EEC)
6. Blood RP, Rynell PW, Johnson PW (2011) Vehicle design influences whole body vibration exposures: effect of the location of the front axle relative to the cab. J Occup Environ Hyg 8 (6):364–374. https://doi.org/10.1080/15459624.2011.583150
7. Sherwood HB, Donze A, Giebe J (1994) Mechanical vibration in ambulance transport. J. Obstet. Gynecol. Neonatal. Nurs. 23(6):457–463. https://doi.org/10.1166/jnn.2012.6247

Real-Time Monitoring of the Posture at the Workplace Using Low Cost Sensors

Marco Tarabini[1(✉)], Marco Marinoni[1], Matteo Mascetti[1],
Pietro Marzaroli[1], Francesco Corti[1], Hermes Giberti[2],
Paolo Mascagni[3], Alberto Villa[4], and Tammy Eger[5]

[1] Dipartimento di Meccanica, Politecnico di Milano, Via La Masa 1, Milan, Italy
marco.tarabini@polimi.it
[2] Dipartimento di Ingegneria Industriale, Università di Pavia, Pavia, Italy
[3] Dipartimento di Medicina del Lavoro, Ospedale di Desio, Desio, Italy
[4] Canali SPA, Sovico, Italy
[5] Centre for Research in Occupational Safety and Health, Laurentian University,
Sudbury, Canada

Abstract. The aim of this paper is to show a method that can be used to monitor the human posture in an industrial environment. The method is based on the fu- sion of the data coming from two different sensors: a time-of-flight camera (Microsoft Kinect V2) and a wearable motion capture system that uses inertial measurement units to identify the body posture (Notch Wearable). The combined use of these two systems overcomes the intrinsic limitations of the two methods, deriving from occlusions and electromagnetic interferences, respectively. First, the algorithms implemented and the calibration of the two measurement systems in a controlled environment are described. Second, the method applied in a workplace to monitor the posture of the workers during different tailoring operations, is explained. The data acquired have been ana- lyzed in the time domain, and used to compute the cumulative probability density function of different body angles. The results are compared to the subjective evaluation of occupational doctors, and used to compute the OCRA index in an auto- mated way, for the assessment of workers exposure to repetitive movements of the upper limbs.

Keywords: Posture · Biomechanical loads · Repetitive tasks · Measurements

1 Introduction

The repetitive tasks, together with other factors like vibrations, impulsive force and awkward posture, have been linked to musculoskeletal disorders in the upper limbs [1, 2]. Due to the high number of causes, so far, a univocal method for assessing the level of risk to develop such disorder has not been defined [3]. The main methods used by ergono- mists are self-reported questionnaires [4] and direct observation [5, 6]. These methods rely on subjective evaluations. For this reason, the results may vary a lot when comparing two different methods with a direct measurement [7]. Moreover, results may vary, thus

S. Bagnara et al. (Eds.): IEA 2018, AISC 820, pp. 678–688, 2019.
https://doi.org/10.1007/978-3-319-96083-8_85

resulting in unreliable findings when different ergonomists are evaluating the same tasks, and even when the same ergonomist is asked to repeat the same evaluation after a few weeks [8].

Motion capture techniques can be used in ergonomics to evaluate working postures. Human motion tracking methods can be divided into three main families: visual based, non-visual based and robot aided methods. The visual based methods use different kinds of cameras to record and analyze human body movements. Non-visual based methods use inertial sensors, magnetometers or acoustic sensors to detect the relative motion with respect to a reference position; the robot-aided techniques use the geometrical configuration of the exoskeleton that is supporting the subject during the movement to identify the limbs position.

Recent studies [9–15] focused on the use of time of flight cameras for motion tracking; one of the most popular systems is the Kinect V2, a time of flight camera produced by Microsoft from 2014 to 2017. The starting aim of this camera was to be connected to an Xbox console to interact with the player through voice recognition and body movements. This system is based on the use of an array of 3 lasers, an infrared camera, an RGB camera, and 4 microphones for voice command interpretation. The proprietary Software Development Kit allows visualizing in real time and saving infrared images, color images, depth images and audio files. Vision algorithms allow the user to obtain a human body reconstruction through a skeleton made up of 25 joints.

Different studies focused on the accuracy of the Microsoft Kinect V2; Caruso et al. [9] studied the use of the Kinect V2 as a vision system in manufacturing applications, in order to detect the path of a robot over an object in its working space. Results confirmed the possibility of using the Kinect V2 instead of industrial 3D cameras, that are more accurate but more expensive. Munaro et al. [10] studied performances of the open source software for multi-camera people tracking, OpenPTrack, from which it is possible to state the superiority of Microsoft Kinect V2 over the other similar low-cost sensors used for human body tracking. In particular, in the literature there are several studies focused on the accuracy of the Kinect V2 in tracking the upper body posture. Giancola et al. [11] studied the use of Kinect V2 for upper body motion tracking. All these studies reported that the measure is influenced by the camera- subject relative position. Problems using a frontal acquisition could derive from the overlapping of arm segments with spine segments. This problem was studied by Yang et al. [12] who proposed a system that uses the data fusion of different Kinect cameras posed in different angulations with respect to the subject, in order to have a 360° global field of view and, through non-linear trilateration, solve the problem of body segments overlapping due to the relative angulation subject-camera in controlled conditions. Several works compared the Kinect performances to those of accurate instruments for human motion tracking: Otte et al. [13] compared the performances of the Kinect with those of a Vicon system. Most of the clinical parameters showed consistency between both systems. The main problem related the use of the Kinect in an industrial environment is the possible presence of occlusions that could lead to partial acquisition and as a consequence, misleading information. Plantard et al. [14] faced the problem of Kinect acquisition in real workplace conditions. The aim of the paper is to propose and test a method to estimate RULA score. The problem of occlusions has been faced following a scheme proposed by Plantard in [15]. Occlusions are detected through comparison with a gold standard constituted by a database of human poses listed in the Method Time Measurement list (MTM), captured by a Vicon system.

Another category of systems for monitoring human posture is based on wearable systems. In this field, there are several studies that use inertial measurement units fixed to the body using straps for the identification of the body kinematics. Performances of IMU based methods have been compared to the vision-based ones by Karatsidis et al. [16], which compared the ground reaction forces and moments during gait. The comparison between the Xsens MVN and a state of art visual inspection system (8 infrared camera OMC system with 53 markers) showed strong correlations for the frontal and sagittal plane moments. Kok et al. [17] used a 17 sensors Xsens MVN to estimate the relative position and orientation of different body segments and the absolute position of the body in space. Results showed a quick convergence of the method and the matching in joint angle estimate from IMU and optical reference system. Koenemann et al. [18] used the Xsens MVN to enable the humanoids imitation of complex whole-body motions in real time. Results showed the system was able to reliably reconstruct full body motions allowing the robot to imitate complex motion sequences. The main issues preventing the use of IMU-based systems in industrial environment is the presence of electromagnetic disturbances, which in some conditions may lead to a biased estimation of the sensor rotation with respect to the vertical axis. This happens, for instance, close to electrical motors or to metallic frames, where the earth magnetic field is perturbed. In this work, we describe a method that uses both the Microsoft Kinect V2 and the Notch systems for human body posture identification at the workplace. By using both system simultaneously, it would be possible to compensate for their intrinsic limitations. We propose an occlusion detection algorithm (both body-self-induced and external occlusions), a frame filtering method and a fusion with wearable sensors to compute the biomechanical loads of workers subjected to repetitive tasks of the upper limbs.

The paper is structured as follows: the proposed method is described in Sect. 2; experimental results are presented in Sect. 3. In Sect. 4, the results are discussed and the conclusions are drawn.

2   Method

This section describes the algorithm used to acquire and merge data from the Kinect V2 and the Notch wearable system and the tests carried out in a real workplace environment.

2.1   Data Acquisition and Data Fusion

As previously explained, the Kinect V2 provides the coordinates of the 25 joints of the skeleton. These information must be pre-processed to avoid artefacts due to occlusions and to identify the body's characteristic angles. The pre-processing is done in three steps:

1. The first step consists in resampling the data. This is necessary to be independent from the Kinect data acquisition rate, which is not constant. The trajectories of every joint (x, y, z coordinates) are FIR filtered and resampled with a rate of 30 Hz, to be consistent with the data acquired by the Notch system.
2. Then, the characteristic body angles are computed from the positions of the 25 joints through simple goniometric Eqs.
3. Finally, the identification of unreliable measurements is based on the comparison between the reference skeleton, derived from a static capture of the worker that is being monitored, and the skeleton acquired in each frame during the working operations.

The reliability of the skeleton is based on the Joint Similarity Index, s, which is computed as follows:

$$s(i,j) = \left[1 - \left(\frac{Dynamic\,Length(i,j)}{Static\,Length(i,j)}\right)\right]^2 \tag{1}$$

Where i is the number of the frame that is being analyzed, and j is the index of each body segment of the skeleton. The Overall Similarity Index, S, is then computed at each time frame as follows:

$$S(i) = \sqrt{\sum_{j=1}^{25} s(i,j)} \tag{2}$$

The similarity index S(i) is close to zero when all the joints of the observed subject are similar to those of the static acquisition and is used to understand if the skeleton is reliable or not. All the acquisitions where S(i) is higher than a threshold are not considered in the analysis. The threshold has been chosen with a statistical analysis as the value that guarantees the maximum number of acquired frames with the minimum decrease of frames quality. The use of this index has been validated in tests performed by subjects moving in a complex scene, verifying that unreliable data were automatically discarded.

The Notch wearable system is able to directly provide the relative angles of the body segment on which it is applied, sampled at 30 Hz. The manufacturer ensures an accuracy in angles measure of less than $3°$; this value has been verified with laboratory tests. Such tests were performed with a mannequin and with a real subject, imposing known movements of upper limbs along a given path, controlled with mechanical elements and measured with the digital spirit level 416 mm (Batavia GmbH, Buitenhuisstraat 2a 7951 SM, Staphorst, Netherlands). In particular, such tests were meant to evaluate the effect of the variability in the identification of the standing position affects the measurement accuracy. The results were compatible with the values declared by the manufacturer, in fact the tests performed with the mannequin had a measurement repeatability less or equal to $1°$; the difference between the tests average and the indication of a digital inclinometer were less than $2°$. In tests performed by a real subject, the differences between nominal value and mean value of different tests of less than $3°$.

The two systems did not acquire data simultaneously and the acquisition did not start at the same time. For this reason, the first step of the data fusion was to align the different waveforms. This was done with a cross-correlation function: the time delay was identified with the time at which such function has the absolute maximum. Then, the body angles were estimated as the arithmetic average between the angles acquired by the notch and the angles computed from the data acquired by the Kinect, if they were determined to be reliable.

2.2 Experimental Tests

The previously described acquisition system was used to monitor the tailoring operations at Canali SPA. Tests were carried out in two days, monitoring seven different works operations, identified by a code number:

- 700: it involved picking up a piece of fabric placed on a stock pile located to the left of the worker and laying it on the template of the press ironing machine, then retrieving and laying it on a stock pile located to the right of the worker
- 701: it involved passing a piece of fabric located to the right of the worker under the sewing machine for the whole length of the piece, then laying it on the stockpile located to the left of the worker
- 722: it involved taking a piece of fabric located to the left of the worker and passing a short corner of it once under the sewing machine, then laying it on the stockpile located to the right of the worker.
- 812: it involved taking a piece of fabric located to the left of the worker and passing a side of it once under the sewing machine, then laying it on the stockpile located to the right of the worker
- 1310: it involved picking a piece of fabric hanging on a rail located to the back of the worker and laying it on the template of the press ironing machine, then retrieving and hanging it on another stock pile located to the back of the worker
- 1311: it involved picking a piece of fabric hanging on a rail located to the back of the worker and laying it on the template of the press ironing machine, then retrieving and hanging it on another stock pile located to the beck of the worker
- C913: it involved taking a piece of fabric located to the left of the worker at passing three short corner of it once under the sewing machine, then laying it on the stockpile located to the right of the worker

Eight workers performed the tests; each of them wore six Notch sensors during day A and seven Notch sensors during day B, as shown in Fig. 1. The steady position was identified by using mechanical bars for ensuring the straightness of the arms. Every operation was performed for at least 15 min. Every session of 15 min was divided into 3 acquisitions of 5 min each to limit the time requested for the single download of the files via Bluetooth for the Notch.

At least two Kinects observed the worker in each test. The position of the Kinects was identified with preliminary tests to avoid as many occlusions, as possible, given by the machines and by the handled fabrics. Kinects were located according to the scheme of Fig. 2.

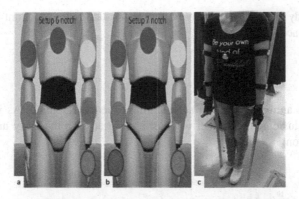

Fig. 1. Notch sensors positions for both days: a, b; and steady position identification setup: c

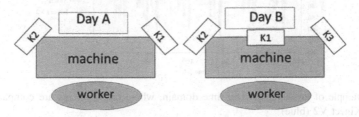

Fig. 2. Setup for Kinect V2 on workstations for both days

For every test, the angles that were evaluated are as follows: shoulder angle: abduction/adduction ($\pm$; α) and flexion/extension ($\pm$; β); elbow angle: flexion/extension ($\pm$; δ) and forearm supination/pronation ($\pm$; ε); wrist angle: flexion/extension ($\pm$; η) and radial/ulnar deviation ($\pm$; ζ). The angles have been evaluated bilaterally and are shown in Fig. 3.

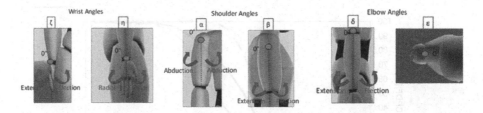

Fig. 3. Body angles evaluated

Videos of the same operations were taken and analyzed by experienced occupational doctors, which provided their evaluation of the maximum and minimum value assumed by the relevant body angles. Their evaluation has been compared to the 10th percentile and the 90th percentile of the average cumulative probability density functions (CPDF)

684   M. Tarabini et al.

of the same body angles acquired by both systems, taking into account the three 5-min sections of the acquisition.

3 Results

Figure 4 shows an example of an acquisition in the time domain. As it is possible to see, due to occlusions, the Kinect was not able to acquire the body angle during the whole acquisition.

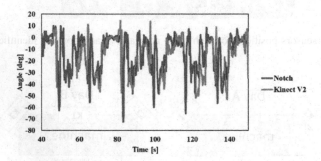

Fig. 4. Example of an acquisition in time domain, where both systems are compared: Notch (red) and Kinect V2 (blue)

The missing data had a relevant impact in the CPDF of the angles computed from both acquisition systems, as it is reported in Fig. 5. However, when the Kinect V2 was able to measure the joints position for more than 90% of the time, the angles measured by both systems were compatible, as it shown in Figs. 6, 7 and 8. The following plots show the average between the measurements performed with the Notch and Kinect systems (black lines). The red lines show the max and min CPDF recorded by one of the two systems.

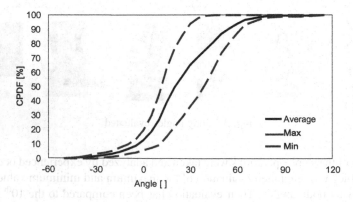

Fig. 5. CPDF of the right Alpha angle during operation 701.

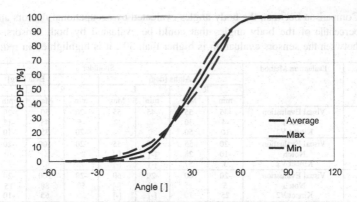

Fig. 6. CPDF of the left Beta angle during operation 701.

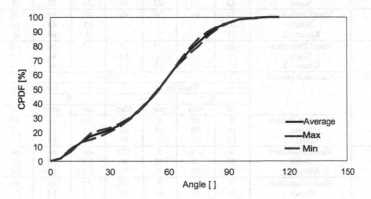

Fig. 7. CPDF of the right Delta angle during operation 700.

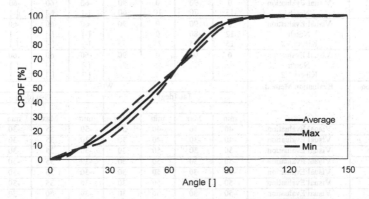

Fig. 8. CPDF of the left Delta angle during operation 700.

Table 1. Comparison between the body angles evaluated by occupational doctors and the 10th and 90th percentile of the body angles that could be evaluated by both sensors. When the difference between the sensors evaluation is higher than 30°, it is highlighted in red.

Operation	Evaluation Method	Shoulder							
		Alpha [deg]				Beta [deg]			
		R		L		R		L	
		min	Max	min	Max	min	Max	min	Max
700	Visual Evaluation	-45	55	-45	55	-20	50	-20	50
	Notch	-4	30	-40	-2	-5	45	-15	40
	Kinect V2	10	50	10	55	-20	35	-10	40
701	Visual Evaluation	-20	55	-20	55	-20	100	-20	100
	Notch	-10	28	[-]	[-]	27	84	30	67
	Kinect V2	3	57	[-]	[-]	-22	47	-10	50
722	Visual Evaluation	-20	60	-20	60	-20	100	-20	100
	Notch	5	35	[-]	[-]	50	88	15	67
	Kinect V2	25	70	[-]	[-]	15	65	-10	95
812	Visual Evaluation	-20	90	-20	90	-20	100	-20	100
	Notch	20	90	55	125	40	93	[-]	[-]
	Kinect V2	0	55	0	105	-15	65	[-]	[-]
1310	Visual Evaluation	-30	45	-30	45	-20	80	-20	80
	Notch	-13	-15	-25	10	20	70	10	65
1311	Visual Evaluation	-30	45	-30	45	-20	80	-20	80
	Notch	10	35	-30	0	15	43	15	48
	Kinect V2	15	80	7	38	-5	60	0	35
C913	Visual Evaluation	-20	60	-20	60	-20	60	-20	60
	Notch	-5	35	[-]	[-]	17	43	20	52
	Kinect V2	47	80	[-]	[-]	47	69	-50	35

Operation	Evaluation Method	Elbow							
		Delta [deg]				Epsilon [deg]			
		R		L		R		L	
		min	Max	min	Max	min	Max	min	Max
700	Visual Evaluation	0	90	0	90	-60	60	-60	60
	Notch	10	80	20	80	[-]	[-]	[-]	[-]
	Kinect V2	13	82	15	85	[-]	[-]	[-]	[-]
701	Visual Evaluation	0	120	0	120	-60	60	-60	60
	Notch	20	74	32	84	[-]	[-]	[-]	[-]
	Kinect V2	31	91	9	70	[-]	[-]	[-]	[-]
722	Visual Evaluation	0	120	0	120	-60	60	-60	60
	Notch	20	67	40	95	[-]	[-]	[-]	[-]
	Kinect V2	20	90	30	85	[-]	[-]	[-]	[-]
812	Visual Evaluation	0	120	0	120	-60	60	-60	60
	Notch	40	90	-40	83	[-]	[-]	[-]	[-]
	Kinect V2	30	97	-20	105	[-]	[-]	[-]	[-]
1310	Visual Evaluation	0	90	0	90	-60	60	-60	60
	Notch	15	70	10	60	[-]	[-]	[-]	[-]
1311	Visual Evaluation	0	90	0	90	-60	60	-60	60
	Notch	12	80	0	35	[-]	[-]	[-]	[-]
	Kinect V2	15	55	25	85	[-]	[-]	[-]	[-]
C913	Visual Evaluation	0	90	0	90	-60	60	-60	60
	Notch	55	78	45	82	[-]	[-]	[-]	[-]
	Kinect V2	15	37	-1	130	[-]	[-]	[-]	[-]

Operation	Evaluation Method	Wrist							
		Tau [deg]				Nu [deg]			
		R		L		R		L	
		min	Max	min	Max	min	Max	min	Max
700	Visual Evaluation	-40	30	-40	30	-30	35	-30	35
701	Visual Evaluation	-40	30	-40	30	-30	35	-30	35
722	Visual Evaluation	-50	30	-50	30	-30	35	-30	35
812	Visual Evaluation	-50	30	-50	30	-30	35	-30	35
1310	Visual Evaluation	-50	30	-50	30	-30	35	-30	35
1311	Visual Evaluation	-50	30	-50	30	-30	35	-30	35
C913	Visual Evaluation	-50	30	-50	30	-30	60	-30	60

Table 1 compares all the measurements performed with the two systems with the subjective angle evaluations performed by an ergonomist. Results show that in general the reliability in the identification of the wrist angles is limited. The subjective evaluation of angles can result in findings that, in particular cases, can be misleading.

4 Discussion and Conclusions

Experimental results showed that both the Kinect and the Notch suffer from intrinsic limitations, the former in presence of occlusions and the latter in presence of electromagnetic interferences. The RMS difference between angles measured by the Kinect and Notch in operative conditions was generally lower than $10°$, and therefore smaller than the discretional margins deriving from the visual inspection performed by skilled operators. The difference between the CPDF can be attributed to occlusions and to the differences between the skeletal models of the two systems. The trunk, for instance is modelled by the Notch as a rigid body, while the Kinect has three nodes on the spine and one node on each shoulder. Whenever the Kinect estimates the position of one joint (instead of measuring it) in more than 10% of time, the measurements are usually unreliable; in the other cases, we decided to estimate the body angles as the arithmetic mean between the ones acquired by the Notch and the ones acquired by the Kinect. Another parameter used to assess the measurement reliability was the difference between the measurements of the Notch and The Kinect; whenever the difference in the computation of the percentiles was larger than the measurement uncertainty, the influence of the false movements was considered important and the body angles are considered unreliable. The next steps for the research will be focused on the adoption of more complex data fusion procedures and on the adoption of alternative methods for the identification of the skeleton using standard cameras.

References

1. Latko WA et al (1999) Cross-sectional study of the relationship between repetitive work and the prevalence of upper limb musculoskeletal disorders. Am J Ind Med 36(2):248–259
2. Occhipinti E (1998) OCRA: a concise index for the assessment of exposure to repetitive movements of the upper limbs. Ergonomics 41
3. David GC (2005) Ergonomic methods for assessing exposure to risk factors for work-related musculoskeletal disorders. Occup Med 55(3):190–199
4. Alexopoulos EC et al (2006) Musculoskeletal disorders in shipyard industry: prevalence, health care use, and absenteeism. BMC Musculoskelet Disord 7(1):88
5. Rafie F et al (2015) Prevalence of upper extremity musculoskeletal disorders in dentists: symptoms and risk factors. J Environ Pub Health 2015
6. Dos REIS, Cunha Diogo et al (2015) Assessment of risk factors of upper-limb musculoskeletal disorders in poultry slaughterhouse. Procedia Manufact 3:4309–4314
7. Spielholz P et al (2001) Comparison of self-report, video observation and direct measurement methods for upper extremity musculoskeletal disorder physical risk factors. Ergonomics 44(6):588–613

688   M. Tarabini et al.

8. Eliasson K et al (2017) Inter-and intra-observer reliability of risk assessment of repetitive work without an explicit method. Appl Ergon 62:1–8
9. Caruso L, Russo R, Savino S (2017) Microsoft Kinect V2 vision system in a manufacturing application. Rob Compu-Integr Manufact 48:174–181
10. Munaro M, Basso F, Menegatti E (2016) OpenPTrack: open source multi-camera calibration and people tracking for RGB-D camera networks. Rob Autonom Syst vol Part B 75:525–538
11. Giancola S, Corti A, Molteni F, Sala R (2016) Motion capture: an evaluation of Kinect V2 body tracking for upper limb motion analysis. In: Wireless mobile communication and healthcare: 6th international conference, Milan, Italy
12. Yang B, Dong H, El Saddik A (2017) Development of a self-calibrated motion capture system by nonlinear trilateration of multiple Kinects v2. IEEE Sens. J. 17(8):2481–2491
13. Otte K, Kayser B, Mansow-Model S, Brandt AU, Verrel J, Schmitz-Huebsch T (2016) Spatial accuracy and reliability of Microsoft Kinect V2 in the assessment of joint movement in comparison to marker-based motion capture (Vicon). In: 20th international congress of parkinson's disease and movement disorders
14. Plantard P, Shum HPH, Le Pierres A-S, Mu F (2017) Validation of an ergonomic assessment method using Kinect data in real workplace conditions. Appl Ergon 65:562–569
15. Plantard P, Shum HPH, Multon F (2017) Filtered pose graph for efficient Kinect pose reconstruction. Multimedia Tools Appl 76:4291–4312
16. Karatsidis A, Bellusci G, Schepers M, de Zee M, Andersen MS, Veltink PH (2017) Net knee moment estimation using exclusively inertial measurement units. In: XXVI congress of the international society of biomechanics, Brisbane, Australia
17. Kok M, Hol JD, Schön TB (2014) An optimization-based approach to human body motion capture using inertial sensors. IFAC Proceedings 47:79–85
18. Koenemann J, Burget F, Bennewitz M (2014) Real-time imitation of human whole-body motions by humanoids. In: 2014 IEEE international conference on robotics and automation (ICRA). IEEE, pp 2806–2812

A Software Toolbox to Improve Time-Efficiency and Reliability of an Observational Risk Assessment Method

Stefano Elio Lenzi[1]([⊠]) [iD], Carlo Emilio Standoli[2] [iD],
Giuseppe Andreoni[2] [iD], Paolo Perego[2] [iD],
and Nicola Francesco Lopomo[1] [iD]

[1] Dipartimento di Ingegneria dell'Informazione, Università degli Studi di
Brescia, Via Branze, 38, 25123 Brescia, Italy
{s.lenzi002, nicola.lopomo}@unibs.it
[2] Dipartimento di Design, Politecnico di Milano, Via Durando 38/A,
20158 Milan, Italy
{carloemilio.standoli, giuseppe.andreoni,
paolo.perego}@polimi.it

Abstract. OCRA is a standard risk assessment method addressing manual handling of low loads at high frequency. This method requires the operator to perform a video analysis checking kind and extension of the movements made by workers. The analyst has to take note about number of performed actions and joint angles amplitude. Often this turn out to be a poorly reliable and time-consuming operation because of the inherent 2D nature of the data. The main goal of this work was to design a software toolbox able to support the operator in collecting, organizing and analyzing the information to obtain the Checklist OCRA index in a more reliable and time-effective way. This toolbox presents three different GUIs to: (1) support the operator in counting the number of technical actions; (2) help the operator in determine the percentage of time in which the worker has an incorrect upper limb posture; (3) automatically perform posture analysis considering real 3D angles data acquired through an IMU-based movement analysis system. Preliminary analysis on reliability was performed on three different operators. Obtained findings confirmed our hypothesis; the automatic analysis, in particular, reduced significantly intra- and inter-operator variability thus making the analysis more objective and reliable. Further evaluations will include structured assessment including several operators with different expertise levels and collecting information about user experience (usability, GUI design, etc.) and overall performance compared to standards (operation time and results accuracy).

Keywords: Ergonomics · Checklist OCRA index · Kinematic analysis
Movement analysis · GUI

© Springer Nature Switzerland AG 2019
S. Bagnara et al. (Eds.): IEA 2018, AISC 820, pp. 689–708, 2019.
https://doi.org/10.1007/978-3-319-96083-8_86

1   Introduction

In Italy, there were about 58.129 complaints for occupational diseases in 2017 [1]; many of them were addressed to the musculoskeletal system (about 61%). This kind of pathologies could rise from a working environment that takes into the wrong account the workers' safety. For this reason, it is necessary and mandatory to take several measures and analyze working conditions so that workers can perform their activities in a controlled and supervised way.

In manual handling tasks, a possible cause of musculoskeletal diseases for the upper limb, is related to the repetition of loading tasks at high frequencies. Several ergonomic investigation methods are available but most of them have limitations for complex tasks.

Among the most widespread approach, the RULA (Rapid Upper Limb Assessment) index [2] can provide evaluation to exposure to risk factors by using diagrams of body postures and three scoring tables. The risk factors considered by the index are: number of movements, static muscle work, force, work postures determined by equipment and furniture and time worked without break. At first, the method requires to analyze separately upper and lower limbs from a kinematic point of view; then, the method provides you reference tables that associates the analyzed effort and musculoskeletal load to a specific score. The same procedure must be done also for the evaluation of force exertion required to accomplish the analyzed tasks, by considering the mass of the handled object and the number of the movement repetitions. The overall score is obtained using a double entry table that combines the score obtained for the upper limb and that one obtained for the lower limb. The main problem when using RULA, is that it is really focused on the evaluation of postures but give less importance to the other risk factors.

A different method was proposed by Moore and Garg [3], focusing on the Carpal Tunnel Syndrome (CTS), and considering force, repetition, posture, recovery time and type of grasp as key factors in the ergonomic investigations of workplaces. This approach underlines how muscle-tendon disorders can be frequently associated to jobs characterized (e.g. required force and repetitiveness). Their "Strain Index" (SI) specifically takes into account 6 multipliers: intensity of exertion, duration of exertion, exertions per minutes, hand/wrist posture, speed of work, duration of task per day. For each multiplier there is a score from 1 to 5 and SI is obtained as the product of all six multipliers. The intensity of an exertion is computed based on the perceived exertion using a scale close to the Borg CR-10 one, whereas the speed of exertion is estimated by the ergonomics team. Limitations with this method are that it is mainly addressed to the distal upper extremity and the relationships between exposure data and multiplier values are not based on mathematical laws. Also, the force is assessed with an observational approach (empirical scale). Other issues are related to its ability only to investigate monotasks job and additional risks factors are not considered.

In the same perspective, OREGE (Outil De Repérage et d' Évaluation des Gestes) method [4] evaluates efforts, joint angular displacement and frequency of movement. In order to estimate the effort, the ergonomist has to evaluate the object mass, the kind of grip, pressures, use of gloves, vibration and couple effect. The effort is computed through a self-assessment scale. Joint angles are estimated starting from the observation of movements made by workers and applying rules contained in the ISO 1005-1,-2,-3,-4. The frequency of the actions is evaluated by observing workers movement and qualitatively establishing if the movements are repeated or not frequently. Starting from the evaluation of each of the three risk factors evaluation, the analyst can put all the scores in a table and summarize the evaluation for each identified action. Each action is evaluated on the basis of a three level scores. Then the overall risk is determined starting from some consideration about the level of risk associated with each task. The main problem of this method is that it requires a combined approach with other methods because it cannot stand alone. Furthermore, it is based on an observational approach that involves workers and their perception of constraints. So, the final assessment is based on the experience of the ergonomist.

Lastly, OCRA (Occupational Repetitive Action) [5] represents a standard ergonomic risk assessment method. It is recognized as the most reliable one, because of its strong scientific basis and it is also suggested as the best method in the third part of the ISO 11228 [6]. This method specifically considers the following risk factors: posture, force and effort, working time and rests, features of the handled objects, vibrations, environmental conditions and frequency of the actions. All these factors are analyzed by the ergonomist and then you can associate a score to them. Postures and frequency are evaluated through the observation of workers in their working environment during their work shift. Force is determined through the Borg CR-10 scale and the other scores are obtained analyzing the work organization.

Summarizing, all these indexes can be considered scientifically well designed and conceived. They are easy to understand and apply, thus to help the ergonomist to quickly perform the risk assessment. However, the common denominator of all these methods relies on that all the kinematics and kinetics related to the working tasks are usually analyzed by observing a recorded video. By definition, this kind of approach is affected by subjective factors, including angles and posture estimation. Furthermore, the use of qualitative and/or semi-quantitative scales for the determination of the perceived effort let the indexes be not reliable and very operator-dependent, presenting low repeatability and high variability in the measurement themselves. Moreover, no specific and dedicated tools that can support analyst in collecting and organizing data are widely used and accepted, so far.

The aim of the present work was to design a software toolbox that could help analyst to collect, organize and analyze the information for the risk assessment of upper limb musculoskeletal diseases. The main hypothesis was that, with the aid of such a tool, the ergonomic analysis can achieve more reliable and objective results, in a time-effective way. Here specifically, without loss of generality, OCRA index was considered because it represents one of the most reliable method for the ergonomic risk assessment, as also suggested by the ISO 11228-3.

2 Methodological Approach

In the perspective to develop a software toolbox to help the ergonomist to conduct the risk assessment, the design requirements represent a fundamental information. For this reason, several hints about theoretical background and key points are hereinafter reported.

2.1 Theoretical Background: ISO 11228-3

ISO 11228-3 reports guideline about the evaluation of tasks involving manual handling of low loads at high frequencies and suggest the OCRA Index as one of the more reliable method. The OCRA index is specifically defined as the ratio between the number of Actual Technical Actions (ATAs) execute by workers and the Reference Technical Actions (RTAs) that is the acceptable number of actions to prevent the occurrence of Upper Limb Work-related Musculoskeletal Disorders (UL-WMSDs), as in Eq. 1:

$$OCRAIndex = nATA / nRTA \qquad (1)$$

RTAs and ATAs must be computed for each upper limb side. ATAs is computed starting by the number of technical actions observed during the work shift, concerning a specific task. Then, it is possible to compute the frequency of actions per minute, through the definition of the ratio between number of actions observed and the time of the observation.

Once nATA is calculated, it is possible to calculate the number of RTAs. A value for each different multiplier is computed using empirical rules described in the international standard. Then all values are multiplied and thence nATA is obtained.

OCRA Checklist index method is a simplified version of the more complex OCRA. It requires to firstly determine, besides all the organisational aspects, the technical action for each upper limb side but without considering the type of action executed. Then postures must be analysed using the thresholds about the joint angles included in the ISO 11228-3. Hand, wrist, elbow and shoulder for each side are analysed using different criteria (Table 1), specifically:

- Shoulder: flexion or abduction at more than 80°;

Table 1. ISO 11228-3 range of motion limits for each considered joint

Joint	Movement	Limits
Shoulder	Flexion/Extension	>80°
Elbow	Pronation/Supination (Dynamic Movement)	>60°
	Flexion/Extension (Dynamic Movement)	>60°
Wrist	Palmar flexion/Dorsal extension	>45°
	Ulnar Deviation	>20°
	Radial Deviation	>15°

- Elbow: pronation/supination or flexion/extension above 60°. In this case only dynamic movement must be considered. If the workers maintain a static posture in which the elbow joint describes an angle above the threshold it is not considered as time in awkward posture;
- Wrist: palmar flexion or dorsal extension higher than 45°, ulnar deviation greater than 20° and radial deviation with a value over 15°.

Concerning the hand risk assessment, the approach considers the percentage of time in which the worker holds in his/her hand an object.

In this way, once the angular displacement of joints from the neutral posture have been evaluated, it is possible to easily estimate the percentage of time in which workers have an incorrect posture during his activity.

Referring to the Checklist OCRA algorithm, it must be noted that the computed percentage of time is cumulative for movements of a specific joint. The condition for considering a time span as "incorrect" for posture, is that at least one of the movements of the specific joint reaches an angle that is beyond the defined threshold.

2.2   Design Requirements for a Dedicated Software Tool

Design required to consider several key points (see Fig. 1). Firstly, the software should be easy to use, with intuitive commands, well-structured sequence of operations according to the norm, and it has to give back real-time feedbacks about the results and concerning the on-going analysis.

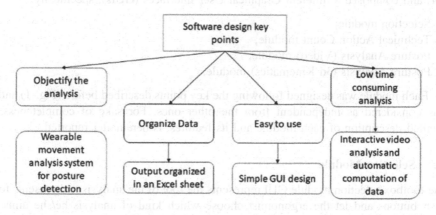

Fig. 1. Software design key points

A further need was identified in improving time efficiency in conducting the risk evaluations. The toolbox, by automatizing several assessment tasks, was thought to allow the ergonomist to perform more analyses at the same time and obtain a wider view about the estimated level of risk.

A third key point in the toolbox design was related to the necessity to organize all the video and processed data in a structured way, including output spreadsheet and reporting files, where all data present their own labels and values, and they can be easily retrieved by the ergonomist for further analysis or synthesis.

The final question concerned what the ergonomist really needed to conduct a more reliable and objective risk assessment and, in particular, how 2d video information can be enhanced. This led to the identification of a method able to easily acquire 3d kinematic data in the workplace. In this perspective, all the movement analysis technologies could be considered, including - for instance - optoelectronic, electromagnetic and inertial-based systems. However, taking into account complexity, costs and environmental constraints, the identified optimal way was to use wearable technology. In particular, IMUs (Inertial Measurement Unit) sensors were chosen, since they let worker be more comfortable and able to walk and complete all tasks without any movement limitations. In fact, considering the actual technological state-of-the-art, the worker had only to wear some compact and light sensors on the surface of his body. As hereinafter reported, by using the kinematic data acquired by IMUs during the trials, it was possible to establish and evaluate the correct angular displacement for each investigated joint without any kind of uncertainty.

3 Software Tool Description

The software toolbox prototype was realized by using Matlab (The Mathworks Inc.) IDE and comprised 4 different Graphical User Interfaces (GUIs), specifically:

- Selection module;
- Technical Action Count module;
- Posture Analysis (Video) module;
- Posture Analysis (3d Kinematics) module.

Each module was designed following the key points described before (Fig. 1) and it was considered as independent from the other ones. For sake of completeness, a detailed description of each module and its features, is hereinafter reported.

3.1 Selection Module

The toolbox selection module GUI represents the root of the analysis and presents four push buttons and let the ergonomist choose which kind of analysis he/she aims to conduct (Fig. 2). For each selection, an additional sub-GUI is opened, implementing the specific features required by the analysis.

3.2 Technical Action Module

The Technical action module (Fig. 3) was implemented with the aim to support ergonomists in counting technical actions and estimate static posture.

This GUI presents three different main parts: (i) video control, (ii) technical action count and frequency, (iii) static posture time and percentage of time.

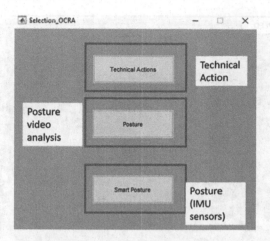

Fig. 2. Selection module GUI

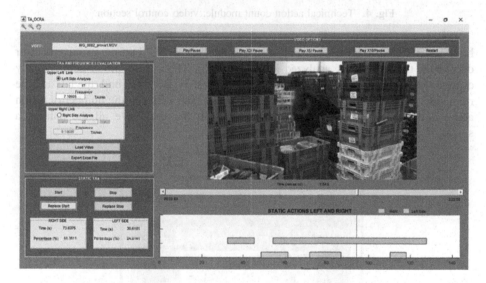

Fig. 3. Technical action count module.

The "video control" panel (Fig. 4) allows the user to choose between 4 different push buttons, thus to control the video framerate. Every speed was thought as lower compared to the real one, thus to allow the user to carefully analyze each technical action. By pushing the play button, the framerate of the video is eight time slower than it really is. The other three buttons increment the video framerate value x2/x5/x10 with respect to the initial framerate.

The "technical action count and frequency" panel allows the user to take notes about the number of technical action performed by the worker. Once the video is loaded with the "Load Video" button and the side ("left upper limb" or "right upper

Fig. 4. Technical action count module: video control section.

limb") is selected, the user has only to push the "+" or "−" buttons to increment or decrement the number of analyzed actions (Fig. 5). Then, the toolbox automatically estimates the frequency of technical actions. This can be done for both the left and the right side. The frequency is simply obtained by the ratio between the number of action and video duration. The "Export File" button let the user save all the data in a spreadsheet file (e.g. *.xls file format).

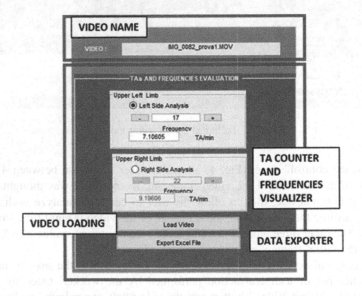

Fig. 5. Technical action count module: technical action count and frequency control panel.

In order to take note about the presence of static action, the ergonomist can interact with the "static action" control panel (Fig. 6). The buttons were designed to tag the instants of time in which the worker begins to assume a static position with the load on his hand and when he/her begins to move once again.

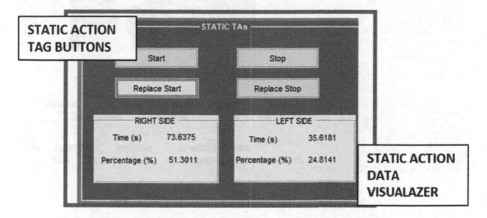

Fig. 6. Technical action count module: static action control panel

All the information about the duration and the percentage of time in which the worker had a static posture are automatically computed and displayed in the panel. By using the "Replace Start" and "Replace Stop" buttons, the user can correct the start and the stop point previously determined. Only static action that last longer than 5 s are considered in the analysis. Furthermore, it is possible to see a graphical representation of results in the static action bar plot (Fig. 7).

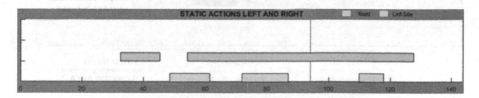

Fig. 7. Technical action count module: static action bar plot.

3.3 Posture Analysis (Video) Module

The posture analysis video module (Fig. 8) was designed to help the ergonomist in conducting the posture assessment by using only the 2d video information. "Video control" panel (Fig. 9) is very similar to the "Technical Action" one, as previously described. In this case, the framerate can be selected considering only two options (x1/x2 the initial framerate). It is also possible to take note about the working area in which the worker is doing his/her activity and see the video information in a dedicated textbox.

698   S. E. Lenzi et al.

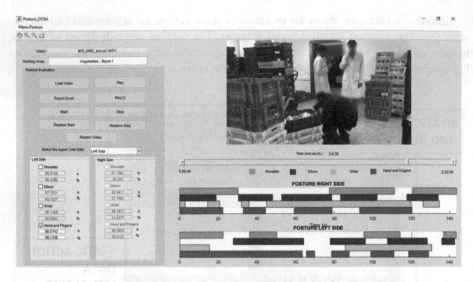

Fig. 8. Posture analysis (video) module.

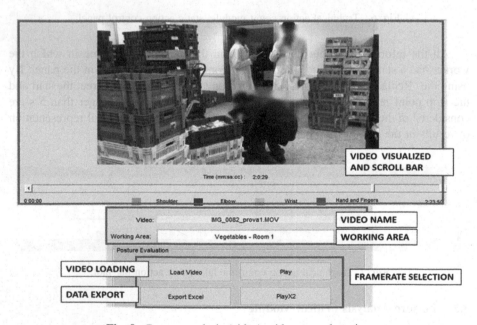

Fig. 9. Posture analysis (video): video control section.

At first the ergonomist has to load the video he wants to analyze and then he can act on the working area field, selecting which limb side and joint he wants to analyze. Only when the side has been selected, he can directly interact with the posture control panel.

The "posture control" panel (Fig. 10) let the user tag the time in which the worker begins to assume an awkward posture (i.e. "Start" button). Then, when the worker's posture goes back to being correct, it is possible to push the "Stop" button. As soon as this button has been pressed, all the information about time and percentage of time the worker presented a wrong posture based on threshold described in the ISO 11228-3. In the "static action bar plot" panel, the ergonomist can exploit also an intuitive visual feedback on the distribution and the duration of different period in which the workers have a non-adequate posture from the ergonomic point of view. There is also a line that is synchronized with the video time frame. In this way the ergonomist can control the exact temporal information.

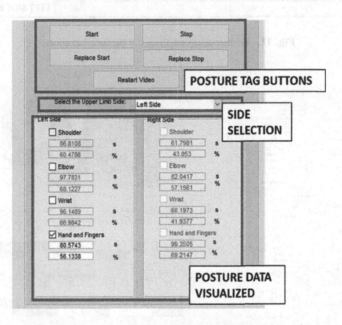

Fig. 10. Posture analysis (video): posture control panel.

Finally, once the analysis is performed, the ergonomist can export all the data in a well-structured spreadsheet file (e.g. *.xls file format) in order to keep trace of the analysis (Fig. 11).

3.4 Posture Analysis (3d Kinematics) Module

This module GUI (Fig. 12) looks very similar to the previous one. The most crucial difference from the "Posture Analysis (video)" module is that durations and percentages of time in which workers present awkward postures are automatically computed, by considering data about the angular kinematics acquire and estimated by means of a movement analysis technology (e.g. inertial sensors).

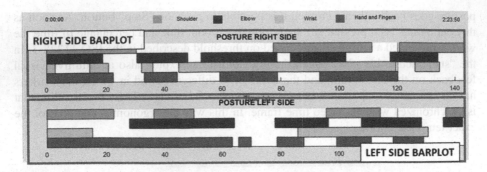

Fig. 11. Posture analysis (video): posture bar-plots.

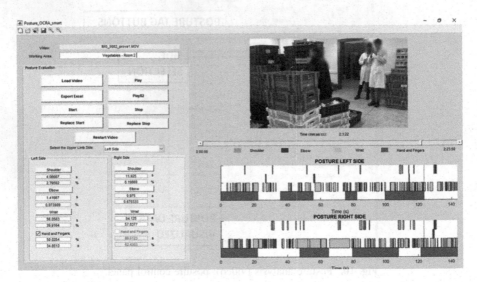

Fig. 12. Posture analysis (3D kinematics)

One of the limitation of this kind of approach can be related to the difficulties in estimating the quality and the type of grip during the working activities (as it can happen by considering only IMU-based system). For this reason, the toolbox allows to perform this analysis through the usual video assessment.

Procedurally, once the GUI has been opened, the user has to load first the video he/her wants to analyze, then fills up the working area box with the required information and finally has to press on the button of each joint in order to import data concerning joint 3d kinematics (Fig. 13). Then the analysis is made in an automatic way by using the algorithm shown in Fig. 14.

Data referring to the angles described by each joint obtained by the elaboration of IMU sensors outputs are loaded. Then a counter ($count_{sign}$) is initialized to keep trace of

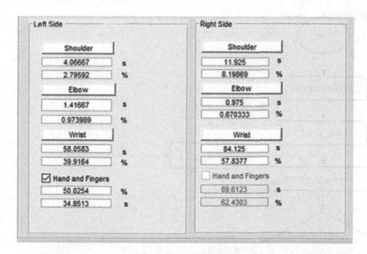

Fig. 13. Load data buttons for each joint and for both sides

the discrete time referring to the sample that is going to be examined. Once the counter has been initialized, the corresponding sample referring to the first joint angle (i_{mov1}) is read. The value is compared with the defined threshold (thr_{mov1}). If the value is above the threshold, a counter will be then incremented ($count_{mov1}$). Subsequently, the next sample contained in the sequence will be read and a new comparison with the threshold is performed. If the sample presents a value below the defined threshold, the algorithm evaluates the sample contained in another plane of the same joint and then the workflow follows the steps, as previously described. The exit condition from this loop is reached when the last sample is analysed.

Then the percentage of time in which the worker presents an incorrect posture for the joint is computed with the following formula (Eq. 2):

$$t\% = \frac{\frac{count_{mov_i}}{f_{sampl}}}{t_{tot}} \times 100\% \qquad (2)$$

where $count_{mov\ i}$ is the counter that counts samples in which the worker has an awkward posture, f_{sampl} is the sampling frequency and t_{tot} is the time of the trial.

For the elbow, the algorithm was adapted because, as noted above, it controls that the angle value is not maintained for more than 5 s. Once the algorithm stops, all the duration and percentage are shown in the dedicate panel and the visual information are loaded in the dedicate graph. Then, pressing the export button, it is possible to download in a structured spreadsheet file (e.g. *.xls file format) all the elaborated data.

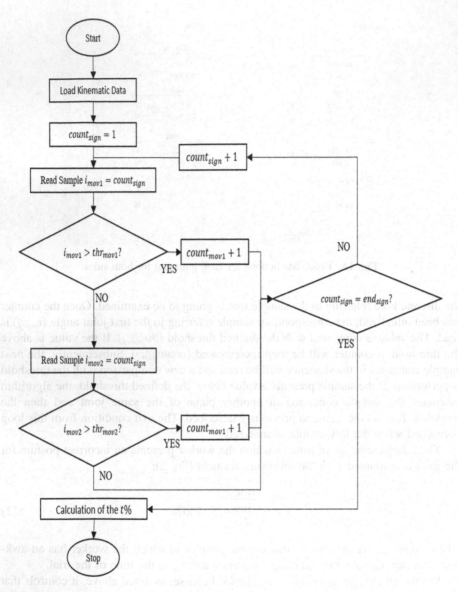

Fig. 14. General scheme of the algorithm for the computation of percentage of time in which the worker has an incorrect posture for each joint.

4 "On Field" Application

We performed several "on field" trials in order to test the effectiveness of using the software toolbox, specifically considering two different tasks: shelfing and cashier's activities.

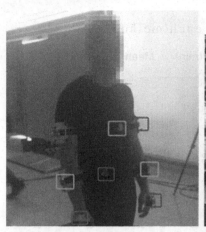

Fig. 15. Cameras setup (right) and sensors placement protocol (left).

To test all the reported features, also 3d kinematics was acquired. In particular, workers wore a set of commercial wearable inertial sensors (Notch system - Notch Interfaces Inc.). Sensors were placed on the workers' body surface as reported in the left part of Fig. 15. Sensors were specifically attached on the following anatomical regions: stern process, left upper arm, right upper arm, left lower arm, right lower arm, left hand, right hand, pelvis. Furthermore, each trial was recorded using two cameras with two distinct point of view. In this way the ergonomist can choose the best perspective from which he can properly conduct his/her analysis. Usually cameras were placed at about one meter from the workers (Fig. 15, right). To test the software, three expert operators were enrolled.

They were instructed to analyze 7 videos randomly chosen. The analysis were made following three different methods: (1) by analyzing the video and annotating results on a sheet, (2) by using the software tool for the video analysis of postures and technical actions, (3) by using the automated tool for postures analysis. In addition, for each joint and for the posture analysis, the operators were instructed to take note of the time spent to complete the evaluation.

Results were analyzed in term of time variation between the different approaches and correlation between obtained assessment. These trials were made in order to, on one hand, establish if the toolbox can simplify and speed up the analysis and, on the other hand - under the hypothesis that the 3d kinematics method can be considered as "gold standard" -, as a way to assess the error made by operators when analyzing the tool with a subjective approach (video analysis) compared with the objective one (i.e. by using IMUs data). Results can be interpreted as a preliminary study on the effectiveness and usability of the presented toolbox.

Starting from the analysis of time (Figs. 16 and 17), we found a reduction of about the 98% passing from the manual approach to the automated one. This result underlined the clear effectiveness of this tool if compared with the manual approach. This can reduce heavily the time spent by the operator in conducting the analysis, letting him do his evaluation quicker and in an objective way.

704   S. E. Lenzi et al.

Mean time (min) for each method

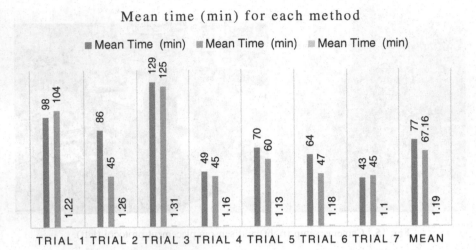

Fig. 16. Mean time (min) for posture detection for each method

Time reduction (%) when using the automatic
tool respect to the manual approach when
analysing postures.

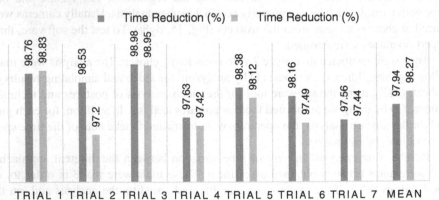

Fig. 17. Time reduction (%) when using the automatic tool respect to the manual approach when analysing postures.

For the analysis of inter-operator variability, it was performed R^2 correlation between assessment outcomes obtained by different operators. Results (Fig. 18) show that error are not coherent and it impossible to clearly interpret all the data in a straightforward way. R^2 values for the shoulder show the highest value ($R^2 > 0.6$). In

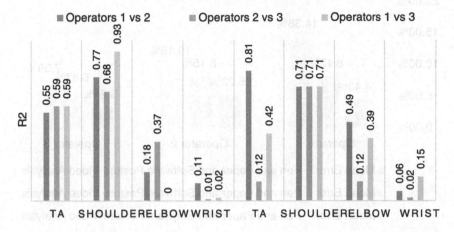

Fig. 18. Inter-operator analysis results: R^2 values for each method and each couple.

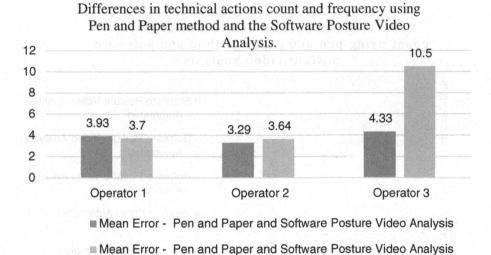

Fig. 19. Differences in technical actions count and frequencies using Pen and Paper method and the Software Posture Video Analysis.

general, there was a fair correlation, that becomes slightly higher when considering technical action count and shoulder evaluation.

Outcomes reported in Figs. 19 and 20 reveal always high values of errors (i.e. the difference with respect to the gold standard). Highest values are obtained for technical

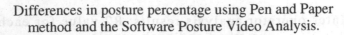

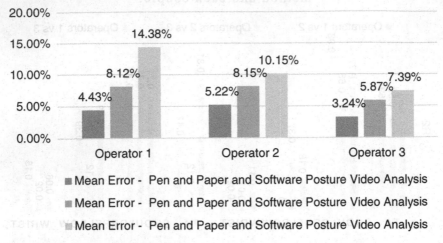

Fig. 20. Differences in posture percentage using Pen and Paper method and the Software Posture Video Analysis.

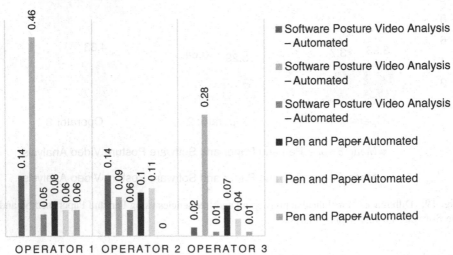

Fig. 21. R^2 values from the comparison between manual methods and the automatic method base on kinematic data gathered from the IMU based movement analysis system

action count when analyzing trials of the operator 3. The errors are generally higher when considering elbow and wrist joint.

In Fig. 21 are shown data resulting from the comparison between manual methods and automated method. All the correlation (R^2) values are very low. The worst results are obtained for the wrist joint.

5   Conclusion

This work reported the design, implementation and preliminary application of a new and promising software toolbox able to improve reliability and efficiency of the ergonomics risk assessment of work activities, specifically considering upper limb and repetitive tasks.

Preliminary trials pointed out that 2d video analysis is not the best method to objectively assess the level of risk associated to working activities. Results confirmed our primary hypothesis for the need to provide the ergonomists with specific tools - indeed as the one described in the paper - that could improve the reliability of the assessment and the objectivity of the analysis.

Inter- and intra-operators' variability analysis seems to suggest that the video analysis of task is strongly operator dependent. Ergonomists, in this study, seem to have different metrics in the evaluation of technical action and postures. Moreover, the same operator, with two different analysis approaches, seems to judge in a different way the observed tasks. Many efforts have to be done in order to suggest new methodologies, tools that could be able to eliminate any kind of error due to the subjectivity of the analysis.

This work can be seen also as a reference and a starting key point for designing more efficient and objective tools able to guide ergonomists in the risk assessment procedure following the Checklist OCRA index method.

Furthermore, this work underlined also that the use of the wearable technology is strongly encouraged in occupational ergonomics, because of its reliability, coherent costs and low intrusiveness with respect to both the worker and the working environment.

Further efforts have to be done in order to test the software in a more complete way.

References

1. INAIL   Opendata   2017.   https://www.inail.it/cs/internet/comunicazione/sala-stampa/comunicati-stampa/com-stampa-open-data-2017.html
2. Mcatamney L, Corlett EN (1993) RULA: a survey method for the investigation of world-related upper limb disorders. Appl Ergon 24(2):91–99
3. Steven Moore J, Garg A, (1995) The strain index: a proposed method to analyze jobs for risk of distal upper extremity disorders. Am Ind Hyg Assoc 56(5):443–458
4. Trask M, Lafaurie AS, Tronchet L, Atain-Kouadio JJ (2000) OREGE: un outil simple d'evaluation des facteurs de risque biomechcaniques de TMS du membre superieur. Note Sci Tech l'INRS 196:1–122

5. Colombini D, Occhipinti E (2015) L'Analisi e la gestione del rischio nel lavoro manuale ripetitivo: Manuale per l'uso del sistema OCRA per la gestione del rischio da sovraccarico biomeccanico in lavori semplici e complessi. Franco Angeli, Milano
6. ISO 11228:2007(E) (2007) Ergonomics - Manual handling 04 01

Musculoskeletal Disorders Among Orthodontists: Risk Factors and Ergonomic Intervention

Rianina D. Borres, John Ulric Lim, Rex Aurelius Robielos[✉],
and Marquin Jose Pacaña

Mapúa University, Manila, Philippines
racrobielos@mapua.edu.ph

Abstract. Musculoskeletal disorders (MSD) prevalence was highest among the orthodontists and oral physicians compared to the general population working in different environments. This study aims to explore ergonomic issues in the orthodontic workplace in small to medium-scale dental clinics in Metro Manila. Assessments were made on the physical workstation set-up of the orthodontic clinic – the design, equipment, tools/instruments, environmental hazards, and the manner in which the workers perform their tasks. The Cornell Musculoskeletal Discomfort Questionnaire (CMDQ) and the National Institute for Occupational Safety and Health (NIOSH) Generic Job Stress Questionnaire were used to gather critical information about MSDs. Anthropometric measurements were taken and used in the Rapid Entire Body Assessment (REBA) to determine the most risky orthodontic tasks. It was found that all orthodontists reported body pains and the highest occurrence had been in the neck (87%), lower back (87%), shoulders (83%), upper back (80%), and forearms (77%). The factors significantly associated with the discomfort scores were age, weekly working hours, workload, and posture. Among the tasks most commonly performed, tooth extraction puts the orthodontists at the highest risk of developing MSDs. To reduce the prevalence of MSDs symptoms among Filipino orthodontists, some of the ergonomic interventions are (1) designing ergonomic workstations to better fit the population, (2) employing certain administrative controls and (3) modifying how tasks are performed.

Keywords: Orthodontics · Musculoskeletal disorders · REBA CDMQ

1 Introduction

Musculoskeletal disorders in the work-place are one of the most reoccurring and highly reported problems by most workers particularly in the field of orthodontics because of their exposure to many ergonomic risk factors that affect their line of work (Shaik et al. 2012; Park et al. 2017). According to Occupational Information Network, orthodontics was ranked as the number one most damaging career to one's health in 2015 due to their workload and job requirements (La Rochelle 2017). Orthodontist cannot avoid prolonged static postures and despite having optimal seating posture there is still little

© Springer Nature Switzerland AG 2019
S. Bagnara et al. (Eds.): IEA 2018, AISC 820, pp. 709–733, 2019.
https://doi.org/10.1007/978-3-319-96083-8_87

movement of the joints and the muscles in the body are contracted for certain periods of time (Shaik et al. 2012). Furthermore, an orthodontist area of work is only small and limited and they are kept in positions with their heads bent, arms distanced from each other, and there is continuous rotation and repetition of positions (Jafari and Yekta-Kooshali 2017). Thus, it is then very likely that these working conditions would lead to certain back, neck, or shoulder pains or worst will result to musculoskeletal disorders (MSDs).

Musculoskeletal disorders (MSDs) are injuries or disorders pertaining to the human body's movement (musculoskeletal system) which consists of muscles, tendons, ligaments, and nerves. According to the Bureau of Labor Statistics in the U.S., MSDs are the single largest category of workplace injuries such as muscle/tendon strain, tendonitis, and carpal tunnel syndrome (Middlesworth 2012). Symptoms of musculoskeletal disorders (MSDs) may include stiff joints, swelling, and recurrent pain which can be caused by several factors such as posture, repetition of activities, poor fitness, and poor work habits. Thus, MSDs can affect daily tasks of workers and considered a very costly problem for most companies and firms.

The field of orthodontics is one of the most demanding professions that requires concentration and high precision (Finsen et al. 1997). It is a branch of specialized dentistry that focuses on patients that have an improper positioning and alignment of teeth as well as in-charge on the reconstruction of the entirety of the faces aesthetic appeal. To improve patient's appearance by properly aligning the crooked teeth, orthodontists commonly use braces, aligners, and retainers in their methods to set and reposition the teeth. Those working in this field usually work several hours a day and are kept in prolonged positions with short intervals in between patients. Thus, this exposes a high risk of physical complications among orthodontists.

Compared to the general population, orthodontists are at greater risk of work-related musculoskeletal disorders. The upper body symptoms of pathological conditions appear more frequently in orthodontist compared to those working in different environments (Kierklo et al. 2011). Musculoskeletal disorders are commonly attributed to numerous risk factors such as prolonged static posture, repetitive movements, suboptimal lighting, poor positioning, mental stress, and physical conditioning (Valanchi and Valanchi 2003). These risk factors are all present in the common work place of orthodontists. Among the most common practices performed by dental professionals, the most notable is the use of vibratory tools, excessive repetition, and maintenance of inadequate static postures for certain periods of time (Morse et al. 2010). These occupational activities may involve a high demand of physical strength and stamina (Dantas and De Lima 2014).

There have been several studies made regarding work-related musculoskeletal disorders among dental health care workers but most studies are mainly focused on countries such as the United States. One such study has reported that approximately 81% of American dentists suffer from neck, shoulder, and lower back pain (La Rochelle 2017). In the Philippines there are only very few studies that have been made regarding the dental industry and none have been able to discuss thoroughly the ergonomic aspect of this profession.

Thus, the general aim of this study is to thoroughly asses the workplace and working practices of orthodontics in the Philippines. An investigation on the prevalence of musculoskeletal discomfort in different body regions shall be made and significant factors influencing the development of musculoskeletal disorders will be identified. The current workplace design, equipment, work environment, and practices are to be evaluated and job posture analysis shall be made on the different tasks performed by orthodontists. Through this study, high risk task shall be identified and recommendations regarding such tasks will be made. Ergonomic intervention is also to be used in formulating recommendations that will be able to address the problems discovered.

2 Methodology

To determine the risk factors affecting musculoskeletal disorders for orthodontists, the Cornell Musculoskeletal Disorder Questionnaire (CDMQ) was distributed to 30 orthodontists working in small and medium scale dental clinics located in Metro Manila. The participants consisted of 18 females and 12 males with age ranging from 29 to 60 years (Mean = 43.1 yrs). The individual characteristic and general job information such as age, gender, height, weight, anthropometry, job tenure, type of task, working hours, shifts, time spent on other jobs, type of dental clinic, and the presence of a dental assistant were also collected.

Then, the CMDQ was used to measure musculoskeletal complaints especially on the three issues identified for each body site. These are (i) the frequency of occurrence of the discomfort for the previous week, (ii) the discomfort level, and (iii) the degree to which the experienced discomfort has interfered with the employee's ability to work. The CMDQ uses a five-point scale rating for the frequency and uses a three-point scale rating for both the discomfort level and interference. Two CDMQ questionnaires were administered to orthodontists; one for the entire body and another one focusing on the hands.

The National Institute for Occupational Safety and Health (NIOSH) Generic Job Stress Questionnaire was also administered to dental practitioners. The five stress factors chosen for this study was perceived to be physical environment, health condition, job requirements, job satisfaction, and non-work activities. Job requirements and job satisfaction had been considered to be psychosocial risk factors. The perceived sound intensity, ambient temperature, local humidity, ventilation and illumination in the work area were all included in the perceived physical environment.

The dental workplace had also been assessed by evaluating the room, dental instruments used, dentist chair, and patient chair. The Rapid Entire Body Assessment was used to assess the current working postures used by dental practitioners in performing different work-related tasks, through direct observation and proper documentation of the orthodontists during work.

Anthropometric measurements of participants were also taken and compared to the existing measurements of the work station. Measurements for both standing and sitting posture were collected then compared to the current measurement. These measurements included height, shoulder height, upper arm, lower arm, elbow height, and waist height.

3 Results

The demographic characteristics of the population under study are given in Table 1. Thirty (30) orthodontists were surveyed and interviewed, who are working in one of the 20 small and medium dental clinics in Metro Manila. The subjects consisted predominantly of females (60%) compared with males (40%) with age ranging from 29 to 60 years with mean age 43.10. The average orthodontist stands 167.78 cm high, and weighs 60.83 kg. The average body mass index (BMI) is 21.86, which is within the normal range. On the average, the orthodontists consume 1.03 cigarettes per day, and reported a total of 8 days sick leave for the past month. Overall, the perceived general health status of orthodontist is relatively fair (score = 47.27).

Table 1. Descriptive statistics of the 30 orthodontists

Variable	% or Mean (N = 30)
Gender	
Female	60.00%
Male	40.00%
Age (mean)	43.10
Height, cm (mean)	167.78
Weight, kg (mean)	60.83
BMI, kg/m^2 (mean)	21.86
General health	
Daily cigarette consumption (mean, total)	1.03, 31
Job accidents (mean)	0
Sick leave (mean, total)	0.27, 8
Perceived health (mean)	47.27
Illnesses (mean, total)	0.27, 8

In terms of job related information, the average work experience of all orthodontists is 13.83 years. As to employment status, 12 orthodontists (40.00%) are sole proprietors of their clinics, 13 orthodontists (43.33%) are partner dentists, and 5 orthodontists (16.67%) are working as employee dentists. It was found that the majority of the surveyed orthodontist work without an assistant (63.33%) while 36.67% work with an assistant. The data also indicated that more than half of the orthodontists (80%) also practice general dentistry while the other 20% are restricted to orthodontics only. Sixty percent (60%) of the 20 clinics visited has one dentist, 30% has 2 dentists, and 10% has 3 dentists (Table 2).

For their work schedule, orthodontists work on the average around 43.33 h per week, with an average of 2.60 h doing overtime work. It can be seen that almost half (46.67%) of them also work on weekends. Two of the orthodontists (6.67%) have other jobs and they spend an average of 12 h per week in those jobs. On the average, orthodontists get 37 appointments per week and each appointment lasts for 20 min. Thirteen (43.33%) of the orthodontists reported that they work without any break, 7

Table 2. Job related information of orthodontists

Variable	% or Mean (N = 30)
Job information	
Tenure (mean)	13.83
Employment status	
Sole proprietor	40.00%
Partner dentist	43.33%
Employee dentist	16.67%
Presence of assistant	
With assistant	36.67%
No assistant	63.33%
Type of practice:	
Orthodontics only	20.00%
General dentistry and orthodontics	80.00%
Number of dentists in a clinic (20 clinics)	
1 dentist	60.00%
2 dentists	30.00%
3 dentists	10.00%
Working hours	
Work hours in a week (mean)	43.33
Overtime per week, hrs (mean)	2.60
Time spent on other jobs, hrs (mean)	0.80
Weekend work	
Working on weekends	46.67%
No work on weekends	53.33%
Number of breaks in a day	
None	43.33%
1	23.33%
2	16.67%
3	6.67%
After every patient	10.00%
Other jobs	
With other jobs	6.67%
No other jobs	93.33%
Dental appointments	
Appointment duration, mins (mean)	20.0
Weekly number of appointments (mean)	37.0
Working position	
Standing	56.67%
Sitting	43.33%

orthodontists (23.33%) reported one break per day, 5 orthodontists (16.67%) reported two breaks per day, 2 orthodontists (6.67%) made three breaks per day and 3 orthodontists (10%) took a break every after patient. In terms of working position, more than half of the orthodontists surveyed (56.67%) work in a standing position while 43.33% prefers to work at a sitting position beside the patient.

Prevalence of Musculoskeletal Pain

All participants complained of more than one musculoskeletal pain one week prior to questionnaire completion. Two orthodontists complained of discomfort in four body regions, the minimum number of painful sites reported by an orthodontist, while one complained of experiencing discomfort in 10 out 12 body sites. The results of CMDQ revealed that the most commonly affected body regions were the upper back (100%), neck (87%), lower back (80%), wrists (80%), forearms (73%), upper arms (73%), and shoulders (70%) (See Fig. 1).

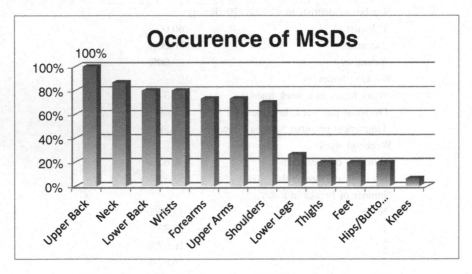

Fig. 1. Frequency of occurrence of MSDs among orthodontists

When the severity of discomfort and interference were taken into account, it was observed that discomfort scores in the right upper extremities slightly exceeded the discomfort scores in their left counterparts. Table 3 shows the discomfort scores which was obtained by getting the product of frequency, discomfort, and interference.

Seventy-eight percent (78%) of the discomforts were categorized as slight while the rest were classified as moderate, and 27.50% slightly interfered with the workers' ability to work while 72.50% has not affected their work at all. Body pains were evidently more prevalent in the upper extremities. Seven out of the 30 participants (23.33%) have been absent from work in the past month because of some kind of musculoskeletal disorder. The major reason for being absent from work in the past month was back pain (57.14%) while for job restriction the main reason was upper arm pain (13.27%), neck pain (12.83%), and shoulder pain (9.07%).

Table 3. Discomfort scores

Body region	Discomfort score	%
Lower back	124	15.31%
Forearms	112	13.83%
Upper arms	108	13.33%
Neck	105	12.96%
Upper back	104	12.84%
Shoulders	104	12.84%
Wrists	102	12.59%
Lower legs	19.5	2.41%
Thighs	9	1.11%
Feet	9	1.11%
Hips/Buttocks	9	1.11%
Knees	4.5	0.56%

Comparison of CMDQ scores between genders showed that there is no significant difference between males and females in terms of the discomfort scores. The same is true among employment status groups, which showed that the discomfort scores do not significantly vary among sole proprietors, partner dentists, and employee dentists. Whether an orthodontist has an assistant or none, there is also no effect on the CMDQ scores. However, those who practice general dentistry in addition to orthodontics have significantly higher CMDQ scores. The same goes for orthodontists who also work on weekends compared with those who do not. Orthodontists who prefer to work at a standing position have significantly higher CMDQ scores, as well as those who do not take breaks during work compared to those who take at least one break.

Results of the analysis of the relationships between the CMDQ scores and the demographics, job information, psychosocial load and perceived environment through correlation test are presented in Table 4. No significant correlations ($p > 0.05$) were found between CMDQ scores and other demographic information (BMI, number of sick leaves for the past month, and daily cigarette consumption). Significant, positive correlations ($p < 0.05$) were found between the CMDQ score prevalence and all the variables related to work (tenure, weekly working hours, weekly overtime, appointment duration, and average number of appointments in a week). Weekly working hours had the strongest, positive correlation with the CMDQ score, followed by the appointment duration. Those who worked overtime also had higher CMDQ scores. MSD symptoms also increase with the number of appointments in a week and the orthodontist's work experience. Older orthodontists are also more predisposed to MSD symptoms, as well as those who have poor health. Height has a moderate, positive relationship with the CMDQ scores, signifying that taller orthodontists are more prone to MSDs, which may be due to bending at greater angles to have a better view of their patient's teeth. Psychosocial load (job requirements and perceived workload and responsibility) as well as perceived environment showed significant correlations.

Focusing on those orthodontists who are restricted to orthodontic tasks alone (no general dentistry practice), the frequency of occurrence of the musculoskeletal

Table 4. The association between CMDQ scores and demographics, job information, psychosocial load, and perceived environment

Parameter	Correlation coefficient, r	p-value
Demographics		
Age	0.478	0.008*
BMI	−0.063	0.740
Height	0.373	0.042*
General health	0.472	0.008*
Number of sick leaves	−0.55	0.771
Daily cigarette consumption	0.274	0.143
Job information		
Tenure	0.374	0.042*
Weekly working hours	0.894	0.000*
Weekly overtime	0.561	0.001*
Appointment duration	0.603	0.000*
Average number of appointments in a week	0.535	0.002*
Psychosocial load		
Job requirements	0.499	0.005*
Workload and responsibility	0.614	0.000*
Perceived environment	0.687	0.000*

*Significant at p < 0.05

complaints differed a bit compared to the previous analysis considering the entire sample population. Members of the subgroup who are restricted to orthodontics alone complained of pain in the upper back (100%), neck (83%), lower back (83%), and wrists (67%). None of them complained of pain in the thighs, feet, and knees. It is important to note that 5 out of the six orthodontists who only focus on orthodontic tasks prefer to work in a sitting position (Fig. 2).

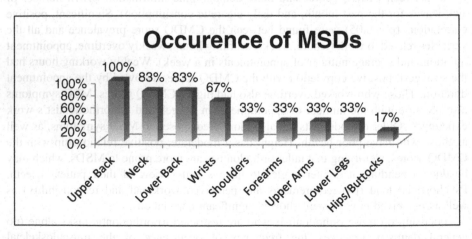

Fig. 2. Frequency of occurrence of MSDs for those who practice orthodontics only

To further analyze the problems being experienced by those who practice orthodontics only, the discomfort scores were also evaluated and it was found that the discomfort scores was highest in the lower back (Table 5). Furthermore, orthodontists who concentrate only on orthodontic tasks experience more pain the upper back, neck, and wrists.

Table 5. Discomfort scores for those who practice orthodontics only

Body region	Discomfort score	%
Lower back	23	21.70%
Upper back	21.5	20.28%
Neck	15	14.15%
Wrists	12	11.32%
Forearms	10	9.43%
Upper arms	10	9.43%
Shoulders	8.5	8.02%
Lower legs	4.5	4.25%
Hips/Buttocks	1.5	1.42%

Correlation analysis was also carried out between the CMDQ scores of orthodontists without general dentistry practice and the different factors considered in this study (bio-demography, job information, psychosocial load and perceived environment). Table 6 shows the results of the correlation tests. Age, general health, tenure, weekly working hours, and perceived environment remained to be significantly correlated with the CMDQ scores.

Table 6. The association between CMDQ scores and demographics, job information, Psychosocial load and perceived environment among Orthodontists who Practice Orthodontics Only

Parameter	Correlation coefficient, r	p-value
Demographics		
Age	0.948	0.004*
BMI	−0.669	0.146
Height	0.767	0.075
General health	0.874	0.023*
Number of sick leaves	−0.483	0.332
Job information		
Tenure	0.955	0.003*
Weekly working hours	0.887	0.019*
Weekly overtime	0.244	0.641
Appointment duration	0.152	0.773
Average number of appointments in a week	0.290	0.577

(*continued*)

Table 6. (*continued*)

Parameter	Correlation coefficient, r	p-value
Psychosocial load		
Job requirements	0.765	0.076
Workload and responsibility	0.053	0.921
Perceived environment	0.978	0.001*

*Significant at p < 0.05

Work Postures and Exposure

One way to analyze job posture is through the use of Rapid Entire Body Analysis (REBA). In this study, the orthodontists were observed in their workplace through two body groups namely: A - neck, upper and lower back and legs; B – shoulders, elbows and wrists. By observing each job posture of the different body parts, the REBA score was obtained based on the basis of their posture angle. Then the REBA score was analyzed to determine the appropriate level of action (Table 7). The REBA score revealed that the four activities (restoration, extraction, cleaning, and brackets installation) did not differ much in posture, except for extraction that needs more force than the others. Orthodontists who prefer to work while standing also has greater leg scores than those who prefer the sitting position, which led to higher REBA scores. Table 8 shows in detail the posture assumed by the orthodontists when performing operations in the different parts of the mouth.

Table 7. REBA action levels

Action level	REBA score	Risk level	Action (including further assessment)
0	1	Negligible	Change not necessary
1	2–3	Low	Change may be necessary
2	4–7	Medium	Further investigation, change soon
3	8–10	High	Investigate and implement change
4	11+	Very high	Implement change now

Table 8. Postures when working in the different parts of the mouth

Location	Body Part	Posture
Right lower jaw teeth	Neck	Neck is bent downwards at an angle greater than 20°
	Trunk	Trunk is bent downwards at an angle less than 20°
	Legs	Legs are at a neutral position (sitting) or weight is shifted from one leg to the other (standing)
	Upper arm	Upper arms abducted and angled at less than 20°, shoulders raised
	Lower arm	Lower arms angled at less than 60°
	Wrist	Wrists are bent at angles less than 15°

(*continued*)

Table 8. (*continued*)

Location	Body Part	Posture
Right upper jaw teeth	Neck	Neck is bent sideways and downwards at an angle greater than 20°
	Trunk	Trunk is bent sideways and downwards at an angle greater than 20°
	Legs	Legs are at a neutral position (sitting) or weight is shifted from one leg to the other (standing)
	Upper arm	Upper arms abducted and angled at less than 20°, shoulders raised
	Lower arm	Lower arms angled at less than 60°
	Wrist	Wrists are bent at angles greater than 15°
Left lower jaw teeth	Neck	Neck is bent downwards at an angle greater than 20°
	Trunk	Trunk is bent downwards at an angle less than 20°
	Legs	Legs are at a neutral position (sitting) or weight is shifted from one leg to the other (standing)
	Upper arm	Upper arms abducted and angled at less than 20°, shoulders raised
	Lower arm	Lower arms angled at less than 60°
	Wrist	Wrists are bent at angles less than 15°
Left upper jaw teeth	Neck	Neck is bent sideways and downwards at an angle greater than 20°
	Trunk	Trunk is bent downwards at an angle less than 20°
	Legs	Legs are at a neutral position (sitting) or weight is shifted from one leg to the other (standing)
	Upper arm	Upper arms abducted and angled at less than 20°, shoulders raised
	Lower arm	Lower arms angled at less than 60°
	Wrist	Wrists are bent at angles greater than 15°
Anterior teeth	Neck	Neck is bent downwards at an angle less than 20°
	Trunk	Trunk is bent downwards at an angle less than 20°
	Legs	Legs are at a neutral position (sitting) or weight is shifted from one leg to the other (standing)
	Upper arm	Upper arms abducted and angled at less than 20°
	Lower arm	Lower arms angled at less than 60°
	Wrist	Wrists are bent at angles less than 15°

Since majority of the orthodontists (80%) also practice general dentistry, REBA was also performed on general dentistry tasks such as restoration and extraction. Tables 9 and 10 show the REBA scores for different orthodontics and general dentistry activities commonly performed in different parts of the mouth. REBA scores for the standing position slightly exceeded those of the sitting position because of higher leg scores.

Table 9. REBA scores for different dentistry activities – sitting position

Sitting Position	Arm and Wrist Score	Neck, Trunk and Leg Score	Activity Score	Final REBA Score
Restoration				
Restoration of the right lower jaw teeth	5	3	1	5
Restoration of the right upper jaw teeth	6	6	1	9
Restoration of the left lower jaw teeth	5	3	1	5
Restoration of the left upper jaw teeth	6	4	1	7
Restoration of anterior teeth	3	2	1	3
Extraction				
Extraction of the right lower jaw teeth	7	4	1	8
Extraction of the right upper jaw teeth	8	7	1	11
Extraction of the left lower jaw teeth	7	4	1	8
Extraction of the left upper jaw teeth	8	5	1	9
Extraction of anterior teeth	5	3	1	5
Cleaning				
Cleaning of the right lower jaw teeth	5	3	1	5
Cleaning of the right upper jaw teeth	6	6	1	9
Cleaning of the left lower jaw teeth	5	3	1	5
Cleaning of the left upper jaw teeth	6	4	1	7
Cleaning of anterior teeth	3	2	1	3
Brackets/Braces Installation				
Brackets installation in the right lower jaw teeth	5	3	1	5
Brackets installation in the right upper jaw teeth	6	6	1	9
Brackets installation in the left lower jaw teeth	5	3	1	5
Brackets installation in the left upper jaw teeth	6	4	1	7
Brackets installation in anterior teeth	3	2	1	3

The final results of job posture analysis using REBA showed that activities done in the right upper jaw teeth while standing (restoration, extraction, cleaning, brackets installation) have very high risk levels (scores higher than 10), necessitating immediate changes. On the other hand, the extraction at sitting position (with a score of 11) had the highest risk level. In addition, other activities such as restoration, cleaning, and brackets installation in the right lower jaw teeth while in the sitting position have also recorded to be with high risk levels (scores of 8–10), as well as the extraction of teeth from other parts of the jaw except the anterior part. For the standing position,

Table 10. REBA scores for different dentistry activities – standing position

Standing Position	Arm and Wrist Score	Neck, Trunk and Leg Score	Activity Score	Final REBA Score
Restoration				
Restoration of the right lower jaw teeth	5	5	1	7
Restoration of the right upper jaw teeth	6	8	1	11
Restoration of the left lower jaw teeth	5	5	1	7
Restoration of the left upper jaw teeth	6	6	1	9
Restoration of anterior teeth	3	4	1	5
Extraction				
Extraction of the right lower jaw teeth	7	6	1	10
Extraction of the right upper jaw teeth	8	9	1	12
Extraction of the left lower jaw teeth	7	6	1	10
Extraction of the left upper jaw teeth	8	7	1	11
Extraction of anterior teeth	5	5	1	7
Cleaning				
Cleaning of the right lower jaw teeth	5	5	1	7
Cleaning of the right upper jaw teeth	6	8	1	11
Cleaning of the left lower jaw teeth	5	5	1	7
Cleaning of the left upper jaw teeth	6	6	1	9
Cleaning of anterior teeth	3	4	1	5
Brackets/Braces Installation				
Brackets installation in the right lower jaw teeth	5	5	1	7
Brackets installation in the right upper jaw teeth	6	8	1	11
Brackets installation in the left lower jaw teeth	5	5	1	7
Brackets installation in the left upper jaw teeth	6	6	1	9
Brackets installation in anterior teeth	3	4	1	5

restoration of the left upper jaw teeth, extraction of the lower jaw teeth, cleaning and brackets installation in the left upper jaw teeth posed a high risk level (scores of 8–10), which needed investigation and interventions. The task that puts the orthodontists at the highest risk for MSDs is tooth extraction.

In summary, the results of REBA identified that 15% of the activities were classified as very high risk, 27.5% were high risk, 50% with average risk and 7.5% with low risk. Orthodontists extracting the right upper jaw teeth had the score corresponding to a very high risk (11–12). Thus, an immediate action must be done in order to improve their posture.

Dentists usually carry out these activities while sitting or standing directly at the patient's right side, or behind the patient's head slightly to the right side. This makes orthodontists assume awkward postures when performing operations in the patient's right upper jaw just to have a good view of the teeth. Correlation analysis revealed that REBA scores have significant, positive correlations with the CMDQ scores, upper extremity MSDs, shoulder pain, upper arm pain, and forearm pain. Issues identified during the observation of the orthodontists while at work are as follows:

a. Working with the neck tilted to one side.
b. Shoulders elevated.
c. Side bending to left or right.
d. Forward bending/overreaching at waist.
e. Shoulders flexed and abducted.
f. Elbows flexed greater than 90°.
g. Wrists flexed/deviated in grasping.
h. Position maintained for 20+ min per patient.
i. Narrow work area (mouth).

Having known the MSD risk levels for each task category, another concern would be the duration of exposure of the workers to these risks. In general, the longer the duration, the greater was the risk for injury. For this, work sampling was performed on the workers to determine the proportion of time each occupational group spends on each activity. Workers were classified into two groups: those who practice both general dentistry and orthodontics and those who are restricted to orthodontics only. The results of the work sampling are shown in Table 11.

Table 11. Proportions of tasks among occupational groups

Activity	General dentistry and orthodontics (n = 24)	Orthodontics only (n = 6)
Restoration	20.00%	0.00%
Extraction	21.67%	0.00%
Cleaning	32.50%	43.33%
Brackets/Braces Installation	24.17%	52.50%
Others	1.67%	4.17%

Orthodontists who also practice general dentistry spent most of their time on cleaning, which needs to be done before every restoration and installation of braces or brackets. For those who practice orthodontics only, more than half (52.50%) of their time is spent on braces/brackets installation and 43.33% on cleaning. Other tasks include check-ups and cleaning the tools, among others.

Anthropometric Data
For the anthropometric data, body dimensions of 18 females and 12 males were measured and obtained the shortest and tallest orthodontist with 150 cm and 183 cm, respectively. The age of the participants raged from 29 to 60 yrs old, while weights varied from 49 to 76 kg. The mean standing height for males was higher than for females at 165 cm and 171 cm, respectively. The anthropometric measurements considered relevant for the study included the standing height, shoulder height, upper arm length, lower arm length, elbow height, waist height, knee height, overhead reach, forward reach, elbow height (sitting), hip breadth, buttock-knee length (sitting), and popliteal height (sitting).

Anthropometric measurements for both standing and sitting posture were collected. The 5th and 95th percentile groups will help in establishing the maximum and minimum levels of work heights. From these data, comfortable heights of dental chairs and operatories can be recommended. Table 12 lists the workers' anthropometric measurements.

Table 12. Anthropometric measurements

Anthropometric measurements (cm)	Female					Male				
	Mean	5th percentile	50th percentile	95th percentile	Std. dev.	Mean	5th percentile	50th percentile	95th percentile	Std. dev.
1. Height	165.34	150.46	160.02	180.29	12.05	171.45	165.10	170.69	177.47	3.98
2. Shoulder height	135.26	123.10	130.91	147.49	9.86	140.26	135.07	139.64	145.19	8.20
3. Upper arm	31.10	28.30	30.10	33.91	2.27	32.25	31.06	32.11	33.38	1.88
4. Lower arm	23.97	21.82	23.20	26.14	1.75	24.86	23.94	24.75	25.73	1.45
5. Elbow height	104.16	94.79	100.81	113.58	7.59	108.01	104.01	107.53	111.81	6.31
6. Waist height	87.63	79.75	84.81	95.55	6.39	90.87	87.50	90.46	94.06	5.31
7. Knee height	47.12	42.88	45.61	51.38	3.43	48.86	47.05	48.65	50.58	2.86
8. Overhead reach	208.01	189.30	201.32	226.82	15.16	215.70	207.71	214.74	223.27	12.61
9. Forward reach	72.75	66.20	70.41	79.33	5.30	75.44	72.64	75.10	78.09	4.41
10. Elbow height	22.06	15.61	21.36	29.42	4.49	24.5	21.22	24.51	27.68	3.91
11. Hip breadth	41.02	37.76	41.21	44.28	2.29	34.4	30.55	35.27	36.81	4.01
12. Buttock-knee length	34.98	31.57	35.66	37.44	2.25	34.65	31.77	35.35	36.72	2.08
13. Popliteal height	47.12	42.88	45.61	51.38	3.43	48.86	47.05	48.65	50.58	2.86

Assessment of the Orthodontic Workplace

Correlation analysis has revealed that the discomfort scores had the second strongest correlation with the physical environment as perceived by the orthodontists. This means that the workstations, tools, equipment, and appliances may not be designed to fit workers, and can lead to problems such as improper work heights, awkward working postures, tools poorly designed for the intended task, and even accidents. With this in mind, we assessed the orthodontic workplace and identified potential risk factors for MSDs existing in the orthodontic clinic. Issues of concern in the orthodontic workplace include the work heights, tools and equipment, and environmental hazards.

Workstation Set-up

With the orthodontists' anthropometric measurements at hand, the workstation set-up can be evaluated by comparing working heights with the orthodontists' body measurements. Points of interest in the workstation assessment were the orthodontist's seat, the patient chair, placement of tools, and the dental operatory in general. Table 13 compares the current measurements of the orthodontist's seat, patient's chair, and certain dimensions of the operatory with the gathered anthropometric data from the respondents.

Table 13. Comparison between orthodontic workplace measurements and anthropometric data

Workplace dimensions	Body dimensions	Anthropometric measurements	Workplace measurement	Assessment
Dentist's seat arm rest height (from base of seat)	Elbow height (from base of seat)	17–29 cm (15° inclination for upper arm)	25 cm (average)	Dentist's seat arm rest is fixed. Too low for 95th percentile and too high for the 5th percentile
Dentist's seat width	Hip breadth	34–44 cm	46 cm (average)	Dentist's seat width is fixed but acceptable.
Dentist's seat height	Popliteal height	43–51 cm	46–50 cm (adjustable)	Does not accommodate 5th percentile (male and female) and 95th percentile (male and female)
Dentist's seat depth	Buttock-knee length	34–43 cm	35 cm (average, fixed)	Acceptable
Patient's head rest height	Elbow height (from base of seat)	40–61 cm (15° inclination for lower arm and 25 cm incisors-head height)	46–108 (adjustable)	Does not accommodate 5th percentile female
Instruments holder distance from dentist	Forward reach	66–78 cm	Adjustable height (64–90 cm) and distance	Acceptable

Most of the workplace dimensions do not accommodate all or a certain portion of the subjects. While most of the appliances are adjustable, these still fail to fit to 90% of the subjects (5th to 95th percentile). Only the dentist's seat width and the instruments holder have acceptable measurements.

Table 14 lists the problems identified in the present orthodontic workstation set-up based on direct observation and interviews with the respondents. This study is geared towards addressing these problems and the evident misfit between the current orthodontic workplace and the user's body measurements.

Associations Between Risk Factors and Musculoskeletal Complaints

To determine which among the aforementioned factors are good predictors of the orthodontists' discomfort scores, the multiple regression technique was performed showing the independent variables that resulted to significant correlations with the discomfort scores. Table 15 shows the output of the multiple regression method.

The R^2 value was equal to 89.75%, indicating 89.75% of the variation in the discomfort score can be explained by age, weekly working hours, workload and

Table 14. Problems observed in present dental workspace

Problem area	Present situation	Body parts affected
Dentist's seat arm	• Dentist's chair has no lumbar, thoracic, or arm support	Shoulders and arms
Instrument table	• Instruments table is not positioned properly within the comfortable distance. Hence, during the operation the orthodontist sometimes struggle to reach the tools placed on the instrument table	Back and shoulders
	• Wrists are bent	
Dental operatory	• Edges of chairs, instrument tables and work surfaces are sharp and uncomfortable	Arms and hands
Dentist's seat	• Dentist's seat is not adjustable	Neck, shoulders, back, arms
Workspace area	• The actual workspace area (patient's mouth) is too narrow. Most dentists do not use lip and cheek retractor to have a better view of the teeth	Neck, shoulders, back, arms, wrists

Table 15. Predictors of MSDs in the whole body

Risk factor	p-value
Age	0.020*
Weekly working hours	0.000*
Workload and responsibility	0.002*
REBA score	0.045*

*Significant at $p < 0.05$

responsibility, and posture (REBA score). These four independent variables come together to form the best predictive model with the highest R^2 value. Hence, out of the 17 factors considered in this study, only four emerged to be significant predictors of total MSDs. Weekly working hours was the strongest risk factor, followed by workload and responsibility, age, and posture (REBA score).

Regression equations for the total CMDQ score is as follows:

$$CMDQ = -162.0 + 0.285\, Age + 2.512\, Weekly\, work\, hours$$
$$+ 1.405\, Workload\, and\, responsibility + 2.58\, REBA\, score$$

Discomfort scores were significantly higher in the upper extremities than the lower extremities, particularly in the back, arms, neck, and shoulders. Analyzing the upper and lower extremities separately, the six factors namely: tenure, weekly working hours, job requirements, workload and responsibility, posture (REBA score), and general dentistry practice are significant predictors of the upper extremity CMDQ scores. Working on weekends significantly influenced the development of MSDs in the lower extremities (Table 16).

726   R. D. Borres et al.

Table 16. Predictors of MSDs in the upper and lower extremities

Risk factor	p-value
Upper extremities: R = 90.40%	
Tenure	0.005*
Weekly working hours	0.000*
Job requirements	0.004*
Workload and responsibility	0.019*
Weekly overtime	0.143
REBA score	0.000*
General dentistry practice	0.041*
Lower extremities: R = 33.54%	
Work on weekends	0.0033*
Perceived physical environment	0.118

*Significant at p < 0.0

The occurrence of MSDs, or the number of body regions affected, was significantly influenced by weekly working hours and cigarette consumption. Further analyses revealed that weekly working hours significantly influence shoulder pain, and upper arm. Working position (sitting or standing) also greatly contribute to wrist and forearm pain. Weight was the single significant predictor for neck pain. The causes for upper back pain were also obtained and were found to be weekend work, daily cigarette consumption, age, BMI, and the presence of an assistant. For upper arm, this was predicted by workload and responsibility, general health, BMI, and tenure. Lastly, the forearm pain can be predicted by workload and responsibility, posture (REBA score), general health, and perceived physical environment.

Analyzing those orthodontists who focus on orthodontic tasks alone, three factors turned out to be significant predictors of CMDQ scores namely: perceived physical environment, weekly working hours, and general health. These were different when compared to the results of the multiple regression of the entire sample population.

Regression equations for the total CMDQ score is as follows:

$$CMDQ = -85.61 + 1.821 \, Perceived \, Physical \, Environment - 0.3214 \, Weekly \, work \, hours + 1.393 \, General \, Health$$

Table 17 shows the significant predictors of the upper extremities and lower extremities analyzed separately. In addition to the physical environment and the weekly overtime, significant predictors also included the number of sick leaves and the presence of an assistant. For the lower extremities, tenure, weight, BMI, and workload were of substantial influence to the CMDQ scores.

Table 17. Predictors of MSDs in the upper and lower extremities among orthodontists without general dentistry practice

Dependent factor	Significant predictors
Upper extremities:	Perceived physical environment
	Weekly overtime
	Number of sick leaves
	Presence of an assistant
Lower extremities:	Tenure
	Weight
	BMI
	Workload and responsibility

4 Discussion

The most common musculoskeletal complaints among the orthodontists who participated were frequently occurring in the upper extremities. They had reported complaints
in several body regions mainly pertaining to upper back (100%), neck (87%), lower
back (80%), wrists (80%), forearms (73%), upper arms (73%), and shoulders (70%).
This coincides with the results of a study conducted in America that reported
approximately 81% of American dentists suffer from neck, shoulder, and lower back
pain (La Rochelle 2017). It is assumed that during dental operations the upper limb
muscles are frequently affected compared to other regions of the body (Rundcrantz
et al. 2010) due to repeated forward positioning of the head and bending of low back
during clinical procedures (Alexopoulos et al. 2004). A total of 23.33% of participants
have reported being absent in the past month due to some form of musculoskeletal pain
within these regions with the major reason being back pain (57.14%). Job restriction
during work had also been reported to be caused by mainly upper arm pain (13.27%) &
neck pain (12.83%). In comparison to dental practitioners who also practice general
dentistry, those who focus on orthodontic related tasks experience more pain in the
back, neck, and wrists.

The Cornell Musculoskeletal Disorder Questionnaire (CMDQ) scores are not
affected by gender, employment status, and presence of an assistant. However, those
who practice general dentistry in addition to focusing on orthodontics have significantly higher CMDQ scores due to an increase in workload. The addition of
orthodontics tasks requires higher levels of precision and concentration (Finsen et al.
1997). For orthodontists who also work on weekends compared with those who do not,
they exhibit higher CMDQ scores as well as orthodontists who do not take breaks
during work. Similarly, in another study it was determined that dentists with more total
work time reported a higher prevalence of problems (Finsen et al. 1997).

The Rapid Entire Body Assessment (REBA) scores demonstrated positive significant relationships with the upper extremity. The task that poses the highest risk levels
were the activities done in the right upper jaw teeth while standing (restoration,
extraction, cleaning, brackets installation). REBA scores for the standing position
exceeded those of the sitting position. The most risky task among the four activities

was tooth extraction due to the force that orthodontists need to apply for a tooth to be pulled out.

Comparing the orthodontists' anthropometric measurements with the current dimensions of the workstation, it was found that most of the workplace dimensions do not accommodate all size groups. In addition, most of the orthodontists' seats are not adjustable; forcing them to assume postures that may increase their musculoskeletal risks. The actual workplace area (patient's mouth) is too narrow and as a result, their necks must be bent in order to have a good view at the patient's mouth.

Out of the all the factors considered in this study, only four emerged to be significant predictors of total MSDs namely: age, weekly working hours, workload and responsibility, and posture (REBA score). Weekly working hours was the strongest risk factor, followed by workload and responsibility, age, and posture (REBA score). Age, weekly working hours, workload, and responsibility all increased with the CMDQ score. This implies that older orthodontists are more prone to musculoskeletal symptoms as well as those who work for extended hours and those who have heavy workload. As anticipated, REBA score was also a significant predictor of the discomfort scores. This verified that incompatibilities between body size and workplace design are significantly related to work-related body pains.

The existing orthodontic clinics studied in this research appear to be inadequately designed to fit the worker's needs. Among the significant predictors of the CMDQ score, posture (REBA score) is the factor that can be most easily controlled by altering the workplace set-up and by adopting appropriate work practices that deal with the correct working positions.

Table 18 lists the recommendations that focused primarily on the improvement of the posture since this is the easiest factor that can be controlled. To counteract the weekly working hours and the workload, proper administrative controls were also suggested.

Table 18. Current vs. proposed orthodontic workstation set-up and instruments

Customer requirement (from QFD)	Design/Ergonomic consideration	Current	Proposed
Comfort and ease of using tools and equipment	Provision of high-end instruments and equipment	No cheek retractors and magnification aids for a better view of the teeth	Use cheek retractors to lessen neck bending and dental loupes (telescopes) so that the orthodontist can sit in an upright position

(continued)

Table 18. (*continued*)

Customer requirement (from QFD)	Design/Ergonomic consideration	Current	Proposed
		Instruments have no padding to reduce contact stress. Some handles are not round. Diameter ranged from 6.5 mm to 11 mm. Weight ranged from 10 to 25 g	Dental instrument diameter should range from 5.6 to 11.5 mm and should have paddings, especially those that are gripped for longer periods of time. Instruments should not exceed 15 g to reduce muscle workload and pinch force. Round handles will reduce muscle force and compression
		Some instruments need sustained pinch grip to operate. Due to obsolescence, a number of instruments have become dull and difficult to use	Choose instruments that are easy to activate and operate. This will reduce the risk for errors. Make sure that the tools are sharp so that the orthodontist will not need to apply additional force
Ease of accessing tools and instruments	Proper positioning of instruments table	During work, orthodontists sometimes struggle to reach tools from the instruments table	The instruments table should be positioned in a manner which allows the orthodontists to maintain a neutral working posture. It should require minimum adjustment and effort to access so as to reduce postural deviation while working. It should be kept within a "comfortable distance" (55–66 cm) and not above shoulder height or below waist height

(*continued*)

Table 18. (*continued*)

Customer requirement (from QFD)	Design/Ergonomic consideration	Current	Proposed
Comfort	Orthodontist's seat height adjustability	Some chairs have fixed heights. For adjustable chairs, height ranged at 46–50 cm	Orthodontists' seat height must be adjustable and should range from 43 cm (5th percentile, female) to 51 cm (95th percentile, male). Foot rests should be provided
Comfort	Back and arm supports for orthodontist's chair	Some chairs have no back and arm supports. For those that do have, arm supports are at an average height of 25 cm	Use chairs that have adjustable arm rests ranging from 17 cm (5th percentile, female) to 29 cm high (95th percentile, male) that allows the arms to be positioned at an angle of 15° from its neutral axis
Ease of accessing tools and instruments	Proper arrangement of instruments	Instruments are not arranged and organized prior to operation; orthodontists sometimes take much time in searching for the appropriate tool	Organize instruments prior to operation. Instruments may be arranged based on the sequence at which the tools are used or may be grouped according to function. Color-coding and labels may make instrument identification easier
Comfort, Orthodontist's safety	Proper administrative controls	43.33% of the orthodontists do not take breaks and only 10% take breaks every after patient. More than half 56.67% of them prefer to work at the standing position	Orthodontists should take breaks in between patients especially when performing highly repetitive tasks. During those breaks, it is recommended that they take strengthening exercises and rest their eyes to reduce eye strain. It is preferred to work at the sitting position to reduce MSDs. Better yet, orthodontists can alternate between the standing and sitting position

(*continued*)

Table 18. (*continued*)

Customer requirement (from QFD)	Design/Ergonomic consideration	Current	Proposed
Comfort	Working posture	Orthodontists were observed to bend their necks, elevate their shoulders, twist their wrists and bend forward while at work. Majority (57%) prefer to work while standing	Orthodontists should maintain an upright position with the arms and wrists at neutral position. It is recommended that they work close to the body to prevent overextension of their arms, neck, and back. Working while in a sitting position is preferable to a standing position

5 Conclusion

Further development of dental ergonomics must take place among Filipino orthodontists to address the risks for the development of musculoskeletal disorders that they encounter in their everyday work. Investigation of musculoskeletal pain sheds light on the prevalence of MSDs in the upper back (100%), neck (87%), lower back (80%), wrists (80%), forearms (73%), upper arms (73%), and shoulders (70%). Among the factors studied, age, weekly working hours, workload and responsibility, and posture were significantly associated with the discomfort scores, while psychosocial factors (job requirements, workload and responsibility) were not, although it exhibited a substantial correlation with MSDs. MSDs were more prevalent in the upper extremities and in the right part of the body compared to their left counterparts.

In addition, the REBA method showed that the major orthodontic tasks (restoration, extraction, cleaning, and brackets/braces installation) exposed the orthodontists to working postures that were highly at risk for MSD. REBA scores for the standing position slightly exceeded those of the sitting position because of higher leg scores. REBA has shown that activities done in the right upper jaw teeth while standing (restoration, extraction, cleaning, brackets installation) have very high risk levels (scores higher than 10), necessitating immediate changes. Among the four tasks, tooth extraction puts the orthodontists at the highest risk of developing MSDs.

The problem lies in fixed-height work surfaces that do not accommodate multiple body sizes of orthodontists. Some seats have no back and arm supports and lack adjustable features, thus failing to accommodate some portion of the study population. Improvement of work-place design must be taken into account and accommodate majority of the participants measurements. While some of the significant factors are uncontrollable, endeavors can be focused on improving risk factors that can be controlled (posture and workload). Given the association of the posture with work-station dimensions, designing ergonomic workstations to better fit the population, employing certain administrative controls and modifying how tasks are performed may be beneficial in reducing the prevalence of MSD symptoms among Filipino orthodontists.

References

Akesson I, Hansson G, Balogh I, Moritz U, Skerfving S (1997) Quantifying work load in neck, shoulders and wrists in female dentists. Int Arch Occup Environ Health 69:461–474

Alexopoulos E et al (2004) Prevalence of musculoskeletal disorders in dentists. BMC Musculoskelet Disord 5:16

Armstong T (2006) Circulatory and local muscle responses to static manual work. PhD dissertation, The University of Michigan, Ann Arbor, Michigan

Bezik J (2012) The different types of dental specialists. http://www.dentistinfairfaxva.com/family-care/the-different-types-of-dental-specialists/. Accessed 10 May 2015

Bureau of Labor Statistics, U.S. Department of Labor (2004) Special Report provided on June 28

Chao E, Opegrand J, Axmear F (2010) Three-dimensional force analysis of finger joints in selected isometric hand functions. J Biomech 9:387–396

Dantas F, de Lima K (2014) The relationship between physical load and musculoskeletal complaints among Brazilian dentists

Fasunloro A, Owtade FJ (2004) Occupational hazards among clinical dental staff. J Contemp Dent Pract 5(2):134–152

Finsen L, et al (1997) Musculoskeletal disorders among dentists and variation in dental work

Gorter RC et al (2000) Burnout and health among Deutch dentist. Eur J Oral Sci 108(4):261–267

Hayess M, Cockrell D, Smith DR (2009) A systematic review of musculoskeletal disorders among dental professionals. Int J Dent Hyg 7(3):159–165. https://doi.org/10.1111/j.1601-5037.2009.00395

Horstman S, Horstman B, Horstman F (2007) Ergonomic risk factors associated with the practice of dental hygiene: a preliminary study. Prof Saf 42:49–53

Jabbar TAA (2008) Musculoskeletal disorders among dentist in Saudi Arabia. Pak Oral Dent J 28(1):135–144

Jafari H, Yekta-Kooshali M (2017) Work-related musculoskeletal disorders in Iranian dentists: a systematic review and meta-analysis

Kierklo A et al (2011) Work-related musculoskeletal disorders among dentists – a questionnaire survey. Ann Agric Environ Med 2011(18):79–84

Kupcinskas L, Petrauskas D (2003) Hepatitis-Mediku Profesineliga. J Stomatologija Suppl1 (1):22

Lalumandier J, McPhee S (2011) Prevalence and risk factors of hand problems and carpal tunnel syndrome among dental hygienists. J Dent Hyg 75:130–134

La Rochelle N (2017) Work-related musculoskeletal disorders among dentists and orthodontists

Lin T et al (2012) Prevalence of and risk factors for musculoskeletal complaints among Taiwanese dentists

Meador H (2003) The biocentric technique: a guide to avoiding occupational pain. J Dent Hyg 67:38–51

Michalak-Turcotte C (2011) Controlling dental hygiene work-related musculoskeletal disorders: the ergonomic process. J Dent Hyg 74:41–48

Moosavi S et al (2015) Ergonomic analysis to study the intensity of MSDs among practicing Indian dentists

Nasl Saraji J, Hosseini MH, Shahtahei SJ, Golbabaei F, GhasemKhani M (2005) Evaluation of ergonomic postures of dental professions by REBA. J Dent 18(1):61–68

Newell T, Kumar S (2004) Prevalence of musculoskeletal disorders among orthodontists in Alberta. Int J Ind Ergon 33(2):99–107

Pandis N, Pandis BD, Pandis V, Eliades T (2007) Occupational hazards in orthodontics: a review of risks and associated pathology. Am J Orthod Dentofac Orthop 132:280–292

Philippine Dental Association (2012) Taking a look back to the history of dental practice in the Philippines. http://www.dentalclinicphilippines.com/news/takingalook-back-to-the-history-of-dental-practice-in-the-philippines/. Accessed 25 Apr 2015

Rabiei M, Shakiba M, Shahreza H, Talebzadeh M (2011) Musculoskeletal disorders in dentists. Int J Occup Hyg

Rempel D, Keir P, Smutz W, Hargens A (2007) Effects of static fingertip loading on carpal tunnel pressure. J Orthop Res 15:422–426

Rolander B, Bellner A (2010) Experience of musculoskeletal disorders, intensity of pain, and general conditions in work-the case of employees in non-private dental clinics in a county in southern Sweeden. Work 17:65–73

Romualdez A et al (2001) The Philippines health system review. Health Systems in Transition, vol 1, no 2

Rundcrantz BL, Johnson B, Moritz U (2010) Occupational cervico-bronchial disorders among dentists. Scand J Soc Med 19(3):174–180

Sachan A et al (2013) Ergonomics, posture and exercises - painfree, prolong orthodontic career. J Orthod Res 1:89–94

Sanders M, Turcotte C (2002) Strategies to reduce work-related musculoskeletal disorders in dental hygienists: two case studies. Hand Ther 15:363–374

Simmer-Beck M, Branson B (2010) An evidence-based review of ergonomic features of dental hygiene instruments. Work 35:477–485. https://doi.org/10.3233/WOR-2010-0984 IOS Press

Valanchi B, Valanchi K (2003) Preventing musculoskeletal disorder in clinical dentistry. J Am Dent Assoc 134(12):1604–1612

Viragi P et al (2013) Occupational hazards in dentistry – knowledge attitudes and practices of dental practitioners in Belgaum city. J Pierre Fauchard Acad 27:90–94

Yee T, Crawford L, Harber P (2005) Work environment of dental hygienists. J Occup Environ Med 47:633–639

Quantifying Vertebral Endplate Degeneration Using the Concavity Index

Menekse Salar Barim[1]([✉]), Richard F. Sesek[2], M. Fehmi Capanoglu[2],
Wei Sun[3], Sean Gallagher[2], Mark C. Schall Jr.[2], and Gerard A. Davis[2]

[1] Oak Ridge Institute for Science and Education (ORISE) Research Fellow,
Cincinnati, OH 45202, USA
mzs0053@auburn.edu
[2] Auburn University, Auburn, AL 36830, USA
[3] Data Scientist, Colaberry Inc., St. Louis, MO, USA

Abstract. A novel morphometric measurement of endplate degradation was compared with qualitative ratings of intervertebral disc degeneration (Pfirrmann Grading) in a double-blinded study to investigate a new, quantitative method for relating disc morphology and bony changes using MR imaging techniques known as the "Concavity Index" (CI). By adding a quantitative measure of vertebral endplate degeneration, the CI could provide further insight into structural changes related to disc breakdown and subsequent low back pain. The continuous nature of the CI may also allow medical professionals to more closely monitor a patient's low back health. T2-weighted MRI scans of the sagittal profile of the lumbar endplates (L2-S1) were collected from 50 subjects (25 females and 25 males) whose ages ranged from 20–40 years. Three trained examiners independently measured the height and the concavity levels of each lumbar vertebrae (L2-S1) as well as assessed the health of the intervertebral discs using Pfirrmann's lumbar disc degeneration grading method. Concavity Indices (CIs) were computed by dividing measured concavity level by disc height (CL/DH). A larger CI was hypothesized to be indicative of spinal degradation and subsequent low back pain. Intra- and inter-rater reliabilities were assessed for both the CI measurements and Pfirrmann's lumbar disc degeneration grades. The categorical intra-observer agreement for Pfirrmann ratings ranged from 26 to 63%. However, the CI, which is a continuous measure, varied by only 2% (average absolute error) among raters. Endplate concavity is indicative of fracturing and damage and is hypothesized to lead to subsequent disc degeneration due to impediment of nutrient flow to the discs themselves. The CI shows promise as a means for potentially quantifying low back health and identifying risk for future low back pain prior to significant disc degeneration.

Keywords: Vertebrae degeneration · MRI · Concavity index · Pfirrmann grading

© Springer Nature Switzerland AG 2019
S. Bagnara et al. (Eds.): IEA 2018, AISC 820, pp. 734–741, 2019.
https://doi.org/10.1007/978-3-319-96083-8_88

1 Introduction

Low back pain (LBP) represents one of the most costly and prevalent musculoskeletal disorders (MSDs). MSDs are the leading cause of disability in the United States and represent 48% of all self-reported chronic medical conditions [1]. LBP is a major health issue affecting millions of people worldwide [2–4].

Despite advances in imaging technology, the etiology of the underlying pain is frequently illusive. Morphological changes related to normal disc aging often appear on MR imaging without any corresponding symptoms. Despite an incomplete understanding of the relationship between physical changes and pain, these MRI-detectible morphological changes show predictive promise and warrant further discussion [5]. Interest in biomechanical models of the spine, particularly detailed knowledge regarding spinal morphometry and the relationships between vertebral segments and corresponding intervertebral discs has been increasing. Several quantitative studies have investigated the external geometry of the vertebrae and adjacent intervertebral discs for different regions of the human spine. In biomechanical models, assumptions have been made assuming the spine acts as a single straight line without considering the volume or the curvature of the spine [6] or incorporating average measurements into models without considering the impact of personal characteristics or differences between individuals [7]. Lakshmanan discovered that the majority of lumbar endplates were concave, while the majority of sacral endplates were flat [8]. In addition to this study, Larsen proposed that concavity was nearly always the result of physical loading [9].

In vertebral motion segment testing, the endplate is typically the first structure to become damaged, and is clearly the weakest link when the spine is loaded in compression [10]. Additionally, it has been shown that compressive loading fractures the endplate before damaging the disc. Increases in concavity related to subsidence can result in vertebral endplate scarring that may impede the flow of nutrients to the intervertebral disc (IVD). It is proposed that increased concavity is related to increased endplate damage and therefore subsequent disc degeneration [10]. In other words, increases in concavity proceed and are related to disc degeneration [11].

The relationship between vertebral disc degeneration and low back pain is well established [12, 13]. This study proposes a novel approach for quantifying vertebral concavity and relating this to disc degeneration. Unlike measures of disc health, which are subjective and discrete, this approach uses a mathematical measure, which quantifies endplate health continuously.

2 Methodology

2.1 Subject Data

MRI scans were collected from fifty subjects (25 male and 25 female) with ages ranging from 20 to 40 (mean 31.1 ± 5.4 years). All Subjects were scanned on a whole body 3T Magnetic Resonance Imaging machine (Siemens Verio open-bore). All imaging was performed in the supine position. Informed consent was obtained from all

736   M. S. Barim et al.

subjects. A T2-weighted image, which provides a comprehensive perception of disc structure and good tissue differentiation, was used for the morphological evaluation of intervertebral discs. MRI was performed on the intervertebral discs from L2-S1. MRI data were obtained using a dedicated abdominal coil. The protocol included the following sequences: Axial Continuous T2-weighted, Sagittal Continuous T2-weighted, and Axial Multi group T2-weighted images with the following parameters; T2-weighted spin-echo (TR – 3440 ms; TE – 41 ms). All MR images were obtained at a 3-mm slice thickness with 385 FoV read & 100% FoV phase.

2.2   Image Assessment

Three, Level-3 certified MRI observers, blinded to each other's measurements, graded lumbar IVDs and adjacent vertebral bodies using both the Pfirrmann Grading system and the CI in a randomized sequence. The Pfirrmann Grading System is widely used and related to the height and health of the IVD. Pfirrmann categorizes IVDs into five grades [14, 15]. Grades I and II have normal disc height and have a healthy structure when compared with other levels. Grades III and IV demonstrate some height changes (becoming narrower) relative to other discs and the disc's structure also begins to change. With Grade V, the distinction between disc nucleus and annulus is completely lost and the disc space has collapsed completely. The grading scale and progression from a healthy disc (Grade I) to a severely compromised disc (Grade V) is illustrated in Fig. 1 [16].

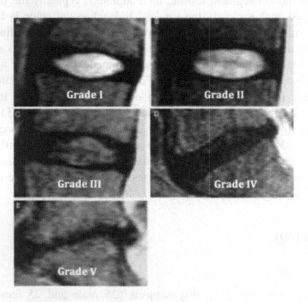

Fig. 1. Pfirrmann grading scores [16]

T2-weighted sagittal MRI scans were used for both Pfirrmann grading and the CI evaluation of lumbar endplates. To calculate the CI, each examiner measured the height and the concavity levels of the lumbar discs (L2–S1). CIs were measured as follows: the superior aspect of vertebral body lengths, which are the distances in the sagittal plane between anterior and posterior borders of vertebral body, were traced first. Then, the perpendicular distance between this line and the vertebral body was measured. This was defined as the concavity level. The concavity level was then divided by the corresponding disc height (CL/DH) (shown in Fig. 2). All measurements were performed using sagittal lumbar spine T2-weighted images and the Osirix software system (OsiriX v8.0.1, 2016).

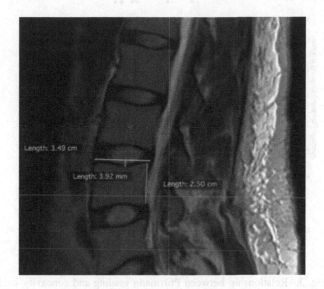

Fig. 2. Measurement of concavity index

3 Results

The absolute agreement between observers was relatively low for Pfirrmann Grading. The highest absolute agreement between any two observers was 63% and the lowest was 26%. Table 1 shows the Cohen's Kappa analysis between observers where "−1" and "+1" represent differences of 1 category.

Table 1. Probability of absolute and near (±1) category inter-observer agreement

Agreement	−1 Category	Absolute agreement	+1 Category	Kappa
Obs 1 vs Obs 2	0.28	0.38	0.32	0.18
Obs 2 vs Obs 3	0.16	0.63	0.17	0.61
Obs 1 vs Obs 3	0.34	0.26	0.36	0.18

High CI agreement (≥ 0.98) between observers suggests that CI measurements are consistent and potentially reliable when compared to Pfirrmann Grading. Table 2 shows the correlation coefficients among CI observers. Each observer's ratings compared against one another and a high level of agreement among observers (r^2 ranging from .963 to .983) was found. The relationship between CI and Pfirrmann is modest with variation in CI at each Pfirrmann level (shown in Fig. 3).

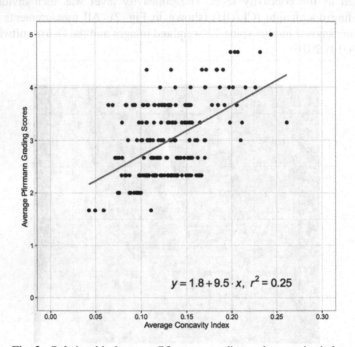

$$y = 1.8 + 9.5 \cdot x, \ r^2 = 0.25$$

Fig. 3. Relationship between Pfirrmann grading and concavity index

A possible explanation for this modest relationship is that the subject pool was a relatively healthy population with few Grade-V discs. Another possible explanation for lack of fit may be a function of observer experience with Pfirrmann grading.

Table 2. Pearson Correlation Coefficient between observers for concavity index

	Observer 1	Observer 2	Observer 3
Observer 1	1.0	0.99	0.99
Observer 2	0.99	1.0	0.98
Observer 3	0.99	0.98	1.0

There appears to be a clear observer effect with respect to Pfirrmann gradings. However, no such observer effect was detected when ICCs were computed for CIs. In fact, ICC model results were similar for one-way and two-way random effects analyses with both yielding ICC values of 0.985 and standard errors of 0.0026 and 0.0027, respectively. The results indicate that ICC is very high, while agreement among Pfirrmann ratings was relatively poor.

4 Discussion

Most inter-rater disagreements for the Pfirrmann scores were within one category and occurred when classifying grade II and III discs. The difference between grades II and III is heavily dependent open disc height which can be difficult to reliable assess visually. This is further complicated since normal disc height is not uniform across all levels in all subjects and disc height often decreased at the L5-S1 compared with other levels, even when disc health appeared to be otherwise healthy (e.g., good color and uniformity). Since the study population was asymptomatic, very few Grade-V discs were observed. Surprisingly, there were also very few Grade-I discs. A broader range of subject ages and symptom statuses would likely provide a wider distribution of Pfirrmann grades and CIs. With a wider distribution of data, relationships may become more clear. Pfirrmann grading relies solely on visual appearance of T2-weighted images. It is possible, however, to more quantitatively determine water and proteoglycan content of discs using MRI signals other than T2-weighted images. This may be used to enhance the Pfirrmann grading system and make the process more objective. The novice Pfirrmann graders in this study might benefit from such an enhancement.

Also, this study did not consider subject symptoms or low back pain. A prospective study including subjects with and without LBP could address this limitation. All of the subjects in this study were young (20 to 40 years of age) and were relatively healthy college students free of LBP. A diverse sampling of subjects from a greater age range and with varying occupational risk factors could address this limitation. Accurate knowledge of normal and degenerative lumbar intervertebral discs is important for medical professionals. Using the CI, medical professionals can potentially make more accurate and early diagnostic interpretations and, subsequently, more precise surgical interventions regarding lumbar vertebrae and intervertebral discs. The CI demonstrated very high agreement despite the lack of medical experience of the research team. Adding objective elements to the Pfirrmann grading system (such as water content evaluation using other MRI signals) could be beneficial for both inexperienced and experienced observers alike.

An ideal classification system for disc degeneration should be simple, easy to apply, discriminatory, and reproducible with good intra-rater reliability [21]. Results of this anatomic study of morphometric measurement for the lumbar vertebrae suggest that the CI method described within has promise for objectively quantifying low back health and possibly predicting future low back pain. The CI allows for relative comparisons because it is a continuous measure rather than an ordinal scale. On the contrary, and consistent with other studies [17], agreement between observers for Pfirrmann grading was relatively low. The Pfirrmann scoring system is simple and easy to use, but its subjective nature lacks the ability to subtly discriminate degradation. The CI, on the other hand, has demonstrated strong intra-rater reliability while being a more objective approach to assessing vertebral health. Together, CI and Pfirrmann paint a more complete picture of intervertebral motion segment health.

5 Conclusion

A novel approach for quantifying vertebral degeneration has been proposed and there appears to be a positive linear relationship between the CI and Pfirrmann grading. The Pfirrmann grading system is widely used and accepted. The CI may provide a complimentary measure capable of predicting disc degeneration and that could be used in conjunction with Pfirrmann grading to provide a more complete assessment of the health of a given spinal motion segment. The CI is easy to apply; requiring limited previous knowledge of MRI scans or low back geometry.

The CI in conjunction with the established Pfirrmann ratings can provide a more complete picture of low back health and could potentially provide a more comprehensive assessment of spinal segment health.

Acknowledgement. This publication was partially supported by Grant # 2T420H008436 from NIOSH. The findings and conclusions in this report are those of the authors and do not necessarily represent the official position of the Centers for Disease Control and Prevention. Use of trade names is for identification only and does not constitute endorsement by the Public Health Service or by the U.S. Department of Health and Human Service.

References

1. BMUS, The Burden of Musculoskeletal Diseases in the United States (2011) United States bone and joint initiative: the burden of musculoskeletal diseases in the United States, prevalence, societal and economic cost, 2nd edn. American Academy of Orthopaedic Surgeons, Rosemont
2. Pope MH, Goh KL, Magnusson ML (2002) Spine ergonomics. Annu Rev Biomed Eng 4:49–68
3. Brooks PM (2006) The burden of musculoskeletal disease – a global perspective. Clin Rheumatol 25(6):778–781 ISSN 0770-3198
4. Woolf AD, Pfleger B (2003) Burden of major musculoskeletal conditions. Bull World Health Organ 81(9):646–656 ISSN 0042-9686
5. Ract I, Meadeb JM, Mercy G, Cueff F, Husson JL, Guillin RA (2015) Review of the value of MRI signs in low back pain. Diagn Interv Imaging 96(3):239–249
6. Merryweather AS, Loertscher MC, Bloswick DS (2009) A revised back compressive force estimation model for ergonomic evaluation of lifting tasks. Work 34(3):263–272
7. Chaffin DB (1969) A computerized biomechanical model-development of and use in studying gross body actions. J Biomech 2(4):429–441
8. Lakshmanan P, Purushothaman B, Dvorak V, Schratt W, Thambiraj S, Boszczyk BM (2012) Sagittal endplate morphology of the lower lumbar spine. Eur Spine J 21:16–164
9. Larsen JL (1985) The posterior surface of the lumbar vertebral bodies, part I. Spine 10 (1):50–58
10. Adams MA (2004) Biomechanics of low back pain. Acupunct Med 22(4):178–188
11. Wang Y, Videman T, Battie MC (2012) Lumbar vertebral endplate lesions: associations with disc degeneration and back history. Spine 1(31):1490–1496
12. Kumar MN, Baklanov A, Chopin D (2001) Correlation between sagittal plane changes and adjacent segment degeneration following lumbar spine fusion. Eur Spine J 10:314–319

13. Kepler CK, Ponnappan RK, Tannoury CA, Risbud MV, Anderson DG (2013) The molecular basis of intervertebral disc degeneration. Spine J 13(3):318–330
14. Rajasekaran S, Babu JN, Arun R, Armstrong BR, Shetty AP, Murugan S (2004) ISSLS prize winner: a study of diffusion in human lumbar discs—a serial magnetic resonance imaging study documenting the influence of the endplate on diffusion in normal and degenerate discs. Spine (PhilaPa 1976) 29(23):2654–2667
15. Marinelli NL, Haughton VM, Munoz A, Anderson PA (2009) T2 relaxation times of intervertebral disc tissue correlated with water content and proteoglycan content. Spine (Phila Pa 1976) 34(5):520–524
16. Pfirrmann CW, Metzdorf A, Zanetti M, Hodler J, Boos N (2001) Magnetic resonance classification of lumbar intervertebral disc degeneration. Spine 26:1873–1878
17. Griffith JF, Wang YX, Antonio GE, Choi KC, Yu A, Ahuja AT et al (2007) Modified Pfirrmann grading system for lumbar intervertebral disc degeneration. Spine 32:708–712

Evaluating the Reliability of MRI-Derived Biomechanically-Relevant Measures

Menekse Salar Barim[1(✉)], Richard F. Sesek[2], M. Fehmi Capanoglu[2], Sean Gallagher[2], Mark C. Schall Jr.[2], and Gerard A. Davis[2]

[1] Oak Ridge Institute for Science and Education (ORISE) Research Fellow, Cincinnati, OH 45202, USA
mzs0053@auburn.edu

[2] Auburn University, Auburn, AL 36830, USA

Abstract. Human geometric dimensions have been estimated and approximated in several ways, most recently using Magnetic Resonance Imaging (MRI) techniques. The reliability of MRI-based measurement of structures has been shown to be relatively high. However, a limitation of reliability evaluations is that they often only compare assessments of the same MRI image (e.g., "slice" of the back); differences are only a function of analyst dexterity (in tracing or measuring the structures). Ideally, a reliability test should compare estimates of biomechanical structures using different scans analyzed by different analysts. This presents a "worst case" scenario and provides a robust test of the process's repeatability. Existing databases of vertebral and intervertebral dimensions tend to be limited with respect to measures of repeatability/reliability with relatively narrow study populations and/or parameters recorded. The objectives of this study were (1) to provide a more accurate data set of lumbar spinal characteristics from 144 Magnetic Resonance Imaging (MRI) scans which were reviewed and measured using the Osirix software program and (2) to assess inter- and intra-rater reliability of the MRI process itself. Reliability for the entire process was evaluated using the aforementioned worst-case scenario of comparing two distinct scans of the same subject with different researchers performing each MRI scan and different researchers performing measurements of the various aspects of vertebral and intervertebral disc dimensions. Geometric dimensions were consistent with measurements obtained in previous MRI-based studies. As expected, larger discrepancies were observed in the "worst case" scenarios (scanners and analysts both different). However, worst case variation was relatively low with 3.6% average absolute difference for anterior endplate measurements, for example, as compared to 2.6% average absolute difference for analysts re-rating their own scans after 1 month. The process for obtaining MRI-derived biomechanical measures appears to be robust.

Keywords: Lumbar vertebrae · Endplate morphology
MRI scan/rescan reliability

© Springer Nature Switzerland AG 2019
S. Bagnara et al. (Eds.): IEA 2018, AISC 820, pp. 742–749, 2019.
https://doi.org/10.1007/978-3-319-96083-8_89

1 Introduction

Human geometric dimensions have been estimated and approximated in several ways, most recently using Magnetic Resonance Imaging (MRI) techniques. The reliability of MRI-based measurement of structures has been shown to be high (e.g., $\geq .90$ ICC) [1, 2]. Low back pain (LBP) is one of the most prevalent and costly health problems experienced by industry [3]. Direct measurements of the spine in multiple planes can provide valuable information about the human vertebrae, particularly for improving subject specific biomechanical models. Research efforts have been made to measure the geometry of the low back using medical imaging techniques and been reported frequently. However, a comprehensive review of the reliability and veracity of the methods themselves has not been studied at the level presented herein. Specifically, a comparison of different scans by operators and reviewed/measured by different analysts has not been conducted on substantive sample size.

Magnetic resonance imaging (MRI) is increasingly used to assess patient lumbar spinal health. MRI has important benefits for imaging the musculoskeletal system [4–6], which provide better visualization of anatomic and pathologic structures, including cartilage, bones and ligaments [4, 7–9]. Morphometric analysis helps to determine the relationships of vertebrae with the anatomical dimensions of low back structures. These morphometric measurements have been questioned by reviewers, specifically, the reliability of the MRI data collection process used here and used previously [10, 11]. While regression relationships to predict low back parameters have been presented, the veracity of their measurement methods has not been adequately studied.

In order to evaluate how precisely these data are collected, a comprehensive scan-rescan study was conducted. Scan-rescan variability is very important because poor reliability of the measurement method itself could call into question the usefulness and accuracy of the regression results.

2 Methodology

2.1 Study Sample Size

MRI scans of the lumbar intervertebral segments (L2/S1) of thirty-six (36) subjects (20 males and 16 females) who were 19 years of age or greater were scanned on a 3T scanner using standardized T2 weighted protocol. Subject demographics (age, gender, height and weight) were obtained. The average age was 23.7 (± 3.1) years for males and the average mean was 25.4 (± 4.7) years for females. Institutional Review Board (IRB) approval and subject informed consent was obtained prior to data collection. MRI data were obtained using a dedicated abdominal coil (Fig. 1). Subjects were placed in a lying position (supine posture) on the scanner, foot support was provided, and they were instructed to keep their body stable (no motion during MRI scans to minimize artifacts).

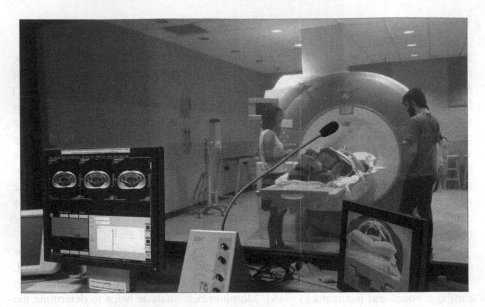

Fig. 1. MRI procedure

2.2 Measurements

MRIs were performed on a 3T unit (Siemens Verio open-bore, Auburn University Research Park, Alabama) using a dedicated abdomen coil. The protocol included the following sequences: Axial Continuous T2-weighted, Sagittal Continuous T2-weighted, and Axial Multi group T2-weighted images with the following parameters; T2-weighted spin-echo (TR 3440 ms; TE 41 ms). All MR images were obtained at a 3-mm slice thickness with 385 FoV read and 100% FoV phase.

Two level-3 MRI certified analysts were provided scans and performed measurements in random order to determine the intra and inter-rater reliability of MRI parameters. In total, 15 parameters were measured. These parameters were measured for the L2/L3, L3/L4, L4/L5, and L5/S1 lumbar regions. Parameters are as follows: Anterior Vertebrae Height (AVH), Posterior Vertebrae Height (PVH), Vertebral Height Index (VHI), Average Height Index (AHI), Sagittal Vertebrae Body Width (SVBW), Sagittal Vertebrae Body Height (SVBH), Height/Weight Index (HWI), Superior Vertebrae Body Length (SVBL) and Inferior Vertebrae Body Length, Length Index (LI) are shown in Fig. 2.

2.3 Repeatability of Measurements

In order to assess the reliability and the repeatability of measurements, two different observers measured all parameters three times with at least one month between repeated measurements of the same scan. Data from two observers and six sets of measurements were compared. In the lumbar MRI scans, there are 50 different slices, which can be chosen to perform measurements. In order to test the reliability, specific image

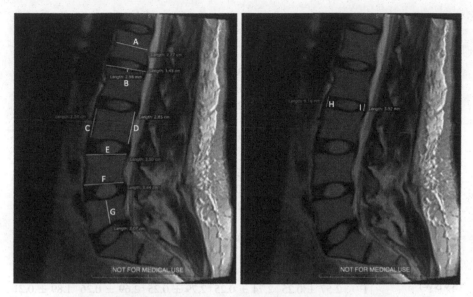

Fig. 2. Sagittal MRI scan with measurements. (A) Sagittal Vertebrae Body Width, (B) Concavity Height, (C) Anterior Vertebral Height, (D) Posterior Vertebral Height, (E) Superior Vertebral Body Length, (F) Inferior Vertebral Body Height, (G) Sagittal Vertebral Body Height, (H) Anterior IVD Height, (I) Posterior IVD Height

slices were not selected prior to measurements. Each observer chose the slice they thought was most appropriate for the measurement in question. The results show that, on average, the same slice was selected 61% of the time and observations were within one slice (3 mm) 90% of the time. This is across all conditions including analysts looking at the same scans and worst-case comparisons of different scans and different analysts. The highest levels of absolute agreement occurred when subjects reanalyzed the same scans. The lowest absolute agreement occurred with different analysts, regardless of scan. Analysts were within 1 slice of each other 78%-85% of the time for these three comparisons and in no case did analysts differ by more than two slices (6 mm).

3 Results

Mean values, standard deviations and range of data for the lower lumbar spine, which were obtained from the MRI measurements are reported. Scheffe tests were done choosing four different scenarios, which are Inter-rater reliability (same scan different observer/analysts), Intra-Best X (observer X measures her/his own scan two different times), Intra-Best Y (observer Y measures her/his own scan two different times) and Worst-Case (different scans observed by different analysts). According to the results, the most different measurements were observed in Worst-Case, which was predicted before the study (Tables 1 and 2).

Table 1. Vertebral body dimensions of lumbar region (L2, L3, L4, L5, S1)

Dimensions	Sex	L2	L3	L4	L5	S1
Anterior Vertebrae Height (AVH)	M	2.66 ± 0.17	2.77 ± 0.19	2.83 ± 0.19	2.89 ± 0.19	3.09 ± 0.19
		(2.09–3.14)	(2.32–3.25)	(2.4–3.42)	(2.29–3.2)	(2.59–3.57)
	F	2.46 ± 0.18	2.6 ± 0.17	2.62 ± 0.19	2.67 ± 0.19	2.98 ± 0.23
		(1.96–2.82)	(2.20–2.91)	(2.05–3.05)	(2.3–3.15)	(2.39–3.37)
	T	2.57 ± 0.21	2.69 ± 0.20	2.74 ± 0.22	2.8 ± 0.22	3.04 ± 0.22
		(1.96–3.14)	(2.20–3.25)	(2.05–3.42)	(2.29–3.2)	(2.39–3.57)
Posterior Vertebrae Height (PVH)	M	2.78 ± 0.19	2.84 ± 0.18	2.73 ± 0.16	2.52 ± 0.19	2.47 ± 0.26
		(2.33–3.18)	(2.3–3.2)	(2.27–3.03)	(2.02–2.83)	(1.55–3.14)
	F	2.57 ± 1.94	2.6 ± 0.17	2.48 ± 0.17	2.26 ± 0.18	2.29 ± 0.21
		(2.02–2.99)	(2.05–2.93)	(2.13–3)	(1.85–2.65)	(1.81–3.26)
	T	2.69 ± 0.22	2.73 ± 0.21	2.62 ± 0.21	2.40 ± 0.22	2.39 ± 0.26
		(2.02–3.18)	(2.05–3.2)	(2.13–3.03)	(1.85–2.83)	(1.55–3.26)
Sagittal Vertebrae Body Width (SVBW)	M	2.90 ± 0.23	3.08 ± 0.22	3.12 ± 0.25	2.98 ± 0.24	2.26 ± 0.26
		(2.29–3.41)	(2.67–3.98)	(2.57–3.74)	(2.32–3.75)	(1.75–2.96)
	F	2.57 ± 0.25	2.74 ± 0.25	2.79 ± 0.25	2.69 ± 0.26	1.89 ± 0.21
		(1.78–3.14)	(2–3.18)	(2.14–3.18)	(2.1–3.32)	(1.43–2.4)
	T	2.75 ± 0.28	2.93 ± 0.29	2.98 ± 0.3	2.86 ± 0.29	2.09 ± 0.3
		(1.78–3.41)	(2–3.98)	(2.14–3.74)	(2.1–3.75)	(1.43–2.96)
Sagittal Vertebrae Body Height (SVBH)	M	2.32 ± 0.23	2.37 ± 0.27	2.41 ± 0.24	2.34 ± 0.25	2.55 ± 0.25
		(1.71–2.71)	(1.72–3.35)	(1.77–2.88)	(1.65–2.76)	(1.87–3.13)
	F	2.24 ± 0.17	2.27 ± 0.16	2.77 ± 0.16	2.15 ± 0.19	2.44 ± 0.21
		(1.85–2.71)	(1.96–2.73)	(1.83–2.66)	(1.63–2.49)	(1.94–2.8)
	T	2.28 ± 0.21	2.33 ± 0.23	2.34 ± 0.22	2.26 ± 0.25	2.50 ± 0.24
		(1.71–2.71)	(1.72–3.35)	(1.77–2.88)	(1.63–2.76)	(1.87–3.13)
Superior Vertebrae Body Length (SVBL)	M	3.08 ± 0.20	3.22 ± 0.19	3.27 ± 0.22	3.28 ± 0.22	3.05 ± 0.24
		(2.38–3.58)	(2.63–3.88)	(2.76–3.84)	(2.83–4)	(2.31–4.25)
	F	2.77 ± 0.25	2.91 ± 0.23	2.97 ± 0.23	2.99 ± 0.22	2.76 ± 0.24
		(1.93–3.3)	(2.18–3.36)	(2.42–3.33)	(2.51–3.37)	(2.31–3.17)
	T	2.94 ± 0.27	3.08 ± 0.26	3.13 ± 0.27	3.15 ± 0.26	2.92 ± 0.28
		(1.93–3.58)	(2.18–3.88)	(2.42–3.84)	(2.51–4)	(2.31–4.25)
Inferior Vertebrae Body Length (IVBL)	M	3.14 ± 0.19	3.22 ± 0.22	3.31 ± 0.22	3.14 ± 0.24	1.94 ± 0.38
		(2.57–3.71)	(2.26–3.76)	(2.92–3.88)	(2.36–4.05)	(1.14–2.99)
	F	2.81 ± 0.25	2.91 ± 0.24	2.99 ± 0.24	2.89 ± 0.28	1.52 ± 0.23
		(1.91–3.3)	(2.21–3.29)	(2.27–3.35)	(2.33–3.44)	(1.03–2.08)
	T	2.99 ± 0.27	3.08 ± 0.28	3.16 ± 0.28	3.03 ± 0.29	1.76 ± 0.38
		(1.91–3.71)	(2.21–3.76)	(2.27–3.88)	(2.33–4.05)	(1.03–2.99)

Table 2. Scheffe's test results for vertebrae dimensions

Dimensions	Contrast coefficient	L2	L3	L4	L5	S1
Anterior Vertebral Height (AVH)	Inter rater	0.7305	0.9963	0.9904	0.9903	0.9944
	Inter Best X	0.9618	0.9974	1.0000	0.9995	0.9943
	Inter Best Y	0.9998	0.7578	0.9607	0.9129	0.9949
	Worst Case	0.8759	0.5398	0.2904	0.1672	0.0830
Posterior Vertebral Height (PVH)	Inter rater	0.0000	0.0000	0.1028	0.0766	0.0496
	Inter Best X	0.9865	0.9943	0.8590	0.9999	1.0000
	Inter Best Y	0.8429	0.5571	0.9992	0.8121	0.9745
	Worst Case	0.0000	0.0000	0.0426	0.0001	0.0127
Sagittal Vertebrae Body Width (SVBW)	Inter rater	0.7059	0.0000	0.0002	0.0973	0.3597
	Inter Best X	0.9997	0.9736	1.0000	0.9912	1.0000
	Inter Best Y	0.9910	0.9601	1.0000	1.0000	1.0000
	Worst Case	0.0463	0.0074	0.0003	0.7799	0.1002
Sagittal Vertebrae Body Height (SVBH)	Inter rater	0.0006	n/a	0.1873	0.9884	0.7720
	Inter Best X	0.9822	n/a	1.0000	1.0000	0.9992
	Inter Best Y	1.0000	n/a	1.0000	1.0000	1.0000
	Worst Case	0.0266	n/a	0.7488	0.9996	0.9863
Superior Vertebrae Body Length (SVBL)	Inter rater	0.0216	0.0000	0.0002	0.0002	0.9887
	Inter Best X	1.0000	0.9836	1.0000	1.0000	1.0000
	Inter Best Y	0.9881	1.0000	0.9994	1.0000	0.9993
	Worst Case	0.0034	0.0000	0.0000	0.0000	0.0321
Inferior Vertebrae Body Length (SVBL)	Inter rater	0.0220	0.0071	0.1949	0.3384	n/a
	Inter Best X	0.9992	0.9998	0.9995	0.9927	n/a
	Inter Best Y	1.0000	1.0000	1.0000	0.9997	n/a
	Worst Case	0.0000	0.0001	0.0143	0.0396	n/a

4   Discussion

Lumbar vertebrae measurements have been performed by a number of studies [12–20]. In all of these studies, the accuracy of the measurement techniques were not reported. Only sample sizes were reported; no measures of repeatability were included. In the present study, the age range was not broad enough to draw conclusions regarding age differences in lumbar spine measurements. However, the number of subjects was sufficient to explore the repeatability of the measurement process itself and to provide accurate information regarding geometric dimensions of vertebrae.

MRI shows great promise for improving biomechanical modelling of the lumbar spine using subject specific information. This study demonstrates the high degree of repeatability associated with MRI as a means of measuring biomechanically relevant structures.

5   Conclusion

The scan matters! There are differences based on the scan taken. Average absolute differences were greatest when different scans were compared. For example, the average absolute difference expected between measures of the same scan for the L2 Anterior Vertebrae Height was 3% (max observed 11%) while the average absolute difference expected for worst case comparisons of the L2 Anterior Vertebrae Height was 4.5% (max observed 20%). This study demonstrates that MRI derived measures are consistent both within observers and across different observers and different scans. Regressions derived from such MRI data sets and used to predict structure sizes should therefore be impacted much more by actual individual differences between subjects than by the measurement error of analysts.

Acknowledgement. This publication was partially supported by Grant #2T420H008436 from NIOSH. The findings and conclusions in this report are those of the authors and do not necessarily represent the official position of the Centers for Disease Control and Prevention. Use of trade names is for identification only and does not constitute endorsement by the Public Health Service or by the U.S. Department of Health and Human Service.

References

1. Tang R, Gungor C, Sesek RF, Foreman KB, Gallagher S, Davis GA (2016) Morphometry of the lower lumbar intervertebral discs and endplates: comparative analyses of new MRI data with previous findings. Eur Spine J 25(12):4116–4131
2. Gungor C, Tang R, Sesek RF, Foreman KB, Gallagher S, Davis GA (2015) Morphological investigation of low back erector spinae muscle: historical data populations. Int J Ind Ergon 49:108–115
3. Zhou SH, McCarthy ID, McGregor AH, Coombs RRH, Hughes SPF (2000) Geometrical dimensions of the lower lumbar vertebrae – analysis of data from digitized CT images. Eur Spine J 9(3):242–248
4. Schibany N, Ba-Ssalamah A, Marlovits S, Mlynarik V, Nobauer-Huhmann IM, Striessnig G, Shodjai-Baghini M, Heinze G, Trattnig S (2005) Impact of high field (3.0 T) magnetic resonance imaging on diagnosis of osteochondral defects in the ankle joint. Eur J Radiol 55:283–288
5. Barr C, Bauer JS, Malfair D, Ma B, Henning TD, Steinbach L, Link TM (2007) MR imaging of the ankle at 3 Tesla and 1.5 Tesla: protocol optimization and application to cartilage, ligament and tendon pathology in cadaver specimens. Eur Radiol 17(6):1518–1528
6. Shapiro MD (2006) MR imaging of the spine at 3T. Magn Reson Imaging Clin N Am 14:97–108
7. Link TM, Sell CA, Masi JN et al (2006) 3.0 vs 1.5 T MRI in the detection of focal cartilage pathology: ROC analysis in an experimental model. Osteoarthr Cartil 14:63–70
8. Phan CM, Matsuura M, Bauer JS et al (2006) Trabecular bone structure of the calcaneus: comparison of MR imaging at 3.0 and 1.5 T with micro-CT as the standard of reference. Radiology 239:488–496
9. Zhao J, Krug R, Xu D, Lu Y, Link TM (2009) MRI of the spine: image quality and normal-neoplastic bone marrow contrast at 3T versus 1.5 T. Am J Roentgenol 192(4):873–880

10. Tang R (2013) Morphometric analysis of the human lower lumbar intervertebral discs and vertebral endplates: experimental approach and regression models. Doctoral Dissertation, Auburn University, Auburn, AL
11. Gungor C (2013) Prediction of the erector spinae muscle lever arm distance for biomechanical models. Doctoral Dissertation, Auburn University, Auburn, AL
12. Berry JL, Moran JM, Berg WS et al (1987) A morphometric study of human lumbar and selected thoracic vertebrae. Spine 12:362–367
13. Einstein S (1983) Lumbar canal morphometry for computerized tomography in spinal stenosis. Spine 8(2):187–191
14. Fang D, Cheung K, Ruan D, Chan F (1994) Computed tomographic osteometry of the Asian lumbar spine. J Spinal Disord 7:307–316
15. Gilad I, Nissan M (1985) Sagittal evaluation of elemental geometrical dimensions of human vertebrae. J Anat 143:115–120
16. Gilad I, Nissan M (1985) Sagittal radiographic measurements of the cervical and lumbar vertebrae in normal adults. Br J Radiol 58:1031–1034
17. Nissan M, Gilad I (1984) The cervical and lumbar vertebrae - an anthropometric model. Eng Med 13:111–114
18. Larsen JL, Smith D (1980) Vertebral body size in lumbar spinal canal stenosis. Acta Radiol Diagn 21:785–788
19. Postacchini F, Ripani M, Carpano S (1983) Morphometry of the lumbar vertebrae – an anatomic study in two Caucasoid ethnic groups. Clin Orthop 172:296–303
20. Van Schaik JJ, Verbiest H, Van Schaik FD (1985) Morphometry of lower lumbar vertebrae as seen on CT scans: newly recognized characteristics. AJR Am J Roentgenol 145(2):327–335

Preventing Back Injury in Caregivers Using Real-Time Posture-Based Feedback

Mohammadhasan Owlia[1,2] (iD), Chloe Ng[3], Kevin Ledda[3],
Megan Kamachi[1,4], Amanda Longfield[5], and Tilak Dutta[1,4(✉)] (iD)

[1] Toronto Rehabilitation Institute, University Health Network,
Toronto, ON, Canada
[2] Department of Mechanical and Industrial Engineering, University of Toronto,
Toronto, ON, Canada
[3] Department of Occupational Science and Occupational Therapy,
University of Toronto, Toronto, ON, Canada
[4] Institute of Biomaterials and Biomedical Engineering, University of Toronto,
Toronto, ON, Canada
tilak.dutta@uhn.ca
[5] Adult Occupational Therapy Program, Saint Elizabeth, ON, Canada

Abstract. Introduction: Poor posture while performing patient handling tasks is a determinant of low back injury in caregiving settings. Traditional training programs relying on lecture-based teaching of body mechanics has shown to be ineffective. The use of real-time feedback during caregiving tasks has potential as an alternative training intervention.

Objectives: This study aimed to investigate the effectiveness of PostureCoach, a device for providing real-time feedback based on spine posture, in reducing the time participants spend in extreme flexion postures during caregiving tasks, as well as the user acceptability of the device as a training tool.

Methods: Eleven novice participants were recruited and divided into intervention (n = 9) and control (n = 2) groups. They were asked to repeat a set of simulated care activities eight times over two separate sessions held on consecutive days. Individuals in intervention group received real-time auditory feedback in some trials when their forward spine flexion exceeded a threshold, while the participants in control did not. Changes in the amount of forward flexion in the lumbar spine was compared between groups and across trials.

Results: PostureCoach reduced the amount of extreme (80th percentile and 95th percentile) spine flexion during caregiving tasks. This reduction indicates that the participants using PostureCoach reduced their risk of back injury compared to the control group.

Conclusions: PostureCoach shows potential as a wearable device to train novice healthcare professionals and home caregivers for back injury prevention.

Keywords: Injury prevention · Posture-based feedback · Family caregivers
Low back injury · Biofeedback

© Springer Nature Switzerland AG 2019
S. Bagnara et al. (Eds.): IEA 2018, AISC 820, pp. 750–758, 2019.
https://doi.org/10.1007/978-3-319-96083-8_90

1   Background

Musculoskeletal (MSK) injuries due to overexertion are more prevalent in healthcare workers compared to most other industries. The injury rate among hospital workers is twice the average among other professions; nursing home workers alone experienced over three times the average of MSK injuries [1]. Although the rate of most occupational injuries has been declining, the rate of MSK injuries among healthcare workers continues to rise [2]. Healthcare workers often have to adopt awkward postures, which is one of several determinants of low back pain reported in healthcare providers at home and in hospitals in Ontario [3]. Among MSK injuries, low back injuries in caregivers were reported to have the highest prevalence worldwide of 40–50% [4], and this is also a major reason for healthcare workers leaving their jobs [5]. Moreover, a national push towards shifting patient care from hospitals to home, with 2.2 million Canadians receiving care at home, has increased the burden on personal support workers and family caregivers [6]. When providing care in the home, space constraints and lack of equipment will result in completing tasks with methods that are far from ideal. Common components of caregiving tasks include manual patient handling, patients transferring and lifting, and carrying heavy objects to assist patients in performing everyday activities. Informal caregivers are often untrained for their responsibilities, making them particularly susceptible to MSK injuries [7].

Attempts to reduce the risk of injury among healthcare workers have been mainly focused on *fitting tasks to the worker* by redesigning tasks [8]. Adjustable working surfaces or mechanical lift devices are examples of this approach. However, many tasks such as bathing and toileting patients in homecare are yet to be restructured. Therefore, it is necessary to acknowledge the importance of educational programs that help to *fit the worker to their tasks* [9]. Traditional training programs relying on lecture-based teaching of body mechanics has shown to be ineffective, especially in long-term behavior change [2]. This highlights the necessity for developing new methods for more effective training.

One of the most important strategies for preventing low back injury is to minimize lumbar spine flexion. This can be done by retraining worker movement patterns to focus on bending from hips and knees since flexion/extension motions dramatically increase risk of back injury and intervertebral disc herniation of the long term [10–12]. Another benefit of maintaining lumbar lordosis during lifting is that in this position, facet joints will be locked together and the load of will be shared between disc and vertebra, and reduce stress on the disc [13–15].

To reduce the risk of injury during caregiving tasks, we have developed PostureCoach, a wearable device for providing real-time feedback based on lumbar spine flexion. This device consists of a pair of MTi-3 (Xsens Technologies, Enschede, Netherlands) inertial measurement units (IMUs) that are mounted on the user's back at mid-thoracic level and at the sacrum using a belt and a vest, as is shown in Fig. 1.

Each IMU combines information from tri-axial accelerometers, gyroscopes and magnetometers to calculate its orientation in space by referencing gravity and earth's magnetic field. The PostureCoach system uses a microcontroller to process the output of IMUs to calculate the angle of spine flexion in the lumbar spine and log data to an

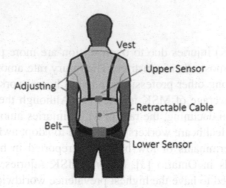

Vest

Upper Sensor

Adjusting

Retractable Cable

Belt

Lower Sensor

Fig. 1. Setup of PostureCoach

SD card. Moreover, it can also provide the user with an instant audio feedback when a pre-set relative angle threshold for forward flexion is exceeded. This would allow the user to instantly perceive their risk of injury and to cue a corrective action.

The purpose of this study is to evaluate the effectiveness of PostureCoach in altering participants' postural performance compared to an educational training video in a sample of inexperienced individuals. We hypothesize that the real-time feedback of PostureCoach will reduce the amount of time spent in extreme spinal flexion among the individuals in the intervention group compared to the control group.

2 Design and Methods

2.1 Participants

In order to represent novice family caregivers, 11 healthy adults with no formal training in caregiving or patient handling were recruited for this study. Participants were randomly assigned to the experimental group (n = 9) or control group (n = 2). The small number of control participants was selected to assess the potential change in novice caregiver spine flexion due to a *learning effect* which our pilot testing indicated was small relative to the *feedback effect* attained by the use of PostureCoach. Participants were recruited from Toronto Rehabilitation Institute (TRI) and students from the University of Toronto. All participants were over the age of 18, and can speak and understand English. They had no formal education in caregiving or patient handling. They had no history of back pain in the last six months and have no musculoskeletal issues related to the spine.

2.2 Instrumentation

This study took place in HomeLab, located in the 12th floor of Toronto Rehabilitation Institute, which consists of a furnished bedroom, living room, bathroom and kitchen. This lab resembles a typical single story house with functioning wiring and plumbing, similar to the ones in which personal support workers and family caregivers perform their daily care-giving activities. One of the members of research team performed the role of patient actor in all trials for consistency. A wheelchair (category 3, 18" × 18" NRG+ , Maple Leaf Wheelchair Mfg Inc., Mississauga, ON) with adjustable armrests, working push brakes and swing out standard footrests was used in this study. Figure 2 shows the setup used for this study.

Fig. 2. Setup of HomeLab for this study

The microcontroller in PostureCoach has been programmed to provide feedback to the participant when pre-set relative angle thresholds for forward bending were exceeded, while saving this angle data to an SD card. This feedback takes the form of audible cues (beeps). Threshold angle for providing feedback was customized for each participant, based on their baseline performance. After the baseline trial, a MATLAB script was used to analyze the data collected by PostureCoach to calculate a personalized threshold. This threshold was set to 10° less than each individual's 90th percentile of flexion during the baseline trial. However, this threshold was not allowed to be less than 20 or more than 45°. For example, if a person's 90th percentile angle value was larger than 45°, the threshold of PostureCoach's feedback was set at 45°. These partially customized threshold limits have been selected to avoid over-prompting, that could discourage future use of the system, by avoiding prompting participants who are already performing well at baseline, and gently encouraging participants who spend the most time in extreme spine flexion to reduce flexion.

2.3 Procedure

The same member of the research team played the role of patient actor for the simulated tasks and a second member of our research team was present to guide the participant through each data collection session. An overview of this study's procedure can be seen in Fig. 3.

Session One. At the beginning of the first session and after obtaining written consent, the research walked the participant through a typical series of caregiving tasks and

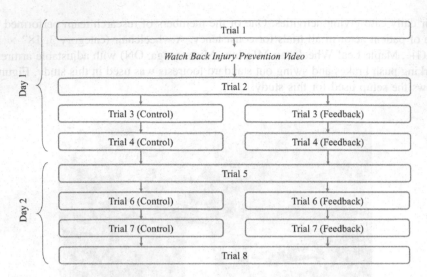

Fig. 3. Schematic of the protocol used for this study

answered related questions. Following introduction, the participant was asked to don PostureCoach, enter HomeLab, where the patient actor is already present, and perform a series of care tasks (*Trial 1*). Then a training video with information about safe patient handling strategies was played for individuals. The video used for training body mechanics and safe patient handling was developed in collaboration with Occupational Health & Wellness team of Saint Elizabeth Healthcare. The points discussed in the video were adapted from key scientific literature [9, 16, 17].

After watching the training video, participants were asked to perform the same simulated care tasks again (*Trial 2*). Afterwards, participants again performed the caregiving tasks twice more, but this time the intervention group received posture-based real-time feedback, while the control group did not. Participants were asked to attempt a toe-touch at certain points in the series of simulated care activities to track changes in spine flexion ability over the two-day study period. Participants were asked to attempt to bend over and touch their toes at a number of points during the study period to provide a maximum spine flexion angle to use for normalizing flexion data to account for differences in sensor placement over the two days.

Session Two. On the next day, participants started the second session by performing the same simulated care activities without feedback, to measure baseline performance during this session. Afterwards participants repeated the care activities again two more times, but with feedback turned on for the intervention group and feedback turned off for those in the control group. After a short break, they again performed the care tasks, without feedback (both groups), in order to determine whether the use of PostureCoach resulted in a short-term behavior change.

Caregiving Activates. The participants were asked to complete the following simulated care activities in each trial on patient actor, in order:

- Wheeling the wheelchair from living room to bedroom
- Transferring the patient from the bed to the wheelchair
- Wheeling the patient to living room and placing the wheelchair near couch
- Transferring the patient to the couch
- Transferring the patient from the couch to wheelchair
- Wheeling the patient from living room to bathroom
- Helping the patient to stand, doff his/her pants, and sit on the toilet
- Helping the patient to stand, don his/her pants, and sit on the wheelchair
- Wheeling the patient back to bedroom and preparing the bed
- Transferring the patient from wheelchair to bed
- Moving wheelchair back to living room.

3 Results and Discussion

The maximum forward flexion values were used to normalize spine flexion data. This normalization helped to eliminate the differences in forward flexion angles measured that were due to variations in sensor placement, as well as individuals' height and flexibility.

For the intervention group, the data from all participants was used to plot a histogram of forward spine flexion for trials 1, 2 and 8 as shown in Fig. 4 (recall that no feedback was given during these three trials). This histogram demonstrates that the individuals in intervention group spent less time with high amounts of spine flexion over the two-day training period during trials when they received no feedback from PostureCoach. These findings suggest that participants had shifted their behaviour to safer movement patterns over the duration of the study. However, it is possible that some of this change in behaviour was the result of the *learning effect* resulting from repeating the same care activities eight times over two days. Therefore, we next

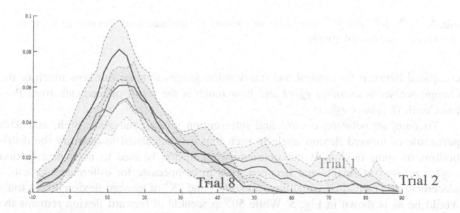

Fig. 4. Average histogram of forward spine flexion in intervention group for Trial 1, Trial 2 and Trial 8 trials along with corresponding 95% confidence intervals

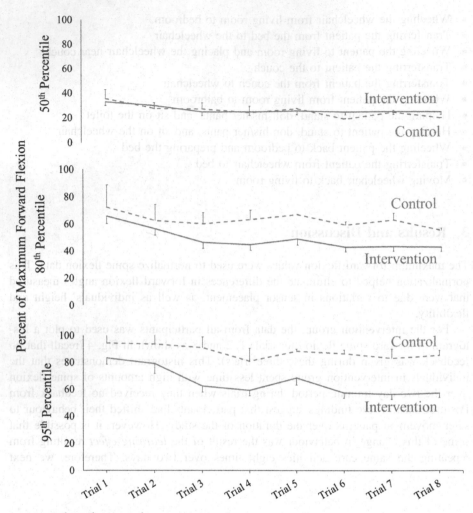

Fig. 5. 50th, 80th and 95th percentile of forward flexion/maximum flexion across trials in intervention and control groups

compared between the control and intervention groups to tease out how much of the change we see is *learning effect* and how much is the result of feedback from PostureCoach (*feedback effect*).

To compare between control and intervention groups, the 50th, 80th, and 95th percentile of forward flexion angle in each trial was calculated to quantify the distributions of spine flexion. If maximum forward flexion is used to normalize flexion angles across participants and between trials to compensate for differences in sensor placement and individual variability, 50th, 80th, and 95th of forward flexion across trials would be as is shown in Fig. 5. While 50th percentile of forward flexion remains the same between groups, the value of 80th and 95th percentile of flexion decrease in intervention group much more than control group as the study goes forward.

Independent t-test revealed that the 80th and 95th percentile of forward flexion in *Trial 8* among individuals in intervention group ($M_{80th} = 42.7$, $SE_{80th} = 3.33$ *and* $M_{95th} = 57.1$, $SE_{95th} = 3.36$) is significantly less than the control group ($M_{80th} = 55.3$, $SE_{80th} = 1.55$ *and* $M_{95th} = 85.0$, $SE_{95th} = 0.80$), with $p = .05$ and $p = .003$ respectively.

Although in this case the sample size is small, given the relatively large effect size (*Cohen's d* = 1.49 *and* 2.93 *respectively*), Type I error rate can be reasonable [18]. This difference could not be observed ($p = .35$) in 50th percentile of flexion ($M_{ctrl.} = 19.9$, $SE_{ctrl.} = 2.00$ *and* $M_{interv.} = 24.97$, $SE_{interv.} = 2.30$).

These results indicate that there were no differences between the two groups at the beginning of study (*Trial 1*), there were statistically significant differences at the end (*Trial 8*). The histogram in Fig. 4 agrees with these results since it shows a shift to the left and toward median values when we compare trial 1 to trail 8. It should be noted that the individuals in the intervention group did not receive any feedback during either of these trials, and their performance is based on their retained behavior change.

4 Conclusion

In this study, we investigate the effectiveness of a real-time posture-coaching device to reduce the risk of back injury in novice caregivers. The results indicated that participants in intervention group using PostureCoach utilized lower values of spine flexion for performing care activities after a two-day training period. This reveals that a wearable device with real-time feedback based on posture can be effective for quickly reducing the risk of back injury among novice individuals. Therefore, PostureCoach has the potential to be integrated into a training program for caregivers, including but not limited to personal support workers as well as informal caregivers.

References

1. Bureau of Labour Statistics (2015) Nonfatal Occupational Injuries and Illnesses Requiring Days Away From Work
2. Fragala G, Bailey LP (2003) Addressing occupational strains and sprains: musculoskeletal injuries in hospitals. Aaohn J 51(6):252–259
3. Workplace Safety and Insurance Board, WSIB By The Numbers: 2015 (2015)
4. Hignett S (1996) Work-related back pain in nurses. J Adv Nurs 23(6):1238–1246
5. Nelson A, Fragala G, Menzel N (2003) Myths and facts about back injuries in nursing: the incidence rate of back injuries among nurses is more than double that among construction workers, perhaps because misperceptions persist about causes and solutions. the first in a two-part series. ajn. Am J Nurs 103(2):32–40
6. Sinha M, Bleakney A.: Receiving care at home. 2014: Statistics Canada = Statistique Canada
7. Reinhard SC et al: Supporting family caregivers in providing care (2008)
8. Grandjean E, Kroemer KH (1997) Fitting the task to the human: a textbook of occupational ergonomics. CRC press, Boca Raton
9. McGill SM (2009) Evolving ergonomics? Ergonomics 52(1):80–86

10. Marras WS et al (1995) Biomechanical risk factors for occupationally related low back disorders. Ergonomics 38(2):377–410
11. Callaghan JP, McGill SM (2001) Intervertebral disc herniation: studies on a porcine model exposed to highly repetitive flexion/extension motion with compressive force. Clin Biomech 16(1):28–37
12. Colloca CJ, Hinrichs RN (2005) The biomechanical and clinical significance of the lumbar erector spinae flexion-relaxation phenomenon: a review of literature. J Manipulative Physiol Ther 28(8):623–631
13. McGill SM (1997) Distribution of tissue loads in the low back during a variety of daily and rehabilitation tasks. J Rehabil Res Develop 34(4):448
14. Skotte J et al (2002) A dynamic 3D biomechanical evaluation of the load on the low back during different patient-handling tasks. J Biomech 35(10):1357–1366
15. Bauer S, Paulus D, Keller E (2015) How do different load cases affect the spinal structures of a well-balanced lumbar spine? a multibody simulation analysis. Int J Innov Res Comput Sci Technol (IJIRCST) 3(5):28–33
16. Burgess-Limerick R (2003) Squat, stoop, or something in between? Int J Ind Ergon 31 (3):143–148
17. McGill SM (1997) The biomechanics of low back injury: implications on current practice in industry and the clinic. J Biomech 30(5):465–475
18. De Winter JC (2013) Using the student's t-test with extremely small sample sizes. Pract Assess Res Eval 18(10)

Equotherapy Center at a Glance for Ergonomic Activity: Epidemiological Profile Versus Therapeutical Practices

Marcelo Dondelli Boaretto[✉], Jullia Maria Rodrigues Zullim,
Bruno Sobral Moreschi, and Maria de Lourdes Santiago Luz

Universidade Estadual de Maringá, Maringá, Brazil
marceloboaretto95@gmail.com, jullia.zullim@gmail.com,
bruno.sobral94@gmail.com, mlsluz@uem.br

Abstract. An ergonomic study at an equotherapy center – method in which a horse is used as the mean to promote the handicap people, positive results according to the individual therapeutical objectives for the rehabilitation of themselves. The study has its primary focus, the equotherapeutical personnel who faces on the therapeutical attendance, constraints by the biomechanical efforts. The demand reflected on identifying, knowledge need and comprehend how to establish the activities done by the professionals. The study was based on the Ergonomic Work Analysis (EWA). On the initial phases of the EWA, a perception questionary was applied in order to characterize. The professional profile, demographic and epidemiological, at the same time with the job description done and its specificity for the professionals. To comprehend the epidemiological profile of the organization, it has been applied the bipolar questionary, resulting in information which brought the activities which cause the major discomfort. Within all the listed activities the major ones detected was "stabilize the practitioners by the horse" and "walking by the horse side." It has been understood during the data collection, the practitioner diversity, whose biomechanical efforts demanded by the professional vary (practitioner versus therapy). It has been observed that major incidences of discomfort and intensity were identified on the shoulder and forearm. The results allowed to correlate the difficult degree with the epidemiological profile of the professionals. Such understanding allows that each professional adequate the appointment agenda, alternating the daily shift between the cases which demand more or less physical and posture efforts.

Keywords: EWA · Equotherapy · Biomechanical efforts

1 Introduction

The article focus on ergonomics, the equotherapy professional who works in the equotherapy center.

The equotherapy method, according to [1, 2], utilize the tridimensional movements of the horses to help the handicap people. The stimulus promoted by the horse, with the attention of the multidisciplinary team ended up by benefiting the equotherapy practice,

S. Bagnara et al. (Eds.): IEA 2018, AISC 820, pp. 759–764, 2019.
https://doi.org/10.1007/978-3-319-96083-8_91

promoting muscle relaxation, modifying the muscle tonus, besides the psychological aspects such as self-confidence, fearless, and sensitivity development. In summary, According [3]: "a therapeutical and educational method utilizing the horse in an interdisciplinary approach, in the health area, education, and equitation, targeting the biopsychosocial development of the handicap people or with special needs."

According to [4], the equotherapy has four types of programs: hippotherapy, education/rehabilitation pre-sport, and para-equestrian. The three first ones were made at the research location. The space in which the center is located there is three types of floor trackings, sand, and paved spaces. According to [5] the horse cadence interferes on the muscle tonus stimulus of the practitioner. Known that, the floor type influences on the horses' cadence and from the individual therapeutical objectives of the practitioner the strategies are done for the floor type the sessions will occur.

1.1 Methodology

The equotherapeuticals activities analysis, it was utilized the ergonomics work analysis (EWA) – the methodology which procedures makes feasible the information flow acquisition from the employees and global observations and activities systematics of the activities, with the objective to comprehend and improvement proposal that minimize the skeletal muscle constraint caused by the job done. The studied center presented an organization of four professionals who work individually on the attendances, but help themselves to exchange knowledge and improvement suggestions with the practitioner.

Within the information input, a perception questionary was applied, intending to comprehend the activities at the glance of the professionals of the area and to become aware of the activities which cause the significant discomfort.

It aims to identify the discomfort regions, and it has been applied to the bipolar questioner of [6] (Fig. 1). The equotherapeuticals assigned within the listed human body parts, which are the discomfort types and its degree of intensity according to the scale previewed established.

Subsequently, after the systematic observations, it was elaborated a follow-up chart aiming to detect and correlate the difficulty level at the appointments, according to the therapeutic perceptions.

It was made a questionary with closed questions and objective enabling to a comparative analysis within the equotherapeuticals group. It was considered, organizational aspects such as working schedule breaks within the attendances and its sequences with the difficult degree and any discomfort presence.

The follow-up chart was structured with general information of the attendance and with specific questions correlating the therapeutical method with the practitioner.

The scale adopted to classify the difficult degree was starting from easy, medium and difficult. This scale was established, based on a compiled of all appointments of the center, taking into consideration the criteria – weight and the practices stability, the height and the holding difficulty (arms position of the equotherapeutical), the horse height, and the equotherapy program, the equitation equipment type – and defined from the comparison degree within the appointments.

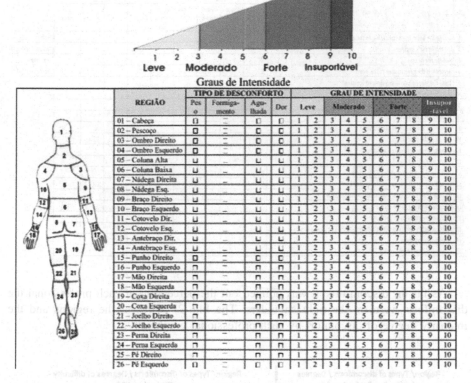

	REGIÃO	TIPO DE DESCONFORTO				GRAU DE INTENSIDADE										
		Pes o	Formiga-mento	Agu-lhada	Dor	Leve		Moderado			Forte			Insupor-tável		
	01 – Cabeça	□	–	□	□	1	2	3	4	5	6	7	8	9	10
	02 – Pescoço	□	–	□	□	1	2	3	4	5	6	7	8	9	10
	03 – Ombro Direito	□	–	□	□	1	2	3	4	5	6	7	8	9	10
	04 – Ombro Esquerdo	□	–	□	□	1	2	3	4	5	6	7	8	9	10
	05 – Coluna Alta	u	–	u	u	1	2	3	4	5	6	7	8	9	10
	06 – Coluna Baixa	u	–	u	u	1	2	3	4	5	6	7	8	9	10
	07 – Nádega Direita	u	–	u	u	1	2	3	4	5	6	7	8	9	10
	08 – Nádega Esq.	u	–	u	u	1	2	3	4	5	6	7	8	9	10
	09 – Braço Direito	u	–	u	u	1	2	3	4	5	6	7	8	9	10
	10 – Braço Esquerdo	u	–	u	u	1	2	3	4	5	6	7	8	9	10
	11 – Cotovelo Dir.	u	–	u	u	1	2	3	4	5	6	7	8	9	10
	12 – Cotovelo Esq.	u	–	u	u	1	2	3	4	5	6	7	8	9	10
	13 – Antebraço Dir.	u	–	u	u	1	2	3	4	5	6	7	8	9	10
	14 – Antebraço Esq.	u	–	u	u	1	2	3	4	5	6	7	8	9	10
	15 – Punho Direito	□	–	□	□	1	2	3	4	5	6	7	8	9	10
	16 – Punho Esquerdo	n	–	n	n	1	2	3	4	5	6	7	8	9	10
	17 – Mão Direita	n	–	n	n	1	2	3	4	5	6	7	8	9	10
	18 – Mão Esquerda	n	–	n	n	1	2	3	4	5	6	7	8	9	10
	19 – Coxa Direita	n	–	n	n	1	2	3	4	5	6	7	8	9	10
	20 – Coxa Esquerda	n	–	n	n	1	2	3	4	5	6	7	8	9	10
	21 – Joelho Direito	n	–	n	n	1	2	3	4	5	6	7	8	9	10
	22 – Joelho Esquerdo	n	–	n	n	1	2	3	4	5	6	7	8	9	10
	23 – Perna Direita	n	–	n	n	1	2	3	4	5	6	7	8	9	10
	24 – Perna Esquerda	n	–	n	n	1	2	3	4	5	6	7	8	9	10
	25 – Pé Direito	n	–	n	n	1	2	3	4	5	6	7	8	9	10
	26 – Pé Esquerdo	□	–	□	□	1	2	3	4	5	6	7	8	9	10

Fig. 1. Bipolar questionary

2 Development

2.1 Perception Questionary

The results presented with the perception questionary with the bipolar questionary, identified the activities the equotherapeuticals execute in a daily basis at the equotherapy center, the timing necessary to complete one cycle of this activity and the execution position (Table 1). Besides, it has been diagnosed the discomfort felt by the professionals during the working journey, presented below in Fig. 2.

It has been possible to perceive some differences with the activities done within the equotherapeutical, including factors such as, execution timing and postures in which they are made. Observed the differentiated glance of each equotherapeutical activities. Observed the personalized attendances, the equotherapeutical execute the activities in different ways, to accomplish the therapeutical objectives, at the end the practitioner evolution.

Table 1. Perception questionary compilation

Nº	Activities	Time						Position				
		Does not execute	Up to 15 min	15 min up to 30 min	30 min up to 45 min	45 min up to 60 min	Greater than 60 min	Standing	Sitting	Walking	Crouching	
1	Guide the horse from the stable to the equestrian track	XX X		X								Professional 01 X
2	Do the horse's hygiene	X	X X					X X		X	X X	Professional 02 X
3	Saddle the horse	X	X X					X X		X	X	Professional 03
4	Approach the practitioner to the horse		XX X					XX		XX X		Professional 04 X
5	Take the wheelchair to the desired location		XX X					XX		XX X		
6	Put the practitioner on the horse		XX X					XX X		X		
7	Put the safety belt on the practitioner		XX X					XX X				
8	Walk beside the horse			XX X				XX		XX X		
9	Guide the moviments of the horse	X		XX				X		X X		
10	Stabilize the practitioner		XX	X				XX X		XX		
11	Change the practitioner's position		XX	X				XX X		XX		
12	Conduct assisted exercise		X	XX				XX X				
13	Remove the practitioner from the horse		XX X					XX X				
14	Lead the practitioner to your responsible		XX X					X		XX X		
15	Update the practitioner's status to the responsible		X	X				X X	X X	X		
16	General cleaning of the center	X		XX				X	X	X X	X X	
17	Plan future equoterapy section		XXX					XX	XX X	XX		
18	Hold monthly meetings with those responsible for practitioners						XX X		XX X			
19	Hold team meetings						XX X	X	XX X	X		
20	Guide the horse to the stable	X	X	X						XX		

As far the discomforts, it was pointed out the perception of each professional the discomforts were on the right shoulders. The Fig. 2, represent the regions and the intensity degree of discomfort by each professional graphically.

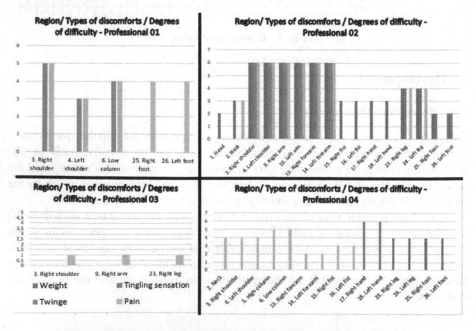

Fig. 2. Compilation graphics of data coming from the perception questionary.

2.2 Follow up Chart

The follow-up chart objective was to identify the practicers profile and the appointments of each equotherapeuticals.

The follow-up chart allowed to define the difficult degree of the appointments for each professional (Fig. 3).

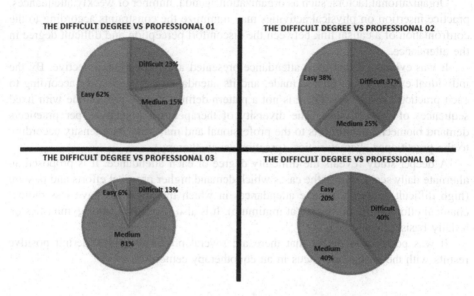

Fig. 3. The difficult degree graphics vs the equotherapeutical professionals of the center.

2.3 Discussion

Starting from the professional results, presented in the graphics Fig. 3, it was possible to establish a causal link between the intensity and the range of the body region with discomfort and the difficult attendance degree.

The professional 01 considers 61,5% of its appointment as level "easy," and if comparing with the data of the perception questionary (Fig. 2), she demonstrated that she defines the degree of discomforts as moderated in a few corporal regions. The professional 02 and professional 04, the perception questionary replies, were the ones that the most demonstrated discomfort in more body region (Fig. 2) and in the request of difficult degree they considered in their appointments level medium to difficult, totaling 62,5% and 80% respectively. Although, the professional 03, considered her appointments (81,3%) as level medium and only 6,3% level easy, answered she does not feel much discomfort and few parts of her body were identified. As a justification, the pain exposure was minimized by the muscle enhancement by the exercises with personal trainer sessions.

3 Final Consideration

It is considered that there are factors identified which are inherent, as individual characteristics, biological profile and professional age, nevertheless the major remaining factors are possible to be modified by action plans to minimize the physical wear of the equotherapeuticals.

Organizational factors, such as organization agenda, number of weekly attendances, practice insertion on physical activities may minimize the constraints according to the confrontation of a causal link between the discomfort perception and difficult degree in the attendance.

It was evidenced that each attendance presented a therapeutical objective. By the individual evaluation, a plan is made, and its attendance strategies vary according to each practicers responses. There is not a pattern defined even a job routine with fixed sequences of the activities. The diversity of therapeutical objectives per practicers demand biomechanical efforts to the professional and may vary its intensity according to the practitioner characteristics. (practice versus therapy).

A deeper study reveals the difficulty degree of each attendance. It is proposed an alternate daily scale, within the cases which demand higher physical efforts and posture (high difficult degree) with the attendances in which the attendees have less biomechanical efforts (difficult degree at minimum). It is also suggested labor gymnastics on a daily basis.

It was possible to detect that there are several proposals which permit positive results with the ergonomic focus in an equotherapy center.

References

1. Queiroz COV Visualização da Semelhança entre os Movimentos Tridimensionais do Andar do Cavalo com o Andar Humano. http://equoterapia.org.br/media/artigos-academicos/documentos/20082221.pdf. Accessed 14 Nov 2017
2. Severo JT (2010) Equoterapia: Equitação, Saúde e Educação. In: SENAC, São Paulo, SP
3. Alves EMR (2009) Prática Em Equoterapia – Uma Abordagem Fisioterápica, São Paulo, SP, Atheneu
4. ANDE-BRASIL. equoterapia.org.br. Accessed 11 Mar 2018
5. Wickert H O Cavalo Como Instrumento Cinesioterapêutico, 1995, Brasília - DF – ANDE BRASIL, Associação Nacional de Equoterapia, Trabalhos Técnicos Científicos. http://equoterapia.org.br/media/artigos-academicos/documentos/24101008.pdf. Accessed 11 Mar 2018
6. Corlett EN, Bishop RP (1976) Ergonomics 19(2): 175–182

Patterns and Predictors of Work-Related Musculoskeletal Disorders Among Commercial Tricycle (Keke Napep) Riders in Nigeria

Echezona Nelson Dominic Ekechukwu[1,2](✉) (iD),
Martins Oshomah Okaku[3], Samson Adaramola[2,4],
and Ifeoma Nmachukwu Onuorah[1,2]

[1] Department of Medical Rehabilitation, Faculty of Health Sciences
and Technology, College of Medicine, University of Nigeria, Enugu, Nigeria
nelson.ekechukwu@unn.edu.ng
[2] Ergonomic Society of Nigeria (ESN), Enugu Region, Enugu, Nigeria
[3] Physiotherapy Department, University of Benin Teaching Hospital,
Benin, Nigeria
[4] Institute of Occupational Health, University of Port Harcourt,
Port Harcourt, Nigeria
http://www.unn.edu.ng/

Abstract. Objective: This study assessed the prevalence, pattern and predictors of Work related Musculoskeletal Disorders (WMSD) among tricycle (keke) ridders in Nigeria.

Method: A total of 384 keke riders participated in this cross-sectional study. WMSD was assessed with Cornwell questionnaire while the demographic, job related, and anthropometric variables were assessed with the use of a self-designed interview guide. Data obtained was analysed descriptively using frequency and percentage, mean and standard deviation and inferentially using chi-square and logistic regression. Level of significance was set at $\alpha = 0.05$.

Results: The mean age of the participants was 38.73 ± 10.55 years with a mean ridding duration 8.50 ± 1.39 h/day. The general WMSD prevalence in this study was 79.9%; and was more prevalent at the low back (60.4%), neck (49.7%) and shoulder (39.8%) regions. There was a significant association between the general WMSD and each of ridding with passengers in the front cabin ($X^2 = 19.156$, $p < 0.001$), vibration ($X^2 = 28.568$, $p = 0.001$), twist while making a turn ($X^2 = 39.874$, $p < 0.001$), awkward sitting posture ($X^2 = 4.243$, $p = 0.039$). Twisting while ridding (OR = 1.006, $p = 0.009$) and job stress (OR = -1.398, $p = 0.01$) were the significant predictors of general WMSD.

Conclusion: There is a high prevalence of WMSD among Nigerian Tricycle ridders. Job stress and twisting while driving are important predictors of WMSD in this group. Ergonomic training programmes are recommended for this population.

Keywords: Predictors · Musculoskeletal disorders · Commercial tricycle ridders

© Springer Nature Switzerland AG 2019
S. Bagnara et al. (Eds.): IEA 2018, AISC 820, pp. 765–777, 2019.
https://doi.org/10.1007/978-3-319-96083-8_92

1 Introduction

Musculoskeletal disorders are injuries and disorders that affect the human body's movement or musculoskeletal system such as muscles, tendons, ligaments, discs etc. Musculoskeletal disorders affect all ages, reoccur most times and frequency increases with age (Ojoawo et al. 2014). It represents the largest categories of work related illness affecting individuals (Shaik et al. 2014). Work related musculoskeletal pain is defined as musculoskeletal disorders that results from a work related event (Tinubu et al. 2010) and/or affects work efficiency. It has been recognized to be most common cause of decreased productivity among the working population (Wang et al. 2017) such as commercial drivers.

Commercial motor cycle popularly known as "Okada" hitherto was the major means of transportation within major cities in Nigeria (Badejo and Salau 2010). However, their operation in these cities have been outlawed because of the increased road traffic accident associated with their operations (Atubi and Gbadamosi 2015). This may have led to the surge in the operation and patronage of commercial tricycle popular known as "Keke Napep". Most commercial tricycle riders in Nigeria accommodate passengers in their front cabin possibly due to poor traffic surveillance as well as the quest to make more returns especially in the present prevailing economic downturn in Nigeria (Marshal and Solomon 2017). In an attempt to accommodate this extra passenger on a seat meant solely for the rider, the rider is forced to adopt an awkward, constrained postures with some degree of deviation of the spine to the opposite direction of the passenger. Also, the job task of tricycle riding requires a great deal of manual handling, exposure vehicular vibrations amplified by a continuously contracted set of postural muscles, repetitive movement of the wrist, contact stress and poor carbin design which are common risk factors of work related musculoskeletal disorder.

Several studies (Ezeukwu et al. 2011; Akinbo et al. 2008; Ojoawo et al. 2014; Ismaila et al. 2011; Akinpelu et al. 2011) on musculoskeletal disorders among Nigerian workers exist. In a study by Ezeukwu et al. (2011), on the prevalence of work related musculoskeletal pain among Timber workers in Enugu metropolis, they found a 12 months' prevalence of 90.1% among their participants with repetitive task as the most reported pro-domestic job factor. Also in a study by Ojoawo et al. (2014), on the Assessment of work related musculoskeletal pain among professional bus drivers in the service of a tertiary institution in Nigeria, they reported work related musculoskeletal disorders prevalence of 77% with prolonged driving reported as the major cause of musculoskeletal disorder. In similar studies by Akinpelu et al. (2011) and Akinbo et al. (2008), similarly high prevalence of work related musculoskeletal disorders were reported among occupational drivers. However, there is a dearth in literatures on the prevalence and predictors of work related musculoskeletal disorders among commercial tricycle riders in Nigeria. Therefore, this study assessed the prevalence, pattern and predictors of work related musculoskeletal disorders among commercial tricycle riders in Nigeria.

2 Method

2.1 Participants

Commercial tricycle riders in Enugu, Nigeria who were recruited using consecutive sampling technique participated in this study with a cross-sectional exploratory design. Only keke riders aged 18 years and above, who had been on the job for at least one year and operated within Enugu Metropolis were included. However, subjects that met the inclusion criteria but have obvious cofounding musculoskeletal deformities such as scoliosis, kyphosis etc. were excluded from the study. The minimum sample size was calculated using the expression of medium effect formula

$$N = n(Z_1 - Z_2)/ES^2$$

Where, N = minimum sample size; n = number of groups = 1; Z_1 = α-confidence interval at $0.05 = 1.96$; Z_2 = β-confidence interval at $0.01 = 0.84$; ES = medium effect = 0.15. A minimum sample size of 348 keke riders were projected to participate in this study.

2.2 Instrument

Cornell Musculoskeletal Discomfort Questionnaire (CMDQ): This questionnaire was developed by Hedge et al. (1999) at Cornell University. It was used to assess MSD among the participants and their adopted measures for preventing and treating their MSD. A study by Afifehzadeh-Kashani et al. (2011) reported that CMDQ has an excellent reliability (Cronbach's alpha Coefficient = 0.986).

Self-Structured Interview Guide: A proforma was developed through a two-stage development process. The first stage involved the formulation of the items and sub-categories while the second stage involved expert review, vetting and assessment of content validity index. Items with scores less than 3 out of a maximum score 5 by the experts were omitted. The final proforma had three sections (A, B, and C). Section A assessed participants' demographic details, this section had 10 items. The second section assessed their job related variables and had 16 items while section C had 10 items and was used to assessed the ergonomic details in their workstations. Thus the proforma had a total of 36 items.

Stadiometer (Manufactured in USA): This was used to assess participant's height with a 0.01 m precision.

Bathroom Weighing Scale (Manufactured in China): This was used to assess the participants' weight to the nearest kilogram (Kg).

Tape Rule (Manufactured in Nigeria): This was used to assess waist and hip circumferences to the nearest centimeter (cm).

2.3   Procedure

Ethical approval was sought and obtained from the Health Research and Ethics Committee of the University of Nigeria Teaching Hospital (UNTH), Ituku-Ozalla, Enugu, Nigeria. The aim, purpose and relevance of the study were explained to keke drivers after being screened for eligibility. The informed consent form was given to the those who volunteered to participate in the study. Demographic were recorded while anthropometric variables such as height, weight, BMI, waist circumference, hip circumference, waist-hip ratio were assessed using standard protocols (Thomas et al. 2015). The outcome measures (Cornell questionnaire and interview guide) were then administered. Some ergonomic concepts such as reach, clearance, awkward and constrained posture, symptoms of musculoskeletal disorders, job stress indicators etc. were explained to the participants using lay terms. The participants response about their seat height, reach envelope, pedal position, exposure to vibration and repetitive movements were elicited using the interview guide.

2.4   Data Analysis

Data obtained from the field was cleaned and analysed using statistical package for social sciences, version 20.0 (SPSS Inc. Chicago, IL, USA). Descriptive statistics of frequency, percentages, charts, mean and standard deviation was used to present participants demographic, anthropometric and work related details. Chi-square test used to assess the association between MSD and other selected variables. Binomial logistic regression model was used to predict the likelihood of the occurrence of MSD. Level of significance was set at $\alpha = 0.05$.

3   Results

3.1   Summary of Participants' Variable

A total of 384 tricycle drivers participated in this study, most of whom were males (99%), married (63.3%) and between 21 and 65 years. Their mean age, BMI and years on the job were 38.73 ± 10.55 years, 26.66 ± 3.75 kg/m^2 and 3.75 ± 1.84 years respectively. The mean working and driving duration of the participants were 12.89 ± 1.65 h/day and 8.50 ± 1.39 h/day respectively as shown in Table 1. Majority of the participant took alcohol (73.2%), carried passengers in the front cabin (97.7%) in order to increase daily revenue (85.7%) even though they were not comfortable with it (50.0%). A good number of the participants responded that they felt the vibration from their tricycle (77.3%) which they described as discomforting (74.7%). Most of the participants used key ignition system (51%) and reported that the job is stressful (86.7%) as shown in Table 2. Majority of the participants reported that their seat height (95.8%) and back support (93.2%) were ideal. Most of the participants twist their spine (90.0%), lean forward (51.8%) and adopted awkward postures (64.6%) while driving as shown in Table 3.

Predictors of WMSDs Among Nigerian Keke Riders   769

Table 1. Summary of participants' variable (N = 384)

Variables	Min.	Max.	Mean	Std. Deviation
Age	21.00	65.00	38.73	10.547
Weight	50.00	117.00	77.76	11.025
Height in metre	1.47	1.88	1.7082	0.06208
Body mass index	18.8189	42.9687	26.662533	3.7487212
Waist circumference	68.00	115.00	87.13	7.987
Hip circumference	84.00	126.00	97.80	7.588
Waist hip ratio	0.6053	1.0748	0.891327	0.0513327
Years of employment as keke driver	0.00	8.00	3.751	1.8384
Total monthly income	10000	74000	41813.54	10715.42
Total working duration	6.00	18.00	12.89	1.650
Driving duration	5.00	13.00	8.50	1.390
Waiting/sitting duration	1.00	7.00	4.40	0.934
Number of body parts affected	0.00	8.00	3.18	2.173

Table 2. Frequency distribution of participant Demographic and Job variables (N = 384)

Variables	Categories	Frequency	Percentage
Other sources of income	Yes	185	48.2
Other jobs other than keke driving	Yes	188	49.0
		196	51.0
Alcohol	Yes	281	73.2
	No	103	26.8
Smoke	Yes	106	27.6
	No	278	72.4
Part time employer	Yes	132	34.4
	No	252	65.6
Drive temporary	Yes	133	34.6
	No	251	65.4
Side job	Yes	123	32.0
	No	261	68.0
Replacement for another driver	Yes	26	6.8
	No	358	93.2
Night shift	Yes	158	41.1
	No	226	58.9
Passenger on front seat	Yes	375	97.7
	No	9	2.3
Reason for carrying passenger in front seat	To make more money	329	85.7
	Just for the fun of it	4	1.0
	For no reason	45	11.7

(*continued*)

770   E. N. D. Ekechukwu et al.

Table 2. (*continued*)

Variables	Categories	Frequency	Percentage
	Don't like sitting alone	2	0.5
	No answer	4	1.0
Vibration	Yes	297	77.3
	No	87	22.7
Comfortable with vibration	Yes	97	25.5
	No	287	74.7
Ignition	Key	196	51.0
	Hand pedal	54	14.1
	Automatic starter	134	34.9
Job is too stressful	Yes	333	86.7
	No	51	13.3

Table 3. Summary of the ergonomic assessment of participants' workstation (N = 284)

Variables	Categories	Frequency	Percentage
Seat height ideal	Yes	368	95.8
	No	16	4.2
Reach envelop ideal	Yes	368	95.8
	No	16	4.2
Back support ideal	Yes	358	93.2
	No	26	6.8
Pedal position ideal	Yes	364	94.8
	No	20	5.2
Twist while driving	Often	130	33.9
	Sometimes	181	47.1
	Never	73	19.0
Forward lean while driving	Yes	199	51.8
	No	185	48.2
Awkward seating	Yes	248	64.6
	No	136	35.4
Stretch or change position while driving	Yes	216	56.3
	No	168	43.8
Comfortable when sitting with passenger	Yes	192	50.0
	No	192	50.0

3.2 Prevalence of WMSD and Health Seeking Behaviour

Approximately eighty percent of the participants had WMSD. The prevalence of the WMSD was higher in the low-back (60.4%) and neck (49.7%) regions of the body. Majority of the participants had WMSD symptoms of moderate intensity (43.2%), that occurred not more than twice within the previous week (44%), that neither made them miss work (57.85) nor become inactive (77.1%) and had not sought medical advice (74.2%) as shown in Table 4. Most of the participants didn't seek for help anywhere (65.4%), had no change in their prognosis (67.7%) and had no knowledge of ergonomics (79.9%). Very few of the participants (12.5%) reported to have received Physiotherapy treatment for their condition as shown in Table 5.

Table 4. Frequency distribution of musculoskeletal disorders among the participants (N = 384)

Variable	Categories	Frequency	Percentage
General MSD Prevalence		307	79.9
Regional MSD prevalence	Neck	191	49.7
	Shoulder	153	39.8
	Upper back	116	30.2
	Elbow	86	22.4
	Lower back	232	60.4
	Hip	39	10.2
	Wrist	50	13.0
	Finger	26	6.8
	Ankle	77	20.1
	Knee	85	22.1
	Upper arm	45	11.7
	Forearm	10	2.6
	Thigh	12	3.1
	Lower leg	26	6.8
	Foot	75	19.5
MSD frequency	1–2 times last week	169	44.0
	3–4 times last week	60	15.6
	Once everyday	67	17.4
	Several times everyday	14	3.6
	Not at all	74	19.3
Pain intensity	Severe	34	6.8
	Moderate	166	43.2
	Mild	120	31.3
	Not at all	74	19.3
Inactivity	Yes	88	22.9
	No	296	77.1
Missed work as a result of pain	Yes	162	42.2
	No	222	57.8
Medical advice	Yes	99	25.8
	No	285	74.2

Table 5. Summary of health seeking behaviour of the participants (N = 284)

Variable	Categories	Frequency	Percentage
Where care was sought	PMV	64	16.7
	Hospital	69	18.0
	Nowhere specifically	251	65.4
Prognosis	Better	123	32.0
	No change	258	67.7
	Worse	3	0.8
Strategy	Yes	69	18.0
	No	315	82.0
Knowledge and awareness	Yes	77	20.1
	No	307	79.9
Physiotherapy treatment received	Yes	48	12.5
	No	336	87.5

Key: PMV = Patent Medicine Vendors

3.3   Association Between WMSD and Selected Participants' Variables

There was a significant association between the general MSD and marital status (X^2 = 30.033, p < 0.0001), job strain (X^2 = 49.709, p < 0.0001), and experiencing discomfort while sitting with passengers in the front cabin (X^2 = 35.884, p < 0.0001). Also, there was a significant association between having an MSD and the practice of accommodating passengers in the driver's cabin (X^2 = 19.156, p < 0.0001), exposure to vibrations while driving (X^2 = 28.568, p < 0.0001) and adopting an awkward sitting posture while driving (X^2 = 4.242, p = 0.039). However, there was no significant association between having an MSD and sex (X^2 = 1.014, p = 0.310) as well as the type of ignition system (X^2 = 4.439, p = 0.109) as shown in Table 6.

Table 6. Association between participants' variables and musculoskeletal Disorders (N = 384)

Variable	X^2 (p-value)		
	General MSD	Neck MSD	Low back MSD
Sex	1.014 (0.310)	1.032 (0.310)	2.648 (0.104)
Marital status	*30.033 (<0.001)	0.156 (0.693)	*42.706 (0.0001)
Passenger on front seat	*19.156 (<0.001)	*9.120 (0.003)	*9.368 (0.002)
Vibration	*28.568 (0.001)	*44.219 (0.0001)	*23.781 (0.0001)
Ignition	4.439 (0.109)	*12.298 (0.002)	*6.277 (0.043)
Stress while working	*49.709 (0.0001)	*18.669 (0.0001)	*33.461 (0.0001)
Twist while driving	*39.874 (0.0001)	*47.424 (0.0001)	*47.323 (0.0001)

(*continued*)

Table 6. (*continued*)

Variable	X^2 (p-value)		
	General MSD	Neck MSD	Low back MSD
Forward lean while driving	3.456 (0.178)	*17.329 (0.0001)	*7.271 (0.026)
Awkward sitting	*4.243 (0.039)	*5.161 (0.023)	*4.000 (0.045)
Comfortable while sitting with passenger	*35.884 (0.0001)	*19.261 (0.0001)	*27.223 (0.0001)

Key: * = significant

3.4 Regression Model for Predicting WMSD Among the Participants

A logistic regression model was performed to ascertain the effects of marital status, highest educational qualification, alcohol consumption, smoking status, total work duration, perceived job stress, and twisting while driving on the likelihood that the participants have MSD at 12 month (period prevalence). The logistic regression model was statistically significant ($X^2 = 16.42$; $p < 0.0001$). The model explained 38.0% of the variance in MSD and correctly classified 75.02% of cases. Participants who were single were 1.6 times more likely to develop MSD than their married counterparts. Also, having a tertiary education decreases the odds of having a WSMD (OR = −1.199, $p < 0.0001$). Other significant predictors of WMSD in this population were job strain (OR = 1.398, $p = 0.01$), and twisting the trunk (OR = 1.006, $p = 0.009$) as shown in Table 7.

Table 7. Regression model for predicting participation among the participants (N = 384)

Variable	Categories	n (N)	Model summary		Predictor model	
			$X^2(P)$	$R^2(C)$	OR	p-value
Marital status	Married	215 (243)	... 16.42 (< 0.001)	... 0.38 (75.0%)	1	–
	Single	92 (141)	...	...	1.617	<0.001*
Highest educational status	None	8 (9)			1	–
	Primary	32 (48)			−1.990	0.093
	Secondary	79 (218)			0.530	0.208
	Tertiary	88 (109)			−1.199	<0.001*
Alcohol	No	77 (103)			1	–
	Yes	230 (281)			−1.64	0.58

(*continued*)

Table 7. (*continued*)

Variable	Categories	n (N)	Model summary		Predictor model	
			$X^2(P)$	$R^2(C)$	OR	p-value
smoke	No	235			1	–
	Yes	72 (106)			0.832	0.06
Total work duration	NA	NA			−0.138	0.097
Stress while working	No	22 (51)			1	–
	Yes	285 (333)			−1.398	0.01*
Twist	Often	114 (130)			1	–
	Sometimes	154 (130)			0.794	0.53
	Never	39 (73)			−1.006	0.009*
Forward lean	No	140 (184)			1	–
	Yes	67 (200)			−0.518	0.66

Key: n $(N)^{\#}$: Number of participants with MSDs (number of participants in each category);
X^2 (p): Chi-square and significance of the model;
R^2 (C): Nagelkerke R Square (degree of classification) by the model;
OR: the odds ratio of each predictor variable;
p-value: significance of each predictor variable;
* = significant.

4   Discussion

About Eight percent of the participants in this study had a WMSD. This implies that about four of every five tricycle riders in Nigeria may have WMSD. Similarly high WMSD prevalence has been reported among commercial drivers (Akinpelu et al. 2011; Ojoawo et al. 2014) and other occupation requiring manual handling task and exertion (Ezeukwu et al. 2011). In the study by Ojoawo et al. (2014) among professional drivers in the university community in Ibadan, 70% of their participants had WMSD while the study by Akinpelu et al. among commercial bus drivers recorded a higher prevalence (89.3%) of WMSD. These statistics are but epidemiological figures, they do not tell the type of suffering these cohorts are exposed to. It is therefore pertinent that training programmes on how to prevent the development of these WMSDs among this population is needed.

In this study, WMSD was more prevalent in the spine (low-back and neck). Similarly high prevalence of WMSD in the spine was reported by Tawiah et al. (2015) among Ghanian Gold miners and also by Malikraj et al. (2011) among Indian welders.

The spine especially the low-back region appears to be the most abused part of the human body. People often seen using their back muscles for lifting (like a crane) due to the lack of knowledge and awareness of proper lifting techniques. This behaviour if not corrected may strain or sprain the back muscles and ligaments respectively and may subsequently results in mechanical injuries to the intervertebral disc structures or increases the odds of MSD of the spine (Orr et al. 2015; Donnally and Dulebohn 2017). Tricycle ridding is performed in a sitting position but requires movements of the upper body parts e.g. reaching and maneuvering, thus activating the muscles of the spine for the stabilization of the upperlimbs. The impact on the spine especially the low back region becomes multiplied when the rider adopts an awkward posture (that about two-third of participants in this study assumed) while driving. Secondly, it may be due to poor back care and health seeking behaviour that was found to be common among the participants. Regular assessment of the musculoskeletal system of this population especially their low-back region is recommended. Other preventing measures such as observing rest breaks, back muscle stretch and exercise are also recommended.

There was a significant association between WMSD and having passenger on the front cabin as well as twisting while driving. This implies that the common practice of having passengers sit with the keke driver in order to make more returns is a contributing factor to the occurrence of WMSD. The body tends to become asymmetrical with an extra person on a seat designed for just the rider due to the displacement of centre of gravity away from the passenger. Frequently working in this awkwardly constrained body posture abnormally exerts the trunk muscles and may result in WMSD. Similar report was found in an Indian study by Shaik et al. (2014). Therefore, Keke ridders should be properly educated on the dangers associated with this practice of having passengers in the front cabin. Also, relevant agencies such as the Federal Road Safety Corp (FRSC) should monitor and discourage this practice.

Also, there was a significant association between MSD and vibration. This implies that being exposed to vibration may increase the chances of developing WMSD. In a systematic review by Moraes et al. (2016) whole body vibration was found to be significantly associate with MSDs among professional truck drivers. Another systematic review by Charles et al. (2017) reported that occupational exposure to vibration is associated with MSDs. The body can be exposed to localized vibrations or whole body vibrations. Localised vibrations can cause changes in muscles, tendons, and joints as well as the nervous system. These effects are collectively called Hand Arm Vibration Syndrome (HAVS) or Raynaud's phenomenon (Johanning 2015). Also, whole body vibration can cause fatigue, loss of balance, headaches etc. and prolong exposure may lead to health disorders especially MSDs (Caffaro et al. 2016). According to Diyana et al. (2017) there are three factors to be considered in decreasing vibration and its effects on vehicle operators which are vehicular factors, environmental factors and operators factors. The vehicular factors include regular servicing and changing of worn out parts while environmental factors entails fixing bad roads and surfaces. Personal factors involves adopting neutral postures and avoiding twists (Diyana et al. 2017).

Job strain was in this study was not only significantly associated with WMSD, it was also a significant predictor of MSD. In a systematic review and meta-analysis of 54 longitudinal studies by Hauke et al. (2011) on the impact of work-related psychosocial stressors on the onset of musculoskeletal disorders in specific body regions, they found psychosocial factors (such as stress and strain) to be independent predictors of the onset of MSDs. There is a dynamic relationship between physical factors at work, psychosocial factors at work, stress symptoms and MSDs (Heuvel 2014). Psychological stress/strain influences musculoskeletal discomfort through the physiological and biological systems, affecting internal tolerances e.g. the adrenaline and noradrenaline released due to Psychological stress may cause increase muscle tightness (physiologic stress) that further increases the risk of MSD (Whysall 2008). Twisting while driving was also a significant predictor of WMSD among the commercial keke riders. Twisting the spine places undue physical stress on the soft tissues of the spine. Therefore, prolonged twisting of the spine results in MSDs. Similar findings have been reported by Lynch et al. (2014) who found twisting of the neck and trunk to be associated with MSD among logging machine operators in United States.

5   Conclusion

Based on the results obtained from this study, the following conclusions were made. First, that there is a high prevalence of WMSD among commercial tricycle riders in Nigeria. Secondly, that vibration, carrying passengers in the front cabin, job stress, twisting are associated with WMSD among commercial tricycle riders. Finally, Job stress and Twisting the spine are important predictors of WMSD among commercial tricycle riders in Nigeria. We therefore recommend ergonomic training programmes for commercial tricycle riders in Nigeria, regular assessment of their musculoskeletal system especially the low-back region and appropriate legislation and enforcement to stop them from having passengers in their front cabin.

References

Afifehzadeh-Kashani H, Choobineh A, Bakand S, Gohari MR, Abbastabar H, Moshtaghi P (2011) Validity and reliability of farsi version of cornell musculoskeletal discomfort questionnaire (CMDQ). Occup Health J 7:1–10

Akinbo SR, Odebiyi DO, Osasan AA (2008) Characteristics of back pain among commercial drivers and motorcyclists' in Lagos, Nigeria. West Afr J Med 27(2):87–91

Akinpelu AO, Oyewole OO, Odole AC, Olukoya RO (2011) Prevalence of musculoskeletal pain and health seeking behaviour among occupational drivers in Ibadan, Nigeria. Afr J Biomed Res 14:89–94

Atubi AO, Gbadamosi KT (2015) Global positioning and socio-economic impact of road traffic accidents in Nigeria: matters arising. Am Int J Contemp Res 5(5):136–146

Badejo B, Salau TI (2010) Transportation and the environment: the Lagos Example. Transportation

Caffaro F, Cremasco MM, Preti C, Cavallo E (2016) Ergonomic analysis of the effects of a telehandler's active suspended cab on whole body vibration level and operator comfort. Int J Ind Ergon 53:19–26

Charles LE, Ma CC, Burchfiel CM, Dong RG (2017) Vibration and ergonomic exposures associated with musculoskeletal disorders of the shoulder and neck. Saf Health Work

Diyana NA, Karuppiah K, Rasdi I, Sambasivam S, Tamrin SBM, Mani KK, Syahira PA, Azmi I (2017) Vibration exposure and work-musculoskeletal disorders among traffic police riders in Malaysia: a review. Ann Trop Med Public Health 10(2):334

Donnally III CJ, Dulebohn SC (2017) Lumbosacral Disc Injuries

Ezeukwu AO, Ugwuoke J, Egwuonwu AV, Abaraogu UO (2011) Prevalence of work-related musculoskeletal pain among timber workers in Enugu metropolis, Nigeria. Cont J

Hauke A, Flintrop J, Brun E, Rugulies R (2011) The impact of work-related psychosocial stressors on the onset of musculoskeletal disorders in specific body regions: a review and meta-analysis of 54 longitudinal studies. Work Stress 25(3):243–256

Hedge A, Morimoto S, Mccrobie D (1999) Effects of keyboard tray geometry on upper body posture and comfort. Ergonomics 42(10):1333–1349

Heuvel S (2014) Psychosocial risk factors for musculoskeletal disorders (MSDs)

Ismaila AS, Victor AU, Adeolu OJ (2011) Int J Occup Saf Ergon 17(1):99–102

Johanning E (2015) Whole-body vibration-related health disorders in occupational medicine–an international comparison. Ergonomics 58(7):1239–1252

Lynch SM, Smidt M, Merrill PD, Sesek RF (2014) Incidence of MSDs and neck and back pain among logging machine operators in the southern US. J Agric Saf Health 20(3):211–218

Malikraj S, Senthil Kumar T, Ganguly AK (2011) Ergonomic intervention on musculoskeletal problems among welders. Int J Adv Eng Technol 2(3):33–35

Marshal I, Solomon ID (2017) Nigeria economy and the politics of recession: a critique. J Adv Econ Finance 2(4):259

Moraes GFDS, Sampaio RF, Silva LF, Souza MAP (2016) Whole-body vibration and musculoskeletal diseases in professional truck drivers. Fisioterapia em Movimento 29 (1):159–172

Ojoawo AO, Onaade O, Adedoyin R, Okonji A (2014) Assessment of work related musculoskeletal pain among professional drivers in the service of a tertiary institution. Am J Health Res. Special Issue: Supplementary Prescribing

Orr RM, Johnston V, Coyle J, Pope R (2015) Reported load carriage injuries of the Australian army soldier. J Occup Rehabil 25(2):316–322

Shaik R, Gotru CK, Swamy CG, Sandeep R (2014) The prevalence of musculoskeletal disorders and their association with risk factors in auto rickshaw drivers - a survey in Guntur City. Int J Physiother 1(1):2–9

Tawiah AK, Oppong-Yeboah B, Idowu Bello A (2015) Work-related musculoskeletal disorders among workers at gold mine industry in Ghana: prevalence and patterns of occurrence

Thomas JR, Silverman S, Nelson J (2015) Research methods in physical activity, 7E. Human kinetics

Tinubu BM, Mbada CE, Oyeyemi AL, Fabunmi AA (2010) Work-related musculoskeletal disorders among nurses in Ibadan, South-west Nigeria: a cross-sectional survey. BMC Musculoskelet Disord 11(1):12

Wang X, Dong XS, Choi SD, Dement J (2017) Work-related musculoskeletal disorders among construction workers in the United States from 1992 to 2014. Occup Environ Med 74(5):374–380

Whysall Z (2008) Link between stress and musculoskeletal disorders

Effectiveness of a Pain Education Programme for Persistent Work-Related Musculoskeletal Pain

Deepak Sharan[1]([⊠]) and Joshua Samuel Rajkumar[2]

[1] Department of Orthopaedics and Rehabilitation, RECOUP Neuromusculoskeletal Rehabilitation Centre, 312, 10th Block, Anjanapura, Bangalore 560108, KA, India
deepak.sharan@recoup.in

[2] Department of Physiotherapy, RECOUP Neuromusculoskeletal Rehabilitation Centre, 312, 10th Block, Anjanapura, Bangalore 560108, KA, India
joshua.samuel@recoup.in

Abstract. Pain education is an important factor in the self-management of work-related musculoskeletal pain and can help in reduction of pain, disability, anxiety, stress, improved physical activity, positive thoughts, and reduced catastrophising of pain. Hence the objective of this study was to investigate the effectiveness of a pain education programme on workers with persistent work-related musculoskeletal pain in improving knowledge, self-efficacy and reduction in their pain. A survey questionnaire was used to evaluate the knowledge and understanding of the workers regarding pain and a questionnaire was used to evaluate their social and psychological status. The primary outcomes used were the McGill Pain Questionnaire and revised Neurophysiology of Pain Questionnaire. The pre-post analysis regarding the effectiveness of the course programme by analysing the outcomes of McGill pain questionnaire, Neurophysiology of pain questionnaire and the programme evaluation score showed a statistically significant difference in the post-scores compared to the pre-scores. This showed that the pain education programme was effective in the reduction of pain, disability and better conceptualisation of pain.

Keywords: Pain education · Work related musculoskeletal pain
Self management

1 Introduction

Pain education is an important factor in the self-management of work-related musculoskeletal pain and can help in reduction of pain, disability, anxiety, stress, improved physical activity, positive thoughts, and reduced catastrophising of pain. As workers with persistent work-related musculoskeletal pain gain knowledge about the neuroscience of pain, they develop increased awareness of pain and develop positive coping strategies to overcome pain. However, clinical application of a pain education programme in the occupational health context is still an understudied area.

2   Objective

To investigate the effectiveness of a pain education programme on workers with persistent work-related musculoskeletal pain in improving knowledge, self-efficacy and reduction in their pain.

3   Methodology

A prospective study was carried out among 40 workers with persistent work-related musculoskeletal pain, who participated in a pain education programme for a period of 8 weeks. A survey questionnaire was used to evaluate the knowledge and understanding of the workers regarding pain and a questionnaire was used to evaluate their social and psychological status. The primary outcomes used were the McGill Pain Questionnaire and revised Neurophysiology of Pain Questionnaire. Homework activities to support the program were also given. Workers were taught the basics of anatomy and physiology, stress, sleep disturbances, pharmacological and non-pharmacological interventions etc., and were taught to assess and reassess their pain and automatic thoughts, advised to modify the diet, exercises best suited to their health condition were taught, ergonomic advises were given, and training in Cognitive Behavioral Therapy was provided. At the end of the education programme, an examination was conducted to assess the workers' knowledge gained from the programme. A follow-up of the enrolled participants was done 1 year after the program to see if the effect has been maintained.

4   Results

The average duration of pain experience of the workers was 7.5 ± 4.90 years. The age group was between 18 to 60 years. The commonest site of pain was knee (56.25%), upper back (36.25%), shoulder (22.65%), neck (20.09%), lower back (20.07%), ankle (7.18%), and wrist and hand (5%). Analysis of survey questionnaire which was taken to evaluate the knowledge, attitude and understanding of pain showed that the knowledge and understanding of pain was very less among the participants, but their attitude towards gaining knowledge about the pain was considerably high, which showed that the participants were willing to learn about pain. The pre-post analysis regarding the effectiveness of the course programme by analysing the outcomes of McGill pain questionnaire, Neurophysiology of pain questionnaire and the programme evaluation score showed a statistically significant difference in the post-scores compared to the pre-scores. This showed that the pain education programme was effective in the reduction of pain, disability and better conceptualisation of pain. There was a significant improvement in self-efficacy in the social and psychological questionnaire. In the follow-up evaluation 90% of the participants were able to be contacted and their respective outcomes were collected, which showed that the positive outcomes were maintained for 80% of the participants.

5   Conclusions

Our study confirmed the effectiveness of a pain education programme on workers with persistent work-related musculoskeletal pain in improving knowledge, self-efficacy and reduction in their pain. The obtained outcomes were long lasting too.

Co-morbidities of Myofascial Low Back Pain Among Information Technology Professionals

Deepak Sharan(✉)

Department of Orthopaedics and Rehabilitation, RECOUP Neuromusculoskeletal
Rehabilitation Centre, 312, 10th Block, Anjanapura, Bangalore 560108, KA,
India
deepak.sharan@recoup.in

Abstract. Co-morbidity can be defined as the presence of one or more disorders (or diseases) in addition to a primary disease or disorder, or the effect of
such additional disorders or diseases. The underlying basis for such studies is
that if there is a presence of two or more diseases simultaneously, they may have
a common origin. Myofascial Pain Syndrome (MPS) of the lower back or
Myofascial Low Back Pain (MLBP) is one of the commonest WRMSD noted
among IT professionals. The aim of this study was to identify the prevalence of
MLBP as a WRMSD and its co-morbidities among IT professionals. The data
was collected from 8500 IT professionals from a single IT company who visited
the on-site clinics situated at their office campuses in 8 cities in an Industrially
Developing Country. All the reports from the year 2005 to 2017 which were
maintained in a database were reviewed. The study participants were predominantly males (78%). The mean age of the male and female subjects were
33.30 ± 5.99 years and 27.38 ± 5.59 years respectively. 44% of the population
used laptops, 42% desktops and 14% both. 48.5% of the participants had
MLBP. MLBP was the third common WRMSD, following MPS of neck and
Thoracic Outlet Syndrome. The present study revealed that MPS of neck,
Thoracic Outlet Syndrome, Fibromyalgia Syndrome and Patellofemoral Pain
Syndrome were found to be co morbid among IT professionals with MLBP.

Keywords: Co-morbidity · Myofascial low back pain · Information
technology

1 Introduction

Work related musculoskeletal disorders (WRMSD) are highly prevalent among
Information Technology (IT) professionals. Various risk factors have been proposed
for WRMSD, but its etiology is still being debated. A newer and different approach to
determine the etiology of WRMSD is the consideration of co-morbidity. Co-morbidity
can be defined as the presence of one or more disorders (or diseases) in addition to a
primary disease or disorder, or the effect of such additional disorders or diseases. The
underlying basis for such studies is that if there is a presence of two or more diseases
simultaneously, they may have a common origin. Co-morbidity of several WRMSD
leads to increment in the work absenteeism. Earlier studies have revealed that co-
morbidity of neck and low back pain affects healthcare utilisation and absenteeism.

© Springer Nature Switzerland AG 2019
S. Bagnara et al. (Eds.): IEA 2018, AISC 820, pp. 781–782, 2019.
https://doi.org/10.1007/978-3-319-96083-8_94

Myofascial Pain Syndrome (MPS) of the lower back or Myofascial Low Back Pain (MLBP) is one of the commonest WRMSD noted among IT professionals. The aim of this study was to identify the prevalence of MLBP as a WRMSD and its co-morbidities among IT professionals.

2 Methodology

The data was collected from 8500 IT professionals from a single IT company who visited the on-site clinics situated at their office campuses in 8 cities in an Industrially Developing Country. All the reports from the year 2005 to 2017 which were maintained in a database were reviewed. The database included key demographic data were collected of the participants including age, gender, computer usage per day, and the type of use (Laptop/Desktop). Data regarding type and intensity of the musculoskeletal problems were collected from the medical records of an Occupational Health/Rehabilitation Physician and a Physical Therapist in the same online database. Employee's feedbacks were also used for evaluating the status of musculoskeletal health of the IT professional. The physician's diagnosis with standardized clinical testing to rule in and rule out the clinically relevant work related musculoskeletal dysfunctions revealed the type and severity of the clinical features. The database was grouped accordingly and the statistical analysis was done. Descriptive statistics were used to describe the age, gender, body area affected and distribution of video display users. Chi square test was used to find the association between the various musculoskeletal discomfort co morbidities and MLBP.

3 Results

The study participants were predominantly males (78%). The mean age of the male and female subjects were 33.30 ± 5.99 years and 27.38 ± 5.59 years respectively. 44% of the population used laptops, 42% desktops and 14% both. 48.5% of the participants had MLBP. MLBP was the third common WRMSD, following MPS of neck and Thoracic Outlet Syndrome. Analysis revealed that there was a significant association between the presence of MLBP and MPS of neck (p < 0.001), MLBP and Thoracic Outlet Syndrome (p < 0.001), MLBP and Fibromyalgia Syndrome (p < 0.001), and between MLBP and Patellofemoral Pain Syndrome (p < 0.001).

4 Conclusions

The present study revealed that MPS of neck, Thoracic Outlet Syndrome, Fibromyalgia Syndrome and Patellofemoral Pain Syndrome were found to be co morbid among IT professionals with MLBP.

Application Study: Biomechanical Overload in Physiotherapists

Deepak Sharan[1(✉)], Joshua Samuel Rajkumar[2],
and Rajarajeshwari Balakrishnan[2]

[1] Department of Orthopaedics and Rehabilitation, RECOUP
Neuromusculoskeletal Rehabilitation Centre, 312, 10th Block, Anjanapura,
Bangalore 560108, KA, India
deepak.sharan@recoup.in
[2] Department of Physiotherapy, RECOUP Neuromusculoskeletal Rehabilitation
Centre, 312, 10th Block, Anjanapura, Bangalore 560108, KA, India
{joshua.samuel,rajarajeshwari.b}@recoup.in

Abstract. Healthcare professionals, especially Physiotherapists (PTs) are highly predisposed to WRMSDs. PTs tasks are generally complex and involve many physical activities that can lead to acute and chronic WRMSD, e.g., shifting the patient from one position to other or while performing manual therapy techniques. The Timing Assessment Computerized System (TACO) method was used to evaluate the risk of WRMSD among PTs and its relationship with musculoskeletal symptoms, exertion and workload. A prospective experimental study was conducted among a group of 55 PTs working in a rehabilitation center. The results of the study showed that 80% of PTs reported pain. The predominant posture causing high risk was "lumbar spine fully bent with operational areas under the knee" (54%). The commonest site of pain was lower back (60%), neck (52%) and upper back (50%). 80% of the participants were in the High Risk/Very High Risk category according to the TACO outcome.

Keywords: Biomechanical overload · Physiotherapists · Musculoskeletal pain

1 Introduction

Work Related Musculoskeletal Disorders (WRMSDs) represent one of the leading causes of occupational injury, disability, absenteeism and incapacity among workers. Healthcare professionals, especially Physiotherapists (PTs) are highly predisposed to WRMSDs. PTs tasks are generally complex and involve many physical activities that can lead to acute and chronic WRMSD, e.g., shifting the patient from one position to other or while performing manual therapy techniques. Existing risk assessment tools fail to appropriately evaluate the nature of task, variable postures involved and the duration of exposure that can predict the risk of a PT developing WRMSD. The Timing Assessment Computerized System (TACO) method was used to evaluate the risk of WRMSD among PTs and its relationship with musculoskeletal symptoms, exertion and workload. TACO was developed by Daniela Colombini, Marco Tasso (Italy) and

Joshua Samuel Rajkumar (India), in 2016. The TACO is a software based programme that calculates risk categories for the development of WRMSD based on working postures and duration. This was an adapted version from the original TACO tool, for specific needs of the PTs tasks.

2 Methodology

A prospective experimental study was conducted among a group of PTs working in a rehabilitation center. 55 PTs with a mean age of 30.45 years (56% female), working 9 h per day, including 1 h of breaks, 6 days a week, were evaluated with the TACO tool based on video recordings of their job tasks, and a self-reported questionnaire included demographics, job details, health status, physical risk factors, Work Style Questionnaire (Short Form), Nordic Musculoskeletal Pain Questionnaire, Borg CR 10 and NASA Task Load Index.

3 Results

The results of the study showed that 80% of PTs reported pain. The predominant posture causing high risk was "lumbar spine fully bent with operational areas under the knee" (54%). The commonest site of pain was lower back (60%), neck (52%) and upper back (50%). 75% of the PTs did not exercise regularly. 65% of the subjects reported an adverse workstyle risk (total score >28). The perceived exertion and workload were also high as 70% of the PTs were in the scores of >15 (Borg CR 10) and >50 (NASA Task Load Index) respectively. 80% of the participants were in the High Risk/Very High Risk category according to the TACO outcome.

4 Conclusions

Awkward postures should be monitored and eliminated by appropriate measures to reduce the risk of WRMSD among PTs. The other risk factors, i.e., psychosocial, exertion and workload, also significantly contribute towards WRMSDs and need to be appropriately addressed.

Why Do Information Technology Professionals Develop Work Related Musculoskeletal Disorders? A Study of Risk Factors

Deepak Sharan[1]([✉]) and Joshua Samuel Rajkumar[2]

[1] Department of Orthopedics and Rehabilitation, RECOUP
Neuromusculoskeletal Rehabilitation Centre, 312, 10th Block, Anjanapura,
Bangalore 560108, KA, India
deepak.sharan@recoup.in
[2] Department of Physiotherapy, RECOUP Neuromusculoskeletal Rehabilitation
Centre, 312, 10th Block, Anjanapura, Bangalore 560108, KA, India
joshua.samuel@recoup.in

Abstract. Work Related Musculoskeletal Disorders (WRMSD) are highly prevalent among Information Technology (IT) professionals. A prospective analysis of 8200 IT Professionals (age 20 to 60 years, mean 33 years, 74% males), in an Industrially Developing Country was conducted. The employees were evaluated by a detailed questionnaire consisting of demographic data, job details, health status, physical risk factors, short-form Work Style Questionnaire and Nordic Musculoskeletal Pain Questionnaire. pain (36%). Musculoskeletal problems increased the fatigue levels as recorded by Borg CR-10 scale. 76% of the employees were diagnosed by an experienced occupational health physician to have a WRMSD, among which 65% of the employees were laptop users. 78% had widespread body pain, 72% neck pain, 63% lower back pain, 52% shoulder pain and others with upper arm, thigh, knee and foot pain. Increasing age, high Body Mass Index, longer working hours, hazardous body postures, static loading, resting elbows and wrists on hard surfaces, and adverse work-style were positively correlated (r < 0.01) with the presence of WRMSD. On the other hand, rest breaks during work, regular exercises and formal ergonomics training were negatively correlated (r < −0.01).

Keywords: Work-related musculoskeletal disorder · Information technology
Risk factors

1 Introduction

Work Related Musculoskeletal Disorders (WRMSD) are highly prevalent among Information Technology (IT) professionals. WRMSD is reported to be influenced by genetic factors, socio-economic factors, environmental factors, lifestyle and individual perceptions, high physical work demands, e.g., hazardous body postures, monotonous work, repetitive arm movements, prolonged standing, work with arms above shoulder height, and heavy lifting. However, the risk factors associated with the development of

WRMSD in IT professionals are not clear. Hence, the objective of this study was to evaluate the risk factors that predispose IT professionals to WRMSD.

2 Methodology

A prospective analysis of 8200 IT Professionals (age 20 to 60 years, mean 33 years, 74% males), in an Industrially Developing Country was conducted. The employees were evaluated by a detailed questionnaire consisting of demographic data, job details, health status, physical risk factors, short-form Work Style Questionnaire and Nordic Musculoskeletal Pain Questionnaire. The data was extracted, and statistical analysis was performed.

3 Results

The results of the study showed that 76% of the employees were diagnosed by an experienced occupational health physician to have a WRMSD. 65% of the employees were laptop users, 30% were desktop users and 5% used both. A total of 54% of the employees worked for at least 5–9 h per day and 46% for 10–14 h per day. Most of the male workers complained of low back and radiating pain in upper or lower limbs, compared to female workers who complained predominantly of neck and shoulder pain. Both the population had eye strain and increased fatigue in common. 78% had widespread body pain, 72% neck pain, 63% lower back pain, 52% shoulder pain and others with upper arm, thigh, knee and foot pain. 78% were diagnosed to have Myofascial Pain Syndrome, followed by Thoracic Outlet Syndrome (36%), Fibromyalgia (35%), Tendinopathies of shoulder, elbow or wrist (21%), Patellofemoral Pain Syndrome (12%) and Type 1 Complex Regional Pain Syndrome (6%). Increasing age, high Body Mass Index, longer working hours, hazardous body postures, static loading, resting elbows and wrists on hard surfaces, and adverse work-style were positively correlated (r < 0.01) with the presence of WRMSD. On the other hand, rest breaks during work, regular exercises and formal ergonomics training were negatively correlated (r < −0.01) with the presence of WRMSDs, as more frequent breaks, regular exercises and prior ergonomics training showed lower prevalence of WRMSD.

4 Conclusions

The presence of co morbidities like joint hypermobility, diabetes, hypothyroidism, hyperuricemia, low bone mineral density, hypovitaminosis D and B12 had a positive influence on the prevalence of WRMSDs in the study population. Other specific factors like work experience, hand dominance, type of computer used also had an influence on the development of WRMSDs. The risk factor analysis gives an insight to the appropriate areas of ergonomic interventions among IT professionals.

The Ergo-UAS System and a New Design Approach: Overview and Validation

Gabriele Caragnano[1,2(✉)] and Roberta Bonfiglioli[3]

[1] Partner PwC Italy, Operations 20149 Milan, Italy
gabriele.caragnano@pwc.com
[2] Direttore Tecnico Fondazione Ergo, 21100 Varese, Italy
[3] Occupational Medicine, Department of Medical and Surgical Sciences,
Occupational Health Unit, Sant' Orsola Malpighi Hospital,
University of Bologna, 40138 Bologna, Italy
roberta.bonfiglioli@unibo.it

Abstract. Musculoskeletal disorders are the second largest contributor to disability worldwide affecting people across the life-course. Risk identification and design of interventions to reduce the rates of work-related musculoskeletal disorders need to be based on valid and reproducible methods. The Ergo-MTM model is a method used for balancing and design purposes based on the definition of standard time that considers at the same time two of the most important issues in the definition of a fair load. The innovative aspect, is the definition of the Ergonomic Factor determined for every workplace accordingly to the biomechanical load coming from the combination of the assigned operations and quantified with EAWS method. The concept is based on recent relevant standards related to the biomechanical load, which is influenced by the sequence, the repetitiveness and frequency of the operations, in addition to the characteristics of every movement. In the paper the methodology of the project of validation through longitudinal epidemiological study of the EAWS system for the assessment and prevention of biomechanical overload will be presented. Fondazione Ergo and University of Bologna are carrying out the study with the scientific contribution of a panel of experts from academic and non-academic institutions and the Bioethics Committee of the University of Bologna.

Keywords: ERGO-UAS system · Ergonomic factor
Longitudinal epidemiological study

1 Introduction

Musculoskeletal conditions are the second largest contributor to disability worldwide affecting people across the life-course. The prevalence and impact increase with age, but younger people are also affected [1]. Physically demanding jobs can contribute to the development of musculoskeletal disorders with at least reasonable evidence of a causal relationship [2]. Risk identification and design of interventions to reduce the rates of work-related musculoskeletal disorders need to be based on valid and reproducible methods.

A recent systematic review concluded for moderate evidence that supplementary breaks, compared to conventional break schedules, are effective in reducing symptom intensity in various body regions. However, better quality studies are needed to allow definitive conclusions on the effectiveness of work organizational or psychosocial interventions to prevent or reduce work-related musculoskeletal disorders [3].

2 Ergo MTM Model

The Ergo-MTM model is a method to calculate the standard time: basic time plus allowances. It is usually used for balancing purposes and for the calculation of the target production quantity. The standard time of a workplace is calculated with the aim of considering at the same time the two most important issues in the definition of a fair workload [4, 5]:

- The basic time according to the internationally recognized MTM scale, which is the outcome of the work method analyzed through the MTM-UAS technique.
- The allowance time, which is calculated in proportion to the biomechanical load determined through the EAWS technique accordingly to the operations sequence and their repetitiveness or frequency.

2.1 Standard Time Calculation Procedure Through the Ergo-MTM Model

In the following example, in order to simplify the calculation, we consider the case (however very common) when Total Basic Time coincides with The Active Time of the workplace.

In this case, the Total Standard Time (Tstd) is determined starting from the Total Basic Time (Tbasic) which is amplified by the Allowance Factor (ALW), as shown in the picture below (Fig. 1):

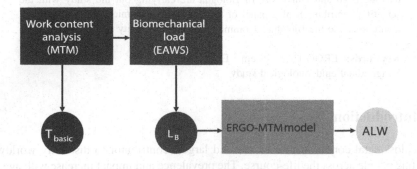

$$T_{std} = T_{basic} \times (1 + ALW)$$

Fig. 1. Standard time calculation procedure with Ergo MTM Model

2.2 Calculation of the Ergonomic Factor

The Ergonomic Factor (and consequently the Allowance factor containing the Work Delay Factor), is determined for every workplace accordingly to the combination of the assigned operations (balancing) and not for every basic movement and quantified with EAWS method.

This need comes from the recent relevant standards related to the biomechanical load, which is influenced by the sequence, the repetitiveness and frequency of the operations, in addition to the characteristics of every movement.

The Allowance Factor is directly linked to the biomechanical load calculation through the Ergo-MTM curve, as shown in the picture below (Fig. 2):

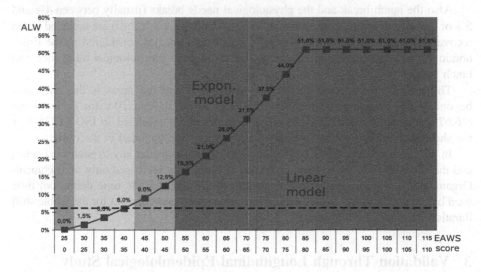

Fig. 2. Ergo MTM curve

This curve has been defined by the International MTM Directorate after an extended experimentation which has involved some of the biggest European car and home-appliances manufacturers.

This experimentation was necessary because this new model, which determines the allowance to compensate the physical efforts considering entire work sequences, is more responding to the recent relevant standards.

The values of the ERGO-MTM curve have been determined in order to exponentially smooth the biomechanical load for rising values of risk score. The growing rate of the curve (inclination) and its positioning in the graph "Allowance Factor vs. EAWS Index" have been calculated in order to:

– give enough recovery time to exit the high-risk zone (red area)
– give enough recovery time to work correctly in the medium-risk zone (yellow area)
– don't assign recover period in the green area

The recovery time is a period of inactivity of one or more body parts usually involved in work activity. The Ergonomic Factor assessment is necessary to give time to be used as recovery time: the higher is the biomechanical load, the longer is the recovery time. The recovery time could be organized in a continue or discrete way:

- Continue recovery time: small period included in each work cycle (takt) and available continuously at regular intervals
- Discrete recovery time: it is usually a break that lasts some cycle: usually at least 8 min break periods (pauses). The discrete recovery time could also be a period during which the worker is performing a task without using the body parts usually involved in the work cycle (e.g. walking, visual control, material replenishment, etc.)

Also the lunch break and the physiological needs breaks (usually between 4% and 5% of the working time, corresponding to 20 min during a 8 h shift) are included in the recovery time; nevertheless the physiological breaks are not considered in the Ergonomic Factor, that has to be considered an addition to physiological allowance and lunch break.

The limit of Ergonomic Factor between yellow and red areas is the last value beyond which the continue recovery time has to be at least 10 s for each minute (16,67%, that in the ERGO-MTM curve becames16,5%), defined in ISO 11228-3 as the shortest time that can be considered a recovery time included in the cycle.

In the green area the ERGO-MTM model does not consider any Ergonomic Factor, and the standard time is the same as the MTM basic time increased only of Technical-Organizational Factor (if present), according to the MTM basic time definition: time used by a trained worker, with medium force and speed and constant for the whole shift duration, without any allowance.

3   Validation Through Longitudinal Epidemiological Study of the EAWS System for the Assessment and Prevention of Biomechanical Overload

The main purpose of the project is to evaluate the capability of the EAWS system to evaluate and prevent biomechanical overload. Fondazione Ergo and University of Bologna carry out the study thanks to the collaboration of a large automotive company and with the scientific contribution of a panel of experts from academic and non-academic institutions. The study protocol has been submitted for evaluation and approved by the Bioethics Committee of the University of Bologna.

3.1   "Predictive Validity" Objective

In a group of workers assigned to manual activities designed on the basis of the results of the EAWS system, the presence and frequency of musculoskeletal symptoms and diseases will be assessed over time.

The study hypothesis consists in demonstrating that workers involved in manual tasks designed according to the above mentioned system do not present a significant increase in the risk of suffering from disorders and/or diseases affecting the musculoskeletal system (predictive validity).

The objective will be pursued:

- by estimating the prevalence and incidence of self-reported pain and disability in the musculoskeletal system by means of a standardized questionnaire
- by estimating the musculoskeletal diseases diagnosed through clinical examination and instrumental findings.

Data from non-exposed workers will be used to calculate the risk of developing musculoskeletal disorders in workers involved in manual task designed on the basis of the results of the EAWS system.

3.2 Project Activities

The various activities that will be carried out consist in a detailed exposure assessment and a clinical evaluation in a randomly selected sample of workers who perform manual and repetitive tasks (assembly lines or workstations) and in a control group of workers employed in the same company who do not perform manual tasks (white collar workers). Monitoring over time of both exposure and frequency of musculoskeletal disorders and/or diseases in the study sample to assess the ability of EAWS to reduce the risk of occurrence of musculoskeletal disorders.

3.3 Expected Benefits and Possible Developments

In the short term, data will provide an overview of the association of musculoskeletal disorders with different levels of biomechanical load assessed by the EAWS system, controlling for personal risk factors. In the medium-long term it will be possible to evaluate the capacity of the EAWS system to reduce the risk of the onset of disorders and/or diseases affecting the musculoskeletal system.

4 Conclusions and Future Perspectives

The Ergo-MTM model is an innovative method used for balancing and design purposes based on the definition of standard time that considers at the same time two of the most important issues in the definition of a fair load. Especially the Ergonomic Factor is determined for every workplace accordingly to the biomechanical load coming from the combination of the assigned operations and quantified with EAWS method. Risk identification and design of interventions to reduce the rates of work-related musculoskeletal disorders need to be based on valid and reproducible methods. In this perspective a project of validation through longitudinal epidemiological study of the EAWS system for the assessment and prevention of biomechanical overload is started, involving Fondazione Ergo and University of Bologna with the contribution of a panel of experts from academic and non-academic institutions and the Bioethics Committee of the University of Bologna.

The project will be carried out on 36 months data collation with two different output: in the short term, data will provide an overview of the association of musculoskeletal disorders with different levels of biomechanical load assessed by the EAWS system, controlling for personal risk factors and in the medium-long term it will be possible to evaluate the capacity of the EAWS system to reduce the risk of the onset of disorders and/or diseases affecting the musculoskeletal system.

References

1. WHO Fact sheet, February 2018. http://www.who.int/mediacentre/factsheets/musculoskeletal/en/
2. Da Costa BR, Vieira ER (2010) Risk factors for work-related musculoskeletal disorders: a systematic review of recent longitudinal studies. Am J Ind Med 53(3):285–323
3. Stock SR, Nicolakakis N, Vézina N et al (2018) Are work organization interventions effective in preventing or reducing work-related musculoskeletal disorders? A systematic review of the literature. Scand J Work Environ Health 44(2):113–133
4. Caragnano G, Lavatelli I (2012) ERGO-MTM model: an integrated approach to set working times based upon standardized working performance and controlled biomechanical load. Work 41(Suppl 1):4422–4427
5. Vitello M, Galante LG, Capoccia M, Caragnano G (2012) Ergonomics and workplace design: application of Ergo-UAS system in fiat group automobiles. Work 41(Suppl 1):4445–4449

Development of a Risk Assessment Procedure for Upper Limbs Based on Combined Use of EAWS 4th Section and OCRA High Precision Checklist

Enrico Occhipinti[1] and Lidia Ghibaudo[2(✉)]

[1] EPM Research Unit and International School, Milan, Italy
epmenrico@tiscali.it
[2] Fiat Chrysler Automobiles - EMEA Region – Manufacturing Planning
and Control – Direct Manpower Analysis and Ergonomics, 10135 Turin, Italy
lidia.ghibaudo@fcagroup.com

Abstract. Italian legislation on the subject of Workers' Health and Safety recalls the ISO standards for the ergonomics analysis and mentions the OCRA method as 'preferred' for risk assessment of upper limbs overload. Over the years, other methods for ergonomics risk assessment not currently mentioned in ISO Standards, were developed. As for example, EAWS method – proposed by International MTM network for ergonomic design and analysis of workstations – is greatly adopted by automotive OEMs because of its integration in the ErgoUAS system that allows to define the sequence of tasks and the time needed to perform related to the biomechanical load. This paper presents the "Cut Off" Project developed to define a risk assessment procedure for upper limbs based on the combined use of EAWS section 4th and OCRA Checklist. "Cut off" are a set of criteria, related to the main risk factors for upper limbs musculoskeletal disorders, characterized by threshold values based on EAWS section 4th. The aim of the procedure is to apply EAWS section 4th to all manual workstations during both design and risk assessment phase and to analyze with OCRA checklist only workstations, characterized by the overcome of one or more thresholds based on EAWS factors. The procedure is going to be constantly monitored and possibly improved by new data collection.

Keywords: EAWS method · OCRA checklist · Risk assessment
Cut off

1 Introduction

Concurrent with the radical changes of manufacturing process, workers' health disease patterns have changed.

From years 2000, an increase in diseases associated with industrial working tasks has been observed. Results of various studies report that, in spite of the increasing mechanization and automation of production processes, yet a major part of work activities, especially related to high quality customers requirement, are manually performed by workers [1]. Musculoskeletal disorders are worldwide considered as the

© Springer Nature Switzerland AG 2019
S. Bagnara et al. (Eds.): IEA 2018, AISC 820, pp. 793–799, 2019.
https://doi.org/10.1007/978-3-319-96083-8_98

principal occupational health problem and cost, as well as the most common cause of occupational injury and disability in industrialized countries [2, 3]. The most well-known MSDs are low back pain and upper limb disorders. The first one is mainly associated with manual material handling while the second one is mainly associated with manual repetitive tasks with combinations of high frequency, force and awkward work postures. Across Europe it is estimated that around 70 million workers suffer of work-related MSDs [3] and data coming from the Sixth European Working Conditions Survey, conducted in 35 countries, report that half of European workers suffer from work-related MSDs of which more than 44% is related to muscular pains at shoulders, neck and/or upper/lower limbs [3].

The level of exposure to physical workload is assessed quantifying the risks factors and the exposure time associated with work-related musculoskeletal disorders. Many methods and tools are available and presented in peer-review literature and consist in observational methods, instrumental or direct methods, self-reports, checklist and other psychophysiological methods.

In design phase as well as in the risk assessment phase, it's fundamental to be effective and to employ resources and time in workstations with critical issues, this should be possible using simple and quick approach as proposed in ISO TR 12295 with Quick Assessment [4].

This paper presents the Cut Off Project, to develop a risk assessment procedure for upper limbs based on the combined use of EAWS 4th section and OCRA Checklist. This procedure should improve and fast the risk assessment phase allowing to focus and deep analyze workstations with critical issues.

2 Materials and Methods

2.1 Risk Assessment Methods

The Italian legislation regarding Health and Safety of Workers, Dlgs.81/08, indicates that ISO Standards 11228 parts 1-2-3 define reference criteria for the ergonomics risk assessment related to manual material handling.

The OCRA index method, mentioned in the ISO standards 11228-3 [5] as 'preferred' method for risk assessment of upper limbs overload, especially as checklist has been widely adopted since years 2000 as a tool for the assessment and mapping of risk due to biomechanical overload of the upper limbs. Over years, OCRA authors have collected suggestions for improving the accuracy of the method and for making the application easier and in 2011 the method was upgraded and the OCRA High Precision was presented [6–8].

Over the years, other methods for ergonomics risk assessment not currently mentioned in ISO Standards, were developed. In particular, the EAWS method – Ergonomic Assessment Work Sheet – published in 2008 by Fondazione Ergo, that has become the main tool for ergonomic design and analysis of workstations for the International MTM (Methods-Time Measurement) network [9, 10].

EAWS worksheet is structured in four macro sections, each dedicated to a specific biomechanical risk factor analysis: body postures (Section 1st), force actions (Section 2nd), manual material handling (Section 3rd) and manual material handling of low load at high frequency (Section 4th). This structure leads to define EAWS as a comprehensive method for assessing the biomechanical overload of whole body and upper limbs classifying in a traffic light coded evaluation (green, yellow, red) the final index as required by Machine Directive. The wide spectrum of analysis makes it particularly suitable for the automotive production, characterized by working tasks in which the biomechanical overload is defined by the simultaneous presence of risk factors such as awkward postures, high frequency upper limbs movements and force actions, exerted by the whole body or by the upper limb.

The EAWS method is greatly used by automotive OEMs, because it's the ergonomic assessment method integrated in the ErgoUAS system. The ErgoUAS system is based on two main components: UAS and EAWS. This system is used especially during the industrialization phase, to balance the line: defining the sequence of tasks and the time needed to perform them related to the biomechanical load [9, 10].

UAS is the methods-time measurement used to assign pre-determined time to elementary tasks in production line batch, called basic time. Each elementary task is characterized by physical parameters in order to assess the biomechanical load of the overall working tasks.

The evaluation of workstation's biomechanical load is performed with EAWS method, according to the biomechanical load an additional amount of time, the rest factor is given to each workstations to assure the physiological recovery time and the optimized biomechanical load [10, 11].

During the production phase, the risk assessment is performed according to the legal requirements and each workstation is analyzed with the OCRA checklist method. So at the starting point of the project the methods were sequentially applied.

2.2 Cut off Project and Scope

The Cut Off project is born on the experimental hypothesis of applying a combined approach based on the use of the EAWS section 4th and OCRA checklist methods for the risk assessment analysis of manual workstations, in the design and production phases, being compliant to legal requirements.

The aim was to define some strictly criteria, called "cut off", related to the main risk factors for upper limbs musculoskeletal disorders and the related threshold value, to define the procedure for the combined use of EAWS section 4th and OCRA Checklist method during the design and risk assessment phases, related to the working tasks and the organizational structure of FCA plants located in EMEA Region [13, 14].

In fact, the project scope is to:

- to apply the EAWS section 4th to all manual workstations during both design and the preliminary risk assessment phase;
- analyze in more detail, by the OCRA checklist, just the workstations, characterized by the overcome of one or more screening thresholds based on EAWS factors, defined cut off.

Since the procedural hypothesis was based on a limited number of comparative analysis deriving from a previous project work, the working group defined a phase of analysis and monitoring based on an experimental data collection.

The experimental phase involved all FCA Italian plants and consisted on the acquisition of upper limbs risk factors - such as frequency, postures of segments and joints and forces - simultaneously analyzed with EAWS section 4th and OCRA checklist method of a great number of workstations. It's worth to underline that the widest spectrum of case studies in terms of working tasks, cycle time and traffic light risk index, was taken into account.

Another crucial aspect was the analysis and the definition of the relationship between the two methods coming from experimental data, in order to define a circumscribed forecast model.

2.3 Data Analysis and Cut off Definition

The first correlation analysis was led on the association between the final score of the two methods, using a simplified statistical approach based on the linear regression model on 260 pairs of available data [13]. Data come from the analysis with both methods of single videotaped tasks. To simplify the analysis and to guarantee a statistically sound number of analyzed tasks with both methods in a short term, the analyzed tasks were extracted FROM the working activity constituting the workstation in this respect, it is worthwhile recalling that, in an automotive production line, workstations are highly influenced by the mix: optional elements, different fuels and motorizations, etc. Consequently, the production mix is a fundamental variable that has an high impact on workstations final score. In other words, the data-points in the figure are to be considered only for the purpose of comparing the two methods and not as risk assessments of workstations.

Being aware of the above simplification and of the different traffic light classification scale of the two methods (Table 1); in the definition of the linear regression equation, OCRA Checklist index was defined as dependent variable and EAWS section 4 index as independent variable (Fig. 1):

Table 1. Traffic light classification scale

	Green zone	Yellow zone	Red zone
OCRA Checklist	0–7.5	7.6–11	>11
EAWS	0–25	25.5–50	>50

$$OCRA\ checklist\ index = A + B * EAWS\ section\ 4\ index$$

The cut off criterion based on the EAWS section 4th score threshold has to allow a high confidence level (99% confidence limits) that, even in the worst SCENARIO, the OCRA Checklist score was at the most in the yellow zone. This means that the criterion (EAWS section 4th score ≥ 40) guarantees a very limited risk (1%) of not intercepting workstations with an OCRA Checklist score above 11. An additional set of criteria,

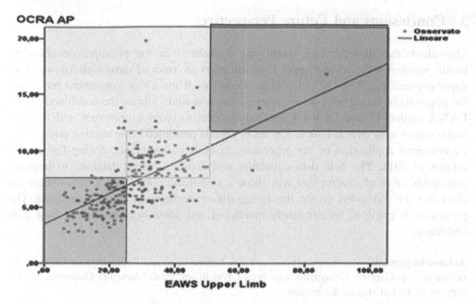

Fig. 1. Linear regression graph between OCRA Checklist and EAWS section 4th

based on the most common risk factors for upper limbs biomechanical overload, was set to reduce even further the possibility of not intercepting in the design phase analysis workstations that OCRA checklist might classify as potentially at risk.

So, the working group defined the following the cut off, concerning the main risk factors, and the related threshold [14]:

- EAWS section 4 score ≥ 40
- frequency of real actions ≥ 30 action/min
- exposure time of unfavorable grips (such as pinch grip; or not power grip or contact grip) $\geq 25\%$ of the cycle time
- exposure time of awkward shoulder posture (arm elevation $>80°$) $\geq 10\%$ of cycle time
- stereotypy in terms of joints or segments engaged posture $>50\%$ of the cycle time or static actions $>50\%$ of the cycle time
- force in two different cases:
  - presence of medium force (approximately force level $\geq 25\%$ of the maximum force exerting by 15th neutral percentile as reported in Force Atlas [11, 12]) for an exposure time $\geq 25\%$ of the time cycle.
  - presence of force peaks (approximately force level $\geq 50\%$ of the maximum force exerting by 15th neutral percentile as reported in Force Atlas [11, 12]) regardless of the exposure time.

3   Conclusions and Future Perspectives

Musculoskeletal disorders are worldwide considered as the principal occupational health problem and cost and upper limb disorders are ones of most well-known. This paper presents the "Cut Off" Project developed to define a risk assessment procedure for upper limbs based on a set of criteria. The procedure allows the combined use of EAWS section 4th and OCRA Checklist in order to employ resources and time on workstations with critical issues. The set of criteria presented is the starting point to the experimental application of the procedure, in FCA Italian plants, during the second quarter of 2018. The new data collection will define a robust database to improve thresholds value of criteria that will show a systematic conservative behavior or the ones that are conceived on the theoretical difference between the two methods. The procedure is going to be constantly monitored and possibly improved by new data collection.

Acknowledgements. Authors want to thank Fondazione Ergo and Politecnico of Turin, in the person of Ing. Gabriele Caragnano, Ing. Ivan Lavatelli and Prof. Maria Pia Cavatorta, for the support in the last step of the project.

References

1. Chau N, Bhattacherjee A, Kunar BM (2009) Relationship between job, lifestyle, age and occupational injuries. Occup Med 59(2):114–119
2. EU-OSHA - European Agency for Safety and Health at Work, OSH in figures: Work-related musculoskeletal disorders in the EU - Facts and figures
3. Taiwo OA, Cantley LF, Slade MD, Pollack KM, Vegso S, Fiellin MG et al (2009) Sex differences in injury patterns among workers in heavy manufacturing. Am J Epidemiol 169(2):161–166
4. ISO TR 12295 (2014) Ergonomics – Application document for International Standards on manual handling (ISO 11228-1, ISO 11228-2 and ISO 11228-3) and evaluation of static working postures (ISO 11226), Geneva, Switzerland
5. ISO 11228-3 (2007) Ergonomics - Manual handling - Handling of low loads at high frequency. ISO, Geneva, Switzerland
6. Colombini D, Occhipinti E, Cerbai M, Battevi N, Placci M (2011) Aggiornamento di procedure e di criteri di applicazione della ChecklistOCRA. Med Lav 102(1)
7. Colombini D, Occhipinti E, Delleman D, Fallentin N, Kilbom A, Grieco A (2001) Exposure assessment of upper limb repetitive movements: a consensus document. In: Karwowski W (ed) International Encyclopaedia of Ergonomics and Human Factors. Taylor & Francis, New York
8. Colombini D, Grieco A, Occhipinti E (1998) Occupational musculoskeletal disorders of the upper limbs due to mechanical overload. Ergonomics 41(9). Special issue
9. Schaub K, Mühlstedt J, Illmann B, Bauer S, Fritzsche L, Wagner T, Bullinger-Hoffmann A (2012) Ergonomic assessment of automotive assembly tasks with digital human modelling and the 'ergonomics assessment worksheet' (EAWS). Int J Hum Factors Model Simul 3(3/4)
10. The European Assembly Worksheet, Schaub K, Caragnano G, Britzke B, Bruder R (2012) Theoretical Issues in Ergonomics, 1–23. iFirst Science

11. Schaub K et al (2010) The European assembly worksheet. In: Mondelo P, Karwowski W, Saarela K, Swuste P, Occhipinti E (eds) Proceedings of the VIII international conference on occupational risk prevention, ORP 2010, 5–7 May 2010, Valencia. ISBN: 978-84-934256-8-5
12. Wakula J et al (2009) Der montagespezifische Kraftatlas. (The assembly specific force atlas) BGIA-Report 3/2009. (Hrsg.) von der Deutschen Gesetzlichen Unfallveriherung (DGUV) (Institute for Occupational Safety and Health of the German Social Accident Insurance (IFA)). Berlin, March 2009, Final report. ISBN 978-3-88383-788-8. ISSN 1869-3491
13. Occhipinti E (2017) La relazione tra EAWS e CK-OCRA AP nella casistica di 275 postazioni FCA: alla ricerca di una procedura di individuazione dei casi EAWS da approfondire con CK OCRA. Internal document
14. Occhipinti E, Caragnano G, Cavatorta MP, Lavatelli I, Ghibaudo L (2018) Progetto di sviluppo procedura di valutazione del rischio arti superiori basata sull'utilizzo combinato di EAWS sez.4 e checklist OCRA Alta Precisione, Internal Document

Author Index

Printed in the United States
By Bookmasters